MATHEMATICAL REVIEW

Area of a circle of radius R $A = \pi R^2$
Circumference of a circle $C = 2\pi R$
Surface area of a sphere $A = 4\pi R^2$
Volume of a sphere $V = \frac{4}{3}\pi R^3$
Area of a triangle $A = \frac{1}{2}bh$
Volume of a circular
    cylinder of length l $V = \pi R^2 l$
Pythagorean Theorem $C^2 = A^2 + B^2$
    $\sin \theta = A/C$
    $\cos \theta = B/C$ $\tan \theta = \dfrac{\sin \theta}{\cos \theta}$
    $\tan \theta = A/B$
Quadratic Equation: Where $ax^2 + bx + c = 0$

$$x = \frac{-b \pm \sqrt{b^2 - 4ac}}{2a}$$

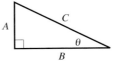

PHYSICAL CONSTANTS

Quantity	Symbol	Value
Gravitation constant	G	$6.672\,59 \times 10^{-11}$ N·m²/kg²
Speed of light in vacuum	c	$2.997\,924\,58 \times 10^{8}$ m/s
Electron charge	e	$1.602\,18 \times 10^{-19}$ C
Planck's Constant	h	$6.626\,076 \times 10^{-34}$ J·s
		$4.135\,669 \times 10^{-15}$ eV·s
Universal gas constant	R	$8.314\,510$ J/mol·K
Avogadro's number	N_A	$6.022\,137 \times 10^{23}$ mol^{-1}
Boltzmann Constant	k_B	$1.380\,66 \times 10^{-23}$ J/K
		$8.617\,39 \times 10^{-5}$ eV/K
Coulomb force constant	k	$8.987\,55 \times 10^{9}$ N·m²/C²
Permittivity of free space $(1/\mu_0 c^2)$	ϵ_0	$8.854\,19 \times 10^{-12}$ C²/N·m²
Permeability of free space	μ_0	$1.256\,64 \times 10^{-6}$ T·m/A
Permeability constant	$\mu_0/4\pi$	10^{-7} T·m/A
Electron mass	m_e	$9.109\,39 \times 10^{-31}$ kg
Electron rest energy	$m_e c^2$	$0.510\,999$ MeV
Electron magnetic moment	μ_e	$9.284\,77 \times 10^{-24}$ J/T
Electron charge/mass ratio	e/m_e	$1.758\,82 \times 10^{11}$ C/kg
Electron Compton wavelength	λ_c	$2.426\,31 \times 10^{-12}$ m
Proton mass	m_p	$1.672\,623 \times 10^{-27}$ kg
		$1.007\,276$ u
Proton rest energy	$m_p c^2$	938.272 MeV
Proton magnetic moment	μ_p	$1.410\,608 \times 10^{-26}$ J/T
Neutron mass	m_n	$1.674\,929 \times 10^{-27}$ kg
		$1.008\,66$ u
Neutron rest energy	$m_n c^2$	939.566 MeV
Neutron magnetic moment	μ_n	$9.662\,37 \times 10^{-27}$ J/T
Bohr magneton	μ_B	$9.274\,015 \times 10^{-24}$ J/T
Stefan-Boltzmann Constant	σ	$5.670\,51 \times 10^{-8}$ W/m²·K⁴
Rydberg constant	R	$1.097\,373 \times 10^{7}$ m^{-1}
Bohr radius	r_1	$5.291\,77 \times 10^{-11}$ m
Faraday constant	F	$9.648\,53 \times 10^{4}$ C/mol

VOLUME ONE
PHYSICS

EUGENE HECHT
Adelphi University

Brooks/Cole Publishing Company
Pacific Grove, California

I(T)P ™ The trademark ITP is used under license.

Brooks/Cole Publishing Company
A Division of Wadsworth, Inc.

Printed in the United States of America

10 9 8 7 6 5 4 3 2 1

The Library of Congress has cataloged the single-volume edition as follows:
Hecht, Eugene.
    Physics / Eugene Hecht.
        p. cm.
    Includes index.
    ISBN 0-534-09114-8
    1. Physics.   I. Title.
QC23.H3917 1994
530—dc20

93–43106
CIP

ISBN 0-534-09130-X

Sponsoring Editor: *Lisa J. Moller*
Project Development Editor: *Suzanne Ewing*
Editorial Assistant: *Beth Wilbur*
Production Editor: *Marjorie Z. Sanders*
Production Assistant: *Tessa A. McGlasson*
Manuscript Editor: *Stuart Kenter*
Permissions Editor: *May Clark*
Interior and Cover Design: *E. Kelly Shoemaker*
Cover Photo: *Tom Skrivan*
Art Coordinator: *Susan Haberkorn*
Interior Illustration: *Precision Graphics, Carl Brown,*
    LM Graphics, Matrix Communications
Photo Coordination and Digital Photo Design: *Larry Molmud*
Photo Researcher: *Stuart Kenter*
Typesetting: *Beacon Graphics*
Cover Printing: *Phoenix Color Corporation*
Printing and Binding: *R.R. Donnelley & Sons Company/Willard*

Credits continue on p. C1.

To Ca, b. w. l.

Preface

Physics is the study of the material Universe—all there *is*. And that's a bold and wonderful agenda. The Universe is incredibly awesome and tantalizingly mysterious, and we, after all, are just beginning to understand it. Almost 3000 years in the making, physics—incomplete as it is—stands as one of the great creations of the human intellect. It's been a privilege and an unending joy to have spent much of my life studying physics, and it is out of gratitude and admiration that this book takes its form. If this work, while insightfully teaching basic physics, transmits a sense of the grandeur, unity, and vitality of the subject, it will have met my primary objectives.

Philosophy and Goals

Designed for a one-year College Physics course, this text assumes the student comes with little more than a rudimentary knowledge of algebra and just the basics of geometry. There is an appendix that will help to develop all the requisite mathematical skills. Whatever trigonometry is required is taught in place, as the need arises. Special attention is always paid to carefully and slowly developing mathematical ideas that might be new to the student—graphs, vectors, logarithms, and so forth.

The central glory of twentieth-century physics is the discovery of an overarching unity in nature. There is revealed an internal simplicity that is well confirmed, even if its complete comprehension is, as yet, just beyond our reach. All matter is composed of myriad identical clones of a small number of fundamental interacting particles. Everything physical is presumably understandable within that context. Thus, to treat the subject as if it were an encyclopedic collection of unrelated ideas is to miss the point of the twentieth century. From the beginning of this text to its end, we will study the unity of natural phenomena, the various manifestations of matter interacting with matter via the fundamental Four Forces. Even as we explore concepts conceived centuries ago, we will bring to bear the perspective of contemporary physics.

Bulk matter is atomic, and to truly comprehend its behavior (mechanical, thermal, electrical, magnetic, acoustical, optical, etc.), we must appreciate how atoms interact to produce all the phenomena of nature. It's no longer appropriate to wait until Chapter 30 and the second semester of College Physics to learn that the objects of our everyday world are composed of atoms. Nor is it necessary to teach the initial 16 chapters of the first semester as if nineteenth-century thought was still valid in all regards—it is not. We have learned a great deal during this century, and that knowledge should inform everything taught, including Classical Physics.

Pacing

The standard College Physics course makes a whole range of challenging intellectual demands on the student, and the experienced instructor knows that the first few weeks of the semester can be crucial. It is then that some students are likely to become overwhelmed. With that in mind, years ago I began to explore the benefits of rearranging the material in kinematics in order to introduce new physical concepts more gradually and allow time for the ideas to be assimilated. Thus, Chapter 2 deals only with speed, displacement, and velocity. The explanations are elaborate, and there are many examples, graphs, and illustrations. The concept of vectors is introduced as it relates to displacement. It is justified, made logically appealing, and very carefully extended and applied to velocity. With the visual aid of multi-frame drawings, the concept of vector addition is next applied to relative velocities. Only then, after the student has presumably worked out dozens of problems and has begun to learn how to learn physics, do we turn to Chapter 3, acceleration. By the time the equations of uniform acceleration are reached, the typical student is much better able to deal with the logical complexities involved.

With the mathematics of trigonometry and rudimentary vector algebra in place, and with a more realistic understanding of the demands of the experience, the serious student is ready to move ahead more rapidly.

A Modern Perspective

The discoveries of this century have radically changed our perception of every aspect of the physical Universe. And yet Classical Physics is often taught as though the last hundred years of revelation had little or no effect on our thinking. This is certainly not the case. Just consider the impact the discovery of the photon has had. If radiant energy is particulate, if the electromagnetic field is quantized, then continuous electromagnetic waves lose their traditional reality, and electric and magnetic field lines become a metaphor for an alternative corpuscular vision.

One of the most far-reaching insights of twentieth-century physics (one still rarely discussed in books at this level) is the theoretical importance of the relationship between symmetry and conservation. This relationship, which culminates in the theory of the electroweak interaction, is a leitmotif that runs throughout the text. Students are usually fascinated to learn that symmetries of space and time manifest themselves in the conservation laws. The relationship is first discussed, albeit briefly, in Chapter 1, with an introduction to the concept of symmetry. It recurs in Chapter 4 with Conservation of Linear Momentum, in Chapter 8 with Conservation of Angular Momentum, in Chapter 9 with Conservation of Energy, and so on all the way up to gauge symmetry in Chapter 33. Further, modern insights are gently introduced whenever they are relevant and as early as possible. Chapter 1, for example, briefly summarizes and considers the interrelationships of the various theoretical formalisms (Classical Physics, Relativity, and Quantum Mechanics). It is important for the student to have an overview of the discipline, to know that this is an ongoing, unending study. Chapter 1 also introduces some of the vocabulary and ideas that will be considered throughout the development. In this way, the concept of quantization will be a familiar one long before it is dealt with formally in Chapter 30. Likewise, atoms, electrons, protons, neutrons, photons, gravitons, and quarks are all at least introduced early on.

The central role played by the Four Forces of nature is an ever-present theme, and an effort is made to see everyday phenomena from this unifying perspective. Students are made aware from the outset that contact forces, friction, cohesion, adhesion, etc. are the result of electromagnetic interaction, although they will have to wait until Chapter 17 to study the details.

Contemporary, Practical Applications

A driving force shaping this text is the belief that students are more easily motivated to learn about what directly affects their lives and concerns. The perspective of this book is therefore largely practical: How do we walk? Why do we get backaches? How does an airplane fly? The text is rich in applications of physics to the life sciences, but it strives to be broadly interdisciplinary, drawing from and reflecting back upon biology, geology, astronomy, architecture, medicine, and meteorology.

Problem Solving and Self-Study

Examples

The text contains a great many example calculations systematically worked out in detail, in order to guide students through the processes of analysis and numerical problem solving. All such examples are followed by a Quick Check, which teaches a wide variety of verification methods and helps to establish the habit of checking one's computations.

Problem-Solving Techniques

Every chapter includes a section of suggestions on problem solving that may also contain approximation techniques, as well as a discussion of the pitfalls and common errors specific to the material at hand. For example, a common error is to compute the average speed of a uniformly accelerating object using $v_{av} = \frac{1}{2}(v_f - v_i)$ rather than $v_{av} = \frac{1}{2}(v_f + v_i)$.

Core Material

Every chapter includes a distilled review of the core material for that chapter.

Discussion Questions

All chapters contain a selection of discussion questions designed to develop and extend the conceptual understanding of the material.

Multiple Choice Questions

A group of roughly 20 multiple choice questions similar to those found on national medical school entrance exams is included in each chapter. Among other types, these include single-concept calculational questions, as well as pointedly probing conceptual questions.

Numerical Problems

An extensive selection of problems, most built on real data and referring to actual situations, is provided at the end of each chapter. An instructor will find a wealth of choices from which to assign homework. The problem sets are arranged in three levels of difficulty and always include a large selection of single-concept problems that explore one idea at a time (from several perspectives) to help students establish a strong foundation of competence and confidence with the basics. Each set of problems is deployed as a carefully developed and integrated pedagogical unit that will carry the student from one level of mastery to the next. All students working diligently should soon be able to do grade-I problems without much trouble. Grade-II problems require a more developed competence, and a student who is comfortable with them should do well in the course. Approximately 15% of the end-of-chapter problems are worked out succinctly at the back of the book in order to encourage and strengthen independent study. These problems are indicated by a boldface numeral. Problem solving, like playing an instrument, can only be mastered by a combination of practice and guiding example. Students should be encouraged to spend several hours a day solving problems on their own, especially those with boldface numerals for which skeletal solutions are provided. Another 10% of the solutions are worked out in a student solutions manual, which is available at the instructor's request to the bookstore. These problems are indicated by an italic numeral.

Organization

A good working text has responsibilities to both the student and the instructor. It should, as much as possible, provide a complete resource to students who are studying on their own at home. At the same time, it must provide an instructor with a broad range of conceptual material from which to fashion a course that meets his or her specific requirements.

It is impossible to lecture in class on every topic covered in a text that deals with the whole of introductory physics. The book has been organized in a way that addresses this problem and should therefore be a more effective teaching instrument. All section headings are coded in three colors: brown, black, and purple. Topics bearing   are intended to be read and understood by students working on their own prior to coming to class. This information can then be assumed as basic communal knowledge, and it thereafter requires little or no classroom time. Topics bearing **black headings** represent the basic material from which the instructor can fashion the specific course. Topics with   are enrichment sections that can be used to enhance and strengthen the primary discourse.

Units

Units play a special role in physics and so represent an issue that requires considerable attention. Without exception, the proper units to be used in physics are those of the Système International (SI). Thus, SI units will be used almost exclusively throughout this text. The commitment to the Système International must, however, be flexible enough to contend with three outstanding challenges: (1) The United States is the only major country in which there still lingers an antique system of units. Consequently, the intuitive experiences of most students are framed in terms of feet, miles, quarts, and pounds. Moreover, the data of everyday life, which must be brought into the discussion, is invariably given in U.S. Customary units. Apparently baseball pitchers hurl balls in miles per hour exclusively. (2) In order to extend our analyses into the domains of other disciplines, we must deal realistically with their idiosyncratic unit preferences. Like it or not, the pressure in the lungs is measured in centimeters of water, and for good reason. (3) There is a vast body of extant scientific data in a variety of units; the well-trained student will need to have access to that treasure.

Thus, this text will gradually, compassionately, and yet unswervingly move toward the total adoption of SI units. Even as it does, it will attempt to deal effectively with the substantive issues considered here.

Art and Photography Program

The graphics and text for this book were conceived and developed from the very start as an integrated whole; every photograph and every drawing is justified in its inclusion. The result is an unparalleled level of pedagogical effectiveness. There are about 1300 fresh, insightful, and uncompromisingly accurate illustrations that make it possible for students to visualize a diversity of physical phenomena. Many of these are multi-frame sequential drawings that allow students to apprehend the temporal unfolding of complex events (e.g., Fig's. 2.27, 2.28, 13.38, and 13.39). There are nearly 500 photographs (many of them overlaid with explanatory graphics) that form an important part of the pedagogical program.

A Complete Ancillary Package

Accompanying the text is a complete ancillary package, which includes the following instructional aids:

Instructor's Solutions Manual
    Includes answers to all discussion questions, answers to all multiple choice questions, and solutions to all problems.

Student Solutions Manual for Selected Problems
    Includes answers to selected odd discussion questions, answers to odd multiple choice questions, and solutions to selected odd problems (approximately 10%) not already solved in the book.

Transparencies
    Includes approximately 100 full-color transparencies illustrating key concepts.

Test Bank
    Includes 1000 multiple choice and short-answer questions.

Electronic Testing (EXP-Test, MathWriter)
    All test items are available in electronic format for IBM PCs and Macintosh computers.

BCX-Physics Software
    An electronic study guide for the Macintosh and IBM PC, consisting of multiple choice and fill-in-the-blank questions with hints, selected from the multiple choice problems and numerical problems in the text.

Acknowledgments

Over the several years during which this work has developed, there have been many people who have been kind enough to share their thoughts about the book with me. Accordingly, I take this opportunity to express my appreciation to:

Reviewers

William Achor
Western Maryland College

Robert Boughton
Bowling Green State University

Michael Browne
University of Idaho

Anthony Buffa
California Polytechnic State University, San Luis Obispo

Gary Buckwalter
Daytona Beach Community College

Lawrence B. Coleman
University of California, Davis

Miles Dresser
Washington State University

David Ernst
Vanderbilt University

Roger Freedman
University of California, Santa Barbara

Sherman Frye
North Virginia Community College

Robert Hallock
University of Massachusetts, Amherst

Fred Inman
Mankato State University

Gordon Jones
College of Charleston

Sanford Kern
Colorado State University

James Kettler
Ohio University Eastern Campus

Paul Lee
Louisiana State University

David Markowitz
University of Connecticut

Robert March
University of Wisconsin

Reviewers (continued)

George Matous
Indiana University of Pennsylvania

Kandula Sastry
University of Massachusetts, Amherst

Marvin Morris
San Jose State University

Jerald Tunheim
Dakota State University

Melvin Oakes
University of Texas, Austin

Lonnie Van Zandt
Purdue University

Kurt Reibel
Ohio State University

Richard Whitlock
University of North Carolina

Answer Checkers

Sol Friedman
University of California, Santa Cruz

Betty Richardson
Cornell University

Petr Gaidarev
Cornell University

Gerardo A. Rodriguez
Cornell University

Chris Keller
University of California, Berkeley

V. K. Saxena
Purdue University

Stefan Koch
Cornell University

Robin Shelton
University of Wisconsin

Leo Krzewina
University of Wisconsin

Bingxi Sun
Cornell University

Wei Liu
Adelphi University

William Tiernan
Trinity College

Scott Milster
University of Wisconsin

Ashok Tripathi
Cornell University

Sangwook Park
Purdue University

Deborah Wolkovitch
Cornell University

In particular, I thank Professor G. N. Rao of Adelphi University for all the delightful time we have spent talking about physics. Anyone wishing to exchange ideas is welcome to write to the author c/o Physics Department, Adelphi University, Garden City, NY 11530.

The team that worked so mightily to produce this book had as its production editor, Marjorie Sanders; permissions editor, May Clark; supplements editor, Audra Silverie; marketing manager, Connie Jirovsky; and advertising manager, Margaret Parks. The design and layout were brilliantly done with an uncompromising commitment to excellence by Kelly Shoemaker; the art was skillfully managed by Susan Haberkorn, and the photos artfully rendered by Larry Molmud. I thank them for bearing with me for these many months. Each can be justly proud; they have produced a beautiful book. I also thank Sue Ewing, whose clear thinking kept the project on course; Beth Wilbur, whose good counsel sustained the effort; Stuart Kenter, whose experience and patience was so valuable; and Lisa Moller, whose devotion, energy, and wisdom guided the book into being.

My appreciation is also extended to Amy Allison, Jennifer Michal, and Jamey Adam for their assistance in the preparation of every aspect of this work. Finally, I nod appreciatively to my good friend, Carolyn Eisen Hecht, for tracking down hundreds of photographs and reading thousands of pages and, of course, for going through all of this, one more time.

Eugene Hecht

Brief Contents

V O L U M E   O N E

1 An Introduction To Physics 1

2 Kinematics: Speed and Velocity 23

3 Kinematics: Acceleration 57

4 Newton's Three Laws: Momentum 89

5 Dynamics: Force and Acceleration 125

6 Equilibrium: Statics 163

7 Gravity, According to Newton 207

8 Rotational Motion 239

9 Energy 277

10 Solids 317

11 Fluids 353

12 Oscillations and Waves 401

13 Sound 441

14 Thermal Properties of Matter 481

15 Heat and Thermal Energy 513

16 Thermodynamics 549

V O L U M E   T W O

17 Electrostatics: Forces 587

18 Electrostatics: Energy 625

19 Direct Current 661

20 Circuits 689

21 Magnetism 721

22 Electromagnetic Induction 763

23 AC and Electronics 795

24 Radiant Energy: Light 833

25 The Propagation of Light: Scattering 863

26 Geometrical Optics and Instruments 893

27 Physical Optics 939

28 Special Relativity 979

29 The Origins of Modern Physics 1015

30 The Evolution of Quantum Theory 1043

31 Quantum Mechanics 1073

32 Nuclear Physics 1101

33 High-Energy Physics 1137

Contents

1 An Introduction to Physics   1

1.1   Law and Theory   2
1.2   The Modern Perspective   3

MEASUREMENT   8
1.3   Length   8
1.4   Mass and Weight   11
1.5   Time   13
1.6   Significant Figures   13

THE LANGUAGE OF PHYSICS   16
1.7   Equations   16
1.8   Graphs   17
1.9   Approximations and Checks   17

Core Material   19
Suggestions on Problem Solving   19
Discussion Questions   20
Multiple Choice Questions   20
Problems   21

**2 Kinematics:
Speed and Velocity   23**

SPEED   24
2.1   Average Speed   24
2.2   Constant Speed   27
2.3   Delta Notation: The Change in
         a Quantity   28
2.4   Instantaneous Speed   29

VELOCITY   31
2.5   The Displacement Vector   32
2.6   Some Vector Algebra   32
2.7   Average Velocity   35
2.8   Instantaneous Velocity   36

RELATIVE MOTION   39
2.9   Velocity with Respect to . . .   40
2.10 Vector Components   44

Core Material   50
Suggestions on Problem Solving   50
Discussion Questions   51
Multiple Choice Questions   52
Problems   53

3 Kinematics: Acceleration   57

THE CONCEPT OF ACCELERATION   58
3.1   Average Acceleration   58
3.2   Instantaneous Acceleration   60

**UNIFORMLY ACCELERATED
MOTION   62**
3.3   The Mean Speed   63
3.4   Integrals and Intervals   65
3.5   The Equations of Constant
         Acceleration   66

FREE-FALL   69
3.6   Acceleration Due to Gravity   71
3.7   All Fall Down: Constant Acceleration   73
3.8   Projectile Motion   75

Core Material   82
Suggestions on Problem Solving   83
Discussion Questions   83
Multiple Choice Questions   84
Problems   85

**4 Newton's Three Laws:
Momentum   89**

THE LAW OF INERTIA   90
4.1   Ballistic Motion   92
4.2   Force   94

THE SECOND LAW   100
4.3   Mass   100

4.4 Linear Momentum   101
4.5 Impulse and Momentum Change   102
4.6 Varying Force   106
   INTERACTION   108
4.7 Rockets   111
4.8 Conservation of Linear Momentum   112

Core Material   117
Suggestions on Problem Solving   117
Discussion Questions   118
Multiple Choice Questions   118
Problems   120

5 Dynamics: Force and Acceleration   125

FORCE, MASS, AND ACCELERATION   126
5.1 Instantaneous Motion   128
5.2 Weight: Gravitational Force   130
5.3 Vertical Accelerations   134
5.4 Effectively Weightless   136
5.5 Coupled Motions   138
CURVILINEAR MOTION   139
5.6 Centripetal Acceleration   140
5.7 Center-Seeking Forces   141
5.8 Circular Motion   143
FRICTION AND MOTION   146
5.9 Static Friction   146
5.10 Moving via Static Friction   150
5.11 Kinetic Friction   152
5.12 Rolling Friction   154
5.13 The Causes of Friction   154

Core Material   156
Suggestions on Problem Solving   157
Discussion Questions   157
Multiple Choice Questions   158
Problems   159

6 Equilibrium: Statics   163

TRANSLATIONAL EQUILIBRIUM   164
6.1 Parallel and Colinear Force Systems   164
6.2 Concurrent Force Systems   167
ROTATIONAL EQUILIBRIUM   173
6.3 Torque   174
6.4 Rigid Bodies   178

EQUILIBRIUM OF EXTENDED BODIES   184
6.5 Center-of-Gravity   184
6.6 Stability and Balance   188
6.7 Structures, Walls, and Arches   191

Core Material   193
Suggestions on Problem Solving   194
Discussion Questions   194
Multiple Choice Questions   195
Problems   197

7 Gravity, According to Newton   207

THE LAW OF UNIVERSAL GRAVITATION   208
7.1 Evolution of the Law   208
7.2 Newton's Confirmation of $1/r^2$ 210
7.3 The Product of the Masses   211
7.4 Gravitational Constant   212
7.5 Terrestrial Gravity   213
7.6 Earth's Density   214
7.7 Gravity of a Sphere   215
7.8 An Imperfect Spinning Earth   216
THE COSMIC FORCE   218
7.9 The Laws of Planetary Motion   218
7.10 Gravity and Kepler's Laws   220
7.11 Satellite Orbits   224
7.12 The Gravitational Field   227
7.13 New Directions   229

Core Material   231
Suggestions on Problem Solving   231
Discussion Questions   231
Multiple Choice Questions   233
Problems   234

8 Rotational Motion 239

THE KINEMATICS OF ROTATION 241
8.1 Angular Speed   243
8.2 Angular Acceleration   246
8.3 Equations of Constant Acceleration   247

THE DYNAMICS OF ROTATION 250
8.4 Rotational Inertia   250
8.5 The Center-of-Mass   254
8.6 Rolling Down an Incline   256
8.7 Angular Momentum   257
8.8 Conservation of Angular
        Momentum   259

Core Material   268
Suggestions on Problem Solving   269
Discussion Questions   269
Multiple Choice Questions   271
Problems   272

9 Energy 277

THE TRANSFER OF ENERGY 278
9.1 Work   278
9.2 Power   286

MECHANICAL ENERGY 288
9.3 Kinetic Energy   288
9.4 Potential Energy   291
9.5 Internal Energy   296

**CONSERVATION OF MECHANICAL
ENERGY 297**
9.6 Mechanical Energy   297
9.7 The Energy Perspective   299
9.8 Collisions   303
9.9 Energy Conservation and Symmetry   308

Core Material   309
Suggestions on Problem Solving   310
Discussion Questions   311
Multiple Choice Questions   312
Problems   313

10 Solids 317

ATOMS AND MATTER   318
10.1 Atomism   319
10.2 Avogadro's Number   319
10.3 Density   320
10.4 The Size of Atoms   322
10.5 The States of Matter   323
10.6 Solids   324

ELASTICITY   327
10.7 Hooke's Law   327
10.8 Stress   332
10.9 Strain   335
10.10 Strength   337
10.11 Fatigue   340

MODULI AND ENERGY   341
10.12 Young's Modulus   341
10.13 Shear and Bulk Moduli   343

Core Material   346
Suggestions on Problem Solving   347
Discussion Questions   347
Multiple Choice Questions   348
Problems   349

11 Fluids 353

FLUID STATICS   354
11.1 Liquids   354
11.2 Gases   354
11.3 Hydrostatic Pressure   355
11.4 Boyle's Law   364
11.5 Pascal's Principle   365
11.6 Buoyant Force   368
11.7 Specific Gravity   369
11.8 Surface Tension   371

FLUID DYNAMICS 374
11.9 Fluid Flow 374
11.10 The Continuity Equation 377
11.11 Bernoulli's Equation 379
11.12 Airfoils 384
11.13 Vortices 385
11.14 Blood Pressure 386

Core Material 389
Suggestions on Problem Solving 390
Discussion Questions 390
Multiple Choice Questions 392
Problems 394

12 Oscillations and Waves 401

HARMONIC MOTION 402
12.1 Simple Harmonic Motion 402
12.2 Elastic Restoring Force 408
12.3 Gravitational Restoring Force:
　　　　Pendulums 412
12.4 Damping, Forcing, and Resonance 415
12.5 Self-Excited Vibrations 418

MECHANICAL WAVES 420
12.6 Wave Characteristics 420
12.7 Transverse Waves: Strings 425
12.8 Compression Waves 429

Core Material 433
Suggestions on Problem Solving 434
Discussion Questions 434
Multiple Choice Questions 435
Problems 436

13 Sound 441

ACOUSTICS 442
13.1 Sound Waves 442
13.2 The Superposition of Waves 444
13.3 Wavefronts and Intensity 447
13.4 The Speed of Sound 449
13.5 Hearing Sound 451
13.6 Intensity-Level 454
**PRODUCTION AND PROPAGATION
OF SOUND 457**
13.7 Beats 458
13.8 Standing Waves 458
13.9 Sound Waves 465
13.10 The Doppler Effect 469

Core Material 473
Suggestions on Problem Solving 474
Discussion Questions 474
Multiple Choice Questions 476
Problems 477

**14 Thermal Properties of
Matter 481**

TEMPERATURE 482
14.1 Thermodynamic Temperature and
　　　　Absolute Zero 484
14.2 The Range of Temperatures 485
THERMAL EXPANSION 487
14.3 Linear Expansion 487
14.4 Volumetric Expansion:
　　　　Solids and Liquids 490
THE GAS LAWS 494
14.5 The Ideal Gas Law 496
14.6 Real Gases: Liquefaction 498
14.7 Phase Diagrams 500
14.8 Kinetic Theory 502

Core Material 507
Suggestions on Problem Solving 507
Discussion Questions 508
Multiple Choice Questions 509
Problems 510

15 Heat and Thermal Energy 513

THERMAL ENERGY 514
15.1 Caloric versus Kinetic 514
15.2 Heat and Temperature 515
15.3 Quantity of Heat 518
15.4 The Mechanical Equivalent of
　　　　Heat 520

15.5   The Zeroth Law   521
15.6   Specific Heat Capacity   521
       CHANGE OF STATE   528
15.7   Melting and Freezing   528
15.8   Vaporization   530
15.9   Boiling   532
       **THE TRANSFER OF
       THERMAL ENERGY   534**
15.10 Radiation   534
15.11 Convection   536
15.12 Conduction   537

       Core Material   540
       Suggestions on Problem Solving   541
       Discussion Questions   541
       Multiple Choice Questions   543
       Problems   545

16 Thermodynamics   549

       **THE FIRST LAW OF
       THERMODYNAMICS   550**
16.1   Conservation of Energy   550
       **CYCLIC PROCESSES: ENGINES
       AND REFRIGERATORS   561**
16.2   The Carnot Cycle   562
16.3   Internal Combustion   566
16.4   Refrigeration Machines and
           Heat Pumps   566
16.5   Absolute Temperature   568
       **THE SECOND LAW OF
       THERMODYNAMICS   570**
16.6   The Clausius Formulation   571
16.7   Entropy   572
16.8   Order and Disorder   576
16.9   Entropy as Probability   577

       Core Material   579
       Suggestions on Problem Solving   580
       Discussion Questions   580
       Multiple Choice Questions   581
       Problems   583

17 Electrostatics: Forces   587

       ELECTROMAGNETIC CHARGE   588
17.1   Positive and Negative Charge   588
17.2   Charge Is Quantized and Conserved   589
17.3   Charging by Rubbing   590
17.4   The Transfer of Charge   591

17.5   Insulators and Conductors   592
       THE ELECTRIC FORCE   595
17.6   Coulomb's Law   595
17.7   Electrostatic Induction   600
       THE ELECTRIC FIELD   602
17.8   Rationalized Units   607
17.9   Field Lines   609
17.10 Conductors and Fields   610
17.11 Gauss's Law   613
17.12 Quantum Field Theory   617

       Core Material   618
       Suggestions on Problem Solving   619
       Discussion Questions   619
       Multiple Choice Questions   621
       Problems   622

18 Electrostatics: Energy   625

       ELECTRIC POTENTIAL   626
18.1   *Electrical*-PE   626
18.2   Electric Potential   626
18.3   Potential in a Uniform Field   631
18.4   Potential of a Point-Charge   632
18.5   Equipotentials   634
18.6   The Potential of Several Charges   635
18.7   Potential and the Distribution of
           Charge   636
18.8   Conservation of Charge   638
       CAPACITANCE   639
18.9   The Parallel-Plate Capacitor   641
18.10 Capacitors in Combination   643
18.11 Energy in Capacitors   648

       Core Material   651
       Suggestions on Problem Solving   651
       Discussion Questions   652
       Multiple Choice Questions   653
       Problems   654

19   Direct Current   661

CURRENTS   662
19.1   The Battery   664
19.2   Electric Fields and the Drift
            Velocity   669

RESISTANCE   671
19.3   Resistivity   673
19.4   Superconductivity   677
19.5   Voltage Drops and Rises   678
19.6   Energy and Power   679

Core Material   681
Suggestions on Problem Solving   682
Discussion Questions   682
Multiple Choice Questions   684
Problems   685

20   Circuits   689

CIRCUIT PRINCIPLES   690
20.1   Sources and Internal Resistance   691
20.2   Resistors in Series and Parallel   693
20.3   Maximum Power and Impedance   698
20.4   Ammeters and Voltmeters   699
20.5   *RC*-Circuits   700

NETWORK ANALYSIS   702
20.6   Kirchhoff's Rules   702

Core Material   707
Suggestions on Problem Solving   707
Discussion Questions   708
Multiple Choice Questions   711
Problems   712

21   Magnetism   721

MAGNETS AND THE MAGNETIC FIELD   722
21.1   Poles   722

21.2   The Magnetic Field   724
21.3   The Earth's Magnetism   726
21.4   Monopoles   727
21.5   Magnetism on an Atomic Level   727

ELECTRODYNAMICS   730
21.6   Currents and Fields   731
21.7   Confirming Ampère's Hypothesis   737
21.8   Calculating *B*-Fields   739

MAGNETIC FORCE   741
21.9   The Trajectory of a Free Particle   743
21.10 Forces on Wires   746

Core Material   752
Suggestions on Problem Solving   752
Discussion Questions   753
Multiple Choice Questions   756
Problems   759

22   Electromagnetic Induction   763

ELECTROMAGNETICALLY INDUCED EMF   764
22.1   Faraday's Induction Law   765
22.2   Lenz's Law   768
22.3   Motional Emf   770
22.4   Induced Magnetic and Electric
            Fields   773

GENERATORS   774
22.5   The AC Generator   774
22.6   The DC Generator   776
22.7   Eddy Currents   778

SELF-INDUCTION   779
22.8   Inductance   779
22.9   The Back-Emf   780
22.10 The *R-L* Circuit: Transients   782
22.11 Energy in the Magnetic Field   783

Core Material   785
Suggestions on Problem Solving   785
Discussion Questions   785
Multiple Choice Questions   788
Problems   790

23   AC and Electronics   795

ALTERNATING CURRENT   796
23.1   AC and Resistance   798
23.2   AC and Inductance   800
23.3   AC and Capacitance   803

L-C-R AC NETWORKS   805
23.4   Series Circuits   805

23.5 The Transformer   812
23.6 Domestic Circuits and Hazards   815

ELECTRONICS   818
23.7 Semiconductors   818
23.8 The *pn*-Junction and Diodes   819
23.9 Transistors   822

Core Material   825
Suggestions on Problem Solving   825
Discussion Questions   826
Multiple Choice Questions   828
Problems   829

24 Radiant Energy: Light   833

THE NATURE OF LIGHT   834
24.1 Waves and Particles   834
24.2 Electromagnetic Waves   835
24.3 Waveforms and Wavefronts   837
24.4 The Speed of Propagation: c   840
24.5 Energy and Irradiance   842
24.6 The Origins of EM Radiation   843
24.7 Energy Quanta   845
24.8 Atoms and Light   846
24.9 Scattering and Absorption   847

THE ELECTROMAGNETIC-PHOTON SPECTRUM   848
24.10 Radiowaves   848
24.11 Microwaves   850
24.12 Infrared   851
24.13 Light   852
24.14 Ultraviolet   854
24.15 X-rays   856
24.16 Gamma Rays   857

Core Material   857
Suggestions on Problem Solving   858
Discussion Questions   858
Multiple Choice Questions   859
Problems   860

25 The Propagation of Light: Scattering   863

SCATTERING   864
25.1 Rayleigh Scattering: Blue Skies   864
25.2 Scattering and Interference   865
25.3 The Transmission of Light through Dense Media   866

REFLECTION   867
25.4 Internal and External Reflection   868
25.5 The Law of Reflection   869
25.6 The Plane Mirror   871

REFRACTION   873
25.7 The Index of Refraction   874
25.8 Snell's Law   876
25.9 Total Internal Reflection   879

THE WORLD OF COLOR   882
25.10 White, Black, and Gray   882
25.11 Colors   883

Core Material   885
Suggestions on Problem Solving   886
Discussion Questions   886
Multiple Choice Questions   887
Problems   888

26 Geometrical Optics and Instruments   893

LENSES   894
26.1 Aspherical Surfaces   895
26.2 Spherical Thin Lenses   897
26.3 Focal Points and Planes   901
26.4 Finite Imagery: Lenses   903
26.5 Magnification   905
26.6 A Single Lens   907
26.7 Thin-Lens Combinations   916

MIRRORS   925
26.8 Aspherical Mirrors   925
26.9 Spherical Mirrors   926
26.10 Finite Imagery: Mirrors   929

Core Material   932
Suggestions on Problem Solving   932
Discussion Questions   933
Multiple Choice Questions   934
Problems   935

27 Physical Optics 939

POLARIZATION 940
27.1 Natural Light 940
27.2 Polarizers 942
27.3 Polarizing Processes 946

INTERFERENCE 950
27.4 Coherence 951
27.5 Young's Experiment 952
27.6 Thin Film Interference 956
27.7 The Michelson Interferometer 959

DIFFRACTION 960
27.8 Single-Slit Diffraction 963
27.9 The Diffraction Grating 965
27.10 Circular Holes and Obstacles 967
27.11 Holography 969

Core Material 971
Suggestions on Problem Solving 972
Discussion Questions 972
Multiple Choice Questions 974
Problems 975

28 Special Relativity 979

BEFORE THE SPECIAL THEORY 980
28.1 The Michelson-Morley Experiment 980

THE SPECIAL THEORY OF RELATIVITY 982
28.2 The Two Postulates 983
28.3 Simultaneity and Time 985
28.4 The Hatter's Watch: Time Dilation 990
28.5 Shrinking Alice 993
28.6 The Twin Effect 995
28.7 Wonderland: Space-Time 997
28.8 Addition of Velocities 998

RELATIVISTIC DYNAMICS 1000
28.9 Relativistic Momentum 1000
28.10 Relativistic Energy 1003

Core Material 1008
Suggestions on Problem Solving 1009
Discussion Questions 1009

Multiple Choice Questions 1011
Problems 1012

29 The Origins of Modern Physics 1015

SUBATOMIC PARTICLES 1016
29.1 The Quantum of Charge 1016
29.2 Cathode Rays: Particles of Charge 1017
29.3 X-Rays 1022
29.4 The Discovery of Radioactivity 1025

THE NUCLEAR ATOM 1028
29.5 Rutherford Scattering 1028
29.6 Atomic Spectra 1031
29.7 The Proton 1032
29.8 The Neutron 1034

Core Material 1036
Suggestions on Problem Solving 1036
Discussion Questions 1037
Multiple Choice Questions 1038
Problems 1039

30 The Evolution of Quantum Theory 1043

THE OLD QUANTUM THEORY 1044
30.1 Blackbody Radiation 1044
30.2 Quantization of Energy: The Photoelectric Effect 1049
30.3 Bremsstrahlung 1054
30.4 The Compton Effect 1055

ATOMIC THEORY 1056
30.5 The Bohr Atom 1057
30.6 Stimulated Emission: The Laser 1062
30.7 Atomic Number 1065

Core Material 1066
Suggestions on Problem Solving 1067
Discussion Questions 1067
Multiple Choice Questions 1069
Problems 1070

31 Quantum Mechanics 1073

THE CONCEPTUAL BASIS OF QUANTUM MECHANICS 1074
31.1 De Broglie Waves 1074

31.2 The Principle of Complementarity 1076
31.3 The Schrödinger Wave Equation 1076

QUANTUM PHYSICS 1081
31.4 Quantum Numbers 1081
31.5 The Zeeman Effect 1082
31.6 Spin 1083
31.7 The Pauli Exclusion Principle 1084
31.8 Electron Shells 1084
31.9 The Uncertainty Principle 1089
31.10 QED and Antimatter 1092

Core Material 1096
Suggestions on Problem Solving 1096
Discussion Questions 1096
Multiple Choice Questions 1098
Problems 1099

32 Nuclear Physics 1101

NUCLEAR STRUCTURE 1102
32.1 Isotopes: Birds of a Feather 1102
32.2 Nuclear Size, Shape, and Spin 1104
32.3 The Nuclear Force 1107
32.4 Nuclear Stability 1109

NUCLEAR TRANSFORMATION 1113
32.5 Radioactive Decay 1113
32.6 The Weak Force 1118
32.7 Gamma Decay 1119
32.8 Half-Life 1119
32.9 Induced Radioactivity 1122
32.10 Fission and Fusion 1123

Core Material 1131
Suggestions on Problem Solving 1132
Discussion Questions 1132
Multiple Choice Questions 1133
Problems 1134

33 High-Energy Physics 1137

ELEMENTARY PARTICLES 1138
33.1 Leptons 1139
33.2 Hadrons 1140

QUANTUM FIELD THEORY 1142
33.3 Gauge Theory 1145
33.4 Quarks 1148
33.5 Quantum Chromodynamics 1150
33.6 The Electroweak Force 1152
33.7 GUTs and Beyond:
       The Creation of the Universe 1156

Core Material 1159
Suggestions on Problem Solving 1159
Discussion Questions 1160
Multiple Choice Questions 1162
Problems 1162

Appendixes: A Mathematical Review A-1

APPENDIX A   ALGEBRA   A-1
A-1 Exponents A-2
A-2 Powers of Ten: Scientific Notation A-3
A-3 Logarithms A-4
A-4 Proportionalities and Equations A-5
A-5 Approximations A-6
A-6 The Change in a Quantity A-7

APPENDIX B   GEOMETRY   A-8

APPENDIX C   TRIGONOMETRY   A-10

APPENDIX D   VECTORS   A-11

APPENDIX E   DIMENSIONS   A-13

Answer Section   AN-1

Credits   C-1

Index   I-1

NOTE: The Table of Contents is color-coded as per p. x in the Preface.

1

An Introduction to Physics

PHYSICS IS THE STUDY of the physical Universe. And though we all know what the words *physical Universe* refer to, it would take volumes to attempt to explain what they mean. The problem is that the basic underlying concepts—matter, space, and time—are difficult if not impossible to define conceptually, a shortcoming we share with the philosophers.

All the material things we are familiar with, from stars to toenails, are structures composed of ever smaller structures. Ultimately, the smallest indivisible subatomic grains of matter are the fundamental *real particles*, which compose all that exists. The primary property of matter is that it is observable; it interacts in ways that result in changes we can detect. **Physics, then, is the study of matter, interaction, and change**.

Physicists try to describe phenomena in the simplest, most precise way, an approach that leads to a language of exact terminology. There are, for instance, concepts in physics that correspond to basic properties of matter (such as mass, charge, momentum, and energy); that describe the location of matter in space and time (such as displacement, velocity, and acceleration); or that pertain to the behavior of matter in bulk (such as heat, current, resistance, and pressure). All of these concepts have very specific scientific meanings.

In addition to conceptual or dictionary-like definitions, physicists rely on *operational definitions* to augment the working language of the discipline. These definitions are procedural, and include actual things: a standard mass, a thermometer, a stream of cesium atoms, and so on. Operational definitions demarcate a quantity by telling us how to measure it. For example, a *second* is the interval of time corresponding to exactly 9 192 631 770 oscillations of a cesium-133 atom. Not being able to conceptually define time in a completely satisfactory way doesn't prevent us from thinking about it, or from being able to measure it (or even from growing old).

The natural sciences, as distinct from the social sciences, have nature as their subject; that is, material phenomena. Physics, the premier natural science, deals with composition, structure, shape, creation, annihilation, motion, light and sound, atoms and molecules, fission and fusion, solids and liquids and gases—with the whole of things physical. Its purview is all that can be observed, all that can be measured.

1.1 Law and Theory

The aim of physics is to understand the natural events that we are party to, to understand the Universe—what it is, how it works, what it is doing, and maybe even why it exists. If that bold agenda is possible at all, it is only because natural phenomena occur in reproducible ways; there are rules and rhythms in the seeming chaos.

The doing of physics usually involves a record of phenomena: observation and the collection of **data**—information objectively perceived. An event is observed, either deliberately or not, and things are recorded, *measured* (How much? How long? How many? How big?). **Physics quantifies; it associates numbers with its concepts**. Within the body of observations, patterns are sought that reveal relationships among the data. A **law** is a description of a relationship in nature that manifests itself in recurring patterns of events. It can be a prescription for how things change, or it might be a statement of how things remain invariant to change.

How does our basic underlying knowledge, our world view, of space and time, of matter interacting with matter, account for the patterns of events? How can we understand what is observed? **Theory** is the explanation of phenomena in terms of more basic natural processes and relationships. To explain phenomena, we draw on intuition and imagination and guess at what is happening. We propose **hypotheses**, and leap beyond what we know to what might be. *A construct of definitions, hypotheses, and laws that explains some observed order in nature is the essence of theory.* A powerful theory allows us to deduce already known laws, as well as to predict new occurrences and relationships that, once tested and confirmed, may become new laws.

All theory is tentative. The formalisms of physics cannot be proven absolutely true, or even absolutely false. Our current world view may be wrong and all the proofs based on it equally wrong, our hypotheses may be in error, our laws may only be approximate. Newtonian Mechanics worked incredibly well for 200 years before it began to deviate even slightly from the most refined observations. Then Albert Einstein proved that the world view on which it was based is wrong; Newtonian Mechanics is a brilliant approximation of a more complete truth.

Physics must be continuously tested against nature. Our ideas must correspond in every detail to the observed workings of the Universe, but even that does not prove them true. The testing never ends. The process goes on—checked and rechecked—predictions, confirmations, discrepancies, refinements, new visions, new laws, new theories, evermore powerful understandings.

At this moment, there is no single significant theory in physics that can be said to be both finished and completely satisfactory. We have a magnificent understanding, but it's only the beginning. The game is wonderfully wide open and the real surprise, as Einstein suggested, is that we can do so much and know so little.

Insofar as the propositions of mathematics refer to reality they are not certain, and insofar as they are certain they do not refer to reality.

ALBERT EINSTEIN

1.2   The Modern Perspective

Physics has evolved to its present state as a result of about 2500 years of effort. Simple early theories were displaced by increasingly more effective later ones, and these, in turn, were subsumed by still more powerful, far-reaching, and complex contemporary treatments. A theory that has functioned well is assumed to possess some measure of truth, and so each new formulation must encompass the wisdom of its predecessors.

Classical Physics

The discipline, as it developed up until the 1920s, is known as Classical Physics. The classical period rests on three theoretical pillars (Fig. 1.1): Newtonian Mechanics, Electromagnetic Theory, and Thermodynamics. Within its boundaries are also several subdisciplines, such as acoustics, wave motion, condensed matter physics, and so on. The completion of Classical Physics was accomplished by Einstein. His Special Theory of Relativity (1905) reformulates our conception of space, time, and motion. His General Theory of Relativity (1915) considers gravity in terms of curved space-time. It subsumes the Special Theory, and contains and transcends Newton's Theory of Gravitation. The General Theory describes things on a grand scale, allowing us to begin to comprehend such cosmic phenomena as gravitational lenses, black holes, pulsars, and the Big Bang creation of the Universe.

Classical Physics produced a wonderful but limited picture of the physical world. Indeed, humanity was able to set foot on the Moon in 1969 thanks largely to Newton's theory. Except for the electronics, which would never have existed without our modern knowledge, the journey to the Moon was a classical trip. Classical Physics represents the basic conceptual material that should be understood if one is to interact effectively with the physical environment on an everyday macroscopic level.

Philosophically, Classical Physics is *deterministic*—it maintains that things can be known with certainty, that immutable law constrains nature to unfold in a totally predictable way. The cosmos is thus viewed as a great machine

Figure 1.1   Classical Physics encompasses Newtonian Mechanics (the study of motion and gravity), Electromagnetic Theory (the synthesis of electricity and magnetism), Thermodynamics (the study of thermal energy), and Relativity. It corresponds to physics prior to the 1920s.

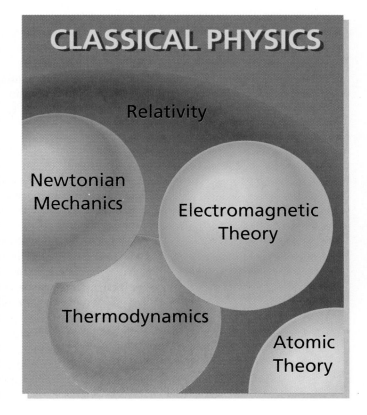

that grinds along according to an intricate internal order. Once that order is known, we can, in principle at least, compute both the future and the past.

Physicists were supremely confident throughout the last decades of the 1800s, an era dominated by breakthroughs in early experimental atomic physics when everyone expected to successfully apply established theory to the atom. Surprisingly, as time went by, the atomic domain seemed to become more and more *inaccessible* to classical analysis.

Modern Physics

The reality of the atom was finally established only in the beginning of the twentieth century, in large measure due to the work of Einstein. Moreover, in direct conflict with classical theory, Einstein showed that light was emitted and absorbed in minute blasts, or *quanta.* These particles of light, the most tenuous form of matter, are called *photons.* Today, the material Universe is considered to be composed of myriads of identical clones of a handful of elementary particles—the *quarks, leptons,* and photons. All of us are basically organized clouds of quarks surrounded by leptons (electrons), absorbing and emitting photons.

Contemporary physics has extended its domain from the subnuclear to the entire Universe. Though it seems presumptuous, Genesis has become a branch of Modern Physics. Particle accelerators can replicate conditions that existed in the first moments of creation. All of this, and more, are the fruits of Quantum Mechanics. We know now that physical processes, although they appear to be continuous on a macroscopic level, are fundamentally discrete. The entities of the microworld exist and move in ways that are distinctly restricted. The continuum of motion is an illusion. *Change occurs in jumps;* the landscape is quantized.

Philosophically, it can be argued that Quantum Mechanics embraces determinism, but it certainly is not the all-knowing determinism of Classical Physics. The Universe may be a great machine, but its tiniest parts are mysteriously elusive. We cannot possibly know everything we would like to know. There is a soft edge to certainty, and the future has yet to be written. Despite its baffling text and strange version of reality, Quantum Mechanics has never once failed an experimental test. It is extremely reliable, though not transparently comprehendible. It is likely true that "no one understands Quantum Mechanics" (Sect. 31.9), although it is equally true that in some wonderful way Quantum Mechanics understands the Universe.

The Four Forces

Matter, whatever it is, interacts. The Universe is laced with a web of interactions that has evolved since the moment of its creation. Contemporary theoretical physics maintains that in the world as it exists today, there are four distinct basic interactions, or forces: gravitational, electromagnetic, strong, and weak. These influence the entire range of observable occurrences (Table 1.1).

When a change in the state of something occurs, we assume a cause and logically presume that the change is effectuated by a force. When an apple falls, or a grasshopper leaps, or a supernova explodes, or a neutron decays, these very different events unfold because of the involvement of the Four Forces. Thus, we define **force** broadly as *the agent of change.*

The **gravitational force** keeps you, the atmosphere, and the seas fixed to the surface of the planet. Though gravity is the weakest of all the interactions, it is also the least selective—it acts between all particles. Because its range is unlimited and because it is only attractive, gravity rules the cosmos on a grand scale. It holds

Light is, in short, the most refined form of matter.

LOUIS DE BROGLIE (1892–1987)
French physicist

To be sure, we say that an area of space is free of matter; we call it empty, if there is nothing present except a gravitational field. However, this is not found in reality, because even far out in the universe there is starlight, and that is matter.

ERWIN SCHRÖDINGER (1887–1961)
Austrian physicist

the Earth in orbit around the Sun, keeps the Sun locked within our galaxy of a hundred thousand million stars, and reaches all across the thousands of millions of galaxies that constitute the Universe. If it were much stronger than it is, it would quickly halt the present expansion of the Universe and send all the galaxies collapsing back down into oblivion (it may even be strong enough, as is, to inevitably do that anyway).

The **electromagnetic force** binds together the smaller things, such as atoms, molecules, trees, buildings, and you. As with the gravitational force, we can perceive its consequences directly in the macroworld (just hold two magnets near one another). The electromagnetic force acting on a microscopic level is responsible for a variety of seemingly different macroscopic forces. It produces contact forces, as for example, between a fist and a punching bag, or a hamburger and teeth. It generates friction and drag, produces adhesion and cohesion, and is responsible for the elastic force. The range of the electromagnetic force is also unlimited, but it can be either attractive or repulsive, and that fact inhibits its influence over great distances. Electromagnetism governs chemistry and biology, rules life and death, and keeps the Earth and everything on it from being crushed by gravity.

This galaxy is one of perhaps one hundred thousand million in the Universe, and it resembles our own Milky Way galaxy. Our galaxy is an island of about one hundred thousand million stars, as well as dust and gas, all pulled together by gravity. The Sun is but one rather ordinary star, around which orbit nine tiny planets, also held in place by the gravitational force.

The **strong force** binds quarks together to form neutrons and protons, and binds these together to form the nuclei of atoms (Fig. 1.2). It's an extremely powerful, exceedingly short-range force whose influence extends about as far as the diameter of a proton. That tiny range accounts for its not obtruding directly into our normal experience, and, as a result, not being discovered until this century. Without it, familiar matter, from planets to puppies, would all disintegrate into a fine quark dust.

The **weak force** is a million times fainter than the strong force and a hundred times shorter in range. Among other things, it transforms one flavor or type of quark into another. Thus, it can change protons into neutrons and is thereby responsible for the slow decay of certain radioactive atoms, such as uranium. A star like the Sun derives its energy from the thermonuclear furnace at its core. There, it "burns" hydrogen into helium, a process that relies on the gradual transformation of protons via the weak force. Hence, no weak force, no sunshine, no life on Earth.

Most physicists believe that the Four Forces are quantized. They do not act as continuous streamers, or invisible tentacles extending from one interacting body

TABLE 1.1 The Four Forces

Force	Couples with	Strength*	Range
Strong	Quarks and particles composed of them	10^4	$\approx 10^{-15}$ meter
Electromagnetic	Electrically charged particles	10^2	Unlimited
Weak	Most particles	10^{-2}	$\approx 10^{-17}$ meter
Gravitational	All particles	10^{-34}	Unlimited

*Strengths listed are the forces (in newtons) between two protons separated by a distance equal to their diameter (≈ 2 femtometers). See Appendix A-2 for scientific notation.

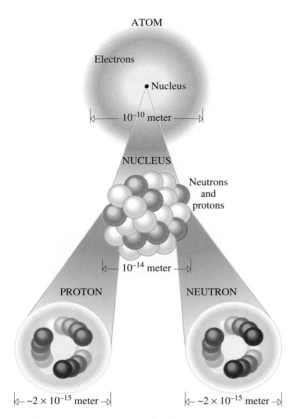

ATOM

Electrons

• Nucleus

|←——— 10^{-10} meter ———→|

NUCLEUS

Neutrons
and
protons

|←—— 10^{-14} meter ——→|

PROTON   NEUTRON

|←– ~2 × 10^{-15} meter –→| |←– ~2 × 10^{-15} meter –→|

Figure 1.2 An atom is composed of a tiny nucleus surrounded by a cloud of fast-moving electrons. The nucleus is a minute cluster of neutrons and protons. Each neutron and proton consists of three quarks.

Figure 1.3 A uniform disc mounted on a central axle displays rotational symmetry.

to the other. Instead, forces are caused by the back-and-forth exchange of a special class of force-mediating particles. These *virtual particles* only exist as transmitters of force and cannot be otherwise observed. This line of thinking has been quite successful, though it remains tantalizingly incomplete and philosophically problematic.

For the moment, the greatest impasse experienced by contemporary physics involves the inability to bring into harmony two fundamental theories: General Relativity (primarily its vision of gravity) and Quantum Mechanics. Relativity is an overview grounded in the analysis of space and time. All theory should be relativistically correct. Quantum Mechanics is an overview grounded in the analysis of energy and momentum, in the way change occurs. All theory should be quantum-mechanically correct. As yet, there is no satisfactory blending of the two. There is no adequate quantum theory of gravity. If researchers succeed, we will be well on the road to what is brashly called the Theory of Everything (Sect. 33.7).

Symmetry

Physics in the last quarter of this century has come to a remarkable revelation: on a fundamental level, nature possesses a range of symmetries that are associated with the most basic laws of physics. By understanding these symmetries, we can produce theories that presumably have the correct internal logical structure. Physicists now believe that for a basic theoretical construct to be right, it must have a subtle mathematical form called *gauge symmetry*. Such an incredible statement could not have been made a hundred years ago, nor could scientists have uttered it with much assurance even three or four decades back. Our confidence derives from an amazing series of discoveries that culminated in 1983.

Gauge symmetry was introduced as an abstract mathematical idea in the 1950s. By the mid-1960s, theoreticians were attempting to use it to produce a theory of the weak force. Gradually, the work began to suggest that the weak and electromagnetic interactions were actually two aspects of the same phenomenon. The treatment predicted the existence of three extraordinarily massive new particles. Although experiments in the 1970s confirmed a series of other predictions based on the theory, no laboratory in the world had equipment powerful enough to create the trio of hypothetical particles. When that feat was finally accomplished with a new generation of colliding-beam machines in 1983, the three particles created were found to have exactly the charges and almost exactly the masses predicted. Gauge symmetry was firmly established. We discuss symmetries and trace the development of these profound ideas throughout this text.

Noether's Principle

There are many kinds of symmetry, the simplest of which is geometrical. Imagine a perfectly uniform disc mounted on a central axle (Fig. 1.3). Suppose you close your eyes and someone rotates the disc. When you open your eyes, everything is as it was before—you can't tell that anything has occurred. *When a change in a physical*

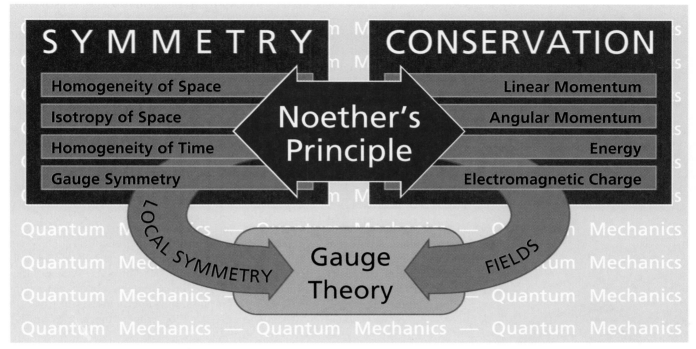

Figure 1.4 A summary of a number of interrelated ideas that we develop throughout this text. Noether's Principle relates symmetry and conservation.

system leaves some aspect of the system unchanged, the system possesses a corresponding symmetry. The existence of a symmetry means that a feature of the system is changeless, or *invariant.* In this instance, we say that the disc possesses rotational symmetry and that the system is invariant under rotation about the central axis. The amount of rotation can be varied smoothly, and the symmetry is said to be *continuous.*

The letter S is invariant under a rotation of 180°, just as a snowflake is invariant under a rotation of 60° (see illustration). In contrast, the letter A is symmetric under a mirror reflection along a vertical centerline. That kind of symmetry is *discontinuous;* that is, it jumps—all or nothing. These are all geometrical symmetries that we can appreciate by just looking at them, but, on a deeper level, any invariance suggests a symmetry.

In nature, there are important invariances associated with more subtle symmetries. The first person to grasp the physical implications of this was Amalie Noether. Among the central conceptual pillars of Classical Physics are three conservation laws, powerful insights that took hundreds of years to develop: Conservation of Linear Momentum, Conservation of Angular Momentum, and Conservation of Energy. Each maintains that a specific quantity is constant at every moment as an isolated system undergoes change. **Noether's Principle** states that *for every continuous symmetry there is a corresponding conservation law and vice versa.* In due course, we will learn how each of the great conservation laws turns out to be a direct consequence of a specific symmetry of space and time (Fig. 1.4). Noether's Principle is more general yet; it also applies to subtler abstract symmetries, such as gauge symmetry. In Chapter 18, we see how this principle relates to the conservation of electromagnetic charge.

Now, suppose we paint the disc in Fig. 1.3 half red and half green (Fig. 1.5). If the viewers happen to be color-blind, nothing is altered, but if they are not, the

60° snowflake symmetry.

Figure 1.5 A uniform disc whose symmetry was broken by being painted.

symmetry is radically reduced—it's *broken*. In nature, symmetries are often *spontaneously broken*. Imagine a perfectly symmetrical ballpoint pen balanced on its tip. We might suppose that if there were no vibrations, winds, or other external influences the pen would remain upright indefinitely. But that's not likely. We know from Quantum Mechanics that things on an atomic level are always changing; the symmetry will spontaneously break, and the pen will fall.

In an analogous way, many physicists believe that the wildly energetic, incredibly dense infant Universe was a place of great symmetry and undifferentiated simplicity. The early Universe was ruled by a single superforce that ultimately broke into the Four Forces as the Universe cooled, calmed, and shattered into complexity (Fig. 33.23).

MEASUREMENT

Physics is founded on experimentation: we ground the science directly to nature through observations that entail the measurement of physical quantities. The basic notions of length, volume, weight, and time were quantified in antiquity and simply carried over into physics. Today, the scientific community follows the *Système International* (SI), a program in which the majority of measurements are based on the following units: for length, the *meter* (m); for mass, the *kilogram* (kg); for time, the *second* (s); for electric current, the *ampere* (A); and for temperature, the *kelvin* (K).

1.3 Length

Length is a distance or extent in space (whatever distance is, or for that matter, whatever space is). Given the limitations of language, it's easier to measure length precisely than to talk about it precisely. Stick out your arm and announce that the distance from elbow to farthest outstretched fingertip is a length hence to be known as a *cubit*. That's the way it was done over 4000 years ago in Egypt and Mesopotamia. *An agreed-upon measure by which we reckon any physical quantity is called a* **unit**. The Great Pyramid was built to cubit specifications, as was Noah's Ark. Of course, you can't build a large structure if everyone's forearms are different. An advanced society must evolve *an unchanging embodiment of each unit to serve as a primary reference, or* **standard**. The black granite master cubit was just such a standard against which cubit sticks in Egypt were regularly checked.

For centuries, primitive body measures were used throughout Europe: the width of a man's thumb is an inch; the length of a sandal is a foot. The rise of modern science painfully exposed the lack of a universally standardized system. In 1790, Thomas Jefferson proposed a decimal scheme for weights and measures, but the U.S. Congress would have nothing to do with it. At the same time in France, the climate of the Revolution was right for innovation. The French adopted a decimal approach for their *Metric System,* dividing all quantities into 10, 100, 1000, and so on, equal parts. The French decided on a new unit of length, the **meter** (or *metre*, from the Greek *metron*, a measure), abbreviated m. It was to be one ten-millionth of the distance from the North Pole to the Equator along a meridian line passing through Paris. However patriotic, errors in the difficult measurement made the whole exercise quite arbitrary. They could have just as well stayed home and used the length of Napoleon's sword instead. Be that as it may, the distance between two lines inscribed in a bar of platinum-iridium alloy became—in 1889—the world's standard meter (Fig. 1.6).

Amalie (Emmy) Noether (1882–1935) was an outstanding mathematician who did most of her work in abstract algebra. After a long struggle she won the right as a woman to lecture, without pay, at Göttingen University in Germany. It was then, in 1918, that she presented the results of an analysis dealing with symmetry that became a guiding principle for contemporary physics. Noether taught at Göttingen until 1933 when she came to the United States after the Nazis learned that she was Jewish and expelled her from Germany.

The **Metric System**, which is the basis of the SI, gradually spread across Europe, although it was shunned by the British for over a century. The United States, in happy distant oblivion, continued to use the dreadful British system and is still trying to make the transition. The system used in the United States is called *U.S. Customary units*. It is based on length, weight, and time expressed in feet, pounds, and seconds, respectively (Table 1.2).

Because a billion (1 000 000 000) in the United States is a thousand times smaller than a European billion (1 000 000 000 000), the common names for numbers beyond a million are rarely used in scientific work. Instead, prefixes (Table 1.3) are added to the units. Greek prefixes like kilo, mega, and giga are multiples; the Latin ones like centi, milli, and micro are subdivisions (see Appendix A-2 for a review of scientific notation).

A related approach that is used, especially where things are on a small scale, is the cgs (centimeter, gram, second) system. Here, a *centimeter* (cm) is a one-hundredth part of a meter (1 cm = 0.01 m) and 1 inch (in.) is defined to be exactly 2.54 cm. The centimeter, gram, and second are SI units, but all the others in the cgs system (the units for force, pressure, and density, for instance) are not.

God taking the measure of the Universe, painted by the poet and artist William Blake (1757–1827).

Example 1.1 How many centimeters are there in 25.00 meters?

Solution: [Given: a length of 25.00 m. Find: the corresponding number of centimeters.] The most systematic way to do all unit conversions is to form a ratio equal to 1, with the unit you want on top and the one you want to replace on the bottom. Since 1 m is exactly 100 cm,

$$1 = \frac{100.0 \text{ cm}}{1.000 \text{ m}}$$

Then multiply the original number by the unit ratio

$$25.00 \text{ m} \times \frac{100.0 \text{ cm}}{1.000 \text{ m}} = \boxed{2500 \text{ cm}}$$

▶ **Quick Check:** There are 100 cm per 1 m, hence 25 m × 100 cm/m = 2500 cm.

Twenty-nine duplicates of the International Prototype Meter were made, and the major industrial nations still preserve their copies (the United States maintains Number 27). After the Nazis conquered France, it became obvious that the world's reliance on a single vulnerable set of standards was foolhardy—far better to redefine the meter in terms of some measurement that could be performed in any well-equipped laboratory. In 1960, the *Système International* was accepted, and the meter became 1 650 763.73 wavelengths of the orange-red light from a krypton-86 lamp.

In the decades that followed, with the advent of stable lasers, the speed of light was measured with extraordinary precision. In 1983, the decision was made to use it as the basis of a redefined meter that would be better than 10 times more precise.

In the scientific world, the kilogram is the basic unit of mass.

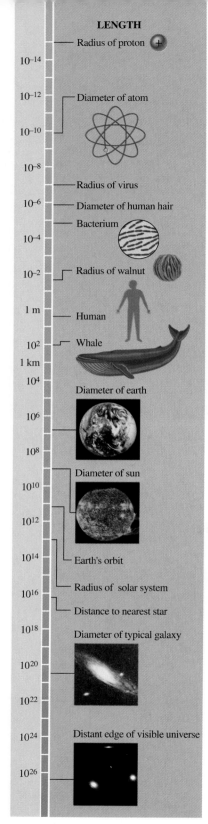

LENGTH

- 10^{-14} — Radius of proton
- 10^{-12} — Diameter of atom
- 10^{-10}
- 10^{-8}
- 10^{-6} — Radius of virus
- — Diameter of human hair
- 10^{-4} — Bacterium
- 10^{-2} — Radius of walnut
- 1 m — Human
- 10^2 — Whale
- 1 km
- 10^4
- — Diameter of earth
- 10^6
- 10^8
- — Diameter of sun
- 10^{10}
- 10^{12}
- 10^{14} — Earth's orbit
- 10^{16} — Radius of solar system
- — Distance to nearest star
- 10^{18} — Diameter of typical galaxy
- 10^{20}
- 10^{22}
- 10^{24} — Distant edge of visible universe
- 10^{26}

Figure 1.6 The sizes (in meters) of objects from protons to the Universe itself.

The speed of light in vacuum (c) is now defined to be exactly 299 792 458 meters per second, which is the best measured value. The meter is, then, *the distance light travels in vacuum during a time of exactly 1/299 792 458 second.* This new standard will never need to be revised. Improvements in measuring time will simply improve the measurement of the meter.

TABLE 1.2   Length and volume equivalents

1 light-year (5.88×10^{12} mi)	$9.460\,55 \times 10^{15}$ m
1 mile (5280 ft)	1.609 344 km
1 yard (yd; 3 ft)	0.914 4 m
1 foot	0.304 8 m
1 inch	2.54 cm
1 meter	3.281 ft
1 meter	39.37 in.
1 meter	1.094 yd
1 gallon U.S. (4 quarts)	$3.785\,412 \times 10^{-3}$ m^3
1 quart (2 pints)	$9.463\,530 \times 10^{-4}$ m^3
1 ounce	$2.957\,353 \times 10^{-5}$ m^3
1 tablespoon (1/2 oz)	$1.478\,676 \times 10^{-5}$ m^3
1 cubic foot	28 320 cm^3
1 cubic foot	0.028 32 m^3
1 cubic meter	1.308 yd^3
1 cubic meter	61 023 in.^{3}

TABLE 1.3   Common unit prefixes

Power of ten	Prefix	Pronunciation	Symbol	Example
10^{18}	exa-	ex'a	E	exajoule, EJ
10^{15}	peta-	pe'ta	P	petasecond, Ps
10^{12}	tera-	ter'a	T	terahertz, THz
10^9	giga-	jig'a[a]	G	gigavolts, GV
10^6	mega-	meg'a	M	megawatt, MW
10^3	kilo-	kil'oe	k	kilogram, kg
10^2	hecto-	hek'toe[b]	h	
10	deka-	dek'a[b]	da	
10^{-1}	deci-	des'i	d	decibel, dB
10^{-2}	centi-	sen'ti	c	centimeter, cm
10^{-3}	milli-	mil'i	m	millimeter, mm
10^{-6}	micro-	my'kroe	μ	microgram, μg
10^{-9}	nano-	nan'oe	n	nanometer, nm
10^{-12}	pico-	pee'koe	p	picofarad, pF
10^{-15}	femto-	fem'toe	f	femtometer, fm
10^{-18}	atto-	at'toe	a	attocoulomb, aC

[a]Although the given pronunciation is supposed to be standard, many physicists use gig'a instead.
[b]The prefixes hecto- and deka- are avoided in physics.

Example 1.2 How many cubic centimeters are in a cubic meter?

Solution: [Given: 1 cubic meter. Find: the number of cubic centimeters.] We know that 1 m = 100 cm exactly, hence a cube 1 m by 1 m by 1 m is 100 cm by 100 cm by 100 cm. From Fig. 1.7 and Appendix B, the volume (V) is given by the length times width times height:

$$V = (100 \text{ cm}) (100 \text{ cm}) (100 \text{ cm}) = \boxed{1\,000\,000 \text{ cm}^3}$$

▶ **Quick Check:** Imagine a big cube made up of 1 cm³ blocks. There will be 100 layers, each 1 cm tall and 100 cm long by 100 cm wide. Each layer contains 100 × 100 or 10 000 little cubes, and there are 100 such layers, yielding 100 × 10 000, which equals 10^6 cm³.

1.4 Mass and Weight

When the French set out to construct the Metric System, they followed the ancient Babylonian example of defining weight in terms of a fixed volume of water. The gram (g), originally a unit of weight, was specified as "the absolute weight of a volume of pure water equal to a cube of one-hundredth part of a meter [that is, a cube one centimeter on a side]."

Historically, weight was the downward force acting on an object at the surface of the Earth, and it was generally assumed to be constant. The fact that the weight of an object varies with location was discovered inadvertently in 1671 by J. Richer, who found that his pendulum clock, when brought from Europe to French Guiana, ran slow, losing $2\frac{1}{2}$ minutes each day. Newton promptly explained this surprising effect by clearly distinguishing between *weight* and *mass,* two ideas that had been around for centuries.

Mass is a property of a sample of matter immersed in the Universe-at-large (p. 92). It is apparently independent of the local distribution of matter and is constant everywhere on the Earth. Indeed, the mass of an object in a laboratory will be measured to be constant no matter where in the Universe the laboratory is. Masses interact gravitationally, and this causes the force we call **weight**. What gives rise to mass is still an unsettled issue.

The weight of an object at the surface of our planet is due to the gravitational interaction between that object and the Earth. Speaking practically, weight arises out of a local gravitational interaction between the sample and any massive object in its immediate vicinity, such as a planet, moon, or asteroid. At a far distance from all such objects, the weight of a sample effectively vanishes. Mass never vanishes. Richer's pendulum bob, its mass constant, was moved to a place where gravity was reduced, and consequently its weight was reduced. Thus, the pendulum was pulled down on each swing with less force, and it swung more slowly.

Ultimately, physicists properly ceased considering weight to be a fundamental property of matter, and in 1889,

Figure 1.7 The areas and volumes of several geometric figures. See Fig. B4 in Appendix B.

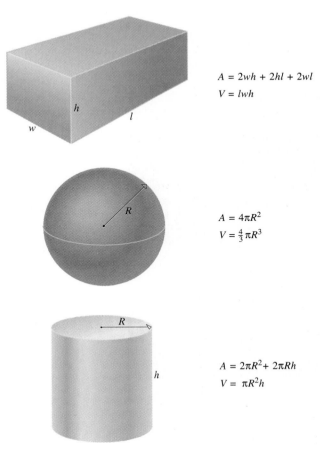

$A = 2wh + 2hl + 2wl$
$V = lwh$

$A = 4\pi R^2$
$V = \frac{4}{3}\pi R^3$

$A = 2\pi R^2 + 2\pi Rh$
$V = \pi R^2 h$

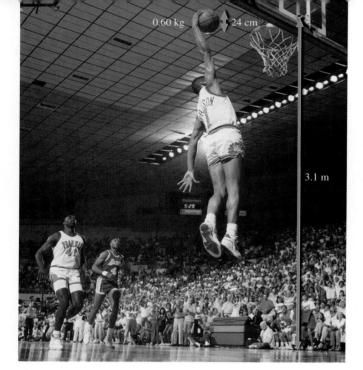

0.60 kg · 24 cm

3.1 m

A basketball must weigh between 20 and 22 ounces. It must be no less than $29\frac{1}{2}$ in. and no more than 30 in. in circumference ($2\pi R$). The basket must be hung 10 ft above the floor. Check that the diameter of the ball is about 24 cm.

the **kilogram** (kg) was redefined as the unit of mass. The standard kilogram is a platinum-iridium cylinder watched over by the International Bureau of Weights and Measures. It's a little thing about the size of a whisky shot glass, 39 mm (1.5 in.) in both height and diameter. By definition, a gram is a one-thousandth part of a kilogram. Once the International Kilogram was agreed upon and copies made, a pan balance could then be used anywhere on the planet to compare the mass of an object with a set of standard masses. Since gravity is the same at each pan, if the balance is level the masses are equally independent of what they might weigh at that location.

Forty nearly identical masses were made in London in 1884. The standard for the United States is Kilogram Number 20. It and a clone, Kilogram Number 4, sit next to each other under sealed glass domes in a basement vault at the National Institute of Standards and Technology (NIST). Kilogram Number 20 is rarely handled, for fear that the slightest trace of metal may be worn away. Even so, it has traveled to Paris on four occasions in its lifetime of over one hundred years. The kilogram is the only remaining unit in the SI that is still based on an actual artifact; this brings special problems. Exposing the mass to air or touching it even with tongs can change it, but if it's to be used, it must be handled, however carefully. Although an atomic standard for mass is probably inevitable, the cylinder survives because it's easy to compare objects of that size to within an accuracy of several parts in a thousand million.

Because weight depends on mass, we explore the physics of converting from one to the other in Chapter 4. Whenever you buy anything by the gram, kilogram, ounce, grain, carat, or pound, that measure must be traceable directly back through a long list of calibrations to the primary standard (Table 1.4).

Mass is a fundamental property of matter that we can attempt to define via three seemingly different approaches: (1) it is that which participates in the gravitational interaction; (2) it is a measure of an object's resistance to a change in motion; and (3) it is a measure of the quantity of matter possessed by an object. So primary is the concept of mass that we return to it many times in this text (Fig. 1.8).

Kilogram Number 20 is the standard mass of the United States. It is a copy of the International Standard Kilogram housed at Sèvres, France.

TABLE 1.4	Weight-mass correspondence on Earth*	
1 gram	0.035 273 40 ounce
1 kilogram	2.204 622 pounds
1 ounce	28.349 52 grams
1 pound	0.453 592 4 kilogram

*Mass and weight are very different features of matter. The values given here are the weights that would be measured for the corresponding masses at the surface of the Earth.

1.5 Time

The rhythms of life—the passing of successive days, the pounding of a heart—suggest a progression, a sequencing of occurrences. Time is a measure of the unfolding of events. Space is revealed by the juxtaposition of matter; time is revealed by the change in that juxtaposition. Whatever time *is,* it can be measured against uniform change. That's often done with some cyclic phenomenon like the rising of the Sun. More than 3000 years ago the Egyptians divided the day and night into 12 equal hours. Babylonian arithmetic had 60 as its number base, and there subsequently developed a tradition of subdividing things into 60 parts. It was probably sometime in the fourteenth century, after the arrival of the mechanical clock, that the hour (h) got divided into 60 minutes (min). The word minute was abbreviated from the medieval Latin for *first minūte part* (which is why the words minute and minūte are spelled the same). Later, when the so-called *second minūte part* could be measured, it was natural to take 60 seconds (s) to a minute.

As a young pre-med undergraduate in 1582, Galileo noticed that a swinging pendulum kept a steady beat and suggested that it might serve to measure time. By the early 1600s, astronomers were using pendulums 39.1 in. long (0.994 m), pushed every now and then by hand, to count out seconds. A tremendous improvement in accuracy was achieved in 1656 when Huygens introduced the pendulum clock, which had an error of only about 10 s a day. Minute hands, which had been rare, became the rule, and clocks were soon fitted with second hands for the first time. By the 1920s, the free pendulum clock was keeping time to better than 0.005 s per day.

The day, high noon to high noon, has been the benchmark of time for millennia, but the Earth's spin rate is slowing down, and the days grow longer. In 1967, the SI unit of time, the *second,* was defined as the interval required for 9 192 631 770 vibrations of the cesium-133 atom measured via an atomic beam clock. At present, such clocks are in error by about 1 second in 3 000 000 years. Their remarkable constancy is a manifestation of the precisely quantized energy states of the cesium atom.

So accurate is modern timekeeping, that the atomic clocks of the NIST at Boulder, Colorado, have been found, as expected from the General Theory of Relativity, to run 15.5 nanoseconds (ns) faster per day than those in Paris and Washington. Einstein showed that time depends on gravity: the city of Boulder is a mile above sea level; at that altitude, time runs a little faster.

Operationally, time is that which is measured by a clock. Conceptually, time is a measure of the rate at which change occurs. Facetiously, time is that which keeps everything from happening all at once (Fig. 1.9).

If we ever communicate with extraterrestrials, they surely won't use meters, kilograms, and seconds, but they will certainly measure length, mass, and time. We should have no trouble making the necessary conversions. Our physics, to the extent that it's correct, must be their physics. And that's what gauge symmetry is all about.

1.6 Significant Figures

Measurement is very different from counting, even though both associate numbers with notions. We can count the number of beans in a jar and know it exactly. But we cannot measure the height of the jar exactly. *There is no such thing as an exact measurement.* Practically, measurements are made to some desired precision that suits

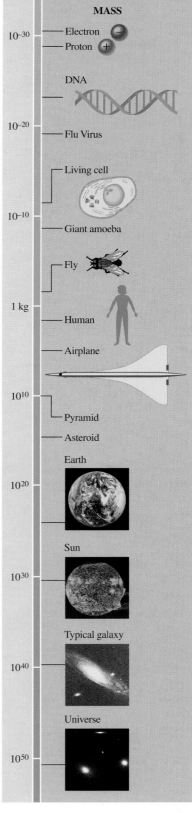

Figure 1.8 Masses (in kg) of objects from the electron to the Universe.

TIME

10^{-23} — Time for light to cross nucleus

10^{-21}

— Period of nuclear vibration

10^{-15} — Period of visible light

10^{-10} — Lifetime of "strange" particles
— Period of microwaves

— Period of AM radio

10^{-6}

— Period of musical middle C

10^{-2}

1 sec — Period of human heartbeat

— College lecture

10^5 — 1 day

10^7 — 1 year

— Life of a human

10^9
— Lifetime of Roman empire
— Lifetime of a mountain range
Age of Earth

10^{13}

10^{15}

Age of Universe

10^{17}

Figure 1.9 Intervals (in seconds) from the time it takes light to traverse a nucleus to the age of the Universe.

the experimenter's purposes (often determined by the limitations of the available instruments).

Figure 1.10 shows a rod whose length is being measured with a ruler. Assuming one end of the ruler and rod are lined up precisely, the object is between 4.1 and 4.2 cm long. Because there are no finer divisions than 1 mm on the scale, fractions of a millimeter will have to be estimated. Doing that, we conclude that the rod is 4.15 cm long. It's inappropriate to say that the length is exactly 4.15 cm. The length is closer to 4.15 cm than to either 4.14 cm or 4.16 cm, which is what we mean when we write 4.15, although it can be stated explicitly as 4.15 ± 0.01. The 4 is certain, as is the 1, but the 5 might be in error by as much as ±1, and trying to arrive at any more figures would be meaningless. A digit in a number is said to be a **significant figure** when it is known with some reliability, like the 4, the 1, and even the 5. When a result is presented, such as 4.15 cm, it can be assumed that all the figures are significant and that the 5 is the least significant. Note that to specify the rod's length exactly would require an infinite number of significant figures.

To three significant figures, the length of the rod is 4.15 cm, or 41.5 mm, or 0.041 5 m, or 0.000 041 5 km. Each of these is equivalent, and each has three significant figures. Moving the decimal point and changing the unit prefix has no effect on the accuracy of the number. Hence, zeros to the left of the 4 only locate the decimal point and are not significant figures. Zeros to the right, however, are significant. If you use a micrometer and measure the rod to have a length of 4.150 cm, there are four significant figures. Needless to say, carelessness with these right-hand zeros can cause ambiguities. For example, the distance to the Sun is 146 million kilometers, but writing it as 146 000 000 km implies 9 significant figures, which is not the case. Only significant figures should be given and the decimal should be located using scientific notation. The distance to the Sun is then 146×10^9 m, or 146×10^6 km, or 14.6×10^7 km, and so forth.

When we conduct a careful measurement, a number of determinations are made and averaged. The deviations are examined, all possible errors analyzed, and a numerical uncertainty is associated with the measurement. Thus, it's found that the mass of an electron is $9.109\,389\,7 \times 10^{-31}$ kg, with an uncertainty of 0.59 *parts per million* (ppm). That latter phrase is translated: $(0.59/1\,000\,000) \times (9.109\,389\,7) = 0.000\,005\,4$. Hence, the mass is $(9.109\,389\,7 \pm 0.000\,005\,4) \times 10^{-31}$ kg.

Processing Significant Figures

Computations frequently involve measured quantities, and we need to establish how to deal with the multiplication, division, addition, and subtraction of significant figures. Consider the area of a rectangle with sides of 10.7**5** cm and 3.5**4** cm (to help keep track of things, the doubtful digits are boldfaced). The area (Appendix B) is the product of length and width, (10.7**5** cm)(3.5**4** cm) = 38.**055 0** cm². As a rule, we *keep only one doubtful digit, and round off the answer accordingly. If the first of the insignificant figures being dropped is 5 or more, raise the last significant figure being retained by 1; if it's less than 5, leave it unchanged.* Here 38.**055 0** cm² is rounded up to 38.1 cm². Rather than computing the number of significant figures in each case, we use this rule:

The result of multiplication and/or division should be rounded off so that it has as many significant figures as the least precise quantity used in the calculation.

This rule is the equivalent of saying that making a calculation cannot improve on the precision of the numbers used. The rule is rough; it can occasionally reduce the precision. For example, it's possible, using the same instrument, to measure a flat plate to have sides of 0.9**1** m, and 1.5**1** m, with 2 and 3 significant figures, respectively. The area is (0.91 m)(1.51 m) = 1.374**1** m^2. According to the rule, that's 1.4 m^2, even though the uncertainty is in the 7 and 1.37 m^2 is a better answer. Regardless of the rare exceptions, we will stick with the rule.

$$
\begin{array}{r}
10.75 \\
3.54 \\
\hline
4300 \\
5\ 375 \\
32\ 25 \\
\hline
38.0550
\end{array}
\qquad
\begin{array}{r}
1.51 \\
0.91 \\
\hline
151 \\
1\ 359 \\
\hline
1.3741
\end{array}
$$

Example 1.3 Someone is 2.00 yd tall. Using the fact that 1 in. is exactly 2.54 cm, how tall is the person in centimeters?

Solution: [Given: height = 2.00 yd, and 1.00 in. = 2.54 cm. Find: height in centimeters.] Note that the data, the height, is given to three significant figures. We have a relationship between inches and centimeters, but the height is in yards. If we had the height in inches, it would be easy to find it in centimeters. So, we compute it in inches. Since there are 3.00 feet per yard,

$$1 = \frac{3.00 \text{ ft}}{1.00 \text{ yd}}$$

The height, call it h, in feet, is

$$h = 2.00 \text{ yd} \frac{3.00 \text{ ft}}{1.00 \text{ yd}} = 6.00 \text{ ft}$$

Similarly, since there are 12.0 inches per foot,

$$1 = \frac{12.0 \text{ in.}}{1.00 \text{ ft}}$$

and $$h = 6.00 \text{ ft} \frac{12.0 \text{ in.}}{1.00 \text{ ft}} = 72.0 \text{ in.}$$

Inasmuch as there are 2.54 centimeters per inch,

$$1 = \frac{2.54 \text{ cm}}{1.00 \text{ in.}}$$

hence $$h = 72.0 \text{ in.} \frac{2.54 \text{ cm}}{1.00 \text{ in.}} = 182.88 \text{ cm}$$

and to three significant figures $\boxed{h = 183 \text{ cm}}$.

▶ **Quick Check:** 1 yd ≈ 0.9 m; 2 yd ≈ 1.8 m ≈ 180 cm.

The significance of digits in addition and subtraction is determined in relation to the location of the decimal point. Suppose a rod of length 140 mm is added to one 2.0 m long, and we want the total length. It might be assumed that (0.140 m) + (2.0 m) = 2.140 m, but we know nothing about the digits in the second and third decimal places of the 2.0******-m rod and cannot know the sum accurate to three decimal places. It makes sense to round off the sum to the least number of decimal places of any number appearing in the summation: the length of the rod is 2.1 m and both the 2 and the 1 are significant figures. As a rule:

$$
\begin{array}{r}
0.140 \\
+2.0 \\
\hline
2.140
\end{array}
$$

> The result of addition and or subtraction should be rounded off so that it has the same number of decimal places (to the right of the decimal point) as the quantity in the calculation having the least number of decimal places.

Thus, (275 s − 270 s) = 5 s and (120 kg − 40.0 kg) = 80 kg.

In time, we will work with several transcendental functions: sines, cosines, tangents, logarithms, and exponentials. All of these can be determined using a scientific calculator; for example, sin 35.1° = 0.575 005 252. Here, the number whose sine is being taken, the *argument* of the function, is 35.1°. The calculator, trying to please, provides

Figure 1.10 This rod measures somewhere between 4.1 cm and 4.2 cm, closest to 4.15 cm, but that last digit is not certain.

A fine micrometer can measure distances in divisions of 0.001 mm.

The sight of day and night, of months and the revolving years, of equinox and solstice has caused the invention of number and bestowed on us the notion of time.

PLATO
Timaeus

us with a full register of figures, most of which are meaningless. What it actually gives us is sin 35.100 000 00°. We shall follow the rule that:

> The values of transcendental functions have the same number of significant figures as their arguments.

Thus, sin 35.**1**° = 0.57**5**. The 35 in the argument is certain, but there is some doubt about the .1°, and so on the right we are certain about the 0.57 and have doubt about the .005. In agreement with the rule, sin 35.0° = 0.574 and sin 35.2° = 0.576.

Because rounding off intermediate numbers within a calculation can produce cumulative errors, *one or two insignificant figures should be carried through intermediate calculations and only the final answers should be properly rounded off.* It's appropriate to run through an entire analysis on a calculator and round off only the final answer.

THE LANGUAGE OF PHYSICS

Physics is done in almost every language on Earth, and yet it has a symbolic dialect of its own that is understandable to all its practitioners. Learning that vocabulary is the most basic task confronting the student. Being able to make the translation from everyday discourse to the language of physics and on to the symbolic statements of mathematics is part of the process of mastering the subject. As you move through this text, *commit to memory the definitions of physical concepts* like displacement, velocity, acceleration, and so on. Physics is progressively elaborated in terms of quantities that have previously been defined: you cannot understand acceleration without remembering velocity, and you cannot understand velocity without remembering displacement. Physics, perhaps more than any other discipline you have studied before, builds on definitions and basic principles.

Once we learn the scientific vocabulary, we can transpose the experiences of everyday life into the carefully defined terminology of physics. For example, How much space is occupied by a big round red balloon 10 meters across? This question translates into, What is the volume of a sphere 10 m in diameter? The scientific words *volume, sphere,* and *diameter* have exact meanings, and there is a specific relationship between them—we know how to calculate the volume of a sphere in terms of its diameter (Fig. 1.7). No such relationship exists for the ordinary words *space, round,* and *across.*

1.7 Equations

Newton was the last great physicist to do his work without routinely representing physical quantities symbolically and writing equations describing their interrelationships (see Appendix A-4). Today, that approach would be unthinkable. We assign each physical quantity a symbol; t for time, a for acceleration, V for volume, R for radius, and so forth. Of course, we try to pick symbols that will be easy to remember, though that's not always possible. Moreover, we have standardized mathematical signs that act as a conceptual shorthand (Table 1.5).

The reason for transforming verbal scientific statements into symbolic ones is twofold. First, they are concise. A sentence such as "The volume of a sphere is equal to four-thirds of the product of pi times the radius cubed" becomes $V = \frac{4}{3}\pi R^3$. Certainly you have to know the definitions of volume, radius, and pi for this equation to make any sense. Once the technique is mastered, though, a whole page of symbols can be read like a story. Second, there exists a mathematical symbolic logic called algebra that's immediately applicable to physics. If you know the volume of a sphere and want to <u>find its</u> radius, the rules of algebra tell us that $3V = 4\pi R^3$, $R^3 = 3V/4\pi$, and $R = \sqrt[3]{3V/4\pi}$. Try thinking that through in sentences! Using algebra, we can easily combine several physical equations to arrive at new relationships that would otherwise be extremely difficult to deduce.

1.8 Graphs

Complex relationships can be represented algebraically, but it's often hard to get a feeling for the interplay of the variables in an equation. That's a great virtue of graphs; we can plot an equation and literally see what happens to one quantity as another varies, and take it all in at a glance. Futhermore, physicists usually have collections of data, and by graphing them they can get a handle on the interdependencies of the measured quantities.

The simplest situation is when one variable (y) depends *directly* on another (x): $y = mx$, where m is a constant. The larger m is, the faster y changes with changes in x. For example, if you get a 10% discount ($m = 0.10$) on the price of any shirt (x) in the store, you save ($y = mx$) dollars. Figure 1.11 is a plot of y versus x, and m is the slope of the line (see Appendix A-4). Because the graph is a straight line, $y = mx$ is called a *linear* equation. There are many physical relationships, such as Ohm's Law and Hooke's Law, that are linear. Here, the line passes through zero so that at $x = 0$, $y = 0$, but that needn't be the case. Figure 1.12 shows a graph of the function $y = mx + b$, where m is again the slope and b is the point where the line now crosses the y-axis when $x = 0$.

Frequently, a quantity depends on the square of some other parameter. Mathematically, that situation appears as $y = kx^2$, where k is a constant. For example, the area (A) of a square whose sides are of length l is given by $A = l^2$ (where $k = 1$). Double the side length and the area of the square increases by a factor of four. A second degree equation of this kind is said to be *quadratic,* and the graph of y versus x is a *parabola* (Fig. 1.13). As we will see, a falling object descending along a straight line drops a distance (y) that is proportional to the time of fall (x) squared. A plot of distance versus time is parabolic. In a very different way, a ball thrown into the air follows an arcing path in space that's essentially parabolic (p. 77). Equations can describe relationships between physical parameters, but they can also represent actual flight paths through space, and both can be graphed.

Another common situation occurs when a quantity (y) depends *inversely* on some other quantity (x): $y = k/x$ or $yx = k$, where k is constant. The bigger x gets, the smaller y becomes, but yx is constant. For example, suppose you are designing a rectangular floor that must have a fixed area (k). The longer (x) the floor is made, the narrower (y) it must be, since $yx = k$. The graph of y versus x (Fig. 1.14) is a hyperbola that gradually approaches zero as x increases. At a constant temperature, a plot of pressure against volume for a variety of gases is hyperbolic.

1.9 Approximations and Checks

There are about 10^{11} galaxies in the Universe. No one has ever counted them, but we have photographed the sky, counted the number of galaxies in a small region, and then estimated the total. Making approximations like that are part of the everyday doing of physics. We approximate how much material we need for an experiment, how much time it takes, how much power to expend, and so forth.

TABLE 1.5	Mathematical symbols
Symbol	**Meaning**
$=$	equal to
$\neq$	not equal to
$\equiv$	defined as
$\pm$	plus or minus
$\mp$	minus or plus
$\parallel$	parallel
$\perp$	perpendicular
$\div$	divide by
$\rightarrow$	approaches
$\approx$	approximately equal to
$<$	less than
$>$	greater than
$\ll$	much less than
$\gg$	much greater than
$\leq$	less than or equal to
$\geq$	greater than or equal to
$|z|$	absolute value of z
$\propto$	proportional to
∞	infinity
$\sum$	sum of

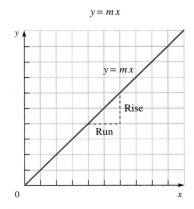

$$y = mx$$

Figure 1.11 A curve is linear when it plots as a straight line, which means that the slope, the rise over the run, is constant and equal to m. This plot passes through the origin when $x = 0$, $y = 0$.

$y = mx + b$

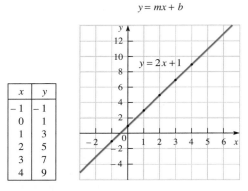

x	y
-1	-1
0	1
1	3
2	5
3	7
4	9

Figure 1.12   Here $y = 2x + 1$, the slope is $+2$, and the y-intercept (b) is 1. When $x = 0$, $y = 1$. Between $x = 0$ and $x = 4$, the rise is 8, the run is 4, and the slope, 8/4, is 2.

$y = kx^2$

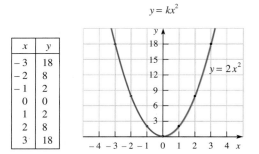

x	y
-3	18
-2	8
-1	2
0	0
1	2
2	8
3	18

Figure 1.13   A plot of $y = kx^2$ (where k is a constant) is a curve that is symmetric about the y-axis called a *parabola*. At $x = 0$, $y = 0$, and the curve passes through the origin. There are many physical relationships that are parabolic.

How accurate an approximation is depends on the accuracy of the data that goes into it and the level of detail we devote to the calculation. For example, if all the people in the United States were placed head-to-foot, how far would they extend? There are about 250×10^6 people, each roughly 2 m tall; that's $\approx 5 \times 10^8$ m, which is greater than the distance to the Moon (3.8×10^8 m). That's not a very good approximation because 2 m is a poor guess for the average height—it's too tall. So what we calculated is an extreme. We know that the exact answer (there probably isn't one, with people being born and dying) is less than 5×10^8 m. We could improve things by assuming there are $\approx 186 \times 10^6$ adults ≈ 1.6 m tall and $\approx 64 \times 10^6$ children averaging ≈ 1 m tall. That yields a distance of $\approx 4 \times 10^8$ m, which can't be too far off—after all, if everyone in the country averaged only 1 m tall, the distance would still be $\approx 2.5 \times 10^8$ m. Notice how looking at the extremes of a situation puts things in perspective.

The volume of gasoline pumped into the tank is directly proportional to the time (t) during which it is pumped. If V is the volume of gas in the tank at any moment, b is the amount in at the start, and m is the constant flow rate, then $V = mt + b$.

A long-exposure image of a small region of the southern sky. Here galaxies appear red and stars green. Clearly, there are a lot of galaxies in the Universe.

It's also common practice to check the solutions of problems by making quick approximations (see Appendix A-5). For example, suppose we had a swimming pool 3.1079 m deep, 4.921 m wide, and 18.689 m long. What's its volume (V)? Since the volume is the length times width times depth (Fig. 1.7):

$$V = (18.689 \text{ m})(4.921 \text{ m})(3.1079 \text{ m}) = 285.8 \text{ m}^3$$

To check this result, round each number off to 1 significant figure and redo the calculation, as follows:

$$V = (2 \times 10^1 \text{ m})(5 \text{ m})(3 \text{ m}) = 3 \times 10^2 \text{ m}^3$$

Because we rounded up more than down, this approximation can be expected to be on the large side. Still, it provides a quick check to show us that 285.8 m³ is roughly the right size.

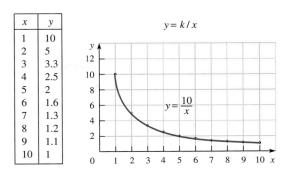

x	y
1	10
2	5
3	3.3
4	2.5
5	2
6	1.6
7	1.3
8	1.2
9	1.1
10	1

Figure 1.14 A plot of $y = k/x$ (where k is constant) is a hyperbola that approaches $y = 0$ as $x \to \infty$. For the gas in an air pump, the pressure times the volume is constant at a fixed temperature. Double the pressure, and you halve the volume.

Core Material

Physics is the study of the physical Universe. **Data** is the objective record of our observations. A **law** is a statement of a recurring pattern of events. **Theory** is the explanation of phenomena.

An agreed-upon measure of a physical quantity is a **unit**. An unchanging embodiment of a unit that serves as a primary reference is a **standard**. The SI is based on the meter, kilogram, and second. U.S. Customary units use the foot, the pound, and the second.

The result of multiplication and/or division should be rounded off so that it has as many **significant figures** as the least precise quantity used in the calculation. The result of addition and/or subtraction should be rounded off so that it has the same number of decimal places as the quantity in the calculation having the least number of decimal places. The values of transcendental functions have the same number of significant figures as their arguments.

Suggestions on Problem Solving

1. Problems are usually posed in nonscientific prose, for example, "A young man runs 20 meters from his front door to the street in 10 seconds. On average, how fast was he going?" We develop the physics to handle this problem in Chapter 2; here, we are only concerned with general principles of problem solving. Clearly, the question is full of extraneous information—it doesn't matter that it's a young man, and it doesn't matter that he goes from door to street. The following is a good way to proceed. (a) Decide on what's happening as far as the physics is concerned and lift that information from the inconsequential background. Object travels 20 m in 10 s—how fast? (b) Translate the statement into the language of physics, as you continue to learn that language. Object travels a *distance* (20 m) in a *time* (10 s) at what *average speed*? (c) Symbolically represent the variables: Object travels a distance ($d = 20$ m) in a time ($t = 10$ s); find average speed (v_{av}). Once the problem has been so restated, write down what was *given* ($d = 20$ m, and $t = 10$ s) and what you must *find* (v_{av}). That completes the crucial translation phase of the solution.

2. Determine a relationship between the given quantities and the one(s) to be found. It is the business of physics to produce such relationships in the form of laws and definitions. For example, a 10.0-meters tall cylinder contains 50.0 cubic meters of water. What is its diameter? First, translate it: A cylinder of height 10.0 m has a volume of 50.0 m³; find its diameter. Notice that the original statement did not contain the crucial word *volume*. Without that conceptual translation we could not proceed. Now write down what is given and what's to be found. *Given:* cylinder, $h = 10.0$ m, $V = 50.0$ m³; *Find: D*. This problem is more one of math than of physics, but the procedures are the same. Next, use an equation that provides a relationship between the quantities (Fig. 1.7), namely, $V = \pi R^2 h$. We want the diameter ($D = 2R$), not the radius. Solve for $R = \sqrt{V/\pi h} = 1.26$ m and double it.

3. As a rule, it's always a good idea to draw a diagram. The process helps to organize your thoughts.

4. Wherever possible, check your calculation. We all make numerical mistakes, it's perfectly normal. The thing you must do is catch them! With the above cylinder, take $R = 1.26$ m and compute something we know, namely V: $V = \pi R^2 h = \pi(1.26 \text{ m})^2(10.0 \text{ m}) = 49.9 \text{ m}^3$—close enough, since we rounded off R to three figures.

5. Show all of your work when making a calculation. Doing so will allow you, and anyone else attempting to understand your efforts at some later date, to follow your logic.

6. When solving a problem involving quantities with units having prefixes such as ms, or kg, or μm, replace the prefixes by scientific notation before carrying out the calculation. Notice that $(1.0 \text{ km})^3 = (1.0 \times 10^3 \text{ m})^3 = 1.0 \times 10^9 \text{ m}^3$, not 1.0×10^9 m.

Discussion Questions

1. There are people who define science as that which scientists do. Why isn't that a reasonable approach?

2. Deliberate hoaxes occur occasionally in science. A good example is the famous Piltdown man (around the turn of the century, a human skull fragment and an apelike jaw were combined and touted as the "missing link"). How do you think such hoaxes get revealed for what they are?

3. Given that a law says that some relationship among physical quantities will be found to *always* exist, how does that transcend experience? Implicit in the statement of law is the understanding that it applies *everywhere*. Comment on the practicalities of this notion.

4. What do you think the French mathematician H. Poincaré meant when he wrote, "Nature can only be governed by obeying her"?

5. What did Einstein mean when he wrote:

The aim of science is, on the one hand, a comprehension, as *complete* as possible, of the connection between the sense experiences in their totality, and, on the other hand, the accomplishment of this aim *by the use of a minimum of primary concepts and relations.*

What's the point of specifying "a minimum of primary concepts"?

6. When W. K. Clifford, more than a century ago, said "scientific thought is the guide to action," what might he have had in mind?

7. Discuss the distinction between *data* and *fact.* When Einstein was confronted by some reporters who excitedly proclaimed that the facts of a new experiment had proven Relativity to be incorrect, he simply replied, "The facts are wrong." Months later, the data was reinterpreted and the theory confirmed. Fact is not immutable. Indeed, a good trial lawyer knows that facts can often be cast within a shadow of doubt.

8. The poet Samuel Taylor Coleridge said that beauty was "unity in variety." How is that applicable to physical theory?

9. Having spoken for an hour or so about his recent theoretical work on elementary particles, Wolfgang Pauli settled down for the usual discussion involving the audience. It was New York, 1958, and the great Niels Bohr was there. After some heated remarks, Bohr rose and said, "We are all agreed that your theory is crazy. The question which divides us is whether it is crazy enough to have a chance of being correct." How would they ultimately find out if it was "crazy enough"?

10. Robert Millikan, the precise experimentalist and Nobel laureate, wrote in 1936, "In science, truth once discovered always remains truth." What can you say about this idea in light of the tentative nature of science? In what way was Millikan naive?

Multiple Choice Questions

1. Which of the following is longest?   (a) 10^4 cm   (b) 10^4 mm   (c) 10^6 μm   (d) 10^9 nm   (e) none of these.

2. Which length is the largest?   (a) 10^1 cm   (b) 10^{-10} m   (c) 1×10^2 mm   (d) 1 m   (e) none of these.

3. Which mass is the smallest?   (a) 10^5 μg   (b) 10^2 g   (c) 1 kg   (d) 10^3 mg   (e) none of these.

4. Given are four masses: (1) 10 mg   (2) 1000 μg   (3) 10^2 kg   (4) 10^{-4} kg. These are ordered in ascending size as   (a) 1, 2, 3, 4   (b) 2, 1, 4, 3   (c) 4, 3, 2, 1   (d) 2, 1, 3, 4   (e) none of these.

5. A day has roughly   (a) 86×10^2 s   (b) 8640 s   (c) 9×10^4 s   (d) 1.44×10^3 s   (e) none of these.

6. A year has roughly   (a) 8.77×10^2 h   (b) 5×10^5 min   (c) 3.7×10^3 days   (d) 32×10^5 s   (e) none of these.

7. A cube 1000 cm on a side has a volume of   (a) 10^2 cm^2 (b) 10^2 cm^3 (c) 10^6 cm^3 (d) 10^9 cm^3 (e) none of these.

8. A rectangular floor is 6.6 m by 12 m. Its area is   (a) 79 m^2 (b) 18.6 m^2 (c) 7.92 m^2 (d) 79.2 m   (e) none of these.

9. A femtosecond is   (a) 10^{-12} s   (b) -15 s   (c) 10^{15} s   (d) 10^{-15} s   (e) none of these.

10. A 20.0-in. bar is   (a) 20.0 cm long   (b) 508 mm long   (c) 51 m long   (d) 2.54/20 cm long   (e) none of these.

11. One pound has an equivalent mass of exactly 453.592 37 g. To four significant figures, that's   (a) 453.5 g   (b) 453.592 3 g   (c) 400.0 g   (d) 453.6 g   (e) none of these.

12. The product of 12.4 m and 2 m should be written as   (a) 24.8 m   (b) 24.8   (c) 25 m^2 (d) 0.2×10^2 m   (e) none of these.

13. The product of 15.0 cm and 5 cm should be written as   (a) 75 cm^2 (b) 7.5×10^1 cm^2 (c) 0.75×10^2 cm^2 (d) 0.8×10^2 cm^2 (e) none of these.

14. The weight of 1 kg on Earth is about   (a) 1 lb   (b) 1000 g   (c) 2-1/4 lb   (d) 0   (e) none of these.

15. If a bag of screws costs 10¢ per pound, a kilogram of them will cost about   (a) 100¢   (b) 22¢   (c) 4.5¢   (d) $22   (e) none of these.

16. If coffee is $12 a kilogram, roughly how much will it be by the pound?   (a) $26   (b) $12   (c) $5.5   (d) 55¢   (e) none of these.

17. A liter is 1000 cm^3, which means that a cube 100 cm on a side has a volume of   (a) 1000 liters   (b) 0.001 m^3 (c) 100 liters   (d) 1000 liters^3 (e) none of these.

18. A kilometer is   (a) just under half a mile   (b) just over half a mile   (c) about 1000 ft   (d) roughly 5280 ft   (e) none of these.

19. A dime is about   (a) 1 cm thick   (b) 1 kg in mass   (c) 1 mm thick   (d) 10^{-3} m in diameter   (e) none of these.

Problems

MEASUREMENT

1. [I] The human brain has about 10 (U.S.) billion nerve cells, more tightly packed than in any other tissue. Write that number out and then put it in scientific notation.

2. [I] It's been estimated that the human brain can store 100 (U.S.) trillion bits of information. Write that number out, put it in scientific notation, and then express it by adding the proper prefix to the word *bits*.

3. [I] Express each of the following quantities in seconds in decimal form: (a) 10.0 ms; (b) 1000 μs; (c) 10.000 ks; (d) 100 Ms; (e) 1000 ns. (Don't worry about significant figures in this problem.)

4. [I] Express each of the following quantities in micrograms: (a) 10.0 mg; (b) 10^4 g; (c) 10.000 kg; (d) 100×10^3 g; (e) 1000 ng. (Don't worry about significant figures.)

5. [I] How many millimeters are in 10.0 km?

6. [I] A red blood cell lives for about four months and travels roughly 1.0×10^3 mi through the body. How far is that in kilometers?

7. [I] How many millimeters are in 3.00 in.?

8. [I] The unit known as an angstrom (Å), exactly equal to 0.1 nm, is still widely used. It's named after the nineteenth century physicist Anders J. Ångstrom (Å is the letter that comes after Z in the Swedish alphabet). Given that an atom is about 1 Å in diameter, how much is that in centimeters?

9. [I] The length of a lightwave is about 5×10^3 Å (see previous problem). How much is that in nanometers?

10. [I] About how many centimeters tall is a stack of 25 dimes?

11. [I] A nickel (5-cent coin) has a mass of about 5 g. How many nickels correspond to a kilogram?

12. [I] It takes your brain about one five-hundredth of a second to recognize a familiar object once the light from that object enters your eye. Express that time interval in milliseconds, microseconds, and nanoseconds, each to one significant figure.

13. [I] How many seconds are there in an exactly 24-h day?

14. [I] Each second the human brain undergoes 1×10^5 different chemical reactions. At that rate, how many will it experience in 10 h?

15. [I] The distance light travels in a year is called a *light-year* (ly). Given that 1.00 ly $= 5.88 \times 10^{12}$ mi, how far is that in meters?

16. [I] By what number would you multiply a given number of centimeters to convert it to millimeters?

17. [I] There are an average of 32 million bacteria on each square inch of the human body. Given a total skin area of 1.7 m^2, how many bacteria are you carrying around (exclude the bacteria that are internal)?

18. [I] How many square centimeters are in a square inch?

19. [I] In a human nose, the total area of the region that detects odors is about 3/4 in.^{2}. Compare that to the sensory organ of a hunting dog, whose nose has an active area of about 65 cm^2.

20. [I] Can the quantity 100.0 kg be expressed in grams in decimal form with the correct number of significant figures? Explain.

21. [I] Express each of the following in grams in decimal form with the correct number of significant figures: (a) 1.00 μg; (b) 0.001 ng; (c) 100.0 mg; (d) 10 000 μg; and (e) 10.000 kg.

22. [I] Determine the number of significant figures for each of the following: (a) 0.002 0; (b) 0.99; (c) 1.75 $\pm$ 0.02; (d) 1.001; (e) 4.44×10^4; and (f) 0.01×10^{34}.

23. [I] Write the speed of light in vacuum (299 792 458 m/s) to one, three, four, and eight significant figures.

24. [I] State the number $\pi = 3.141\,592\,65\ldots$ to one, three, four, and five significant figures.

25. [I] The electromagnetic charge of an electron is $1.602\,177\,33 \times 10^{-19}$ coulombs. If the uncertainty is 0.30 ppm, which figures are in doubt and by how much?

26. [I] One of the most important numbers in physics is Planck's constant: $6.626\,075\,5 \times 10^{-34}$ kg·m^2/s $\pm$ 0.60 ppm. (a) Write Planck's constant out to four significant figures. (b) Write it out to two significant figures. (c) Write it out to one significant figure. (d) Write it out to five significant figures.

27. [I] To 4 significant figures, how many square meters correspond to a square foot?

28. [I] Liters are sometimes used for liquid and gas measurement, although the preferred unit is the cubic meter. How much is 1.000 liter in m^3?

29. [I] The radius of the Moon ($R_{\mathbb{C}}$) is 1.738×10^3 km. Assume the Moon is a sphere and compute its volume to two significant figures.

30. [II] A GM diesel locomotive pulling 40 to 50 loaded freight cars at 70 mi/h uses 1.0 gallon of fuel every 632 yd. How many meters can it travel on 10 gallons of fuel?

31. [II] By what number would you multiply a given number of centimeters to convert it into inches? Give the answer to 4 significant figures and show your work.

32. [II] A Boeing 747 jumbo jet carrying 385 people while cruising at 39 000 ft travels 280 yd on a gallon of aviation fuel. How many gallons will it need to travel 2000 km?

33. [II] Most people lose about 45 hairs per day out of a typical headful of 125 000. Suppose each hair averages 10 cm long. If you placed a year's lost hairs end-to-end, how far would they extend?

34. [II] What's the SI equivalent of 1.000 in.^{2}?

35. [II] Express the equivalent of 1.000 ft^2 in SI units.

36. [II] In an average lifetime, a human inhales roughly 5.0×10^5 yd^3 of air. How many cubic meters is that?

37. [II] One cup is equivalent to 237 milliliters. How much is that in the preferred SI unit of cubic meters?

38. [II] Roughly what size cube of water would have a mass of 1.0 kg?

39. [II] Each hour a large man sheds about 6×10^5 particles of skin. In a year, that amounts to about 1.5 lb (on Earth). Roughly how much is the mass of each such particle? What mass of skin will be shed in 50 years of adulthood?

40. [II] We are advised to keep our total blood cholesterol level at less than 200. What that actually means is 200 mg of cholesterol per deciliter of blood. Physicists totally shun the *deci* prefix and avoid the liter as well, so convert this quantity to SI units.

41. [II] Compute the sum of the following quantities: 0.066 m, 1.132 m, 200.1 m, 5.3 m, and 1600.22 m.

42. [II] Add the following quantities: 0.10 ms, 20.2 s, 6.33 s, 18 μs, and 200.55 ms.

43. [II] Add the following quantities: 1.00 g, 1.00 mg, 1.00 kg, and 1.00 μg.

44. [II] Compute the circumferences of circles having diameters of   (a) 5.42 μm,   (b) 0.5420 nm, (c) 5.420 000 mm,   (d) 542.0 m,   (e) 0.542 km.

45. [II] What is the product of the following quantities: 0.002 l g, 655.1 kg, and 4.41 μg?

46. [III] A *pipe* is 2 *hogsheads* and a hogshead is 63 gallons (or 2 barrels). Given that 1 U.S. gallon is 231 in.^{3}, what's the SI equivalent of a pipe to three significant figures?*

***Answers** to all odd-numbered Multiple Choice Questions and all odd-numbered Problems are in the Answers in the back of the book. Numbers in boldface indicate that a **solution** is provided in the Answers in the back of the book. Numbers in italic indicate that a **solution** is provided in the Student Solutions Manual.

2

Kinematics:
Speed and
Velocity

EVERY ATOM IN EVERY solid object, including the human body, is jiggling wildly. The Earth is spinning and revolving around a star that is itself whirling along with the galaxy. The entire Universe is expanding. Everything is in motion. When anything happens, when an event occurs, there is change, and change on a fundamental level is predicated on motion. The idea of motion is so basic that it has occupied human thought for millennia, and we still struggle with the subtleties of space and time. Physics *quantifies*—it associates numbers with concepts in a way that renders them *measurable*. In particular, the branch of physics that describes motion, that

Study for Dynamic Force of a Cyclist I by the Italian futurist Umberto Boccioni (1882–1916). The artist attempts to capture the feeling of motion; the physicist attempts to quantify motion and thereby measure it.

defines the needed ideas and develops their interrelationships *with no concern for what is causing the motion,* is known as **kinematics**. This chapter develops the concepts of speed, displacement, velocity, and relative velocity. Once these ideas are in place, we go on to study changes in velocity—acceleration (Chapter 3). Thus, armed with kinematics (the vocabulary of motion), we can begin to understand dynamics—what drives the motion, what drives the Universe.

SPEED

The faster a thing moves, the farther it goes in any interval of time. We have meter sticks to determine *distance* and clocks for *time;* we only need to relate these two fundamental ideas to quantify speed, to measure "how fast." Among the oldest existing thoughts on the subject are those of Aristotle (384–322 B.C.) and he, like his fellow Greeks of the era, specified speed as the distance traveled in a given amount of time. That's the way it was treated for well over 1000 years—one traveled with a speed of "so many miles in so many hours."

2.1   Average Speed

Today, we define the **average speed** *as the distance traveled divided by the time it took to do the traveling,* or

$$\text{Average speed} = \frac{\text{distance traveled}}{\text{time elapsed}}$$

Dividing by the time in specific units—seconds or minutes or hours—yields the *distance covered per unit time.* The Greeks never performed that division, apparently because they could not bring themselves philosophically to divide the unlike quantities of space and time. The division creates a new concept: a single quantity with a single numerical value and its own unit.

The above definition can be expressed symbolically as

$$v_{\text{av}} = \frac{l}{t} \tag{2.1}$$

where v_{av} is the average speed,* l the length of the path traveled, and t the time it took to travel it (Fig. 2.1). The units of speed are always those of *length over time* (kilometers per hour, meters per second, and so on). An automobile traversing a total along-the-road distance of 30 km in 2.0 h has an average speed of 15 km/h. Evidently, it could have stopped a few times on the way and then sped up and still covered the 30 km in the same 2.0 h—that's why we call this concept *average* speed.

A wooden horse on a carousel can swing around on its pedestal through 40 m in 20 s, and though it is forever coming back to the same spot in space, it still travels with an average speed of (40 m)/(20 s) = 2.0 m/s. *Speed is independent of the direction of motion,* and like all such quantities that have nothing to do with spatial orientation (such as length, temperature, time, mass, density, charge, and volume), it has only a *scale* or size and no associated direction. Accordingly, it's called a **scalar quantity**.

Figure 2.1 The rocket travels a distance measured along its actual path through space. After a time *t*, having gone a distance *l*, its average speed is $v_{\text{av}} = l/t$.

*We do not use s for speed because it has often been associated with the displacement in "space" (see p. 32). The symbol v derives from the word velocity (see p. 31).

Example 2.1 The Moon moves in a nearly circular orbit around the Earth with an average radius R of 3.84×10^8 m. If it takes 27.3 days for it to complete one revolution, determine its average orbital speed in m/s.

Solution: [Given: $R = 3.84 \times 10^8$ m and a once-around time of $t = 27.3$ days. Find: v_{av}.] We compute the motion over one complete orbit, which takes 27.3 days (that is, 27.3 d $\times$ 24 h/d $\times$ 60 min/h $\times$ 60 s/min = 2.359×10^6 s). The Moon's orbital path has an overall length equal to the circumference, namely $2\pi R$ (Appendix B). From Eq. (2.1), $v_{av} = l/t = 2\pi R/t$, or

$$v_{av} = \frac{2\pi(3.84 \times 10^8 \text{ m})}{2.359 \times 10^6 \text{ s}} = \boxed{1.02 \times 10^3 \text{ m/s}}$$

▶ **Quick Check:** *Always make sure your answer is the correct order-of-magnitude* (Appendix A-5). The quantity $2\pi \approx 6$ and the circumference is about $6 \times 4 \times 10^8$ m $\approx 24 \times 10^8$ m; that value divided by the time (about 24×10^5 s) yields 10^3 m/s.

Example 2.2 A student drives from home to school at an average speed of 25.3 km/h. If it takes 4.72 h to get to her destination, how far did she travel?

Solution: [Given: $v_{av} = 25.3$ km/h and $t = 4.72$ h. Find: l.] We are provided the average speed and time, and are asked to find distance traveled. Equation (2.1) contains all of these quantities, and solving it for distance we get

$$l = v_{av}t = (25.3 \text{ km/h})(4.72 \text{ h})$$

Since the least number of significant figures is three, $\boxed{l = 119 \text{ km}}$.

▶ **Quick Check:** Recalculate one of the given terms using l, say v_{av}: $v_{av} = l/t = (119.4 \text{ km})/(4.72 \text{ h}) = 25.297$ km/h = 25.3 km/h.

Table 2.1 shows some typical speeds and Table 2.2 compares mi/h with several other unit representations. An average speed in one set of units can be expressed in any other set of units. To convert, we use multiplicative factors equal to 1 so that the

TABLE 2.1	Typical speeds	
Speed, m/s	**Motion**	**Speed, mi/h**
300 000 000	Light, radio waves, X-rays, microwaves (in vacuum)	669 600 000
210 000	Earth-Sun travel around the galaxy	481 000
29 600	Earth around the Sun	66 600
1 000	Moon around the Earth	2300
980	SR-71 reconnaissance jet	2200
333	Sound (in air)	750
267	Commercial jet airliner	600
62	Commercial automobile (max.)	140
37	Falcon in a dive	82
29	Running cheetah	65
10	100-yard dash (max.)	22
9	Porpoise swimming	20
5	Flying bee	12
4	Human running	10
2	Human swimming	4.5
0.01	Walking ant	0.03
0.000 045	Swimming sperm	0.000 1

TABLE 2.2 A rough comparison of speeds in different units			
mi/h	**ft/s**	**km/h**	**m/s**
0	0	0	0
10	15	16	4
20	29	32	9
30	44	48	13
40	59	64	18
50	73	80	22
60	88	97	27
70	103	113	31
80	117	129	36
90	132	145	40
100	147	161	45
200	293	322	89
500	734	805	224
1000	1467	1609	447

TABLE 2.3 Speed conversion table*

m/s	km/h	mi/h	ft/s	⊶
1	0.2778	0.4470	0.3048	**m/s**
3.600	1	1.609	1.097	**km/h**
2.237	0.6214	1	0.6818	**mi/h**
3.281	0.9113	1.467	1	**ft/s**

*The units being converted head up the columns. The units into which they are transformed via multiplication are on the right of each row. Example: To go from km/h (second column) to mi/h (third row), multiply by 0.6214. Thus, 1.00 km/h = (1.00)(0.6214) mi/h.

size of the physical quantity never changes. For example, to change 15 km/h into m/s, first convert kilometers to meters by multiplying by (1000 m/1 km), which equals 1 exactly:

$$\left(\frac{15 \text{ km}}{1 \text{ h}}\right)\left(\frac{1000 \text{ m}}{1 \text{ km}}\right) = \frac{15\,000 \text{ m}}{1 \text{ h}}$$

To convert this expression to m/s, multiply by whatever it takes to transform the hours into seconds, namely (1 h/3600 s) = 1:

$$\left(\frac{15\,000 \text{ m}}{1 \text{ h}}\right)\left(\frac{1 \text{ h}}{3600 \text{ s}}\right) = 4.2 \text{ m/s}$$

Converting from one system of units to another is a practical necessity, especially in the United States, where quantities of interest are not always given in SI units. To go from U.S. Customary units to SI, or vice versa, requires the knowledge of a numerical constant that relates the two systems for each quantity involved. For example, for distance 1.000 in. = 2.540 cm, or 1.0000 mi = 1.6093 km. Thus, the conversion of 60 mi/h to SI is

$$60 \text{ mi/h} = \left(\frac{60 \text{ mi}}{1 \text{ h}}\right)\left(\frac{1.609 \text{ km}}{1 \text{ mi}}\right)\left(\frac{1000 \text{ m}}{1 \text{ km}}\right)\left(\frac{1 \text{ h}}{3600 \text{ s}}\right) = 27 \text{ m/s}$$

The three terms multiplying 60 mi/h can be compressed into a single conversion factor; that is, (0.4470 m/s) = (1.000 mi/h), (see Table 2.3). *It is not recommended that such specific conversion factors be memorized;* 1.000 mi ≈ 1.609 km should do nicely.

Example 2.3 Roger Bannister was the first person to run the mile in less than 4 min. He accomplished this feat in a time of 3.00 min 59.4 s in 1954. Compute his average speed in both miles per hour and meters per second.

Solution: [Given: l = 1.00 mi and t = 3.00 min 59.4 s. Find: v_{av}.] First, convert the time interval into some fraction of an hour because we want the answer in mi/h. (59.4 s)/(60 s/1 min) = 0.990 min and adding this to 3.00 min yields a total time of 3.990 min, or (3.990 min)/(60 min/1 h) = 0.0665 h. From Eq. (2.1)

$$v_{av} = \frac{l}{t} = \frac{1.00 \text{ mi}}{0.0665 \text{ h}} = \boxed{15.0 \text{ mi/h}}$$

Using Table 2.3 to convert the units, we obtain

$$v_{av} = (15.04 \text{ mi/h})\left(0.4470 \frac{\text{m/s}}{\text{mi/h}}\right) = \boxed{6.72 \text{ m/s}}$$

▶ **Quick Check:** 1.00 mi = 1.6093 km, hence v_{av} = (1.6093 km)/(3.00 min × 60 s/1 min + 59.4 s) = 6.72 m/s. Or (1 mi)/(4 min) = 15 mi/h.

Example 2.4 The 1980 speed record for human-powered vehicles was set on a measured 200-m run by a sleek machine called Vector. Pedaling back-to-back, its two drivers averaged 62.92 mi/h. This awkward mix of units is the way the data appeared in an article reporting the event. Determine the speed in m/s and then compute how long the record run lasted.

Solution: [Given: $l = 200$ m and $v_{av} = 62.92$ mi/h. Find: v_{av} in m/s and t.] Changing miles to feet, $v_{av} = (62.92 \text{ mi/h})(5280 \text{ ft/mi}) = 3.322 \times 10^5$ ft/h. Since 1.00 m $= 3.281$ ft, $v_{av} = (3.322 \times 10^5 \text{ ft/h})/(3.281 \text{ ft/}$ m$) = 10.12 \times 10^4$ m/h. To convert hours to seconds, divide by (3600 s/h); so $(10.12 \times 10^4 \text{ m/h})/(3600 \text{ s/h}) = \boxed{28.1 \text{ m/s}}$. Equation (2.1) relates l, v_{av}, and t, hence

$$t = \frac{l}{v_{av}} = \frac{200 \text{ m}}{28.1 \text{ m/s}} = \boxed{7.12 \text{ s}}$$

▶ **Quick Check:** 200 m $= (200 \text{ m})(6.214 \times 10^{-4} \text{ mi/}$ m$) = 0.124$ mi. $t = (0.124 \text{ mi})/(62.92 \text{ mi/h}) = 1.98 \times 10^{-3}$ h $= 7.1$ s. Or, with $v_{av} \approx 60$ mi/h $= 88$ ft/s ≈ 29 m/s; $(200 \text{ m})/(29 \text{ m/s}) = 7$ s.

Example 2.5 A 600-km cross-country automobile race is won by a team of two drivers, each of whom had the wheel for half the distance of the trip. If one averaged 60 km/h and the other 20 km/h, what was their overall average speed?

Solution: Here two distances and average speeds are involved, so introduce a subscript notation that will keep track of things. [Given: total length $l = 600$ km; distances traveled by drivers 1 and 2, $l_1 = l_2 = 300$ km (half the total distance); average speeds $(v_{av})_1 = 60$ km/h and $(v_{av})_2 = 20$ km/h. Find: v_{av}.] The overall average speed, as stated in Eq. (2.1) is the total length l divided by the total time t. We have l but need $t = t_1 + t_2$. Problems will often be encountered that deal with portions of a total distance or time—*the sum of the parts equals the whole* is one more equation to be used in the analysis. The individual times can be computed from the individual journeys using Eq. (2.1), as

$$t_1 = \frac{l_1}{(v_{av})_1} = \frac{300 \text{ km}}{60 \text{ km/h}} = 5.0 \text{ h}$$

and

$$t_2 = \frac{l_2}{(v_{av})_2} = \frac{300 \text{ km}}{20 \text{ km/h}} = 15 \text{ h}$$

The total time elapsed is 20 h, and therefore the average speed for the total trip is

$$v_{av} = \frac{l}{t} = \frac{600 \text{ km}}{20 \text{ h}} = \boxed{30 \text{ km/h}}$$

Notice that this is *not* equal to the average of the two average speeds, which is

$$\frac{(v_{av})_1 + (v_{av})_2}{2} = \frac{60 \text{ km/h} + 20 \text{ km/h}}{2} = 40 \text{ km/h}$$

The two are not the same because one person drove three times longer than the other and has a correspondingly greater influence on v_{av}. If that bothers you, imagine two rooms, one containing 100 basketball players all 6 ft tall and the other containing 2 kids each 3 ft tall. Find the average height of the people in each room and then find the average height of all the people. If you got 4.5 ft, you missed the point.

▶ **Quick Check:** Since the second driver traveled three times longer, that person's speed must be three times more weighty in the average. One person drives for 5 h out of 20 h, and the other for 15 h out of 20 h, hence

$$v_{av} = \frac{5(60 \text{ km/h}) + 15(20 \text{ km/h})}{20} = \frac{120 \text{ km/h}}{4}$$

and

$$v_{av} = 30 \text{ km/h}$$

2.2 Constant Speed

All of us have traveled in automobiles and have some intuitive sense of what **constant** or **uniform speed** is—just keep the speedometer locked at, say, 55 km/h and you have it. The scholars of the Middle Ages produced the first workable definition: *Uniform speed corresponds to the traversal of equal distances in equal intervals of time—of any fixed duration, however small* (Fig. 2.2). Galileo, 300 years later, was

l

*v*_{av}

t

Figure 2.2 An object traveling at a constant speed covers equal distances in equal intervals of time.

A modern speedometer in the United States shows speed in both mph (miles per hour) and kph (kilometers per hour).

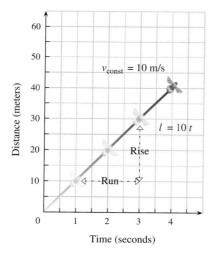

Figure 2.3 Distance-time graph of a bumblebee traveling at a constant speed (v_{const}). Take the distance to be the vertical axis and the time to be the horizontal axis. Note that the angle (although not the slope) of the line depends on the scale used to plot the curve.

very careful to stress that word *any* in his definition. If the interval is large, a body moving alternately faster and then slower can still have the same average speed over each such interval of time and so seem to be moving with a constant speed.

If v_{const} is the constant speed with which some object moves, then that is clearly also its average speed, and Eq. (2.1) becomes $v_{const} = l/t$. The path length traveled in a given time at a fixed speed is:

$$l = v_{const}t \qquad (2.2)$$

When the speed is constant (or if we know the average speed), the distance traversed during any time period can be determined.

Figure 2.3 is a **distance-time graph** for a bumblebee flying at a constant speed of 10 m/s. Starting at $t = 0$ and $l = 0$, the bee traverses 10 m during each second of motion; its *speed* or *rate-of-change of distance* is constant, and so the plot is a straight line. The *slope* of the line is the ratio of its *rise* to its *run* (the vertical distance between any two points on the line divided by their horizontal separation—see Appendix A-4). For the two points shown in the figure, the rise is (30 m − 10 m) = 20 m, and the run is (3.0 s − 1.0 s) = 2.0 s. Thus, the slope is 20 m/2.0 s = 10 m/s. *The slope of the straight line numerically equals the value of the constant speed:* a faster object produces a greater slope (Fig. 2.4).

Another important plot is the **speed-time graph**, which in this case of constant speed is simple. At each of the 1st, 2nd, 3rd, 4th, and so on, seconds, the speed of the bee is always 10 m/s. Hence, the curve is a horizontal straight line (Fig. 2.5). The rectangular area contained under the line within, say, the first 3.0 s of flight equals the height (that is, the speed, 10 m/s), multiplied by the length (that is, the time, 3.0 s), which yields 30 m, the distance. The bee traveled 30 m in those 3.0 s. *The area under the speed-time curve bounded by any two moments in time equals the distance traveled during that interval.* This statement may not seem like much of a revelation in this instance, but it applies to all speed-time graphs. In fact, it was in order to determine just such areas that integral calculus was devised (p. 65).

2.3   Delta Notation: The Change in a Quantity

Imagine now that you decide to measure the average speed of your new Ferrari (Fig. 2.6). As you pass some arbitrary initial point, P_i, on the road, your watch reads an initial time, t_i, of 12:04. Just at that moment, the odometer shows an initial distance, l_i, of 16 275.50 km. A far-off tree marks the final point of the course, P_f. As the car passes it, the odometer reads a final distance, l_f, of 16 285.50 km and the clock a final time, t_f, of 12:14. Here, $l_f - l_i = 10$ km and $t_f - t_i = 10$ min.

The Greek capital letter delta, Δ, placed before the symbol for some quantity means *the change in that quantity*. Thus, Δl (read delta "el") is the *change in the distance* or the length traversed: $\Delta l = l_f - l_i$. Similarly, Δt (read delta "tee") is the *change in time,* or the duration elapsed: $\Delta t = t_f - t_i$. Consequently,

$$v_{av} = \frac{\Delta l}{\Delta t} = \frac{l_f - l_i}{t_f - t_i} \qquad (2.3)$$

Figure 2.4 For a uniformly moving object, the distance-time graph is a straight line, and its slope is the speed. The greater the slope, the greater the speed.

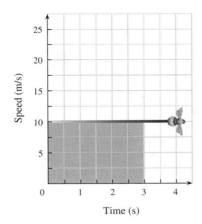

Figure 2.5 Plotting speed along the vertical axis and time along the horizontal axis gives us a speed-time graph. The area under the curve between any two times is the height (that is, the speed) multiplied by the length (that is, the time), and that result equals the distance traveled.

In this case, where $\Delta l = 10$ km and $\Delta t = 10$ min or 1/6 h, $v_{av} = 60$ km/h. This form of Eq. (2.1) is useful when the data are presented in terms of initial and final values, as opposed to overall lengths and times. Had you used a stopwatch and a resettable odometer or "tripmeter," you could have started the run at $t_i = 0$ and $l_i = 0$. In that instance, $\Delta l = l_f = l = 10$ km, the whole distance, and $\Delta t = t_f = t = 10$ min, the whole time. Eq. (2.3) becomes identical to Eq. (2.1).

2.4 Instantaneous Speed

Scholars of the mid-1300s dealt successfully with the idea of uniform speed. But when they tried to take the next logical step, to define the speed at any moment, they failed. They lacked the mathematical imagery of motion that Newton would create centuries later—the calculus of change. While in a car or an airplane, if you ask "How fast am I going NOW?" you want the *instantaneous speed*—the speed at that very moment. Today, that's usually read from a speedometer. Still, what are we really talking about?

The bumblebee flying in Figs. 2.3 and 2.5 has a constant speed, and it does not matter when you yell NOW, its speed at that moment (at any moment) would be 10 m/s. More realistically, Fig. 2.7a depicts a nonlinear distance-time plot. The bee leaps into the air at $t = 0$, lands on a flower after flying 2.0 m, rests, and takes off again at $t = 3.0$ s. It flies past your nose at $t = 5.0$ s, and alights on a branch at $t = 7.0$ s, ending the journey. Its average speed over any portion of the trip is the slope of the line from the beginning to the end of that portion. In Fig. 2.7b, for the interval from say $t = 2.0$ s to $t = 7.0$ s, $v_{av} = \Delta l/\Delta t = 8.0$ m/5.0 s $= 1.6$ m/s.

How do we specify the speed at some instant—NOW—like the one when the bee flew past the point P on the tip of your nose at $t = 5.0$ s? To that end, evaluate v_{av} using as an interval Δl, a small test course that extends along the flight path from a little before to a little after point P (Fig. 2.8a). Because it takes a time Δt for

Figure 2.6 Measuring the average speed of a car over a run from point P_i to point P_f. The distance traveled is $l_f - l_i = 10$ km and the time elapsed is $t_f - t_i = 10$ min. The average speed is: $v_{av} = 10$ km/10 min $= 60$ km/h.

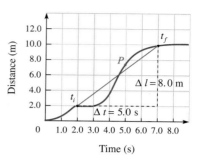

Figure 2.7 (a) The distance-time curve for a bee. The bee flew to a flower at $t = 2.0$ s, rested until $t = 3.0$ s, and then flew past your nose at point P, alighting on a branch at $t = 7.0$ s.

(b) The average speed of the bee during the interval from $t = 2.0$ s to $t = 7.0$ s is $v_{av} = \Delta l/\Delta t = 8.0$ m/ 5.0 s = 1.6 m/s.

the bee to traverse this Δl, the average speed is again $\Delta l/\Delta t$. Our scheme will become increasingly better as the length interval Δl straddling P is narrowed, and that narrowing will occur if we reduce the time interval of the measurement (Fig. 2.8b). For example, suppose the bee flew from 0.000 005 m before P to 0.000 005 m beyond P in 0.000 002 s. Its average speed over that tiny distance is equal to 0.000 010 m divided by 0.000 002 s, or 5 m/s. This is the average speed measured during a small but finite time N-O-W as opposed to an instantaneous NOW. It is not necessarily the speed exactly at P, but it is very close to it. By further shrinking Δt, and thereby Δl, we can get as close as we like to the speed NOW.

There is a geometrical way to envision the instantaneous speed of the bee at P. As Δt is made ever smaller, Δl also shrinks, and $\Delta l/\Delta t$ (the average speed) approaches the slope of the distance-time curve at P. The slope of any curve at any point is the slope of the *tangent* to the curve at that point (Fig. 2.9a). The slope precisely at P is the instantaneous speed at P, just as the slope anywhere along the curve of Fig. 2.3 is the constant instantaneous speed 10 m/s. From Fig. 2.9b, we see that the bee flew by your nose at 2.9 m/s.

The central notion here is that Δl will become infinitesimally small as Δt becomes infinitesimally small, and yet their ratio, $\Delta l/\Delta t$, will approach a finite limiting value: a tiny number divided by another tiny number need not itself be tiny. This limiting value approached by $\Delta l/\Delta t$ as $\Delta t \to 0$ (that is, as Δt gets infinitesimally small) is called the **instantaneous speed**, v. Mathematically that's written as:

$$v = \lim_{\Delta t \to 0} \left[\frac{\Delta l}{\Delta t} \right] \qquad (2.4)$$

A 0.30-caliber bullet "frozen" in an exposure lasting less than a millionth of a second. During that time, the bullet certainly moved, but not enough to blur its image. It had already cut the card in half before the picture was taken. The average speed of the bullet during the exposure is a practical value for its instantaneous speed at the "moment" shown.

The instantaneous speed is the limiting value of the average speed ($\Delta l/\Delta t$) determined as the interval over which the averaging takes place (Δt) approaches zero. Henceforth, when we talk about "speed," this concept is what we have in mind. Notice that $\Delta l/\Delta t$ is to be determined as Δt *approaches*, rather than *reaches*, zero. It would make no sense to ask for the average speed over a zero time interval; nothing moves a real distance in no time, and the definition of average speed falls apart when $\Delta t = 0$. This subtlety was a point of confusion for centuries. It was to solve the very problem of instantaneous speed that Newton devised the calculus. Indeed,

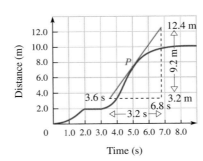

Figure 2.8 To find the instantaneous speed of the bee at point P, determine the average speed over a tiny interval straddling P. Then shrink Δl until v_{av} remains constant at the instantaneous speed, v.

Figure 2.9 (a) As Δt shrinks, Δl shrinks and the line from t_i to t_f approaches the tangent to the curve at P. In this case, both t_i and t_f approach $t = 5.0$ s. (b) The slope of the tangent at P is $v = 9.2$ m/ 3.2 s = 2.9 m/s, and this value is the instantaneous speed at $t = 5.0$ s.

Eq. (2.4) serves as the definition of the *derivative*, written as $v = (dl/dt)$. Speed is the derivative of l with respect to t. Not to worry, we will not be using calculus. For us, it will suffice to think of v as v_{av} taken over a tiny interval Δt centered on NOW, where Δt is small enough so that v is constant within it.

VELOCITY

To describe the motion of an object completely, we must specify both the rate of travel (that is, the speed) and its direction. The single concept that at once embraces both speed and direction is called **velocity**. Although we have managed thus far with just the notion of speed, more complex circumstances will require us to consider direction as well. A situation in which several motions occur simultaneously in different directions can only be analyzed using velocities. For example, a boy running east at 5 m/s throws a ball north at 10 m/s with respect to him. In what direction and at what speed relative to the ground will the ball travel? We now develop the mathematics needed to treat all such directional quantities (velocity, acceleration, force, momentum, and so on), the mathematics of *vectors* (Appendix D).

On Newton, Leibniz and Limits.
Neither man [the two founders of the calculus] succeeded in doing more with the limit concept proper than confusing himself, his contemporaries, and even his successors. At one point in his Principles, *Newton does state the correct version of the notion of instantaneous rate of change but apparently he did not recognize this fact, for in later writings he gave poorer explanations of the logic of his procedure.*

MORRIS KLINE (1953)
Mathematics in Western Culture

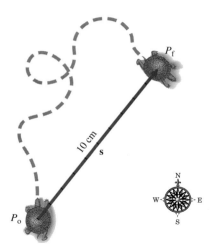

Figure 2.10 The displacement vector **s** of a turtle that meandered from point P_o to point P_f. The vector quantity **s** tells us where the turtle ended up in relation to where it started.

2.5 The Displacement Vector

Suppose we put a little turtle on a piece of paper at some point P_o and leave for a while. On our return, the turtle is discovered at a new spot, P_f. Nothing is known about the route it took, but we do know the turtle has been displaced. This *displacement is designated as the straight-line shift in position from P_o to P_f specifying both its length and direction*. In Fig. 2.10, the meandering reptile ended up 10 cm northeast of where it started. Any concept such as this that can be stated completely only if we provide both its magnitude (size) and direction is known as a **vector quantity**. It's standard in textbooks to designate vector quantities by boldfaced letters, reminding us that we are dealing with something other than a scalar.

To be precise, **the displacement of an object from a starting point P_o is the vector extending from that point to the position where the object is finally**. Displacement is often of considerable interest, as is the case with the displacement of a football from the person who threw it. It is common practice to represent the *displacement vector* by **s**, a symbol that comes from the idea of the space between two points.

In Fig. 2.10, the vector **s** can be represented pictorially by an arrow 10 cm long pointing northeast. And if 10 cm is inconveniently long to draw, simply scale it down to 10 "anythings" and just label it 10 cm. The *length* of the arrow representing a vector corresponds to what is called the **magnitude** (the positive numerical value) of that vector, and it is symbolized by an italic light typeface letter. Thus, the magnitude of the displacement vector **s** is s, which is a scalar specified by a single number, in this case equal to 10 cm.

Imagine now that the turtle takes off on a new trip across the page, this time traveling in a straight line east for 3 cm and thereby undergoing its first displacement $\mathbf{s}_1$, arriving at point A in Fig. 2.11. After an idle moment, it turns 90-degrees left, going from A straight north for 4 cm to point P_f, thereby undergoing a displacement $\mathbf{s}_2$. The *resultant* displacement with respect to P_o is the vector **s**, which is independent of the actual path taken in getting to P_f. Its magnitude can be computed from the right triangle using the famous theorem of Pythagoras (Appendix B), $s = \sqrt{3^2 + 4^2} = \sqrt{25}$. Fortunately, the turtle walked two sides of a 3-4-5 right triangle, so s, the hypotenuse, comes out neatly as 5 cm long—P_f is 5 cm from P_o in a direction somewhat east of north.

There is a considerable difference between *distance* and *displacement*. For instance, imagine yourself throwing a football to someone running in zigzags. He started at your side and is now far down the field, dodging the opposition. You certainly might worry about the total distance traveled by the runner; surely he is considering it. But you are more concerned with the moment-by-moment *displacement* of the potential receiver—with the straight-line arrow from you to him.

2.6 Some Vector Algebra

We have not completely defined what a vector is until we establish some rules for its behavior. One such needed operation is **vector addition**. We do not have a complete idea of what numbers, or scalars, are until we know how to manipulate them, and the same is true with vectors.

Referring to Fig. 2.11, we can envision somehow adding the intermediate displacements ($\mathbf{s}_1 + \mathbf{s}_2$) to arrive at the total displacement **s**, but it's evident that we cannot simply add in the usual algebraic manner. The total path-length traveled is $s_1 + s_2 = 7$ cm, while the magnitude of the total displacement is $s = 5$ cm, and obviously $s \neq s_1 + s_2$. The illustration itself provides a general rule for adding vectors known as the **tip-to-tail method**. *Two vectors can be added together by arranging them such that the tail of the second is at the tip of the first; the sum of these, also*

Figure 2.11 The total displacement **s** of a turtle that traveled from P_o to A to P_f via displacements $\mathbf{s}_1$ and $\mathbf{s}_2$. The total displacement equals the vector sum of the individual displacements: $\mathbf{s} = \mathbf{s}_1 + \mathbf{s}_2$.

called the **resultant vector**, *extends from the tail of the first to the tip of the second.* In the process of adding two vectors, the sum will not change if either one (or both) is moved around, provided it is kept parallel to its original direction.

Notice that had the turtle walked north first and then east (inverting the order of the displacements), it would still have arrived at P_f, which means that $\mathbf{s} = \mathbf{s}_1 + \mathbf{s}_2 = \mathbf{s}_2 + \mathbf{s}_1$. Just as with ordinary scalars, the order of addition is irrelevant, and you can put either tip to either tail when adding two vectors.

Figure 2.12 makes the same point, this time with vectors that form an oblique triangle. It suggests another technique for graphically adding two vectors. Beginning with the tip-to-tail scheme and realizing that the order of addition is irrelevant, we can construct the resultant vector in two ways that, taken together, form a parallelogram. In other words, draw the two vectors we wish to add, $\mathbf{s}_1$ and $\mathbf{s}_2$, emerging from a *common origin*. Using these as adjacent sides, construct a parallelogram. *The diagonal emerging from the origin is then the resultant of the two.* This procedure is called the **parallelogram method**.

Once we have a rule for adding two vectors, we have a rule for adding any number of vectors. Thus in Fig. 2.13, we can add the four vectors, $\mathbf{A} + \mathbf{B} + \mathbf{D} + \mathbf{F}$,

(a)

(b)

(c)

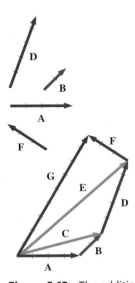

Figure 2.12 (a) Two vectors that are to be added in two ways. (b) The tip-to-tail addition of two vectors ($\mathbf{s}_1 + \mathbf{s}_2 = \mathbf{s}_2 + \mathbf{s}_1 = \mathbf{s}$). (c) The parallelogram formed with $\mathbf{s}_1$ and $\mathbf{s}_2$ as its sides.

Figure 2.13 The addition of several vectors ($\mathbf{A} + \mathbf{B} + \mathbf{D} + \mathbf{F}$) via the tip-to-tail method. Here $\mathbf{C} = \mathbf{A} + \mathbf{B}$, $\mathbf{E} = \mathbf{C} + \mathbf{D}$, and $\mathbf{G} = \mathbf{E} + \mathbf{F} = \mathbf{A} + \mathbf{B} + \mathbf{D} + \mathbf{F}$. Knowing how to add two vectors allows us to add any number of vectors.

two at a time, following the tip-to-tail method. Accordingly, **A** plus **B** is **C**, to which we add **D**, such that **C** + **D** = **E**, to which we add **F**, such that **E** + **F** = **G**, so that **A** + **B** + **D** + **F** = **G**. Evidently, it is not really necessary to bother with the intermediate sums of **C** and **E**. The figure shows that we could have just as well tipped-to-tailed **A**, **B**, **D**, and **F** directly and, for that matter, we could have done it in any order at all.

Example 2.6 A treasure map was torn into six pieces by its maker, a clever one-eyed gentleman with a parrot. The first piece shows a large recognizable tree and simply reads "start here and go." Although in no particular order, each of the other five fragments reads, in turn: "3 m−east," "5 m−north," "5 m−southeast," "2.5 m−south," and "6.5 m−west." Find the treasure.

Solution: Each of the five instructions is a displacement vector and vector addition is independent of order. It follows from Fig. 2.14 that the treasure is buried 1 m due south of the old tree or, if you like, the resultant displacement vector has a magnitude of 1 m and points south. Try this example with cut matchsticks; so long as their lengths and orientations remain constant, the resultant will be the same for a fixed starting point independent of the order in which the sticks are put down.

Figure 2.14 The five vectors in (a) can be added in any order to find the treasure. In (b), we start at the tree and first move east 3 m. In (c), that 3-m displacement is last, but we still arrive at the same spot where the treasure is buried.

Vectors that are either parallel or antiparallel (that is, oppositely directed) are added in the usual tip-to-tail way, and the results are simple and useful (Fig. 2.15). In the case of two parallel vectors, s_1 and s_2, the magnitude of the resultant, s, equals the sum of the magnitudes of the scalar distances.

Suppose now that we add to a vector **s** an identical vector **s**. The resultant of (**s** + **s**) lies in the same direction as either constituent **s** and is twice as long as either one. It would seem reasonable from the usual procedures of algebra to write the resultant as 2**s** = **s** + **s**, which defines the process of **multiplication of a vector by a scalar**. Accordingly, if **s** is a displacement of 10 cm−east, 6**s** is a displacement of

60 cm−east. *Multiplying a vector by a positive scalar multiplies the vector's magnitude in the usual way, leaving its direction unchanged.*

If two vectors are antiparallel (Fig. 2.15b), the magnitude of the resultant equals the difference of the individual magnitudes, $s_1 - s_2$. *Vectors that are oppositely directed subtract algebraically.* A displacement of 10 m−east, followed by a displacement of 10 m−west, yields a total displacement of zero. If we interpret the symbol $(-\mathbf{s})$ to be a vector equal in magnitude but antiparallel to $\mathbf{s}$, then the sum $\mathbf{s} + (-\mathbf{s}) = \mathbf{s} - \mathbf{s} = 0$, so the two cancel one another in the usual algebraic way. Accordingly, **vector subtraction** can be defined as follows: any vector $\mathbf{B}$ can be subtracted from any vector $\mathbf{A}$ by reversing its direction to form $-\mathbf{B}$ and then adding them in the usual tip-to-tail way (Fig. 2.16).

Multiplying a vector by a negative scalar multiplies the vector's magnitude in the usual way but also reverses its direction. If $\mathbf{s}$ is a displacement of 10 cm−east, $-6\mathbf{s}$ is a displacement of 60 cm−west. Note that the magnitude is always positive.

Each of these operations (vector addition, subtraction, and multiplication by a scalar) has its direct application in our description of the physical Universe.

2.7   Average Velocity

Speed is a scalar that tells us "how fast" without dealing with "in what direction." But more often than not, direction is an essential aspect of motion. There is a difference between the command "March at 2 km/h" and "March at 2 km/h–south, toward the cliff." The culminating concept of this chapter is *instantaneous velocity*, and to understand it we have to examine what happens to the average velocity vector as $\Delta t \to 0$. As we shall see, the instantaneous velocity is the vector whose direction at any moment is the direction of motion and whose magnitude is the speed.

Back to our bumblebee, which in a time t now flies from P_o to P_f in an arc shown in Fig. 2.17a. We see from Eq. (2.1) that the bee's average speed is $v_{av} = l/t$, and, with this expression as a guide, we define the **average velocity** to be the *vector formed by the ratio of the displacement, measured from the starting point of the motion, over the time elapsed*, or

$$\mathbf{v}_{av} = \frac{\mathbf{s}}{t} \tag{2.5}$$

Since t is a positive scalar, multiplying $\mathbf{s}$ by $1/t$ has no effect on its direction; $\mathbf{v}_{av}$ is parallel to $\mathbf{s}$. Average velocity is a necessary idea on the way to considering instantaneous velocity, although by itself $\mathbf{v}_{av}$ is of limited usefulness—a racecar roaring around a track will have zero displacement and zero average velocity whenever it returns to its starting point, no matter how fast it is going.

Figure 2.15   (a) The sum of two parallel vectors. Vectors that point in the same direction can be added like scalars in that $\mathbf{s} = \mathbf{s}_1 + \mathbf{s}_2$. (b) The sum of two antiparallel vectors.

Example 2.7   A pulse from a laser travels at a constant speed in a straight line reaching a detector 3×10^4 m due north in 100 μs. Calculate the average velocity of the beam.

Solution: [Given: speed is constant; $\mathbf{s} = 3 \times 10^4$ m−north; $t = 100 \times 10^{-6}$ s. Find: $\mathbf{v}_{av}$.] The beam travels with a constant speed in a straight line, so it has a constant velocity. This constant velocity is also equal to its

average velocity, namely

$$\mathbf{v}_{av} = \frac{\mathbf{s}}{t} = \frac{3 \times 10^4 \text{ m−north}}{100 \times 10^{-6} \text{ s}} = \boxed{3 \times 10^8 \text{ m/s−north}}$$

The displacement is due north, and so the average velocity is due north.

▶ **Quick Check:** $v_{av}t = s = (3 \times 10^8 \text{ m/s})(10^{-4} \text{ s}) = 3 \times 10^4$ m.

(a)

(b)

(c)

(d)

Figure 2.16 The sum and differences of two vectors. In (a) we see vectors **A** and **B** and their negatives −**A** and −**B**; (b) shows **A** − **B**; and (c) shows −**A** − **B** and **A** + **B**; (d) adds −**A** and **B** to get **B** − **A**. Notice that −**B** + **A** is the negative of −**A** + **B**.

Before the idea of instantaneous speed was introduced (p. 29) we studied the average speed in terms of intervals, $\Delta l/\Delta t$. The same thinking applies to the average velocity where we now focus on a portion of the flight of length Δl. As the bee in Fig. 2.18 flies the segment of its path from P_i to P_f in a time Δt, the vector **s**, measured from P_o, changes as it tracks the bee instant-by-instant. The displacement with respect to P_o initially, $\mathbf{s}_i$, plus the change in the displacement, $\Delta\mathbf{s}$, equals the final displacement with respect to P_o, $\mathbf{s}_f$ (Fig. 2.19). The symbolic form is $(\mathbf{s}_i + \Delta\mathbf{s} = \mathbf{s}_f)$ and, therefore, $\Delta\mathbf{s} = \mathbf{s}_f - \mathbf{s}_i$, which is reminiscent of Eq. (2.3). Note that $\Delta\mathbf{s}$ is also the displacement of the bee at P_f with respect to its starting point P_i.

Using the definition as stated in Eq. (2.5), we can now express the average velocity over the curved route from P_i to P_f as

$$\mathbf{v}_{av} = \frac{\Delta\mathbf{s}}{\Delta t} = \frac{\mathbf{s}_f - \mathbf{s}_i}{t_f - t_i} \tag{2.6}$$

Notice that when P_i and P_f are coincident, $\Delta\mathbf{s} = 0$ and $\mathbf{v}_{av} = 0$.

2.8 Instantaneous Velocity

Earlier, we transformed the notion of average speed into instantaneous speed by causing the time interval over which the averaging was happening to shrink toward zero. We now apply the same limiting process to the average velocity, Eq. (2.6), to get the *instantaneous velocity* **v** (or the velocity, for short).

Figure 2.17 The distance traveled (*l*) and the displacement (**s**) in the two different cases of (a) motion along a curve, and (b) motion along a straight line. For the special case of straight-line motion in a fixed direction, the magnitude of the average velocity *s/t* equals the average speed, *l/t*. That is generally not the case.

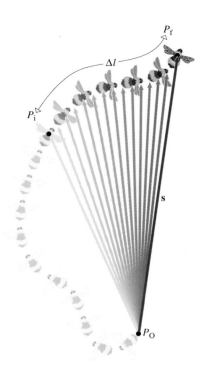

Figure 2.18 The bumblebee flies from P_i to P_f always tracked by the vector **s**. The distance it travels in the time Δt is Δl.

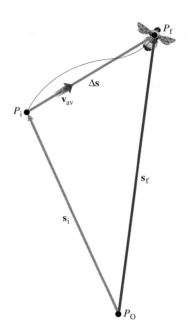

Figure 2.19 The change in the displacement of the bumblebee from P_i to P_f occurring in a time Δt is Δ**s**. Notice that **v**$_{av}$ is in the direction of Δ**s**.

Figure 2.20 The bee flies from point P_i to point P_f. In successive parts of the figure, from (a) to (d), less and less time passes, and point P_f comes closer and closer to P_i. As Δt shrinks, Δs approaches Δl, and the magnitude of **v** approaches v.

The bee in Fig. 2.19 flies from P_i to P_f in a time Δt, experiencing a displacement in the process, Δ**s**. As the time, Δt, allowed for the bee to fly shrinks, the insect cannot travel as far; P_f moves nearer to P_i and Δ**s** diminishes as well (Fig. 2.20). Accordingly, we define the **instantaneous velocity** *at any point* (P_i) *as the limit* (*as* $\Delta t \to 0$) *of the ratio of the displacement of the object from that starting point over the time interval,* or

$$\mathbf{v} = \lim_{\Delta t \to 0} \left[\frac{\Delta \mathbf{s}}{\Delta t} \right] \tag{2.7}$$

Now that we have **v** mathematically, let's see what it is physically. There are two crucial insights here: first, as P_f approaches P_i, the magnitude of Δ**s**, that is Δs, gradually shrinks and the change in path-length, Δl, shrinks right along with it (Fig. 2.20). For extremely small values of Δt, Δs and Δl become increasingly indistinguishable. We can conclude that as $\Delta t \to 0$, the two merge and become equal, which means that in the limit $\Delta s/\Delta t = \Delta l/\Delta t$; *the magnitude of the instantaneous velocity vector,* **v**, *equals the instantaneous speed* v. That is why we used the symbol v for speed in the first place—*speed is the magnitude of the velocity.*

The second important quality of instantaneous velocity as defined by Eq. (2.7), is its direction. As the vector Δ**s** diminishes in size (Fig. 2.20) with decreasing Δt, its end point glides along the curve toward P_i. As $\Delta t \to 0$, P_f moves closer and closer to sitting on top of P_i and $\Delta\mathbf{s}$, however small, becomes tangent to the path at P_i. Since $\mathbf{v}$ is defined by the limit of this shrinking process, *the velocity vector at any point is tangent to the path, pointing in the direction of motion* (Fig. 2.21). The magnitude of the velocity vector is the speed, and its direction is always tangent to the path in space.

Example 2.8   A particle-beam pulse fired due north, 45° up from the horizon strikes a distant orbiting satellite. If the beam travels in a straight line at a constant speed of 2.5×10^8 m/s and it takes 1.0 ms to reach the target, what is the displacement of the pulse from the source at the moment it hits the satellite?

Solution: [Given: $\mathbf{v} = 2.5 \times 10^8$ m/s–north-45° up, and $t = 1.0$ ms. Find: $\mathbf{s}$.] The instantaneous velocity of

the beam is constant, hence

$$\mathbf{s} = \mathbf{v}t = (2.5 \times 10^8 \text{ m/s–north-45° up})(1.0 \times 10^{-3} \text{ s})$$

and so $\boxed{\mathbf{s} = 2.5 \times 10^5 \text{ m–north-45° up}}$

▶ **Quick Check:** because the motion is in a straight line $s = vt$; $v = 1/4 \times 10^9$ m/s, $t = 10^{-3}$ s, $s = 1/4 \times 10^6$ m.

Example 2.9   A large clock mounted on a wall in a railroad station has a second hand 1.0 m long. Assuming that the hand sweeps smoothly around, find the velocity of a point on its very tip at exactly 15 s after noon.

Solution: [Given: the point moves around on a circle of radius $r = 1.0$ m, its speed is uniform, and it traverses the circumference in 60 s. Find: $\mathbf{v}$ at $t = 45$ s.] Most of what is given is not stated explicitly—you need to recognize the implied. Nature is rarely explicit. *Draw a diagram* (Fig. 2.22). We have to find the point's instantaneous velocity—speed and direction—when the second hand is horizontal. The total distance swept out by the very tip of the hand in one revolution is $2\pi r = 2(3.14)1.0$ m $= 6.28$ m. Because the motion is uniform, the point's average speed equals its instantaneous speed, thus

$$v = \frac{l}{t} = \frac{6.28 \text{ m}}{60 \text{ s}} = 0.10 \text{ m/s}$$

The velocity 15 s after noon is then $\boxed{0.10 \text{ m/s–straight downward}}$, while 15 s before noon it's 0.10 m/s–straight upward.

Figure 2.22   A clockface at 15 s after 12:00. At the instant shown, the point on the very end of the second hand is moving downward with a velocity **v**. At 15 s before noon, the hand will be pointing at 9 and **v** will be directed straight upward.

▶ **Quick Check:** The velocity is always tangent to the circle (that is, perpendicular to the radius), pointing in a direction determined by the clockwise motion. $2\pi R \approx 6$ m, $v \approx 6$ m/60 s ≈ 0.1 m/s.

Instantaneous velocity cannot be measured with the same accuracy that it can be thought about. In the end, Δt will be as tiny as you can measure it, but finite, meaning that we will always be restricted to measuring average velocity. This restriction holds even if we measure so rapidly there is no practical distinction between average velocity and the ideal of instantaneous velocity.

One-Dimensional Motion

An especially simple graphical relationship exists between displacement and velocity when the motion is confined to one dimension. To illustrate, suppose an object moves on a straight line (Fig. 2.23) along the x-axis. Let the magnitude of the displacement from the origin O be s_x (Fig. 2.24). This object is already moving with a velocity of 0.6 m/s in the positive x-direction at $t = 0$, and it continues at that rate (constant slope) until $t = 9$ s, when the speed slowly begins to decrease (Fig. 2.24). At $t = 11.5$ s, the tangent to the displacement-time curve has zero slope and the speed is zero. Because we are plotting displacement from O and not the total distance traveled, s_x can decrease and the curve will have a negative slope whenever the object moves back toward O. At $t = 19.6$ s, the object momentarily stops and thereafter moves in the positive x-direction.

The velocity is always in the direction of $\Delta\mathbf{s}_x$ which is the direction of motion. When the slope of Fig. 2.24 is positive, $\mathbf{v}$ points away from the origin, and when the slope is negative, $\mathbf{v}$ points toward the origin. Thus, since we know the direction of motion, we can talk about velocity and not just speed. *For 1-dimensional motion, the slope of the displacement-time curve at every point in time is the velocity at that instant.* Fig. 2.25 is a plot of the instantaneous velocity determined from the tangents to the curve in Fig. 2.24.

RELATIVE MOTION

Most people believe that they can know with certainty that either a thing is moving or it is not; they think that motion is absolute. But the Earth is orbiting the Sun (at about 66 000 mi/h), and the Sun is whirling along with the galaxy (at about 480 000 mi/h), and the whole Universe is expanding. Nothing is at rest; besides, and even more importantly, there is no way to measure the difference between rest and uniform motion. There is no physical difference between rest and uniform motion. That's why it feels as though you are at rest at this very moment, sitting reading this book.

We learn from the Special Theory of Relativity that motion is relative, not absolute. Motion is always measured with respect to something. We usually hold the Earth to be at "rest," and think about objects moving relative to it. Even so, there are countless situations not quite that simple: two things moving with respect to the Earth may surely be moving relative to each other, and that latter motion might even

Figure 2.21 The velocity is everywhere tangent to the path, pointing in the direction of motion.

The plane moving at a constant speed is an inertial reference frame. The rockets are traveling at relatively low speeds with respect to it. By comparison, the speed of the rockets with respect to the ground is much higher.

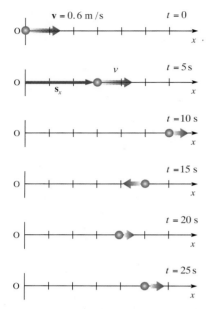

Figure 2.23 The motion of an object in one dimension at successive 5-s intervals. Notice that sometime between $t = 10$ s and $t = 15$ s, the object stops and reverses direction. It can't move along a line and change direction without stopping, if only for an instant.

Figure 2.24 A plot of the magnitude of the displacement versus time for that object as it moves along its straight path.

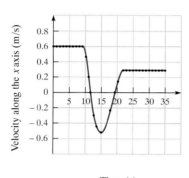

Figure 2.25 The velocity along the x-axis determined each second from the slope of the curve of Fig. 2.24. At $t = 11.5$ s and $t = 19.6$ s, the slope of the displacement versus time curve is zero and the velocity is zero.

be the more important one. A front-runner in a race should be as concerned about her velocity relative to the person in second place as with her velocity relative to the finish line. A pilot landing a plane on the deck of a moving aircraft carrier has relatively similar concerns.

Consider a laboratory that is either at rest or moving with a constant velocity. Such a system is called an *inertial frame of reference* or an *inertial system*. In the remainder of this chapter, we study uniformly moving objects traveling with respect to inertial systems. In most cases of motion occurring on the surface of the Earth, the planet itself (that is, the ground) serves as just such an inertial frame (Fig. 2.26).

The central idea in relative motion is *velocity*. To combine the effects of several simultaneous motions, we need only to add the corresponding velocity vectors. The procedure is an elegant example of the power of the vector formulation.

2.9 Velocity with Respect to . . .

Consider the journey of a little turtle meandering eastward along the moving ruler of Fig. 2.27. At the last moment, frozen in the diagram, the *r*uler has been displaced with respect to the *E*arth by $\Delta\mathbf{s}_{RE}$. The *t*urtle, in turn, has undergone a displacement with respect to the end of the *r*uler of $\Delta\mathbf{s}_{TR}$. The total displacement of the *t*urtle with respect to the *E*arth (that is, with respect to the starting point P_o fixed on the ground) is given by the vector sum

$$\Delta\mathbf{s}_{TE} = \Delta\mathbf{s}_{TR} + \Delta\mathbf{s}_{RE} \qquad (2.8)$$

An observer sitting stationary on the Earth would see the turtle travel from P_o to P_f in some time interval Δt. Consequently, if we divide both sides of the equation by Δt and take the limit as $\Delta t \to 0$, we get, via Eq. (2.7), the corresponding velocity expression, namely

$$\mathbf{v}_{TE} = \mathbf{v}_{TR} + \mathbf{v}_{RE} \qquad (2.9)$$

If, for example, the velocity of the *t*urtle with respect to the *r*uler is 0.01 m/s–east, and the velocity of the *r*uler with respect to the *E*arth is 0.50 m/s–east, then $\mathbf{v}_{TE} = $ 0.51 m/s–east. The velocity vectors add to give the resultant, just as the displacement vectors do. Remember that the equation is really quite general even though the subscripts may change; it applies as well to *t*urtles, *r*ulers, and *E*arths as to *t*rains, *r*ubberbands, and *e*lephants.

The order of the subscripts in Eq. (2.9) is crucial to the logic, even though the order of the vector addition is not. When setting up expressions such as this one, arrange the subscripts so that the first and last on one side of the equation are the same as the first and last on the other side. It is convenient to imagine that identical letters cancel when they come one after the other in the sequence of subscripts on the right. For instance, the vector addition effectively removes the R subscripts in Eq. (2.9)—the intermediate role of the moving ruler is subsumed.

Suppose that the turtle in Fig. 2.27 turns around and begins heading *west* at a speed with respect to the ruler of 0.01 m/s. Equation (2.9) can still be used. In fact, it would apply regardless of the directions of the two vectors. In the present instance, $\mathbf{v}_{TR}$ and $\mathbf{v}_{RE}$ are antiparallel. The resulting speed is the *difference* between their speeds, and subtracting the smaller from the larger yields $\mathbf{v}_{TE} = $ 0.49 m/s–east. The turtle approaches the left edge of the ruler but, because v_{RE} is greater than v_{TR}, it is nonetheless carried east of the fixed point P_f in Fig. 2.28.

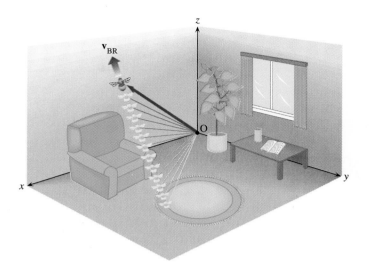

Figure 2.26 A bee flying a straight path in a room. The room is fixed to the ground and represents an inertial reference frame. A vector from the origin, O, to the bee tracks the insect's motion. The velocity of the bee with respect to the room is $\mathbf{v}_{BR}$. If the room happens to be in a mobile home, the bee's motion with respect to the ground can get rather complicated.

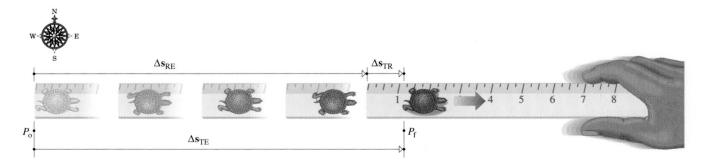

Figure 2.27 The displacement of a turtle walking on a moving ruler. The turtle walks to the right with respect to the ruler, while the ruler moves right with respect to the Earth. As a result, the little traveler ends up a rather large distance Δs_{TE}, from the point on Earth, P_o, where it started.

Figure 2.28 With respect to the Earth, the turtle travels east even though it's heading toward the west end of the ruler. The situation is much like that of a person walking to the back of a jet airliner while flying at 600 mi/h. If the walk takes 1 min, the plane travels 10 mi in the process.

On the other hand, if the turtle began moving west at P_f with a constant speed equal to that of the ruler (0.50 m/s), it would again march to the left end of the ruler and yet still be trudging along "in place" at P_f. Its speed relative to the Earth would be zero. This is what occurs when someone aboard a train pulling out of a station runs to the rear of the car trying to linger another moment face-to-face with a friend standing on the platform. It is the classic treadmill scheme well known to rodents and cardiologists.

Example 2.10 A big, black dog saunters down the aisle of an Amtrak Pullman at a velocity of 5.00 km/h–east with respect to the car. The train, all the while, is traveling at 10.00 km/h–east. A resident flea heading rumpward along the hound's back moves at 0.01 km/h–west with respect to the dog. Find the velocity of the flea with respect to the Earth.

Solution: [Given: v_{DT} = 5.00 km/h–east; v_{TE} = 10.00 km/h–east; v_{FD} = 0.01 km/h–west. Find: v_{FE}.] Extending Eq. (2.9) to a three-vector sum and arranging the first and last subscripts on the right to be F and E, respectively, we get

$$v_{FE} = v_{FD} + v_{DT} + v_{TE} \qquad (2.10)$$

Notice that the pairs of Ds and Ts will effectively can-

cel, yielding the desired resultant. In particular, v_{FE} = (0.01 km/h–west) + (5.00 km/h–east) + (10.00 km/h–east). Because parallel and antiparallel vectors can be dealt with algebraically, we simply add the speeds of similarly directed velocities and subtract those of oppositely directed ones. The easterly speed is greater, and so we subtract 0.01 km/h–west from 15.00 km/h–east to get $\boxed{v_{FE} = 14.99 \text{ km/h–east}}$.

▶ **Quick Check:** Even though the flea is running due west, it is going east with respect to the ground—still, it is getting right to the end it wants. Both the dog and the train are going east, so $v_{FE} \approx$ 15 km/s is reasonable. Some keen-eyed observer outside at rest watching this circus will see the train, the dog, and the flea all moving east at different speeds.

Now, for a slight variation, imagine the midnight mail train carrying a safe full of gold heading due north at $v_{TE} = 20$ km/h relative to the Earth. A robber, riding hard at $\mathbf{v}_{RE} = 25$ km/h−north next to the caboose, is about to leap onto the train. What is his speed relative to the train (v_{RT})? His speed is his main concern because he wants to land on the train, not the ground. Assuredly, unless he misses, he couldn't care less about either $\mathbf{v}_{RE}$ or $\mathbf{v}_{TE}$ individually.

We want $\mathbf{v}_{RT}$ and can construct an appropriate vector equation for it, namely

$$\mathbf{v}_{RT} = \mathbf{v}_{RE} + \mathbf{v}_{ET} \qquad (2.11)$$

which is fine except we don't have $\mathbf{v}_{ET}$. Notice that *reversing the order of the subscripts reverses the direction of the vector*. For example, $\mathbf{v}_{ET} = -\mathbf{v}_{TE}$; that is, the motion of the train with respect to the sheriff standing still on the Earth is 20 km/h−north (that is, $\mathbf{v}_{TE}$), but to the robber, once he makes his jump onto the caboose, the sheriff will be receding evermore *southward* at 20 km/h from him (that is, $\mathbf{v}_{ET}$). Therefore Eq. (2.11) can be rewritten in terms of vectors whose values we already know, namely

$$\mathbf{v}_{RT} = \mathbf{v}_{RE} + (-\mathbf{v}_{TE})$$
$$\mathbf{v}_{RT} = (25 \text{ km/h−north}) + (-20 \text{ km/h−north}) \qquad (2.12)$$

This gives the robber a velocity just before he leaps of 5.0 km/h−north with respect to the train. The analysis applies as well to spaceships docking, airplanes midair-refueling, relay runners baton-passing, and police cars hotly pursuing.

Example 2.11 Two knights with lances and banners are about to joust. Sir John the Slow gallops south at 5.0 km/h, while Sir Peter the Fast races north at 25 km/h. At what speed do they close the gap between one another?

Solution: [Given: $\mathbf{v}_{JE} = (5.0$ km/h−south); $\mathbf{v}_{PE} = (25$ km/h−north). Find: $\mathbf{v}_{JP}$.]

$$\mathbf{v}_{JP} = \mathbf{v}_{JE} + \mathbf{v}_{EP}$$

We don't have $\mathbf{v}_{EP}$, but it equals $(-\mathbf{v}_{PE})$, and we have that. Thus,

$$\mathbf{v}_{JP} = \mathbf{v}_{JE} - \mathbf{v}_{PE}$$
$$\mathbf{v}_{JP} = (5.0 \text{ km/h−south}) - (25 \text{ km/h−north})$$

The minus sign means that, if we do the vector addition graphically, we must draw the second vector reversed; that is, pointing south. The resultant is then

$$\boxed{(30 \text{ km/h−south})}$$

▶ **Quick Check:** Peter sees John coming south at 30 km/h and John sees Peter ($\mathbf{v}_{PJ}$) galloping north toward him at 30 km/h, where $\mathbf{v}_{JP} = -\mathbf{v}_{PJ}$. The distance between them is being covered from both ends at a net speed of 30 km/h.

Figure 2.29 shows an ant scampering with a velocity of $\mathbf{v}_{AP}$ east across a piece of paper that is simultaneously being pulled north at a velocity of $\mathbf{v}_{PE}$. The ant, as seen from above by an observer at rest on the Earth, has a diagonal velocity of $\mathbf{v}_{AE}$. In exactly the same way, we can think of this situation in terms of a stream rushing north ($\mathbf{v}_{SE}$) being crossed by a boat bearing due east ($\mathbf{v}_{BS}$) with respect to the stream. The resultant motion of the boat with respect to the inertial frame of the Earth (that is, the shore) is diagonal ($\mathbf{v}_{BE}$)—the stream carries the boat somewhat northward even though its prow is always heading east.

The same vector relationships in Fig. 2.29 will exist for many situations. For example, imagine a jet plane heading east through the air ($\mathbf{v}_{JA}$) which, in turn, is

Figure 2.29 An ant walking across a sheet of paper that is itself being moved at a speed v_{PE}. The bug is carried along with the paper so that it moves northeast with respect to the Earth.

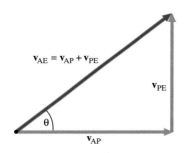

flowing north as a strong wind ($\mathbf{v}_{AE}$) with respect to the Earth's surface, as in Fig. 2.30a. If the air speed of the jet is a low 120 km/h and the wind speed a constant 50.0 km/h, the Pythagorean Theorem gives a hypotenuse of $v_{JE} = 130$ km/h. The plane has a ground speed in excess of its air speed. The pilot must get the plane to its destination, in spite of winds, even if it means pointing the aircraft more toward the south in order not to be carried off course by a sustained wind blowing north. As shown in Fig. 2.30b, the southerly component of the plane's velocity must cancel the northerly velocity of the wind, leaving only an easterly motion of the craft.

The angle of flight, θ, can be determined via the definition of the tangent function for right triangles (Appendix C), as

$$\tan \theta = \frac{\text{opposite side}}{\text{adjacent side}}$$

The tangent of either acute angle of a right triangle is the ratio of the side opposite the angle to the side adjacent to that angle. In the case of Fig. 2.30a, (50.0 km/h)/(120 km/h) = 0.416 7 = $\tan \theta$. We want the angle (θ) whose tangent is known, and that angle is obtained using the *inverse* or *arc tangent* function; $\tan^{-1}(0.417) = \theta = 22.6°$, which here is north of east. On a calculator, the sequence is [0] [·] [4] [1] [6] [7] [tan^{-1}].

2.10 Vector Components

We know how to add two vectors to form a third *resultant* vector that produces the same physical effects as the original two. The reverse process is also possible and also useful. Figure 2.31 shows two perpendicular vectors (acting along the x and y directions) $\mathbf{A}_x$ and $\mathbf{A}_y$. Using the parallelogram method (p. 33), it is evident that $\mathbf{A} = \mathbf{A}_x + \mathbf{A}_y$. The same diagram can be interpreted as showing that vector $\mathbf{A}$ may be *resolved into two perpendicular components:* $\mathbf{A}_x$ and $\mathbf{A}_y$. Given a specific value of $\mathbf{A}$ (that is, given its magnitude, A, and direction, θ), it is possible to measure these two components graphically. It will often be more convenient and accurate, however, to use the sine and cosine trigonometric functions instead. If you know any two sides, or any side and either of the acute angles of a right triangle, everything else about it can be determined easily.

Note (see Appendix C) that, for a right triangle, the sine of either acute angle is the ratio formed of the side opposite that angle over the hypotenuse:

$$\sin \theta = \frac{\text{opposite side}}{\text{hypotenuse}}$$

(a)

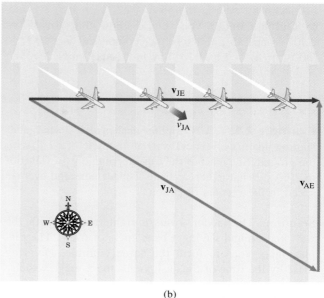

(b)

Figure 2.30   (a) A plane heading east across a wind blowing north. The plane is carried N of E at an angle θ. (b) To cross the air stream due east, the plane must point south-of-east such that its southerly velocity component effectively cancels the northerly velocity of the wind. Someone hitting a golf ball in a crosswind has the same problem as the pilot. When you are moving across a medium that's moving, you can't head straight toward your destination.

Similarly, the cosine of either acute angle of a right triangle is the ratio formed of the side adjacent to that angle over the hypotenuse:

$$\cos \theta = \frac{\text{adjacent side}}{\text{hypotenuse}}$$

The triangle in Fig. 2.31 has sides equal to the lengths (or magnitudes) of the vectors $\mathbf{A}_x$, $\mathbf{A}_y$, and $\mathbf{A}$. Accordingly, the appropriate ratios of these are

$$\sin \theta = A_y/A \quad \text{and} \quad \cos \theta = A_x/A$$

Multiplying both sides of both these equations by A yields

$$A_y = A \sin \theta \quad \text{and} \quad A_x = A \cos \theta \quad (2.13)$$

These then, are the scalar x- and y-components of $\mathbf{A}$.

Example 2.12   An aircraft in a 45.0° dive is traveling at a constant 800 km/h. Determine its speed in the vertical direction—that is, the rate at which the plane is losing altitude.

Solution: The plane's *constant* velocity vector $\mathbf{v}$ points downward at an angle θ. [Given: $v = 800$ km/h and $\theta = 45.0°$. Find: v_y.] Equation (2.13) provides the needed relationships between any vector and its two perpendicular components.

Therefore

$$v_y = v \sin \theta = (800 \text{ km/h}) \sin 45.0°$$

Because both acute angles of this velocity triangle are equal (both are 45°), it is isosceles and the two sides, v_x and v_y, are equal. The plane's horizontal and vertical speeds are therefore both $\boxed{566 \text{ km/h}}$. As the aircraft advances horizontally, it simultaneously descends at the same speed vertically.

(continued)

(continued)
▶ **Quick Check:** You should know that sin 45° = 0.707 = cos 45°, and so either component is ≈(0.7) × (800 km/h) ≈ 5.6 × 10² km/h. *Always make a rough* check *of the numerical computation to be sure you have pushed the correct calculator keys:* [4] [5] [·] [0] [sin] [×] [8] [0] [0] [=].

Example 2.13 A firefighter dashes up a 26-m ladder leaning against a vertical wall at an angle of 67.4° with the ground. She reaches the top, stepping onto a flat roof in 20.0 s. What is her vertical displacement and average speed in the vertical direction?

Solution: The ladder forms a right triangle with the wall and ground having a base angle of 67.4°. The displacement vectors form the same triangle. Dividing these by the time of the dash yields the average velocity vectors, which form a similar triangle. [Given: s = 26 m, t = 20.0 s and θ = 67.4°. Find: s_y the vertical displacement and the average vertical speed.] From Eq. (2.13)

$$s_y = s \sin \theta = (26 \text{ m})(\sin 67.4°) = (26 \text{ m})(0.923)$$

and $\boxed{s_y = 24 \text{ m}}$. The average vertical speed is the vertical distance traveled divided by the time, (24 m)/ (20.0 s) = $\boxed{1.2 \text{ m/s}}$. Similarly, the average speed along the ladder is (26 m)/(20.0 s) = 1.3 m/s while the average horizontal speed is $s_x/t = (s \cos \theta)/t$ = (10 m)/ (20.0 s) = 0.50 m/s.

▶ **Quick Check:** The displacement triangle is a special whole-number right triangle (Appendix B), a 5-12-13. Its sides are 10-24-26. The same must be true for the velocity triangle.

Now let's develop the most powerful analytic scheme for summing any number of vectors. Figure 2.32a shows two arbitrary vectors **A** and **B**, which we wish to add. The usual graphical tip-to-tail technique is illustrated in Fig. 2.32b, and the resultant vector **C** is constructed in Fig. 2.32c. Notice that each of the components of the resultant are equal to the sums of the components of the constituents. In other words,

$$C_x = A_x + B_x \qquad \text{and} \qquad C_y = A_y + B_y \qquad (2.14)$$

Find the scalar x-components of the vectors to be summed, give them proper signs (Fig. 2.33), and then add them algebraically. That total is the scalar x-component of the resultant and the same procedure applies to the y-component. To reconstruct **C** analytically, use the Pythagorean Theorem

$$C = \sqrt{C_x^2 + C_y^2} \qquad (2.15)$$

to compute the magnitude of the resultant. The most convenient way to find its orientation is to use

$$\theta = \tan^{-1} \frac{|C_y|}{|C_x|} \qquad (2.16)$$

where $|C_y|$ and $|C_x|$ are the positive values of the components (Fig. 2.34), as discussed in detail in Appendix D.

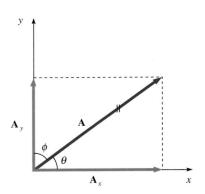

Figure 2.31 Any vector **A** can be thought of as equivalent to its two perpendicular components A_x and A_y. Notice that $A_x = A \cos \theta$ and $A_y = A \sin \theta$, but if angle ϕ is used, $A_x = A \sin \phi$ and $A_y = A \cos \phi$. Be careful with those angles.

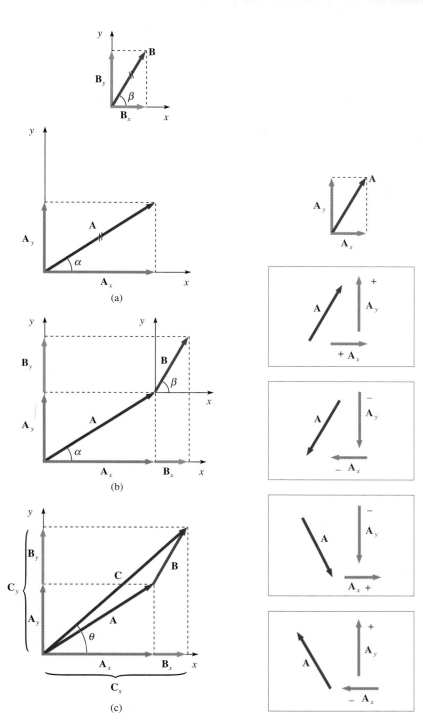

Figure 2.32 The summation of two vectors using their perpendicular components. (a) Those x- and y-components are $A_x = A \cos \alpha$, $A_y = A \sin \alpha$, $B_x = B \cos \beta$, and $B_y = B \sin \beta$. In (b) we place **A** and **B** tip-to-tail. In (c) we see how this arrangement is equivalent to adding their components in the x- and y-directions: $C_x = A_x + B_x$ and $C_y = A_y + B_y$.

Figure 2.33 The vector **A** has components in the x- and y-directions of $\mathbf{A}_x$ and $\mathbf{A}_y$. Here we see the signs that must be associated with the scalar quantities A_x and A_y if they are to be correctly added algebraically to the scalar components of other vectors.

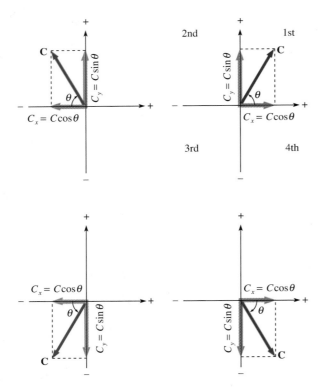

Figure 2.34 In the process of adding any two vectors **A** and **B** to get **C** use Eq. (2.14) to determine C_x and C_y. Make sure to associate the proper signs with A_x and B_x, and A_y and B_y before adding them (see Fig. 2.33). The signs of C_x and C_y determine in which quadrant **C** will be and also determine θ. Thus, if C_x is negative and C_y is positive, **C** is in the 2nd quadrant.

Example 2.14 Two vectors lie in the x-y plane, as shown in Fig. 2.32. Vector **A** is 10 units long and points upward 30° above the positive x-axis. Vector **B** is 15 units long and points 45° above the positive x-axis. What is the resultant of the two vectors?

Solution: [Given: **A** and **B**. Find: **C** = **A** + **B**.] From Eq. (2.13) $A_x = A \cos \alpha = 10 \cos 30° = 8.66$ and $A_y = A \sin \alpha = 10 \sin 30° = 5.00$; $B_x = B \cos \beta = 15 \cos 45° = 10.61$ and $B_y = B \sin \beta = 15 \sin 45° = 10.61$. Hence, $C_x = 8.66 + 10.61 = 19.27$ and $C_y = 5.00 + 10.61 = 15.61$. Using Eq. (2.15),

$$C = \sqrt{C_x^2 + C_y^2} = \sqrt{(19.27)^2 + (15.61)^2} = 24.8$$

To two significant figures, $\boxed{C = 25 \text{ units}}$. To find the angle θ that **C** makes with the x-axis, we use Eq. (2.16):

$$\tan \theta = \frac{|C_y|}{|C_x|} = \frac{15.61}{19.27} = 0.8101$$

and $\tan^{-1} 0.8101 = \theta = 39°$. Both **A** and **B** point above the x-axis and so **C** is $\boxed{39°}$ above it too.

▶ **Quick Check:** A_x and B_x are positive as are A_y and B_y, so C_x and C_y must be positive; that is, $90° \geq \theta \geq 0°$. From the parallelogram method, **C** should lie between **A** and **B** and be greater than either, and it is. The smaller the angle between **A** and **B** (45° − 30°), the more C approaches $A + B$ as a maximum. Thus, $C = 20$ units is reasonable.

Example 2.15 A blimp cruising with a constant air speed of 180 km/h heads due north against a steady 100 km/h wind blowing out of the northwest. Compute its ground speed, its actual direction of flight, and the distance traveled after 3.00 h of cruising.

Solution: This problem concerns relative velocity. [Given: $\mathbf{v}_{BA} = 180$ km/h–north; $\mathbf{v}_{AG} = 100$ km/h–southeast; $t = 3.00$ h. Find: $\mathbf{v}_{BG}$ and s.] Inasmuch as we need $\mathbf{v}_{BG}$, set up the appropriate expression for it, using Eq. (2.9) as a guide:

(continued)

(continued)

$$\mathbf{v}_{BG} = \mathbf{v}_{BA} + \mathbf{v}_{AG}$$

Figure 2.35 represents this equation graphically. To carry out the addition, we will first resolve $\mathbf{v}_{BA}$ and $\mathbf{v}_{AG}$ into their respective x- and y-components, which is easily done for $\mathbf{v}_{BA}$; it points along the positive y-axis and is the entire y-component—it has no x-component. The scalar x-component of $\mathbf{v}_{AG}$ is just $v_{AG} \cos 45°$ and the y-component is $-v_{AG} \sin 45°$. The minus sign is used here because the vector points downward in the negative y-direction. It has to be subtracted from v_{BA}, which points in the positive y-direction. For the moment, let's represent vectors pointing either left in the negative x-direction (west), or down in the negative y-direction (south), as negative. Similarly, those pointing either right in the positive x-direction (east), or up in the positive y-direction (north) are positive. In that way the algebraic subtraction of oppositely directed vectors takes place systematically as one adds all the components in a given direction. With the signs taken care of, every numerical value is entered as a positive number.

The scalar y-component of $\mathbf{v}_{BG}$ is then

$$v_{BA} - v_{AG} \sin 45° = 180 \text{ km/h} - (100 \text{ km/h})(0.707)$$
$$= 109 \text{ km/h}$$

while the scalar x-component is

$$v_{AG} \cos 45° = (100 \text{ km/h})(0.707) = 70.7 \text{ km/h}$$

The magnitude of $\mathbf{v}_{BG}$ can now be gotten from the Pythagorean Theorem:

$$v_{BG} = \sqrt{(109 \text{ km/h})^2 + (70.7 \text{ km/h})^2} = \boxed{130 \text{ km/h}}$$

The angle $\mathbf{v}_{BG}$ makes with the x-axis (or east-west line) is θ where

$$\tan \theta = \frac{109 \text{ km/h}}{70.7 \text{ km/h}} = 1.54$$

and so $\tan^{-1}(1.54) = \boxed{\theta = 57.0° \text{ N of E}}$.

In 3.00 h, the blimp travels a distance

$$s = v_{BG}t = (130 \text{ km/h})(3.00 \text{ h}) = 390 \text{ km}$$

▶ **Quick Check:** There are many problems that can be checked using the Law of Sines (Appendix C). This law maintains that for any triangle the ratio of any one side to the sine of the angle opposite that side equals the ratio of any other side to the sine of the angle opposite it. For the triangle in Fig. 2.35, $v_{BG}/\sin 45° = 184$ m/s and this quantity should equal $v_{AG}/\sin (90° - 57°) = 184$ m/s, and all is well.

Figure 2.35 (a) A blimp heads north through the air with a velocity $\mathbf{v}_{BA}$ while the air is moving with a velocity $\mathbf{v}_{AG}$ with respect to the ground. Thus, the blimp is blown off course and travels northeast with a velocity $\mathbf{v}_{BG}$ with respect to the ground. (b) To find its velocity $\mathbf{v}_{BG}$, vectors $\mathbf{v}_{BA}$ and $\mathbf{v}_{AG}$ are resolved into their components along the x- and y-axes. (c) All the components along each axis are added, yielding two perpendicular vectors that are themselves the x- and y-components of the resultant $\mathbf{v}_{BG}$.

Core Material

Average speed is *the distance traversed in one unit of time,* or

$$v_{av} = \frac{l}{t} \qquad [2.1]$$

When the speed is constant

$$l = v_{const}t \qquad [2.2]$$

The slope of the distance-time curve at any point is the speed at that point. The area under the speed-time curve (bounded by any two moments in time) equals the distance traveled during that interval (p. 28). The **instantaneous speed** is

$$v = \lim_{\Delta t \to 0} \left[\frac{\Delta l}{\Delta t} \right] \qquad [2.4]$$

Any physical quantity that can be specified completely only if both its magnitude and direction are provided is known as a **vector quantity**. Two vectors can be added by arranging them such that the tail of the second is at the tip of the first; the sum of these, also called the **resultant vector**, extends from the tail of the first to the tip of the second (p. 32). The **average velocity** is

$$\mathbf{v}_{av} = \frac{\mathbf{s}}{t} \qquad [2.5]$$

and the **instantaneous velocity** is

$$\mathbf{v} = \lim_{\Delta t \to 0} \left[\frac{\Delta \mathbf{s}}{\Delta t} \right] \qquad [2.7]$$

Figure 2.26 illustrates the concept of relative motion. In this case, the velocity of the turtle with respect to the Earth is

$$\mathbf{v}_{TE} = \mathbf{v}_{TR} + \mathbf{v}_{RE} \qquad [2.9]$$

The scalar *x*- and *y*-components of any vector **A** are

$$A_y = A \sin \theta \qquad \text{and} \qquad A_x = A \cos \theta \qquad [2.13]$$

If **C** is the sum of the two vectors **A** and **B**, then

$$C_x = A_x + B_x \qquad \text{and} \qquad C_y = A_y + B_y \qquad [2.14]$$

and

$$C = \sqrt{C_x^2 + C_y^2} \qquad [2.15]$$

where

$$\theta = \tan^{-1} \frac{|C_y|}{|C_x|} \qquad [2.16]$$

Suggestions on Problem Solving

1. Reread the Suggestions on p. 19. Translate the problem into the language of physics and then express it symbolically. If one idea, like distance, has several values, call them distance 1 and distance 2, and so on. When several distances or times are involved, determine their relationship to the total distance or time. List what is *given* and what is to be *found*. This list is the best summary of the exercise. It will suggest which equations will be useful in the solution.

2. Read the problem several times and determine the broad subject to which it pertains—the subject may be obvious from the location in a chapter, but it might not be so obvious in the uncategorized real world or even on a final exam or MCAT. Determine the subgrouping of ideas to which the problem most clearly belongs, such as average speed, instantaneous speed, displacement, and so on. Look for key ideas and phrases that distinguish one group of notions from another, such as references to direction. These indications tell you that you are dealing with displacement or velocity. Watch for expressions like *straight road, steady motion, constant speed,* and so forth—they provide vital information.

3. Draw sketches that illustrate the problem being analyzed. Put into these all the known information and, where appropriate, draw in the vectors to represent directional quantities.

4. For most simple single-concept problems there will be one unknown that can be determined from one formula. Write down the equation that contains the desired unknown (and only that one unknown) and solve for it algebraically in terms of the remaining knowns. Problems involving two unknowns require solving two equations simultaneously (Appendix A-4). Remember you will need the same number of equations as there are unknowns to be found.

5. **Unless otherwise required, work all problems in SI units.** Convert the given numerical information into appropriate units *before* substituting into equations. It's best to convert as soon as possible to avoid accidentally using the wrong values (10 cm is *not* 10 m). Carry the units for each term through the whole calculation as a double check that you haven't messed up the algebra: an answer of 10 km/h for how *far* the chicken traveled suggests something is amiss; an answer of just 10 tells you a good deal less and is incomplete. **Always** include units with your numerical answers.

6. Always try to check your answers by recomputing the problem in a different way.

7. Watch out for the number of significant figures. If you travel 10.2 m (three significant figures) in 2.1 s (two significant figures), your average speed is 4.9 m/s (two significant figures) and *not* 4.857 142 9 m/s.

8. To foster some independence from the calculator, a number of problems are given with "nice" numbers so they are easier to do by hand than by machine. Avoid using the calculator to divide 100 by 2, wondering all the while why you bothered. Some important national examinations do not allow calculators to be used. If you plan to take such exams, you must know what the curves of the sine, cosine, and tangent functions look like. Also know their values for 0°, 30°, 45°, 60°, and 90°. Become familiar with the 3-4-5 and 5-12-13 right triangles.

Discussion Questions

1. Suppose that the average speed over some time interval is zero. Is it possible for the average speed over a still smaller segment of that interval to be nonzero? Suppose that the average velocity over some time interval is zero. Is it possible for the average velocity over a still smaller segment of that interval to be nonzero?

2. If the tangent at a given point on the distance-time graph of some object is horizontal, what is the object's instantaneous speed at that moment? What does it mean if the tangent to the graph is vertical? Can that actually occur?

3. Is it possible for a plot of *l* versus *t* to have a negative slope at any point? Explain.

4. Is it possible to travel from one place to another with some average speed without ever having had an instantaneous speed equal to it somewhere along the journey? Explain.

5. Imagine that you leave the base of a mountain at 9:00 A.M. and ascend along a path reaching the top at 11:00 A.M. You sleep there and descend along the same path starting at 9:00 A.M. the next day. Will there be some point along the trail that you reach at exactly the same time you were there the day before, regardless of the variations in your speeds during the two journeys?

6. Can the change in displacement of an object in some interval of time exceed the distance traveled in that time?

7. Is it possible at any moment to be moving with a constant velocity and not a constant speed? How about the reverse of this situation—a constant speed but not a constant velocity?

8. When we talk about displacement vectors for travel on our planet, we assume a flat Earth, or at least trips short enough so it's approximately flat. Just for fun, suppose you were standing at the North Pole and walked on the surface (no burrowing) 10 km–south, 20 km–east, and 10 km–north—where would you end up?

9. Figure Q9 is a plot of the total distance traveled by a dog running in a field. Describe the motion. What can be said about the dog's displacement at any time? What was the total distance traveled? When was he moving the fastest?

Figure Q9

10. Is it possible during a given interval for a graph of *distance* (*l*) *versus time* for the same object to be different from a graph of *the magnitude of the displacement* (*s*) *versus time?* Explain.

11. Suppose we have a vector representing some physical quantity. Will the numerical value of the magnitude change if we express it in different units? Will its orientation change?

12. After reaching cruising altitude, a jet can fly along at a constant 750 km/h, over the "shortest path" from New York to Tokyo. What dynamical quantity would you say was constant for that portion of the flight? Explain.

13. Figure Q13 is a plot of the displacement of a mouse running through a straight, narrow tunnel, the entrance to which is taken as the origin. Describe the mouse's motion in general terms throughout its journey. Where and when did it start moving? Where did it finish? Did it ever rest?

Figure Q13

14. A plane dives in a straight line at a fixed angle. With the Sun right above it, the plane's shadow moves across the ground at a constant speed. What can you say about the plane's rate of descent?

15. If two vectors have magnitudes of 4 units and 6 units, what are the largest and the smallest possible resultants? Can the vector **A** have a component with a magnitude that is larger than *A*? If two vectors are to produce a resultant equal to zero, what can you say about them?

16. Suppose we multiply a vector **M** representing some physical quantity, by a scalar *n* representing yet another physical quantity. Will *n***M** have the same units as **M**?

17. If the magnitude of the displacement of a body is given by $s = (At^2 + Bt)/D(C + t)$ where A, B, C, and D are constants, (a) determine the displacement at $t = 0$; (b) find the approximate value of s when t is very much larger than C—that is, when $t \gg C$. What is the approximate displacement when $t \ll C$?

18. Determine the resultant of the two vectors in Fig. Q18 where $A_1 = A_2 = A = 25$. Show that the resultant has a magnitude of $2A \cos 30°$. Now try the last part again, this time rotating the coordinate system along which you resolve the vectors so that the *y*-axis bisects the 60° angle. Notice how much easier the problem is now!

Figure Q18

Multiple Choice Questions

1. If $[L]$ represents the dimensions of length and $[T]$ that of time, then the dimensions of speed are (a) $[L + T]$ (b) $[T/L]$ (c) $[L/T^2]$ (d) $[L/T]$ (e) none of these.
2. When converting a given number of kilometers per hour to kilometers per second, you can expect the resulting number to be (a) always smaller (b) the same (c) sometimes smaller (d) never smaller (e) none of these.
3. How far does light travel in one year moving at 3×10^8 m/s? (a) 3×10^8 m (b) 9.5×10^{15} m (c) 3×10^8 km (d) 9.5×10^8 km (e) none of these.
4. A speed of 300 m/s is equivalent to (a) 0.300 km/h (b) 984 ft/s (c) 134 mi/h (d) 83.3 km/h (e) none of these.
5. A speed of 60 mi/h is equivalent to (a) 44 ft/s (b) 22 m/s (c) 88 ft/s (d) 11 km/h (e) none of these.
6. The three-person crew of *Apollo 11* made the 3.8×10^5-km journey to the Moon in exactly 3 days with an average speed of (a) 5.3×10^3 km/d (b) 53×10^3 km/h (c) 5.3×10^3 km/s (d) 5.3×10^3 km/h (e) none of these.
7. If a body moves at a constant speed v along a closed path, its average speed in comparison to v is (a) greater (b) less (c) the same (d) sometimes greater, sometimes less (e) none of these.
8. On a distance-time graph, the slope at any point is (a) the distance traveled (b) the time elapsed (c) the instantaneous speed (d) the average speed (e) none of these.
9. During a journey over which the speed varies, the average speed in comparison to the maximum attained speed is always (a) one-half of it (b) greater than it (c) less than it (d) equal to it (e) none of these.
10. A horse on a continuously and uniformly turning carousel is carried through three rotations every 6 min. Its average speed over each and any 18-min interval is (a) a constant (b) undetermined and we can not say anything about it (c) always different (d) may be different (e) none of these.
11. The *Mariner 2* spacecraft lifted off from Cape Canaveral on August 27, 1962. After a successful flight, it passed out of tracking range at about 87 million kilometers (54×10^6 mi) on January 4, 1963. The magnitude of its average velocity up until that point was roughly (a) not enough information to calculate (b) 1.7×10^4 km/h (c) equal to the average speed (d) 28×10^3 km/h (e) none of these.
12. When is the resultant that arises when **A** is subtracted from **B** equal to that when **B** is subtracted from **A**? (a) always (b) never (c) not enough information (d) only when $A = B$ (e) none of these.
13. If, just as its fuel ran out, the speed of a World War II German V-2 rocket was 5633 km/h–straight up, what was the velocity of the Earth with respect to the rocket at that moment? (a) zero (b) 5633 km/h–downward (c) 5633 km/h–upward (d) 11 266 km/h (e) none of these.
14. Two vectors of magnitude 8.0 and 3.0 can be added so as to produce a third vector of magnitude (a) 12 (b) 15 (c) 8.0 (d) 3.0 (e) none of these.
15. A vector 10 units long pointing northeast is added to a vector 24 units long pointing northwest. The magnitude of the resultant is (a) 26 units (b) 14 units (c) 34 units (d) 0 units (e) not enough information.
16. The inhabitants of the mythical planet Mongo measure distance in units of "glongs," one of which is the length of their leader's breathing tube. A Mongoian displacement vector 90.0 glongs long points south from the truffle tree to the methane fountain, while a vector 120 glongs long points west from the fountain to the main reflector. The displacement from tree to reflector is (a) 150 glongs, 49° S of W (b) 150 m, 37° S of W (c) 150 glongs, 37° S of W (d) 80.0 glongs, 41° S of E (e) not enough information.
17. In general, if a vector **A** is to be added to a vector **B** the magnitude of the resultant when $A \geq B$ must be between (a) $2A$ and B (b) A and $A + B$ (c) $A + B$ and $A - B$ (d) $B - A$ and B (e) none of these.
18. Figure MC18 is a portion of a distance-time curve for a jogger. At $t = 14$ s the instantaneous speed is about (a) 0.50 m/s (b) 0.27 m/s (c) 6.5 m/s (d) 0.46 m/s (e) none of these.

Figure MC18

Problems

SPEED

1. [I] In the next 30.0 s, the Earth will travel roughly 885 km (550 mi) along its orbit around the Sun. Compute its average orbital speed in km/s.

2. [I] You know how you close your eyes when you sneeze? Suppose you are driving along at a constant 96.5 km/h (60 mi/h) and you experience a 1.00-s-long, eyes-closed, giant sneeze. How many meters does the car travel while you are out of control?

3. [I] Light travels in vacuum at a fixed speed of roughly 2.998×10^8 m/s and its speed in air is only negligibly (0.03%) slower than that.   (a) How long does it take light to traverse 1.0 ft?   (b) That means that when you look at something 1000 m away, you are seeing it as it was _____ second(s) back in time.

4. [I] If an alien power straight out of a comic book were to cause the Sun to vanish *now,* we would still be bathed in sunshine for the next 8.3 min. We would see our star blazing in the sky as usual for all that time. Using that information and taking the speed of light to be roughly 3.0×10^8 m/s, compute the average Earth–Sun distance.

5. [I] Referring to Table 2.3 (p. 26), derive each conversion factor in the first column.

6. [I] Referring to Table 2.3 (p. 26), derive each conversion factor in the second column.

7. [I] Figure P7 is a plot of the speed of a cat versus time. How far did the cat travel during the third second of its journey? What were its maximum and minimum speeds? When, if ever, was its speed nonzero and constant?

Figure P7

8. [I] The Earth rotates once around its spin axis in 23 h 56 min and its equatorial diameter is 1.276×10^7 m (7927 mi). At what speed would you be traveling with respect to the stationary stars while sitting still in Mbandaka, Zaire, on the Equator?

9. [I] A 30-km trip is negotiated in two equal-length segments. The first, on highways taking 15 min; the second, through heavy traffic and taking 45 min.

Compute   (a) the average speed along each leg, and (b) the average speed over the whole trip.

10. [I] The SR-71 strategic reconnaissance aircraft, the *Blackbird,* set a world speed record by flying from London to Los Angeles (5463 mi) in 3 h 47 min 36 s. (a) Compute its average speed in m/s.   (b) It recaptured the 1000-km closed-circuit-course record (previously held by the Russian MIG-25 *Foxbat*) at an average speed of 2092 mi/h. How long in time was that flight?

11. [I] Use Fig. 2.5 (p. 29) to calculate the distance traveled by the bee (whose speed-time graph is plotted) in the time interval from 1.33 s to 2.83 s.

12. [I] The driver of a car sets a tripmeter to zero and starts a stopwatch while en route along a highway. The following are a series of pairs of readings taken during the run: 6.0 min, 2.75 km; 30 min, 13.8 km; 45 min, 20.6 km; 1.0 hr, 27.5 km; 1.00 h 15 min, 34.4 km; 1.00 h 30 min, 41.3 km. Draw a distance-time graph and describe the car's speed.

13. [I] Figure P13 shows the distance traveled versus time curve for a toy car. What was the toy car's average speed during the time interval from $t = 2.0$ s to 8.0 s?

Figure P13

14. [II] Suppose you fire a rifle bullet (1600 m/s) in a shooting gallery and hear the gong on the target ring 0.731 s later. Taking the speed of sound to be 330 m/s and assuming the bullet travels straight downrange at a constant speed, how far away is the target?

15. [II] The speed of sound in air (at 0°C) is a constant 3.3×10^2 m/s, while the speed of light is about 3.0×10^8 m/s. If you see a flash of lightning and then 5.0 s later hear its roll of thunder, how far away did the bolt strike?

16. [II] The evil Dr. X secretly leaves Space Port L4 in a warship capable of traveling at an average speed of $(v_{av})_X$. Two hours later her escape is noticed and our hero blasts off after her at a speed that will effect rendezvous in

6.4 hours. Write an expression for his speed, $(v_{av})_H$, in terms of hers, $(v_{av})_X$.

17. [II] The bad guys come roaring onto a highway at 100 km/h headed for the Mexican border 300 km away. The cops, in hot pursuit, arrive at the highway entrance one-half hour later. What must be their minimum average speed if they are to intercept the crooks this side of the border?

18. [III] Having traveled halfway to the end of a journey at an average speed of 15 km/h, how fast must you traverse the rest of the trip in order to average 20 km/h?

19. [III] A swimmer crosses a channel at an average speed of 10 km/h and returns at half that rate. What was his overall average speed for the round trip?

20. [III] If the equation describing the motion of a rocket in SI units is $l = 10 + 5t^2$, find its average speed during the first 5.0 s of flight. (Hint: At a time $(t + \Delta t)$, the object is at $(l + \Delta l)$.) Write an expression for the instantaneous speed of the rocket [see Eq. (2.4)].

21. [III] Given the equation describing the motion of a falling object $l = Ct^2$ (where C is a constant), show that its average speed during the interval from t to $(t + \Delta t)$ is

$$v_{av} = 2Ct + C\Delta t$$

(Hint: At a time $(t + \Delta t)$, the object is at $(l + \Delta l)$.) Now determine an equation for the instantaneous speed from Eq. (2.4).

VELOCITY

22. [I] A mouse runs straight north 1.414 m, stops, turns right through 90°, and runs another 1.414 m due east. Through what distance is it displaced?

23. [I] A green frog with a body temperature of 18°C at rest at the edge of a 1.0 m high table jumps off and lands on the ground 1.0 m out from the edge. Determine the length of its displacement vector.

24. [I] A toy electric train runs along a straight length of track. Its displacement versus time curve is shown in Fig. P24. Is the train's velocity ever constant, and if so, when? What is its instantaneous velocity at $t = 2.0$ s and at $t = 6.5$ s? Did it change direction, and if so when?

Figure P24

25. [I] A robot on the planet Mongo leaves its storage closet and heads due east for 9.0 km. It then turns south for 12 km and stops. Neglecting the curvature of the planet, what is the magnitude of the robot's displacement?

26. [I] A jogger in the city runs 4 blocks north, 2 blocks east, 1 block south, 4 blocks west, 1 block north, 1 block east, and collapses. Determine the magnitude of the jogger's displacement.

27. [I] A stationary wooden horse on a merry-go-round is carried along a 31.4-m circumference each time the machine turns once around. Determine the magnitude of the horse's displacement for half a turn.

28. [I] Draw the vector **A** 10 units long pointing due east and graphically add to it the vector **B** 15 units long pointing northeast at 45°. Measure both the magnitude of the resultant and its direction.

29. [I] A boy scout troop marches 10 km east, 5.0 km south, 4.0 km west, 3.0 km south, 6.0 km west, and 8.0 km north. What is their total displacement from their starting point? How far have they marched?

30. [I] Graphically determine **C**, which is the resultant of **A** = 20 units–30° S of E and **B** = 10 units–30° N of E.

31. [I] Someone runs up four zigzagging flights of stairs, each one 10 m long and rising 6.0 m high. At the end of the climb, the person stands directly above the starting point. Determine the runner's total displacement.

32. [I] A marble, on the first shot of a tournament game in 1942 in Brooklyn, was displaced by **s** equal to 3.0 m in a direction 60° N of E. On the next shot, it was displaced from its new location by −4.0 **s**. Where did it end up?

33. [I] Graphically add **A** (5.0 units–east) to **B** (7.1 units–45° N of W) to get **C**.

34. [I] Graphically subtract **A** (5.0 units–east) from **B** (7.1 units–45° N of E) to get **C**.

35. [I] A bumblebee flew 43 m along a twisting path only to land on a flower 3.0 m due south of the point on its hive from which it started. If the entire journey took 10 s, what was its average speed and average velocity?

36. [I] A driver in a cross-country race heads due east, attaining a speed of 30 km/h in 10 s. Maintaining that speed for another 30.0 min, he then makes a hard 90° left turn. After 50.0 min at that speed and heading, he veers right and slows down, moving northeast at 20 km/h for the next 40.0 min, at which point he crosses the finish line. What is the car's velocity at each of the following times: (a) 40 s, (b) 30.0 min, (c) 50 min, and (d) 100 min?

37. [II] A cannonball fired from a gun located 20 m away from a castle rises high into the air in a smooth arc and sails down, crashing into the wall 60 m up from the ground. Determine the projectile's displacement.

38. [II] Figure P38 shows the displacement-time graph of a

Figure P38

gerbil running inside a straight length of clear plastic tubing. (a) How far did it travel during the interval from 15 s to 20 s? (b) What distance did it traverse in the first 35 s? (c) What is its average speed during the first 35 s? (d) What is its instantaneous speed at $t = 20$ s?

39. [II] Refer to Fig. P38 and assume the gerbil starts out heading north. (a) What is his velocity at $t = 12$ s; at $t = 38$ s? (b) What is his velocity at 42.5 s?

40. [II] An athlete dashes past the starting point in a straight-line run at a fairly constant speed of 5 m/s for the first 2 s and then at 10 m/s for the next 3 s. At that moment she abruptly stops, turns around, waits 1 s, and runs back to the start with a constant speed in 3 s. Draw a graph of the magnitude of her displacement versus time.

41. [II] Ideally, a bullet fired straight up at 300 m/s will constantly slow down as it rises to a height of 4588 m in 30.6 s, at which point it will come to a midair stop. It then plummets back down again (and overlooking friction losses) and will speed up, reaching the gun barrel at the original 300 m/s after falling 30.6 s. Determine its average speed and average velocity on both the upward leg and on the round trip.

42. [II] Referring to the previous problem: (a) What is the velocity of the bullet just as it leaves the gun? (b) What is the bullet's velocity at $t = 30.6$ s; at $t = 61.2$ s?

43. [II] A chicken is resting at a location 3.0 m north of a stationary farmer who weighs 185 lb. The chicken then meanders to a new location 3.0 m east of the farmer in a time of 2.0 s. Compute the bird's average velocity during its little journey.

44. [II] A trolley travels along a straight run of track and Fig. P44 is a plot of its velocity versus time. Approximately how far did it travel in the first 3.0 s of its journey? How far from its starting point is it at $t = 6.0$ s?

Figure P44

45. [III] A cannon is fired due north with an elevation of 45°. The projectile arcs into the air and then descends, crashing to the ground 600 m downrange 3 s later. With the Sun directly overhead, the projectile's shadow races along the flat stretch of Earth at a fairly constant speed. (a) Compute the instantaneous velocity of the shadow at the moment of impact. (b) What was the average velocity of the shell for the entire flight?

46. [III] An observer on a golf course, at 2:00 in the afternoon, stands 60 m west of a player who drives a ball due north down the fairway. If the ball lands 2.0 s later, 156 m from the observer, what was its average velocity?

RELATIVE MOTION

47. [I] Two runners at either end of a 1000-m straight track each jogs toward the other at a constant 5.00 m/s. How long will it take before they meet?

48. [I] The jet stream is a narrow current of air that flows from west to east in the stratosphere above the temperate zone. Suppose a passenger plane capable of cruising at an air speed of 965 km/h (600 mi/h) rode the jet stream on a day when that current remained at a constant 483 km/h (300 mi/h) with respect to the Earth. Find the plane's ground speed.

49. [I] Suppose that a fly were to go back and forth, essentially without a pause, from one runner to the other in Problem 47 at an average speed of 10 m/s. How far will it have traveled in total by the time the athletes meet? (Hint: How long is the fly in the air?)

50. [I] A deckhand on a ship steaming north walks toward the rear of the vessel at 5.00 km/h carrying a horizontal wooden plank. A ladybug on the plank scampers away from the human at 0.01 km/h–south. If the ship cuts through the calm sea at 15.00 km/h: (a) What is the speed of the ladybug with respect to the ship? (b) What is the speed of the plank with respect to the Earth? (c) What is the velocity of the bug with respect to the shore?

51. [I] Determine the two acute angles of an exactly 5-12-13 right triangle accurate to two decimal places (4 significant figures).

52. [I] A pack of hounds running at 15.0 m/s is 50.0 m behind a mechanical rabbit, itself sailing along at 10.0 m/s. (a) How long will it take before the rabbit is caught? (b) How far will it travel before being overtaken?

53. [I] A rope tied to the top of a flagpole makes an angle of 45° with the ground. If a circus performer walks up the $10\sqrt{2}$-m rope, what is her altitude at the top?

54. [I] What is the ratio of the sides of a right triangle whose acute angles are 30° and 60°? *This ratio is a good one to remember.* Which side is opposite the 30° angle?

55. [I] A telescope pointed directly at the top of a distant flagpole makes an angle of 25° with the ground. If the scope is low to the ground and 25 m from the base of the pole, how tall is the pole?

56. [I] A Roman catapult fires a boulder with a launch speed of 20 m/s at an angle of 60° with the ground. What are the projectile's horizontal and vertical speeds? (See the previous problem.)

57. [I] A skier is sailing straight down a 3000-ft-tall mountain with a 60.0° slope at 85 km/h. At what rate is his altitude decreasing?

58. [I] A pintail duck is flying northwest at 10 m/s. (a) At what rate is it progressing north? (b) What is the magnitude of its westward velocity component?

59. [I] A bullet is fired from a 1905 9-mm Luger pointed up at 32.0° with respect to the horizontal. If the muzzle speed—that is, the speed at which the projectile leaves the gun—is 300 m/s, find the magnitudes of: (a) the

horizontal, and (b) the vertical, components of the bullet's velocity. (c) Check your results using the Pythagorean Theorem.

60. [II] A jet fighter heading directly north toward a flying target drone fires an air-to-air missile at it from 1000 m away. The drone has a constant speed (due north) with respect to the ground of 700 km/h. The fighter maintains a ground speed of 800 km/h, and the missile travels at 900 km/h with respect to it. (a) What is the speed of the plane with respect to the drone? (b) What is the ground speed of the missile? (c) How long will it take the missile to overtake the target?

61. [II] At the moment this sentence was written, I was in New York at a latitude of just about 41° N. Using Fig. P61, compute the circumference of the circle (the *parallel* of latitude) swept out by a point at that location as the planet rotates. Take the Earth to be a sphere of 6400 km radius turning once in 24 h. At what speed was I moving about the spin axis while sitting here at "rest"? Where are you and how fast are you moving?

Figure P61

62. [II] A ferryboat has a still-water speed of 10 km/h. Its helmsman steers a course due north straight across a river that runs with a current of 5.0 km/h–east. If the trip lasts 2.0 h, how wide is the river and where does the ferry dock?

63. [II] A hawk 50 m above ground sees a mouse directly below running due north at 2.0 m/s. If it reacts immediately, at what angle and speed must the hawk dive in a straight line, keeping a constant velocity, to intercept its prey in 5.0 s? Incidentally, the mouse escaped by jumping in a hole.

64. [II] A ship, in an old black-and-white war movie, is steaming 45° S of W at 30 km/h. At that moment 50 km due south of it, an enemy submarine heading 30° W of N, spots it. At what speed must a torpedo from the sub travel if it's to strike the ship? Assume that the sub fires straight ahead from its forward tubes, and overlook the fact that this problem is a bit unrealistic. (Hint: What can be said about both westerly displacements?)

65. [III] At times of flood, the Colorado River reaches a speed of 30 mi/h near the Lava Falls. Suppose that you wanted to cross it there perpendicularly, in a motorboat capable of traveling at 30 mi/h in still water. (a) Is that possible? Explain. (b) If on a quieter day the water flows at 20 mi/h, at what angle would you head the boat to cut directly across the river? (c) At what speed (in SI units) with respect to the shore would you be traveling?

66. [III] A train is traveling east along a straight run of track at 60 km/h (37 mi/h). Inside, two children 2.0 m apart are playing catch directly across the aisle. The kid wearing a Grateful Dead T-shirt throws the ball horizontally north. The ball crosses the train and is caught in 1.0 s by her little brother. Ignoring any effects of gravity or friction, (a) find the ball's velocity with respect to the little brother. (b) What is the velocity of the ball as seen by someone outside standing still?

67. [III] A cruise ship heads 30° W of N at 20 km/h in a still sea. Someone in a sweatsuit dashes across the deck traveling 60° E of N at 10 km/h. What is the jogger's velocity with respect to the Earth?

68. [III] A ship bearing 60° N of E is making 25 km/h while negotiating a 10-km/h current heading 30° N of E. (a) What is the velocity of the ship with respect to the Earth? (b) How fast is it traveling northward?

69. [III] The Kennedy Space Center on the east coast of Florida is located at a latitude of about 28.5°. Using Problem 61, we can show that the Space Center is moving eastward at 1470 km/h as the Earth rotates. (a) Suppose a rocket is fired so it arcs over and travels horizontally westward at 1470 km/h. What's its speed with respect to both the center and the surface of the Earth? (Incidentally, it's the former speed that determines the orbit of a satellite.) (b) What would be the rocket's speed with respect to the center of the Earth if it were launched at 1470 km/h eastward? (c) Which way do you think we usually fire rockets from Kennedy? (d) Why, overlooking the weather, was Florida rather than Maine originally chosen for the Kennedy Space Center?*

*Answers to all odd-numbered Multiple Choice Questions and all odd-numbered Problems are in the Answers in the back of the book. Numbers in boldface indicate that a solution is provided in the Answers in the back of the book. Numbers in italic indicate that a solution is provided in the Student Solutions Manual.

3

Kinematics:
Acceleration

VARIATIONS IN MOTION ARE commonplace. Change is the rule rather than the exception. We travel from one place to the next, speed up, slow down, turn, pause, and move again. How do we quantify this kind of complex motion? In the previous chapter, we learned about velocity, the time-rate-of-change of displacement, but were careful to avoid such complexities. This chapter goes a step further and evolves the concept of *acceleration,* the time-rate-of-change of velocity. To keep things manageable, the discussion will be limited to objects undergoing constant accelerations. But even confining ourselves in this manner will allow us to develop a set of equations that describes the motion of a great range of occurrences, from braking automobiles to falling apples.

THE CONCEPT OF ACCELERATION

Aristotle hinted at the idea of acceleration, although he never grasped it clearly. Not long afterwards, the philosopher-physicist Strato (ca. 300 B.C.) rightly suggested that a body is accelerating when *equal increments of distance are traversed in shorter and shorter times.* An alternative formulation appeared during the twelfth century in Europe: *acceleration occurs when, during equal periods of time, a body travels greater and greater distances* (Fig. 3.1). What is missing in both of these definitions is the realization that velocity is a distinct concept that may vary from moment to moment. This point was first appreciated in the fourteenth century.

The change in velocity must be part of our definition of acceleration, but it's only a part. The *rate* at which change occurs is crucial. Waiting in a car, at rest on the entrance ramp to a highway, we know what is needed. The car must move from an initial speed $v_i = 0$ to a final speed $v_f = 88.5$ km/h (that is, 55 mi/h) to merge into a gap in the flowing ribbon of traffic. Clearly, the change in velocity ($\Delta \mathbf{v}$) must occur in a specific amount of time (Δt)—it cannot take 10 h, or 10 min, but more like 10 s. **Acceleration is the time-rate-of-change of velocity**.

3.1   Average Acceleration

The **average acceleration** ($\mathbf{a}_{av}$) of a body is defined as *the ratio of the change in its velocity over the time elapsed in the process,* thus

$$\text{Average acceleration} = \frac{\text{change in velocity}}{\text{time elapsed}}$$

$$\mathbf{a}_{av} = \frac{\Delta \mathbf{v}}{\Delta t} = \frac{\mathbf{v}_f - \mathbf{v}_i}{t_f - t_i} \tag{3.1}$$

Figure 3.1 Uniform acceleration. Notice how the distances traveled in equal time intervals increase. As we'll see, those distances go as the odd integers 1, 3, 5, 7, ...

A vector divided by a scalar is a vector; both velocity and acceleration are vector quantities. Something accelerates in a specific direction—it changes its velocity in that direction.

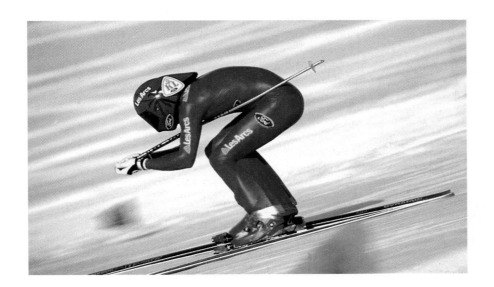

Speed skiing can take a person from 0 to 60 mi/h in 3 s and up to 100 mi/h in 6 s. The men's world record is 139 mi/h. The special rubber suit, helmet, and leg fins have been wind-tunnel tested to cut air resistance.

Acceleration is a two-part concept that arises because of a change in the *direction* of the velocity or because of a change in the *magnitude* of the velocity (that is, speed) or both. A racehorse running faster and faster on a straightaway is accelerating because its speed is changing. A merry-go-round stallion revolving at a constant speed is also accelerating, because the direction of its motion (the direction its nose is pointing) is forever changing. A colt chasing butterflies in a field is likely to be accelerating because of changes in both speed and direction.

Whenever an object moves along a curved path, its velocity vector changes direction and its acceleration will not be parallel to the direction of motion. By contrast, much of this chapter will deal with the simpler situation of rectilinear motion, where velocities and accelerations occur along the same straight line. Even so, there are a great many practical situations that meet this condition (a falling rock is just one of them). Accordingly, $\mathbf{v}_f$ and $\mathbf{v}_i$ are either parallel or antiparallel and $(\mathbf{v}_f - \mathbf{v}_i)$ is in the direction of one or the other. Putting it slightly differently, $\mathbf{v}_i$, $\mathbf{v}_f$, and $\mathbf{a}_{av}$ are *colinear;* that is, they all act along the same line. As we have seen, colinear vectors can be treated algebraically by assigning them appropriate signs. *Let's take the direction in which the object first moves to be positive.* Hence, if $\mathbf{v}_f$ is in the same direction as $\mathbf{v}_i$ and is larger than it, $\mathbf{a}_{av}$ is also in that direction (the time difference $t_f - t_i$ must be a positive number). When the speed increases in the positive direction, it corresponds to a positive acceleration.

By contrast, if $\mathbf{v}_f$ turns out to be smaller than $\mathbf{v}_i$ (or oppositely directed), $\mathbf{a}_{av}$ is negative. A decrease in speed in the positive direction is a negative acceleration, often called a *deceleration.* Many people, however, avoid that term, not wanting to suggest that two different ideas are at work. Regardless of whether velocity is increasing or decreasing, its time-rate-of-change is the acceleration, be it positive or negative.

With an appropriate sign convention, Eq. (3.1) can be written as a scalar, thus

$$a_{av} = \frac{\Delta v}{\Delta t} = \frac{v_f - v_i}{t_f - t_i} \tag{3.2}$$

In the same way, Eq. (2.3) was formulated. The dimensions of a_{av} are those of *speed over time* or, equivalently, *distance over time, over time*—meters per second, per second (m/s²); feet per second, per second (ft/s²), and so on. The former is the preferred SI unit of acceleration, and it is read "meters per second squared" (although (m/s)/s is more informative—an object changes speed by so many meters per second, every second).

Example 3.1 A yellow messenger robot is traveling at 1.00 m/s in a straight line along a ramp in a spaceship. If it speeds up to 2.50 m/s in a time of 0.50 s, what is the magnitude of its average acceleration?

Solution: [Given: $v_i = 1.00$ m/s, $v_f = 2.50$ m/s, and $\Delta t = 0.50$ s. Find: a_{av}]. From the definition

$$a_{av} = \frac{\Delta v}{\Delta t} = \frac{(2.50 \text{ m/s} - 1.00 \text{ m/s})}{0.50 \text{ s}}$$

and

$$a_{av} = \boxed{3.0 \text{ m/s}^2}$$

▶ **Quick Check:** $\Delta v = a_{av}\Delta t = (3.0 \text{ m/s}^2)(0.50 \text{ s}) = 1.5$ m/s.

Example 3.2 A red, 12-cylinder Jaguar XJ12L sedan (getting a dreadful 13 miles per gallon) can travel a straight line from a dead stop to 48.3 km/h (30.0 mi/h) in 3.80 s. It takes an additional 3.00 s to reach 80.5 km/h (50.0 mi/h) and 16.7 more seconds to attain 161 km/h (100 mi/h). What is the magnitude of its average acceleration in m/s² during each of these intervals?

Solution: We take the direction in which the car is traveling as positive. [Given: for the three intervals $v_{i1} = 0$, $v_{f1} = 48.3$ km/h, $\Delta t_1 = 3.80$ s; $v_{i2} = 48.3$ km/h, $v_{f2} = 80.5$ km/h, $\Delta t_2 = 3.00$ s; $v_{i3} = 80.5$ km/h, $v_{f3} = 161$ km/h, $\Delta t_3 = 16.7$ s. Find: the corresponding values of a_{av}.] From the definition of a_{av} we'll need Δv over Δt. The three successive values of $\Delta v = v_f - v_i$ are 48.3 km/h, 32.2 km/h, and 80.5 km/h. To convert these to m/s, multiply each by [(1000 m/km)/(3600 s/h)], which equals 0.2778 according to Table 2.3 (p. 26), thereby yielding 13.42 m/s, 8.945 m/s, and 22.36 m/s. During the first 3.80 s

$$(a_{av})_1 = \frac{13.42 \text{ m/s}}{3.80 \text{ s}} = 3.53 \text{ m/s}^2$$

The car moves faster and faster, increasing its speed on the average 3.53 m/s each second. During the next 3.00 s

$$(a_{av})_2 = \frac{8.945 \text{ m/s}}{3.00 \text{ s}} = 2.98 \text{ m/s}^2$$

And over the last 16.7 s

$$(a_{av})_3 = \frac{22.36 \text{ m/s}}{16.7 \text{ s}} = 1.34 \text{ m/s}^2$$

▶ **Quick Check:** As expected, the vehicle, encountering increasing air resistance, has a decreasing acceleration. From Table 2.2 (p. 25), we can get some fast rough answers: 30 mi/h ≈ 13 m/s, 50 mi/h ≈ 22 m/s, and 100 mi/h ≈ 45 m/s. Hence, (13 m/s)/4 s = 3 m/s², (22 m/s − 13 m/s)/3 s = 3 m/s², and (45 m/s − 22 m/s)/17 s = 1.4 s.

3.2 Instantaneous Acceleration

The people who made the great advances in kinematics in the early fourteenth century were scientists at Oxford with marvelous names like Bradwardine, Dumbleton, Heytesbury, and Swineshead. From their work, we understand that constant or **uniform acceleration** happens *when equal changes in speed occur during equal intervals of time, of any duration.*

For example, if an object at rest increases its speed precisely by 5 m/s during each 1 s of travel, the magnitude of its acceleration will be constant at 5 m/s^2. Because the motion is along a straight line, we graph velocity versus time (Fig. 3.2a), with the initial direction of motion positive. The plot is a diagonal *straight line,* where $v = 0$ at $t = 0$. At the end of successive 1-s intervals, each point is plotted 5 m/s higher than the previous point, and so the graph has to be a straight line, rising at a constant rate. The slope, the rise (Δv) over the run (Δt), is the same everywhere; it's constant at 5 m/s over 1 s and *equal to the scalar value of the average acceleration.* The greater the constant acceleration, the greater the slope of the line. If the speed was *decreasing* ($\Delta v < 0$), the rise over the run would be *negative.* A negative slope, one progressing downward, corresponds to a negative acceleration (Fig. 3.2b). By contrast, a constant speed corresponds to a horizontal line, a slope of zero, and no acceleration (Fig. 2.5, p. 29).

Figure 3.3 is a plot of the velocity-time curve for the Jaguar XJ12L in the previous example. Although it is fairly straight for the first 3 or 4 s, the curve leans over more and more, leveling off as time goes on. Its slope (that is, the slope of the tangent at any point) decreases, slowly at first but more so as time progresses—the car's speed is increasing, but ever more slowly. Evidently, it cannot accelerate forever because of increasing air drag. The Jaguar will never hit 1000 km/h, even with the gas pedal floored for a week.

Again, because the motion varies in time, we need to consider what is happening at each moment. It's not very meaningful to ask about the vehicle's average acceleration without specifying the interval over which we want it. Furthermore, the larger that time interval, the more obscured the details of the motion become. Thus, the car ends up with an average acceleration of 1.90 m/s^2 for the whole run. That single number is poor in information, and has less meaning than the graph that makes it clear there is a range of acceleration. What we need is the average acceleration at *any moment* over a *tiny interval.* The limiting value approached by the average acceleration—Eq. (3.1)—as the time interval (over which the averaging occurs) shrinks toward zero is known as the **instantaneous acceleration**, and is expressed as

$$\mathbf{a} = \lim_{\Delta t \to 0} \left[\frac{\Delta \mathbf{v}}{\Delta t} \right] \qquad (3.3)$$

The limiting value has the same form as Eq. (2.4). This expression represents the moment-by-moment value of the acceleration of an object. *The slope of the tangent*

(a)

(b)

Figure 3.2 Graphs of the motions of two objects moving along straight lines. The slope of the velocity-time plot is the average acceleration over the interval. In (a), the slope is positive and constant, as is the acceleration. In (b), the slope is constant and negative. The body decelerates to $v = 0$ and continues to accelerate in the negative direction such that, after $t = 4$ s, it has a negative velocity.

Figure 3.3 The speed-time graph for a classic Jaguar XJ12L sedan traveling along a straight line.

to the velocity-time curve at any moment is the instantaneous acceleration at that moment. The Jaguar's velocity-time curve shows a gradual decrease in slope, and therefore a gradual decrease in acceleration, as time goes on.

UNIFORMLY ACCELERATED MOTION

Although in the real world, acceleration is rarely if ever constant, there are many circumstances where it can be treated that way, at least over some limited time interval. The first few seconds of an automobile's acceleration is one such case (Fig. 3.3). Without calculus, we have to be content considering only situations in which **a** may be taken to be constant, and therefore equal to $\mathbf{a}_{av}$. Thus, no matter how large Δt is

$$\mathbf{a} = \frac{\Delta \mathbf{v}}{\Delta t} = \frac{\mathbf{v}_f - \mathbf{v}_i}{t_f - t_i} \tag{3.4}$$

Continuing to restrict the discussion to straight-line motion, we can treat the acceleration as a scalar. Furthermore, if we start our clocks so that $t_i = 0$, the final time t_f is the entire trip-time, which we represent simply as t. Hence, Eq. (3.4) becomes

$$a = \frac{v_f - v_i}{t} \tag{3.5}$$

Multiplying both sides by t and shifting v_i over to the other side, leaving v_f to stand alone, we arrive at one of the equations governing *uniformly accelerated motion,* namely

$$v_f = v_i + at \tag{3.6}$$

Four quantities are interrelated by Eq. (3.6) and given any three, the remaining one can be calculated. *Taking the direction of initial motion as positive, numerical values entered into Eq. (3.6), or any other equation in this chapter, can be positive or negative.* For example, if **a** is in the negative direction opposing the initial motion, the numerical value of a must be negative when put in Eq. (3.6), which will ensure that $v_f < v_i$.

Example 3.3 A red Jaguar XJ12L can screech to a straight-line stop from 96.54 km/h (60.00 mi/h) in about 3.7 s. Compute the magnitude of its acceleration, assuming it to be constant.

Solution: [Given: $v_i = 96.54$ km/h, $v_f = 0$, and $t = 3.7$ s. Find: a.] We have v_i, v_f, and t and need a, which suggests Eq. (3.6), or Eq. (3.5). In either case, we will require the initial speed in m/s: $v_i = (96.54$ km/h$) \times (1000$ m/1 km$)(1$ h/3600 s$) = 26.82$ m/s. Since

$$a = \frac{v_f - v_i}{t}$$

we have

$$a = \frac{0 - 26.82 \text{ m/s}}{3.7 \text{ s}} = \boxed{-7.2 \text{ m/s}^2}$$

▶ **Quick Check:** 60 mi/h = 88 ft/s and $a = -23.78$ ft/s^2, multiplying by 0.304 8 m/ft, we get -7.2 m/s^2.

Example 3.4 A bicyclist pedaling along a straight road at 25.0 km/h uniformly accelerates at 3.00 m/s^2 for 3.00 s. Find her final speed.

Solution: The words *uniformly accelerating* immediately classify this as a problem for which $a =$ constant and to which Eq. (3.6) applies. We select the direction of motion as positive. [Given: $v_i = +25.0$ km/h, $a = +3.00$ m/s^2, and $t = 3.00$ s. Find: v_f.] Put everything in the same SI units:

$$v_i = \frac{(25.0 \text{ km/h})(1000 \text{ m/km})}{3600 \text{ s/h}} = 6.94 \text{ m/s}$$

Hence
$$v_f = v_i + at$$
$$v_f = (6.94 \text{ m/s}) + (3.00 \text{ m/s}^2)(3.00 \text{ s})$$

and to three significant figures, $\boxed{v_f = 15.9 \text{ m/s}}$.

▶ **Quick Check:** $\Delta v / \Delta t = (15.9 \text{ m/s} - 6.94 \text{ m/s})/3 \text{ s} = 3 \text{ m/s}^2$.

3.3 The Mean Speed

Perhaps the most important contribution to the theory of motion that came from the Middle Ages was the Mean-Speed Theorem—a gift from Oxford, around 1335. Having successfully defined uniform acceleration, scholars next wanted to relate a to the distance traveled. They knew that the magnitude of the displacement s traversed in rectilinear motion at some average speed was given by $s = v_{av}t$ (though all of this work was done long before any algebraic expressions were ever written out). Could it then be possible to determine an *average speed* that would produce the same displacement in the same time interval as would the constant acceleration in question? (See Fig. 3.4.) In other words, does an average value of the speed exist for an object that is uniformly accelerating along a straight line from an initial speed v_i to a final speed v_f? The answer is yes, and that **mean speed** is given by

$$v_{av} = \tfrac{1}{2}(v_i + v_f) \tag{3.7}$$

Figure 3.4 A body that accelerates uniformly from v_i to v_f over a certain straight-line distance will cover exactly the same distance in the same time traveling at a fixed speed of v_{av}, known as the *mean speed*.

Today, we often contend that a constant acceleration produces a uniformly changing speed, and so the average value is midway between the initial and final speeds (Fig. 3.5). There were proofs like that given in the fourteenth century, but let's look at a different approach that is quite fruitful. Around 1360, Nicolas Oresme used a primitive speed-time graph to assert that the area under the sloping straight line representing uniform acceleration corresponded to the magnitude of the displacement (Fig. 3.6). We considered something similar in Chapter 2.

The area in question has two distinct pieces and can be thought of as the sum of the areas of the rectangle (base t times height v_i) and the triangle (one-half the base t times the altitude $v_f - v_i$). The total area is therefore given by

$$v_i t + \tfrac{1}{2}(v_f - v_i)t$$

or
$$(v_i - \tfrac{1}{2}v_i + \tfrac{1}{2}v_f)t = \tfrac{1}{2}(v_i + v_f)t$$

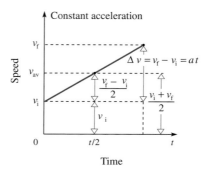

Figure 3.5 For a body that is uniformly accelerating, the mean-speed, v_{av}, is the height of the midpoint of the straight-line speed-time graph.

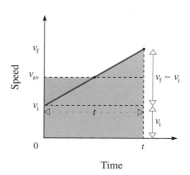

Figure 3.6 The area under the speed-time curve is the distance traveled. The area under the curve (that is, the tilted line) equals the area under the v_{av} line. In other words, $v_{av}t = v_i t + \frac{1}{2}(v_f - v_i)t$.

Setting this equal to the magnitude of the displacement (s) experienced during the uniform accelerated motion, we arrive at what was called the *Mean-Speed Theorem*:

$$s = \tfrac{1}{2}(v_i + v_f)t \qquad (3.8)$$

This equation is simply $s = v_{av}t$ with v_{av} given by Eq. (3.7). Notice that the mean speed is attained at the journey's halfway point in time (Fig. 3.5) and not at the halfway point in space. Also remember that Eq. (2.1) is always true by definition, whereas Eqs. (3.7) and (3.8) are true only when a is constant.

Example 3.5 A bullet fired from a 38-caliber handgun with a 6.00-in. barrel attains a muzzle speed of 330 m/s. Assuming a constant acceleration, how much time does it take the bullet to travel down the barrel?

Solution: This is a case of constant acceleration. [Given: $v_i = 0$, $v_f = 330$ m/s, $s = 6.00$ in. Find: t.] First, put everything in SI units: (6.00 in.)(2.54 cm/in.)/(100 cm/m) = 0.152 m. Because all these quantities are related by $s = \frac{1}{2}(v_i + v_f)t$ we have

$$s = 0.152 \text{ m} = \tfrac{1}{2}(0 + 330 \text{ m/s})t$$

and $t = 2(0.152 \text{ m})/(330 \text{ m/s}) = \boxed{9.21 \times 10^{-4} \text{ s}}$.

▶ **Quick Check:** In U.S. Customary units, $\frac{1}{2}$ ft $= \frac{1}{2}(1083 \text{ ft/s})t$; $t = 0.92$ ms.

Example 3.6 That classic 12-cylinder Jaguar can go from rest to 48.3 km/h (30 mi/h) in 3.8 s, accelerating uniformly (in a straight line) at a rate of 3.54 m/s². Find its average speed and determine how much of a runway it will use in the process.

Solution: The key words here are *accelerating uniformly*, which means a is constant. Take the direction of motion as positive. [Given: $v_i = 0$, $v_f = +48.3$ km/h, $t = 3.8$ s and $a = +3.54$ m/s². Find: v_{av} and s.] First we convert v_f to m/s, so

(continued)

(continued)

$$v_f = \frac{(48.3 \text{ km/h})(1000 \text{ m/km})}{3600 \text{ s/h}} = 13.4 \text{ m/s}$$

Equation (3.7) gives us

$$v_{av} = \tfrac{1}{2}(v_i + v_f) = \tfrac{1}{2}(0 + 13.4 \text{ m/s}) = \boxed{6.7 \text{ m/s}}$$

which equals 15 mi/h. The distance traveled [Eq. (3.8)] is that average speed multiplied by the time of travel, or

$$s = (6.7 \text{ m/s})(3.8 \text{ s}) = \boxed{25 \text{ m}}$$

or about 84 ft.

▶ **Quick Check:** From Table 2.3, $v_f = (48.3) \times (0.2778)$ m/s $= 13.4$ m/s. Since 60 mi/h $= 88$ ft/s, 30 mi/h $= 44$ ft/s; hence, $v_{av} = 22$ ft/s $= 6.7$ m/s.

3.4 Integrals and Intervals

Oresme's area-under-the-curve analysis introduced the use of graphs to represent motion and served as an early cornerstone for the integral calculus of Newton (1665–1666) and Leibniz (1673–1676).

Thus far, we have computed areas under curves, or portions of curves, that were simple straight lines. Now, we want to generalize the technique to include nonuniform motion, such as that represented in Fig. 3.7. Since the object depicted in Fig. 3.7a is moving along a straight path, we can call the diagram a velocity-time curve. We would like to be certain that the area under it, bounded by the times t_i and t_f, is the scalar value of the displacement of the body. In this case, where the velocity is always positive, the magnitude of the displacement equals the distance traveled. Divide the region into several subintervals, each corresponding to a small rectangular area. During any one of these finite subintervals, the body will travel a finite distance.

Let's look at the second subinterval for which the distance traveled is Δs_2. We fix the height of the rectangle at a value of speed equal to $(v_{av})_2$ such that $\Delta s_2 = (v_{av})_2 \Delta t_2$. In so doing, we make its area exactly equal to Δs_2, the appropriate subinterval displacement. If we add up all the subareas, the sum equals the total distance traveled by the body, but we still don't know if that is exactly the area *under the curve*.

The tops of all the rectangles can be made to more closely coincide with the curve by increasing the number of subintervals, making each one correspondingly narrower. If we allow the number of subintervals to approach infinity while the

(a) Time

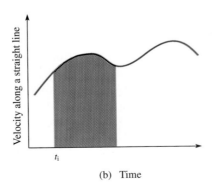

(b) Time

Figure 3.7 (a) The velocity-time curve for an object moving in a straight path at a speed that changes from moment to moment. The second area element is $\Delta s_2 = (v_{av})_2 \Delta t_2$, and the total area under the curve between t_i and t_f is the distance traveled in that time interval. (b) As each time interval Δt shrinks, and the number of such intervals increases, the uppermost points form the curve. The total area of the rectangles approaches the area under the curve.

width of each one (that is, Δt) approaches zero, the tops of the rectangles will blend smoothly into the curve. Those topmost speeds on the rectangles are the various v_{av} values and, in the limit as $\Delta t \rightarrow 0$, they become the instantaneous speeds [Eq. (2.4)] that correspond to points on the curve. Thus, the area under the curve is the distance traveled (Fig. 3.7b).

Isaac Beeckman in 1618 divided time into "indivisible moments," and areas into infinitesimal rectangles, long before the work of Newton and Leibniz. The scheme for adding up an infinity of infinitesimals is the heart of integral calculus. For us, it will suffice to draw the velocity-time curves on graph paper and count the little boxes. Keep in mind that if the object moves in the negative direction (as in Fig. P44, p. 55), the area will then be negative, and any such contribution must be subtracted from the total positive area. An object that moves outward, stops, turns around, and comes all the way back, will have zero displacement and, properly, zero net area under its velocity-time curve.

Example 3.7 Figure 3.8 depicts a speed-time curve for a toy compressed-gas rocket car that has enough fuel for 3.0 s of full power. Approximately how far did the toy car travel between the second and fourth seconds?

Solution: We want the area under the curve between $t = 2.0$ s and $t = 4.0$ s. In the interval from 2.0 to 3.0 s, the area is roughly that of a triangle above a rectangle: $\frac{1}{2}(30 \text{ m/s} - 20 \text{ m/s})1.0 \text{ s} + (20 \text{ m/s})1.0 \text{ s} = 25$ m. Between 3.0 and 4.0 s, the speed is fairly constant at 30 m/s and the area is that of a rectangle, about $(30 \text{ m/s})1.0 \text{ s} = 30$ m. Therefore, the total distance is $\boxed{\approx 55 \text{ m}}$.

▶ **Quick Check:** The average speed is >25 m/s and <30 m/s so for 2 s, 55 m is reasonable.

Figure 3.8 The speed-time curve for a toy car traveling in a straight line. The acceleration is positive and nearly constant for the first 3 s.

3.5 The Equations of Constant Acceleration

Usually, acceleration changes with time, and the calculus is necessary to describe the motion. Nonetheless, many situations occur with uniform, or almost uniform, acceleration, and we already have the kinematics to deal with those. To that end, we will develop five constant-a equations that interrelate the five motion variables (v_i, v_f, a, s, t). These derive from the two basic definitions of speed and acceleration, and so there are only two independent variables that can be solved for. First, from the definition of acceleration, we have

$$v_f = v_i + at \qquad [3.6]$$

The average speed is

$$v_{av} = \frac{1}{2}(v_i + v_f) \qquad [3.7]$$

although it does not shed any new light on the relationship among v_i, v_f, a, s, and t. The Mean-Speed Theorem gives us

$$s = v_{av}t = \tfrac{1}{2}(v_i + v_f)t \qquad\qquad [3.8]$$

It would be useful to have an expression for s that is not explicitly dependent on v_f but is written in terms of v_i, t, and a. Often, objects begin their motion from rest, so that v_i (equal to zero) is known. Furthermore, although s and t are easy to measure, v_f is frequently not available because it's much more difficult to determine directly. Replace v_f in Eq. (3.8) by Eq. (3.6):

$$s = \tfrac{1}{2}[v_i + (v_i + at)]t = \tfrac{1}{2}(2v_it + at^2)$$

and
$$s = v_it + \tfrac{1}{2}at^2 \qquad\qquad (3.9)$$

The first term (v_it) is the scalar value of the body's displacement as it coasts along at its initial speed, even if it is not accelerating $(a = 0)$. To the first term is added another displacement $(\tfrac{1}{2}at^2)$ traversed as the speed increases beyond v_i. If a is negative, the term $(\tfrac{1}{2}at^2)$ is also negative, and the object is slowing down—it will not travel as far as (v_it). In fact, since **s** is the displacement, it's possible for the negative acceleration to carry the object back to where it started such that $s = 0$, or even to where $s < 0$. Using Eq. (3.9) is equivalent to using both Eqs. (3.6) and (3.8), and that's sometimes more convenient.

Example 3.8 According to *Road & Track* magazine, a Lotus Esprit super-coupe can travel 100 ft from rest in 3.30 s. Assume the acceleration is constant, as it nearly is, and determine it in m/s^2.

Solution: This is a uniform-acceleration problem for which we currently have four applicable equations. Take the direction of motion as positive. [Given: $v_i = 0$, $s = +100$ ft, and $t = 3.30$ s. Find: a.] This list (v_i, s, t, and a) suggests Eq. (3.9), which contains all, and only, these quantities. Hence:

$$s = v_it + \tfrac{1}{2}at^2$$

$$(100 \text{ ft})(0.304\,8 \text{ m/ft}) = 0 + \tfrac{1}{2}a(3.30 \text{ s})^2$$

Multiplying both sides by 2 and dividing by $(3.30 \text{ s})^2$ yields

$$a = \frac{60.96 \text{ m}}{(3.30 \text{ s})^2} = 5.598 \text{ m/s}^2$$

and to three figures, $\boxed{a = 5.60 \text{ m/s}^2}$. Note that s and a are both positive quantities.

▶ **Quick Check:** Using Eq. (3.6), $v_f = v_i + at = 0 + (5.6 \text{ m/s}^2)(3.3 \text{ s}) = 18.5 \text{ m/s}$, so the average speed is 9.25 m/s, which in 3.30 s means a distance traveled of 30.5 m = 100 ft.

The last of the five expressions we want relates the displacement, speeds, and acceleration *independent of time*. First, solve Eq. (3.6) for the time, $t = (v_f - v_i)/a$.

Then substitute this along with v_{av} from Eq. (3.7) into Eq. (3.8) so that t no longer appears explicitly:

$$s = \left(\frac{v_i + v_f}{2}\right)\left(\frac{v_f - v_i}{a}\right) = \frac{v_f^2 - v_i^2}{2a}$$

Multiplying both sides by $2a$ and adding v_i^2 to both sides yields

$$v_f^2 = v_i^2 + 2as \qquad (3.10)$$

Whenever we have a problem concerning uniform acceleration in which time does not explicitly appear, Eq. (3.10) will likely be the key formula in the analysis.

Example 3.9 The fastest animal sprinter is the cheetah, reaching speeds in excess of 113 km/h (70 mi/h). These animals have been observed to bound from a standing start to 72 km/h in 2.0 s. Compute the cheetah's maximum acceleration, assuming it to be constant. What minimum distance is required for the cheetah to go from rest to 17.9 m/s (40 mi/h)?

Solution: [Given: for the first part, $v_i = 0$, $v_f = 72$ km/h, $t = 2.0$ s. Find: a. In the second part, $v_i = 0$, $v_f = 17.9$ m/s, and a is assumed constant. Find: s.] Eq. (3.6) is perfect for the first part: $v_f = 72$ km/h $= 20$ m/s

$$a = \frac{v_f - v_i}{t} = \frac{20 \text{ m/s}}{2.0 \text{ s}} = \boxed{10 \text{ m/s}^2}$$

The second part calls for s without specifying the corresponding time and that suggests Eq. (3.10). Using $v_f = 17.9$ m/s and rearranging ($v_f^2 = v_i^2 + 2as$) where $v_i = 0$, we get

$$s = \frac{v_f^2}{2a} = \frac{(17.9 \text{ m/s})^2}{2(10 \text{ m/s}^2)} = \boxed{16 \text{ m}}$$

The cheetah can hit 17.9 m/s from a dead stop in a mere 16 m.

▶ **Quick Check:** $v_{av} = \frac{1}{2} 72$ km/h $= 10$ m/s; hence, it takes a distance $s = (10 \text{ m/s})(2.0 \text{ s}) = 20$ m to reach 20 m/s, and so 16 m for a top speed of 17.9 m/s is reasonable.

Suppose you are driving a car and decide to stop. Assuming you apply steady pressure to properly working brakes that do not lock, and assuming good road conditions, the vehicle should experience a uniform negative acceleration. How far can you expect to travel before coming to rest? According to Eq. (3.10), realizing that $v_f = 0$, we have

$$0 = v_i^2 + 2as$$

and

$$-v_i^2 = 2as$$

Thus, the *stopping distance* is

$$s = -\frac{v_i^2}{2a} \qquad (3.11)$$

Because a is a negative number, the presence of the minus sign ensures that s will be positive. You brake, and the car continues to be displaced a distance s in the positive direction of the initial motion.

The stopping distance depends on v_i *squared*—double the driving speed, and you will need four times the space to stop. The stopping distance also depends on the negative acceleration of your car (the condition and type of its brakes and tires), as well as the road surface. Under good conditions, a superb automobile like a Lamborghini has a negative acceleration of around -9.3 m/s^2, whereas a less than superb compact might slow down at around -6.6 m/s^2. If you brake while traveling 36 m/s (80 mi/h) in that compact on a sunny day and manage not to skid, you will coast for 98 m before the thing stops. (Which is one good reason to never drive that fast.)

Our description overlooks the fact that ordinarily it takes some time (t_R) for the driver to react. That reaction time might be as little as 0.2 s, but it's likely to be more. Thus, if you are traveling at a speed v_i, a distance $s = v_i t_R$ will be traversed before you even activate the brakes. Adding this term to Eq. (3.11) means that stopping will take a distance of

$$s = v_i t_R - \frac{v_i^2}{2a} \qquad (3.12)$$

which should make the point that tailgating is a bad idea (Fig. 3.9).

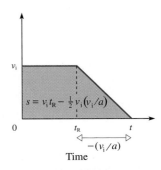

Figure 3.9 The speed-time curve for a driver who, while traveling at v_i, waits a time t_R before slamming on the brakes. The area under the curve is the distance traveled, s. The slope of the curve becomes negative once the brakes engage at $t = t_R$.

Example 3.10 It is a typical sunny day. Assume your reaction time is 0.50 s and the acceleration of your car is -8.2 m/s^2. What is the no-skid stopping distance when the car's speed is initially 26.8 m/s (60 mi/h)?

Solution: [Given: $t_R = 0.50$ s, $a = -8.2$ m/s^2, $v_i = 26.8$ m/s. Find: s.] This specialized problem relates to the analysis that produced Eq. (3.12):

$$s = v_i t_R - \frac{v_i^2}{2a}$$

$$s = (26.8 \text{ m/s})(0.50 \text{ s}) - \frac{(26.8 \text{ m/s})^2}{2(-8.2 \text{ m/s}^2)}$$

and $s = 13.4 \text{ m} + 43.8 \text{ m} = \boxed{57 \text{ m}} \approx 188 \text{ ft}$

▶ **Quick Check:** From Eq. (3.6), the time to decelerate is $(-26.8 \text{ m/s})/(-8.2 \text{ m/s}^2) = 3.3$ s, and from Eq. (3.9) the distance traveled with the brakes engaged is 43.8 m.

FREE-FALL

Probably the most important situation where acceleration is effectively constant is that of free-fall. Drop a heavy object at the Earth's surface and it will descend in such a way that Eqs. (3.6) through (3.10) describe the motion quite accurately.

A falling object ordinarily moves through the surrounding air, so the essential nature of the motion can be obscured by friction. We now know that, in a vacuum, all bodies uniformly accelerate downward at the same rate, regardless of their weight. This crucial insight reveals the hidden nature of gravity, but the realization of the role played by gravity was a long time coming. Aristotle created a theory of the Universe that dominated the science of Muslims and Christians alike, right through the Renaissance. He maintained that a vacuum could not exist (p. 90), and so objects always fell through some medium.

This Aristotelian notion produced the misleading conclusion that all falling bodies descend at different rates proportional to their individual weights, which is a fair picture of what happens to extremely light objects, such as feathers and snowflakes, in air. It applies, as well, for heavy bodies descending through thick media, such as oil or honey, but it misses the central point. For centuries, people wrongly thought that a stone 10 times heavier than another would fall through air 10 times faster—even today many believe this to be the case.

The person who contributed most effectively to the modern understanding of falling bodies was Galileo Galilei, the hero of free-fall. Galileo was born in Italy in 1564. He was vigorous and robust, built short and solid with wavy red hair, a man who fathered three more children than perhaps was proper for a lifelong bachelor. He was an arrogant, argumentative, self-assured spokesman of modernity.

Galileo really was not saying anything new when he asserted that, **in a vacuum, where there is no air resistance, all bodies fall at the same rate**. Others (John Philoponus in the sixth century, for instance) had guessed as much before. By contrast, when an object falls, moving ever more rapidly through the air, the resistance to its motion, the *drag,* also increases. (Riding on a motorcycle gives a sense of how formidable the effect can be—at 100 mi/h the force exerted on a 1.0 ft^2 perpendicular area is about 25 lb, and it increases to 100 lb at 200 mi/h.) The greater the rate of descent, the greater the resistance, as more air must be pushed aside per second (roughly speaking, for small objects at low speeds, resistance goes up with v; for large objects at high speeds, it goes up with v^2). Finally, a balance is reached where no further increase in falling speed can occur. This point is called the **terminal speed**, and it depends on the shape, surface, and weight of the object (see Table 3.1).

A sky diver, after dropping about 620 m (in the thick atmosphere below an altitude of 3000 m or so) with arms and legs extended spread-eagle, will reach a top speed of roughly 200 km/h (120 mi/h).* In a head-down dive, the same person will have a terminal speed up around 300 km/h (185 mi/h). With an open parachute, drag increases tremendously, and the terminal speed drops to around 30 km/h (20 mi/h). On the other hand, a compact heavy object, a smooth stone or an aerial bomb, might drop 150 m to 200 m before even beginning to deviate from free-fall.

All else being constant, terminal speed decreases as the size of an object decreases. The terminal speed of a BB ($\approx$9 m/s) is lower than that of a cannonball ($\approx$250 m/s). The terminal speed of a child is less than that of an adult. This fact accounts, in part, for small animals and insects surviving falls that are immense in proportion to their size. Air friction spares us from being hurt by raindrops and hailstones, which ordinarily strike the Earth at only about 25 km/h to 30 km/h, but might otherwise fall at several hundreds of kilometers per hour (Problem 49, p. 87).

Light objects with a lot of surface area have low terminal speeds that are reached quickly even in air, which is why they seem to conform to Aristotle's notion of weight-dependent fall (why a piece of paper drops slower than a penny). But crumple the paper into a ball, cutting drag without changing weight, and it will fall together with the penny. Galileo knew that for all falling bodies, air resistance would "render the motion uniform," provided the descent was long enough. His ideas were confirmed some 30 years later, when Robert Boyle pumped the air out of a long cylinder and, for the first time, dropped various objects in a vacuum. No amount of theorizing is as convincing as seeing a feather and a lead ball come slamming down together.

TABLE 3.1	Approximate terminal speeds
Object	**Speed (m/s)**
Fluffy feather	0.4
Sheet of paper (flat)	0.5
Snowflake	1
Parachutist	7
Penny	9
No. 6 shot	9
Mouse	13
Sky diver spread-eagle (reached in 10 to 12 s)	58
Bullet (high caliber)	100
Large rock	200
Cannonball	250

*Five seconds is the record for a 310-ft free-fall onto an air bag. The stunt man, who reached 80 mi/h and burst the bag on impact, walked away smiling.

The terminal speed of each person in this position is about 200 km/h. Notice how this flat, belly-down orientation is stable. Throw some playing cards into the air. Do they fall edge-first?

3.6 Acceleration Due to Gravity

Aristotle implied that falling bodies accelerate, but his remarks were ambiguous. His successor Strato, however, was perfectly clear: *objects in free-fall accelerate.* Rainwater running off a roof begins its descent in a continuous stream that ultimately breaks up into a flutter of separate splashes. "This could never happen," Strato maintained, "unless the water was falling more swiftly through each successive distance than it had through the earlier ones." The droplets do accelerate, and the ones falling longer move faster and pull away from the slower ones behind them. On some level, you already knew that falling bodies accelerate—although you might volunteer to catch a baseball dropped from a few meters, you would surely decline if it was to come down from the top of the Sears Tower in Chicago (443 m). The ball would then descend at a tremendous rate, which could only mean that it accelerated.

Leonardo da Vinci maintained that a body in free-fall moves at a speed proportional to its time of flight, which is equivalent to saying that a body accelerates uniformly, since, by Eq. (3.6), when $v_i = 0$, $v_f = at$. But it was Galileo who first designed and carried out an experiment to confirm that the acceleration is indeed constant. Two practical obstacles made it impossible for him to simply drop an object, measure v_f at several values of t, and then compute a for each. The first was that free-fall happens rapidly and clocks in his era were extremely crude. He surmounted this problem by slowing things down—instead of dropping balls, he rolled

On free-fall. For all things that fall through the water and thin air, these things must needs quicken their fall in proportion to their weights, just because the body of water and the thin nature of air cannot check each thing equally, but give place more quickly when overcome by heavier bodies. But, on the other hand, the empty void cannot on any side, at any time, support anything, but rather, as its own nature desires, it continues to give place; wherefore all things must needs be borne on through the calm void, moving at equal rate with unequal weights.

LUCRETIUS (ca. 94–55 B.C.)
On the Nature of Things

THE RECORD FREE-FALL

Lieutenant I. M. Chisov of the former Soviet Union was flying his *Ilyushin 4* on a bitter cold day in January, 1942, when it was attacked by 12 German *Messerschmitts*. Convinced that he had no chance of surviving if he stayed with his badly battered plane, Chisov bailed out at 21 980 ft. With the fighters still buzzing around, Chisov cleverly decided to fall freely out of the arena. It was his plan not to open his chute until he was down to only about 1000 ft above the ground. Unfortunately, he lost consciousness en route. As luck would have it, he crashed at the edge of a steep ravine covered with 3 ft of snow. Hitting at about 120 mi/h, he plowed along its slope until he came to rest at the bottom. Chisov awoke 20 minutes later, bruised and sore, but miraculously he had suffered only a concussion of the spine and a fractured pelvis. Three and one-half months later he was back at work as a flight instructor.

A cannon ball weighing one or two hundred pounds, or even more, will not reach the ground by as much as a span ahead of a musket ball weighing only half a pound, provided both are dropped from a height of 200 cubits.

GALILEO GALILEI

Regarding terminal speeds. *You can drop a mouse down a thousand-yard mine shaft and, on arriving at the bottom, it gets a slight shock and walks away. A rat is killed, a man is broken, a horse splashes.*

J. B. S. HALDANE
British geneticist, 1892–1964

them along inclined planes (Fig. 3.10). He argued that if a was constant for each incline, it would continue to be constant, even when the track was vertical and the ball fell freely.

The second difficulty Galileo encountered was that he had no way of measuring the instantaneous final speed at the end of each run. To handle this obstacle, he reformulated the problem, replacing v_f by quantities that were directly observable. Thus, he arrived at a verbal statement of Eq. (3.9), $s = v_i t + \frac{1}{2} a t^2$. *If the acceleration was constant, starting from rest ($v_i = 0$), the distance traveled by the ball would depend on the time squared ($s \propto t^2$).* Doubling the time a ball was allowed to move would quadruple the distance it traveled, and that situation would be an easy one to observe. Galileo's results were remarkably good (even though he overlooked the fact that the balls were rolling, not sliding, which decreases the constant acceleration—see Fig. 8.16). "In such experiments," he wrote, "repeated a full hundred times, we always found that the spaces traversed were to each other as the squares of the times, and this was true for all inclinations of the plane."

In the absence of air friction, all bodies, regardless of their weight, fall with the same uniform acceleration. Well, not quite. The acceleration is the same for all bodies, but it is not exactly uniform. Newton pointed out that the acceleration gradually decreases with increasing height above the Earth (Eq. 7.3). At an altitude of 1000 m, the acceleration is about 0.03% less than it is at the planet's surface. Over the small distances objects fall in the laboratory, this altitude dependence is insignificant, and it's not surprising that Galileo missed it.

The acceleration due to gravity is represented by its own symbol, g, traditionally taken to be the net *measured* downward acceleration (Table 3.2). In nearby space there is no ambiguity, but *on* Earth that acceleration does not depend only on gravity. There is a small negative contribution (see Fig. 7.7) arising from the spin of the planet; objects at rest at the surface (a rock in your hand, for example) are revolving around the center of the Earth. If the planet stopped spinning, g would

Figure 3.10 To see if the acceleration of a falling body was constant, Galileo rolled balls down inclined planes of increasing tilt. As the angle increased, the acceleration increased, but for each position it was constant. He argued that at a tilt of 90° the object would free-fall, also with a constant acceleration.

TABLE 3.2 Acceleration due to gravity*		
	g	
Location	m/s^2	ft/s^2
Equator	9.780	32.09
Panama	9.782	32.09
Honolulu	9.789	32.12
Key West	9.790	32.12
Austin	9.793	32.13
Tokyo	9.798	32.15
San Francisco	9.800	32.15
Princeton	9.802	32.16
New York	9.803	32.16
Chicago	9.803	32.16
Munich	9.807	32.18
Winnipeg	9.810	32.18
Leningrad	9.819	32.22
Arctic Red River	9.824	32.23
North Pole	9.832	32.26

*The measured values of g are actually due to both gravity and the rotation of the Earth.

Galileo Galilei at the age of about 60, drawn by his contemporary Ottavio Leoni. Galileo was born in 1564, just three days before Michelangelo died, and within months of Shakespeare's birth.

increase slightly everywhere but at the poles. Because the effect is small (<0.4 %), it's usually ignored. We can generally take g to be a constant equal to its average value of 9.806 65 m/s^2 (32.2 ft/s^2). Galileo never did determine g—he did not have a clock accurate enough for the purpose. C. Huygens, who invented the pendulum clock in 1656, was the first to measure g. He showed that it could be adduced from the swing of a pendulum using only a ruler and a good timepiece (see Sect. 12.3).

3.7 All Fall Down: Constant Acceleration

If an object is moving freely near the surface of the Earth, its gravitational acceleration is along a vertical line and is assumed to be constant. All the equations of uniformly accelerated motion—Eqs. (3.6), (3.7), (3.8), (3.9), and (3.10)—apply where $a = g$. **In free-fall the acceleration is always g and always straight downward regardless of the motion**.

Example 3.11 A salmon is dropped by a hovering eagle. How far will the fish fall in 2.5 s before the bird catches it again? Neglect air drag.

Solution: The initial motion is down, so take down as positive. [Given: $a = g = +9.81$ m/s^2, $v_i = 0$, and $t = 2.5$ s. Find: the distance fallen, s.] We have a, v_i, and t, and need s, which suggests Eq. (3.9). Thus

$$s = v_i t + \tfrac{1}{2} g t^2 = 0 + \tfrac{1}{2}(+9.81 \text{ m/s}^2)(2.5 \text{ s})^2$$

The fish falls a distance $s = +30.6$ m or to two figures,

$$\boxed{s = 31 \text{ m}}.$$

▶ **Quick Check:** Its final speed is $v_f = v_i + gt = 24.5$ m/s, and since $v_f^2 = v_i^2 + 2gs$, $s = (24.5 \text{ m/s})^2/2g = 30.6$ m.

Example 3.12 A ball is thrown straight down from the roof of a dormitory at 10.0 m/s. If the building is 100 m tall, at what speed will the ball hit the ground? How long will the trip take?

Solution: Since the initial motion is down, take that as the positive direction. [Given: $v_i = +10.0$ m/s, $a = g = +9.81$ m/s^2, and $s = +100$ m. Find: v_f and t.] Using Eq. (3.10), we have

$$v_f^2 = v_i^2 + 2gs = (+10.0 \text{ m/s})^2 + 2(+9.81 \text{ m/s}^2)(100 \text{ m})$$

and $\boxed{v_f = 45.4 \text{ m/s}}$. The time can most easily be gotten

from Eq. (3.6), $v_f = v_i + at$ where

$$t = \frac{v_f - v_i}{g} = \frac{45.4 \text{ m/s} - 10.0 \text{ m/s}}{+9.81 \text{ m/s}^2} = \boxed{3.61 \text{ s}}$$

▶ **Quick Check:** Using $s = v_i t + \frac{1}{2}at^2$, $\frac{1}{2}(9.81$ m/s$^2)t^2 + (10.0$ m/s$)t - 100$ m $= 0$. From the quadratic equation (Appendix A-4)

$$t = \frac{-10.0 \text{ m/s} \pm 45.4 \text{ m/s}}{9.81 \text{ m/s}^2}$$

ignoring the negative solution $t = 3.6$ s.

If a ball is thrown upward, its speed continuously diminishes at a rate of -9.81 m/s per second as it climbs. Always acting downward, g has the effect of increasing the speed of descending objects and decreasing the speed of ascending objects. The rising ball, moving slower and slower, will momentarily stop (the acceleration is still g) and then plunge downward faster and faster. After all, a body cannot move in opposite directions along a straight line without stopping first. The height of that stopping-point, the maximum or **peak altitude** (s_p), is the vertical equivalent of the stopping distance for a uniformly decelerating car, Eq. (3.11)—the distance it takes to go from v_i to $v_f = 0$.

The up-and-down journey is symmetrical in space and time, around the peak altitude (Fig. 3.11). Ignoring air friction, the object will be moving at the same speed when at the same altitude, whether it's on the way up or down. Moveover, the speeds will be equal for any amount of time before and after the peak altitude is reached. The ascent unfolds as if it were a movie of the descent run backward. And the descent is exactly the same as if the object were dropped from the peak altitude. At high projectile speeds, air friction limits the motion causing the trajectory to be lower and asymmetrical (see p. 93). For example, typical antiaircraft fire rises to less than about 1500 m (see Example 3.13).

For a projectile shot into the air, at any instant on the way up or down, s is the scalar value of the displacement: *it is the distance the object is from the starting point.* It is not the total distance traversed to get there. When a projectile is thrown straight up, reaches peak altitude, and comes halfway back down, the displacement ($\frac{1}{2}s_p$) at that moment is one third of the distance traveled.

Example 3.13 A .32-caliber bullet fired from a revolver with a 3-in. long barrel will have a relatively low muzzle speed of about 200 m/s. If it's shot straight up, neglecting air resistance, (a) What is the peak height the bullet will reach? (b) How fast will it be moving when it returns to the height of the gun? (c) How long will the whole trip take?

Solution: Select up as the positive direction. [Given: $v_i = +200$ m/s. Find: (a) s_p; (b) final speed; (c) total

time t_T.] For the upward leg of the trip use Eq. (3.10), where $a = g$, to get the height at which the speed goes to zero ($v_f = 0$). Thus

$$0 = v_i^2 + 2gs_p$$

and

$$s_p = -\frac{v_i^2}{2g}$$

which is essentially Eq. (3.11). The bullet is decelerating, g is downward, and so we enter it as a negative number.

(continued)

(continued)

Hence the answer to part (a) is

$$s_p = -\frac{(200 \text{ m/s})^2}{2(-9.81 \text{ m/s}^2)} = \boxed{+2.04 \times 10^3 \text{ m}}$$

(b) Because the flight is symmetrical, the bullet ideally returns to the height of the gun at a speed of $\boxed{200 \text{ m/s}}$. Table 3.1 tells us that the actual speed is closer to 100 m/s, which is still very dangerous. (c) The time for the trip *up* can be computed using Eqs. (3.6), (3.8), or (3.9)—Eq. (3.6) is the easiest. Accordingly

$$v_f = v_i + gt$$
$$0 = 200 \text{ m/s} + (-9.81 \text{ m/s}^2)t_p$$

and $t_p = -(200 \text{ m/s})/(-9.80 \text{ m/s}^2) = 20.4$ s, which is half the total flight time and thus $\boxed{t_T = 40.8 \text{ s}}$.

▶ **Quick Check:** Compute the downtrip time starting with Eq. (3.9), $s = v_i t + \frac{1}{2}gt^2$. Now $v_i = 0$, $g = -9.80 \text{ m/s}^2$, and the displacement is -2040 m measured from the top of the trajectory downward to the initial height of firing. Hence

$$t_p = \left(\frac{2s}{g}\right)^{1/2} = \left[\frac{2(-2040 \text{ m})}{-9.81 \text{ m/s}^2}\right]^{1/2} = 20.4 \text{ s}$$

Alternatively, factor Eq. (3.9) and set $s = 0$; $t(v_i + \frac{1}{2}gt) = 0$, hence $t = 0$ and $(v_i + \frac{1}{2}gt) = 0$, and so $t = t_T = -2v_i/g = 40.8$ s.

Example 3.14 A ball is hurled straight up at a speed of 15.0 m/s, leaving the hand of the thrower 2.00 m above ground. Compute the times and the ball's speeds when it passes an observer sitting at a window in line with the throw 10.0 m above the point of release.

Solution: Because nothing about ground level is to be computed, the height of the hand at launch is taken as the origin. Take *up* as positive throughout; $a = g$, which is always downward and a negative number. [Given: $v_i = +15.0$ m/s, $a = -9.81 \text{ m/s}^2$, and $s = +10.0$ m. Find: the *two* times and the *two* final speeds.] We could use $s = v_i t + \frac{1}{2}at^2$ to find t directly, which yields a quadratic equation (Problem 74). Instead we will find v_f first, via $v_f^2 = v_i^2 + 2as$. Consequently

$$v_f^2 = (15.0 \text{ m/s})^2 + 2(-9.81 \text{ m/s}^2)(+10.0 \text{ m})$$
$$v_f^2 = 28.9 \text{ m}^2/\text{s}^2$$

and $\boxed{v_f = \pm 5.37 \text{ m/s}}$

The ball passes the window traveling at +5.37 m/s on the way up and −5.37 m/s on the way down. The corresponding times (t_u and t_d) can be gotten from $v_f = v_i + at$. On the way up

$$+5.37 \text{ m/s} = +15.0 \text{ m/s} + (-9.81 \text{ m/s}^2)t_u$$
$$\boxed{t_u = 0.982 \text{ s}}$$

On the way down, $v_f = -5.37$ m/s, and so

$$-5.37 \text{ m/s} = +15.0 \text{ m/s} + (-9.81 \text{ m/s}^2)t_d$$
$$\boxed{t_d = 2.08 \text{ s}}$$

Notice that the peak altitude was reached when $v_f = 0 = v_i + (-9.81 \text{ m/s}^2)t_p$; that is, when $t_p = 1.53$ s. Hence, the ball passed the window on the way up 0.55 s earlier than t_p and on the way down, 0.55 s later. This symmetry around the peak can be useful in problem solving.

▶ **Quick Check:** $s = v_i t + \frac{1}{2}at^2 = (15.0 \text{ m/s}) \times (0.982 \text{ s}) + \frac{1}{2}(-9.81 \text{ m/s}^2)(0.982 \text{ s})^2 = 10.0 \text{ m}$.

3.8 Projectile Motion

Suppose that we throw an object into the air at some angle other than straight up. Galileo recognized that the projectile's unpowered movement can be imagined as if it were two simultaneous *independent* motions: a horizontal flight (ideally at constant speed) and a vertical gravity fall (ideally at constant acceleration). The gravitational interaction causes a vertical acceleration, but no such influence exists horizontally, and the horizontal acceleration (overlooking air drag) is zero. We will

Speeds both upward and downward

(ft/s)　(m/s)

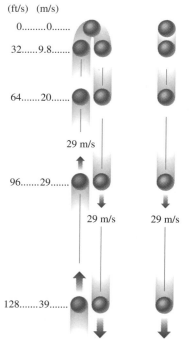

29 m/s

29 m/s　29 m/s

Figure 3.11 A ball thrown straight up at some speed—in this case 39 m/s (128 ft/s)—will (neglecting air friction) return to its launch height at that same speed. In fact, the speeds moving up and down are equal at equal heights. In this instance, the ball takes 4.0 s to reach maximum altitude and 4.0 s to get back down.

At the surface of the Earth, in situations where air friction is negligible, objects fall with the same acceleration regardless of their weights.

study the forces at work in this situation in the next chapter; here, we concentrate on the kinematics, the description of the motion.

Imagine a ball thrown horizontally. There is then no initial component of the velocity in the vertical direction ($v_{iV} = 0$). The ball sails straight off horizontally at a *constant speed* equal to its initial speed, v_{iH}. But on Earth, heavy objects fall freely downward with a constant acceleration. Accordingly, the projectile will continuously descend, faster and faster, as it progresses laterally, sweeping out a smooth arc that curves increasingly downward (Fig. 3.12).

The shape of the arc traveled by a projectile can be approximated by forming an equation relating its horizontal and vertical displacements, an equation that is independent of time since the path is fixed, even if the object moving along it is not. In the case of a projectile launched with an initial horizontal speed (v_{iH}), which is assumed constant, the horizontal distance traveled is proportional to the time of flight ($s_H = v_{iH}t$). The vertical free-fall distance ($s_V = \frac{1}{2}gt^2$) varies with the time squared. These two distance equations can be combined to produce an expression for s_V in terms of s_H, which is independent of time. Rearranging the first expression to get $t = s_H/v_{iH}$ and substituting this into the second yields

$$s_V = \left(\frac{g}{2v_{iH}^2}\right)s_H^2 \qquad (3.13)$$

Figure 3.12 The faster each cannonball is fired, the farther it will go, but all fall at the same rate and hit the water after the same flight times. The horizontal speed, v_{iH}, is constant, provided air friction is negligible. The vertical speed increases at the rate g.

The term in parentheses is constant for any given launch and that means that Eq. (3.13) is the equation of a *parabola*—one space variable proportional to the square of another space variable (Fig. 3.13). To the extent that both the curvature of the Earth and friction can be neglected, the path of a ballistic projectile is parabolic.

Figure 3.13 The parabola can be visualized as the intersection of a plane and a cone. The plane is tilted parallel to the edge of the cone.

Sparks fly into the air, arcing along parabolic paths.

Example 3.15 A youngster hurls a rock horizontally at a speed of 10 m/s from a bridge 50 m above a river. Ignoring air resistance: (a) How long will it take for the rock to hit the water? (b) What is the velocity of the rock just before it lands? (c) How far from the bridge will it strike?

Solution: [Given: The initial horizontal speed $v_{iH} = 10$ m/s and $s_V = 50$ m. Find: (a) t; (b) the final velocity $\mathbf{v}_f = \mathbf{v}_{fH} + \mathbf{v}_{fV}$, and (c) s_H.] (a) The time it takes to hit the water is just the free-fall time given by Eq. (3.9), $s = v_i t + \frac{1}{2}at^2$, applied exclusively to the vertical motion as if it alone were occurring. With $a = g$ and no initial *vertical* speed, either up or down, this equation becomes

$$s_V = \tfrac{1}{2}gt^2$$

Taking down as positive

$$(50 \text{ m}) = \tfrac{1}{2}(9.81 \text{ m/s}^2)t^2$$

and
$$t = \sqrt{\frac{2(50 \text{ m})}{(9.81 \text{ m/s}^2)}} = 3.19 \text{ s} = \boxed{3.2 \text{ s}}$$

(b) The velocity at which the rock hits the water (Fig. 3.14) is the vector resultant of its final horizontal, $\mathbf{v}_{fH}$, and vertical, $\mathbf{v}_{fV}$, motions. We know that the horizontal speed is constant at $v_{iH} = 10$ m/s and so must now find $\mathbf{v}_{fV}$. Equation (3.6), with no initial vertical speed, yields

$$v_{fV} = gt = (9.81 \text{ m/s}^2)(3.19 \text{ s}) = 31.3 \text{ m/s}$$

The resultant speed (p. 46) is then

$$v_f = \sqrt{v_{fH}^2 + v_{fV}^2} = \sqrt{1079} \text{ m/s} = 33 \text{ m/s}$$

Notice that this quantity is larger than the largest of the two components, as it should be. From Fig. 3.14, we see that the angle, θ, made by $\mathbf{v}_f$ and the horizontal is such that

$$\tan \theta = \frac{v_{fV}}{v_{fH}} = \frac{31.3 \text{ m/s}}{10 \text{ m/s}} = 3.13$$

and therefore $\theta = 72°$ below the horizontal. We could use either $\sin \theta$ or $\cos \theta$ instead of $\tan \theta$ but that would involve using v_f, and if an error was made in computing

Figure 3.14 A projectile thrown horizontally from a bridge hits the water with a velocity $\mathbf{v}_f$ that has both horizontal, $\mathbf{v}_{fH}$, and vertical $\mathbf{v}_{fV}$, components. With negligible air-friction losses, the initial horizontal speed v_{iH} equals v_{fH}.

v_f, it would make θ wrong, too. It's best, if all else is equal, to use quantities that are least likely to be in error. Thus, $\boxed{v_f = 33 \text{ m/s–}72° \text{ below horizontal}}$. (c) Because the rock's horizontal speed is constant and because it is in the air for a time t

$$s_H = v_{iH}t = (10 \text{ m/s})(3.2 \text{ s}) = \boxed{32 \text{ m}}$$

▶ **Quick Check:** The average vertical speed is $\frac{1}{2}v_{fV} = 15.65$ m/s; the time-of-flight is then $t = 50$ m/(15.65 m/s) $= 3.19$ s. Observe that v_{fV} is about 3 times v_{fH}, so a fairly large value of θ is reasonable.

Now for the more general case. Imagine a cork popping out of a champagne bottle into the air at some angle θ with an initial speed v_i, as depicted in Fig. 3.15a. (This could just as well be a leaping frog, an athlete executing a broad jump, or a kicked football—any unpowered flight.) We deal with this situation practically

$$v_{iV} = v_i \sin \theta$$

$$v_{iH} = v_i \cos \theta$$

Figure 3.15 (a) Launching a projectile into the air at an angle θ. (b) Notice how the vertical component of the velocity varies along the trajectory. A longstanding record was set in Reno, Nevada on July 4, 1981, when a cork from a champagne bottle was "fired" 105 ft 9 in. Because of air friction, corks are launched at 40° to cover more distance than height.

by realizing that the motion unfolds as if it were two independent superimposed motions—one vertical, with an initial speed of

$$v_{iV} = v_i \sin \theta$$

the other horizontal, with an initial speed of

$$v_{iH} = v_i \cos \theta$$

That's equivalent to saying that **s**, the displacement vector drawn from the point of launch to the projectile at any moment, has a scalar horizontal component of

$$s_H = v_{iH}t$$

since $a_H = 0$, and a scalar vertical component of

$$s_V = v_{iV}t + \tfrac{1}{2}gt^2$$

since $a_V = g$.

As before, when dealing with the scalar components of vectors, we must select certain directions to be positive so that the components along and opposite to those directions can be added algebraically, preserving the directional nature of the process. Accordingly, g can be either a positive or negative number (± 9.81 m/s^2). If the initial motion has an upward component, we might choose *up* to be positive, in which case v_{iV} is positive, but g, because it is downward as always, is then negative.

The cork rises (in the positive direction) accelerating at -9.81 m/s^2, slowing down under the influence of a downward acceleration, until it reaches its peak altitude. At that point, its vertical speed is zero although it continues uninterruptedly to move horizontally (Fig. 3.15b). Thereafter, the cork descends along an arc, as ever under the influence of gravity. Accelerating all the while, it returns to the launch elevation at very nearly the speed with which it was first fired. During the descent (because *up* is positive), the velocity increases in the negative direction.*

The equation for s_V describes the trip down as well as the trip up; you need not change signs midway through the journey. We are describing the *displacement* from the launch point, *not* the distance traveled. On the way up, it increases; on the way down, it decreases. When the projectile drops back to the launch height, s_V is zero. If it should fall into a hole or over a cliff so that it lands beneath launch level, s_V would become negative (being in the downward direction).

Provided friction losses are negligible, the path is a parabola, symmetrical around the peak altitude. But if you ran along at a constant speed staying directly under the cork, it would have no horizontal velocity with respect to you; it would be seen simply to rise up, stop at peak altitude, and come straight down. The time that it takes (t_p) to reach peak altitude (s_p) is again obtained from Eq. (3.6), where the final vertical speed is zero and, hence, $0 = v_{iV} + gt_p$. Accordingly, $t_p = -v_{iV}/g$ (which is simple to derive and should not be memorized). When *up* is positive so that v_{iV} is positive, g will be negative and vice versa. In any event, *the time is always positive*.

In the simplest case, the journey ends when the projectile returns to the height at which it was launched. Then, because of the symmetry, twice the time-to-peak equals the **total-flight time** t_T:

$$t_T = -\frac{2v_{iV}}{g} = -\frac{2v_i \sin \theta}{g} = \frac{2v_i \sin \theta}{9.8 \text{ m/s}^2} \tag{3.14}$$

The **peak height** attained is

$$s_p = -\frac{v_{iV}^2}{2g} = -\frac{(v_i \sin \theta)^2}{2g} = \frac{(v_i \sin \theta)^2}{2(9.8 \text{ m/s}^2)} \tag{3.15}$$

which applies as well to basketballs or water fountains. As long as *up* is positive, $g = -9.8$ m/s^2 and s_p is positive.

Example 3.16 The person in the gondola shown in Fig. 3.16 throws a ball at 40.0° to the horizon at 10.0 m/s. If the ball is launched at a height of 100 m, where will it land? Ignore air friction.

Solution: [Given: $v_i = 10.0$ m/s, $\theta = 40.0°$, and the launch height is 100 m. Find: s_H.] We need the total time of flight, but this path is not symmetrical and Eq. (3.14) does not apply. As a rule, we can get the total-flight time by finding how long the vertical motion lasts. With the launch point as the origin

$$s_V = v_{iV}t + \tfrac{1}{2}gt^2$$

and if we take up positive, the journey ends when $s_V = -100$ m, at which point $t = t_T$. Using $v_{iV} = (10.0 \text{ m/s}) \sin 40.0° = 6.428$ m/s in the expression for s_V, we get

(continued)

*When the Palestine Liberation Organization left Beirut, Lebanon, in 1982, they fired so many shots into the air to celebrate that at least half a dozen people were seriously wounded by falling bullets.

(*continued*)
$$-100 \text{ m} = (6.428 \text{ m/s})t_\text{T} + \tfrac{1}{2}(-9.807 \text{ m/s}^2)t_\text{T}^2$$

To solve for t_T, rearrange this equation so it has the standard form of the quadratic equation (Appendix A-4), namely

$$(4.904 \text{ m/s}^2)t_\text{T}^2 + (-6.428 \text{ m/s})t_\text{T} + (-100 \text{ m}) = 0$$

whereupon t_T then equals

$$\frac{6.428 \text{ m/s} \pm \sqrt{(-6.428 \text{ m/s})^2 - 4(4.904 \text{ m/s}^2)(-100 \text{ m})}}{2(4.904 \text{ m/s})}$$

There is only one positive solution, namely, $t_\text{T} = 5.22$ s. Hence, the ball hits the ground at a horizontal distance from the launch point equal to

$$s_\text{H} = v_{\text{iH}}t_\text{T} = (10.0 \text{ m/s})(\cos 40.0°)(5.22 \text{ s}) = \boxed{40.0 \text{ m}}$$

▶ **Quick Check:** The time it takes to get back down to the launch height is $2t_\text{p} = 1.31$ s. At that point, the ball is moving down at 6.428 m/s and if the answer is right, it will take $(5.22 \text{ s} - 1.31 \text{ s}) = 3.91$ s to descend the 100 m. Hence, $s_\text{V} = v_{\text{iV}}t + \tfrac{1}{2}gt^2 = (6.428 \text{ m/s}) \times (3.91 \text{ s}) + \tfrac{1}{2}(9.81 \text{ m/s}^2)(3.91 \text{ s})^2 = 100$ m and all's well.

Figure 3.16 A ball is thrown at 40.0° from a gondola 100 m above the ground. Where will it land?

Range s_R *is the total horizontal distance that a projectile travels returning to the same height from which it was launched.* The only way this notion makes sense is if the object is fired in some direction upward. Again *up* is positive and g is negative. Since v_{iH} is constant

$$s_\text{R} = v_{\text{iH}}t_\text{T} = (v_\text{i} \cos \theta)t_\text{T}$$

Using Eq. (3.14) for the total-flight time, the range becomes

$$s_\text{R} = -\frac{2v_\text{i}^2}{g} \cos \theta \sin \theta = \frac{2v_\text{i}^2}{9.8 \text{ m/s}^2} \cos \theta \sin \theta \qquad (3.16)$$

This horizontal displacement is nonzero and positive (to the right of the launch point) when $\theta < 90°$ whereupon both $\sin \theta$ and $\cos \theta$ are positive. Notice that when $\theta = 90°$, the projectile goes straight up and down, $\cos 90° = 0$, and $s_\text{R} = 0$; moreover, when $\theta = 0°$, $s_\text{R} = 0$. Neglecting aerodynamic effects, maximum range occurs when $(\cos \theta \sin \theta)$ has its maximum value. A simple plot of that product reveals that it reaches a maximum when $\theta = 45°$. Altering the launch angle by the same amounts, either increasing or decreasing it from 45°, produces the same range (Fig. 3.17). *The projected horizontal distance is the same for each member of any pair of firing angles that add up to 90°.*

Figure 3.17 Ideally, the maximum range of an artillery shell corresponds to a firing angle of 45°. Notice that if we change the angle by going up or down some amount from 45°, the range is the same. For example, firing at 45° + 15° or 45° − 15° yields the same range.

Example 3.17 A baseball recoiling from a bat soars into the air at an angle of 40.0° above the ground traveling at 45.7 m/s (150 ft/s). Calculate its range ignoring aerodynamic effects.

Solution: [Given: $\theta = 40.0°$, and $v_i = 45.7$ m/s. Find: s_R.] It is not necessary to memorize Eq. (3.16) when it's so easily derived from $s_R = v_{iH}(2t_p)$. In any event

$$s_R = -\frac{2v_i^2}{g} \cos \theta \sin \theta$$

$$s_R = \frac{2(45.7 \text{ m/s})^2}{9.81 \text{ m/s}^2} \cos 40.0° \sin 40.0°$$

or $\boxed{s_R = 210 \text{ m}}$ (688 ft). In reality, a baseball experiences both drag and lift. At the speed given, it's not likely to go much farther than about 500 ft. Moreover, maximum range is actually attained when the ball is hit at 40° rather than 45°.

▶ **Quick Check:** $t_T = 2t_p = -2v_{iV}/g = -2(45.7 \text{ m/s}) \times \sin 40°/9.8 \text{ m/s}^2 = 6.0 \text{ s}$; $s_R = v_{iH}t_T = (45.7 \text{ m/s}) \cos 40° \times (6.0 \text{ s}) = 210 \text{ m}$.

Air friction (p. 93) has a great effect on projectiles, especially on those that are light and fast. Accordingly, cannon fire can travel much farther than bullets. For example, the 210-mm (diameter) howitzer (used by Iraq in the 1991 Gulf War) has the greatest range of all field artillery, namely 57 000 m. That is more than ten times the range of a high-speed rifle bullet.

Core Material

The **average acceleration** is defined as:

$$\mathbf{a}_{av} = \frac{\Delta \mathbf{v}}{\Delta t} = \frac{\mathbf{v}_f - \mathbf{v}_i}{t_f - t_i} \qquad [3.1]$$

As $\Delta t \to 0$, this ratio approaches the instantaneous acceleration

$$\mathbf{a} = \lim_{\Delta t \to 0} \left[\frac{\Delta \mathbf{v}}{\Delta t} \right] \qquad [3.3]$$

Restricting the treatment to *straight-line motion at constant acceleration*, there are five useful equations:

$$v_f = v_i + at \qquad [3.6]$$

$$v_{av} = \tfrac{1}{2}(v_i + v_f) \qquad [3.7]$$

$$s = \tfrac{1}{2}(v_i + v_f)t \qquad [3.8]$$

$$s = v_i t + \tfrac{1}{2}at^2 \qquad [3.9]$$

and
$$v_f^2 = v_i^2 + 2as \qquad [3.10]$$

Objects in free-fall at the surface of the Earth (for which air friction is inconsequential) descend with a nearly *uniform acceleration* (g) equal to 9.81 m/s² or 32.2 ft/s². The above set of equations can be applied to projectiles (p. 80) where the constant-speed horizontal motion and the constant-acceleration vertical motion occur independently. Such quantities as the time to peak altitude (t_p), the total time of flight (t_T), and the range (s_R) are important ideas that can be easily calculated when friction is negligible.

Suggestions on Problem Solving

1. Watch units. When making conversions anticipate whether the answer will be larger or smaller than the original number *before* doing the calculation. Never mix different sets of units in the same problem, let alone the same equation. Be careful not to take 32 as the numerical value of g in a problem using SI units.

2. Be consistent with the signs of the vector quantities (**s**, **v** and **a**), particularly in free-fall and projectile problems. Take the initial direction of motion as positive and stick with it throughout the problem. This approach is not always necessary, but in the beginning it is certainly advisable. *Draw diagrams.*

3. Consider Eq. (3.9), which varies with t and t^2 both. When t is given, it's easy to find s, v, or a. When t is the unknown to be determined, the most straightforward approach is to use the quadratic equation. Rearranging Eq. (3.9) yields $\frac{1}{2}at^2 + v_it - s = 0$, which has the general form $At^2 + Bt + C = 0$, where $A = \frac{1}{2}a$, $B = v_i$, and $C = -s$. The solution is then $t = (-B \pm \sqrt{B^2 - 4AC})/2A$ (see Appendix A-4). The quadratic equation can be avoided by using combinations of other equations: Eq. (3.10) to find v_f, and then Eq. (3.6) or (3.8) to get t. The wise choice of equations will save time and effort.

4. Think about your answers. Often an error will produce a result that is unrealistically large or small. If you think the problem through so that you can anticipate the size, direction, sign, and so on, most computational errors will be spotted. We all make computational errors; not all of us find them. A reindeer falling off the roof of a one-story building is not going to land on the ground at 5000 km/h. Given the real-world physical situation of a problem, ask yourself if your answer is reasonable.

5. Check your answers. The best way to do that (although it may not always be practical) is to recalculate the answer using a different approach. At least go over your calculations several times.

6. When a problem requests the distance traveled during a certain time interval, such as the fifth second of motion, this is not the same as asking how far the body moved in five seconds. The fifth second is the one that begins at $t = 4$ s and ends at $t = 5$ s.

7. It is not really necessary to remember the whole set of equations for projectile motion: t_p is easily derived from the definition of a, where $v_V = 0$; t_T is just $2t_p$ and s_R is simply v_Ht_T. The fact that $v_V = 0$ (and $v_H \neq 0$) at the peak altitude is important and should always be kept in mind.

8. The Mean-Speed Theorem applies to straight-line motion in one direction, and so is appropriate for dealing with the uniformly accelerated vertical component of a projectile's movement. *Do not apply it to the overall motion* (as for example, in an attempt to determine the total final speed v_f knowing s and t)— the projectile travels along an arc and $l \neq s$.

Discussion Questions

1. Does a car's speedometer measure a vector or a scalar? How about the odometer? What does it measure? Is it possible for the car to accelerate and yet have the speedometer read a constant value? Is it possible that at some moment two airplanes have the same acceleration and different speeds? The same speeds and different accelerations? Can either plane at a given instant have a large acceleration and a small velocity? A small acceleration and a large velocity? Explain.

2. Is an ant stuck to a glob of honey fixed on a spinning phonograph record accelerating (a) with respect to the record? (b) with respect to the room? Explain.

3. Galileo used a "thought experiment" from the Middle Ages to justify the conclusion that all things fall at the same rate in vacuum. See if you can reconstruct it from the following: consider three identical objects (three balls of clay, for instance). How would they fall separately in vacuum? Stick two together. How would the pair of clay masses fall now? Explain.

4. Overlooking the planet's own motion, why is it likely that everything moving for any appreciable distance on the surface of the Earth is accelerating?

5. An airplane in a dive increases its speed every second by one-tenth its original value. What, if anything, can you say about its acceleration? Is it possible for a rocket to have an acceleration due east and a velocity due west? Explain.

6. Can the average acceleration of a body over a finite time ever equal the instantaneous acceleration for more than an instant during the journey? Can it ever not equal the instantaneous acceleration for at least an instant? Explain.

7. Two identical cars are in a drag race. The first starts and continues to uniformly accelerate up to the finish line some time later. The second car follows the first out of the starting box 1 s afterward traveling exactly as it did. Will the cars be the same distance apart throughout the race? If not, when will they be closest to each other? How much time will there be between their successive crossings of the finish line?

8. Is it accurate to say that the total distance traveled is the product of the time and the average speed, even though the instantaneous speed varies from moment to moment?

9. A rocket can accelerate at an increasing rate as its fuel is consumed and it gets lighter and lighter. If the rocket starts from

rest, what can we say about the *average speed* as compared with the *final speed?*

10. Explain why the distance traversed by a body during its tenth second of fall is greater than the distance covered during its first second—if, in fact, it is. What's the average speed of an apple during its first second of fall from a tree?

11. Suppose you are in a free-falling elevator and you hold your keys motionless right in front of your face and then let go. What will happen to them? Explain.

12. What is the average speed and acceleration of a banana at maximum altitude when hurled straight up at the surface of the Earth? A pea is fired straight up through a drinking straw by an undergrad in the cafeteria. When is it moving at its greatest speed? Imagine that you have taken a motion picture of a ball fired straight up, reaching maximum altitude and then falling straight down. What would it look like run backwards? Explain.

13. Consider a free-falling sky diver descending straight down. Make a graph of v versus t showing (a) the straight-line terminal speed v_T; (b) the zero-drag free-fall with $v_i = 0$; and (c) your approximation of the actual speed-time descent curve.

14. If you throw a ball straight up next to something whose height you already know (like a building) and time the round trip, you can determine g. How would you proceed?

15. To get a feeling for how extremes of a variable affect an equation, suppose that the acceleration of a body is given by $a = [(At^2 - Bt)/(t + C)D] + Et^2/(t - C)D$ where A, B, C, D, and E are constants: (a) Determine the value of a when $t = 0$. (b) Find the value of a when $t \gg C$. (c) Find the value of a when $t \ll C$.

16. Imagine an object thrown straight upward into the air. Ignoring friction, draw a rough plot of the vertical velocity versus time. Does the velocity ever become negative, and if so, when? What can you say about the object's acceleration?

17. A portion of the flight of a toy rocket that travels straight up from the surface of the Earth is depicted in Fig. Q17. Describe the motion as completely as you can.

Figure Q17

18. A football is kicked into the air at some angle θ and lands down field. Ignoring air friction, draw a rough plot of its speed versus time. Is the slope of the curve the acceleration? Explain.

Multiple Choice Questions

1. The expression $s = vt$ is applicable when   (a) the speed is constant   (b) the acceleration is constant   (c) the distance is constant   (d) the acceleration is linear   (e) none of these.

2. A body moving with an acceleration having a constant magnitude must experience a change in   (a) velocity   (b) speed   (c) acceleration   (d) weight   (e) none of these.

3. The term "10 m/s²–south" specifies   (a) displacement   (b) speed   (c) acceleration   (d) velocity   (e) none of these.

4. Which of the following pairs of concepts cannot both simultaneously be constant and nonzero for a given body?   (a) the speed and velocity   (b) the distance and displacement   (c) the magnitude of the acceleration and the acceleration   (d) the velocity and acceleration   (e) none of these.

5. A mouse runs along a straight narrow tunnel. If its velocity-time curve is a straight line parallel to the time axis, then the acceleration is   (a) a nonzero constant   (b) zero   (c) varying linearly   (d) quadratic   (e) none of these.

6. Which one of the following must be held constant if *uniformly accelerated motion* is likely to occur? The   (a) speed   (b) direction   (c) acceleration   (d) velocity   (e) none of these.

7. The speed of a body traveling in a straight line with a constant positive acceleration increases linearly with   (a) distance   (b) time   (c) displacement   (d) distance squared   (e) none of these.

8. If [L] represents the dimension of length and [T] that of time, then the dimensions of acceleration are   (a) $[L + T^2]$ (b) $[L/T]$ (c) $[L^2/T]$ (d) $[L/T^2]$ (e) none of these.

9. At any moment the slope of the velocity-time graph of a body moving along a straight line is its   (a) displacement   (b) distance traveled   (c) speed   (d) time   (e) none of these.

10. Taking the direction of the initial motion to be positive, the slope of the velocity-time curve of a body moving in a straight line is negative when the   (a) acceleration is in the opposite direction to the initial motion   (b) acceleration increases the speed   (c) acceleration decreases in time   (d) acceleration is constant and in the same direction as the initial motion   (e) none of these.

11. If the displacement of a body is a quadratic function of time, the body is moving with   (a) a uniform acceleration   (b) a nonconstant acceleration   (c) a uniform speed   (d) a uniform velocity   (e) none of these.

12. The expression $s = v_i t + \frac{1}{2}at^2$ is applicable only when   (a) v is constant   (b) t is constant   (c) a is constant   (d) s is constant   (e) none of these.

13. A driver traveling at 20 km/h abruptly stops the car in 3 m. Later, while moving at 40 km/h, the driver again stops the car with the same deceleration, bringing it to a halt after   (a) 6 m   (b) 9 m   (c) 60 m   (d) 12 m   (e) none of these.

14. Two statues of Kermit the Frog, one made of aluminum and one of brass, are the same size, although the former is 3.2 times lighter than the latter. Both are dropped at the

same moment from the same height of 2 m. They hit the ground   (a) at nearly the same time with very different speeds   (b) at very different times with nearly the same speeds   (c) at nearly the same times with nearly the same speeds   (d) at very different times with very different speeds   (e) none of these.

15. A ballast bag of sand dropped from a hot air balloon hits the ground at a certain speed and the craft slowly rises and soon comes to a stop. If a second identical bag is then dropped and it hits the ground twice as fast as the first, how high was the balloon when it was dropped in comparison to when the first bag fell?   (a) 1/2 as high   (b) 2 times as high   (c) 4 times as high   (d) 8 times as high   (e) none of these.

16. The average speed of a coconut during a 2-s fall from a tree, starting at rest, is   (a) 19.6 m/s   (b) 9.8 m/s^2 (c) 39.2 m/s   (d) 9.8 m/s   (e) none of these.

17. An alien space traveler exploring Earth records that his phasor pistol, dropped from a high cliff, fell a distance of 1 glong in a time of 1 tock. How far will it fall in two tocks?   (a) 1.5 glongs   (b) 2 glongs   (c) 3 glongs   (d) 4 glongs   (e) none of these.

18. A body moving with a uniform acceleration in a straight line has an acceleration-time curve (a vertical against t horizontal) that is   (a) a vertical straight line   (b) a horizontal straight line   (c) a vertical parabola   (d) unable to be plotted   (e) none of these.

Problems

THE CONCEPT OF ACCELERATION

1. [I] A rocket lifts off its launch pad and travels straight up attaining a speed of 100 m/s in 10 s. Calculate its average acceleration.

2. [I] A canvasback duck heading south at 50 km/h at 2:02 A.M. is spotted at 2:06 A.M. still traveling south but at 40 km/h. Calculate its average acceleration over that interval—magnitude and direction.

3. [I] An android on guard duty in front of the Institute of Robotics is heading due south at 1:07 P.M. at a speed of 10 m/s when it receives a command to alter course. At 1:09 P.M. it is recorded to be moving at 10 m/s due north. Compute its average acceleration over that interval—magnitude and direction.

4. [I] At time $t = 0$, a body located at $s = 0$ has a speed of $v = 0$ and an acceleration of 1 m/s^2. It travels in a straight line maintaining that constant acceleration for 4 s and then immediately ceases accelerating. That condition lasts for 1 s at which point it decelerates at 2 m/s^2 for 2 s and then again ceases accelerating. Draw graphs of a versus t, and v versus t.

5. [I] A finalist in the Soap Box Derby starts with a push down a long hill at an initial speed of 1.0 m/s. At the bottom, 1.0 min 2.0 s later, it reaches a speed of 15.0 m/s. Find its average acceleration.

6. [I] A car goes from rest to 10 km/h in 10 s; at the end of 20 s it's moving at 20 km/h; at the end of 30 s it reaches 30 km/h. What can you say about its acceleration? Make a graph of a.

7. [I] A piston-engine dragster set a world record by starting from rest and hitting a top speed of 244 mi/h over a measured track of 440 yd. Compute its average acceleration in m/s^2.

8. [I] A VW Rabbit can go from rest to 80.5 km/h (50.0 mi/h) in a modest 8.20 s. Assuming its acceleration is precisely constant (which, of course, it isn't), how long will it take to speed up from 48.3 km/h to 64.4 km/h?

9. [I] What is the instantaneous acceleration of the object whose motion is depicted in Fig. 3.2a (p. 61) at a time of 1.58 s?

10. [I] Figure P10 is a velocity-time graph for a test car on a straight track. The test car initially moved backward in the

Figure P10

negative x-direction at 20 m/s. It slowed, came to a stop, and then moved off in the positive x-direction at $t = 2.0$ s. What was its average acceleration during each of the time intervals 0 to 0.5 s, 1.5 s to 2.0 s, and 2.0 s to 2.5 s?

11. [II] Referring back to Fig. P10, what was the car's average acceleration during the interval from $t = 0$ to $t = 3.0$ s?

12. [II] Figure P12 shows the speed-time curves of three cyclists. Describe their motions and compute their average accelerations over the entire interval shown.

Figure P12

13. [II] Using Fig. 3.3 (p. 61), graphically determine the car's approximate acceleration at 3.8 s into the run.

14. [II] A motorboat starting from a dead stop accelerates at a constant 2.0 m/s² for 3.0 s, then very rapidly roars up to 4.0 m/s² and holds it constant for 4.0 s. What is its approximate average acceleration over the first 5.0 s of motion?

15. [II] Suppose that in the previous problem, after holding at 4.0 m/s² for 4.0 s, the boat next decelerates to a stop 20 s later at a rate of 1.1 m/s².   (a) What is its average acceleration 27 s after the start of the run?   (b) What is its instantaneous acceleration 10 s into the trip?

16. [II] A rocket accelerates straight up at 10 m/s² toward a helicopter that is descending uniformly at 5.0 m/s². What is the relative acceleration of the missile with respect to the aircraft as seen by the pilot who is about to bail out?

17. [II] Superman slams head-on into a locomotive speeding along at 60 km/h, bringing it smoothly to rest in an amazing 1/1000 s and saving Lois Lane, who was tied to the tracks. Calculate the constant deceleration of the train.

18. [III] Two motorcycle stuntpersons are driving directly toward one another, each having started at rest and each accelerating at a constant 5.5 m/s². At what speed will they be approaching each other 2.0 s into this lunacy?

19. [III] The speed of a flying saucer in ascent mode-4 as reported by an alleged eyewitness (who got it straight from the talkative alien navigator) is given by the expression $v = At + Bt^2$. Amazingly, A and B are constants in SI units but the little green navigator would not reveal their values. Determine the instantaneous acceleration of the craft. (Hint: at $(t + \Delta t)$ the speed is $(v + \Delta v)$.)

UNIFORMLY ACCELERATED MOTION

20. [I] Ask a friend to hold his or her thumb and forefinger parallel to each other in a horizontal plane. The fingers should be about an inch apart. Now you hold a 1-ft ruler vertically in the gap just above and between these fingers so it can be dropped between them. Have your friend look at the ruler and catch it when you, without warning, let it fall. Calculate the corresponding response time. Now position a dollar bill vertically so that Washington's face is between your accomplice's fingers—is it likely to be caught when dropped?

21. [I] A locust, extending its hind legs over a distance of 4 cm, leaves the ground at a speed of 340 cm/s. Determine its acceleration, presuming it to be constant.

22. [I] A kangaroo can jump straight up about 2.5 m—what is its takeoff speed?

23. [I] A wayward robot, R2D3, is moving along at 1.5 m/s when it suddenly shifts gears and roars off, accelerating uniformly at 1.0 m/s² straight toward a wall 10 m away. At what speed will it crash into the wall?

24. [I] Regarding Problem 17, Lois was a mere 10 cm down the tracks from the point where Superman struck the train. How far away from her did the "Man of Steel" finally stop the engine? (Thank goodness he arrived in time!)

25. [I] A Jaguar in an auto accident in England in 1960 left the longest recorded skid marks on a public road: an incredible 290 m long. As we will see later, the friction force between the tires and the pavement varies with speed, producing a deceleration that increases as the speed

decreases. Assuming an average acceleration of −3.9 m/s² (that is, −0.4 g), calculate the Jag's speed when the brakes locked.

26. [I] A modern supertanker is gigantic: 1200 to 1300 ft long with a 200-ft beam. Fully loaded, it chugs along at about 16 knots (30 km/h or 18 mi/h). It can take 20 min to bring such a monster to a full stop. Calculate the corresponding deceleration in m/s² and determine the stopping distance.

27. [I] A bullet traveling at 300 m/s slams into a block of moist clay, coming to rest with a fairly uniform acceleration after penetrating 5 cm. Calculate its acceleration.

28. [I] The velocity-time graph for a rocket is shown in Fig. Q17 (p. 84). Roughly what was its peak altitude?

29. [II] A driver traveling at 60 km/h sees a chicken dash out onto the road and slams on the brakes. Accelerating at −7 m/s², the car stops just in time 23.3 m down the road. What was the driver's reaction time?

30. [II] The longest passenger liner ever built was the *France*, at 66 348 tons and 315.5 m long. Suppose its bow passes the edge of a pier at a speed of 2.50 m/s while the ship is accelerating uniformly at 0.01 m/s². At what speed will the stern of the vessel pass the pier?

31. [II] A swimmer stroking along at a very fast 2.2 m/s ceases all body movement and uniformly coasts to a dead stop in 10 m. Determine how far she moved during her third second of unpowered drift.

32. [II] A little electric car having a maximum speed of 40.0 km/h can speed up uniformly at any rate from 1.00 to 4.00 m/s² and slow down uniformly at any rate from 0.00 to −6.00 m/s². What is the shortest time in which such a vehicle can traverse a distance of 1.00 km starting and ending at rest?

33. [II] Two trains heading straight for each other on the same track are 250 m apart when their engineers see each other and hit the brakes. The Express, heading west at 96 km/h, slows down, accelerating at an average of −4 m/s², while the eastbound Flyer, traveling at 110 km/h, slows down, accelerating at an average of −3 m/s². Will they collide?

34. [II] The drivers of two cars in a demolition derby are at rest 100 m apart. A clock on a billboard reads 12:17:00 at the moment they begin heading straight toward each other. If both are accelerating at a constant 2.5 m/s², at what time will they collide?

35. [II] A rocket-launching device contains several solid-propellant missiles that are successively fired horizontally at 1-s intervals.   (a) What is the separation of the first and second missiles just at the moment the second one is fired, if each has an initial speed of 60 m/s and a constant acceleration of 20 m/s² lasting for 10 s?   (b) What is the separation of the first and second just as the third is launched?

36. [II] Referring to the previous problem, what is the relative acceleration of the first missile with respect to the second one once both are in the air and accelerating? What will be their separation 6 s into the flight?

37. [II] Imagine that you are driving toward an intersection at a speed v_i just as the light changes from green to yellow. Assuming a response time of 0.6 s and an acceleration of −6.9 m/s², write an expression for the smallest distance (s_S) from the corner in which you could stop in time. How much is that if you are traveling 35 km/h?

38. [II] Considering the previous problem, it should be clear that the yellow light might reasonably be set for a time t_Y, which is long enough for a car to traverse the distance equal to both s_S and the width of the intersection s_I. Assuming a constant speed v_i equal to the legal limit, write an equation for t_Y, which is independent of s_S.

39. [II] The driver of a pink Cadillac traveling at a constant 60 mi/h in a 55 mi/h zone is being chased by the law. The police car is 20 m behind the perpetrator when it too reaches 60 mi/h, and at that moment the officer floors the gas pedal. If her car roars up to the rear of the Cadillac 2.0 s later, what was her acceleration, assuming it to be constant?

40. [II] With Problems 37 and 38 in mind, how long should the yellow light stay lit if we assume a driver-response time of 0.6 s, an acceleration of -6.9 m/s^2, a speed of 35 km/h, and an intersection 25 m wide? Which of the several contributing aspects requires the greatest time?

41. [II] In Problem 39, how far does the police car travel in the process of closing the Cadillac's 20-m lead?

42. [III] Having taken a nap under a tree only 20 m from the finish line, Rabbit wakes up to find Turtle 19.5 m beyond him, grinding along at 1/4 m/s. If the bewildered hare can accelerate at 9 m/s^2 up to his top speed of 18 m/s (40 mi/h) and sustain that speed, will he win?

43. [III] Superman is jogging alongside the railroad tracks on the outskirts of Metropolis, at 100 km/h. He overtakes the caboose of a 500-m-long freight train traveling at 50 km/h. At that moment he begins to accelerate at $+10$ m/s^2. How far will the train have traveled before Superman passes the locomotive?

44. [III] A motorcycle cop, parked at the side of a highway reading a magazine, is passed by a woman in a red Ferrari 308 GTS doing 90.0 km/h. After a few attempts to get his cycle started, the officer roars off 2.00 s later. At what average rate must he accelerate if 110 km/h is his top speed and he is to catch her just at the state line 2.00 km away?

FREE-FALL

45. [I] At what speed would you hit the floor if you stepped off a chair 0.50 m high? Ignore friction. Express your answer in m/s, ft/s, and mi/h.

46. [I] If a stone dropped (not thrown) from a bridge takes 3.7 s to hit the water, how high is the rock-dropper? Ignore friction.

47. [I] Ignoring air friction, how fast will an object be moving and how far will it have fallen after dropping from rest for 1.0 s, 2.0 s, 5.0 s, and 10 s?

48. [I] A cannonball is fired straight up at a rather modest speed of 9.8 m/s. Compute its maximum altitude and the time it takes to reach that height (ignoring air friction).

49. [I] Calculate the speed at which a hailstone, falling from 3.00×10^4 ft out of a cumulonimbus cloud, would strike the ground, presuming air friction is negligible (which it certainly is not). Give your answer in mi/h and m/s.

50. [I] A circus performer juggling while standing on a platform 15.0 m high tosses a ball directly upward into the air at a speed of 5.0 m/s. If it leaves his hand 1.0 m above the platform, what is the ball's maximum altitude? If the juggler misses the ball, at what speed will it hit the floor? Ignore air friction.

51. [I] Draw a curve of s in meters versus t in seconds for a free-falling body dropped from rest. Restrict the analysis to the first second of fall, and make up a table of values of t, t^2, and s at 1/10 s intervals. Draw a curve of s versus t^2. Explain your findings.

52. [I] A shoe is flung into the air such that at the end of 2.0 s it is at its maximum altitude, moving at 6.0 m/s. How far away from the thrower will it be when it returns to the height it was tossed from? Ignore air friction.

53. [I] Show that the range of a projectile can be expressed as

$$s_R = -\frac{2v_{iH}v_{iV}}{g}$$

Ignore air friction.

54. [I] A lit firecracker is hurled upward from a height of 2.00 m at a speed of 20.0 m/s at an angle above the ground of 60°. If it explodes 2.00 s later, how high above the ground is the blast? Ignore air friction.

55. [I] Suppose you point a rifle horizontally directly at the center of a paper target 100 m away from you. If the muzzle speed of the bullet is 1000 m/s, where will it strike the target? Assume aerodynamic effects are negligible.

56. [I] A rifle must always be pointed slightly above the target—the bullet travels an arc. The peak height it reaches (at midrange) above the horizontal straight line from rifle to target is called the *mid-range trajectory*, and the smaller that is, the better. A 243 Winchester with a muzzle speed of 3500 ft/s has a mid-range trajectory of 4.7 in. when zeroed in on a target 300 yd away. Ignoring air effects, at what angular elevation must the rifle be pointed to hit that target? A bullet like this will lose about 30 to 40% of its initial speed over such a trip, so the above approximation is tolerable, but not great.

57. [I] A raw egg is thrown horizontally straight out of the open window of a fraternity house. If its initial speed is 20 m/s and it hits ground 2.0 s later, at what height was it launched? At such low speeds air friction is negligible.

58. [I] A golfer wishes to chip a shot into a hole 50 m away on flat level ground. If the ball sails off at 45°, what speed must it have initially? Ignore aerodynamic effects.

59. [I] Check the dimensions of both sides of the equation

$$v_i s_V^2 = t v_i^2 \cos \theta + \tfrac{1}{2} g v_i t^2$$

to see if they are the same; if not, the equation is wrong. This technique was introduced by the French mathematician and physicist, J. B. J. Fourier, around 1822. Is it possible that the above equation is correct?

60. [I] The acceleration due to gravity on the surface of the Moon is about $g/6$. If you can throw a ball straight up to a height of 25 m on Earth, how high would it reach on the Moon when launched at the same speed? Ignore the minor effects of air friction.

61. [II] A young kid with a huge baseball cap is playing catch with himself by throwing a ball straight up. How fast does he throw it if the ball comes back to his hands a second later? At low speeds air friction is negligible.

62. [II] An arrow is launched vertically upward from a crossbow at 98.1 m/s. Ignoring air friction, what is its instantaneous speed at the end of 10.0 s of flight? What is its average speed up to that moment? How high has it

risen? What is its instantaneous acceleration 4.20 s into the flight?

63. [II] A lit firecracker is shot straight up into the air at a speed of 50.00 m/s. How high is it above ground level 5.000 s later when it explodes? How fast is it moving when it blows up? How far has it fallen, if at all, from its maximum height? Ignore drag and take g equal to 9.800 m/s^2.

64. [II] According to their literature, the observation deck of the World Trade Center in New York is 1377 ft above ground. Ignoring air friction, how long will it take to free-fall that far? Take $g = 9.81 \text{ m/s}^2$. Considering the discussion in the text, is it reasonable to ignore air drag in this problem? Explain.

65. [II] A human being performing a vertical jump generally squats and then springs upward, accelerating with feet touching the Earth through a distance s_a. Once fully extended, the jumper leaves the ground and glides upward, decelerating until the feet are a maximum height s_{max} off the floor. Assuming the jumping acceleration a to be constant, derive an expression for it in terms of s_a and s_{max}. Ignore friction.

66. [II] A small rocket is launched vertically, attaining a maximum speed at burnout of 1.0×10^2 m/s and thereafter coasting straight up to a maximum altitude of 1510 m. Assuming the rocket accelerated uniformly while the engine was on, how long did it fire and how high was it at engine cutoff? Ignore air friction.

67. [II] Using Fig. 3.17 (p. 82) and extending it where necessary, show that an angle less than 45° with respect to the ground will result in a greater horizontal displacement when the landing point is lower than the launch point. Neglect friction.

68. [II] A flea jumps into the air and lands about 8.0 in. away, having risen to an altitude of about 130 times its own height (that's comparable to you jumping 650 ft up). Assuming a 45° launch, compute the flea's takeoff speed. Make use of the fact that $2 \sin \theta \cos \theta = \sin 2\theta$ and ignore air friction.

69. [II] Two diving platforms 10 m high terminate just at the edge of each end of a swimming pool 30 m long. How fast must two clowns run straight off their respective boards if they are to collide at the surface of the water midpool? Ignore friction.

70. [II] A golf ball hit with a 7-iron soars into the air at 40.0° with a speed of 54.86 m/s (that is, 180 ft/s). Overlooking the effect of the atmosphere on the ball, determine (a) its range and (b) when it will strike the ground.

71. [II] A silver dollar is thrown downward at an angle of 60.0° below the horizontal from a bridge 50.0 m above a river. If its initial speed is 40.0 m/s, where and at what speed will it strike the water? Ignore the effects of the air.

72. [II] Salmon, swimming to their spawning grounds, leap over all sorts of obstacles. The unofficial salmon-altitude record is an amazing 3.45 m (that is, 11 ft 4 in.) jump. Assuming the fish took off at 45.0°, what was its speed on emerging from the water? Ignore friction.

73. [II] Show that at any time t, where friction is negligible, the velocity of a projectile (launched at an angle θ with a speed v_i) makes an angle θ_t with the horizontal given by the expression

$$\theta_t = \tan^{-1}\left(\tan \theta + \frac{gt}{v_i \cos \theta} \right)$$

where $g < 0$.

74. [II] The illustrative Example 3.14 deferred from using the equation $s = v_i t + \frac{1}{2}at^2$ because it required the use of the quadratic formula. Carry out that calculation, solving it for the two values of t. Notice the symmetry around the time of peak altitude.

75. [III] A bag of sand dropped by a would-be assassin from the roof of a building just misses Tough Tony, a gangster 2 m tall. The missile traverses the height of Tough Tony in 0.20 s, landing with a thud at his feet. How high was the building? Ignore friction.

76. [III] Imagine that someone dropped a firecracker off the roof of a building and heard it explode exactly 10 s later. Ignoring air friction, taking $g = 9.81 \text{ m/s}^2$, and using 330 m/s as the speed of sound, calculate how far the cherry bomb had fallen at the very moment it blew up.

77. [III] A burning firecracker is tossed into the air at an angle of 60° up from the horizon. If it leaves the hand of the hurler at a speed of 30 m/s, how long should the fuse be set to burn if the explosion is to occur 20 m away? Ignoring friction, just set up the equation for t.

78. [III] A baseball is hit as it comes in, 1.30 m over the plate. The blast sends it off at an angle of 30° above the horizontal with a speed of 45.0 m/s. The outfield fence is 100 m away and 11.3 m high. Ignoring aerodynamic effects, will the ball clear the fence?

4

Newton's Three Laws: Momentum

THIS CHAPTER AND THE next are about Newton's Three Laws—
(1) the Law of Inertia; (2) the relationship between force and the accompanying change in motion; (3) and the Law of Interaction. These insights initiated
the study of how motion changes, a discipline called **dynamics**. Traditionally, there are two different ways to develop this material: One is through the
concept of momentum and how momentum is changed by force, which we
will concentrate on in this chapter; the other uses acceleration and how it is
produced by force, discussed in Chapter 5. Superficially, the two seem
equally insightful, but they are not. Newton's own formulation was in terms

of momentum, and we now know from Einstein's Special Theory of Relativity that Newton's Second Law is perfectly correct as he gave it. However, the alternative formulation in terms of acceleration ($F = ma$) is actually only a low-speed approximation—a very useful one, but an approximation nonetheless.

We concentrate first on momentum because it has been learned in this century that momentum is one of the premier ideas of physics. When there is no net external force acting on a system, its linear momentum remains unchanged and we have the all-important law of Conservation of Linear Momentum. We now know that this powerful principle arises from the homogeneity of space itself (p. 116). As we shall see, in all of physics there are only a handful of these wonderful conservation laws, and each is related to some fundamental symmetry of the Universe.

Though their perspectives are different, Chapters 4 and 5 both use Newton's laws to develop *dynamics;* that is, what happens when momentum changes. Chapter 6 uses those same laws to elaborate *statics;* that is, what happens when momentum remains constant.

THE LAW OF INERTIA

Sir Isaac Newton, 1642–1727.

Isaac Newton was born on Christmas day, 1642, within a year of Galileo's death. Newton was a short man, nearsighted, already silver-gray in his thirties, inattentive to personal appearance, incredibly forgetful, and probably virginal. This nervous, hypochondriacal, sensitive, pious, vulnerable individual was one of the greatest geniuses to ever live. Among Newton's many achievements was a theory of motion that withstood every test for over 200 years.

The first of Newton's Three Laws was actually formulated much earlier by Galileo. The **Law of Inertia** states that *a body once set in motion and thereafter undisturbed will continue in uniform motion forever, all by itself.* Some scholars thought along similar lines in Galileo's day, but this concept was completely opposed to the prevailing Aristotelian view. To Aristotle and his followers, any motion of a heavy object other than free-fall required the exertion of a force. They argued—wrongly—that without an impelling force, there can be no sustained motion. End that force, and the motion in progress spontaneously ceases. Aristotle's position appears to agree with experience, but it is wrong. We live on a planet where gravity and friction obscure what is really going on. So much so that it took 18 centuries before Galileo saw to the heart of the matter.

The experiment that led Galileo to the Law of Inertia used two inclined planes set end-to-end, one tilted down, the other up (Fig. 4.1). A ball was rolled, from rest, down the first incline. It swept past the bottom and ascended the second incline, gradually losing speed until it momentarily came to rest. Regardless of the tilt of that second surface, the ball always stopped ascending at a vertical height a little lower than that from which it was first released. Rather than overlook that slight difference, Galileo searched for its cause. He polished the ball and smoothed the track, and finally concluded that the loss was due to friction between the track and the ball. Friction was a crucial piece to the puzzle, a piece Aristotle never appreciated.

Next, Galileo did something quite clever: He gradually reduced the angle of the second plane so that the ball rolled farther and farther before coming to rest. Then he asked the right question: What would happen if the second incline were made *perfectly frictionless* and *perfectly horizontal?* The ball would roll on forever, neither speeding up nor slowing down, with "a motion which was uniform and perpetual." His imagination had gone beyond the limitations of the experiment, from

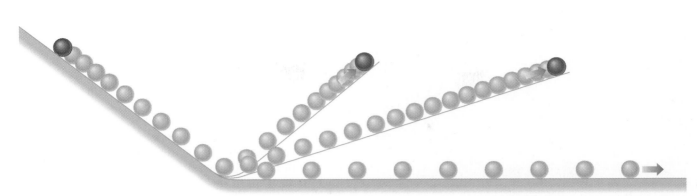

which he could never have totally removed friction. We ordinarily must keep pushing objects to maintain them in motion because of friction (p. 146), not because of the fundamental nature of the process. Had Aristotle lived aboard a spaceship, the Law of Inertia would have been obvious.

Despite the brilliant insight, Galileo soon got lost worrying about how an infinite horizontal plane could exist on a spherical Earth. In the end, he erroneously decided that the unimpeded motion of objects had to be in great circles, in this case around the planet. The law was soon properly linearized by a professional soldier turned philosopher and mathematician—René Descartes (pronounced Day-cart). Guided by conjecture rather than experiment, Descartes maintained, in 1644, "that every body which moves tends to continue its movement in a straight line."

In 1687, Isaac Newton set out the "axioms or laws of motion" in his great work, *Mathematical Principles of Natural Philosophy* (nowadays often called the *Principia* from the original Latin title, and pronounced prin-sip-´pee-uh). His First Law was the Law of Inertia: **Every body continues in a state of rest or of uniform motion in a straight line except insofar as it is compelled to change that state by forces impressed upon it**—and he gave Galileo credit for the insight.

A new equivalence is drawn here between rest and uniform motion. Altering either requires a force, but once either state is established, it persists forever in the absence of force. Rest and uniform motion are only "relatively distinguished," as Newton put it. For example, a coin sits at "rest" in your open hand even though the car you are in is speeding along at a constant 50 km/h. Someone standing outside would see the car, you, and the coin streak past, but the Law of Inertia holds from both perspectives; that is, it holds in both inertial frames. Measured with regard to your hand, the coin's speed is zero and will remain so, just as it will remain 50 km/h with respect to the outside observer, so long as no applied force alters the motion.

The *Apollo* astronauts, in 1969, shut down their engines and coasted in unpowered flight for much of the way to the Moon. At this very instant, *Pioneer 10* is hurtling toward the stars at tens of thousands of kilometers per hour, its rockets burned out decades ago. The famous trick of whisking a tablecloth out from under a setting of dishes is a grandiose First-Law gesture. And the purpose of seat belts becomes clear when the body in motion tending to stay in motion after slamming on the brakes is yours.

Be aware that the Law of Inertia, the first basic insight into the behavior of nature that we have come upon so far, is an idealization. There is no place in this Universe where an object can be completely free of external influences, and the idea of infinite straight-line motion is unrealistic, especially in a cosmos cluttered with

Figure 4.1 A ball rolled down an inclined plane always climbs back to nearly the same height at which it was released before it stops. Galileo thus argued that if the second plane was horizontal and frictionless, the ball would roll forever.

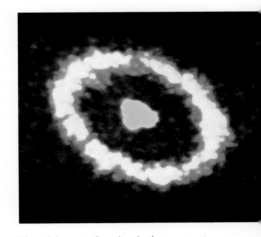

The pink central region is the remnant of a star that was seen to explode in 1987 (supernova 1987A). This 1990 photo shows a yellow ring of glowing gas about 12×10^{12} mi across. It was blown off by the star some 20 000 years ago and is still hurtling out into space in accord with Newton's First Law. The central debris of the supernova is moving outward much faster, and will overtake and tear apart the ring within the next 100 years.

The First Law: A body in motion tends to stay in motion.

galaxies. So, we will not be able to confirm directly the full implications of the First Law, namely that an object free of external influences will move with a constant velocity forever. Still, the Law of Inertia allows us to understand a tremendous range of observed phenomena—and that's its real power.

Newton felt that **inertia**—that is, *the resistance to a change in motion*—was an intrinsic property of matter itself, independent of the environment. Yet, within a few decades of his publication of the Law of Inertia, that opinion was challenged by the philosopher George Berkeley. Many today believe, like Berkeley, that inertia arises from the interaction (p. 208) of all matter in the Universe, an idea that is called Mach's Principle. As we will see, this concept is quite appealing. Still, there are other possibilities. Physicists have certainly not yet resolved the fundamental question of the origin of inertia.

4.1 Ballistic Motion

The unpowered flight of an object launched into the air at the Earth's surface is a fine example of the primary role played by the Law of Inertia. Moreover, it demonstrates how a body is caused to deviate from its straight-line, First-Law inertial path by forces applied to it, in this case gravity acting downward and frictional drag acting tangent to the path of the motion.

The first person to comprehend ballistics was Galileo, precisely because he knew about both free-fall and the Law of Inertia. Galileo understood that an unpowered projectile moves in a way that is the result of the superposition of two independent motions: a straight-line inertial flight in the direction of launch (ideally at constant speed) and a vertical gravity fall (ideally at constant acceleration). The Law of Inertia maintains that left on its own, a ball should sail straight off in the direction thrown, at a *constant speed* equal to its initial speed, forever (and this phenomenon is exactly what would happen in empty space in the absence of air friction and gravity). On a sizeable body such as the Earth, gravity will exert a substantial downward force. From the First Law, we know that such a force will alter the motion, in this instance turning that straight path into a curved trajectory (p. 75). Newton showed that on a spherical planet, with gravity pulling down toward its center, a dragless ballistic trajectory should actually be elliptical. (The parabolic path discussed in Chapter 3 was a flat-Earth approximation, excellent for short ranges.)

If we fire a cannonball horizontally and at that very instant drop a pebble from the same height, *the ball and pebble will hit the ground at very nearly the same moment*—at least to the degree that air effects are negligible. Upon emerging from the cannon, the ball is in free-fall and it immediately accelerates downward at g. Despite the fact that the ball may land quite a distance away, under the influence of gravity it will drop at 9.8 m/s², in step with the descent of the free-falling pebble. Figure 4.2 shows how the ball's flight path at any instant t is composed of a horizontal inertial displacement, $s_H = v_{iH}t$, from which the projectile continuously free-falls a distance $s_V = \frac{1}{2}gt^2$. You can do a similar experiment by placing a dime at the

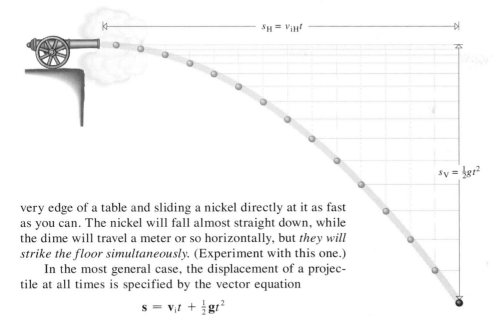

$$s_H = v_{iH}t$$

$$s_V = \tfrac{1}{2}gt^2$$

Figure 4.2 Without friction, a projectile fired horizontally falls as it progresses, sustaining its constant horizontal inertial motion while experiencing a vertical acceleration of *g*.

very edge of a table and sliding a nickel directly at it as fast as you can. The nickel will fall almost straight down, while the dime will travel a meter or so horizontally, but *they will strike the floor simultaneously.* (Experiment with this one.)

In the most general case, the displacement of a projectile at all times is specified by the vector equation

$$\mathbf{s} = \mathbf{v}_i t + \tfrac{1}{2}\mathbf{g}t^2$$

wherein **g** is the vector acceleration due to gravity: 9.81 m/s² pointing downward. The equation provides the displacement from the point of liftoff at each and every moment. Notice that the first term is the straight-line inertial path, whereas the second is the downward free-fall contribution. Without gravity, the ball in Fig. 4.3 would have followed the straight line, being displaced in a time *t* by an amount $\mathbf{v}_i t$. With gravity, the ball falls continuously, and is further displaced by an amount $\tfrac{1}{2}\mathbf{g}t^2$ downward. The vector sum of these two displacements is **s**.

The concern is sometimes voiced that the effect of gravity on a projectile might be different because of the lateral motion, but it isn't. If we perform the free-fall experiment inside a uniformly moving train, the results are unaltered. A dropped bundle of keys does not slam into the rear wall of the car, but instead falls straight down as usual (Fig. 4.4a). To someone outside at "rest" looking in, the keys would have the constant horizontal motion of the train in addition to the downward fall, and so would be seen to move along a parabolic arc (Fig. 4.4b).

Although we will not be able to analyze aerodynamic effects, people concerned with accurate ballistics have to. They consider the complexities of air friction, lift (Sect. 11.12), and wind (Fig. 4.5). In their own way, so do people trying to catch a "high fly" or hit a golf ball to a distant green. Since the retarding force generated on a fast projectile by the air (the **drag**) is proportional to the *square of the speed,* that force becomes increasingly problematic as speeds go up, usually in an attempt to extend the range. The analysis of projectiles in Chapter 3 is fine for cannoneers on the Moon where there is no atmosphere, but on Earth it fails noticeably when we go beyond a gentle game of catch.

A well-struck baseball, in the air for a rather long time, can lose as much as half its initial speed due to aerodynamic drag, and so travel only a bit more than half as far as it would have, had there been no friction. A basketball is

The flight of exploding mortar shells, drawn by Leonardo da Vinci about 500 years ago. Leonardo knew that on a flat Earth the trajectory would ideally be parabolic.

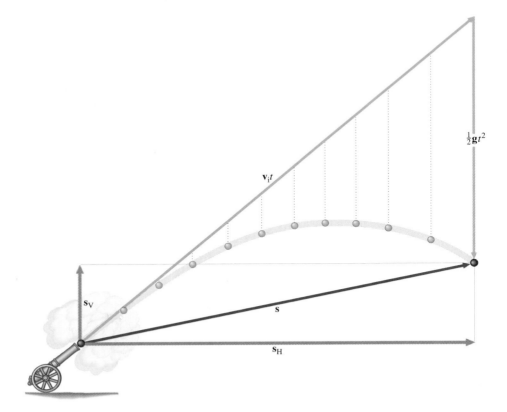

Figure 4.3 At every instant, the motion of the ball is the composite of a straight-line inertial path ($\mathbf{v}_i t$) in the direction of launch, and a vertical free-fall ($\frac{1}{2}\mathbf{g}t^2$) drop down from it.

Were it not for the accidental impediment of the air, I verily believe that, if at the time of the ball's going out of the piece [that is, a cannon] another were let fall from the same height directly downwards, they would both come to the ground at the same instant.

GALILEO

If you insist upon a precise definition of force, you will never get it!

RICHARD FEYNMAN
Nobel Laureate (1963)

usually thrown at a fairly low speed (around 7.6 m/s, or 25 ft/s), which keeps drag down, but the ball's light weight and large surface area nonetheless make air resistance appreciable. The drag force results in a deceleration (≈ 1 m/s^2) tangent to the flight path. A player must compensate for that by launching the ball with a speed about 5% greater than would be necessary if the court were in vacuum. By contrast, a rifle bullet fired at 0.6 km/s will experience a good deal of drag. Although, if unimpeded, it would have a tremendous maximum range of about 40 km according to Eq. (3.16, p. 81), the bullet is not likely to travel much beyond 4 km (Fig. 4.5).

4.2 Force

The First Law tells us that a body at rest will remain at rest and a body in uniform motion will remain in uniform motion *except insofar as it is compelled to change that state by a force impressed upon it.* But what is a force? Force is often naively thought of as pushes and pulls (reasonably enough, we tend to go back to our physiological experiences as a starting point), but we have to be more precise than that. The sensation of force is as subjective as the sensations of taste, smell, and color. To say that force is your muscle sensation of force is simplistic. If we were ghost-like and viewed the world remotely without being able to feel, what would we say about force? Forces are not directly observable. What is perceived is the effect, the change produced by a force.

(a)

(b)

Figure 4.4 An experiment performed inside a uniformly moving train. (a) As seen from inside, the dropped keys fall straight down. (b) As seen from outside, the keys, tossed straight up, move along a parabola.

Horizontal displacement

In the beginning of the *Principia,* Newton defined force as "an action exerted upon a body, in order to change its state, either of rest or of moving uniformly. . . ." **Force is the agent of change**. In the particular case of dynamics, it is *that which alters motion.* In the Universe as it exists today, there are four distinct fundamental forces: gravitational, electromagnetic, strong, and weak. In one way or another, all of physics is concerned with the Four Forces.*

Figure 4.5 Actual and ideal paths of a relatively slow cannonball in free flight through the air, and a high-speed projectile like a rifle bullet with a muzzle speed of 0.6 km. To see the effect, hurl something lightweight, like a wad of tissue paper. Use a piece about an inch or two square and loosely balled up.

*The concept of force or interaction has taken on a broadened meaning in contemporary physics. Thus, for example, the weak force mediates the transmutation of quarks and the decay of the neutron; force not only changes motion, it transforms matter.

Figure 4.6 A spring scale being used to measure force. Shown are several situations that look different but are equivalent as far as the scale is concerned.

An apple of medium size weighs about 1 N.

As with most scientific concepts, we try to define force in a way that allows us to measure it. The easiest procedure to follow is to take a standard mass (p. 11), a kilogram, hang it on a spring scale and call whatever the arrow points to "so many units of force." After all, ordinary weight is the downward gravitational force. With a beam balance, we can divide a kilogram mass into fractional pieces that could then be used to calibrate the spring scale. A 0.453 592 37-kg piece, at a place where $g = 9.806\,65$ m/s², has a weight equal to 1 lb. Once appropriately labeled, the spring scale then provides a convenient, reproducible way to measure forces (Fig. 4.6). Unfortunately, this scheme uses gravity and has little to do with Newton's definition of force as the changer of motion.

Later in the chapter, when we formally define the unit of force, the *newton* (N), in the SI system, it will be by way of the change in motion of an object, independent of gravity. For now, 1 N is the weight on Earth of a mass of about 0.10 kg. In terms of the outmoded pound, 1 N = 0.224 809 lb. From now on, we will treat force in newtons, keeping in mind that 1 N $\approx \frac{1}{4}$ lb—the weight of a medium-sized apple.

Example 4.1 The *Apollo Command Module* on the way to the Moon was *effectively weightless* (p. 136) after being placed in what was called a parking orbit around the Earth. On the other hand, its mass (5896 kg) was unchanged by the ride into space. Compute the craft's weight on the launch pad in both pounds and newtons.

Solution: [Given: $m_A = 5896$ kg. Find: its Earth-weight.] Because a mass of 0.453 59 kg weighs 1.000 0 lb on Earth, the *Apollo* weighed

$$\frac{5896 \text{ kg}}{0.453\,59 \text{ kg/lb}} = \boxed{13.00 \times 10^3 \text{ lb}}$$

Because 1 N = 0.224 809 lb, the inverse of this is 4.448 22 N = 1 lb, and the *Apollo* weighed

$$(13.00 \times 10^3 \text{ lb})(4.448\,22 \text{ N/lb}) = \boxed{57.83 \text{ kN}}$$

▶ **Quick Check:** As we will see [Eq. (5.5)], the proportionality constant between weight and mass is g; that is, the weight of an object divided by its mass equals 9.8 m/s^2 = $(57.83 \times 10^3 \text{ N})/(5896 \text{ kg})$.

Force as a Vector

Hold a book upright, placing a hand flat on either side (Fig. 4.7). Now press your hands together exerting an equal force on each cover. Regardless of how great the force applied by each hand, as long as both are equal, the book remains at rest. It seems that the *net applied force* is zero—no change in the motion, no net force according to the First Law. Two oppositely directed forces acting on the same object work against each other, partially or totally canceling one another. Moreover, several forces applied to a point on a body act as if they were a single force. The man on the left in Fig. 4.8 experiences one large force pulling him backward even though he has four individual adversaries. With your eyes closed, there is no way to tell how many people are pulling against you in a tug-of-war; you experience the one resultant force.

If we say "push on the person next to you with a force of 10 N," it could be done in a number of different ways: up, down, right, left, and so on. The statement is ambiguous. Force is specified completely only when both its magnitude and direction are given. Like displacement and velocity, *force is a vector quantity*. We draw this conclusion from observations. And *it's the net force, the resultant of all the forces acting on a body, that is manifested in a change in motion.*

Figure 4.7 Force is a vector quantity, which means that a force to the left tends to cancel a force to the right. If the two forces on the book are equal, it remains at rest.

Figure 4.8 The four forces to the right (which all pass through a single point in the ring) act as a single net force equal to their vector sum.

Example 4.2 Determine the resultant force exerted on the elephant by the two clowns in Fig. 4.9.

Solution: [Given: $F_1 = 300$ N and $F_2 = 400$ N. Find: the resultant **F**.] We want the vector sum

$$\mathbf{F} = \mathbf{F}_1 + \mathbf{F}_2$$

and since the forces form a 3-4-5 right triangle $\boxed{F = 500 \text{ N}}$. To find the angle **F** makes with the west-to-east axis, we use

$$\theta = \tan^{-1} \frac{300 \text{ N}}{400 \text{ N}} = \boxed{36.9° \text{ north of east}}$$

▶ **Quick Check:** $(500 \text{ N}) \cos 36.9° = 400 \text{ N}$.

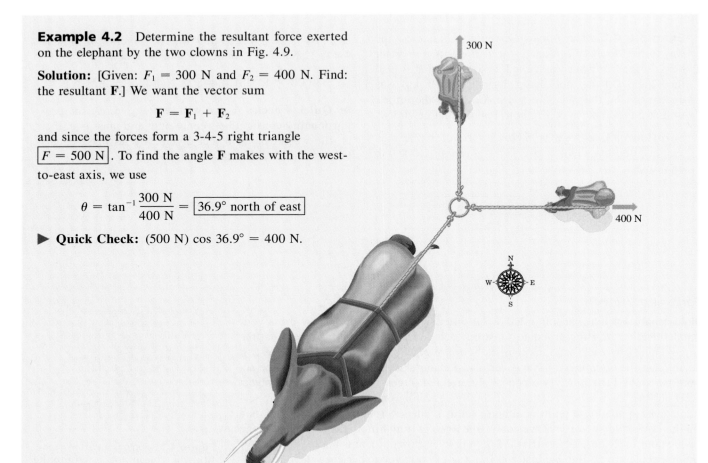

300 N

400 N

Figure 4.9 Each clown pulls on the ring with a force as shown. The elephant feels a net force pulling him north of east.

Example 4.3 Determine the net force exerted on the ring by the three people in Fig. 4.10.

Solution: [Given: $F_1 = 707$ N, $F_2 = 500$ N, and $F_3 = 966$ N. Find: $\mathbf{F} = \mathbf{F}_1 + \mathbf{F}_2 + \mathbf{F}_3$.] Resolve the three vectors into their components to get the resultant (see Appendix D). The choice of the two perpendicular axes along which to do the resolving is arbitrary, but often one set yields simpler computations, and it's usually along one of the force vectors. Accordingly, resolve the forces along and perpendicular to the y-axis down which $\mathbf{F}_3$ acts, as in Fig. 4.10b. The y-component of the resul-

tant is then

$$F_y = F_{1y} + F_{2y} - F_3 = F_1 \cos 30° + F_2 \cos 45° - F_3$$

where F_3 is entered with a negative sign because it points in the negative y-direction. Hence

$$F_y = (707 \text{ N})(0.866) + (500 \text{ N})(0.707) - (966 \text{ N})$$

and so to 3 significant figures

$$F_y = 0$$

(continued)

(continued)

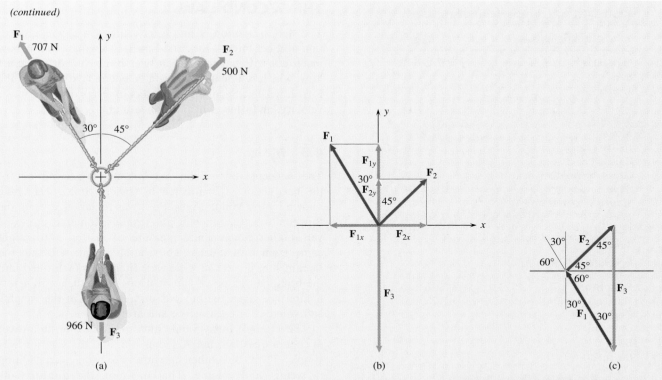

(a)

(b)

(c)

Figure 4.10 (a) Three forces passing through a point are applied in different directions. (b) The forces are resolved in the two perpendicular directions *x* and *y*. (c) Because the net force is zero, the three force vectors form a closed triangle.

Similarly, in the *x*-direction

$$F_x = F_{2x} - F_{1x} = F_2 \sin 45° - F_1 \sin 30°$$

and therefore

$$F_x = (500 \text{ N})(0.707) - (707 \text{ N})(0.500) = 0$$

The people heave and strain, but if the situation began with the ring at rest, it remains at rest because the net force acting on it is zero!

▶ **Quick Check:** If the three vectors actually add up to zero, they must form a closed triangle, as in Fig. 4.10c. The angles inside the triangle do add up to 180°, which is a good sign. To be sure things are right, we can use the Law of Sines:

$$\frac{707}{\sin 45°} = \frac{500}{\sin 30°} = \frac{966}{\sin 105°} = 10^3$$

Everyday ideas about force are often erroneous. That only living creatures can exert force is a commonly held (invalid) belief. Moreover, there are people who wrongly insist that only self-propelled things, living or not, are force makers. Surely a 5000-kg safe resting on your foot exerts a force. The gravitational interaction with the planet pulls the safe downward with a crushing force (its weight), and both Earth and safe are inanimate.

TABLE 4.1   A comparison of masses

Approximate masses	kg
Our galaxy	2×10^{41}
Sun	2×10^{30}
Jupiter	2×10^{27}
Earth	6×10^{24}
Mercury	3×10^{23}
Moon	7×10^{22}
Indian Ocean	10^{20}
Lake Erie	10^{15}
Great Pyramid at Giza	10^{10}
Super tanker	10^{8}
Saturn V rocket	3×10^{6}
Boeing 747	3.5×10^{5}
Blue whale	1.4×10^{5}
Tank	5×10^{4}
Elephant, African	6×10^{3}
Car	10^{3}
Polar bear	3×10^{2}
Human—adult	0.7×10^{2}
Dog	10^{1}
Chicken	3
Book	10^{0}
Baseball	1.5×10^{-1}
Golf ball	4.6×10^{-2}
Hummingbird, Helena's	2×10^{-3}
Paperclip	0.5×10^{-3}
House spider	10^{-4}
Ant	10^{-5}
Giant amoeba	10^{-8}
Pencil mark	10^{-10}
Bacterium	10^{-13} to 10^{-15}
DNA molecule	10^{-20}
Insulin molecule	10^{-23}
Oxygen atom	3×10^{-26}
Proton	2×10^{-27}
Electron	9×10^{-31}

THE SECOND LAW

The state of motion of an object changes with the application of a net force—the First Law tells us that. But by how much does it change? The answer is in Newton's Second Law, which is a quantified restatement and extension of the First Law. The Second Law speaks in terms of force, mass, velocity, and time, which are measurable.

4.3   Mass

The concepts of weight and volume, conceived out of the necessities of commerce in ancient times, were long known by the Middle Ages. So it was remarkable when the thirteenth-century theologian Aegidius Romanus proposed yet another fundamental measure of matter. His insight came from a quandary: The church maintained that the sacramental bread and wine were literally transformed into the body and blood of Christ. How could such a transformation be, given the obvious discrepancies in the weights and volumes of each before and after?

Few people today would look to physics for an answer to such a question, but Aegidius did, and his solution was simple: There is another measure of substance in addition to weight and volume. That measure, the **quantity-of-matter**, indicates *how much matter* there is in a material object. However unlikely, this was the origin of one of the most important ideas in physics, the concept of **mass** (*m*). Although the innovation of Aegidius had no effect on the course of theology, it did emerge as a useful, though undefined, concept in medieval science. Almost 400 years later, Newton used the terms *mass* and *quantity-of-matter* interchangeably, after providing a rather unsatisfactory definition of his own.

Mass manifests itself in two seemingly distinct ways: gravitationally and inertially. We now know from the General Theory of Relativity that these are equivalent properties of matter, that mass is mass, that there is only one kind of mass. In fact, if inertia actually is the result of the gravitational interaction of any one object with all other objects, this conclusion is quite natural. No matter what mass is conceptually (and science is certainly not finished trying to define it), we can measure it *relative to a standard kilogram*. We need only counterpoise any object against an appropriate selection of standard masses on a simple beam balance. Gravity presumably acts equally on each side, and so once balanced, the masses are equal as well. In effect, this definition of mass is an operational one that scrupulously avoids any conceptual description of what is being measured. The mass of an object is whatever we measure with a beam balance, but that still requires gravity (Table 4.1).

If you were to go off into the distant reaches of space, far from anything else, your weight would all but vanish—without gravity, there is no weight. In outer space, there is no such thing as standing *on* a scale; you would simply float above it while it reads zero. Other than that dramatic weight loss, nothing else about you would change in the journey. You would certainly look the same, even if you were overweight. Moreover, the force it takes to push you is the same whether you are hovering weightlessly in space or standing on slick ice in Brooklyn. That's the crucial point and it reveals a need for a concept, associated with matter, that remains constant—namely, *mass*.

4.4 Linear Momentum

The realization that speed alone fails to give us some essential quality of the motion of an object goes back to the Middle Ages. A Parisian, Jean Buridan, working around 1330, elaborated on an idea that had been around for centuries. When a body is thrown, it continues to move because it supposedly captured some of the original push, or, as he called it, impetus. That explanation was naive, but it led, 300 years later, to the Law of Inertia.

Buridan's most important contribution came from his attempt to understand why an iron ball of the same size as a wooden one would travel much farther when both were launched at the same speed. Or, equivalently, why you would rather be hit by a firefly traveling at 60 km/h than by a fire engine at the same speed. Buridan reasoned that the crucial concept was the product of the quantity-of-matter and the speed. The "drive," the "motive power," the "impetus"—all those ideas that convey the sense of something unalterably plowing along—cannot be described by speed alone. The moving body possesses more than speed (the firefly, after all, has the same speed as the fire engine). The essential measure of motion was a new basic quantity, the product of *mass* and *speed*: (*mv*).

By the time Galileo's work matured, his *impeto* had an inertial quality related to what he called *momento* (weight times velocity). Descartes spoke of the measure of motion as the *quantity-of-motion,* equating it to the product of "matter" and speed. When Sir Isaac Newton set out the central concept of his theory—the measure of motion—he defined the quantity-of-motion, or the **momentum (p)**, as *the product of the object's mass and velocity:*

$$\mathbf{p} = m\mathbf{v} \tag{4.1}$$

and because velocity is a vector quantity, so too is momentum. The direction of **p** is the direction of **v**; the magnitude of **p** is *mv* with SI units of kg·m/s. When **p** is constant, there is straight-line motion, and so it's more precisely called *linear momentum.*

A critical feature of momentum is that oppositely directed momenta can cancel, which is a key quality of vectors. If a train travels with a velocity of 20 km/h–east, and you jog along on its flatcars with a velocity of 20 km/h–west, you can run right off the end, motionless with respect to the ground. You will drop *straight down* as if you had stepped off a chair in your living room. Your horizontal velocity with respect to the Earth up there in midair will be zero, and your momentum will be zero as well. The momentum imparted in the eastward direction by the train will be canceled by the westward momentum gained in the running. Like velocity, linear momentum is relative. The momentum of an object moving at **v** with respect to some observer is *mv* with respect to that observer. The momentum of a pilot flying at 800 km/h is zero with respect to the plane.

The modern physicist may rightfully be proud of his spectacular achievements in science and technology. However, he should always be aware that the foundations of his imposing edifice, the basic notions of his discipline, such as the concept of mass, are entangled with serious uncertainties and perplexing difficulties that have as yet not been resolved.

MAX JAMMER (1961)

Example 4.4 Rich Gossage set a fastball record by hurling a 0.14-kg baseball at a speed of 46.3 m/s (that is, 153 ft/s). What was the magnitude of the ball's momentum as it left his hand? Here everything is measured with respect to the Earth.

Solution: [Given: $m = 0.14$ kg and $v_B = 46.3$ m/s. Find: p_B.] The momentum is provided by the scalar form of Eq. (4.1):

$$p_B = m_B v_B = (0.14 \text{ kg})(46.3 \text{ m/s}) = \boxed{6.5 \text{ kg·m/s}}$$

▶ **Quick Check:** A baseball thrown at a casual 7.2 m/s has a momentum of $(0.14 \text{ kg})(7.2 \text{ m/s}) = 1$ kg·m/s, so the above momentum at a speed ≈ 6.4 times faster is correct.

4.5 Impulse and Momentum Change

> The change of motion is ever proportional to the motive force impressed; and is made in the direction of the right [straight] line in which that force is impressed.

This translation from the Latin is the Second Law as Newton stated it. It is not the version you see in textbooks because the language is archaic. In spite of that, it is worth the effort to appreciate Newton's thinking because it's much more reasonable than most modern versions. We know from the First Law that force is the changer of motion—if there is no applied force, there is no change in straight-line uniform motion.

When a force is applied to a body, there will be a resulting proportional change in its motion. Moreover, the measure of that motion is the quantity-of-motion, the momentum. In fact, when Newton spoke about "the motion," he meant the momentum. Thus, in the definition of quantity-of-motion, he wrote, ". . . in a body double in quantity [that is, mass], with equal velocity, the motion is double." In other words, if you double m, keeping v constant, then p doubles. The Second Law therefore states that an applied force (F) produces a proportional change in the momentum of the body (Δp): $F \propto \Delta p$. Since momentum is relative, the significant quantity is not p, which depends on the motion of the observer, but Δp, which does not.

Example 4.5 A 0.149-kg baseball traveling at 28 m/s due south approaches a waiting batter. The ball is hit and momentarily crushed to almost half its original size in a collision with the bat lasting a few milliseconds. The ball springs back, sailing away at 46 m/s due north. Determine the magnitudes of its initial and final momenta and the change in its momentum.

Solution: [Given: $m_B = 0.149$ kg, $\mathbf{v}_i = 28$ m/s–south, $\mathbf{v}_f = 46$ m/s–north. Find: p_i, p_f, and Δp.]

$$p_i = m_B v_i = (0.149 \text{ kg})(28 \text{ m/s}) = \boxed{4.2 \text{ kg·m/s}}$$

Similarly

$$p_f = m_B v_f = (0.149 \text{ kg})(46 \text{ m/s}) = \boxed{6.9 \text{ kg·m/s}}$$

Since

$$\Delta \mathbf{p} = \mathbf{p}_f - \mathbf{p}_i$$

we can either visualize the subtraction vectorially, reversing $\mathbf{p}_i$ and adding it tip-to-tail to $\mathbf{p}_f$, or let north be positive and subtract scalar values, whereupon

$$\Delta p = (+6.9 \text{ kg·m/s}) - (-4.2 \text{ kg·m/s})$$

and

$$\Delta p = +11 \text{ kg·m/s}$$

Either way, $\boxed{\Delta \mathbf{p} = 11 \text{ kg·m/s–north}}$.

▶ **Quick Check:** $p_i/p_f = 28/46 = 4.2/6.9 = 0.61$, so the numbers look right. The ball has had its direction reversed, as if a momentum of 11 kg·m/s–north was added to its original 4.2 kg·m/s–south by the impact. This kind of large change always occurs when the motion of an object is turned around in a collision.

We know from experience that time is involved here, too. If several people push on a stalled car, the longer they push, the more swiftly it will roll away, and the greater will be Δp. Interestingly, there is no explicit statement about time in the original version of the Second Law. But it is clear from Newton's discussion that he was speaking about the change in motion *per unit time*. Double the force exerted on a body and the resulting change in momentum will double (assuming the forces act for the same time); triple it, and the change in momentum will triple. *The force applied equals the corresponding change in momentum per unit time.* Thus, if a certain change in motion (Δp) is to occur quickly, the applied force must be large. (Have twice as many people push, and the car will get up to speed in half the time.) To get a desired momentum change, the force applied and the interval of time over which it must be applied (Δt) vary inversely: $F \propto 1/\Delta t$. Bring to bear a large force and that force need be applied for only a short time.

If Δt is substantial, it is certainly possible that the force might vary during that interval (someone pushing behind the car could get tired). So, to be precise, we should either let Δt approach zero and work with the instantaneous force or let Δt be appreciable and use the average force over the interval (as we used v_{av} and a_{av} earlier). For the moment, we will do the latter, which will be sufficient for many practical situations, including those where the force is constant and therefore always equal to F_{av}. Accordingly, a modern version of the Second Law reads:

The force of the impact crushes a tennis ball. The racket and ball are in contact for a duration known as the *collision time*.

The average force exerted on a body equals the resulting change in momentum divided by the time elapsed in the process.

Newton was quite aware of the directional nature of the phenomenon—the change in motion is in the direction of the force impressed, which is something we can deal with rather nicely using vectors. Hence, $\Delta \mathbf{p}$ is parallel to $\mathbf{F}_{av}$, and putting it all together symbolically, we have

$$\mathbf{F}_{av} = \frac{\Delta \mathbf{p}}{\Delta t} = \frac{\Delta(m\mathbf{v})}{\Delta t} \qquad (4.2)$$

The **average net force equals the time-rate-of-change of momentum**.

Example 4.6 A rocket fires its engine, exerting an average force of 1000 N for 40 s. By how much will its momentum change?

Solution: [Given: $F_{av} = 1000$ N and $\Delta t = 40$ s. Find: Δp.] The momentum change occurs in the direction of the force. Hence, we can use the scalar form of Eq. (4.2), namely

$$F_{av} = \frac{\Delta p}{\Delta t}$$

Thus $\Delta p = F_{av}\Delta t = (1000 \text{ N})(40 \text{ s})$

and so $\boxed{\Delta p = 40 \times 10^3 \text{ N·s}}$

▶ **Quick Check:** 1 kN × 40 s = 40 kN·s.

The impulse exerted by the punch changes the momentum of the recipient's head. The brain, which is surrounded by a liquid, slams into the skull. The gloves are heavily padded to increase Δt thereby decreasing F_{av}, but it often doesn't help.

Equation (4.2) suggests a dynamic means of quantifying force since we can measure m, v, and t. Consequently, we now define 1 *newton* (N) to be the amount of force that will produce a change in the momentum of *any* object equal to 1 kg·m/s in 1 s. That is

$$1 \text{ N} = 1 \text{ kg·m/s}^2$$

Newton's early studies had been done around the same time as important work by Hooke, Wren, Wallis, and Huygens was accomplished on collisions. Billiards was very popular at the time, and there was a lot of interest in momentary impacts. In such cases, neither $\mathbf{F}_{av}$ nor Δt is known very well, but the change in momentum is easily determined. Consequently, it is useful to combine the force and time into a single notion that equals the known momentum change. That single concept ($\mathbf{F}_{av} \Delta t$) is called the **impulse** of the force, and an alternative statement of the Second Law becomes

$$\text{Impulse} = \text{change in momentum}$$

$$\mathbf{F}_{av} \Delta t = \Delta \mathbf{p} \qquad (4.3)$$

A given impulse will produce a specific change in momentum no matter what the mass or speed of the recipient body. An object originally at rest will take off in the direction of the net applied force, acquiring a momentum $\Delta(m\mathbf{v}) = \mathbf{F}_{av} \Delta t$, which is what happens when you throw a dart, discharge a syringe, or strike a golf ball. As long as the club is in contact with the ball, applying force, there is an ongoing gain in momentum in the direction of the force (Table 4.2). Once the ball leaves the club, it flies off under the First Law, forceless and changeless. The longer the barrel of a gun, or the greater the pitcher's throwing motion—the larger will be the time (Δt) during which the propelling force acts and the more the change in momentum (Δp) of the projectile (Fig. 4.11).

Force applied to a body already in motion may either increase or decrease its momentum depending on whether $\mathbf{F}_{av}$ acts parallel or antiparallel to the initial ve-

Figure 4.11 The pitcher exerts a force on the ball over as long a distance and for as long a time as possible. The greater is $F\Delta t$, the larger is Δp, and the higher the release speed. Here the ball is shown every 1/100th of a second. The greater the distance between images of the ball, the greater the speed.

TABLE 4.2 Typical parameters for balls hit from rest by the appropriate object

Ball	Mass (kg)	Speed imparted (m/s)	Impact Time (ms)
Baseball	0.149	39	1.25
Football (punt)	0.415	28	8
Golf ball (drive)	0.047	69	1
Handball (serve)	0.061	23	12.5
Soccer ball (kick)	0.425	26	8
Tennis ball (serve)	0.058	51	4

locity ($\mathbf{v}_f = \mathbf{v}_i + \Delta\mathbf{v}$). To slow the *Lunar Excursion Module* as it plunged toward the Moon's surface in 1969, a downward-pointing retrorocket was fired that exerted an upward force on the craft, thereby decreasing its downward momentum and speed.

Example 4.7 On September 12, 1966, a *Gemini* spacecraft piloted by astronauts Conrad and Gorden met and docked with an orbiting *Agena* launch vehicle. With plenty of fuel left in the spacecraft, NASA decided to determine the mass of the *Agena*. While coupled, *Gemini*'s motor was fired, exerting a constant thrust of 890 N for 7.0 s. As a result of that little nudge, the *Gemini-Agena* sped up by 0.93 m/s. Assuming *Gemini*'s mass as a constant 34×10^2 kg, compute the mass of the *Agena*.

Solution: [Given: $F_{av} = 890$ N, $\Delta t = 7.0$ s, $\Delta v = +0.93$ m/s, and $m_G = 34 \times 10^2$ kg. Find: m_A.] This calculation will be inexact to the degree that both the

(continued)

(continued)

thrust and m_G are assumed constant. The basic relationship is

$$\mathbf{F}_{av}\,\Delta t = \Delta\mathbf{p} \qquad [4.3]$$

Because m is constant

$$\Delta\mathbf{p} = \Delta(m\mathbf{v}) = m\,\Delta\mathbf{v}$$

where

$$m = m_A + m_G$$

Since the direction of motion is of no concern, we can use a scalar form

$$F_{av}\,\Delta t = (m_A + m_G)\Delta v$$

The rocket's impulse produces a well-defined *change* in speed, regardless of the initial speed. Substituting in the

numbers yields

$$(890\ \text{N})(7.0\ \text{s}) = (34 \times 10^2\ \text{kg} + m_A)(0.93\ \text{m/s})$$

Solving for m_A

$$m_A = \frac{(890\ \text{N})(7.0\ \text{s})}{(0.93\ \text{m/s})} - (34 \times 10^2\ \text{kg})$$

and $m_A = 3299$ kg, or to two figures, $\boxed{33 \times 10^2\ \text{kg}}$.

▶ **Quick Check:** Assume the answer is correct and calculate the force. $F_{av} = (67 \times 10^2\ \text{kg})(0.93\ \text{m/s})/(7.0\ \text{s}) = 890$ N, as given.

It follows from Eq. (4.3) that for a given change in momentum (either an increase or a decrease), the longer the force acts (Δt), the smaller $\mathbf{F}_{av}$ will have to be. Your hands provide the stopping force for an incoming ball when you catch it; that is, when you change $\mathbf{p}_i$ to $\mathbf{p}_f = 0$. If the ball is soft and deforms as it comes to rest, the stopping time will be large and the stopping force will be comparatively small. Even with the same incoming momentum, so that $\Delta\mathbf{p}$ is the same when the balls are brought to rest, a soft ball will take longer to stop and so require less force to catch than a hard ball. Furthermore, if you wear a padded mitt (which will extend Δt) and draw the glove back toward you just as the ball strikes it, the stopping force will be still smaller. Automobile bumpers that compress on impact extend the stopping time (Δt) and thereby decrease the force of collision even though the car undergoes the same change in momentum.

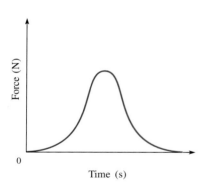

Figure 4.12 The force-time graph for a body undergoing an impact. The force rises, peaks, and falls to zero.

4.6 Varying Force

Force may change from moment to moment (the force on your feet as you walk around certainly does). To obtain a complete picture, we must envision a concept that varies in time. At each and every moment, there is a value of the **instantaneous force**; that is, *the limit approached by the average force as the time interval (over which the averaging is taking place) shrinks to zero*:

$$\mathbf{F} = \lim_{\Delta t \to 0}\left[\frac{\Delta\mathbf{p}}{\Delta t}\right] \qquad (4.4)$$

Push on something for a moment. The force exerted initially builds, the flesh on your hand compresses and distorts, the force peaks, and then decreases to zero. That's what happens when you shove a book or hurl a shot put. Similarly, a quick blow with a hammer or the blast of an explosive charge will usually produce a force that even more rapidly rises, peaks, and falls off to zero.

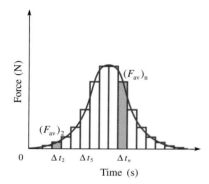

Figure 4.13 The area under the force-time graph is the net impulse exerted, and that equals the net change in momentum imparted to the body in question.

Figure 4.12 is a force-time graph for an impact of some sort, a plot of the instantaneous force versus time. The width of the curve and the details of its shape will vary from one situation to another, but the area under the curve is the net impulse imparted to the body. To quantify the impulse, divide the region under the curve into a number of subintervals, each with a little rectangular area whose base is an interval of time: $\Delta t_1, \Delta t_2, \Delta t_3$, and so forth (Fig. 4.13). Next, fix the height of each rectangle at a value equal to the average force exerted over that particular time interval. Thus, for example, the area of the second rectangle is $(F_{av})_2 \Delta t_2$, which corresponds to the *impulse* associated with that subinterval. The sum of all the little subareas is exactly equal to the total impulse exerted by the force.

If the number of subintervals is made increasingly large while the width of each approaches zero ($\Delta t \to 0$), the tops of the rectangles will blend ever more smoothly into the curve. In other words, the value of the average force for each subinterval will approach the value of the instantaneous force and these, in turn, are the corresponding points on the curve. Just as the area under any portion of the speed-time curve is the distance traveled during a given period, **the area under the force-time curve is the impulse exerted during that interval and it therefore equals the resulting change in momentum**.

Example 4.8 A golfer's club hits a 47.0-g golf ball from rest to a speed of 60.0 m/s in a collision lasting 1.00 ms. The force on the ball rises to a peak value of F_{max} and then drops to zero as it leaves the club. Compute a rough value for this maximum force by approximating the force-time curve, with a triangle of altitude F_{max}.

Solution: [Given: $m = 47.0 \times 10^{-3}$ kg, $v_i = 0$, $v_f = 60.0$ m/s, and $\Delta t = 1.00 \times 10^{-3}$ s. Find: F_{max}.] First we will find the change in momentum, and that equals the net impulse. The net impulse, in turn, equals the area under the force-time curve, which can be approximated by a triangle with an altitude of F_{max}. Thus

$$F_{av} \Delta t = \Delta(mv) = (47.0 \times 10^{-3} \text{ kg})(60.0 \text{ m/s})$$

and the impulse is 2.82 kg·m/s, or 2.82 N·s. Representing the force-time curve by a triangle of altitude F_{max} and base Δt, the area encompassed is $\frac{1}{2}\Delta t(F_{max})$, which equals the impulse; that is

$$2.82 \text{ N·s} = \tfrac{1}{2}(1.00 \times 10^{-3} \text{ s})F_{max}$$

Hence, $\boxed{F_{max} = 5.64 \text{ kN}}$ or 1.27×10^3 lb.

▶ **Quick Check:** Whenever the force-time curve is approximated by a triangle, $F_{av} \Delta t = \frac{1}{2}(\Delta t)F_{max}$ and the peak force is twice the average force. As a consequence, F_{max} may break bones, even though F_{av} is not large enough to do damage. $F_{av} = (2.82 \text{ N·s})/(1.00 \text{ ms})$ and $2F_{av} = 5.6 \times 10^3 \text{ N} = F_{max}$.

A real force-time curve can have a complex shape (Fig. 4.14). A spaceship with a variable-thrust engine can generate a force-time curve with many bumps and wiggles, but the area encompassed in a given time interval will always be equal to the resulting change in momentum. If the ship turns 180° around, the same engine can fire in the opposite direction, against the motion. That would generate a negative impulse, a negative change in momentum, and a decrease in speed. The total impulse is, then, the area of the force-time curve above the axis minus the area below the axis. If those two portions happen to be equal, the net change in momentum will be zero. When you throw a punch, your fist starts with $p_i = 0$ and ends with $p_f = 0$. It accelerates up to a maximum speed and then decelerates to zero when your arm is extended. Given this fact, a karate blow is always aimed at a point inside the target so that it makes contact when p is maximum.

When a car crashes into a brick wall, its front end deforms as it slows to a crushing stop. Today's average production vehicle compresses roughly 1 inch for every mile per hour of speed just prior to impact; smaller cars typically collapse a

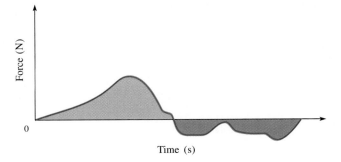

Figure 4.14 Instantaneous force versus time. The total impulse is the net area under the curve; take the area above the axis as positive and that below as negative. Oppositely directed forces produce impulses with opposite signs and adding the areas above and below the axis takes this opposition into account.

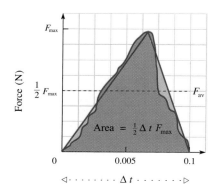

Figure 4.15 A triangular approximation to the area under the force-time curve.

little less than this, larger cars a little more. The process is similar to a struck baseball, except the car does not spring back to its original shape.

If we assume that the collision-deceleration is fairly uniform, the crush-distance s_c divided by $v_{av} = \frac{1}{2}(v_i + v_f)$, with $v_f = 0$, is the impact time. Thus

$$\Delta t = \frac{s_c}{v_{av}} = \frac{2s_c}{v_i}$$

Since tests show that s_c is proportional to v_i, we conclude that the impact time for cars of comparable stiffness should be independent of the speed of impact. A typical head-on, brick-wall collision lasts around 100 milliseconds. By the way, a driver would not be much better off hitting an oncoming car. In fact, crashing head-on into an identical car traveling toward you at the same speed is effectively the same as hitting a stationary brick wall. To appreciate that, just imagine a large movable sheet of steel mounted on wheels. Now, if both cars plow into this steel sheet on opposite sides, the autos will be quite destroyed, but the sheet will remain in the middle of the melee unmoved, as if it were unmovable, just like the brick wall.

We can expect the force-time curve for a car colliding with a stationary barrier to have a base width of about 100 ms. An 1800-kg car impacting at 26.8 m/s (60 mi/h) will be crushed about 1.5 m (60 in.). The engine will end up in the passenger compartment as the remainder of the car moves forward, past it. The momentum change of the car is $\Delta p = 48.2 \times 10^3$ kg·m/s, which equals the area under the force-time curve in Fig. 4.15. If we again approximate that curve by a triangle, the peak force exerted on the car is approximately 0.96×10^6 N, or over 100 tons.

Example 4.9   A 70-kg passenger riding in a typical automobile is involved in a 17.9 m/s (40 mi/h) head-on collision with a concrete barrier. Taking the stopping time as 100 ms, compute the average force exerted by the seat belt and shoulder strap on the person.

Solution: [Given: $m_p = 70$ kg; taking the initial direction of motion as positive, $\Delta v = -17.9$ m/s, $\Delta t = 0.100$ s. Find: F_{av}.] This collision problem involves mass, speed, time, and force, and suggests impulse-momentum, that is

$$F_{av} \Delta t = m_p \Delta v$$

$$F_{av} = \frac{(70 \text{ kg})(-17.9 \text{ m/s})}{0.100 \text{ s}}$$

Hence, $\boxed{F_{av} = -1.3 \times 10^4 \text{ N}}$ or -2.8×10^3 lb. The minus sign means the force is in the opposite direction to the initial motion.

▶ **Quick Check:** Assume the force is correct and calculate the time; $\Delta t = (70 \text{ kg})(-17.9 \text{ m/s})/(-1.3 \times 10^4 \text{ N}) = 0.096 \text{ s} = 0.10 \text{ s}$.

INTERACTION

Newton's Third Law completes the logical picture of the concept of force. A change of motion arises out of a net impressed force. Whenever we talked about a force applied to a body, it was an **external force**, one whose source was external to the body. Newton defined force as any action that causes an object to change its momentum. But there is some circular logic here. By definition, whenever a body's momentum is altered, the cause must be a force acting upon it. Something is still missing. What ties the concept of force back to the observable Universe? What links the cause of the force to the effect, the change of momentum?

An isolated body obeys the Law of Inertia. It cannot alter its own uniform straight-line motion because that requires some outside intervention via one or more external entities. We can assume that when two bodies interact, *both* will be affected by the encounter; the motions of both will be altered, both will experience an external force. Two identical bodies moving toward each other at the same speed must experience the same alteration of motion on impact, and therefore must exert the same force on each other.

Newton had other things to draw insight from, in addition to billiards. Experiments on gases, for example, fashionable when he was a young man, demonstrated that when a gas is compressed within a container by a piston pushing down on it, there is a backward reaction—the piston tends to pop back up. The "spring of the air" was a common descriptive phrase coming out of those studies and, indeed, a spring was a perfect image to use—push on it and it pushes back on you; pull it in one direction and it pulls you in the opposite direction.

Newton provided the final link in the credo of the Three Laws. In his own words:

> To every action there is always opposed an equal reaction: or the mutual actions of two bodies upon each other are always equal, and directed to contrary parts.

The interaction of two bodies always occurs via a force and an equal-magnitude oppositely directed counterforce. There is no such thing as a single force exerted by some active entity on a passive one; there is only **interaction**. A hand cannot simply push on a chair—the chair must always push back on the hand. Entities interact, one upon the other: *force is a thing of pairs.* Whenever we see some object deviate from the Law of Inertia, we must be able to find another interacting with it.

Consider, for example, a balloon being pressed into a wall. You could argue that the balloon is squashed by the pusher alone, without any help from the wall; that the person's hand is doing the pushing, not the wall. Yet the balloon is flattened on both sides just as it would be if the wall itself had moved to where the balloon was being held fixed in hand and proceeded to squash it. If you came along at precisely that

Isaac Newton's death mask (a plaster casting of his face), made soon after he died at about 1:00 A.M. on March 20, 1727.

An object offers as much resistance to the air as the air does to the object. You may see that the beating of an eagle's wings against the air supports its heavy body in the highest and rarest atmosphere . . . .
LEONARDO DA VINCI (1452–1519)
Codex Atlanticus

These firefighters are struggling to control a blasting hose. The water is propelled forward with great force and it reacts back with an equal-magnitude force.

F_{WB} F_{BW} F_{BH} F_{HB}

Figure 4.16 A balloon pressed against a wall. F_{HB} and F_{BH} are an interaction pair, as are F_{WB} and F_{BW}. All forces are the result of interactions and all come in pairs comprising equal and opposite influences.

moment and saw a wall, a flattened balloon, and a hand, you could not possibly tell whether the wall or the hand had moved. The hand pushes on the balloon (Fig. 4.16) with a force F_{HB}, and the balloon pushes back with an equal and opposite force F_{BH}. The balloon, acting as if it were an extension of the hand, transmits the force that is applied to it to the wall ($F_{BW} = F_{HB}$). The wall, in turn, reacts with an equal-magnitude, oppositely directed counterforce F_{WB}. The balloon is squashed in on both sides by the forces *applied to it*, F_{WB} and F_{HB}.

Interaction between entities means *two equal-magnitude, oppositely directed forces (one on each participant) forming the pair*. Here, each interaction pair comprises two forces with reverse-ordered subscripts. **The forces of such a pair are always in opposite directions and never act on the same body.** Thus, F_{HB} and F_{BH} are an interaction pair; F_{HB} is exerted on the balloon while F_{BH} is exerted on the hand. Two external horizontal forces are acting on the balloon, F_{WB} and F_{HB}. These are not an interaction pair even though they happen to be equal in magnitude and opposite in direction. Hence, as far as the motion of the balloon as a whole is concerned, the two external horizontal forces cancel, yielding a net applied force of zero. Momentum remains unchanged—the balloon stays put at rest, squashed up against the wall.

A chisel struck by a mallet transmits a force to a piece of wood, just as the balloon transmitted a force from the hand to the wall. One layer of atoms pushes (electromagnetically) on the next layer down to the end of the chisel. During such a process some distortion occurs, even in a hard steel chisel. By contrast, the distortion of the balloon was obvious. In the real world, a perfectly rigid body does not exist. Even a flea deforms the rock it lands on (granted, it's only by a miniscule amount, but to some extent the rock *is* squashed). The shifting of atoms within the rock and the subsequent change in their electrical interaction supports the flea.

Example 4.10 One tough killer bee has a mass of 0.10 g. Traveling due south at 5.0 m/s, it crashes into the windshield of a 10-ton truck heading due north at 60 km/h. The bee bounces off the window at 35 m/s–north (with respect to the ground) after an impact lasting 0.090 s. (a) Compute the average force exerted by the bee on the truck. (b) What was the change in momentum of the truck as a result of the collision? (c) The bee started out moving south and ended up moving north, which means it must have come to rest, with respect to the Earth, during the collision. Does that mean the truck, with which it was in contact, also stopped for that instant?

Solution: [Given: $m_b = 0.10$ g, $v_{bi} = -5.0$ m/s, $v_{bf} = +35$ m/s, the weight of the truck is 10 tons or 20 000 lb, $v_T = +60$ km/h, and $t = 0.090$ s. Find: (a) the average force exerted by the bee on the truck. Because there are so many subscripts, let's use pointed brackets to indicate average value and write it as $\langle F_{bT} \rangle$; and (b) Δp_T.] Take north as the positive direction. (a) We know the change

in velocity of the bee, and so can compute its change in momentum and the corresponding force exerted on it. The force it exerts on the truck is just equal and opposite to that. The truck exerts a force on the bee $\langle F_{Tb} \rangle$, which alters the bee's momentum such that

$$\langle F_{Tb} \rangle \Delta t = \Delta p_b = m_b \Delta v_b$$

Here, $\Delta v_b = (+35 \text{ m/s}) - (-5.0 \text{ m/s}) = +40$ m/s. The fact that the bee was turned around adds to the change in velocity and momentum. The bee experiences a force pointing north that first brings it to rest and then reverses its direction. Hence

$$\langle F_{Tb} \rangle \Delta t = m_b(40 \text{ m/s})$$

and

$$\langle F_{Tb} \rangle = \frac{(0.10 \times 10^{-3} \text{ kg})(40 \text{ m/s})}{0.090 \text{ s}} = 0.044 \text{ N}$$

The bee exerts an equal and opposite force

$$\boxed{\langle F_{bT} \rangle = 4.4 \times 10^{-2} \text{ N–south}}$$

(continued)

(continued)

(b) The impulse acting on the bee was the same in magnitude and opposite in direction to the impulse acting on the truck, and so $\Delta\mathbf{p}_T = -\Delta\mathbf{p}_b$. Accordingly, $\Delta p_b = m_b\,\Delta v_b$, hence

$$\Delta p_b = (0.10 \times 10^{-3}\ \text{kg})(40\ \text{m/s})$$

and $\qquad \Delta p_b = 4.0 \times 10^{-3}\ \text{kg}\cdot\text{m/s}$

The bee's initial momentum points south; its momentum *change* is positive and points north ($\mathbf{p}_{bi} + \Delta\mathbf{p}_b = \mathbf{p}_{bf}$). Hence, the truck's momentum change is

$$\boxed{4.0 \times 10^{-3}\ \text{kg}\cdot\text{m/s–south}}$$

(c) The bee crashed into the window moving southward at a relative speed of 78 km/h. The window deformed ever so slightly, slowing the bee and bringing it to rest for an instant with respect to the ground. At that moment, the bee was moving southward at 60 km/h with respect to the truck. The bit of glass in contact with the bee did stop for an instant with respect to the ground, but the truck kept right on rolling.

▶ **Quick Check:** As seen by the driver, the bee comes in with a speed of (5.0 m/s + 16.66 m/s) and bounces off with a speed of (35 m/s − 16.66 m/s), giving a net momentum change of [(+21.66 m/s) − (−18.33 m/s)] × (0.10 × 10⁻³ kg) = 4.0 × 10⁻³ kg·m/s and, dividing by 0.090 s, $\langle F_{Tb} \rangle = 0.044$ N.

An X-ray photo of a 45-caliber bullet fired from a pistol. In effect, the gun pushes on the bullet and the bullet pushes back on the gun. Some of that reaction goes into operating the slide exposing the barrel and ejecting the spent shell.

4.7 Rockets

Imagine yourself on roller skates holding a bag of oranges. Now throw one of them due north, and away you go due south. You are thrust backward just as a gun recoils backward when it fires a projectile. You push on the orange in the forward direction during the throw, and it pushes back with the same impulse on you, and back you go. Hurl several oranges and you have an orange-rocket, because that's exactly the way a rocket works. Rather than throwing a few large objects slowly, a rocket engine hurls out a tremendous number of tiny high-speed objects—molecules. For example, during launch, the solid-fuel boosters for the Space Shuttle expel 8.5 tons of fiery exhaust each second. The engines blast exhaust downward and the escaping gas, in turn, pushes up on the rocket (interaction). That's why rockets can be used in space. *Rockets are not propelled by pushing against either the ground or the atmosphere,*

A jet plane sucks in air and blasts out a high-speed exhaust that drives the craft forward. The engines push the exhaust backward, and the exhaust pushes the engines forward.

On flying rockets in space. That Professor Goddard . . . does not know the relation of action to reaction, and of the need to have something better than a vacuum against which to react—to say that would be absurd. Of course, he only seems to lack the knowledge ladled out daily in high schools.

(The *New York Times*, Jan. 13, 1920)

Further investigation and experimentation have confirmed the findings of Isaac Newton in the 17th century, and it is now definitely established that a rocket can function in a vacuum as well as in an atmosphere. The Times *regrets the error.*

(The *New York Times*, July 17, 1969)

Octopuses and squid can accelerate rapidly by squirting a stream of water backward, just like a jet plane. The animal pushes the water stream, the stream pushes back on the animal.

though that erroneous opinion is widely held even today. In fact, a *New York Times'* editorial of 1920 advised Robert H. Goddard, the American who launched the first liquid-fuel rocket, to give up any thoughts of space travel. After all, even a school-boy knows that rockets obviously cannot fly in space because a vacuum is devoid of anything to push on.

Example 4.11   A small rocket engine expels 10.0 kg of exhaust gas per second. If these molecules are ejected at an average speed of 600 m/s, what is the thrust of the engine?

Solution: [Given: for the exhaust gas, $v_i = 0$, $v_f = 600$ m/s, and 10.0 kg are ejected per 1.00 s. Find: the thrust, F_{av}, exerted by the motor.] We know that each second, 10.0 kg of exhaust gas experience an increase in speed of $\Delta v = 600$ m/s. Hence, there's an average force exerted on the gas given by

$$F_{av} = \frac{m\Delta v}{\Delta t} = \frac{(10.0 \text{ kg})(600 \text{ m/s})}{1.00 \text{ s}}$$

and
$$F_{av} = 6.00 \times 10^3 \text{ kg·m/s}$$

The engine pushes the gas backward with a force of $\boxed{F_{av} = 6.0 \times 10^3 \text{ N}}$; that is, 1.3×10^3 lb, and the gas exerts an equal reaction force forward on the rocket.

▶ **Quick Check:** Assume the force is correct and calculate the mass: $m = (F_{av}) \Delta t / \Delta v = 10$ kg.

4.8   Conservation of Linear Momentum

One of the great guiding ideas of physics is the law of Conservation of Linear Momentum: **When the resultant of all the external forces acting on a system is zero, the linear momentum of the system remains constant**. Descartes, long before Newton had refined dynamics, wrote of his Creator of the Universe:

He set in motion in many different ways the parts of matter when He created them, and since He maintained them with the same behavior and with the same laws as He laid upon them in their creation, He conserves continually in this matter an equal quantity-of-motion.

In other words, the total momentum of the Universe persists unchanged and will continue to be preserved forever.

In the process of throwing a bouquet or catching a cigar, a force is applied by the hand to the projectile. Simultaneously, there is an equal and opposite force back on the hand. The *impulse* exerted on the cigar, bringing it to rest, is matched by an equal and oppositely directed impulse on the hand that catches it. The interaction times are the same, and the forces are equal in magnitude and opposite in direction. Because interacting objects exert such forces on each other, the resulting momentum changes must be equal and opposite.

A particle or a group of particles constitutes a *system.* Whenever forces act to change the location of material objects, it's wise to have a clear idea of the extent of the system being dealt with. That's true now when we are concerned with changes in momentum resulting from the application of force, and it will be true later when we deal with changes in energy. Forces whose sources are within the system are said to be **internal forces** (Fig. 4.17a). Hence, *the total momentum of a system of interacting masses must remain unaltered, provided that no net external force is applied.* Just try pushing a car off the road while sitting inside it (p. 128).

The momenta of individual interacting members of a system may certainly change, but each change is accompanied by an equal and opposite change in momentum of the interaction partner. If the whole of our Universe is taken as the system, there are no *external forces* whose sources lie outside the system, and the total momentum must be conserved forever (Fig. 4.17b). If a system is not isolated, and external forces do act on it (Fig. 4.17c), its momentum will change, say, by an

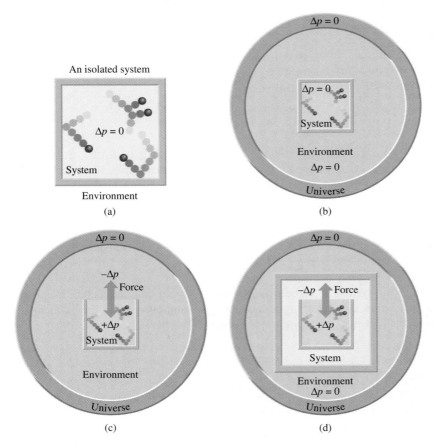

Figure 4.17 (a) A system isolated from the environment conserves momentum; that is, $\Delta p = 0$. (b) Here, we see an isolated system within the Universe. In all cases, Δp for the Universe is zero. (c) When the system communicates with its environment, it may experience a net momentum change, $+\Delta p$, but the environment will then experience an equal and opposite change, $-\Delta p$. (d) If the system is enlarged and yet isolated, its parts can experience momentum changes, but its net change will be zero.

amount $+\Delta p$. But the momentum of the communicating environment will then change by $-\Delta p$. If that system is part of a still larger isolated system (Fig. 4.17d), then the change in momentum of everything going on inside the large system will be zero, as if it were a little universe unto itself.

Figure 4.18, an example of how this phenomenon works, shows two people and a small moon, all floating in deep space and all at relative rest. Let's first take the three bodies to be the system. The total initial momentum is zero. The moon is subsequently thrown back and forth, momentum is imparted to the individual members of the system, but the forces are all internal and the net momentum is always zero, as it was at the outset. If we now take the moon as the system, its momentum changes every time someone applies an external force to it. By contrast, take the moving moon and the person on the right to be the system in Fig. 4.18b and thereafter. There is initially a net momentum of 10 kg·m/s to the right, and it will remain constant (Fig. 4.18c and d) unless the person on the left, who is not in the system, exerts an external force.

Example 4.12   According to published figures, a bullet fired from a 9-mm Luger pistol weighs 124 grains, has a mass of 8.0 g, and a muzzle speed of 1155 ft/s. If the mass of the gun is 0.90 kg, what is its recoil speed when the gun is fired horizontally? Ignore the escaping gases.

Solution: [Given: $m_B = 8.0$ g, $m_G = 0.90$ kg, $v_{Bi} = v_{Gi} = 0$, $v_{Bf} = 1155$ ft/s. Find: v_{Gf}.] The muzzle speed in SI is (1155 ft/s)(0.304 8 m/ft) = 352 m/s. Taking the gun and bullet as the system, there are no external horizontal forces, and so the initial momentum must equal the final momentum, therefore

$$m_B v_{Bi} + m_G v_{Gi} = 0 = m_B v_{Bf} + m_G v_{Gf}$$

hence

$$v_{Gf} = -\frac{m_B}{m_G} v_{Bf} = -\frac{0.008\,0 \text{ kg}}{0.90 \text{ kg}} (352 \text{ m/s})$$

and $\boxed{v_{Gf} = -3.1 \text{ m/s}}$. The minus sign means that the gun moves in the opposite direction to the bullet.

▶ **Quick Check:** $p_{Gf} = p_{Bf}$, the bullet is roughly 100 times less massive than the gun, and its speed is roughly 100 times greater.

The planet often plays a subtle part in experiments on Earth, since it is usually connected to what is happening in some way. For example, if while hovering motionlessly ($p_i = 0$) far off in space you throw a ball, then your final momentum will be equal in magnitude and opposite in direction to the projectile's momentum. On Earth, things can be quite different. The ball gets thrown, but you usually just stand there. In the process, you brace yourself by pushing on the ground with a force (F_{YG}) and the ground pushes back on you with an equal and opposite force (F_{GY}). If you and the ball are the system, F_{GY} is an *external* force acting on that system and momentum is properly not conserved. If, on the other hand, we take the ball, you, and Earth as the system, F_{GY} and F_{YG} are internal forces and momentum is conserved—the Earth must move with a momentum equal and opposite to the ball's momentum (an effect that is much too small to measure directly).

Example 4.13   Imagine two masses, m_1 and m_2, resting on a horizontal frictionless surface. The masses are tied together with a string so that they compress a spring located between them. Now cut the string and let the masses fly apart. (a) Write an expression for m_1 in terms of m_2 and the final recoil speeds v_1 and v_2. (b) What is the expression when m_2 is a standard mass of 1 kg?

(continued)

(continued)
Solution: [Given: m_1, m_2, and the final speeds v_{f1} and v_{f2}. Find: m_1.] There being no external horizontal forces, momentum is conserved, and since it was zero initially

$$0 = m_1 v_{f1} + m_2 v_{f2}$$

and $m_1 = -m_2 v_{f2}/v_{f1}$, When $m_2 = 1$ kg

$$m_1 = -\left(\frac{v_{f2}}{v_{f1}}\right)(1\text{ kg})$$

This procedure can serve as an operational definition of mass that does not require gravity. Given a standard mass (m_2), we can determine the mass of any object without ever defining mass conceptually.

▶ **Quick Check:** v_{f2}/v_{f1} is unitless, so the units are fine. When $m_1 > m_2$, we expect $v_{f2} > v_{f1}$, and the equation agrees with that.

Figure 4.18 Take the two astronauts playing catch with a little moon as the system. There will then be no external forces acting, and momentum will be conserved. (a) Everything starts out at rest with a net momentum of zero: $p = 0$. (b) After the moonlet is thrown, the net momentum of the system is still zero. (c) And it's zero after the moonlet is caught and (d) thrown again. Alternatively, take the moon and/or either astronaut as the system. External forces will now be applied, and the momentum of this less inclusive system will change. Note that we do not know how much momentum the moonlet has in (c).

No experiment can prove that the law of Conservation of Linear Momentum is true, although observations confirm it all the time: There are no places in this Universe free of outside influences. Still, subatomic particles appear to conserve momentum scrupulously. Indeed, Conservation of Momentum is more fundamental

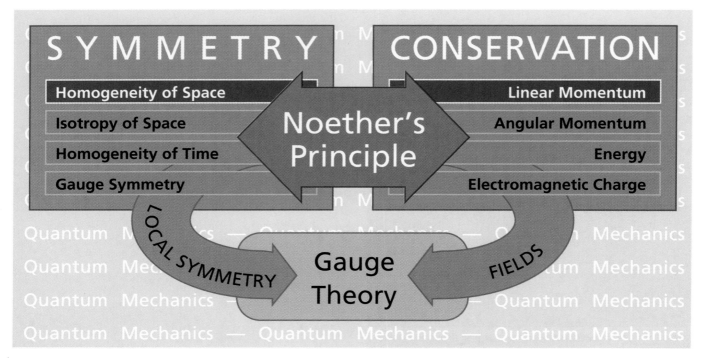

Figure 4.19 Until now we examined symmetry, conservation, and Noether's Principle in general terms. Here, we learn that the homogeneity of space gives rise to Conservation of Linear Momentum.

than the Third Law. That's true because interactions between separated objects take time to propagate, and there can be an appreciable delay between the observed action and the observable reaction. Modern Quantum Field Theory (Ch. 33) maintains that all the basic interactions (gravitational, electromagnetic, strong, and weak) arise from the exchange of momentum-carrying particles that travel at finite speeds. Although momentum is conserved at all times, we may have to wait a while after the action for the reaction.

Linear Momentum and Symmetry

Recall the discussion of symmetry and conservation in Chapter 1. When a change in a physical system leaves it essentially unchanged, that system possesses a corresponding symmetry. Figure 1.4 (p. 6) showed a disc that appears uniform and could therefore be rotated to a new position without producing any observable change—the disc possesses a symmetry. In the same way, we believe that the Universe is homogeneous as regards its governing physical principles. Empty space is the same everywhere. The same laws of physics apply everywhere, and the fundamental constants (for example, the speed of light, the quantum of charge, the gravitational constant) are independent of position. We could relocate our laboratory anywhere in the Universe and the laws of physics, the way things happen, will not change—the Universe possesses a spatial symmetry; there is an invariance in its space-time behavior.

Noether's Principle maintains that for every conservation law there is a corresponding symmetry, and vice versa. In particular, Noether showed that Conservation of Linear Momentum is a consequence of the invariance of nature with respect to place. Insofar as no conceivable experiment can distinguish between one point in empty space and another (on the basis of variations in physical behavior), space is homogeneous and linear momentum must be conserved (Fig. 4.19). Noether's proof of this is beyond our means, but we can show the interrelationship.

Assume that the physical Universe is *not* homogeneous, that the laws of physics are *not* the same everywhere. Now hold two identical objects at separate distant points (A and B) in space. Suppose that they attract each other either electrically, magnetically, or gravitationally. Given the "inhomogeneity," assume the force law is different at the two locations. Then the force on object A due to object B will be different from the force on object B due to object A. These forces are not an equal-and-opposite interaction pair (which conflicts with the Third Law). Suppose the objects are now released with a total initial momentum of zero. They will immediately travel toward each other, increasing in momentum but at different rates, thereby violating Conservation of Linear Momentum. The converse is as follows: space is homogeneous, the force laws are the same everywhere, the two forces are equal in magnitude and opposite in direction, the change in momentum of the two objects is equal in magnitude and opposite in direction, and linear momentum is conserved throughout the Universe.

Momentum is a vector in space, and it is conserved because of the homogeneity of space. That homogeneity is manifest in the equal-and-opposite nature of interaction. Hence, wherever there is a $\Delta\mathbf{p}$, there must, somewhere else in space, be an equal-magnitude, oppositely directed $\Delta\mathbf{p}$.

Core Material

Newton's First Law is the **Law of Inertia**: Every body continues in a state of rest, or in uniform motion in a straight line except insofar as it is compelled to change that state by forces impressed upon it. Force is the agent of change. The *weight* of an object on Earth is the downward force arising from the gravitational interaction between the object and the planet. The SI unit of force, the **newton** (N), is approximately equal on Earth to the weight of a mass of 0.10 kg, or $1 \text{ N} \approx \frac{1}{4}$ lb. It is the *net external force* acting on a body that changes the motion (p. 97).

The measure of motion in Newton's theory is **momentum (p)**, defined as

$$\mathbf{p} = m\mathbf{v} \qquad [4.1]$$

Newton's Second Law says that

$$\mathbf{F}_{av} = \frac{\Delta\mathbf{p}}{\Delta t} = \frac{\Delta(m\mathbf{v})}{\Delta t} \qquad [4.2]$$

and

$$1 \text{ N} = 1 \text{ kg} \cdot \text{m/s}^2$$

Alternatively, *the change in the momentum of a body equals the net applied impulse,* thus

$$\mathbf{F}_{av}\,\Delta t = \Delta\mathbf{p} \qquad [4.3]$$

The **Third Law** maintains that two objects interact such that for every force exerted on one, there is an equal and oppositely directed force exerted on the other.

The law of **Conservation of Linear Momentum** states that *when the resultant of all the external forces acting on a system is zero, the momentum of the system remains constant* (p. 112). Conservation of Linear Momentum arises from the homogeneity of empty space.

Suggestions on Problem Solving

1. Begin problem solving with a sketch. Watch out for the signs. Take the direction of initial motion as positive and stick with it for the whole analysis. Put the signs in your diagram at the outset.

2. When adding force vectors (Appendix D), keep in mind that the x- and y-axes may not be the simplest choice of directions along which to resolve your vectors. Watch for symmetries that tell you force components are equal or perhaps cancel. Two equal and oppositely directed vectors cancel each other no matter what the angle of their common line-of-action is.

3. Do not mix up 32.2 ft/s² and 9.81 m/s²—that is, be sure to use the proper numerical value of g for any given set of units in a problem. Always work in SI units. Most data tabulated

for common consumption in this country are still in U. S. Customary units—you should learn how to deal with them.

4. Think in terms of Conservation of Momentum whenever you have and want masses and speeds before and after some interaction. Newton's laws will usually provide an alternative means of solving such problems, and either approach can be used to check the other.

5. The more experience you have, the further along you will go in the analysis before substituting in numbers. It makes no sense to multiply by some number only to find that you have to divide by it later on. Also, do not forget to square the quantity when you have a problem involving t^2 or v^2.

Discussion Questions

1. In a science-fiction TV show several years ago, the hero spent the entire 40 min fretting about running out of fuel before he could pilot his spaceship back to base. Time and again the camera cut from his worried face to the blasting rockets, to the trembling fuel gauge and back to his pained expression as the engines roared and the fuel diminished. What should he have done and why would you say the writers were Aristotelians?

2. A car stopped at a red light is hit in the rear by another car. Describe what will happen to the two drivers as a result of the impact. What kind of injuries might each driver sustain?

3. Some seat belts have a movable chest strap that locks whenever the brakes are hit sharply while the car is in motion. The belt takeup spool has a tooth wheel on it that is engaged by the top of a little pendulum whenever the latter swings forward. Describe how the First Law rules this arrangement.

4. If, while having lunch aboard a plane flying at 600 km/h, you happen to drop a pea right above your plate, where will it land? Explain. Is that why first class is in the front of the plane?

5. Suppose a hunter fires a blast of No. 6 shot straight up into the air. Given that the pellets have a terminal speed of about 9 m/s, describe the flight of a typical shot and comment on the danger to the hunter. The terminal speed of a .22-caliber bullet (nose first) is about 67 m/s. What does that suggest about shooting into the air anywhere near people? (See the footnote on p. 80 and comment on the physics.)

6. The bombardier in the World War II B-17 shown in Fig. Q6 has evidently just dropped a "stick" of bombs. Explain why they form a nice neat column in the picture.

Figure Q6

7. During the First World War (1918), the Germans surprised the French by bombarding Paris from 115 km away with a super-long-range gun, the *Kaiser Wilhelm Geschütz.* The 120-kg shells, fired with a muzzle speed of 2 km/s, took 3.5 min to reach the city. They were launched at an elevation of 52° and climbed to a height of 40 km. The Germans had found the technique of extending the range of cannon fire quite by accident. What had they discovered that made the bombardment possible?

8. If you dive into the water and land flat on your chest in what is referred to in some circles as a "belly whop," you are likely to remember it for a while. Use the Second Law to explain what happens.

9. Hold a ball high in your right hand and position your left hand a few feet directly below it. Now walk with a constant velocity and, somewhere along the way, drop the ball without breaking stride. Where will it land? Where will the ball land if, just as you drop it, you begin to decelerate? Try it.

10. Suppose you are riding a bike at a constant leisurely rate. Is it possible to bounce a ball (thrown straight downward) off the ground and catch it coming straight up, on the move? Draw a diagram of the ball's trajectory as seen by a stationary bystander as you ride past.

11. It is fairly easy to catch a 20-N steel weight dropped from a height of a $\frac{1}{4}$ m or so. But if you attempt to catch the weight while resting your hand flat on a table, you are likely to be injured. Explain.

12. Why should a batter pay more attention to follow-through when hitting a softball as opposed to a hardball? Every now and then a player will hit a ball so hard that the bat itself shatters. Explain how this happens in terms of Newton's laws. Drop a ball and watch it hit the floor and rebound back upward. According to the First Law, there must have been an externally applied upward force on the ball. What was it? Explain.

13. Which would be more difficult to throw at a given speed while hovering in space: a "weightless" tennis ball or a "weightless" bowling ball? Explain. Can astronauts floating in orbit tell which objects around them would be heavy or light on Earth even though everything is effectively weightless? How? Suppose that while working out in space an astronaut accidentally hits her thumb with a "weightless" hammer. Will it hurt?

14. A human being can survive a feet-first impact at speeds up to roughly 12 m/s (27 mi/h) on concrete; 15 m/s (34 mi/h) on soil; and 34 m/s (75 mi/h) on water—explain the spread in these values.

15. There is a classic demonstration in which a heavy mass is suspended from a fairly fine thread and another length of the same thread is attached to its underside and allowed to hang beneath it. Someone then gives a sharp tug downward on the dangling thread. Which string will break and why? What would happen if the tug were gradual and strong? Explain.

16. Which stings the hand more (if either)—catching a ball in flight while running toward it or away from it? Explain.

17. What would happen to two astronauts if, while floating stationary with respect to one another, they played catch with a baseball?

Multiple Choice Questions

1. If we represent the dimensions of mass, length, and time by M, L, and T, respectively, then the dimensions of impulse are   (a) $[ML/T]$ (b) $[ML^2/T^2]$ (c) $[ML/T^2]$ (d) $[LT/M]$ (e) none of these.

2. The blue team in the Pentagon's annual charity tug-of-war is being dragged toward the losing line by a superior red team. The eight blue pullers are each hauling horizontally with 400 N, but the eight red players are each tugging with 460 N. You can save the day for the blue team by joining them and pulling with a force: (a) of 400 N   (b) in excess of 480 N   (c) of 460 N   (d) of greater than 400 N but less than 460 N   (e) none of these.

3. Two identical rockets are dropped simultaneously from a plane. One is faulty and does not ignite at all; the other fires horizontally for 1 ms and then shuts off.   (a) They hit the ground at about the same time.   (b) The faulty one hits the ground first.   (c) The one that fired sped up and so hits the ground first.   (d) Not enough information is given to reach a conclusion.   (e) None of these.

4. Figure MC4 shows a ball rolling down a curved incline. Which diagram depicts the resulting motion correctly? (a), (b), (c), (d), or (e) none of these.

(a)

(b)

(c)

(d)

Figure MC4

5. Suppose that while walking in a straight line at a uniform speed, you throw a ball straight up about 2 or 3 m without breaking stride. Where will it land?
(a) behind you   (b) in front of you   (c) right back in your hand   (d) to the right of you   (e) none of these. Try it!

6. Figure MC6 depicts a spiral tube lying at rest horizontally on a table, as viewed from above. A ball is rolled into one end of the tube and emerges from the other end. Ignoring friction, which diagram correctly represents the motion: (a), (b), (c), (d), or (e) none of these.

7. Suppose a projectile's speed and mass are both doubled. Its momentum will then be   (a) doubled   (b) unchanged   (c) quadrupled   (d) none of these.

8. A 1.0-kg body at rest experiences a force of 5.0 N exerted in the positive *x*-direction for 2.0 s, followed by a force of

(a)   (b)   (c)   (d)

Figure MC6

10 N exerted in the negative *x*-direction for 1.0 s. Its resulting speed will be   (a) +10 m/s   (b) −10 m/s   (c) +20 m/s   (d) −20 m/s   (e) none of these.

9. Refer to the force-time curve for an object of constant mass *m* in Fig. MC9. When was the object at rest?   (a) 0 s to 1 s   (b) 4 s to 5 s   (c) 2 s to 3 s   (d) 3 s to 4 s   (e) not enough information given.

Figure MC9

10. When did the object in Fig. MC9 have a constant velocity?
(a) 3 s to 4 s   (b) 2 s to 3 s   (c) 4 s to 5 s   (d) 5 s to 6 s
(e) not enough information given.

11. Which interval in Fig. MC9 corresponds to the greatest change in the speed of the body?   (a) 0 s to 1 s   (b) 1 s to 2 s   (c) 2 s to 3 s   (d) 3 s to 4 s   (e) 5 s to 6 s.

12. During which time interval in Fig. MC9 did the body decelerate?   (a) 0 s to 1 s   (b) 2 s to 3 s   (c) 3 s to 4 s (d) 5 s to 6 s   (e) none of these.

13. Light carries momentum so when a beam impinges on a surface, it will exert a force on that surface. If the light is reflected rather than absorbed, the force will be   (a) less   (b) greater   (c) equal either way   (d) not enough information given   (e) none of these.

14. A car traveling at a certain speed undergoes a head-on collision with an identical vehicle traveling at the same speed, and both come to rest in 10 ms. Now suppose, instead, that the car crashes into a brick wall and again comes to a rest in 10 ms. From the perspective of the original driver, which would be less dangerous?   (a) the two-car crash   (b) the car-wall crash   (c) both are identical   (d) not enough information   (e) none of these.

15. A garbage truck crashes head-on into a Volkswagen and the two come to rest in a cloud of flies. Which experiences the greater impact force?   (a) the truck   (b) the Volkswagen   (c) both experience the same force   (d) not enough information given   (e) none of these.

16. Two equal-mass bullets traveling with the same speed strike a target. One of the bullets is rubber and bounces off: the other is metal and penetrates, coming to rest in the target. Which exerts the greater impulse on the target?

(a) the rubber bullet   (b) the metal bullet   (c) both exert the same   (d) not enough information   (e) none of these.

17. An open railroad car filled with coal is coasting frictionlessly. A girl on board starts throwing the coal horizontally backward straight off the car, one chunk at a time. The car   (a) speeds up   (b) slows down   (c) first speeds up and then slows down   (d) travels at constant speed   (e) none of the above.

18. A tank car coasting frictionlessly horizontally along the rails has a leak in its bottom and dribbles several thousand gallons of water onto the roadbed. In the process it   (a) speeds up   (b) slows down   (c) gains momentum   (d) loses momentum   (e) none of the above.

19. A bomb hanging from a string explodes into pieces of different sizes and shapes. After the explosion   (a) the vector momentum of each piece is identical   (b) the total momentum is increased   (c) the momentum of all the pieces, exhaust, and smoke adds up to zero   (d) not enough information to comment   (e) none of the above.

20. A can of whipped cream floating in space develops a hole in the bottom from which it squirts backward a mess of gas and cream at a constant speed with respect to the can. The can thereupon   (a) accelerates forward throughout the squirting   (b) moves forward at a constant speed   (c) remains at rest   (d) first speeds up, and then slows down when the gas runs out   (e) none of the above.

Problems

THE LAW OF INERTIA

1. [I] A youngster sitting in a bus gently throws a ball straight up at a speed of 1.00 m/s. The bus is traveling at a constant rate of 10 m/s and it passes someone sitting on a fence. With what velocity does that person outside see the ball moving when it reaches maximum altitude?

2. [I] The hero in a Western movie is standing on top of a train that is moving due north at 30 km/h, while a pack of bad guys chases in hot pursuit. Cleverly, the hero tosses a lit bomb straight up at a speed of 9.8 m/s. When and where will it land (neglecting air friction)?

3. [I] Marc Antony, stretched out on a couch, is waiting, mouth upward and open. Cleopatra, carrying a bunch of grapes, is dashing across the palace floor straight toward him at a speed of 2.213 6 m/s. She holds one grape 1.000 0 m above his face. How far from him should she release her projectile if it is to land in the middle of his mouth? Neglect air friction and take the acceleration of gravity at the palace to be 9.800 0 m/s^2.

4. [I] A girl running at a constant speed of 2.0 m/s in a straight line throws a ball directly upward at 4.9 m/s. How far will she travel before the ball drops back into her hands? Ignore air friction.

5. [I] When a 12-bore shotgun using No. 6 shot is fired straight up, the pellets attain a height of about 110 m. How does this situation compare with the ideal no-friction approximation, given that the muzzle speed is 405 m/s? Explain the difference.

6. [I] If you use the same setup as in Problem 5 but fire horizontally from a height of 1.22 m (with no air drag), what maximum horizontal distance should the pellets reach? (In actuality, they travel about 85 m.)

7. [I] What is the ideal maximum range you can expect from the shotgun in Problem 5? Maximum range is more nearly 200 m and is attained by firing at a reduced elevation of about 33°. Explain why this might be the case.

8. [I] Calculate your Earth-weight in newtons and then determine your mass in kilograms. What is the combined mass of $\frac{1}{2}$ lb of hard salami, 3 oz of Swiss cheese, 1 oz of mustard, all on a 4-oz roll "fresh this morning" on Earth?

9. [I] What is the force exerted on the wall by the person in Fig. 4.6a (p. 96)?

10. [I] What is the net force exerted on the scale in Fig. 4.6a?

11. [I] What is the net force exerted on the person on the left by the person on the right in Fig. 4.6d?

12. [I] There are five forces acting on the truck in Fig. P12. What is the net external force on it?

Figure P12

13. [I] The two ropes attached to the hook in Fig. P13 are pulled on with forces of 100 N and 200 N. What size single force acting in what direction would produce the same effect?

Figure P13

14. [I] The scale in Fig. P14 is being pulled on via three ropes. What net force does the scale read?

15. [I] What is the net force acting on the ring in Fig. P15?

16. [I] What is the net force acting on the ring in Fig. P16?

17. [II] A baseball is hurled at 30.7 m/s at an angle up from the ground of exactly 45°. Neglecting air friction, compute

Figure P14

Figure P15

Figure P16

(a) its peak altitude, and (b) how far downrange it will be when it returns to the height at which it was thrown. In actuality, because of air friction, the ball will rise to only about 20 m and return to launch level less than 70 m downrange, descending more steeply than it was thrown by about 10° (see Fig. 4.5, p. 95).

18. [II] According to the manufacturer, a 256 Winchester Magnum cartridge shot by a revolver with an 8-in. barrel results in a bullet with a muzzle speed of 2200 ft/s. Fired horizontally, that bullet drops 0.90 in. after traveling 50 yd, 4.4 in. after 100 yd, and 10.9 in. after 150 yd. Compare these figures with those obtained theoretically, neglecting aerodynamic effects. Make a plot of both sets of values. What's happening?

19. [II] Here's a classic: Suppose a conservationist points a dart gun directly at a monkey in a tree. Show that if the monkey drops from its branch just when it sees the gun go off, it will still get hit.

20. [II] Write expressions for the net horizontal and vertical forces acting on the block in Fig. P20.

Figure P20

21. [II] What is the net force acting on the ring in Fig. P21? Try to solve this one in an elegant way.

Figure P21

22. [II] Determine the net force acting on the ring in Fig. P22. Again, try to do this one elegantly.

23. [II] A ring is fixed via a harness to a block of stone and three ropes are attached to the ring. Each rope is to be pulled in the same horizontal plane with a force of 2 kN. How should they be arranged so as to produce a net force due east of 4 kN on the block?

24. [III] Figure P24 shows three people pulling down via ropes, each with 200 N, on the top of a mast 20 m tall. If

Figure P22

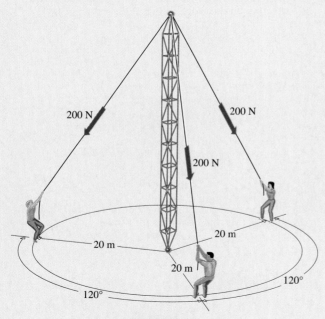

Figure P24

they stand at equal distances of 20 m from the base of the mast along lines 120° apart, what is the net force they exert on the mast?

25. [III] Four firefighters hold a square net, one at each corner. Each person exerts a force of 200 N whose line-of-action passes through a point just below the center of the net and makes an angle with the vertical of 60°. What is the net force the firefighters exert on the net?

THE SECOND LAW
INTERACTIONS

26. [I] Compute the mass of your head in kilograms, given that its Earth-weight in pounds is *very roughly* equal to (0.028 F_W + 6.35), where F_W is your total body weight in pounds.

27. [I] From how high (in meters) must a car fall if it's to have the same momentum it would have driving at 60 mi/h?

28. [I] What is the momentum of the Earth ($m_{\oplus}$ = 5.975 × 10^{27} g) as it moves through space if its orbital speed is about 6.66 × 10^4 mi/h? (SI units please.)

29. [I] A 1.0-kg wad of clay is slammed straight into a wall at a speed of 10 m/s. If the clay sticks in place, what was the impulse that acted on it via the wall?

30. [I] A dried pea fired from a plastic drinking straw has a mass of 0.50 g. If the force exerted on the pea is an average 0.070 lb over the 0.10-s flight through the straw, at what speed will it emerge?

31. [I] A 1.0-kg body initially traveling in the positive *x*-direction at 10 m/s is acted upon for 2.0 s by a force in that same direction of 20 N. It then experiences a force acting in the negative direction for 20 s equal to 2.0 N. Draw a force-time curve and determine the final momentum.

32. [I] A 10-kg asteroid is traveling through space at 2 m/s toward a spaceship. To avoid a collision, an astronaut, with the help of a backpack rocket, exerts a force of 20 N on the asteroid in the opposite direction for 5 s.   (a) Find the asteroid's final speed.   (b) How long should she have pushed in order to just stop the asteroid?

33. [I] A rocket engine blasts out 1000 kg of exhaust gas at an average speed of 2 km/s every second. Calculate the resulting average thrust developed by the engine.

34. [I] Figure P34 shows the force-time curves for two idealized blows delivered during a martial arts contest. Which of the two is likely to correspond to a boxer's punch as opposed to a karate chop? Compute an approximate value for the corresponding impulse in each case. Which is more likely to break bones, and why?

Figure P34

35. [I] If a 0.061-kg handball goes from rest to 20 m/s during a serve and the player feels an average force of 100 N, how long did the impact last?

36. [I] A 1-kg hammer slams down on a nail at 5 m/s and bounces off at 1 m/s. If the impact lasts 1 ms, what average force is exerted on the nail?

37. [I] A 50-kg person at the southernmost end of a 150-kg rowboat at rest in the water walks to the northern end. If, at a given instant, her speed with respect to the water is 10 m/s, what is the velocity of the boat with respect to the water at that moment?

38. [I] Water squirting from a hose emerges with a speed of 10 m/s at a rate of 100 kg/s. Compute the average reaction force exerted on the hose. Have you ever seen firefighters working to control a large hose?

39. [I] A 60-kg cosmonaut jumps from her 5000-kg spaceship in order to meet her copilot who is floating at rest a few hundred meters away. If she sails toward him at a speed of 10 m/s, what is the resulting motion of the ship?

40. [II] A golf ball with a mass of 47.0 g can be blasted from rest to a speed of 70.0 m/s during the impact with a clubhead. Taking that impact to last only about 1.00 ms, calculate the change in momentum of the ball and the average force applied to it.

41. [II] A bullet is fired from a handgun with a barrel 24.0 cm long. Its muzzle speed is 350 m/s and its mass is 6.00 g. Compute the average force exerted on the bullet by the expanding gas in the barrel.

42. [II] During a particular rocket-powered sled run, Colonel J. P. Stapp of the U.S. Air Force decelerated from a speed of 940 ft/s to a dead stop in 1.40 s. Assuming he weighed 175 lb, compute the average force, in newtons, exerted on him. Incidentally, the primary dangers in this sort of test are things like blood vessels tearing away from the lungs or the heart, and even retinal detachment, via the First Law.

43. [II] The human heart pumps about 2 ounces of blood into the aorta on each stroke, which lasts roughly 0.1 s. During that time, the pulse of blood is accelerated from rest to about 50 cm/s. Compute the corresponding average propulsion force exerted on the blood by the heart.

44. [II] Suppose a car stopped on the road is hit from behind by a bus so that it accelerates up to 4.47 m/s (10 mi/h) in 0.10 s. If the driver of the car has a mass of 50 kg, and her front-seat passenger has a mass of 80 kg, what average force will the seat have to exert on them? (Car seats have been known to collapse under this sort of treatment, though they obviously shouldn't.)

45. [II] In Problem 44, how much force must the driver's neck exert to keep her head in line with her body during the collision, if her head weighs 44.5 N (10 lb)? Think about "whiplash."

46. [II] A person can just survive a full-body collision (either to the front, back, or side) at roughly 9 m/s (20 mi/h) with an impact time of approximately 10 ms. At greater speeds or shorter times, fatal brain damage will likely occur. Could someone survive a fall from 4.0 m landing flat on his back on soft soil, so that he decelerates to rest through a distance of 10 cm (that's the total compression of body and soil)? If his mass is 70 kg, what's the impulse exerted on his body by the ground? Assume the deceleration is constant.

47. [II] Suppose that a person of mass m jumps off a ladder at a height s_h and lands on the ground without bouncing. If the total compression of the body and the soil during impact is s_c, and if the deceleration is assumed constant, show that the force exerted by the ground is given by

$$F_{av} = mg\left(\frac{s_h}{s_c}\right)$$

Notice that bending the knees extends s_c considerably, decreasing F_{av} accordingly.

48. [II] During a 40 mi/h car crash, a 50-kg passenger comes to rest in 100 ms. What is the area under the force-time curve for that person? Assuming a seat belt was worn, which would smooth out the curve so there were no devastating peaks; what average force would be experienced?

49. [II] When a person jumps and lands stiff-legged on the heels of the feet, a considerable force can be exerted on the long leg bones. The greatest stress occurs in the tibia, or shin bone, a bit above the ankle where the bone has its smallest cross section. If a force in excess of about 50 000 N is applied upward on the heel, the tibia will probably fracture. Keeping in mind the results of Problem 47, what is the minimum height of a fall above which a 60-kg person is likely to suffer a tibial fracture? Assume the body decelerates uniformly through a distance of 1.0 cm and the landing occurs squarely on both feet.

50. [II] An ion engine is a rocket designed to produce small thrusts for very long periods of time. The SERT 2 engine develops 30 mN of thrust by expelling mercury ions at speeds of around 22 000 m/s (about 50 000 mi/h). How many kilograms of exhaust emerge per second?

51. [II] A modern nuclear aircraft carrier weighs in at around 90 000 tons and is capable of traveling in excess of 30 knots (15.4 m/s). Suppose one of these giants plowed into a pier at 30 knots, coming to rest in 0.50 min, what average force would it exert in the process?

52. [II] A beam consisting of around 1.0×10^{15} electrons per second, each of mass 9.1×10^{-31} kg, paints pictures on the face of a TV tube. Assuming that the electrons are absorbed by the screen and exert an average net force on it of 1.0×10^{-7} N, what is their average speed at impact?

53. [II] A bullet fired into wet clay will decelerate fairly uniformly. If a 10-g slug hits a block of clay at 200 m/s and comes to rest in 20 cm, what average force does it exert on the block?

54. [II] Imagine a car involved in a head-on crash. The driver, whose mass is m, is to be brought uniformly to rest within the passenger compartment by compressing an inflated air bag through a distance s_c. Write an expression for the average force exerted on the bag in terms of m, v_i, and s_c. Compute that average force for a 60 mi/h collision, where the driver's mass is 60 kg and the allowed stopping distance in the bag is 30 cm. Assume the car deforms only negligibly.

55. [II] Suppose a 6.00-g bullet traveling at 100 m/s strikes a bulletproof vest and comes to rest in about 600 μs. What average force will it impart to the happy wearer?

56. [II] A 2785-lb Triumph TR-8 traveling south at 40 km/h (25 mi/h) crashes head-on into a 3840-lb Checker cab moving north at 90 km/h (56 mi/h). If the two cars remain tangled together but free to coast, describe their motion immediately after the collision. What is their final speed?

57. [II] A railroad flatcar having a mass of 10 000 kg is coasting along at 20 m/s. As it passes under a bridge, 10 men (having an average mass of 90 kg) drop straight down onto the car. What is its speed as it emerges with its new passengers from beneath the bridge?

58. [II] A 90-kg astronaut floating out in space is carrying a 1.0-kg TV camera and a 10-kg battery pack. He's drifting toward his ship but, in order to get back faster, he hurls the camera out into space at 15 m/s and then throws the battery at 10 m/s in the same direction. What's the resulting increase in his speed after each throw?

59. [II] An ice skater (with a mass of 55.0 kg) throws a snowball while standing at rest. The ball has a mass of 200 g and moves straight out with a horizontal speed of 20.0 km/h. Neglecting friction, and assuming the skate blades are parallel to the direction of the throw, describe the skater's resulting motion in detail.

60. [II] A 4265-lb Jaguar XJ12L requires 164 ft of runway as a minimum stopping distance from a speed of 60.0 mi/h. Assuming its deceleration is uniform, compute the average stopping force exerted on the car.

61. [II] An F-14A jet weighing 70 000 lb has a typical lift-off distance of 3000 ft and a corresponding lift-off speed of 135 knots (69.5 m/s). Assuming its acceleration is constant, calculate the net average thrust of its engines in newtons.

62. [III] Navy jets are hurled off the deck of a modern carrier by a combination of catapult and engine thrust. The catapult (the C-7, for example) has a 250-ft stroke (that is, runway) and will yank a 70 000-lb F-14A jet, with its engines developing a net average thrust of 16 000 lb, from rest to a speed of 200 ft/s in just 2.4 s.   (a) Compute the average force exerted by the catapult.   (b) Is the acceleration of the plane uniform? (Hint: Try several of the equations for constant a.)

63. [III] During a parachute jump over Alaska in 1955, a United States trooper jumped from a C-119 at 1200 feet, but his chute failed to open. He was found flat on his back at the bottom of a $3\frac{1}{2}$-ft deep crater in the snow, alive and with only an incomplete fracture of the clavicle. Compute the average force that acted on him as he plowed into the snow. Assume the deceleration was constant, take his mass to be 90 kg and his terminal speed to be 120 mi/h.

64. [III] A typical tubular aluminum arrow 70 cm long has a mass of 25.0 g. The force exerted by a bow varies in a fairly complicated fashion from the initial draw force of, say, 175 N (39 lb) to zero, as the arrow leaves the string. Assume a full draw (70 cm) and a launch time of 16.0 ms. If the arrow when fired straight up reaches a maximum

The catapult of a modern aircraft carrier hurls planes off the deck.

height of 143.3 m above the point of launch, then
(a) what is the average force exerted on it by the bow?
(b) What's the area under the force-time curve?   (c) What impulse is applied to the shooter?

65. [III] Two astronauts floating at rest with respect to their ship in space decide to play catch with a 0.500-kg asteroid. Neil (whose mass is 100 kg) heaves the asteroid at 20.0 m/s toward Sally (whose mass is 50.0 kg). She catches it and heaves it back at 20.0 m/s (with respect to the ship). Before Neil catches it a second time, how fast is each person moving, and in what direction?

5

Dynamics: Force and Acceleration

WHEN ALL THE EXTERNAL forces exerted on a system act in such
a way that there is no change in momentum, no acceleration, the system is in
equilibrium—there is a balance, the study of which is called *statics* (p. 163).
Alternatively, the discipline dealing with the nonequilibrium behavior of objects is *dynamics*. A body experiencing a net force will accelerate in the direction of that force. It is the relationship between various kinds of forces
and the resulting accelerations that will be examined in this chapter.

Figure 5.1 The application of a net force F_{av} for a time Δt results in a change in momentum and an average acceleration, $a_{av} = (v_f - v_i)/\Delta t$.

FORCE, MASS, AND ACCELERATION

There is an alternative formulation of Newton's Second Law that, although less general, is very useful. Recall that the average force on an object equals its time-rate-of-change of momentum (Fig. 5.1), expressed as

$$\mathbf{F}_{av} = \frac{\Delta \mathbf{p}}{\Delta t} = \frac{\Delta(m\mathbf{v})}{\Delta t} \qquad [4.2]$$

The change in its momentum $\Delta \mathbf{p} = (\mathbf{p}_f - \mathbf{p}_i)$ equals $(m_f\mathbf{v}_f - m_i\mathbf{v}_i)$, but *if we take the mass of the object to be constant* $m_i = m_f = m$, then $\Delta \mathbf{p} = m(\mathbf{v}_f - \mathbf{v}_i) = m\,\Delta\mathbf{v}$. Consequently, Eq. (4.2) becomes

$$\mathbf{F}_{av} = \frac{m\,\Delta \mathbf{v}}{\Delta t} \qquad (5.1)$$

With Eq. (3.4), $\mathbf{a}_{av} = \Delta\mathbf{v}/\Delta t$, for the average acceleration, the average force becomes

$$\mathbf{F}_{av} = m\mathbf{a}_{av} \qquad (5.2)$$

An average force of 1.0 N will cause a mass of 1.0 kg to accelerate at an average rate of 1.0 m/s², and as we saw before (p. 104), 1 N = 1 kg·m/s². Table 5.1 lists the units of force in the several systems that are likely to be encountered.

The force in Fig. 5.1 is applied to a rope and not directly to the object. A rope, chain, cable, or tendon transmits the applied force to the object via electromagnetic interactions between its intervening atoms. Force is a vector quantity, it acts along a line, and the line of the taut rope is the **line-of-action** of the force. Because that force tends to stretch the rope, it is known as a **tensile force**. Imagine that we cut the rope and splice in a spring scale of negligible mass. The scale will read out the magnitude of the applied tensile force. That scalar quantity is called the **tension** (see Fig. 4.6, p. 96).

Equation (5.2) provides a means of computing average quantities when you know the overall change in motion but not the moment-by-moment details. If we can measure the initial and final speeds, we can compute the average accelerating forces acting on baseballs, cars, rockets, whatever.

Example 5.1 A baseball (0.142 kg) leaving a pitcher's hand at a speed of 45.7 m/s is clocked using gun-radar. If the throw lasted 0.12 s, what was the magnitude of the average force exerted on the ball?

Solution: [Given: $m = 0.142$ kg, $v_i = 0$, $v_f = 45.7$ m/s, and $\Delta t = 0.12$ s. Find: F_{av}.] First determine the average acceleration using

$$a_{av} = \frac{\Delta v}{\Delta t} = \frac{+45.7 \text{ m/s}}{0.12 \text{ s}} = 380.8 \text{ m/s}^2$$

From Eq. (5.2)

$$F_{av} = (0.142 \text{ kg})(380.8 \text{ m/s}^2) = \boxed{54 \text{ N}}$$

This quantity is about 12 lb, which is reasonable given that it corresponds to an average force (see Fig. 4.12).

▶ **Quick Check:** The impulse, $F_{av}\,\Delta t$, is about 6.5 N·s, as is Δp, so it looks okay.

TABLE 5.1   Units of force, mass, and acceleration in various systems

	SI	cgs*	U.S. Customary units
Force	$1 \, N \equiv 1 \, kg \cdot m/s^2$	$1 \, dyne \equiv 1 \, g \cdot cm/s^2$	$1 \, lb$
Mass	$1 \, kg$	$1 \, g$	$1 \, slug^{\dagger} \equiv 1 \, lb \cdot s^2/ft$
Acceleration	$1 \, m/s^2$	$1 \, cm/s^2$	$1 \, ft/s^2$

*The centimeter-gram-second system, which we will avoid, uses the dyne as the unit of force.
$1 \, N = 10^5$ dynes $= 0.225$ lb.
†Although the slug is still used in engineering as the unit of mass, we will make no use of it.

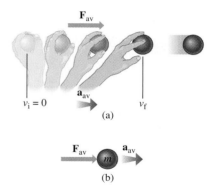

(a)

(b)

Figure 5.2   A ball being launched out in space where the only force acting on it is the pushing hand. (a) The average force accelerates the ball from $v_i = 0$ to v_f at an average rate $\mathbf{a}_{av}$. (b) Displayed is a free-body diagram of the hand pushing the ball. Notice how $\mathbf{a}_{av}$ is in the direction of $\mathbf{F}_{av}$, which in this case is in the direction of motion.

As we progress in our analysis, we will encounter situations in which several forces act on a body in different directions. Even though things can get complicated, there is a wonderful conceptual device called a **free-body diagram** that will help. To create one, isolate the object of concern from the rest of the system; that is, (1) remove anything in contact with it and (2) replace each such source of interaction, be it a pushing hand or a pulling string, by a force vector. Following these procedures leaves us with a picture of the object all by itself with a bunch of force vectors acting on it. Figure 5.2a shows a ball being launched in space, where the only force acting arises from the interaction with the thrower. The free-body diagram (Fig. 5.2b) is the simplest possible one, namely a single force, accelerating a single mass.

Example 5.2   An old Checker cab weighing 17.08 kN cruises along a level road at 35.8 m/s. The driver lets the car coast in neutral, and air drag decelerates it at a non-constant rate down to 22.4 m/s in 24 s. (a) Calculate the average deceleration during the period. (b) Determine the average retarding force acting on the cab.

Solution: [Given: $F_W = 17.08$ kN, $v_i = 35.8$ m/s, $v_f = 22.4$ m/s, and $\Delta t = 24$ s. Find: a_{av} and F_{av}.] Figure 5.3a shows the cab moving to the right, with an initial speed of 35.8 m/s. Ignoring gravity, Figs. 5.3b and c are the free-body diagrams showing the average drag force acting to the left. It opposes the motion and slows the vehicle. (a) From the definition

$$a_{av} = \frac{\Delta v}{\Delta t} = \frac{(22.4 \text{ m/s} - 35.8 \text{ m/s})}{24 \text{ s}} = -0.558 \text{ m/s}^2$$

(a)

(b)

(c)

Figure 5.3   (a) An old cab coasting down from a speed v_i to a speed v_f. Air drag produces an average force $\mathbf{F}_{av}$ and the cab decelerates. Overlooking vertical forces for the moment, (b) is a free-body diagram of the cab. Notice how $\mathbf{a}_{av}$ is in the same direction as $\mathbf{F}_{av}$ but opposite to the motion. (c) When we begin to do a problem, the free-body diagram can be simple.

(continued)

(continued)

or $\boxed{-0.56 \text{ m/s}^2}$. (b) Consequently

$$F_{av} = ma_{av} = (F_W/g)a_{av} = (1741.7 \text{ kg})(-0.558 \text{ m/s}^2)$$

and so $F_{av} = -971.9$ N or $\boxed{-0.97 \text{ kN}}$. The minus sign indicates deceleration.

▶ **Quick Check:** From Eq. (4.3), $F_{av}\Delta t = \Delta p$, (0.97 kN)(24 s) = 23.3 kN·s, while $\Delta p = (1742 \text{ kg}) \times$ (13.4 m/s) = 23.3×10^3 kg·m/s.

5.1 Instantaneous Motion

It is possible to go from average force to instantaneous force in the following way. Return to Eq. (5.1), $\mathbf{F}_{av} = m\,\Delta\mathbf{v}/\Delta t$, and take the limit of both sides as $\Delta t \to 0$. On the left, the average force determined over an infinitesimal interval is the *instantaneous force* (**F**), via Eq. (4.4). On the right, we have the product of the constant mass (*m*) and the *instantaneous acceleration* (**a**), via Eq. (3.3). At each and every moment, the *net force applied to a body equals the mass of the body multiplied by the resulting acceleration,* that is

$$\mathbf{F} = m\mathbf{a} \qquad (5.3)$$

Evidently, the net applied force **F** and the resulting acceleration **a** are always parallel.

Remember that **F** represents the *net externally applied force.* It's helpful to write the equation as

$$\Sigma\mathbf{F} = m\mathbf{a} \qquad (5.4)$$

where Σ (Greek capital sigma) stands for "the sum of" and reminds us that we are dealing with the vector sum of all the forces acting on the body. The equation is true in all directions, and **F** can be resolved into components along any three perpendicular axes, although we restrict ourselves to the simpler case of motion in a plane. The ball in Fig. 5.4 moves in the *xy*-plane, so $\Sigma F_x = ma_x$ and $\Sigma F_y = ma_y$.

If, while in space (Fig. 5.5), we push on a floating weightless $\frac{1}{2}$-kg mass with a constant 1-N force, it will immediately accelerate in the direction of the force, at a rate of 2 m/s² *all the while the force acts.* Similarly, if we push on a 100-kg refrigerator (Fig. 5.6) here on Earth, with a horizontal force of 100 N (and if all other influences such as friction are negligible), away it will go with an acceleration of 1.00 m/s².

When the sum of the external horizontal forces acting on an object is zero (that is, $\Sigma F_H = 0 = ma_H$), as in Fig. 5.7, there will be no horizontal acceleration ($a_H = 0$). The two people and the "fridge" are not necessarily at rest (they may be on a moving sidewalk); they are just not accelerating.

A system may have several interacting parts that push and pull on each other with *internal* forces that do not affect the overall motion. The reason you cannot grab your belt, pull up, and fly, is that you and the belt are the system and the forces exerted are all internal. Your hand pulls up on your belt and your belt pulls down on your hand—that's an interaction pair. Meanwhile, your belt pulls up on your waist and your waist pulls down on your belt—that's an interaction pair. And all the forces are numerically equal. Hence, the net force on the belt is zero, and the net force on you is zero—you go nowhere. You have to get out and push if you want to

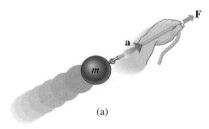

Figure 5.4 (a) An object accelerating in a straight line under the influence of a force **F**. (b) The motion can be thought of as occurring along two perpendicular axes. Newton's Second Law applies independently along both axes.

Figure 5.5 Several examples of how **F** = *m***a** works. In each case the force—at the moment shown— produces a proportional instantaneous acceleration. For as long as the force is applied, the body accelerates in the direction of **F**.

Figure 5.6 (a) If the only horizontal force acting on a body of mass *m* is *F*, then it will accelerate horizontally such that *F* = *ma*. This motion is independent of the vertical forces that we'll deal with shortly. (b) The free-body diagram shows the net horizontal force.

Figure 5.7 (a) Two people apply equal, oppositely directed forces to an object. Because force is a vector quantity, these two forces cancel each other and ΣF_H = 0, which means there cannot be any horizontal acceleration. (b) The free-body diagram makes the same point more directly.

move a stalled car: inside, you push forward on the dashboard and it pushes you backward against the seat. Your back pushes to the rear of the car with the same force your hands push forward with—the net force you exert on the car is zero, and the net force it exerts on you is zero.

Example 5.3 The person in Fig. 5.8 pulls a loaded wagon having a total mass of 100 kg. A force of 100 N is applied through the handle at 30.0°. Ignoring friction, compute the horizontal force on the wagon and the re- sulting horizontal acceleration.

Solution: [Given: *m* = 100 kg, *F* = 100 N, and *θ* = 30.0°. Find: F_H and a_H.] The free-body diagram (Fig. 5.8b) shows **F**, acting up along the handle, resolved into two components, one vertical and one horizontal.

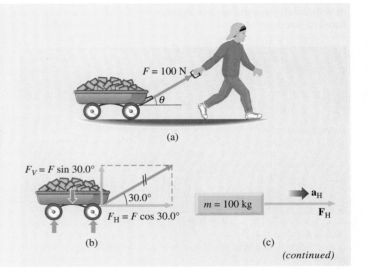

Figure 5.8 (a) The wagon experiences a force directed upward along the handle at an angle *θ*. (b) Its horizontal component F_H is the only horizontal force, and it will deter- mine the horizontal acceleration. (c) Ignoring the vertical forces for the moment, the free-body diagram is simple.

(continued)

(continued)

The guiding equation is

$$\overset{+}{\rightarrow}\Sigma F_H = ma_H$$

where, because the motion is to the right, we make that the positive direction (the arrow and + sign so indicate). Only one force produces the acceleration and it's the horizontal component of **F**. That is,

$$F_H = F \cos\theta = F \cos 30.0° = \boxed{+86.6 \text{ N}}$$

Thus $\overset{+}{\rightarrow}\Sigma F_H = +86.6 \text{ N} = ma_H$

and $\boxed{a_H = +0.866 \text{ m/s}^2}$

▶ **Quick Check:** The total 100-N force acting horizontally would produce an acceleration of 1.00 m/s², so the magnitude of this result is appropriate.

$F = ma$ is one of the most famous equations in science. To the modern physicist it's Newton's hallmark, his signature. Yet it does not appear in the *Principia*—and with good reason. This restricted formulation of the Second Law is the later work of Jacob Hermann (1716), refined by the Swiss mathematician Leonhard Euler. It's a tribute to Newton's insight that his analysis, framed in terms of momentum, carries over naturally into the more precise Theory of Relativity. The formula $F = ma$ is the low-speed approximation of a more accurate velocity-dependent expression. Still, Eq. (5.3) applies at ordinary speeds (much lower than the speed of light), and we will make good use of it.

5.2 Weight: Gravitational Force

Weight, anywhere near the Earth, is the downward force toward the center of the planet experienced by an object as a result of the Earth-object gravitational interaction. Because the Earth is revolving, a scale in your bathroom is actually moving in a circle and therefore accelerating (p. 140). Let's call the "weight" it reads, the **effective weight**. Effective weight is slightly lower than the true weight as defined above, but the differences are usually negligible.

The Second Law provides a relationship between the concepts of weight ($\mathbf{F}_W$) and mass (m). Because the object's Earth-weight is pulling down, causing it to accelerate at g, $\mathbf{F} = m\mathbf{a}$ becomes

$$\mathbf{F}_W = m\mathbf{g} \tag{5.5}$$

that is, the more mass, the more weight. Here, **g** is again the gravitational acceleration vector pointing downward. The weight of an object acting down numerically equals the force that must be exerted up on it (for example, by the floor or a scale) in order to keep the object from accelerating because of gravity (p. 71). The weight on Earth of a 2.0-kg chicken is just (2.0 kg)(9.8 m/s²) = 20 N. Conversely, an apple weighing 1 N on our planet has a mass of (1 N)/(9.8 m/s²) = 0.1 kg. That apple, anywhere in the Universe, has a mass of 0.1 kg, although its weight depends on the local gravity.

We can demonstrate some of these ideas using a quasipractical example. Suppose you meet a 54.4-kN (12 000-lb) bull elephant wearing ice skates and stranded in the middle of a frozen pond, and you decide to push him to

By pushing back on a fixed block the runner causes the block to push forward on her. It is this force that initially accelerates the athlete in the forward direction.

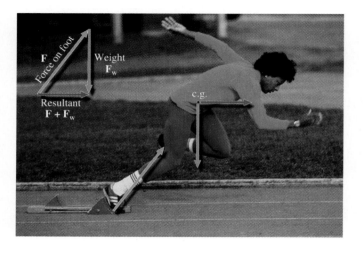

shore. Neglecting friction, how hard must you shove to uniformly accelerate the pachyderm from rest up to 6.7 m/s (15 mi/h) in 10 s? The required acceleration ($a = \Delta v/\Delta t$) is 0.67 m/s², which is not very much. Thus

$$F = ma = \frac{F_W}{g} a$$

and

$$F = \frac{(54.4 \times 10^3 \text{ N})}{9.81 \text{ m/s}^2}(0.67 \text{ m/s}^2) = 3.7 \text{ kN}$$

or 0.84×10^3 lb. Well, forget that idea. If you were even more ambitious and wanted the elephant to accelerate over the ice at $a = 9.8$ m/s² $= g$, a would cancel g in the equation and $F = F_W$. You would have to push with a force equal to the elephant's weight. That's true on a frictionless ice pond or out in space far from any celestial body where gravity is negligible and the object is essentially weightless—the needed force depends on the mass and the acceleration. The moral of this tale is that you *can* get squashed out in space between two colliding weightless elephants.

Weights on Ropes

Let's now examine the forces acting in the system depicted in Fig. 5.9a. A 1-kg brass mass on a rope hangs from a hook. Suppose we want to study the forces on both the rope and brass mass. Accordingly, draw a free-body diagram of each object (Figs. 5.9b and c). Doing so will allow us to write equations for those objects and thereby determine the forces. Gravity tugs on every atom of a body, but it will suffice (for the moment) to draw the overall weight vector acting straight down from the center of each object. The weight of the rope is F_{Wr}, and the weight of the brass mass is F_{Wb}. Gravity pulls down on the brass with F_{Wb}, and the brass pulls down on the rope (F_{br}). The rope, in turn, pulls up on the brass with an equal and opposite interaction (F_{rb}). The hook pulls up on the rope (F_{hr}), and the rope pulls back down on the hook (F_{rh})—another interaction pair ($F_{hr} = F_{rh}$). Applying the Second Law in the vertical direction to the brass object (Fig. 5.9b) and taking up as positive, we have

$$^{+\uparrow}\Sigma F_{Vb} = F_{rb} - F_{Wb} = m_b a_b = 0$$

and $F_{rb} = F_{Wb}$. Three forces act on the rope and, since it is not accelerating,

$$^{+\uparrow}\Sigma F_{Vr} = F_{hr} - F_{Wr} - F_{br} = 0$$

And with $F_{rb} = F_{Wb}$

$$F_{hr} = F_{Wr} + F_{br} = F_{Wr} + F_{Wb}$$

Not surprisingly, the hook (pulling on the rope) holds up all the weight. The tension at the top of the rope (F_{hr}) is greater than the tension on the bottom (F_{br}) by an amount equal to the weight of the rope. From now on, we shall deal only with ropes having negligible weight compared to the loads they support. In that case, *the tension (F_T) at any two points in a rope will be the same (provided no tangential forces act on the rope between those two points)*. Henceforth we can assume that **the tensile force exerted by any rope has a scalar value F_T and is directed along the rope's length toward its center.**

The Physics of Standing Still

The force exerted by a surface on an object is called the **reaction force**. It may have both a tangential component, known as the **force of friction** (**F**$_f$) and a perpendicular component (that is, normal to the surface) known as the **normal force** (**F**$_N$).

Figure 5.9 (a) A hook supporting a rope and a 1-kg brass mass. Suppose we wish to study the forces acting on both the mass and the rope. In that case, draw a free-body diagram of each object of interest, (b) and (c). The weights act down from the middles of the mass (**F**$_{Wb}$) and the rope (**F**$_{Wr}$). All the forces are colinear, and the Second Law applies to each separate object.

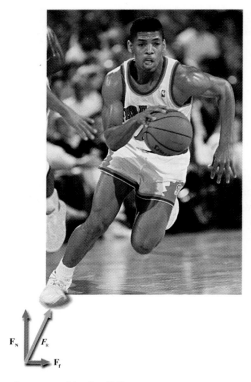

F_N F_R F_f

The floor pushes up and to the left with a reaction force F_R. It has components F_f (a friction force keeping his foot from sliding) and F_N (the normal force equal to his weight).

Both these so-called *contact forces* are due to the electromagnetic interaction between atoms of the surface and atoms of the object when they come into proximity. Let's first deal with the normal force, leaving friction for later. Because we live on a planet where gravity is appreciable, objects most often rest on horizontal surfaces and F_N usually points straight up (Fig. 5.10). That need not be the case: F_N must be perpendicular to the surface, but the latter could be tilted in any direction (a Ping-Pong ball flattened against a vertical paddle experiences a horizontal normal force).

Weight acts down, it behaves as an *external force* acting *on* an object. In effect, the Earth pulls down on the object and the object pulls up on the Earth—that's an interaction pair. As a rule, we do not go plummeting down to the center of the planet, so there must be some other force stopping us. Consider the person in Fig. 5.10. Her weight causes her feet to press on the ground and the ground pushes back on her—that's also an interaction pair. This contact-force pair, which is basically electromagnetic, is the same phenomenon that keeps you from walking through walls.

Call the force exerted perpendicularly on a supporting surface (such as a floor, pillar, or roadway) the **load**. The weight of a brick resting on a floor numerically equals the downward load *on the floor,* and the equal-and-opposite force *on the brick* exerted by the floor is the upward normal force. If you lift up slightly on the brick, decreasing the load, then the normal force will be reduced accordingly. If instead you step onto the brick, the normal force will numerically equal the combined weight, the total load.

When you stand still, the net vertical force *on* you (weight down, normal force up) equals zero. The Second Law then demands that you continue to remain at rest in the vertical direction. If, for structural reasons, the floor cannot exert an upward force equal to your weight, the nonzero net force acting on you would be down, and down you would accelerate through the floor, just as if you were trying to stand on a paper box or walk on water (Table 5.2).

Floors do not usually collapse underfoot. They exactly match your weight with a reaction force, and one might wonder how they manage that. Materials, no matter how rigid, give under the influence of a body pushing on them, regardless of how lightly. When two objects exert a contact force on each other, atoms in both must be

It is the same for me with mechanics as it is with languages. I understand the mathematical laws, but the simplest technical reality demanding perception is harder to me than to the biggest blockheads.

KARL MARX (1818–1883)

TABLE 5.2 **The physiological effects of acceleration**		
Acceleration	**Body orientation**	**Effect**
$2g$	Upright parallel to a	Walking becomes strenuous
$3g$	Upright parallel to a	Walking is impossible
$4g$–$6g$	Upright parallel to a	Progressive dimming of vision due to decrease of blood to retina, ultimate blackout
$9g$–$12g$	Reclining perpendicular to a	Chest pain, fatigue, some loss of peripheral vision, but one is still conscious and can move hands and fingers

somewhat displaced by the interaction. The greater the load, the more a floor stretches, building up a counterforce via springlike atomic interactions (that are electromagnetic) within the floor. The material ceases to distort any further when the upward reaction of the floor exactly cancels the load on it. Wooden floors often betray the sagging by squeaking as one board rubs past another, but all floors sag underfoot.

The Inclined Plane

Although the force of gravity acts straight down, when an object is supported by an inclined plane (Fig. 5.11), components of its weight exist both parallel ($\mathbf{F}_{W\parallel}$) and perpendicular ($\mathbf{F}_{W\perp}$) to the surface. As you can see in Fig. 5.11b, the only possible motion is along the incline, and it is $F_{W\parallel}$ that drives that motion. As we will learn presently, $F_{W\perp}$ influences friction and will be of interest as well. The angle between the line-of-action of the weight vector ($\mathbf{F}_W$) and the normal to the incline equals θ, the incline angle. Consequently

$$F_{W\parallel} = F_W \sin \theta \qquad (5.6)$$

and

$$F_{W\perp} = F_W \cos \theta \qquad (5.7)$$

$F_W \sin \theta$ pushes the object down the incline, and $F_W \cos \theta$ pushes it into the surface. The Second Law applies independently in each perpendicular direction. As shown in the free-body diagram (Fig. 5.11b), there is a force $F_{W\parallel}$ that will accelerate the object down the slope; $\Sigma F_\parallel = ma_\parallel$, and we can compute $a_\parallel$. There is no acceleration perpendicular to the slope; $\Sigma F_\perp = ma_\perp = 0$, and $F_{W\perp}$ must be balanced by an equal and oppositely directed reaction force F_N.

Figure 5.10 The weight of an object is supported by an equal normal force exerted upward by the floor. The free-body diagram shows the two forces acting on the standing person. The normal force is actually distributed over both feet.

Example 5.4 The 50.0-kg skier in Fig. 5.11 coasts along the surface of a snow-covered slope (assumed to be frictionless) tilted at an angle of 30.0°. Compute (a) the magnitude of the normal force acting on her; (b) the magnitude of the force tending to drive her down the inclined plane; and (c) the resulting acceleration ignoring air drag.

Solution: [Given: $m = 50.0$ kg and $\theta = 30.0°$. Find: (a) F_N; (b) the force parallel to the incline; and (c) $a_\parallel$.] The component of the skier's weight pressing her into the surface is

$$F_{W\perp} = mg \cos \theta = (50.0 \text{ kg})(9.81 \text{ m/s}^2) \cos 30.0°$$

and so $\qquad F_{W\perp} = 425$ N

We know that the skier does not leave the surface of the incline, so $a_\perp = 0$. Taking up (normal to the surface) as positive, the sum of the forces perpendicular to the incline is zero, thus

$$+\nwarrow \Sigma F_\perp = 0 = F_N + (-F_{W\perp})$$

$F_{W\perp}$ points into the surface in the negative direction— we enter it with a minus sign. Hence, $\boxed{F_N = 425 \text{ N}}$.

(b) The magnitude of the driving force down the incline is

$$F_{W\parallel} = mg \sin \theta = (50.0 \text{ kg})(9.81 \text{ m/s}^2) \sin 30.0°$$

and $\qquad \boxed{F_{W\parallel} = 245 \text{ N}}$

(c) To compute the down-plane acceleration (taking the direction of motion down the slope as positive), we write

$$+\swarrow \Sigma F_\parallel = ma_\parallel = F_{W\parallel}$$

and, hence

$$a_\parallel = \frac{(mg \sin \theta)}{m} = g \sin 30.0° \qquad (5.8)$$

or $\boxed{a_\parallel = \frac{1}{2}g}$. This result is independent of the mass, and applies to any body sliding down a frictionless incline at $\theta = 30°$.

▶ **Quick Check:** We can think of the inclined plane as a kind of "gravity reducer." With $\sin 30° = \frac{1}{2}$ in Eq. (5.6), the body behaves as if it were in free-fall (down the slope) at $\frac{1}{2}g$. Since the skier's weight is 490 N, in this tilted world it's driven down by $\frac{1}{2}$ 490 N. A normal force of 425 N at a small incline is also reasonable.

(a)

(b)

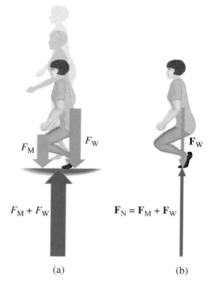

(a)   (b)

The roof of this building in the Northeast is constructed as a strongly tilted inclined plane to allow snow to slide off rather than pile up and load the structure.

Figure 5.11 A body on an inclined plane. (a) The component of weight acting down the inclined plane drives the body downhill. The component of weight acting perpendicular to the plane is matched by the normal force. (b) Applying the Second Law down the incline yields $\Sigma F_{\parallel} = F_{W\parallel} = ma_{\parallel}$.

Figure 5.12 (a) When an individual jumps, two forces act *on* the person, F_W down and the normal force F_N, acting up. (b) The floor experiences a net downward force of ($F_M + F_W$), and it exerts an upward equal and opposite reaction F_N. The net force on the jumper is F_M upward.

5.3 Vertical Accelerations

Have you ever seen $\approx 1\frac{1}{2}$-year-old kids (who have not yet learned to jump) trying desperately to leap into the air? What they do not know is that a net external vertical force must be applied if a mass is to accelerate upward. The Third Law is the key: *push down on the floor and it will push up on you*. Push with your leg muscles downward (F_M) on the floor (Fig. 5.12a). The total load is then ($F_M + F_W$), down. The upward reaction, the normal force (F_N), is then numerically equal to ($F_M + F_W$). But only two forces act *on you*, F_W down and F_N up. F_M is a force exerted by you *on the floor* (Fig. 5.12b). The sum of the vertical forces on you equals your mass times your acceleration. Taking up as positive

$$^{+\uparrow}\Sigma F = ma = (-F_W) + (F_N) = -F_W + (F_M + F_W) = F_M$$

which is a net upward force, and up you go ($a \neq 0$). The harder you push down, the greater the resulting acceleration. Skeptics are encouraged to leap up from a bathroom scale to see it read ($F_W + F_M$) during the accelerated ascent.

Example 5.5 A graduate student (having a mass of 40 kg) is standing inside an elevator on a scale that reads in newtons (Fig. 5.13). (a) If the elevator is at rest, what does the scale read? (b) Now suppose the elevator accelerates upward at one-half the acceleration of gravity ($a = \frac{1}{2}g$)—what does the scale read?

Solution: [Given: $m = 40$ kg, $a = \frac{1}{2}g$. Find: (a) F_W when $a = 0$; (b) F_N when $a = 0.50g$.] (a) Take up as positive. When $a = 0$

$$^{+\uparrow}\Sigma F_V = F_N + (-F_W) = 0$$

The directions of the vectors are taken care of with the signs, so all quantities are entered as positive numbers. Since $F_W = mg$, the load read by the scale is

$$F_W = F_N = (40 \text{ kg})(9.8 \text{ m/s}^2) = \boxed{0.39 \text{ kN}}$$

(Fig. 5.13b), and this is also the case even when the elevator moves at a constant speed. (b) When the elevator accelerates upward at a, it must carry the student up with it (Fig. 5.13c). Taking up as positive

$$^{+\uparrow}\Sigma F_V = F_N + (-F_W) = ma$$
$$F_N - mg = ma$$
and
$$F_N = ma + mg = m(a + g)$$

The floor must push up on the passenger with a total force F_N that equals the sum of her original weight (mg) plus an upward accelerating force (ma); this sum equals the load, and it is the force the scale will read. Since $a = 0.50g$

$$F_N = m(1.5 \text{ } g) = (40 \text{ kg})(1.5)(9.8 \text{ m/s}^2) = 0.59 \text{ kN}$$

The scale reads an "effective weight" of $\boxed{0.59 \text{ kN}}$. Everything within this accelerating environment behaves as if gravity were increased. That's why packages always seem heavier during the brief upward accelerating phase of an elevator ride.

▶ **Quick Check:** Her weight is around 10×40 or 400 N, and with an increase of $\frac{1}{2}g$ the effective weight should be roughly 600 N, so 0.59 kN is reasonable.

(b) (c)

Figure 5.13 (a) A student standing in an elevator on a bathroom scale. (b) When $a = 0$, the normal force equals the weight, and the scale reads F_W. (c) When the elevator accelerates up at a, the normal force increases beyond mg by ma. The scale reads $F_W + ma$.

(a)

A 12-cylinder Jaguar can accelerate from rest at roughly $g/3$. The driver is then pushed forward by the seat in proportion to the acceleration (regardless of the speed). Much the same thing happens to spacebound astronauts. The space shuttle crews are accelerated skyward by an up-pointing normal force. In reaction, they press down against their reclining couches with a peak force three times as much as they felt under normal gravity. Yet even that was a pleasant change from the $6g$ accelerations experienced by the *Apollo* crews during their Saturn-rocket lift-offs to the Moon (Table 5.2, p. 132).

Example 5.6   Jane, a 50-kg undergrad, is rescued from a burning building by a police helicopter (Fig. 5.14a). She hangs at the end of a rope dangling beneath the chopper. (a) Compute the tension in the rope as read by the spring scale when the craft is hovering. (b) Suppose the helicopter accelerates straight downward with respect to the ground at 3.0 m/s². What is the tension in the rope now?

Solution: [Given: $m_J = 50$ kg; a_J is Jane's acceleration with respect to the ground. Find: the tension F_T (a) when $a_J = 0$ and (b) when $a_J = 3.0$ m/s².] (a) Figure 5.14b is Jane's free-body diagram where $a_J = 0$. The tensile force $\mathbf{F}_T$ (that is, the pull of the rope) acts up, while her weight acts down. It doesn't matter, but since the motion will be down later, take down as positive, and so

$$+\downarrow \Sigma F_V = (-F_T) + F_W = m_J a_J = 0$$

Accordingly, the tension equals her weight, $F_T = F_W =$ $m_J g = (50$ kg$)(9.8$ m/s²$)$, and $\boxed{F_T = 0.49 \text{ kN}}$. (b) When the chopper drops downward at a rate less than g, she drops with the same acceleration. The magnitude of the net force *on* her (Fig. 5.13c) equals her mass times that acceleration

$$+\downarrow \Sigma F_V = (-F_T) + F_W = m_J a_J$$

and $$F_T = F_W - m_J a_J$$

Hence

$$F_T = 0.49 \text{ kN} - (50 \text{ kg})(3.0 \text{ m/s}^2) = \boxed{0.34 \text{ kN}}$$

▶ **Quick Check:** The chopper accelerates down at roughly $g/3$. That puts her in a $\frac{1}{3}$-reduced gravity environment. Her scale-measured "effective weight" is reduced by $\frac{1}{3}$, hence the tension should be about $\frac{2}{3}gm \approx \frac{2}{3}500$ N ≈ 0.33 kN.

(a)   (b)   (c)

Figure 5.14 (a) The spring scale reads the tension in the rope produced by weight of the dangling student. (b) When there's no acceleration, the sum of the forces acting on her is zero, and $\mathbf{F}_T = \mathbf{F}_W$. (c) When the person accelerates down, $\mathbf{F}_T < \mathbf{F}_W$.

5.4 Effectively Weightless

Nothing in this Universe of $\approx 10^{22}$ stars is gravityless. Yet a body falling freely behaves as if it were without weight, and so we say it is *effectively weightless*. Hold

This diver, high in the air over Barcelona at the 1992 Olympic games, is effectively weightless in free-fall.

Figure 5.15 Floating around within a free-falling elevator after pushing off from its floor. The elevator moves downward more rapidly than the person who slowed his descent by pushing on the floor.

your palm facing skyward, and place a bunch of keys in it, keeping your hand open. Now drop your arm, accelerating it at 9.8 m/s^2. The keys descend at this rate all by themselves—let them fall freely and simply have your hand precede them down. At the point where your hand reaches $a = g$, you will no longer feel the keys pressing on your palm; your hand will fall away as the keys fall, and the keys will become effectively weightless.

Now imagine that you are in an elevator falling freely, accelerating at g. Not only would the keys again hover "weightlessly" in your hand, but if you were standing on a scale it would read zero. The floor, dropping out from beneath your feet, would exert no upward response unless you pushed down on it with some muscle-generated force (Fig. 5.15). To be precise, the gravitational force still acts; in fact, *it's the only force that does act.* According to our definition of weight, an object in free-fall is not weightless; $F_W = mg$ is not zero. However, if we define the **effective weight** of an object as *the force it exerts on a scale,* then bodies in free-fall have an effective weight of zero, which is what the popular literature calls "zero-g" or "zero gravity." As we will see, your bathroom scale is also accelerating downward, so it too reads effective weight.

Suppose that our elevator is now hurled upward, at first accelerating at a rate a and then let loose to continue up in free unpowered flight. If you stepped off a chair inside the elevator during the powered ascent, you would accelerate down at g while the floor accelerated up to meet you at a. The combined result would be a relative acceleration of you with respect to the floor of $(g + a)$. Once you alight on the floor, you would have an effective weight of $m(g + a)$. During the unpowered portion of the ascent, our elevator-turned-rocketship will continue to climb, slowing down as it rises because of the ever-present tug of gravity. Step off the chair now and you will accelerate Earthward at g as usual, but so will the elevator, even as it continues to climb. Both passenger and craft are in free-fall although heading upward, and you again hover effectively weightless in midair right where you left the chair.

The *Apollo* astronauts floated around in their spaceship on the way to and from the Moon. They were not gravityless, or weightless, just effectively weightless in free-fall. The National Aeronautic and Space Administration (NASA) routinely flies a padded research plane called the "Vomit Comet" (see photo, p. 141) in arcs so that trainee astronauts can float effectively weightless. Gravity acts uniformly on all parts of the body, and so in free-fall one feels nothing—no perception of endless descent—just effortless floating.

5.5 Coupled Motions

Figure 5.16 shows several different situations in which two masses are attached together by an unstretchable rope. The pulleys are weightless and frictionless, so there are no tangential forces, and the tensions are therefore constant throughout each rope. For the moment, the surfaces are frictionless, as well. Suppose the motion takes place in the direction shown by the arrow in each case. The leading mass m_1 pulls the connecting rope along, and the rope pulls the trailing mass m_2 along. The trailing mass can never overtake and slacken the rope, nor can it lag behind, accelerating slower than the rope. Each mass must accelerate at the same rate. With two masses, we can write two coupled Second-Law equations and solve for two unknowns, usually F_T and a, although any two parameters could be unknown.

If there is any ambiguity about how the system moves, *guess at the direction of the overall motion and make that positive,* even if it means down is positive for one part and up for another. If you guess wrong, the values of the unknowns will just come out negative.

Figure 5.16 Three examples of coupled arrangements where two masses are attached together by a rope. The free-body diagram for every mass is drawn. In each of the three examples the same tension acts on both masses, and both accelerate at the same rate.

(a)

(b)

(c)

Example 5.7 Marylou ($m_M = 50$ kg) and her boyfriend, Doug ($m_D = 70$ kg), are tied together by a rope of negligible mass. She is standing on a frictionless horizontal sheet of wet ice when Doug accidentally steps off a cliff (Fig. 5.17a). Assume the unlikely possibility that the tree limb is frictionless and that her length of rope is horizontal. Determine (a) the tension in the rope, and (b) the accelerations of the lovers.

Solution: [Given: $m_M = 50$ kg and $m_D = 70$ kg. Find: F_T, a_M and a_D.] First, draw a free-body diagram of each person, as in Fig. 5.17b and c. As long as the rope does not stretch, $a_D = a_M = a$. Draw a curved arrow in the direction of motion of the system from Marylou over the branch to Doug. For him, down is positive, while for her, to the right is positive. There are two unknowns, a and F_T, and we have to construct two equations. Marylou

(continued)

(continued)

(a)

(b)

(c)

Figure 5.17 (a) Marylou sliding along the ice, after Doug falls off the cliff. Free-body diagrams of (b) Marylou, and (c) Doug.

and the Second Law (for the horizontal motion) yield

$$\overset{+}{\rightarrow}\Sigma F_{HM} = F_T = m_M a \qquad (5.9)$$

while Doug's vertical contribution provides

$$\overset{+\downarrow}{}\Sigma F_{VD} = F_{WD} - F_T = m_D a$$

We solve these two equations simultaneously (Appendix A-4) for either unknown. Let's do it for a first. Substituting $F_T = m_M a$ from Eq. (5.9) into the vertical summation leads to

$$F_{WD} - m_M a = m_D a$$

We do not have Doug's weight, but Eq. (5.5) lets us write it in terms of his mass: $F_{WD} = m_D g$, hence

$$m_D g = m_D a + m_M a$$

and

$$a = \frac{m_D}{(m_D + m_M)} g$$

Putting this expression into Eq. (5.9) produces the de-

sired equation for the tension, namely

$$F_T = \frac{m_M m_D}{(m_D + m_M)} g$$

The numerical values are $\boxed{a = 0.58\ g}$ and $\boxed{F_T = 0.29\ \text{kN}}$. The tension is the magnitude of the force accelerating dear Marylou. If the rope were to be cut, the tension would drop to zero, and she would coast at a fixed speed. Doug would descend in free-fall.

▶ **Quick Check:** When $m_D \gg m_M$, $(m_D + m_M) \approx m_D$, $a \approx g$, which means Doug is approximately in free-fall, and that's reasonable. Moreover, $F_T \approx m_M g$, and that too is to be expected. When $m_M \gg m_D$, $a \approx 0$ and $F_T \approx F_{WD}$, as it would be if Doug were tied to a massive boulder. There is then no acceleration and he just hangs over the edge. Both equations produce sensible results at the extremes, suggesting that they may be correct. Moreover, both have correct units.

CURVILINEAR MOTION

The central problem of Newton's *Principia,* indeed the central question of the age, was "How do the planets move?" The motion of the heavens had long been one of the great issues of Western science, philosophy, and theology. It was Newton who finally welded dynamics and astronomy into a mathematical understanding that swept away Aristotelian scholarship as if that genius had been little more than an amusing schoolboy. At the heart of the Newtonian synthesis is the concept of a center-seeking gravitational force holding the planets in their orbits. We introduce one of the crucial ideas here, leaving a discussion of the role of gravity for Chapter 7.

5.6   Centripetal Acceleration

Newton probably came to the correct understanding of motion along a curve in the early 1680s, no doubt with the inadvertent help of his chief antagonist, the brilliant, argumentative, lascivious Robert Hooke. The common wisdom, then and now (though it is wrong!), was that revolving objects are thrown radially outward by a centrifugal force. Even so, back in 1585, G. B. Benedetti, while considering the motion of a rock in a sling, rightly pointed out that an object revolving in a circle and suddenly unleashed sails off in a straight line *tangent* to the curve at the point of release. (David knew as much intuitively. Perhaps Goliath did, too.)

Whirl a ball tied to a string in a circle. Its velocity continuously changes as it is whipped around, even at a constant speed, because the direction of **v** keeps changing. That means there must be an acceleration and, by the Second Law, a force causing that acceleration. The hand pulls the string inward toward the center of the motion, and the string continuously pulls the ball off its straight-line inertial course. A center-seeking or **centripetal force** must be exerted if any object is to move in a curved path. Remove the inward centripetal force and the motion instantly becomes straight-line inertial, and that's tangential, *not* outwardly radial.

We can derive an expression for the *centripetal acceleration,* $\mathbf{a}_C$, for an object that moves uniformly in a circle (Fig. 5.18). The body travels from *A* to *B* in a time interval Δt, and as it does, the radius *r* sweeps through an angle θ. The velocity $\mathbf{v}_i$ at *A* subsequently changes direction, becoming **v** at *B*. Since both these vectors are perpendicular to *r*, $\mathbf{v}_i$ must also sweep through an angle θ as it moves into **v**, and that is shown in the velocity triangle of Fig. 5.18b. The latter diagram shows that the initial velocity plus the change $\Delta\mathbf{v}$ equals the final velocity ($\mathbf{v} = \mathbf{v}_i + \Delta\mathbf{v}$). Inasmuch as $v = v_i$, since the speed is constant, the velocity triangle is isosceles, just as is triangle *ABO*. Indeed, these two are similar triangles, both having the same vertex angle θ. It follows that the ratios of any two of their corresponding sides are equal, thus

$$\frac{\Delta v}{v} = \frac{\overline{AB}}{r}$$

The side $\overline{AB}$ is the magnitude of the displacement Δs, as the body moves from *A* to *B*, and so

$$\Delta v = \frac{v}{r}\Delta s$$

Dividing by Δt gives us the average acceleration ($\Delta v/\Delta t$) in terms of the magnitude of the average velocity ($\Delta s/\Delta t$), hence

$$\frac{\Delta v}{\Delta t} = \frac{v}{r}\frac{\Delta s}{\Delta t}$$

In the limit as Δt approaches zero, (Fig. 2.20, p. 37), *A* comes ever closer to *B*, $\Delta s/\Delta t$ becomes *v*, by way of Eq. (2.7), and $\Delta v/\Delta t$ becomes *a*, by way of Eq. (3.3). This particular instantaneous acceleration is called the **centripetal acceleration**. For a body in circular motion, a_C is given by

$$a_C = \frac{v^2}{r} \tag{5.10}$$

where the units are m/s^2.

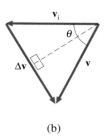

(a)

(b)

Figure 5.18   The geometry of circular motion. In this simple derivation the speed is constant, but that need not be the case. At any instant, $a_C = v^2/r$.

The direction of $\mathbf{a}_C$ is the direction of $(\Delta\mathbf{v}/\Delta t)$ in the limit as $\Delta t \to 0$. There are several ways to see that $\Delta\mathbf{v}$ points inward perpendicular to Δs—for instance, in the two similar triangles, $\overline{OA}$ is perpendicular to $\mathbf{v}_i$, $\overline{OB}$ is perpendicular to $\mathbf{v}$, and therefore Δs must be perpendicular to $\Delta\mathbf{v}$. Alternatively, the dashed angle-bisector in Fig. 5.18b is parallel to Δs and also perpendicular to $\Delta\mathbf{v}$. Since $\mathbf{v}_{av} = \Delta\mathbf{s}/\Delta t$, the average acceleration $(\Delta\mathbf{v}/\Delta t)$ must always be perpendicular to the average velocity. In the limit as $\Delta t \to 0$, the instantaneous acceleration $\mathbf{a}_C$ is everywhere perpendicular to the instantaneous velocity $\mathbf{v}$, which is always tangent to the path; $\mathbf{a}_C$ *points radially inward toward the center.* It's also common practice to refer to a_C as the *radial acceleration.* Interestingly, Huygens was the first to derive Eq. (5.10) but he, like many others, thought it was outward, away from the center.

Example 5.8   A beetle standing on the edge of a 12-in. recording of "Sergeant Pepper's Lonely Hearts Club Band" is whirling around at $33\text{-}\frac{1}{3}$ rotations per minute. Compute the magnitude of the creature's centripetal acceleration.

Solution: [Given: $r = 6.0$ in. $= 0.152$ m and a rotation rate of $33\text{-}\frac{1}{3}$ rpm. Find: a_C.] There is only one expression for a_C, and that is

$$a_C = \frac{v^2}{r} \qquad [5.10]$$

so v must be determined first. The total distance traveled in 1 min is $33\text{-}\frac{1}{3} \times 2\pi r = 31.9$ m; hence $v = (31.9\text{ m})/(60\text{ s}) = 0.532$ m/s. Consequently

$$a_C = \frac{(0.532\text{ m/s})^2}{(0.152\text{ m})} = \boxed{1.9\text{ m/s}^2}$$

▶ **Quick Check:** At a speed of 1 m/s and a radius of 1 m, a_C would be 1 m/s². Here, v is more than one-half that and r about one-sixth, so the answer should be a bit more than $(6/4)(1\text{ m/s}^2) \approx 3/2\text{ m/s}^2$.

Equation (5.10) applies to paths that are not completely circular (Fig. 5.19) provided that, at any instant, we use for r the radius of the circular segment being traveled, the *radius of curvature.*

5.7 Center-Seeking Forces

It follows from the Second Law that if a body of mass m is accelerating, it must be experiencing a net force given by $F = ma$. Any object moving in a circle must therefore be restrained by an inwardly directed centripetal force $\mathbf{F}_C$, with a magnitude of

$$F_C = ma_C = \frac{mv^2}{r} \qquad (5.11)$$

This is the force tugging the body off its otherwise straight-line inertial course. And the tighter the curve (the smaller is r), the more the path deviates from straight-line motion and the greater the needed force (Fig. 5.20). Centripetal force is not a new physical interaction, it is simply the name given to any force directed toward the center of the motion. For the Moon revolving about the Earth, $\mathbf{F}_C$ is gravitational. For a car rounding a turn on a flat highway, $\mathbf{F}_C$ is frictional. There are trains derailed all over the world simply because they are driven too fast around curves; Amtrak's Silver Meteor was derailed in December 1991 by an engineer oblivious to Eq. (5.11).

It was Newton who first showed that the bulge of the Earth (about 20 km at the equator) was due to the planet's rotation. Any piece of spinning Earth tends to fly

By flying along large parabolas in the sky, the so-called Vomit Comet can provide 40 s of effective weightlessness.

Figure 5.19 The centripetal acceleration of a car on a curved road at the moment when the radius of curvature is *r*.

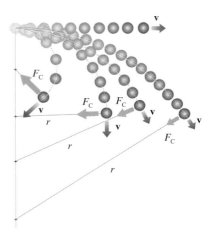

Figure 5.20 It takes no force for an object to move along a straight line at a uniform speed. Such a path can be thought of as a circle having an infinite radius. If the object is made to follow an arc of finite radius *r*, a centripetal force must be applied. The smaller the radius, the tighter the curve, and the greater is the force needed to pull the object away from its inertial path.

off tangentially and the planet essentially stretches out—the elastic forces holding it together provide the needed centripetal force. Failing that (given a high enough *v*), the dirtball would rip apart. It's the equatorial region that stretches out because *v* is greatest there.

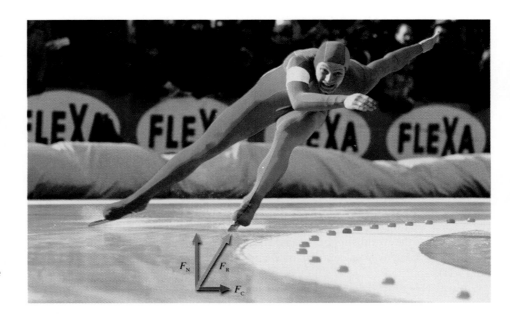

The reaction force F_R acting up along the line of the skate blade has two components: F_N (which is vertical and equal to the weight, and supports the skater) and F_C (which is the needed centripetal force).

Example 5.9 Compute the centripetal acceleration of an object located on the Earth's equator. The period of rotation, or time to spin once around, is 23 h 56 min 4 s (not 24 h; in fact, it's increasing even as you read this—see Problem 8.41) and the equatorial radius is approximately 6378 km. Let g_0 represent the local value of the acceleration due to gravity for a nonspinning Earth. Given that $g_0 = 9.81$ m/s^2 at the equator, at what rate would a stone fall there, and what would be its weight on a scale compared to its nonrotating-Earth weight?

Solution: [Given: $T = 23$ h 56 min 4 s, or 86 164 s, and $r_\oplus = 6378$ km. Find: a_C.] First, compute v:

$$v = \frac{(2\pi r_\oplus)}{T} = 465.09 \text{ m/s}$$

and so

$$a_C = \frac{v^2}{r_\oplus} = \boxed{3.391 \times 10^{-2} \text{ m/s}^2}$$

Any object at rest on the equator is already accelerating downward at 0.033 9 m/s^2, because it is circling around with the planet. A stone would fall with an acceleration of $g_0 - a_C = \boxed{9.78 \text{ m/s}^2}$. Its measured weight would be $\boxed{99.65\%}$ of its nonrotating-Earth weight (see Fig. 7.6, p. 216).

▶ **Quick Check:** Look at Table 3.2, p. 73.

5.8 Circular Motion

It is possible to bank a roadway so that the normal force provides the centripetal force rather than relying on friction to provide it. Figure 5.21 shows how the normal force $\mathbf{F}_N$ (perpendicular to the track's surface) acting on a runner has components both vertical (supporting the weight) and horizontal (providing the centripetal force). With no vertical acceleration

$$^{+\uparrow}\Sigma F_V = F_N \cos\theta - F_W = 0$$

Notice that $F_N > F_W$. There is a horizontal or radial acceleration, and so

$$^+_\rightarrow\Sigma F_H = ma_H$$

Figure 5.21 On a banked road the normal force F_N, which is always perpendicular to the road, is not aligned with the weight F_W, which is always straight down. The normal force has a radial component, $F_N \sin\theta$, that can be made equal to F_C.

(a)

(b)

A banked luge track. The athlete takes a position that will minimize air resistance.

with $a_H = a_C$ $\qquad\qquad F_N \sin \theta = F_C = \dfrac{mv^2}{r}$

The first equation contains $\cos \theta$ and the second, $\sin \theta$. One way to solve for θ is to combine the two into a single expression in terms of $\tan \theta$. Accordingly, write the first equation as $F_N \cos \theta = F_W = mg$ and divide the second one by it, thus

$$\frac{F_N \sin \theta}{F_N \cos \theta} = \frac{mv^2/r}{mg}$$

and

$$\tan \theta = \frac{v^2}{gr} \qquad\qquad (5.12)$$

This expression gives the proper banking angle for any one speed, and it applies equally well to cars on roads, trains on tracks, and even to birds and planes banking during turns in level flight. Any car—Eq. (5.12) is independent of mass—traveling at the posted speed can safely make the tilted turn even if the road is slippery, which is the reason highway ramps are banked. It follows from Eq. (5.12) that the smaller the radius and the greater the speed, the larger is $\tan \theta$ and, hence, the greater the banking angle. Motorcyclists and tobogganers are forever careening at high speeds around incredibly tilted tracks.

Example 5.10 A circular track with a 20-m radius is to be banked at an angle θ appropriate for a "4.0-min mile." Compute θ.

Solution: [Given: $r = 20$ m and v corresponding to a speed of 1 mi per 4 min. Find: θ.] Using Eq. (5.12), we fix θ for a 4.0-min mile; that is, a 240-s (1.609×10^3 m) run, for which

$$v = \frac{1.609 \times 10^3 \text{ m}}{240 \text{ s}} = 6.71 \text{ m/s}$$

Hence

$$\tan \theta = \frac{v^2}{gr} = \frac{(6.71 \text{ m/s})^2}{(9.81 \text{ m/s}^2)(20 \text{ m})} = 0.229$$

and

$$\boxed{\theta = 13°}$$

▶ **Quick Check:** The runner will lean into the curve balancing automatically and quite comfortably at a 13° tilt: $v = \sqrt{gr \tan \theta} = \sqrt{(9.8 \text{ m/s}^2)(20 \text{ m})(\tan 13°)} = 6.7$ m/s.

Figure 5.22 A frog being whirled around in a horizontal circle. The bottom of the bucket exerts a normal force that is the centripetal force.

Imagine a frog sitting in the bottom of a bucket that is being whirled in a horizontal plane (Fig. 5.22). The frog, left to its own, would move in a straight tangential line, but the bottom of the bucket, constrained to a circular path, advances and continuously intercepts that line. The bucket is forever nudging its passenger out of the way, thereby exerting an inward centripetal force on it. This is a normal force exerted *on the frog* by the bucket. The frog, in turn, exerts a radially outward reaction force *on the bucket*.

Suppose this situation were happening in deep space, where there is no appreciable gravity. We could adjust the radius and speed so that $a_C = v^2/r = 9.81$ m/s², whereupon the "frogonaut" would experience a normal force equal to its Earth-weight, as though it were sitting on the ground back at the old mudhole. It has been suggested that great doughnut-shaped, spinning space stations (where "up" is radially inward toward the central axis) could provide a more natural long-term environment for humans who were used to living only on the planet.

The *centrifuge* is a practical device using the same frog-in-the-bucket principle. The idea is simple—for a sample at rest, the sedimentation rate of particles suspended in a liquid (like blood corpuscles in serum) is proportional to g. The centrifuge revolves the sample, creating a simulated gravity of as much as 500 000g at perhaps 60 000 rpm. Particles denser than the surrounding liquid end up in the bottom of a test tube, much as the frog ended up pressing on the bottom of the bucket.

A driver colliding with the door in a turning car (Fig. 5.23) experiences the same thing as a cell in a centrifuge or a frog in a pail. The wall of the car intercepts the driver's linear motion, forcing her radially inward toward the center of the turn. The wall pushes on her, and she pushes back on the wall—with no friction from the seat, there is only one force acting horizontally *on* her and it's the normal force of the wall, which equals $F_C = ma_C$. For ordinary turns, there is usually enough friction via the seats to supply F_C and keep people from crashing into doors.

An amusement-park ride that makes use of gravity to supply centripetal force.

Some interesting physics occurs when we take the rope and bucket and swing the frog in a vertical plane (Fig. 5.24). Now gravity and the tension in the rope work alternately with and against each other to supply the centripetal force. When the rope is at some arbitrary angle θ, the combined weight of the frog and bucket (total mass m) is the vector $\mathbf{F_W} = m\mathbf{g}$. Always acting down, it can be resolved into two components, one radial $F_W \cos \theta$, the other tangential, $F_W \sin \theta$. The weight provides the only tangential force. Because there is a radial acceleration (a_C), there must be a net radial force inward toward O, namely F_C, which we take as positive. This centripetal force must be provided by the only two physical forces acting, the tensile force and the weight, hence

$$F_C = F_T - F_W \cos \theta$$

and since $F_C = ma_C$
$$a_C = \frac{F_T - mg \cos \theta}{m} = \frac{v^2}{r}$$

Solving for the tension,
$$F_T = m\left(\frac{v^2}{r} + g \cos \theta\right)$$

At the bottom of the circle $\theta = 0$, $\cos \theta = 1$ and
$$F_T = m\left(\frac{v^2}{r} + g\right)$$

At the top, $\theta = 180°$, $\cos \theta = -1$ and
$$F_T = m\left(\frac{v^2}{r} - g\right)$$

The fact that the bucket is tied to a rope rather than a rigid rod has not been introduced into the analysis. That is why this last equation allows negative solutions when $g > v^2/r$ at the top of the path. Then the bucket would be whirling too slowly to keep the rope taut, which is fine for a supporting rod that can push or pull on the bucket but not for a rope that can only pull. When $g > v^2/r$ or $v < \sqrt{rg}$, the weight will exceed the needed centripetal force, the rope will slacken, and the bucket and frog will drop out of their circular orbit into a free-fall parabola that never reaches the top. Notice that the tension can be zero if the system is moving such that $v^2/r = g$ at the very top. At that moment, the only force acting is gravity and the system is in free-fall. That's the fun part of those amusement-park rides where the

F_C

Figure 5.23 A car moving along a curve intercepts the straight-line inertial path of the driver, exerting an inward centripetal force on her.

car takes you upside down at the top of a loop. When $v > \sqrt{rg}$, the bucket and its passenger are actually pulled down by the rope with a force greater than in free-fall. If the frog was on a scale on the bottom of the bucket, it would have an effective weight upward, even while it was upside down at the top of the circle. That is why the frog does not drop out of the bucket—the bucket is accelerating down faster than the frog can fall.

FRICTION AND MOTION

Our common experience is that objects in motion do not often stay in motion, despite the unrefuted wisdom of the First Law. In practice we move, or try to move, one thing against another, and there are interactions that resist the motion. A force of this sort that *opposes an impending or actual motion* is said to be a *frictional force*. There are several different mechanisms that retard movement, and all produce what is broadly called **friction**. When friction impedes a motion that is in progress, it's called *kinetic friction;* when it prevents motion from occurring altogether, it's called *static friction*.

Friction arises via the electromagnetic interaction between atoms that can be in any bulk state—solid, liquid, or gas. *Liquid-gas friction* saves us from being pelted by high-speed raindrops; it allows the wind to make the sea choppy and retards the rising bubbles in a glass of beer. *Solid-liquid friction* slows the flow of petroleum in pipelines and blood in vessels; it helps to make bullets fairly useless under water and oil slicks fairly deadly on highways. The friction of the ocean tides as the waters rub across the land decreases the planet's spin and increases the length of the day. *Solid-gas friction* hinders the flow of air in pumps, pipes, and lungs; it slows skaters, cyclists, baseballs, and fire engines (at 70 mi/h a typical car expends 70% of its fuel just pushing air out of its way); it burns up meteors, threatens returning spaceships, and is the unflagging joy of parachutists. *Solid-solid friction* stops your car whenever you use your brakes, and it dissipates about 20% of the engine's power; without it, you could neither walk nor hold up your socks; it keeps cloth, carpets, ropes, baskets, and wicker furniture from unraveling; it allows us to write with pencils and turn pages. Friction is the great two-faced force of nature, at once the bane and the backbone of modern terrestrial technology.

Our present concern is with dry friction and its influence on motion. We limit the discussion to solids sitting, sliding, and rolling on other solids. Even so, our knowledge is rather rudimentary, the analysis is incomplete, and the experimental data are often imprecise and difficult to reproduce. Dry friction is dominated by the presence of contaminants that make the behavior erratic and somewhat unpredictable.

5.9 Static Friction

Figure 5.25 shows a block of weight F_W on a table that supports it with a normal force F_N. A small horizontal force F is applied to the right—the spring scale registers that force, but the block does not accelerate—it does not move at all. The credo of the Three Laws explains what is happening: Because $a = 0$, the net force is zero, and there must be a force to the left equal to F. That force acting parallel to the surface, resisting the motion, is the force of **static friction**, F_f. Increase F and if the object remains at rest, F_f must also have increased. If F is made larger and larger, the block will ultimately break loose and move when F just exceeds the maximum possible value of the static friction $F_f(\text{max})$. When there is no motion, the static friction force is equal and opposite to the driving force F, and that equality continues right up to $F_f(\text{max})$, just as a rope can react with any size tension right up to the moment when it breaks.

Three basic insights regarding $F_f(\text{max})$ come from simple observations (Leonardo da Vinci was aware of them in the mid-fifteenth century). (1) $F_f(\text{max})$ *is proportional to the normal force*, F_N: $F_f(\text{max}) \propto F_N$. Push on an empty chair, first gently, then harder and harder until it slides. Now have someone sit in it and try again. With the added load, $F_f(\text{max})$ is increased tremendously. (2) The same block in Fig. 5.25 sitting on another surface will experience an $F_f(\text{max})$ proportional to F_N, but it will likely be different. Some surfaces are more slick than others. The statement $F_f(\text{max}) \propto F_N$ can be turned into an equality (Appendix A-4) using a constant of proportionality, namely μ_s. This **coefficient of static friction** *depends on the two materials in contact.*

We get around our theoretical limitations by compressing the details of what is happening on an atomic level into a single coefficient that, because it cannot yet be calculated from theory, must be determined experimentally (Table 5.3). Accordingly

$$F_f(\text{max}) = \mu_s F_N \qquad (5.13)$$

A 40-N dictionary on a surface where $\mu_s = 0.3$ will require a minimum horizontal force of 12 N to overcome static friction and make it slide.

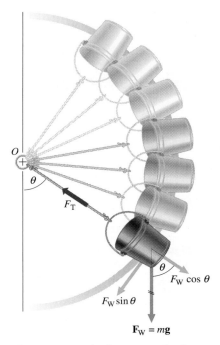

Figure 5.24 The forces on a bucket revolving in a vertical circle. The tension in the rope depends on θ and the speed at that instant v. If you do this with a bucket full of water, make sure that $v > \sqrt{rg}$ at the top ($\theta = 180°$), or you're likely to get wet.

TABLE 5.3   Approximate friction coefficients*

Material	μ_s	μ_k
Steel on ice	0.1	0.05
Steel on steel—dry	0.6	0.4
Steel on steel—greased	0.1	0.05
Rope on wood	0.5	0.3
Teflon on steel	0.04	0.04
Shoes on ice	0.1	0.05
Climbing boots on rock	1.0	0.8
Leather-soled shoes on carpet	0.6	0.5
Leather-soled shoes on wood	0.3	0.2
Rubber-soled shoes on wood	0.9	0.7
Auto tires on dry concrete	1.0	0.7–0.8
Auto tires on wet concrete	0.7	0.5
Auto tires on icy concrete	0.3	0.02
Rubber on asphalt	0.60	0.40
Teflon on Teflon	0.04	0.04
Wood on wood	0.5	0.3
Ice on ice	0.05–0.15	0.02
Glass on glass	0.9	0.4

*The first column lists values of various coefficients of static friction. The second gives the corresponding values of the kinetic coefficients of friction, a concept that will be discussed presently.

An inverted airfoil on the back of this race car pushes down on the rear to increase tire traction without increasing weight.

Figure 5.25   (a) A horizontal force *F* acting through a spring scale is applied to a motionless block of weight F_W. Static friction opposes the impending motion with a force F_f that can be as much as $F_f(\text{max}) = \mu_s F_N$. (b) The free-body diagram indicates that no horizontal acceleration will occur until $F > F_f$.

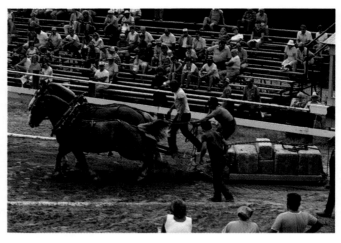

Teams of horses drag heavy loads of concrete across the ground. They strain to overcome static friction.

Example 5.11   A high point of a country fair is the horse-pulling contest. Teams of horses compete to haul the largest weight using a flat-bottomed sled. How much force must two horses exert to get the sled moving if the total load is 50 kN (or 11×10^3 lb) and the coefficient of static friction is 0.60?

Solution: [Given: $F_N = 50$ kN and $\mu_s = 0.60$. Find: $F_f(\text{max})$.] Any horizontal force in excess of $F_f(\text{max})$ will set the sled moving, thus

$$F_f(\text{max}) = \mu_s F_N = 0.60(50 \text{ kN})$$

and

$$\boxed{F_f(\text{max}) = 30 \text{ kN}}$$

▶ **Quick Check:** The force to make it start to slide is 60% of the load, or 30 kN.

The values in Table 5.3 should not be taken literally—μ_s for, say, glass on glass is given as 0.9, but what is really meant is glass on contaminants on glass. Unless special precautions are taken, oxides, dust, oils, and so forth, will be present on the surfaces and will generally reduce the friction. Nonetheless, in most practical situations, we deal with materials covered with foreign matter, so the table is useful, even if a little misleading.

(3) The third insight is the surprising observation that $F_f(\text{max})$ *is independent of the apparent size of the contact area between the two solid surfaces*. The dictionary mentioned earlier can be resting on either its narrow edge or its broad face, and the maximum-static-friction force will be the same.

Example 5.12   A climber stands on the rock face of a mountain. The soles and heels of her boots have a high static-friction coefficient equal to 1.0. (a) What is the steepest slope she can stand on without slipping? (b) Assuming her pants have a static-friction coefficient for rock of 0.3, what happens if she sits down to rest?

(continued)

(continued)

Solution: [Given: $\mu_s = 1.0$. Find: the corresponding maximum slope θ_{max}.] We need to formulate an expression for θ in terms of known quantities. Figure 5.26 shows the climber standing still on the mountain. Her weight, acting straight downward, has two components: $F_W \cos \theta$ perpendicular to the tilted surface, and $F_W \sin \theta$ parallel to the surface. It is $F_W \sin \theta$ that causes her to slide down the slope and it is this component that is resisted by the static friction. Because she is standing still, $a_\perp = 0$ and

$$+\nearrow \Sigma F_\perp = 0 = F_N - F_W \cos \theta$$

or

$$F_N = F_W \cos \theta$$

Static friction always acts parallel to the surface, opposing impending motion. Inasmuch as $a_\parallel = 0$

$$+\nwarrow \Sigma F_\parallel = 0 = F_f - F_W \sin \theta$$

or

$$F_f = F_W \sin \theta$$

To combine the two functions of θ into one tangent expression, divide F_f by F_N, yielding

$$\frac{F_f}{F_N} = \tan \theta$$

This ratio has its maximum value when $F_f = F_f(max)$, whereupon $\theta = \theta_{max}$, thus

$$\frac{F_f(max)}{F_N} = \tan \theta_{max} = \mu_s$$

and

$$\theta_{max} = \tan^{-1} \mu_s \qquad (5.14)$$

Measuring the maximum tilt angle at which a body starts to slide down an incline is a convenient way to determine μ_s. With $\mu_s = 1$, $\boxed{\theta = 45°}$. If she were to sit, $\tan (0.3) = \boxed{17°}$, and she would slide down any hill tilted at more than 17°.

▶ **Quick Check:** Both sides of Eq. (5.14) are unitless, which is okay. At the extremes, when $\mu_s = 0$, $\theta_{max} = 0$; when $\mu_s = \infty$, $\theta_{max} = 90°$, and that makes sense.

(a)

(b)

Figure 5.26 (a) A climber at rest on an inclined surface, where friction opposes her impending downward motion. (b) The free-body diagram is similar to Fig. 5.25b except that it's tilted.

Both the proportionality between F_f and F_N, and the area independence of F_f, are working principles rather than fundamental relationships. They generally hold true, and even when they are off, it's usually by no more than about 10%. As we will see, however, there are a few exceptions. Similarly, the idea that μ_s is always constant is not exactly true either—it can change as the contact time increases. Have you ever tried to unscrew a nut that has been in place for a few years or move a vase that has been sitting on a painted surface for several months?

Example 5.13   A washing machine in a wooden crate has a total mass of 100 kg. It is to be dragged across an oak floor by tugging on a rope (Fig. 5.27), making an angle of 30° with the horizontal. What minimum force will be needed to get the thing moving? Will it be more or less when $\theta = 0$?

Solution: [Given: $m = 100$ kg, $\theta = 30°$, and from Table 5.3, $\mu_s = 0.5$. Find: F.] The central equation is $F_f(\text{max}) = \mu_s F_N$, but here $F_N \neq F_W$. We will use the Second Law to determine F_N and also to explicitly involve the driving force F in the equations. F has two components (one parallel and one perpendicular to the direction of impending motion): $F \cos \theta$ and $F \sin \theta$. Since the crate is at rest, taking vertically up as positive yields

$$^{+\uparrow}\Sigma F_V = 0 = F_N + F \sin \theta - F_W$$

and $F_N = F_W - F \sin \theta$. Taking the direction of impending motion horizontally as positive, we get

$$^{+}_{\rightarrow}\Sigma F_H = 0 = F \cos \theta - F_f$$

and $F_f = F \cos \theta$. The minimum force (F) to move the crate obeys the equation $F \cos \theta = F_f(\text{max})$. Furthermore

$$F_f(\text{max}) = \mu_s F_N = \mu_s(F_W - F \sin \theta)$$

and setting these last two equations for $F_f(\text{max})$ equal to each other yields

Figure 5.27   (a) The washing machine crate experiences an applied force **F** that both lifts upward, lightening the load, and pulls forward to overcome static friction. (b) The normal force equals the load, which now is ($F_W - F_V$). The crate will move when $F_H > F_f(\text{max})$, where $F_H = F \cos \theta$.

$$F \cos \theta = \mu_s F_W - \mu_s F \sin \theta$$

and

$$F = \frac{\mu_s mg}{\cos \theta + \mu_s \sin \theta}$$

Thus

$$F = \frac{490.3 \text{ N}}{(0.866 + 0.250)} = \boxed{0.4 \text{ kN}}$$

When $\theta = 0$, $F = \mu_s F_W = 490$ N, or $\boxed{0.5 \text{ kN}}$.

▶ **Quick Check:** If F was horizontal, then $\theta = 0$, $\cos 0 = 1$, $\sin 0 = 0$, and $F = \mu_s mg$, which equals $\mu_s F_N$, and that suggests that the equation for F might be right.

5.10   Moving via Static Friction

We walk by cleverly arranging for an external force to propel us. Newton's Third Law is at the center of the process—push backward on the floor and it will react with a forward force that propels you. But only if there is friction can you push tangentially backward on the floor. And that push cannot exceed $F_f(\text{max})$ or your foot will slide out from under you. Friction, opposing the backward motion of your foot, propels you forward—*friction is the driving force* (Fig. 5.28).

It takes just a few ounces of force to overcome air drag and keep someone walking at a constant ($a = 0$) pace. To see that, have a person walk in place and as she stands there marking time, push on her with a continuous force—once started, it takes very little effort to keep up a constant forward motion. When you walk at a uniform rate, you gently push backward, but your feet do not move horizontally while in contact with the ground. That's why you make neat footprints trotting on

Figure 5.28 (a) The motion of a running foot resembles that of a point on a uniformly rolling wheel. (b) At the instant shown, the red dot is moving backward at a velocity with respect to the center of the wheel of $\mathbf{v}_{DO}$. This is equal in magnitude and opposite in direction to the wheel's velocity with respect to the Earth, $\mathbf{v}_{OE}$. Hence $\mathbf{v}_{DE} = \mathbf{v}_{DO} + \mathbf{v}_{OE} = 0$ and the dot is at rest, at that instant, with respect to the Earth.

A skater pushes down, to the side, and back, transverse to the skate's rolling axis. The ground pushes forward on the skate, and that frictional reaction propels the person.

The bicycle moves forward when the rear tire pushes back on the ground via friction. The ground pushes the tire, and hence the bike, forward.

damp sand. While traveling at a constant speed, a runner's foot actually moves through space with a motion that closely matches that of a rolling wheel (Fig. 5.28), and *static* friction propels both!

Imagine a dot of red paint on the rim of a uniformly rolling wheel. As the dot comes around, it will reside for an instant on the forward-most portion of the wheel.

It gradually moves down and *back*. At the instant the dot hits the ground, it will be moving backward with respect to the center at the same speed the entire wheel is moving forward. With respect to the ground, just as the dot comes down and touches, it will be motionless.

Example 5.14   The weight distribution of a Datsun 510 with a driver is 55% in front and 45% on the rear wheels. With $\mu_s = 0.9$ for tires-on-road, what is the maximum acceleration this rear-wheel-drive vehicle can attain? It actually goes from 0 to 30 mi/h in 4.1 s. How does that compare?

Solution: [Given: 45% F_W on the rear wheels and $\mu_s = 0.9$. Find: a_{max}.] The back two tires propel the car, and the force they exert on the ground determines the motion. The total load on both drive wheels is 45% mg. The maximum acceleration occurs at the point just before slipping, when

$$F_f(\text{max}) = \mu_s(0.45)mg = ma_{max}$$

and

$$a_{max} = 0.45(0.9)g$$

or

$$\boxed{a_{max} = 0.4g}$$

The car actually accelerates at $a = \Delta v/\Delta t = (13.4 \text{ m/s})/(4.1 \text{ s}) = 3.27 \text{ m/s}^2 = \boxed{0.3g}$, which is a bit less than the predicted maximum and suggests that a more powerful engine could improve the acceleration.

▶ **Quick Check:** All ordinary cars accelerate at a fraction of g, so the units are right and the order-of-magnitude is okay, too.

A car traveling at a constant speed leaves unsmeared tire marks in damp sand because there is no horizontal motion of the tread with respect to the ground. Ideally, the region in contact with the road does not push forward or backward; it does that only when braking or accelerating. Provided there is no slipping, it's the static-friction force between tires and road that propels a car, even when it is in motion. It's the same static-friction force that accelerates it and that supplies the centripetal force on turns (although there is a small amount of slipping that occurs during each one of these maneuvers). Skidding with the brakes locked and the tires motionless or attempting to accelerate so rapidly that the wheels spin (burning rubber), results in something different that is called *kinetic friction*.

5.11   Kinetic Friction

Please look at Fig. 5.25, where we pulled on a body at rest to observe its static-friction force. As the driving force increased, F_f increased until it reached a maximum value of $F_f(\text{max})$. Applying a still greater force will cause the body to break loose and move in the direction of the applied force. The **kinetic-friction force**, *the retarding force exerted on a sliding body in contact with a surface, is equal and opposite to the driving force that is needed to maintain the body in uniform motion.* Once again, experiments show that the three principles of friction apply, this time with a *coefficient of kinetic friction* μ_k, where

$$F_f = \mu_k F_N \tag{5.15}$$

Table 5.3 lists experimentally determined values of μ_k. For example, the force F needed to push a 600-N (135-lb) person wearing leather-soled shoes across a wood floor at a constant speed ($\Sigma F = ma = 0$) is $F = F_f = \mu_k F_N = (0.2)(600 \text{ N}) = 30 \text{ N}$. Table 5.3 makes it clear that static friction is usually greater than kinetic friction (a fact apparent to anyone dragging a heavy mass).

The skier moves ahead by pushing back on the ground with the pole. The ground pushes forward on the pole, which transmits the force to the person.

In practice, kinetic friction decreases as the sliding speed increases. Drivers know from experience that they must ease off the brakes if they are to bring an automobile to a smooth stop—the kinetic friction on the brake shoes increases as the wheels slow, and after a few jolting stops, the new driver learns to lighten the pressure appropriately. Thus, while μ_k for a sliding tire is roughly 0.8 at 8 km/h, it is likely to be less than about 0.5 at 130 km/h, and that can be crucial if your car is skidding. The best way to stop a car is to engage the brakes so the wheels are just on the threshold of locking up but are still turning. This technique will ensure a maximum retardation via *static* friction, which is greater than the kinetic friction of skidding rubber. The process can be done manually—you can feel the wheels locking—but it takes practice, especially if it is to be accomplished under stress. That's why computer-operated antilocking brakes are a good idea.

On friction. Generally it is easier to further the motion of a moving body than to move a body at rest.

THEMISTIUS (390–320 B.C.)
Physica

Example 5.15 A camp is on a mountainside that slopes at 30° (Fig. 5.29). Someone is hired to pull a 200-kg sled up the slippery incline for which the sled-slope coefficient of kinetic friction is 0.1. If the person can exert a continuous force parallel to the incline of 200 N, compute the acceleration of the sled.

Solution: [Given: $\theta = 30°$, $m = 200$ kg, and $\mu_k = 0.1$. Find: a.] We will need F_f and therefore F_N. From the figure, we have

$$+\nwarrow \Sigma F_\perp = 0 = F_N - F_W \cos \theta$$

and again $F_N = F_W \cos \theta$. Taking the direction of motion as that of the applied force **F**, friction must oppose it, and so

$$+\nearrow \Sigma F_\parallel = F - F_W \sin \theta - F_f = ma$$

Hence $ma = F - mg \sin \theta - \mu_k mg \cos \theta$

and

$$ma = 200 \text{ N} - 981 \text{ N} - 170 \text{ N} = -951 \text{ N}$$

STOP!

The component of the weight down the plane (981 N) far exceeds the upward driving force (200 N)—the sled slides downhill and a is negative. That's fine, but we must reverse F_f because it must oppose the motion (Fig. 5.30b)

$$ma = 200 \text{ N} - 981 \text{ N} + 170 \text{ N}$$

and $\boxed{a = -3 \text{ m/s}^2}$. The sled cannot be hauled uphill; the person can only slow the acceleration downhill.

▶ **Quick Check:** Since $\sin \theta = \frac{1}{2}$, the force down the hill ($mg \sin \theta$) is $\frac{1}{2}(200 \text{ kg})g = (100 \text{ kg})g = 981$ N. The normal force is $mg \cos \theta = 1699$ N, and the coefficient of friction is 0.1, so $F_f = 170$ N.

(a)

(b)

Figure 5.29 (a) A person pulling upward on a sled with a force $F = 200$ N. Since $F_W \sin \theta$ exceeds F, the sled slides downhill, but you might not have noticed that at the start of the analysis.

5.12   Rolling Friction

Wheels were used in transportation 5000 years ago, and it has been obvious for all that time that it is easier to roll a load than slide it. Ideally, if a rigid wheel is rolled along a flat inflexible surface and nothing stops the motion, it should go on forever. That does not happen, and we call the operative retarding force **rolling friction**. Even though the primary mechanism is quite different, the effect can be described approximately in the same form as were static and kinetic friction, thus

$$F_f = \mu_r F_N \tag{5.16}$$

where μ_r is the *coefficient of rolling friction*. The friction force is equal in magnitude to the force required to keep the object rolling at a uniform speed.

The effect arises from the deformation of both the wheel and the surface, which creates an obstacle to free rolling that must be overcome by the exertion of a force. When the surfaces are hard, there is little deformation, as when steel wheels roll on steel tracks, whereupon μ_r is only about 0.001. By contrast, a flexible rubber tire on concrete will have a coefficient around 0.01 to 0.02 (with radials lower than crossply). The coefficient for tires actually varies with speed, increasing as much as 15% between 0 and 120 km/h. However small μ_r is, the average auto at 80 km/h uses about 30% of its power to overcome rolling friction. At low speeds it exceeds air drag. That situation changes at around 55 km/h (35 mi/h), depending on the shape of the car, after which air drag predominates. The admonition to maintain proper tire pressure translates into lower rolling friction and better fuel economy.

Example 5.16   A driver in an aerodynamically trim, 1200-kg sports car traveling at 20 km/h shifts into neutral and coasts to a stop uniformly under the influence of rolling friction. Assuming an average $\mu_r = 0.013$, compute the stopping distance.

Solution: [Given: $m = 1200$ kg, $v_i = 20$ km/h, $v_f = 0$, and $\mu_r = 0.013$. Find: s.] Equation (3.11, p. 68) provides the stopping distance: $s = -v_i^2/2a$, but we need a. Notice that $F = ma = F_f = \mu_r mg$, and so $a = \mu_r g$. If

we take the direction of motion as positive, a (which is a deceleration) must be entered as a negative quantity. Hence, $s = v_i^2/2\mu_r g$, and

$$s = \frac{(5.56 \text{ m/s})^2}{2(0.013)(9.81 \text{ m/s}^2)} = \boxed{0.12 \text{ km}}$$

▶ **Quick Check:** The time it takes to stop is $v_i/a = 43.6$ s, and at an average speed of $\frac{1}{2}(5.56$ m/s$)$, $s = 121$ m.

5.13   The Causes of Friction

In 1734, John Desaguliers (an acquaintance of Newton's) rightly concluded that *adhesion* played a role in producing friction. He took two freshly cut pieces of lead, pressed them together by hand and found that it could take as much as 50 lb to pull them apart.

The attraction between the atoms and molecules that holds solid matter together is electromagnetic in nature. Without it, there would be no such thing as textbooks, noses, or teacups, nor would there be any friction. But only when the atoms are close to one another will they experience an appreciable attractive force; the effect is short-ranged. Such forces become negligible at distances equivalent to only about 4 or 5 atomic diameters (each of which is $\approx 1 \times 10^{-10}$ m to 2×10^{-10} m). Thus, if a substantial amount of friction is to exist, the interacting surfaces must be in close contact.

Common objects may seem smooth to the eye and touch, but on a microscopic level they are jagged and rough. Even a polished surface is a craggy affair with features about 3×10^{-8} m across, each hill corresponding to perhaps 100 atomic diameters. When two ordinary surfaces are in contact, they only touch at a few high spots. Any material will deform if the *force per unit area,* or *pressure,* on it is great enough. With the load supported on a few tiny prominences, each experiences a crushing pressure and subsequently flattens out. The contact area thereby increases, decreasing the pressure until the material squashes no further. Still, only a small fraction of the total surface area is in contact, while the remainder is separated by gaps of 10 to 50 atomic diameters. Clearly, friction should be proportional to this *actual* contact area, not the apparent contact area.

Suppose a book is oriented to give the greatest apparent contact area with a surface (Fig. 5.30). Resting on a flat face may mean more contact points, but each will experience less pressure and will deform less. The overall effect is that the actual total contact area is the same, independent of the apparent area. That situation is equivalent to the third principle of friction. The actual contact area depends on the physical properties of the materials. Moreover, the interatomic forces, and thereby the adhesion and the friction, also depend on the kinds of atoms involved, which is equivalent to the second principle of friction.

Static friction arises out of the need to rip apart the areas of bonded contact. Two clean pieces of metal, such as gold, pressed together, will literally fuse at the contact points, a process known as *cold welding.* To move the two surfaces, the welds must be torn. On the other hand, keeping the surfaces separated reduces adhesion and friction—that's what oil and grease do, and why baby powder protects the skin against chafing (Fig. 5.31).

Figure 5.30 (a) When the load is spread out over a large area, the pressure is low, and there are many small contact areas. (b) When the load is concentrated, the pressure is greater and each place of contact is larger, but there are fewer of them. Thus, the total contact area is the same in (a) as in (b). These contact areas literally weld together and must be pulled apart.

Burned rubber due to rapid deceleration. A tire's surface heats up and the rubber burns, leaving marks on the road as the vehicle accelerates. These tracks were left by the two tires on the right side of a truck. Variations show that the driver pumped the brakes.

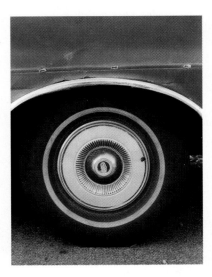

The flattening of the tire contributes to rolling friction. Every time the wheel turns, energy is lost, reshaping the rubber, and the tire heats up.

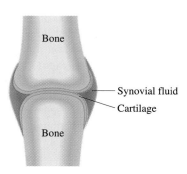

Figure 5.31 A human bone joint lubricated with synovial fluid. As the bones move, the contact area shifts, squeezing fluid out of the porous cartilage. The result is a remarkably small coefficient of friction of about 0.000 3.

Gauge blocks are precise, smooth steel blocks that are used to calibrate measuring tools. When they are squeezed together they interact so strongly that they stick to each other. The adhesive force at work here is the same force that gives rise to friction.

Adhesion is lessened if the two surfaces are already moving past one another, and so usually $\mu_s > \mu_k$. When no relative motion occurs, the surfaces enmesh, coming into close contact. When there is sliding, the surfaces ride across each other mostly on the high prominences. By contrast, in rolling friction the welds are peeled apart rather than sheared or ripped, and that takes less effort. Moreover, during sliding, surface layers of dirt, oxides, and grease will be cut through and good contacts can be made. This action does not happen with rolling objects, where adhesion is still lower and $\mu_k > \mu_r$.

Core Material

The net force (**F**) applied to a body equals the mass (m) of the body multiplied by the resulting instantaneous acceleration (**a**):

$$\mathbf{F} = m\mathbf{a} \qquad [5.3]$$

The weight of an object is

$$\mathbf{F}_W = m\mathbf{g} \qquad [5.5]$$

The force exerted by a surface perpendicular to itself and equal to the load supported is the normal force (F_N).

When the only force acting on an object is gravity, the object is in free-fall, and it is **effectively weightless**.

A centripetal force must be exerted on an object if it is to move in a curved path. The *centripetal acceleration*, a_C, which occurs when any object uniformly moves in a circle, is

$$a_C = \frac{v^2}{r} \qquad [5.10]$$

Such an object must experience an inwardly directed centripetal force given by

$$F_C = ma_C = \frac{mv^2}{r} \qquad [5.11]$$

The maximum static friction force is

$$F_f(\max) = \mu_s F_N \qquad [5.13]$$

where μ_s is the *coefficient of static friction*. For surfaces in relative motion, kinetic friction is given by

$$F_f = \mu_k F_N \qquad [5.15]$$

where μ_k is the *coefficient of kinetic friction*. For the friction encountered when one object rolls across another

$$F_f = \mu_r F_N \qquad [5.16]$$

where μ_r is the *coefficient of rolling friction*.

Suggestions on Problem Solving

1. Draw a diagram. Guess at the direction of the overall motion of the system (actual or impending) and call that positive. If a complicated system is moving clockwise, make that the positive direction and draw a big arrow in that direction from one end of the system to the other. Part of the system may be moving up, part right, and part down. Provided each portion follows the overall clockwise motion, up, right, and down are positive for the corresponding portions of the system. (See, for example, Fig. P22.)

2. Isolate each object of interest via its own free-body diagram. Then apply $\Sigma F = ma$ to each body, remembering that if the body is at rest or moving at a constant velocity, $a = 0$. Apply the Second Law in the direction of actual or impending motion and (if you need normal forces) perpendicular to it. Each such application will provide an independent equation, and the number of equations must equal the number of unknowns.

3. If forces act that are neither perpendicular nor parallel to the actual or impending motion, they must be resolved into components in those directions. The motions in these two directions are independent. For an inclined plane, we apply $\Sigma F = ma$ both down the incline and perpendicular to it—the former tells us about the acceleration, the latter about the normal force.

4. The tensions acting at each end of a rope are equal and opposite (provided no tangential forces act). Thus, if two masses are connected by a rope, the same tension will act in opposite directions on the two bodies.

5. When doing centripetal force problems we often encounter revolving objects for which we must find the speed. Thus, if a body turns at 1000 rotations per minute, a point on its surface a distance r from the axis moves in 1 min through 1000 circular paths, each equal to a circumference of $2\pi r$, for a total distance of $1000(2\pi r)$. The speed is then $v = (1000 \text{ min}^{-1})(2\pi r)/(60 \text{ s/min})$.

6. Don't forget that $F_W = mg$, which means that if you know either the weight or the mass, you can calculate the other.

7. For a body at rest, make sure that the net force can overcome the static friction before you go on to calculate the effects of kinetic friction.

Discussion Questions

1. Can we shoot a hole through a silver dollar floating out in space where the bullet is weightless? An old circus trick is to fire a bullet at a target while holding a pure silk handkerchief so it hangs freely directly in the line of fire. The gun is shot, the target hit, but no hole is made in the handkerchief. Explain.

2. Describe the process of swimming in terms of Newton's laws. Swimmers sometimes use rubber fins on their feet. How do the fins work? While floating in space, can you propel yourself by moving your hands and feet as if swimming? Explain how one moves a boat by rowing.

3. Explain *in terms of Newton's laws* each of the following: How does a rug get cleaned by beating it? What happens when we shake down a mercury thermometer used for measuring body temperatures? How does a hummingbird, a helicopter, or a bumblebee hover in midair? What makes an unsealed, air-filled balloon fly around when it's released? What are the external forces acting and their interrelationships when an automobile is traveling on a level road at a constant speed?

4. A weightlifter and a barbell are both on a large floor scale. While the weightlifter lifts and holds the barbell, will the scale ever momentarily exceed and/or be less than the combined weight of the two bodies? Explain.

5. Can an object ever move in a direction other than that of the net applied force? Can it accelerate in a direction different from that of the force?

6. Two equal-weight monkeys each hang from the ends of a rope passing over a weightless, frictionless pulley. If one accelerates up the rope, what happens to the other?

7. A truck carrying a load of caged canaries pulls onto the weighing scales at a highway toll station. While no one was looking, the driver banged on the side of the cab so the birds would fly around inside. What happens to the weight of the load?

8. A 160-lb gymnast doing a giant swing on a high bar experiences a force directed inward toward the bar of about 120 lb at the top and roughly 790 lb at the bottom. Explain what is happening.

9. Knowing that the mass of muscle tissue increases more rapidly than does its strength, provide a reasonable physical explanation for the fact that dogs and deer run faster than people.

10. What physical mechanism accounts for the "stickiness" of adhesive tape? Attach a piece of tape a few inches long to a solid horizontal surface leaving a little tab to hold on to. Now try to pull the tape off horizontally (that is, shear it off). Next, peel it off with a vertical force. Compare the results and discuss the significance as it applies to rolling friction.

11. Ice skates melt the ice beneath them so that they ride on a layer of water—provided it's not too cold. What role does that process play in skating? Can you skate on dry glass? Why are skis waxed? Suppose you wanted to slide a 4-kN block of marble into a narrow space—why might it be advantageous to slip several ice cubes under the block? There is a rumor that mausoleum workers sometimes use a handful of expendable marbles to move heavy coffins into narrow vaults. Explain the physics behind such a procedure.

12. Why is it advisable to avoid rapid acceleration, "deceleration," and sharp cornering while driving a car on a rainy day? The main purpose of the treads on an automobile tire is to provide a space into which water on the road can be squeezed. Why is that a good idea? What is the purpose of spikes on golf shoes and cleats on football shoes?

13. Take a toy car and wedge the rear wheels so they lock. Now roll it down an inclined plane, front first—what happens and why? Try it again with the front wheels locked. How should the brakes on a real car be engaged? Many cars are biased in the opposite way, which accounts for some of the broadside sluing on icy roads. Why is it recommended to pump the brakes when a car goes into a skid while making a rapid stop?

14. Draw a picture of a horse pulling a wagon over a smooth road. What are all the forces acting on the horse? On the wagon?

15. What is the smallest force with which you can raise an object on Earth into the air? Explain your ideas in detail.

16. What will happen when the string holding the wrench handle in Fig. Q16 burns through?

Figure Q16

Multiple Choice Questions

1. If L stands for length, T for time, and M for mass, the dimensions of force are (a) $[ML^2]$ (b) $[ML/T]$ (c) $[ML/T^2]$ (d) $[LT/M]$ (e) none of these.

2. One *newton* is the force (a) needed to move a 1-kg mass at 1 m/s (b) needed to accelerate a 1-kg mass at 1 m/s^2 (c) needed to accelerate a 1-g mass at 1 cm/s^2 (d) equal to the Earth-weight of a 1-kg mass (e) none of these.

3. Two forces, one 10 N and the other 40 N, cannot be applied to a body so that they produce the same acceleration as a single force of (a) 35 N (b) 50 N (c) 40 N (d) 10 N (e) none of these.

4. The weight of a TV star on Earth is 686 N. Out in deep space his mass is (a) 686 kg (b) 6723 kg (c) 0 kg (d) 70 kg (e) none of these.

5. The mass of an astronaut on a planet where gravity is 10 times greater than Earth's gravity is (a) 10 times smaller (b) 10 times larger (c) $10g$ times greater (d) $10g$ times smaller (e) none of these.

6. Is it possible to devise a technique to push on a table without it pushing back on you? (a) Yes, out in space. (b) Yes, if someone else also pushes on it. (c) A table never pushes in the first place. (d) No. (e) None of these.

7. If a nonzero constant net horizontal force is acting on a body sitting at rest on a frictionless table, the body will (a) sometimes accelerate (b) always move off at a constant speed (c) always accelerate at a constant rate (d) accelerate whenever the force exceeds its weight (e) none of these.

8. If (with no friction) a force F results in an acceleration a when acting on a mass m, then tripling the mass and increasing the force sixfold will result in an acceleration of (a) a (b) $a/2$ (c) $2a$ (d) $a/6$ (e) none of these.

9. A bubble level can be used as an accelerometer. If the level is accelerated due east while in normal operating position aligned east-west, the bubble will (a) move west (b) move east (c) move north (d) remain at rest (e) none of these. Try it.

10. A 250-lb man holding a 30-lb bag of potatoes is standing on a scale in an amusement park. He heaves the bag straight up into the air, and before it leaves his hands, a card pops out of a slot with his weight and fortune. It reads

(a) 250 lb (b) 280 lb (c) less than 250 lb (d) more than 280 lb (e) none of these.

11. Why does it take more force to start walking than to continue walking? (a) It doesn't. (b) Because kinetic friction is less than static friction. (c) Because air drag at walking speeds is much less than the needed inertial force, ma. (d) People get tired walking and prefer to stand still. (e) None of these.

12. Imagine that you are standing on a cardboard box that just supports you. What would happen to it if you jumped into the air? It would (a) collapse (b) be unaffected (c) spring up as well (d) move sidewise (e) none of these.

13. Someone weighing 600 N jumps onto a bathroom scale. In the process, the scale will read (a) a constant 600 N (b) 0 and then rise to 600 N (c) 0 then rise to over 600 N and stay there (d) 0 then rise to over 600 N, then drop down to 600 N (e) none of these.

14. Why do thin people with skinny legs make the best runners? (a) They tend to be strong. (b) Thin legs are longer, so people with them can take longer strides. (c) They have less massive legs to accelerate. (d) Heavy people just don't like to run. (e) None of these.

15. Imagine a flat, lightweight wheeled cart that is low to the ground and has well-oiled bearings. What will happen to it if, while standing at rest on it, you begin to walk along its length? It will (a) remain stationary (b) advance along with you (c) not enough information to say (d) move in the opposite direction (e) none of these.

16. With the previous question in mind, what would happen if you approached the cart, stepped onto it, and walked its length at a constant speed? It would (a) remain nearly stationary (b) advance along with you (c) move rapidly in the opposite direction (d) move forward then backward (e) none of these.

17. For an object moving along a curved path at a constant speed (a) its acceleration is perpendicular to its instantaneous velocity (b) its velocity is constant (c) its acceleration vector is constant (d) its acceleration is zero (e) none of these.

18. Mica is a crystalline material that can be sheared along atomic planes. It is the only material that can be prepared

in a form that is smooth on a molecular level over sizable areas. Place two sheets of mica such that an area A on both are in contact under a 1-kg load, then measure F_f(max) for mica on mica. If that load is then doubled, the maximum static friction force will likely (a) remain unchanged (b) double (c) drop to zero (d) quadruple (e) none of these.

19. With the previous question in mind, suppose the overlapping area of the two sheets is made $2A$. The maximum friction force will thereby (a) remain unchanged (b) decrease by half (c) increase proportionately (d) quadruple (e) none of these.

20. A man is walking with a cane, which at some instant makes an angle θ with the ground. The reaction force on the cane is (a) zero (b) $\mathbf{F}_N$ (c) $-(\mathbf{F}_N + \mathbf{F}_f)$ (d) $+(\mathbf{F}_N + \mathbf{F}_f)$ (e) none of these.

Problems

FORCE, MASS, AND ACCELERATION

1. [I] A person pushes a shopping cart with a force of 100 N acting down at 45°. If the cart travels at a constant speed of 1.0 m/s, what is the value of the combined retarding force acting horizontally on the cart due to friction and air drag?

2. [I] Two bar magnets of mass 1.0 kg and 2.0 kg have their like poles pressed together so that there is a repulsive interaction between them. In a frictionless environment the two fly apart when released such that the more massive one has an initial acceleration of 10.0 m/s^2 due north. What is the initial acceleration of the other magnet?

3. [I] What is the smallest force needed to lift a 0.50-kg bullfrog up from the ground and under what circumstances would that answer be applicable?

4. [I] A 100-kg gentleman standing on slippery grass is pulled by his two rather unruly children. One tugs him with 50 N north, toward the ice cream stand, while the other hauls with 120 N toward the bathroom, due east. Overlooking friction, compute Pop's resulting acceleration.

5. [I] During training on Earth, an astronaut is accelerated upward at $0.50g$. If her mass is m, what is her effective weight?

6. [I] The gravitational acceleration on the surface of Mercury is 0.38 times its value on Earth. What is the weight of a 1.0-kg mass on that planet?

7. [I] A youngster of mass 40.0 kg jumps into the air with an average vertical acceleration of 5.00 m/s^2. Determine the average force exerted by the ground on the jumper during the leap, and compare it to the youngster's weight.

8. [I] A maintenance robot 1.70 m tall, weighing 752 N when constructed on Earth, is found floating 400 m from an orbiting power station. The human crew decides to tie a light rope to it and haul it aboard. How much force must be exerted if the robot is to be uniformly accelerated for 1.00 s at 9.81 m/s^2?

9. [I] An *F-14* twin engine supersonic fighter weighs 69 000 lb. Its two Pratt and Whitney TF30-P-412 engines can each deliver a peak thrust, using afterburners, of 20 200 lb. Compute its maximum acceleration (in terms of g) during horizontal flight (neglect air friction).

10. [I] A rope of mass 10.0 kg is tied to a 1.00×10^3-kg signal relay in intergalactic space. The device is to be returned for repairs as rapidly as possible by a relay-tender android capable of exerting a maximum sustained force on the rope of 10.0 kN. Compute the resulting acceleration.

11. [II] A spread-eagled sky diver whose mass is 65.0 kg is falling straight down at a speed in excess of 150 km/h

with an acceleration of 2.00 m/s^2. Compute the total friction force due to air drag acting at that moment.

12. [II] A parachutist of mass m lands with legs bent, coming to rest with an upwardly directed average acceleration of $4.0g$. What is the average force exerted on him by the ground?

13. [II] Someone of mass 100 kg is standing on top of a steep cliff with only an old rope that he knows will support no more than 500 N. His plan is to slide down the rope using friction to keep from falling freely. At what minimum rate can he accelerate down the rope in order not to break it?

14. [II] The ballistocardiograph is a device used to assess the pumping action of the human heart. A patient lies on a horizontal 2.0-kg platform "frictionlessly" suspended on air-bearings. The rush of blood pumped in one direction will be accompanied by a counterforce on the body and table in the opposite direction. The resulting acceleration of the platform is recorded by an extremely sensitive accelerometer able to measure values as small as 10^{-5} m/s^2. For a young, healthy adult the acceleration over an interval of 0.10 s during the pumping cycle may be as great as 0.06 m/s^2. (a) Compute the force exerted by the heart if the patient has a mass of 70 kg. (b) What is the corresponding change in momentum?

15. [II] A pink pom-pom of mass m hangs on a chain from the rearview mirror of a Corvette L82. That machine can go from rest to 26.8 m/s (60 mi/h) in 6.8 s. The pom-pom serves as a pendulum accelerometer. Assuming a constant maximum acceleration, compute the angle the pom-pom makes with the vertical.

16. [II] On its first flight in 1981, the *Columbia* spaceship was part of a 4.5×10^6-lb, 18-story high launch assembly that developed a total thrust of about 6.4×10^6 lb. (a) What was its initial acceleration at full power? (b) According to newspaper reports, it was traveling at 33.5 m/s (75 mi/h) 6.0 s after blast-off when it cleared the 347-ft support tower. Are these observations consistent? What average acceleration does that correspond to? (c) Compare these values of acceleration and explain any differences.

17. [II] A tug pulls two small barges (of mass $m_1 = 4.00 \times 10^3$ kg and $m_2 = 3.50 \times 10^3$ kg) tied together, one behind the other. The tug exerts 1.00 kN on the line to the first barge (m_1) and accelerates at 0.100 m/s^2. Compute the tension in the two ropes and, knowing that friction opposes the motion, determine the friction on each barge, assuming it's the same for both.

18. [II] Refer to Fig. P18. Compute the acceleration of the 1.3×10^3-N weight. Neglect friction and the mass of the

Figure P18

Figure P22

rope and pulleys. Now remove the 1.8×10^3-N weight, replace it by a constant downward tug of 1.8×10^3 N acting on the end of the rope, and recalculate the acceleration of the first weight.

19. [II] A 2000-kg car in neutral at the top of a 20° inclined driveway 20 m long slips its parking brake and rolls down. At what speed will it hit the garage door at the bottom of the incline? Neglect all retarding forces.

20. [II] A rescue helicopter lifts two people from the sea on an essentially weightless rope. Jamey (100 kg) hangs 15 m below Amy (50.0 kg), who is 5.0 m below the aircraft. What is the tension (a) on the topmost end of the rope, and (b) at its middle while the helicopter hovers? (c) Compute those answers again, this time with the aircraft accelerating upward at 9.8 m/s².

21. [II] The all-time champion jumper is the flea. This tiny pest can attain a range of up to 12 in. or about 200 times its own body length. (That's the same as you jumping five city blocks.) Assuming a 45° lift-off angle, a push-off time of 1.0 ms, and a mass of 4.5×10^{-7} kg, compute (a) the initial acceleration of the flea, assuming it to be constant, and (b) the force it exerts on the floor.

22. [II] In 1784, George Atwood published the description of a device for "diluting" the effect of gravity, thereby facilitating the accurate measurement of g. Figure P22 shows the apparatus: two masses tied together by a length of essentially massless rope slung over an essentially massless free-turning pulley. With $m_2 > m_1$, prove that both masses accelerate at a rate of

$$a = \frac{(m_2 - m_1)}{(m_2 + m_1)} g$$

Show that the tension in the rope is

$$F_T = \frac{2m_1 m_2}{(m_2 + m_1)} g$$

If $m_2 = 2m_1$, what is a? When is a equal to zero? When $m_2 \gg m_1$, find a.

23. [III] Figure P23 shows three masses attached via massless ropes over weightless, frictionless pulleys on a frictionless surface. Compute (a) the tension in the ropes and (b) the acceleration of the system. Draw all appropriate free-body diagrams. (c) Discuss your results in terms of internal and external forces.

Figure P23

24. [III] Two identical blocks, each of mass 10.0 kg, are to be used in an experiment on a frictionless surface. The first is held at rest on a 20.0° inclined plane 10.0 m from the second, which is at rest at the foot of the plane. The one descends the incline, slams into and sticks to the second, and they sail off together horizontally. Calculate their final speed.

25. [III] A physicist on planet Mongo is using a device equivalent to Atwood's machine to measure the Mongoian gravitational acceleration g_M. He fixes one of the two 0.25-kg masses at each end of the rope. While both are at rest, he places a 0.025-kg gronch (a toadlike creature) on one of the masses. That body and its wart-covered passenger descend 0.50 m before the gronch hops off. The body continues traveling downward another 1.2 m in the next 3.0 s. Compute g_M. See Problem 22.

CURVILINEAR MOTION

26. [I] Determine the acceleration of a kid on a bike traveling at a constant 10.0 m/s around a flat circular track of radius 200 m.

27. [I] A 25-g pebble is stuck in the tread of a 28-in. diameter tire. If the tire can exert an inward radial friction force of up to 20 N on the pebble, how fast will the pebble be traveling with respect to the center of the wheel when it flies out tangentially?

28. [I] A 10.0-kg mass is tied to a 3/16-in. Manila line, which has a breaking strength of 1.80 kN. What is the maximum speed the mass can have if it is whirled around in a horizontal circle with a 1.0-m radius and the rope is not to break?

29. [I] A baseball player rounds second base in an arc with a radius of curvature of 4.88 m at a speed of 6.1 m/s. If he weighs 845 N, what is the centripetal force that must be acting on him? Notice how this limits the tightness of the turn. What provides the centripetal force?

30. [I] Compute the Earth's centripetal acceleration toward the Sun. Take the time it takes to go once around as 365 d and the radius on average as 1.50×10^8 km.

31. [I] A hammer thrower at a track-and-field meet whirls around at a rate of 2.0 revolutions per second, revolving a 16-lb ball at the end of a cable that gives it a 6.0-ft effective radius. Compute the inward force that must be exerted on the ball.

32. [I] A youngster on a carousel horse 5.0 m from the center revolves at a constant rate, once around in 15.0 s. What is her acceleration?

33. [I] A supersonic jet diving at 290 m/s pulls out into a circular loop of radius R. If the craft is designed to withstand forces accompanying accelerations of up to $9.0\,g$, compute the minimum value of R.

34. [I] A test tube in a centrifuge is pivoted so that it swings out horizontally as the machine builds up speed. If the bottom of the tube is 150 mm from the central spin axis, and if the machine hits 50 000 revolutions per minute, what would be the centripetal force exerted on a giant amoeba of mass 1.0×10^{-8} kg at the bottom of the tube?

35. [I] A test tube 100 mm long is held rigidly at 30° with respect to the vertical in a centrifuge with its top lip 5.0 cm from the central spin axis of the machine. If it rotates at 40 000 revolutions per minute, what is the centripetal acceleration of a cell at the bottom of the tube?

Figure P36

36. [II] Drop a string of length L through the hole in a vertically held spool of thread (Fig. P36). Tie 10 paper clips to each end and twirl the string so that one bunch of clips moves in a horizontal circle at a speed v, while the other hangs vertically. Neglecting friction, (a) write an expression for the distance d as a function of v and L where $L = (r + d)$. (b) Write an expression for g in terms of v and r, the radius of the circle.

37. [II] A front-loading clothes washer has a horizontal drum that is thoroughly perforated with small holes. Assuming it to spin dry at 1 rotation per second, have a radius of 40 cm, and contain a 4.5-kg wet Teddy bear, what maximum force is exerted by the wall on the bear? What happens to the water?

38. [II] A circular automobile racetrack is banked at an angle θ such that no friction between road and tires is required when a car travels at 30.0 m/s. If the radius of the track is 400 m, determine θ.

39. [II] A skier, with a speed v and a mass m, comes down a slope that has the shape of a vertical segment of a circle (of radius r), ending in a tangential, flat, horizontal run. Write an expression in terms of v, m, r, and g for the normal force exerted by the snow on the skis at the bottom just before the skier leaves the circular portion.

40. [II] A 1000-kg car traveling on a road that runs straight up a hill reaches the rounded crest at 10.0 m/s. If the hill at that point has a radius of curvature (in a vertical plane) of 50 m, what is the effective weight of the car at the instant it is horizontal at the very peak?

41. [II] At a given instant, someone strapped into a roller coaster car hangs upside down at the very top of the circle (of radius 25.0 m) while executing a so-called loop-the-loop. At what speed must he be traveling if at that moment the force exerted by his body on the seat is half his actual weight?

42. [II] Suppose you wish to whirl a pail full of water in a vertical circle without spilling any of its contents. If your arm is 0.90 m long (shoulder to fist) and the distance from the handle to the surface of the water is 20.0 cm, what minimum speed is required?

43. [II] A cylindrically shaped space station 1500 m in diameter is to revolve about its central symmetry axis to provide a simulated 1.0-g environment at the periphery. (a) Compute the necessary spin rate. (b) How would "g" vary with altitude up from the floor (which is the inside curved wall of the cylinder)?

44. [II] A stunt pilot flying an old biplane climbs in a vertical circular loop. While upside down, the force acting upward normally on the seat is $\frac{1}{3}$ of her usual weight. If the plane is traveling at that moment at 300 km/h, what is the radius of the loop? (A World War I pilot did this little trick without fastening his seat belt and—you guessed it—fell out. What can you say about his speed at the top of the loop?)

45. [III] Take the Earth to be a perfect sphere of diameter 1.274×10^7 m. If an object has a weight of 100 N while on a scale at the south pole, how much will it weigh at the equator? Take the equatorial spin speed to be $v = 465$ m/s.

46. [III] Design a carnival ride on which standing passengers are pressed against the inside curved wall of a rotating vertical cylinder. It is to turn at most at $\frac{1}{2}$ revolution per second. Assuming a minimum coefficient of friction of 0.20 between clothing and wall, what diameter should the ride have if we can safely make the floor drop away when it reaches running speed?

FRICTION

47. [I] A dog weighing 300 N harnessed to a sled can exert a maximum horizontal force of 160 N without slipping. What is the coefficient of static friction between the dog's foot pads and the road?

48. [I] Assume that your mass is 70.0 kg and that you are wearing leather-soled shoes on a wooden floor. Now walk over to a wall and push horizontally on it. How much force can you exert before you start sliding away?

49. [I] What is the maximum acceleration attainable by a 4-wheel-drive vehicle with a tires-on-the-road static coefficient of μ_s?

50. [I] Mass m_1 sits on top of mass m_2, which is pulled along at a constant speed by a horizontal force F. If $m_1 = 10.0$ kg, $m_2 = 5.0$ kg, and μ_k for all surfaces is 0.30, find F.

51. [I] Someone wearing leather shoes is standing in the middle of a wooden plank. One end of the board is gradually raised until it makes an angle of 17° with the floor, at which point the person begins to slide down the incline. Compute the coefficient of static friction.

52. [I] A 30.0-kg youngster is dragged around the living room floor at a constant speed (giggling all the while) via a 60-N horizontal force. What was the appropriate pants-carpet friction coefficient?

53. [I] A wooden crate containing old $1000 bank notes has a total mass of 50.0 kg. It is transported on a flatbed truck to a facility to be burned. If the coefficient of static friction between the bed and the crate is 0.3 and the truck begins to climb a 20° incline at a constant speed, will the crate begin to slide?

54. [I] A garbage can partly filled with coal ash weighs 100 N, and it takes a force of 40 N to drag it down to the street at a uniform speed. How much force will it take to drag a can full of ash weighing 150 N?

55. [I] A 60.0-kg woman and a 20.0-kg child are both standing still in a moving passenger car when the engineer hits the brakes decelerating the train at −1.0 m/s². What must be the appropriate coefficients of friction between floor and shoes if neither of them is to slip?

56. [II] What is the steepest incline that can be climbed at a constant speed by a 4-wheel-drive vehicle having a tires-on-the-road coefficient of static friction of 0.90?

57. [II] A crate is being transported on a flatbed truck. The coefficient of static friction between the crate and the bed is 0.50. What is the minimum stopping distance if the truck, traveling at 50.0 km/h, is to decelerate uniformly and the crate is not to slide forward on the bed?

58. [II] A yellow Triumph TR8 automobile with a total weight, including the driver, of 2850 lb is turning a flat curve of radius 200 m at 30 m/s. (a) What is the minimum value of μ_s that must exist between the tires and the road if no skidding is to occur? (b) Will this be possible on icy concrete?

59. [II] Assume an even weight distribution on all four tires of a car with 4-wheel antilock brakes. What is the minimum stopping time from 27 m/s if the coefficients of static and kinetic friction are 0.9 and 0.8, respectively?

60. [II] Place a book flat on a table and press down on it with your hand. Now suppose the hand-book and table-book values of μ_k are 0.50 and 0.40, respectively; the book's mass is 1.0 kg and your downward push on it is 10 N. How much horizontal force is needed to keep the book moving at a constant speed if the hand is stationary?

61. [II] A skier on a 4.0° inclined, snow-covered run skied downhill at a constant speed. Compute the coefficient of friction between the waxed skis and snow on that day when the temperature was around 0°C. Why is it that if the temperature were to drop to −10°C, μ_k would rise to around 0.22? Neglect air friction.

62. [II] A youngster shoots a bottle cap up a 20° inclined board at 2.0 m/s. The cap slides in a straight line, slowing to 1.0 m/s after traveling some distance. If $\mu_k = 0.4$, find that distance.

63. [II] A 100-kg bale of dried hay falls off a truck traveling on a level road at 88.0 km/h. It lands flat on the blacktop and skids 100 m before coming to rest. Assuming a uniform deceleration, compute the coefficient of kinetic friction.

64. [II] A 100-kg trunk loaded with old books is to be slid across a floor by a young woman who exerts a force of 300 N down and forward at 30° with the horizontal. If $\mu_k = 0.4$ and $\mu_s = 0.5$, compute the resulting acceleration.

65. [II] Suppose the woman in Problem 64 puts aside some of the books so that the mass of the load is 50 kg. (a) What will the acceleration be now? (b) A bit annoyed, she squirts some oil under the trunk so that $\mu_s = 0.4$ and $\mu_k = 0.3$. What now?

66. [II] One of two 200-kg crates is dragged along the floor at a constant speed by a horizontal force of 200 N. How much force would it take to pull them both at some constant speed if (a) they are tied together one behind the other? (b) the second is stacked on top of the first?

67. [III] What is the steepest incline that can be climbed at a constant speed by a rear-wheel-drive auto having a tires-on-the-road μ_s of 0.9 and 57% of its weight on the front axle?

68. [III] Determine the force (F) needed to keep the blocks depicted in Fig. P68 moving at a constant speed—m_2 to the right and m_1 to the left. Assume both the frictionless pulley and the rope to be massless: m_1 has a weight of 5.0 N; m_2 has a weight of 10 N; and $\mu_k = 0.40$ for all surfaces.

Figure P68

69. [III] A greasy flatbed truck is carrying a crate weighing 2.0 kN. The truck accelerates uniformly from rest to a speed of 30 km/h in a distance of 30.0 m. In that time, the crate slides 1.0 m back toward the end of the truck. Compute the coefficient of friction between bed and box.

6

Equilibrium: Statics

WE NOW FOCUS ON a special case where, even though two or more external forces act on a body, their effects cancel and there is no change in motion. Such a body is said to be in **equilibrium**—it may or may not be moving, but its motion is unchanging (**a** = 0). The associated discipline is called **statics**. In Chapter 29, we will learn that absolute rest is meaningless. Still, it is convenient for planet-bound observers to consider a thing to be at rest when it is not moving with respect to the Earth (the Golden Gate Bridge is, in that manner of speaking, at rest). An object at rest is in **static equilibrium**.

In practical terms, statics is the study of structures—the analysis of the

Tibia

Achilles tendon

$F_N = F_W$

(a)

F_W

$2.5\,F_W$

$3.5\,F_W$

(b)

Figure 6.1   (a) The foot on tiptoe. Ligaments hold the bones together, and tendons attach the muscles to the bone. Think of the foot pivoted at the ankle joint, with the Achilles tendon pulling upward. (b) The foot can be modeled, and, as we'll see presently, the forces in the tibia and tendon can be computed.

forces acting on and within skyscrapers, towers, wrenches, crutches, and even people. Structurally, the human body shares remarkable similarities with high-rise buildings and trestle bridges. Moreover, about 500 years ago, Leonardo da Vinci discovered that the bones and muscles of vertebrates form a system of levers. We move most of our body parts (fingers, arms, neck, back, legs, feet, and so on) using muscles to pivot one bone against another. For example, when the calf muscles contract, the Achilles tendon pulls up on the heel bone (Fig. 6.1). The foot pivots in the ankle joint, forcing the toes down, and up you go. As we will see, this mechanism is identical to a wide range of systems, from the crowbar to the seesaw. Among other things, knowing the statics of the human body allows us to design efficient sports gear, enhance the techniques of physical therapy, better repair broken bones and damaged teeth, and make improved prostheses.

There are two governing principles in statics, both of which derive from Newton's Second Law: the *sum-of-the-forces equals zero,* and the *sum-of-the-torques equals zero* (p. 174). The first, by itself, can deal with situations where the lines-of-action of all the forces on a body intersect at a single point and are said to be *concurrent.* The second principle must be used when the forces are not concurrent. Concurrent-force systems are easier to analyze, and so after some preliminaries we will focus on them, develop technique, and go on to nonconcurrent systems.

TRANSLATIONAL EQUILIBRIUM

When every part of a body travels with the same speed in the same fixed direction, we have *translational motion,* as compared to *rotational motion,* where the speeds are different and the direction changes from moment to moment. *Translational equilibrium* corresponds to straight-line motion along a fixed direction at a constant speed (which, in this chapter, will be taken to be zero).

If the lines-of-action of all the forces acting on a body lie in a single plane, the system of forces is *coplanar.* Many practical arrangements of cables, pulleys, muscles, and machines correspond to coplanar force systems. Noncoplanar systems are also commonplace (in fact, you are probably sitting on one—the four forces supporting the legs of a chair are not coplanar). Forces not confined to a common plane require more effort to analyze, and so this treatment is limited to the study of *equilibrium under coplanar concurrent forces.*

It follows from Newton's Second Law [Eq. (5.4)] that translational equilibrium (**a** = 0) occurs when

$$\Sigma \mathbf{F} = 0 \qquad (6.1)$$

and this is called the **First Condition of Equilibrium**. If we resolve all the forces acting on a body into their components along *any* two perpendicular axes (for example, horizontal and vertical), the equivalent scalar statement is

$$\Sigma F_H = 0 \qquad \text{and} \qquad \Sigma F_V = 0 \qquad (6.2)$$

These analytic tools are all we will require to deal with concurrent-force systems. What we need now is to learn how they are applied in practice.

6.1   Parallel and Colinear Force Systems

Figure 6.2a shows a pot of begonias resting on an essentially weightless pedestal. The Earth-plant gravitational interaction causes the begonias to have a weight

These people have reached terminal speed. The weight of each diver acting down exactly equals the drag force acting up, or $\Sigma F_V = 0$. Hence, $a = 0$ and each person is in translational equilibrium even though he or she is falling at about 160 km/h.

$F_{Wb} = 300$ N, and the several resulting forces that arise in response are depicted in the diagram. If the potted plant were weightless, those forces would be zero. As is, they all act along the same line and are said to be **colinear**.

As shown in the free-body diagram, only two forces act on the flowers. Given that the begonias are in static equilibrium ($a = 0$), and taking *up* as positive, we have

$$^{+\uparrow}\Sigma F_V = (+F_{PB}) + (-F_{Wb}) = 0$$

and

$$F_{PB} = F_{Wb}$$

The same logic for the pedestal yields

$$^{+\uparrow}\Sigma F_V = (+F_{GP}) + (-F_{BP}) = 0$$

and

$$F_{GP} = F_{BP}$$

Hence $F_{GP} = F_{BP} = F_{PB} = F_{Wb} = 300$ N. The ground holds up the begonias.

When two colinear forces acting on a body point toward each other, the body is under compression—the forces on the pedestal tend to compress or squash it. An object in tension pulls, an object in compression pushes. The external forces that support a system are called **reaction forces**: Here, the ground exerts a vertical reaction force on the pedestal, the normal force.

Let's take things one step further and add to a system of vertical forces a horizontal force (Fig. 6.3). Each tower of the World Trade Center weighs about 1.2×10^9 N (1.4×10^5 tons). Suppose a horizontal wind averaging 50 mi/h blows against a face of one of the towers, having an area of roughly 1.63×10^4 m^2 (1.75×10^5 ft^2). From measurements, we know that such a wind produces a pressure of around 3×10^2 N/m^2 (6 lb/ft^2) and would result in a total horizontal force on the tower of 5×10^6 N ($\approx 1 \times 10^6$ lb). It follows from $\Sigma F_H = 0$ that the ground must exert a horizontal reaction force of 5×10^6 N on the tower in the opposite direction to the wind, or else the building would begin to slide across town (Fig. 6.3b). So a system can experience both horizontal and vertical reaction forces.

Most buildings are constructed on a frame of wood, reinforced concrete, or steel, and that structure carries the loads of weight and wind down to the ground. Skyscrapers were once made rigid with heavy masonry walls, but today they are constructed on a lighter frame, reinforced with steel stiffening members. The human body is constructed on a similar frame of bones, ligaments, and muscles. Loads in the upper body are carried by the spine to the pelvis and then down

Figure 6.2 (a) A 300-N pot of begonias rests on an essentially weightless pedestal. The gravitational force on the pot gives rise to two interaction pairs ($F_{BP} = -F_{PB}$ and $F_{PG} = -F_{GP}$). (b) These both equal 300 N because the pedestal's weight is negligible.

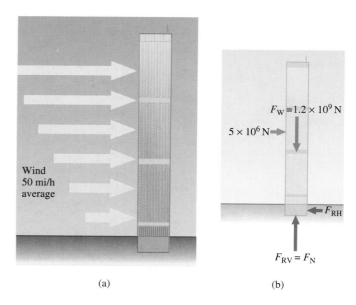

Figure 6.3 The force exerted by wind increases with height and is a major consideration with buildings over 20 or 30 stories. About 10% of the structural weight of a high-rise building goes into wind bracing. (a) One of the 1350-ft-tall towers of the World Trade Center in New York City. A wind with an average speed of 50 mi/h blows against one face and pushes with 5×10^6 N. (b) The free-body diagram shows the needed reaction forces.

through the legs to the ground. The bones of the skeleton are primarily compressive members that carry both body weight and the forces exerted by the muscles. The skeletal muscles are attached at both ends via tendons to different bones. These muscles can only contract, pulling on the bones and putting them in compression.

Ropes and Pulleys

Return to Fig. 5.9, which depicts a brass mass attached to an essentially weightless rope hanging from a ceiling hook. The mass, the rope, and the hook are separately in equilibrium and, for each, $\Sigma F_V = 0$. The hook holds up the rope, and the rope holds up the mass. A single length of rope supporting a load is called a *hanger*. **When there is only one hanger, the tension in it equals the load:** $F_T = F_{Wb}$.

If two lengths of light rope support a load, such as the 300-N pot in Fig. 6.4, the tension in each length is 150 N and the force on each hook is 150 N. If three ropes equally share the 300-N load, the tension in each is 100 N and the force on each ceiling hook to which a rope is attached is 100 N. In each case, the tension in every hanger supporting the load is equal. Figure 6.5 incorporates an essentially weightless, frictionless pulley that distributes the load equally on the two rope segments; other than that, the situation is identical to Fig. 6.4. We apply a force of 150 N, and the system delivers a force of 300 N. This arrangement is a *force-multiplier*, one of the most important simple machines ever devised. It's often set up in a horizontal plane (anchored to a tree to haul heavy loads along the ground, for example).

Figure 6.6 indicates how any number of pulleys can be used to multiply the applied force; in this case by a factor of 4. Several pulleys can be mounted next to, or just below, one another (Fig. 6.7). This ancient device is still used today extensively in applications such as elevators, scaffolds, tow trucks, cranes, and pile drivers. A typical derrick can hoist a 25-ton load into the air via pulleys with 10 or 11 lengths of steel cable sharing the effort.

Each of the pulleys in Fig. 6.8 serves to change the direction of the applied force rather than act as a multiplier. A pulley used in this way is a convenient means of

Figure 6.4 (a) A load supported by two ropes. Since the system is in equilibrium, $\Sigma F_V = 0$. (b) The net upward force ($2F_T$) equals the net downward force (F_W). Here you count the number of ropes, just as you would count the number of legs supporting a table.

Figure 6.5 (a) This pulley arrangement acts as a force multiplier. Count the number of ropes supporting the load. In this case, there are two, and the system applies twice the force applied to it. The hand lifts with 150 N and the system lifts with 300 N. (b) The tension in each supporting length of rope is 150 N. The total upward force is therefore 300 N.

Figure 6.6 (a) When four ropes support the load in equilibrium, each carries an equal tension of $\frac{1}{4}$ that load. (b) Here, the tension is 100 N and the hand need only apply 100 N to hold the load, or a tiny bit more to start it accelerating upward.

applying a constant force (equal to the hanging weight) to a system, whether it's an accelerating mass in a physics lab or a broken leg in a hospital.

6.2 Concurrent Force Systems

A system acted on by concurrent forces (forces that all pass through a single point) is in equilibrium when

$$\Sigma F_x = 0 \qquad \text{and} \qquad \Sigma F_y = 0$$

where x and y are any two perpendicular directions. There are therefore two independent equations, and no more than two unknowns can be determined. Each force vector in the plane of the system corresponds to two quantities: either a magnitude and the angle at which it acts, or any two of its perpendicular components. In general, two equations will be needed to determine a force vector. In practice, however, we often know the line-of-action of the force and require only the magnitude. The nice thing about ropes, cables, and muscles is that they always act in tension, and the forces applied to and by them act along them.

Figure 6.7 (a) This configuration is equivalent to that of Fig. 6.6. (b) Four ropes support the load, and the hand need only pull down with 100 N.

Figure 6.8 It's been common practice since the fourteenth century to use ropes and pulleys to put broken limbs in traction. Shown here is a Thomas leg splint with a Pearson attachment.

Example 6.1 The engine hanging motionlessly in Fig. 6.9 weighs 800 N. If θ is 20.0°, compute (a) the tension in each length of rope and (b) the horizontal force tending to pull out the support pins. Use Table 6.1

Figure 6.9 (a) The engine is supported by a system of ropes, and all the forces acting on that arrangement are concurrent. (b), (c), (d), and (e) are free-body diagrams. (f) Because $\Sigma\mathbf{F} = 0$, the three forces must form a closed triangle such that the resultant is zero.

(continued)

(continued)

to pick an appropriate rope for the job. To avoid failure from high momentary forces (sudden jerks, for example), it's customary to use ropes at no more than about one-sixth their breaking strength.

Solution: [Given: $F_W = 800$ N and $\theta = 20.0°$. Find: tensions F_{T1}, F_{T2}, F_{T3}, and the pull-out force.] The system consists of the three ropes and the load—its free-body diagram is Fig. 6.9b. The force vectors acting on the system—F_{T2}, F_{T3}, and F_W—are concurrent, and pass through the knot. A nice place to start is with F_{T1}; the engine is supported by a single hanger, so the tension in that rope equals the load

$$\boxed{F_{T1} = F_W = 800 \text{ N}}$$

Because of the symmetry, the tensions on either segment of the V-shaped support rope must be equal ($F_{T2} = F_{T3}$). Whatever happens, happens identically to both segments. Let's not take advantage of that fact now, but just press on with the analysis step by step. We need F_{T2} and F_{T3}, and could use either Fig. 6.9b for the whole system or Fig. 6.9c for just the knot. Each must be in equilibrium, and each contains the two forces we want. Take up and to the right as positive. The sum-of-the-horizontal-forces equals zero on the knot, and so

$$\overset{+}{\to}\Sigma F_H = F_{T2} \cos\theta - F_{T3}\cos\theta = 0$$

confirming that $F_{T2} = F_{T3}$. The sum-of-the-vertical-forces equals zero on the knot, and so

$$\overset{+}{\uparrow}\Sigma F_V = F_{T2}\sin\theta + F_{T2}\sin\theta - F_W = 0$$

Hence $\qquad 2F_{T2}\sin 20.0° = 800$ N

and $\boxed{F_{T2} = 1.17 \text{ kN}}$. Note that the tension in either supporting rope is greater than the load.

At each wall pin (Fig. 6.9d), the horizontal force is

$$F_{T2}\cos 20.0° = \boxed{1.10 \text{ kN}}$$

Since $F_{T1} = 800$ N, and $F_{T2} = 1.17$ kN, none of the 3/16-in. ropes will suffice. The braided nylon and Dacron™ 1/4-in. lines will do nicely.

▶ **Quick Check:** Make a sketch of the three force vectors acting at the knot. They must add up to zero and form a closed triangle, and they do (Fig. 6.9f).

TABLE 6.1 Various line strengths

| Diameter (inches) | Breaking strength* | | | | | |
	Manila hemp (kN)	Nylon filament (kN)	Dacron™ filament (kN)	Nylon braid (kN)	Dacron™ braid (kN)	Cable (kN)
3/16	1.8	4.0	4.0	4.6	4.6	19
1/4	2.4	6.6	6.7	7.1	8.2	31
5/16	4.0	9.6	10.3	12.0	12.8	44
3/8	5.4	14.9	15.4	16.9	18.4	64
7/16	7.0	20.0	20.0	23.1	25.0	78
1/2	10.6	27.1	28.3	32.0	32.7	101

*As a rule, working loads should not exceed one-sixth of the breaking strength for ropes and one-fifth for cable.

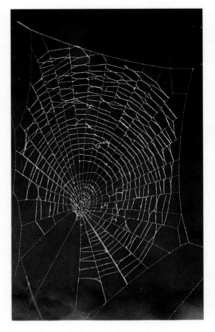

A spider web is a total tension structure. Every fiber is in tension.

Notice in Fig. 6.9 that, if F_W has any nonzero value, θ cannot equal zero because sin 0 equals zero, and an undeflected rope exerts no upward force. The rope must sag as it interacts with the object being supported. In fact, the rope will sag under its own weight. No amount of horizontal force can pull the rope so that it straightens and hangs totally horizontal.

An interesting application of statics for the treatment of congenital hip dislocation. The spring scale reads the tension applied to each leg. The little turnbuckle (on the left of the scale) allows the tension to be adjusted as needed. The double-Y-shaped frame keeps the legs apart and allows $\Sigma \mathbf{F} = 0$.

Example 6.2 Determine the angle θ in the arrangement shown in Fig. 6.10. Assume the pulley is weightless and frictionless.

Solution: [Given: the angle is 20.0°, the leg weight 150 N, and the counterforce 200 N. Find: θ.] With no friction on the weightless pulley, the tension in the rope on the right is uniform throughout. Draw the free-body diagram of the ring (Fig. 6.10b). It should be evident that $F_{T2} = 200$ N, $F_{T1} = 150$ N, and, because θ may not equal 20°, F_{T3} may not equal F_{T2}. At the ring

$$\overset{+}{\rightarrow} \Sigma F_{\mathrm{H}} = (200 \text{ N}) \cos 20.0° - F_{T3} \cos \theta = 0$$

and

$$\overset{+\uparrow}{} \Sigma F_{\mathrm{V}} = (200 \text{ N}) \sin 20.0° + F_{T3} \sin \theta - 150 \text{ N} = 0$$

Hence, $F_{T3} \cos \theta = 187.9$ N and $F_{T3} \sin \theta = 81.60$ N. These are the horizontal and vertical components of F_{T3}, and if we divide the second by the first, the tension cancels. We get

$$\frac{\sin \theta}{\cos \theta} = \tan \theta = 0.434\,274$$

and

$$\boxed{\theta = 23.5°}$$

▶ **Quick Check:** $F_{T3} = (187.9 \text{ N})/\cos \theta = 204.9$ N, and $F_{T3} = (81.60 \text{ N})/\sin \theta = 204.9$ N. Figure 6.10c is the force triangle, and it correctly has a zero resultant.

Figure 6.10 (a) A concurrent force system. (b) The free-body diagram of the ring showing the three forces that meet at its center. (c) The force triangle corresponding to equilibrium.

This person is acted upon by five forces. His gravitational interaction with the Earth produces a downward force F_W. The wall (F_1), table (F_2), and chair (F_3 and F_4) each exert normal forces on him. Since he's in equilibrium, the sum-of-the-forces *on him* equals zero.

Bars, Bones, and Trusses

Many modern structures, from billboards to bridges, are supported by a rigid framework of slender, short, interconnected members called *bars* (which are usually made of steel, aluminum, or wood). We will restrict our treatment to rigid structural members that are light compared to the supported loads. These are attached together by smooth pins (bolts or rivets) at their ends at what are called *joints*. All loads are applied at the joints. This construction has the effect of allowing *only forces that act along the length of each bar*. The bars are either in pure tension or compression, and no bending occurs. To prove that contention, imagine a free-body diagram of a typical member, say bar 1 in Fig. 6.11. Bar 1 is at rest, and if it is to stay at rest, the net force acting on each of its two ends must be equal in magnitude, oppositely directed, and colinear.

Example 6.3 Two lightweight, pinned bars support a 300-N sign (Fig. 6.11). If the angle θ is 30.0°, compute the magnitudes of the forces acting on each member and determine if they are in tension or compression.

Solution: [Given: $F_W = 300$ N and $\theta = 30.0°$. Find: the forces on the bars, F_1 and F_2.] The weight of the sign pulls downward on the joint, and that force must be countered entirely by bar 2 because bar 1 is horizontal. *The internal forces can only act along the line of either pinned member.* As shown in Fig. 6.11c, the free-body diagram of the pin supporting the sign, bar 2 pulls upward and to the left on the joint and is in tension. Bar 2 exerts a force to the left (tending to squash bar 1). This force must be countered by bar 1, which is in compression. If, in your diagram, you put the forces in the wrong directions, they will simply come out negative, so don't worry about that. Thus, for the pin, where the forces are concurrent

$$+\uparrow\Sigma F_V = F_2 \sin 30.0° - F_W = 0$$

and

$$\overset{+}{\rightarrow}\Sigma F_H = F_1 - F_2 \cos 30.0° = 0$$

(continued)

(continued)

Hence, $F_2 = (300 \text{ N})/\sin 30.0° = \boxed{600 \text{ N}}$ and $F_1 = (600 \text{ N}) \cos 30.0° = \boxed{520 \text{ N}}$; both are positive, so they are correct as drawn.

▶ **Quick Check:** Since the system is in equilibrium, the force vectors must add up to zero and so form a closed triangle. Hence, F_2^2 must equal $F_1^2 + F_W^2$, which it does (Fig. 6.11d).

(a)

(b)

(c)

Figure 6.11 (a) A system of concurrent forces. (b) The free-body diagram of bar 1. The only way it can be in equilibrium is if the forces on its ends are equal, oppositely directed, and lie along the axis of the bar. The same is true for bar 2. (c) The free-body diagram of the pin, showing the three forces that intersect at that point. (d) The equilibrium force triangle.

Figure 6.12 (a) The three-bar frame is stable. Push on it, and it will not collapse. The bars cannot stretch or compress to allow the three-bar frame to fold up. (b) The four-bar frame is unstable. Push on that frame, load it with a horizontal force, and it will collapse (c).

A simple *truss* is a structural framework of members joined together at their ends in triangular patterns lying in a plane. Trusses are used in aircraft, space stations, antennas, TV towers, bridges, and buildings. Three-sided component frames are the rule because they are inherently rigid (Fig. 6.12). The truss, using lightweight components of moderate length, provides strength and great space-spanning capability.

In the early days of the railroads, a variety of truss bridges were developed. These were economical, easy to build at remote sites, and fairly rigid under heavy loads. In the United States, where iron was expensive, timber beams were used in compression, with thin iron rods as the tension members. Figure 6.13 shows one of these truss bridges, and it immediately brings to mind the musculoskeletal system of quadrupedal vertebrates, with the spine spanning the space between the shoulder girdle and the pelvis.

A clutter of steel trusses supporting an overhead railroad track. Today, high-strength bolts have replaced the rivets that can be seen here.

Figure 6.13 (a) An old railroad trusswork bridge made of timber and braced with iron tension rods. (b) Vertebrates, such as this lizard, are supported by a similar system of bone in compression, crisscrossed by muscles in tension.

In humans and other animals, the compression members—vertebrae and ribs—are made of bone. The tension members are the intercostal muscles that crisscross between the ribs, pulling them together. Similarly, if the bridge were to sag appreciably under a load, the vertical beams would have to spread apart at their lower ends, an action prevented by the tension crossbars that pull the beams together. The spine-rib-muscle structure is the bridge supporting the weight of the trunk of the body. The four legs are the compression columns on which the entire load rests. So effective is the musculoskeletal system that camels, elephants, donkeys, and horses can easily carry additional loads on their backs.

ROTATIONAL EQUILIBRIUM

The equal-arm balance was already in widespread use well over 4000 years ago. It is not surprising, then, that Aristotle, in his book *Mechanics,* attempted to analyze the practical problem represented by balances, levers, and seesaws. When equal downward forces are applied to each end of a centrally pivoted rod, the system remains balanced horizontally. The beam does not revolve (or, more generally, there is no *change* in its rotational motion), and we say that it is in *rotational equilibrium.* The word *equilibrium* comes from the Latin words *aequus,* meaning even or equal, and *libra,* meaning balance.

The general problem of the lever, where the weights and arm lengths are different on each side of the pivot, was treated (though not altogether satisfactorily) by Aristotle. The job was completed by that genius of Greek science, Archimedes of Syracuse (287 B.C.–212 B.C.). Their rather rudimentary *Law of the Lever* states that

Archimedes, who was kinsman and a friend of King Hieron of Syracuse, wrote to him that with any given force it was possible to move any given weight, and emboldened, as we were told, by the strength of his demonstration, he declared that, if there were another world, and he could go to it, he could move this. Hieron was astonished, and begged him to put his proposition into execution, and show him some great weight moved by a slight force. Archimedes therefore fixed upon a three-master merchantman of the royal fleet, which had been dragged ashore by the great labors of many men, and after putting on board many passengers and the customary freight, he seated himself at a distance from her, and without any great effort, but quietly setting in motion with his hand a system of compound pulleys, drew her towards him smoothly and evenly, as though she were gliding through the water.

PLUTARCH (ca. 46–120 A.D.)
Life of Marcellus

Figure 6.14 Unequal parallel forces acting at unequal distances from the pivot. The system is in rotational equilibrium.

unequal forces, acting perpendicularly on a pivoted bar, balance each other, provided $F_1r_1 = F_2r_2$ (Fig. 6.14). One must consider the sizes of the forces, as well as the distances at which they act from the pivot—that much can easily be confirmed experimentally.

Example 6.4 A boy (mass 30 kg) wishes to play on a centrally pivoted seesaw with his dog Irving (mass 10 kg). When the dog sits 3.0 m from the pivot, where must the boy sit if the 6.5-m-long board is to be balanced horizontally?

Solution: [Given: $m_B = 30$ kg, $m_D = 10$ kg, the dog-pivot distance $r_D = 3.0$ m, and the extraneous board length is 6.5 m. Find: the boy-pivot distance r_B.] The forces are the weights $F_{WB} = m_B g$ and $F_{WD} = m_D g$; therefore, from the Law of the Lever

$$(m_B g)r_B = (m_D g)r_D$$

Hence
$$r_B = \frac{(10 \text{ kg})(3.0 \text{ m})}{(30 \text{ kg})} = 1.0 \text{ m}$$

The boy must sit 1.0 m from the pivot.

▶ **Quick Check:** $m_B r_B = m_D r_D = 30$ kg·m.

The Roman balance, or steelyard (Fig. 6.15), is a practical example of the unequal-arm balance. The steelyard has been used for at least 2000 years to measure heavy objects by counterpoising them against small standard weights. You have probably been weighed in a physician's office on a platform scale that is essentially a steelyard. Extending the idea, da Vinci realized that the human body achieved its mobility via various kinds of levers. Both the human foot (Fig. 6.1a) and the steelyard are levers pivoted at a point between the two forces. da Vinci was also aware of the significance of the *moment of a force,* a concept called the *torque,* which is central to the modern treatment of statics.

6.3 Torque

Figure 6.16 shows situations in which a force **F** acts on a wrench along different directions and at different points. The axis about which rotation will occur runs down the center of the bolt through point O perpendicular to the plane of the diagram. The point of application of the force is marked by the end of the *position vector* **r** drawn from O. In each case, the ability of **F** to produce a rotation of the wrench will be different. It is this turning effect, or *moment,* that da Vinci appreciated as crucial. When the line-of-action of the force passes through the pivot, no rotation will occur at all (Fig. 6.16a). You cannot close a door by pushing on its edge *toward* the

Figure 6.15 An unequal-arm balance, or steelyard. The small weight, the *poise*, is moved down the arm until a balance is reached with the heavy object being weighed. The balanced steelyard is a system of separated forces whose lines-of-action are parallel and therefore nonconcurrent.

hinge. By contrast, the force applied at the very end of the wrench, where the line-of-action is far from O, as in Fig. 6.16c, produces the greatest twist. In fact, it's precisely for that reason that the wrench was invented.

The **moment-arm** *of the force with respect to the axis passing through O is defined as the perpendicular distance ($r_\perp$) drawn from O to the line-of-action of* **F**. In each part of Fig. 6.16, notice that, although the magnitude of the force is unchanged, the moment-arm increases from $r_\perp = 0$ in (a), to $r_\perp = r \sin \theta$ in (b), to $r_\perp = r$ in (c). The **moment of the force** *about O is defined as the product of F and the moment-arm.* We symbolize the moment of the force by the Greek letter *tau* (τ) and particularize it with respect to point O as τ_0. Thus

$$\tau_0 = r_\perp F$$

This quantity is called the **torque** (from the Latin *torquere*, to twist). The term "moment of the force" is also used. The torque is a measure of the twist produced by a force about a particular axis that may be located *anywhere*. Torque has the dimensions of force multiplied by distance and the SI units of N·m.

In Fig. 6.17a, the line-of-action of the force vector **F** makes an angle of θ with the line-of-action of the position vector **r**. Here, $r_\perp = r \sin \theta$, and so

$$\tau_0 = Fr_\perp = Fr \sin \theta$$

Alternatively, resolve **F** into its components (**F**$_\parallel$, parallel to **r**, and **F**$_\perp$, perpendicular to **r**). **F**$_\parallel$ passes through O, its moment-arm is zero, and its torque about O is also zero. It follows that, with respect to O, the torque produced by **F**$_\perp$ must equal the

Figure 6.16 The amount of twist, or torque, one can get from a wrench depends on how big the force is, in what direction it acts, and where it is applied. (a) No torque is exerted when the force goes through the pivot point O. (b) Here, torque about O exists. It can be increased (c) by increasing θ or decreased (d) by decreasing r.

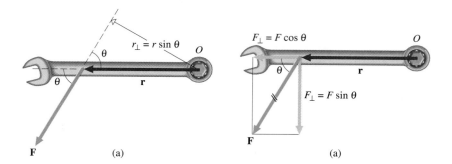

Figure 6.17 The torque can be thought of as $Fr_\perp$ or $F_\perp r$, since $Fr_\perp = F_\perp r = Fr \sin \theta$.

torque produced by **F** as a whole. Since $F_\perp$ (which equals $F \sin \theta$) has a moment-arm of r, its torque about O is

$$\tau_0 = rF_\perp = rF \sin \theta \qquad (6.3)$$

which is the same result we obtained earlier. The torque equals either $r_\perp F$ or $rF_\perp$, and we use whichever form is the more convenient.

Example 6.5 The exercise illustrated in Fig. 6.18 consists of swinging the lower leg out until it's horizontal and then lowering it, all the while supporting a mass m. Write an expression for the torque about O as a function of θ, m, g, and the knee-heel length r. For the time being, ignore the mass of the leg.

Solution: [Given: m, r, and θ. Find: τ_0 where O is at the knee joint.] By definition

$$\tau_0 = r_\perp F$$

The downward force is the weight mg. The moment-arm is the perpendicular distance from O to the line-of-action of the force, that is

$$r_\perp = r \sin \theta$$

where θ is the angle between **r** and **F**. Hence

$$\boxed{\tau_0 = mgr \sin \theta}$$

Figure 6.18 The weight exerts a torque about O that is countered by an equal and opposite muscle-generated torque.

▶ **Quick Check:** The torque varies with $\sin \theta$; reasonably, it's zero when $\theta = 0$ and maximum, $\tau_0(\text{max}) = mgr$, when $\theta = 90°$.

When you look down on the diagram in Fig. 6.17, you can see that the force rotates the wrench counterclockwise about O. Some other force could act on the wrench in a different direction to produce an opposing clockwise twist (as when two people arm wrestle, for instance). The resulting twist produced by opposing torques would then be the sum of the two: $\Sigma\tau_0$, taking one positive and the other negative.

Example 6.6   At the moment depicted in Fig. 6.19, the hand exerts a 200-N force at the end of a pivoted 1.0-m-long lever. The return spring fixed to the lever's midpoint pulls back on it with a horizontal force of 80 N. What is the net torque acting on the lever about the pivot?

Solution: [Given: referring to Fig. 6.19b, $F_h = 200$ N, $r_h = 1.0$ m, $F_s = 80$ N, and $r_s = 0.50$ m. Find: τ_0 where O is at the pivot.] The moment-arm of the hand-force (the perpendicular distance from O to the line-of-action of F_h) is just r_h, while the moment-arm of the spring-force is $r_s \cos 60° = (0.50 \text{ m})(0.50) = 0.25$ m. Arbitrarily taking clockwise as positive, we have

$$\circlearrowleft + \Sigma\tau_0 = r_h F_h + (-r_s F_s \cos 60°)$$

and so

$$\circlearrowleft + \Sigma\tau_0 = (1.0 \text{ m})(200 \text{ N}) - (0.50 \text{ m})(80 \text{ N})(0.50)$$

and

$$\boxed{\circlearrowleft + \Sigma\tau_0 = +0.18 \text{ kN·m}}$$

▶ **Quick Check:** The angle between r_s and F_s is 150°; hence, the torque F_s contributes is $-r_s F_s \sin 150°$, which equals $-r_s F_s \cos 60°$.

(a)

(b)

Figure 6.19   (a) A pivoted bar with two forces acting on it. (b) The forces and their moment-arms about point O.

Inasmuch as both **F** and **r** are vectors, the torque has a vector interpretation that includes a meaningful direction. It is customary to envision the torque vector, $\boldsymbol{\tau}$, lying along the axis of the rotation it tends to cause. That is, it passes through O and is *perpendicular to both* **r** *and* **F**. The sense of the vector, which way it points, is suggested by the *right-hand-screw rule*. When the naturally curved fingers of the right hand point in the direction of the force (Fig. 6.20), the outstretched thumb points in the direction of $\boldsymbol{\tau}$. In Fig. 6.16, the wrench tends to rotate counterclockwise about O under the influence of the torque, and the thumb and $\boldsymbol{\tau}$ point up out of the plane of the paper.

This description can be summarized using the mathematical idea of the **cross product**, which is a procedure for multiplying two coplanar vectors in such a way as to produce a third vector. Accordingly, define **C** as the cross product (also called the vector product) of **A** and **B**, such that

$$\mathbf{C} = \mathbf{A} \times \mathbf{B} \tag{6.4a}$$

which is read: **C** equals **A** cross **B**. The symbol × stands for a set of specific requirements: First, **C** is to be perpendicular to both **A** and **B** (that is, perpendicular to the plane containing **A** and **B**). Second, the magnitude of **C** is given by

$$C = AB \sin \theta \tag{6.4b}$$

Figure 6.20   The right-hand-screw rule. An ordinary screw advances when it's turned in the direction in which the fingers on the right hand point. You will probably use this diagram more often to figure out how to turn screws than find torques.

Figure 6.21   To find the direction of the cross product **A** × **B**, curl the fingers of the right hand from **A** through the smallest angle to **B**. The upwardly extended thumb then points in the direction of **C** = **A** × **B**.

Here, θ is the *smaller* of the two angles between **A** and **B**—each vector can be slid along its respective line-of-action until the two are tail-to-tail (Fig. 6.21) and separated by θ. Third, the direction or sense of **C** is determined by placing the fingers of the right hand along the first vector (**A**) in such a way that the fingers can close naturally through θ as they curl toward the second vector (**B**). The thumb then points in the direction of **C**, and the procedure is an embodiment of the right-hand-screw rule.

If **A** × **B** points upward, **B** × **A** points downward; that is, (**A** × **B**) = −(**B** × **A**). It follows, then, from Eq. (6.3) that, with respect to O, the scalar value of the torque is $\tau_0 = rF \sin \theta$, and

$$\boldsymbol{\tau}_0 = \mathbf{r} \times \mathbf{F} \qquad (6.5)$$

This equation says it all.

Example 6.7   A bicycle handbrake works by squeezing the metal rim of the wheel between two rubber pads. Suppose that the maximum inward force on each brake pad is 50 N and the metal-rubber kinetic coefficient of friction is 0.70. Taking the rim-to-axle distance as 30 cm, determine the maximum torque produced by a set of brakes about the axis of rotation.

Solution: [Given: $F_N = 50\ N$, $\mu_k = 0.70$, and $r = 0.30$ m. Find: τ_0, taking O at the axle.] The friction force F_f applied to the rim by each pad (Fig. 6.22) is

$$F_f = \mu_k F_N = 0.70(50\ N) = 35\ N$$

Since the friction is tangential (**r** is $\perp$ to $\mathbf{F}_f$), the moment-arm equals 0.30 m. Hence, for each pad

$$\tau_0 = (0.30\ m)(35\ N) = 10.5\ N \cdot m$$

and so the total torque is $\boxed{21\ N \cdot m}$, opposing the motion of the wheel.

▶ **Quick Check:** If the wheel is seen revolving clockwise as in Fig. 6.22, **r** lies along a vertical line, $\mathbf{F}_f$

Figure 6.22   The torque about O exerted by a brake on a bicycle wheel.

is horizontal, $\theta = 90°$, and $\mathbf{r} \times \mathbf{F}_f = \boldsymbol{\tau}_0$ extends as shown. The twist of $\boldsymbol{\tau}_0$ (via the right-hand-screw rule) is counterclockwise against the rotation.

6.4 Rigid Bodies

Real bodies deform to some extent under the influence of applied forces, but many change so little that we can take them to be totally rigid for all practical purposes. Assume such a body is made up of a large number of separate, stationary, interacting particles (atoms or molecules). If the body is in equilibrium under the action of several externally applied forces, all of its constituent particles are also in equilibrium. The net force (F) acting on any one particle must be zero. Now, choose some point O—anywhere you like—and determine the resultant torque about O due to the net force acting on that arbitrary particle; that is, $\tau_0 = \mathbf{r} \times \mathbf{F}$. But here, $\mathbf{F} = 0$, so the torque about O is zero, too. That's true for each and every particle in the body, so the net torque on the body due to all the internal and external forces is zero. Because of the co-linear equal-and-opposite nature of the internal interaction pairs holding the body together, the torque about any point due to each pair must cancel. Consequently, the sum of all the torques due to the internal forces must be zero. Thus **the sum-of-the-torques about any point due to all the externally applied forces acting on a rigid body in equilibrium must be zero**. Here we have only the suggestion of a derivation, but it makes the point that this powerful insight is based squarely on Newton's laws and is not some new physical principle.

The fact that

[*translational equilibrium*] $\quad\quad \Sigma\mathbf{F} = 0 \quad\quad$ [6.1]

is the First Condition of Equilibrium. In addition, for a rigid body

[*rotational equilibrium*] $\quad\quad \Sigma\tau_0 = 0 \quad\quad$ (6.6)

which is known as the **Second Condition of Equilibrium**. Note that the point O about which the torques are taken is arbitrary.

When the forces acting on a body pass through a common point, the First Condition suffices for equilibrium. (The Second Condition is then automatically true because the torque produced by a system of *concurrent forces* equals the torque produced by its resultant, which is zero.) *When dealing with concurrent forces, there will be only two independent equations (either sum-of-the-forces or sum-of-the-torques), and no more than two unknowns can be determined.* Figures 6.9, 6.10, and 6.11 depict concurrent force systems, and although we solved those problems using the sum-of-the-forces, we could have used the sum-of-the-torques instead.

The wind blowing to the right on the sail produces a clockwise torque that must be balanced by a net counter-clockwise torque produced by the sailors.

Example 6.8 Figure 6.23 recalls Example 6.3, but this time let's solve for the forces acting on bar 2, using the Second Condition of Equilibrium.

Solution: [Given: $F_W = 300$ N, $\overline{AC} = 1.00$ m, and $\theta = 30.0°$. Find: the force on bar 2, that is, F_2.] We can use either free-body diagram—Fig. 6.23b or 6.23c—since both contain F_2 and the load. Let's select Fig. 6.23c. The torques can be taken about any point, so we should choose cleverly. $\mathbf{F}_1$ passes through A, and since we do not need to find F_1, taking torques about A will exclude it from the equation. The moment-arm of $\mathbf{F}_2$ about point A is the perpendicular distance from A to the line-of-action of $\mathbf{F}_2$, and it equals $(1.00 \text{ m}) \sin 30.0°$. Similarly, the moment-arm of the load about A is the perpendicular distance from A to the line-of-action of the 300-N load, and it equals 1.00 m. Taking clockwise as positive, the sum-of-the-torques about A equals zero, thus

$$\curvearrowleft + \Sigma\tau_A = 0 = (300 \text{ N})(1.00 \text{ m}) - F_2(1.00 \text{ m})\sin 30.0°$$

$$F_2(0.500 \text{ m}) = 300 \text{ N·m}$$

and $\quad\quad \boxed{F_2 = 600 \text{ N}}$

▶ **Quick Check:** F_2 is the same as in Example 6.3.

(continued)

(continued)

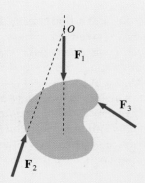

Figure 6.23 (a) The two bars fixed to the wall support the 300-N load. (b) The bars can be isolated in a free-body diagram, in which case forces F_1, F_2, and the load are in equilibrium. (c) The pin at point C is also in equilibrium under the same three forces.

Example 6.9 Prove that any three nonparallel, coplanar forces resulting in equilibrium must be concurrent.

Solution: Suppose the three forces acting are $\mathbf{F}_1$, $\mathbf{F}_2$, and $\mathbf{F}_3$, where $\mathbf{F}_1 + \mathbf{F}_2 + \mathbf{F}_3 = 0$ (Fig. 6.24). Label as O the point of intersection of $\mathbf{F}_1$ and $\mathbf{F}_2$. Now, if $\mathbf{F}_3$ does not pass through O, it will have a moment-arm with respect to it and therefore a torque about it, $\tau_0 \neq 0$. But this possibility violates the Second Condition of Equilibrium. Hence, $\mathbf{F}_3$ must pass through O.

Figure 6.24 Shown are three nonparallel, coplanar forces acting on a body. If the body is in equilibrium under the influence of such forces, then the forces must meet at a point.

Nonconcurrent Forces

When two equal but antiparallel forces act at different points on a body (Fig. 6.25), equilibrium will not be established despite the fact that $\Sigma\mathbf{F} = 0$. There will be a net torque, and the body will revolve increasingly rapidly—it will accelerate. Although the body is in translational equilibrium, it will only be in rotational equilibrium when $\Sigma\tau = 0$. *A nonconcurrent system of any number of parallel separated forces in equilibrium can give rise to only two independent equations (either one force and one torque, or two torque), and only two unknowns can be determined.*

Figure 6.25 (a) Two equal, oppositely directed forces acting at different points on a body do not produce equilibrium even though $\Sigma\mathbf{F} = 0$. There is a torque, and the system cannot be in rotational equilibrium. (b) The block will rotate without translating.

Example 6.10   The steelyard in Fig. 6.15 consists of a nonuniform horizontal bar that is balanced without any added weights when supported at point O. If the weights F_{W1} and F_{W2} are attached, and the system is balanced as shown, write an expression for L_2 in terms of L_1, F_{W1}, and F_{W2}. What is the magnitude of the reaction force F_R supporting the system?

Solution: [Given: F_{W1}, F_{W2}, and the moment-arm L_1. Find: the moment-arm L_2 and the reaction F_R.] The forces are nonconcurrent, and we will have to apply both conditions of equilibrium. The sum-of-the-torques about *any* point must be zero, but not all such points are equally convenient. It would be advantageous not to have to deal with the reaction force in the first part of the problem. Since F_R passes through O, it has no torque around O and consequently does not enter the equation, which therefore contains only one unknown. Weight 2

tends to swing the system clockwise around O, whereas weight 1 tends to rotate it counterclockwise. Hence, with clockwise as positive, we have

$$\circlearrowleft{+}\Sigma\tau_0 = 0 = F_{W2}L_2 - F_{W1}L_1$$

and

$$\boxed{L_2 = \frac{F_{W1}L_1}{F_{W2}}}$$

As for the reaction force

$${+}{\uparrow}\Sigma F_V = F_R - F_{W1} - F_{W2} = 0$$

and, not surprisingly

$$\boxed{F_R = F_{W1} + F_{W2}}$$

▶ **Quick Check:** The problem can be redone using the Law of the Lever, which is a special case of Eq. (6.6). Note that the units check.

Figure 6.26 illustrates a tilted variation of the previous example. The moment-arms are $r_{1\perp} = r_1 \cos 30°$ and $r_{2\perp} = r_2 \sin 60°$; hence, taking clockwise as positive, equilibrium occurs when

$$\circlearrowleft{+}\Sigma\tau_0 = 0 = F_{W2}r_2 \sin 60° - F_{W1}r_1 \cos 30°$$

The steelyard and the seesaw, the crowbar and the scissor (Fig. 6.27), and the human foot are all the same kind of simple machine—a lever of the class of levers that have the fulcrum, or pivot, located *between* the two forces acting on it.

When a system of nonconcurrent forces (that are not all parallel and perpendicular) produces equilibrium, there will be three independent equations, from which no more than three unknowns may be determined.

Figure 6.26 (a) The seesaw. (b) The torque around O due to the weight F_{W1} is counterclockwise. The torque around O due to the weight F_{W2} is clockwise.

Figure 6.27 Two tools that are the same kind of lever as the seesaw. Both have fulcrums or pivot points between the applied force and the force exerted by the device. Shown are the forces acting on the tools.

Example 6.11 A 400-N sign is suspended from the end of a 2.0-m-long, very light horizontal bar, which is attached to a wall via a pivot (Fig. 6.28). The bar is held up by a rope attached one-quarter of the way from its end, making an angle of 50°. Find the reaction force on the pivot and select the lightest possible support rope from Table 6.1.

Solution: [Given: $F_W = 400$ N, $L = 2.0$ m, and F_T acts 0.50 m from the end point E at 50°. Find: F_T and the force on the pivot $\mathbf{F_R}$.] The force acting at O is not at a joint. Therefore, this is not a simple case in which the bar experiences forces only along its length. Moreover, it

might seem that we have two unknowns, but including F_T there are actually three: $\mathbf{F_R}$ is equivalent to F_{RH} and F_{RV}, or F_R and θ—two unknowns. First draw the free-body diagram of the bar, Fig. 6.28b with $\mathbf{F_R}$ in any direction you like: since $\mathbf{F_T}$ is up and to the left, it's clear $\mathbf{F_R}$ is to the right. Moreover, $\mathbf{F_R}$ must act downward if $\Sigma\tau_0$ is to equal zero. Had you drawn it incorrectly, it would simply turn out negative when you solved for it. Now, to determine F_T, we focus on point P so that $\mathbf{F_R}$ is absent from the torque equation

$$(+\Sigma\tau_P = F_W(2.0 \text{ m}) - (F_T \sin 50°)(1.50 \text{ m}) = 0$$

Here, only the vertical component of the tension, $F_T \times$

(continued)

(continued)

sin 50°, contributes to the torque because the horizontal one goes through P. And solving this yields $F_T = 0.70$ kN. The lightest possible rope on which to hang the sign is 3/16-in. hemp, but since that material tends to rot in damp weather, do not stand under the sign.

We have two remaining unknowns, $\mathbf{F_R}$ and θ, and thus require two equations. Let's first try using two torque equations about points O and E

$$\curvearrowleft + \Sigma\tau_O = F_W(0.50 \text{ m}) - (F_R \sin\theta)(1.50 \text{ m}) = 0$$

$$\curvearrowleft + \Sigma\tau_E = (F_T \sin 50°)(0.50 \text{ m}) - (F_R \sin\theta)(2.0 \text{ m}) = 0$$

STOP!

It's clear that these two equations cannot be solved for F_R and θ because they are not independent. We have determined that $F_T = F_W(2.0 \text{ m})/(\sin 50°)(1.50 \text{ m})$, and as soon as we substitute that expression in the second torque equation, it becomes identical to the first equation.

Taking the torques about a point off the line of the bar (as, for example, where the rope attaches to the wall) would give us the needed independent equation (see Problem 40). Applying the torque equation more than once requires care. By contrast, the force equations are far more straightforward, and it's advisable to use them in preference wherever appropriate. From Fig. 6.28c

$$+\uparrow\Sigma F_V = F_T \sin 50° - F_W - F_{RV} = 0$$

$$\xrightarrow{+}\Sigma F_H = F_{RH} - F_T \cos 50° = 0$$

or $F_{RV} = 0.13$ kN and $F_{RH} = 0.45$ kN. Therefore

$$F_R = \sqrt{(0.13 \text{ kN})^2 + (0.45 \text{ kN})^2} = \boxed{0.47 \text{ kN}}$$

The angle that $\mathbf{F_R}$ makes with the horizontal is then

$$\theta = \tan^{-1}\left(\frac{F_{RV}}{F_{RH}}\right) = \boxed{17°}$$

▶ **Quick Check:** Taking torques about O, $F_{RV} = (400 \text{ N})(0.50 \text{ m})/(1.50 \text{ m}) = 0.13$ kN and about A, $F_{RH} = (400 \text{ N})(2.0 \text{ m})/(1.50 \text{ m})(\tan 50°) = 0.45$ kN.

Figure 6.28 (a) The problem is to calculate the reaction at P and the tension in the rope from O to A. Unlike Fig. 6.11, the forces are nonconcurrent, and we cannot learn much from the free-body diagrams of points O or E. (b) Since the bar must be in equilibrium and all the unknown forces act on it, we draw a free-body diagram of the bar. (c) The forces resolved into horizontal and vertical components.

If you want to use three torque equations, the points about which they are taken must not lie on the same line (as do P, O, and E in Example 6.11). If you want to use one force and two torque equations, the points about which the latter are taken must not lie on a line perpendicular to the force-summation axis.

Figure 6.29 The weight of the arm is distributed along its entire length. Nonetheless, it can be represented by a single force **F**$_W$ acting at a single point, the *c.g.*

EQUILIBRIUM OF EXTENDED BODIES

On planet Earth, gravity imparts weight, and weight affects equilibrium. You, right there, are an extended body probably sitting in equilibrium. How do we deal with gravity acting on you?

6.5 Center-of-Gravity

Any finite object can be thought of as composed of a very large number of constituent point-masses. Each atom within you experiences a downward gravitational force, all of which can be considered parallel, and all of which combine to produce a single resultant force, the *weight* of the body. Besides knowing the net weight, the Second Condition of Equilibrium suggests that we must also know where on the body **F**$_W$ can be understood to act. Consequently, we define the **center-of-gravity** (*c.g.*) of an object to be *that point where the total weight,* **F**$_W$*, can be imagined to act.* In other words, having a single force **F**$_W$ act at the *c.g.* produces exactly the same mechanical result as having gravity act on all the point-masses that constitute the body. Hence, the torque about any point arising from the weight of a distribution of mass is exactly the same as if the net weight acted only at the *c.g.* Gravity acts on every atom of the outstreched arm (Fig. 6.29), and yet the effect is exactly equivalent to having the total weight act at the *c.g.* located near the elbow, about 28 cm from the shoulder joint. The torque produced at the shoulder is the weight of the arm multiplied by the distance from the joint to the *c.g.*

The weight of the balanced athlete in the photograph is straight down, and if we think of it as acting through a single point, the *c.g.* must be somewhere on the line-of-action of the normal force. Since the line-of-action of the weight passes through the *c.g., the force of gravity generates no torque about the center-of-gravity.* A body suspended from its *c.g.* is in equilibrium in any orientation whatsoever and will stay where it is put. *The c.g. is the balance point.* If a body of any shape is freely suspended from *any* point, it will so orient itself in reaching equilibrium that the line-of-action of **F**$_W$ will pass through the support point (Fig. 6.30), thus producing a no-torque situation. In that case, **F**$_R$ = −**F**$_W$ and the line-of-action of **F**$_R$, which is vertical, passes through the *c.g.* To find the center-of-gravity experimentally, one need only suspend the object in turn from two different points. The lines-of-action of the two successive reaction forces will intersect at the *c.g.*

For simplicity, suppose we have a flat body made up of a large number of tiny masses: $m_1, m_2, m_3, \ldots$, as shown in Fig. 6.31. We can compute the location of its center-of-gravity in the following way. Each particle experiences a torque (due to its weight) about the *arbitrary* point *O*. To find the moment-arm for each of these tiny torques, extend the line-of-action of the weight of each particle and drop a perpendicular to it from *O*—that *x*-distance is the moment-arm. The torques have the form $F_{W1}x_1$, $F_{W2}x_2$, and so forth. The total weight-generated torque about *O* produced by all the individual particles, taking clockwise as positive, is

$$F_{W1}x_1 + F_{W2}x_2 + \cdots = \Sigma F_W x$$

Because the human body is flexible, the location of the *c.g.* is not fixed. In equilibrium, however, it must be located on the line-of-action of the normal force.

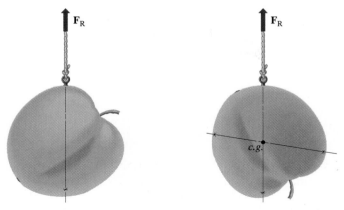

Figure 6.30 The *c.g.* of an object can be located by suspending it from several different points. The *c.g.* is always on the line-of-action of the force supporting the object.

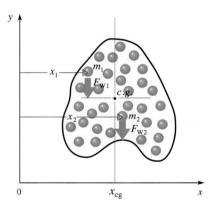

Figure 6.31 The sum of all the torques about any point 0 caused by the weights of the point-masses constituting a body is equal to the torque produced by the total weight of the body acting at the *c.g.* Measured along the *x*-axis, the *c.g.* is at x_{cg}. Measured along the *y*-axis, it's at y_{cg}.

It follows from the definition of the *c.g.* that the total weight, written as

$$F_{W1} + F_{W2} + \cdots = \Sigma F_W$$

must produce that same torque when it acts as a single force at the *c.g.*; that is

$$(\Sigma F_W)x_{cg} = \Sigma F_W x$$

and

$$x_{cg} = \frac{\Sigma F_W x}{\Sigma F_W} \qquad (6.7)$$

Take each point-mass, multiply its weight by its distance from the $x = 0$ axis, add them, divide by the total weight, and you obtain x_{cg}, measured from the axis no matter where the origin was chosen to be. If the object and the coordinate system are both rotated 90° so that the *y*-axis is horizontal, we get an expression of exactly the same form for y_{cg}.

The c.g. *of a regularly shaped body of uniform composition lies at its geometric center.* This statement applies to spheres, cubes, cylinders, rings, and so forth. Envision the object in Fig. 6.31 to be a sphere and slide it over so that the *y*-axis, which is quite arbitrary, passes through its geometric center. Then, by symmetry, $\Sigma F_W x = 0$, and Eq. (6.7) yields $x_{cg} = 0$; the *c.g.* is indeed *on* the central symmetry axis. When a body is composed of several parts, the overall *c.g.* can be located by treating each part as a point-mass located at its own *c.g.*, acted upon by its own weight.

Example 6.12 For a man of mass m and height h, the *c.g.* of the lower leg and foot, and the upper leg are shown in Fig. 6.32 along with the corresponding weights. Find the *c.g.* of the whole leg in the extended position measured from the sole of the foot. This kind of information is important in physical therapy.

Solution: [Given: $F_{W1} = 0.059mg$, $x_1 = 0.19h$, $F_{W2} = 0.097mg$, and $x_2 = 0.42h$. Find: x_{cg}.] Treating the two

(continued)

(continued)

leg parts as point-masses located at their respective *c.g.*s, we find that

$$x_{cg} = \frac{F_{W1}x_1 + F_{W2}x_2}{F_{W1} + F_{W2}}$$

$$x_{cg} = \frac{(0.059mg)(0.19h) + (0.097mg)(0.42h)}{0.097mg + 0.059mg}$$

thus $$x_{cg} = \frac{0.052mgh}{0.156mg} = \boxed{0.33\,h}$$

▶ **Quick Check:** The torque about the heel due to the whole leg is $(0.33h)(0.156mg) = (0.05mgh)$; the torque due to the separate parts $\approx (0.06mg \times 0.2h) + (0.1mg \times 0.4h) \approx 0.05mgh$ and they *are* equal.

Figure 6.32 The locations of the various *c.g.*s for several parts of a man's leg.

Bear in mind that the *c.g.* of an object need not reside within the space occupied by material. The center-of-gravity of a doughnut is at the center of the hole.

Example 6.13 The gadget depicted in Fig. 6.33 consists of a 30-N metal block 3.0 cm thick, epoxied to a 16-cm-long rod weighing 50 N, on whose end is affixed a 6.0-cm-diameter sphere weighing 10 N. The whole thing is to be hung on a thin wire so that it stays horizontal. Find the point where the wire must be attached.

Solution: [Given: $F_{W1} = 30$ N, thickness 0.030 m; $F_{W2} = 50$ N, length 0.16 m, and $F_{W3} = 10$ N, diameter 0.060 m. Find: x_{cg}.] Each piece can be envisioned as replaced by a point-mass located at its geometric center. Taking $x = 0$ at the very left end, the individual *c.g.*s are at $x_1 = 0.015$ m, $x_2 = 0.11$ m, and $x_3 = 0.22$ m for block, rod, and sphere, respectively. Hence

$$x_{cg} =$$
$$\frac{(30\ \text{N})(0.015\ \text{m}) + (50\ \text{N})(0.11\ \text{m}) + (10\ \text{N})(0.22\ \text{m})}{30\ \text{N} + 50\ \text{N} + 10\ \text{N}}$$

Figure 6.33 The *c.g.* of a composite object can be found by representing each constituent body by a point-mass at its *c.g.* For example, the sphere here weighs 10 N and can be replaced by a 10-N point at its center.

and $$\boxed{x_{cg} = 0.091\ \text{m}}$$

The *c.g.* is 9.1 cm to the right of the left end.

▶ **Quick Check:** The net torque ↺+ about the *c.g.* is $(30\ \text{N})(7.555\ \text{cm}) - (50\ \text{N})(1.944\ \text{cm}) - (10\ \text{N}) \times (12.944\ \text{cm}) \approx 0$.

For a complicated inhomogeneous body, like that of a human being, it's far easier to determine the *c.g.* in the laboratory (Problem 63) than it is to calculate it directly. Moreover, the location of the *c.g.* varies with posture and limb position. For a person standing at attention, it lies on a vertical centerline about 3 cm forward

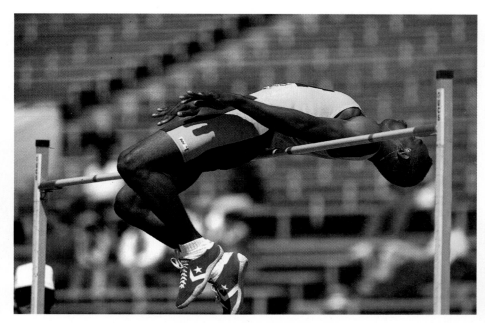

The Fosbury flop. The jumper extends the head and legs downward so the *c.g.* actually passes below the bar. As we'll see, raising the *c.g.* takes energy, and having it slip under the bar means the bar can be higher.

This mechanical arm is driven by two hydraulic pistons that can both push and pull, so each does the work of a pair of muscles.

of the ankle joint and up around the height of the second sacral vertebra, or roughly a few inches behind the navel. The *c.g.* also varies from person to person—in men with large shoulders, it's usually a little higher than in women with large pelvises. The location of the body's *c.g.* is of interest to everyone who stands or walks, but it's of particular concern to dancers and athletes. For example, Dick Fosbury devised a backward high-jump technique in which his body, face upward, arcs over the bar, while his *c.g.* passes beneath it. How high the *c.g.* is raised determines the amount of effort that will have to be expended in the jump (p. 291).

Example 6.14 The biceps muscle (Fig. 6.34) is connected from the shoulder (scapula) to the radius bone at a point 5.0 cm from the elbow. Its contractions flex the arm. Taking the mass of the hand and forearm together to be 5.5% of the total body mass (with a *c.g.* as shown), compute the force exerted by the biceps of a 70-kg man holding a 2.0-kg sphere. Assume the force of the biceps acts vertically.

Solution: [Given: the mass of the arm $m_a = 5.5\% \times$ (70 kg), $m_s = 2.0$ kg, and all the appropriate distances. Find: F_B, the force exerted by the biceps.] Assuming F_B

acts vertically, take the torque about the elbow joint to avoid dealing with the force exerted by the humerus (Fig. 6.34b), whereupon

$$\curvearrowleft + \Sigma\tau_0 = 0 = m_s g(0.34 \text{ m}) + m_a g(0.16 \text{ m}) - F_B(0.050 \text{ m})$$

Accordingly,

$$F_B = g\frac{(2.0 \text{ kg})(0.34 \text{ m}) + (3.85 \text{ kg})(0.16 \text{ m})}{(0.050 \text{ m})}$$

and $\boxed{F_B = 2.5 \times 10^2 \text{ N}}$, or about 57 lb.

(continued)

(continued)

▶ **Quick Check:** Rounding off the numbers, the clockwise torque is roughly $g(2 \text{ kg})(0.3 \text{ m}) + g(4 \text{ kg}) \times (0.2 \text{ m}) \approx 1.4g \approx 14 \text{ m·N}$, and dividing by 0.05 m yields $F_B \approx 0.28 \text{ kN}$.

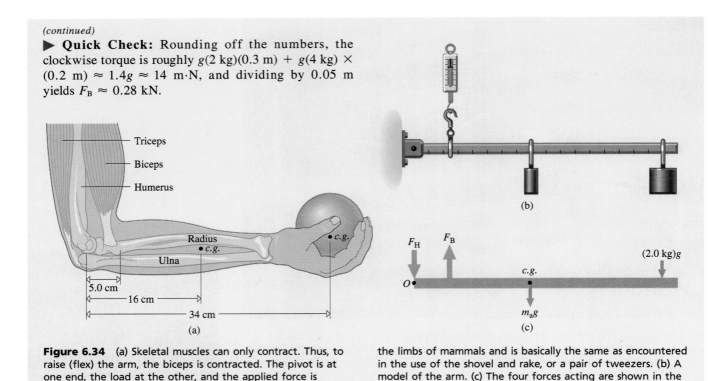

Figure 6.34 (a) Skeletal muscles can only contract. Thus, to raise (flex) the arm, the biceps is contracted. The pivot is at one end, the load at the other, and the applied force is between the two. This kind of end-pivoted lever is common in the limbs of mammals and is basically the same as encountered in the use of the shovel and rake, or a pair of tweezers. (b) A model of the arm. (c) The four forces acting are shown in the free-body diagram.

6.6 Stability and Balance

As bipedal beings, we cope with the problem of stability every day. The common wisdom holds that stability is about objects not toppling over, and that's a good place to start. When a thing falls over, the driving force is gravity and the pivotal concept is the *c.g.* Out in space, where everything is essentially weightless, there's no need to worry about objects tipping over.

An object that is supported against gravity is in equilibrium when $\Sigma \mathbf{F} = 0$ and $\Sigma \tau = 0$, but we can distinguish three degrees of equilibrium. *A body is said to be in* **stable equilibrium** *if, after receiving a small momentary displacement, it returns to its original condition.* Each object in Fig. 6.35 is in stable equilibrium, because once slightly tilted, the resulting torque produced by its weight acting through the *c.g.* tends to rotate the body about the pivot, back to its original orientation. Notice that in each such case the initial displacement causes a raising of the *c.g.*, which, when released, tends naturally to fall back downward.

A body is said to be in **unstable equilibrium** *if, after receiving a small momentary displacement, it moves farther away from its original condition.* A cone balanced on its point is in equilibrium when the normal force and the weight are colinear, $\mathbf{F}_N = -\mathbf{F}_W$ and $\Sigma \tau = 0$. But if the slightest tilt disrupts that condition, $\mathbf{F}_W$ produces a net

Figure 6.35 Bodies such as these are in stable equilibrium because they fall back to their original orientations after a large displacement.

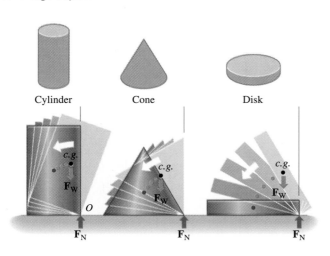

torque about the pivot point and the cone rotates from the vertical. Again the *c.g.* descends, but this time the fall carries it away from the original location.

A body is said to be in **neutral equilibrium** *if, after receiving a small momentary displacement, it remains in equilibrium, tending neither to move away any farther from, nor to return to, its original condition.* Each of the objects in Fig. 6.36 has its *c.g.* directly over the point or line-of-contact with the surface, and this situation holds when it is rolled into some new position. The sum-of-the-torques produced by $\mathbf{F_N}$ and $\mathbf{F_W}$ is always zero. Moreover, the height of the *c.g.* remains unchanged.

As a rule, *an object supported at a point above the c.g. will be stable, while one supported at a point below the c.g. will be unstable.* A cane hanging from its crook on a finger is stable, but one balanced upright on a fingertip is unstable.

Figure 6.36 These objects are in neutral equilibrium. A small shift of the *c.g.* leaves them unchanged—they stay put, with the *c.g.* always above the point or line of support.

Example 6.15 The 0.25-kg toy man on the platform in Fig. 6.37 holds a rod of negligible weight. Considering him alone, his *c.g.* is 8.00 cm up from his feet. The two counterweights, each of mass 0.50 kg, hang on a horizontal level 12.0 cm beneath the top of the platform. Are the counterweights, the toy man, and the rod in stable equilibrium?

Solution: [Given: $m_1 = 0.25$ kg, $m_2 = m_3 = 0.50$ kg, and taking $y = 0$ at the level of the counterweights, $y_1 = 20.0$ cm, $y_2 = y_3 = 0$. Find: the *c.g.* for the total system.] From Eq. (6.7)

$$y_{cg} = \frac{m_1 g y_1 + m_2 g y_2 + m_3 g y_3}{m_1 g + m_2 g + m_3 g} = \frac{m_1 y_1}{m_1 + m_2 + m_3}$$

and
$$y_{cg} = \frac{(0.25 \text{ kg})(0.20 \text{ m})}{1.25 \text{ kg}} = \boxed{0.040 \text{ m}}$$

The system's *c.g.* is 4.0 cm up from the *x*-axis and 8.0 cm below the point of support on the platform, and so it is stable.

▶ **Quick Check:** The toy man has 1/5th of the mass of the system and therefore must be 4 times farther above the *c.g.*, (20 cm × 4/5 = 16 cm) than are the counterweights beneath the *c.g.*, (20 cm × 1/5 = 4.0 cm).

(a)

(b)

Figure 6.37 (a) The two counterweights shift the *c.g.* well below the platform. (b) Side view. Any shift of the little man raises the *c.g.*, giving F_W a moment-arm and causing a torque that drives the system back to its original orientation.

Shenyang Acrobatic Troupe. The combined *c.g.* of the two horizontal men and the pole lies on a vertical line passing through the bottom of the pole. That line then passes between the feet of the man on whose shoulder the pole rests.

Figure 6.38 An object is stable when its *c.g.* is somewhere above the area of support. Someone on all fours is far more stable than someone standing upright.

Figure 6.39 (a) Each World Trade Center tower weighs $\approx 1.2 \times 10^9$ N and has a face area of 1.6×10^4 m². Take the wind to be uniform at 160 km/h (100 mi/h), providing a pressure of 1.2 kN/m² (25 lb/ft²) so the total wind load is 19×10^6 N. (b) At the point where the building begins to tilt, the reaction force is on its leeward edge. The tower is actually well anchored in the ground.

An object resting on a horizontal surface will overturn once its *c.g.* passes beyond the farthest extent of its base of support (whereupon $\mathbf{F}_N$ and $\mathbf{F}_W$ are no longer colinear). Lean forward, hands at your sides, until your *c.g.* is just above the line of your toes—any farther, and over you go. We can imagine an *area of support* bounded by the perimeter of lines connecting the outermost contact points, as shown in Fig. 6.38. *So long as the c.g. is directly above some point within this area of support, the system will be stable.*

Four things affect the ease or difficulty with which the *c.g.* will pass beyond its support: the weight of the object, the height of the *c.g.*, its location within the area of support, and the size of that area. A light object will overturn more easily than a heavy one, all else being equal. So an empty paper cup will blow over in the wind before a full one. On a lurching bus, we instinctively increase the area of support by assuming a wider than normal stance. As a rule, a low *c.g.* in the middle of a large area is stable. The boxer crouches, the sumo wrestler squats, and the football guard gets down on all fours, each in order to lower the *c.g.*

High-rise buildings can experience tremendous wind loads and are therefore anchored to deep foundations. In a strong hurricane, each tower of the World Trade Center swings as much as 7 ft on either side of vertical. Let's make a rough calculation to see if its weight alone would keep the tower from toppling around its bottom leeward edge (Fig. 6.39). Assuming the wind load is uniform over the face of the tower, it can be taken to act at the center (where the *c.g.* ought to be) with a moment-arm of 206 m. The wind produces a torque of (19×10^6 N $\times$ 206 m = 3.9×10^9 N·m) clockwise about the back edge. On the other hand, the weight produces a counterclockwise torque of (1.2×10^9 N $\times$ 18 m = 2.3×10^{10} N·m), and the tower stays put.

When we bipeds walk, our *c.g.* must remain somewhere over the one supporting foot in contact with the floor, and that requires us to do a little side-to-side wiggling. Quadrupeds such as dogs and horses can walk slowly with three feet always in touch with the ground, which maintains a large triangular area of support, even though the animal has small paws or hooves. We are the creatures with the big feet. In fact, the *c.g.* of the upper portion of the human body resides well above the support structure of the hips, and that's not a particularly good design. Birds, which have been

(a)

(b)

bipedal a lot longer than us, are far better suited to that purpose, having their *c.g.* hanging *below* the hip line (Fig. 6.40). It's like comparing a Model-T and a modern sports car (Fig. 6.41). One teeters above the axles, the other hangs below them. Despite Hollywood, we are not likely to see a terrestrial robot designed in the likeness of a human—it would require too much circuitry just to keep the thing in balance as it walked.

Sᴛᴀʙɪʟɪᴛʏ ᴀɴᴅ Bᴀʟᴀɴᴄᴇ

According to Walter Reed Army Medical Center, between 1975 and 1977, 3 young men were killed and 12 severely injured by falling soda vending machines at U.S. army bases. The victims had apparently deliberately tipped the machines, which, when fully loaded, weigh between 800 and 1000 pounds. Since that study was published, another 48 victims (of whom 11 died) were reported. All were male, ranging in age from 10 to 33; 7 were civilians.

6.7 Structures, Walls, and Arches

Children build towers of wooden blocks that invariably come tumbling down. That's not because of some structural failure (the blocks themselves do not crumble). Toy towers topple because they are unstable—the weight of the blocks, instead of holding them together, acts to pull them apart. Much of the ancient stone architecture that still survives was similarly constructed by stacking blocks without any mortar or cement between them. The primary function of mortar is to bring the layers of blocks into good contact, so that the compressive loads are distributed uniformly over the whole surface of the joint. Mortar has comparatively little strength in tension and is not used to bond the blocks together, a point well illustrated every time an automobile crashes through a brick wall. What holds together the cathedral of Notre Dame in Paris is the weight of its parts and the friction between them.

We can expect a stone wall to compress slightly under the combined action of its own weight and whatever load it's carrying, such as the weight of a roof. (A typical modern building 300 m tall compresses its supporting columns by 2 or 3 cm. When the resultant, the so-called *thrust*, acts down the center of the wall, the force spreads out evenly, and the compression is nearly uniform. If the load shifts, the thrust is made to move closer to the wall's edge, and things distort (Fig. 6.42). Once the line-of-action of the thrust passes beyond what is known as the "middle third" of the wall, the masonry experiences compression on that side and tension on the other. The result is usually a crack in the joint where the mortar tears loose. We see that kind of thing all the time in modern buildings. If the condition becomes extreme, and the line-of-action moves beyond the surface of the wall, beyond the area of support, the wall topples over.

The character of a presteel building was determined to a large extent by the kind of roof that it supported. The simplest approach to roofing was to lay a beam from wall to wall. But even King Solomon could not buy roof beams much longer than about 8 m, hewn from the "cedar-trees out of Lebanon." The best solution to spanning a space with stone was the arch. The downward load (the weight of a

Figure 6.40 Birds are stable on two legs because those legs pivot at the pelvis above the *c.g.*, which is low because of the massive breast bone that anchors the wing muscles. In flight, the *c.g.* is supported by, and located under, the wings.

Figure 6.41 Modern automobiles have their *c.g.*s as low as possible. When the *c.g.* is high above the axles, the vehicle is unstable. Large trailer trucks, foolishly loaded, are constantly toppling over in high winds. Many roads are particularly hazardous in this regard and are often closed to commercial traffic whenever the wind comes up.

Figure 6.42 The "middle third" rule maintains that an elastic block will be entirely in compression, provided the load remains in the middle third. Masonry structures are weak in tension and are designed to keep the load in the middle third.

Figure 6.43 The arch channels the load down to the abutments. In the process, each block is compressed in two directions, which is fine since stone has great strength in compression. A masonry arch can readily span 60 m or so. Such arches are also very durable. Many Roman arches from antiquity still stand in excellent condition.

Figure 6.44 The weights of the various parts of the human body are supported by the musculoskeletal system. The hanging arms are held in place by tendons and muscles in the shoulders. That load is carried back to the spine, which transmits the weight of the entire upper body to the pelvis. The bone arch of the pelvis splits the downward force, directing it through the legs to the ground.

bridge or building) squeezes the separate wedge-shaped stones against each other, redirecting the force (Fig. 6.43). In effect, the thrust flows down along the arch-ring and out into the abutments at the base. Each stone acts mainly in compression, where it is strongest. If all goes well, the thrust lines will remain within the arch-ring no matter how the load shifts. If not, cracks will develop just as in the wall, but there is a significant difference. Arches are, by their very nature, stable (without crumbling, the wedge-shaped stones cannot move past each other).

Today, concrete arched highway overpasses are a common sight. In the longest arch bridges, spanning upwards of 1700 ft, steel truss frames replace blocks of stone. The human pelvis (Fig. 6.44) is a bone arch that carries the load of the spine,

A flying buttress of the cathedral of Notre Dame de Paris. The buttress is a half-arch that keeps the walls of the building from moving outward.

dividing it in two and channeling it down to the legs. Not surprisingly, the pelvis would work even better if we walked on all fours.

One of the great inventions of the Middle Ages (done by trial and error long before the development of structural analysis) was the Gothic arch that crowned the cathedrals of Europe. By spanning the same space with a high pointed arch, the horizontal reaction forces were greatly reduced and the support walls made much thinner. The thrust flowed more nearly vertically and stayed within a thinner supporting wall, which could then be ornamented with splendid windows. Where the thrust line might stray out of the bounds of a thin wall (as in a strong wind), a flying buttress channeled the force safely downward (see photo on p. 192).

The pointed arch channels the load straight down to the ground. There is no sideways outward thrusting, as there is with the circular arch. As a result, the walls can be much lighter. That allowed for the soaring architecture of the Gothic cathedral.

Core Material

For *coplanar concurrent forces,* translational equilibrium occurs when (p. 164)

$$\Sigma \mathbf{F} = 0 \qquad [6.1]$$

which is the First Condition of Equilibrium. A system acted upon by forces that all pass through a single point is in equilibrium when

$$\Sigma F_H = 0 \quad \text{and} \quad \Sigma F_V = 0 \qquad [6.2]$$

The **torque** about O is defined as

$$\tau_0 = F r_\perp = F_\perp r = Fr \sin \theta \qquad [6.3]$$

The torque vector $\boldsymbol{\tau}_0$ is *perpendicular to both* $\mathbf{r}$ *and* $\mathbf{F}$, and

$$\boldsymbol{\tau}_0 = \mathbf{r} \times \mathbf{F} \qquad [6.5]$$

For a rigid body in equilibrium, the sum of the torques about any point O must be zero, that is

$$\Sigma \tau_0 = 0 \qquad [6.6]$$

which is the Second Condition of Equilibrium. When we deal with concurrent forces, there will be only two independent equations (either sum-of-the-forces *or* sum-of-the-torques) and no more than two unknowns can be determined. A nonconcurrent system of any number of parallel, separated forces in equilibrium can give rise to only two independent equations (either one force and one torque, or two torque), and only two unknowns can be determined. When a system of nonconcurrent forces (those that are not all parallel and perpendicular) produces equilibrium,

Figure 6.45 An experimental crane boom. The pulley redirects the tension force in the rope (F_{T2}) without altering it. The vertical length of chain is a hanger, and the tension in it (F_{T1}) equals the load (F_W) it supports. The boom is made of a compression spring scale, and it reads the force (F_B) tending to squash it. The bottom of the boom is pivoted on a pin; therefore, the forces (F_B) at its ends act inward along the scale's length. This arrangement is a concurrent force system.

there will be three independent equations, from which no more than three unknowns may be determined.

Figure 6.45 serves as a summary of some of the conceptual details that are involved in analyzing a statics problem.

The **center-of-gravity** *(c.g.) of an object is that point where* *the total weight of an object can be imagined to act* (p. 184), thus the equation

$$x_{cg} = \frac{\Sigma F_W x}{\Sigma F_W} \qquad [6.7]$$

Suggestions on Problem Solving

1. When all the forces acting on a body pass through the same point, we solve the problem by applying the sum-of-the-forces equation (in each of two perpendicular directions if necessary). When the forces are nonconcurrent, at least one application of the sum-of-the-torques equation will be required. Use Eq. (6.6) sparingly to avoid nonindependent torque equations.

2. Always draw a picture of the entire system. Then draw a free-body diagram of each segment of the setup to be analyzed. Guess at the directions of the internal forces, and if you are wrong, they will turn out negative. In time, you should be able to guess correctly by recognizing what the directions of the forces and torques must be to sustain equilibrium. If you are to solve for the force in a given rope or bar, part of that member must be included in one of these diagrams. You cannot solve for a quantity unless it appears in one of the equations. If possible, start the analysis in the simplest place where there is only one unknown. Then move on to the more complicated regions, using what you have just learned.

3. Consider the geometrical symmetries of a given setup. Such symmetries usually mean that the internal forces will be symmetrical if the load is symmetrical, as in Figs. 6.9 and P15. If the arrangement is asymmetrical, the internal forces are also likely to be different (Fig. P11).

4. When finding torques, consider the point about which they will be computed. Avoid the unknown forces you are not interested in by taking torques about points through which those forces pass. By computing the net torque about point *P* in Fig. 6.28, you preclude the unknown reaction force from entering the equation and can immediately solve for the tension.

5. When dealing with pulleys, determine how many lengths or segments of a single continuous rope support the load. A set of pulleys may be fixed to a load and move with it; if so, determine the number of rope segments supporting it (as in Fig. 6.6), because they support the load. If the rope is continuous and there are no tangential forces on it, the tension is the same at every point in it. If there are *N* supporting rope segments, the tension in each is 1/*N*th of the load. When several different ropes support the load (Fig. P14), each may have a different tension in it.

6. To determine the signs in a sum-of-the-torques equation, assign a direction (clockwise or counterclockwise) as positive. Then imagine an axis perpendicular to your drawing running through the point about which you are taking the torques. For each force, determine how it, acting alone, would rotate the system about that axis. For example, in Fig. 6.28, taking ⟨+Στ_p, the weight produces a positive torque, and the tension, a negative torque.

Discussion Questions

1. A body is in equilibrium under the influence of three equal-magnitude nonparallel forces. Determine their relative orientations. Can concurrent forces of 30 N, 40 N, and 80 N produce equilibrium?

2. A rock thrown straight up stops at its peak altitude. Is it in equilibrium at the instant it comes to rest? A ball thrown into a wall on a spaceship deforms and comes to rest for an instant before bouncing back, off the wall. Is it in static equilibrium at that moment? Once the ball springs back to shape and leaves the wall (assuming zero gravity), is it in equilibrium?

3. What is the tension in the rope on the right in Fig. Q3a? Each of the pulley systems in Fig. Q3b is supposed to be in equilibrium. Redraw any of them that may be in error.

4. Imagine a hammock hung by lengths of rope between two trees. Are the ropes more or less likely to break when someone sits in the hammock if it is initially pulled tight or allowed to sag? Explain.

5. Figure Q5 shows a forearm and hand exerting a downward force on a book. Explain the physics of what's happening. How would it be different if the arm were supporting the book? Try it yourself, checking your own muscles.

6. Why is it easier to hold a long stick horizontally at its midpoint rather than at its end? Position a stick horizontally so it

(a)

(b)

Figure Q3

Figure Q5

rests on your two outstretched index fingers held about 2 ft apart. Now slowly move your fingers toward one another. Observe the movement of the stick over your fingers: Why does it alternate, sliding first over one finger and then the other? What is the significance of the point where your fingers finally meet?

7. Why do people lean forward when they ascend an inclined ramp? Why do you think pregnant women often suffer backaches in their third trimester? Why does a swimmer or a sprinter crouched at a starting line lean way forward?

8. About 100 years ago, there was a popular toy called a tumbler (Fig. Q8). When toppled, the toy immediately righted itself. The modern equivalents are great inflatable plastic things that kids punch. How do such toys work?

Figure Q8

9. Stand, feet flat on the ground, with your back and heels against a wall. Try to touch your toes. Explain why you cannot do it. Now face the edge of an open door. Place one foot on either side of the door, so that both feet extend somewhat beyond the edge on which you are resting your nose. Now try to stand up on your toes. Explain what is happening.

10. Try to balance an ordinary kitchen bowl on the eraser end of a pencil, first at its middle on the inside, and then on the bottom outside of the bowl. Explain the difference.

11. Because insects are so light, it would seem that they would be blown over by even a slight wind. What anatomical difference between them and mammals prevents this from happening?

12. Describe the equilibrium state of each configuration shown in Fig. Q12.

Figure Q12

13. What, if anything, can you say about the location of the *c.g.* of the chair in Fig. Q13?

Figure Q13

14. Why is it more strenuous to do situps with your hands behind your head than with them out in front of you?

15. It's common during demonstrations of karate for the practitioner to strike several pine boards held tightly together, one behind the other. It often happens that the back board will break in two, while the front one, which receives the blow, is little more than cracked. Explain.

16. How would you make a pair of "loaded" dice? Explain.

17. Imagine that you are standing straight, hands at sides, holding a weight in each hand. Draw a free-body diagram of your arms and legs and describe the forces acting on them. Are they in tension or compression?

18. Is a truck loaded with several tons of feathers more likely to tip over in a high wind than one carrying the same weight of bricks? Explain. Occasionally, one sees a narrow van filled with cargo carrying a heavy load on its roof. Why is that dangerous, and how can the risk be minimized? Explain the significance of the narrowness of the van.

Multiple Choice Questions

1. Ten scales, each weighing 10 N, are in equilibrium, stacked one on top of the other. What are the readings of the top and bottom scales, and what is the reaction force exerted by the ground on the stack?   (a) 0 N, 100 N, 100 N (b) 10 N, 90 N, 100 N   (c) 10 N, 100 N, 100 N (d) 0 N, 90 N, 100 N   (e) none of these.

2. An 800-N acrobat holds a 40-N chicken in one hand while his 480-N assistant Jane sits on his shoulders holding a 4.0-N box of cigars. The acrobat stands motionlessly on one foot on a 36-N bathroom scale. The scale reads   (a) 1360 N   (b) 1324 N   (c) 662 N   (d) 680 N   (e) none of these.

3. A woman holds a 20-N shovel containing 16 N of dirt at rest horizontally. She grips it with one hand at the end of the handle and one at the middle. The net vertical force she exerts on the shovel is   (a) not enough information to know   (b) 20 N upward   (c) zero   (d) 36 N upward   (e) none of these.

4. The essential aspect of a body in equilibrium is that it has   (a) zero speed   (b) zero mass   (c) zero momentum   (d) zero acceleration   (e) none of these.

5. It is possible to apply forces of 6 N and 10 N to a body in such a way as to counter the effect of a single additional force of   (a) 18 N   (b) 20 N   (c) 7 N   (d) 3 N   (e) none of these.

6. Two forces, each of 100 N acting at a point such that they are 120° apart, are equivalent to a single force of   (a) 100 N   (b) zero   (c) 200 N   (d) 86.6 N   (e) none of these.

7. No matter what the setup is, the tension in a rope used to support a load is always   (a) less than the load   (b) equal to the load   (c) zero   (d) more than the load   (e) none of these.

8. The lightweight pivoted bar in Fig. MC8 will be in rotational equilibrium when a 200-N force acts   (a) down at *D*   (b) up at *B*   (c) down at *E* or up at *C*   (d) up at *C* only   (e) none of these.

Figure MC8

9. An egg balanced on an end is   (a) in stable equilibrium   (b) in unstable equilibrium   (c) in neutral equilibrium   (d) not in equilibrium   (e) none of these.

10. Given a body acted upon by several forces, if the sum of those forces is exactly zero, the body   (a) may be in both translational and rotational equilibrium   (b) must be in both translational and rotational equilibrium   (c) cannot be in both translational and rotational equilibrium   (d) not enough information to come to any conclusion   (e) none of these.

11. In the photo of Chinese acrobats on p. 190, the combined *c.g.* of the pole and the two suspended people   (a) is on a vertical line passing through the third man's left shoulder   (b) is to the left of the third man   (c) is to the right of the third man   (d) is on a horizontal line passing through the third man's shoulder   (e) none of these.

12. If an object is divided into two parts by slicing it vertically straight through its *c.g.*, the two pieces   (a) must have the same weight   (b) must have different weights   (c) can have different weights   (d) must have the same mass   (e) none of these.

13. Two astronauts push on either end of a satellite floating in space. If the craft is then in translational and rotational equilibrium, it is required that   (a) both forces must pass through the satellite's *c.g.*   (b) the two forces be equal, coplanar, and pass through the *c.g.*   (c) there be two forces, they need not be equal, but they must be antiparallel   (d) the forces must be equal, colinear, antiparallel, and pass through the *c.g.*   (e) none of these.

14. A light rope is looped over a weightless pulley and tied to a 50-N rock. The other end is held by a 50-N monkey who climbs up the rope a bit and then stops. The tension in the rope at that moment is   (a) 100 N   (b) zero   (c) 50 N   (d) 25 N   (e) none of these.

15. In analyzing the condition of rotational equilibrium, one takes the sum of the torques about an axis   (a) which must pass through the body's *c.g.*   (b) through which the line-of-action of all the forces must pass   (c) passing through the center of the body   (d) located anywhere   (e) none of these.

16. Someone places a 6.0-m-long steel rod on a rock so that one end is under a baby moose weighing 2.0 kN. The person pushes down on the other end of the rod with a force of 400 N, and the moose is held in the air at rest. The rock was   (a) 1.0 m from the moose   (b) 5.0 m from the moose   (c) 1.0 m from the person   (d) 6.0 m from the person   (e) none of these.

17. A 100-N load hangs from the ceiling on a rope. Someone pushes on the load with a continuing horizontal force, and it moves off to the side a little, coming to rest at a bit of an angle with the vertical. As a result, the tension in the rope   (a) is unchanged   (b) is zero   (c) is decreased   (d) is increased   (e) none of these.

18. A 10-m length of rope, weighing 10 N per meter, is hanging straight down from a ceiling hook. The tensions in the rope at the free end, at 5.0 m up from the free end, and at the hook are, respectively   (a) 0, 100 N, and 100 N   (b) 0, 50 N, and 100 N   (c) 100 N, 100 N, and 100 N   (d) 100 N, 50 N, and 0   (e) none of these.

19. In order to hold the weight in Fig. MC19 in equilibrium, the hand and the ceiling exert forces, respectively, of   (a) 100 N and 300 N   (b) 300 N and 100 N   (c) 100 N and 400 N   (d) 300 N and 600 N   (e) none of these.

Figure MC19

Problems

1. [I] A 200-N chandelier is hung from a ceiling hook by a chain weighing 50 N.   (a) What force is exerted on the hook?   (b) What force acts on the bottom of the chain?
2. [I] What are the tension forces acting on each of the weightless ropes in Fig. P2?

Figure P2

3. [I] A 20-m length of chain weighing 2.0 N per meter is hung vertically from one end on a hook.   (a) What is the tension three-quarters of the way up?   (b) What is the tension 1.0 m from the top and bottom?
4. [I] Using a ruler and a compass, determine if forces of 11 N, 16 N, and 20 N acting at a point can result in equilibrium. Explain your answer.
5. [I] A 10-kg child and a 100-kg adult both stand upright and barefooted at rest on a smooth platform sloping at some angle θ. Who is more likely to slide down the incline?
6. [I] A 200-N child stands at rest on an inclined metal slide tilted at an angle of 24° up from the ground. Compute the friction force on her shoes. What minimum value of μ_s must exist?
7. [I] The 10-kg block in Fig. P7 is held at rest by the four ropes shown. If the tension in the ropes on the right and top are each 98 N, what is the tension in the remaining two ropes?

Figure P7

8. [I] The orthodontal wire brace in Fig. P8 makes an angle of 80.0° with the perpendicular to the protruding tooth. If the tension in the wire is 10.0 N, what force is exerted on the tooth by the brace?

Figure P8

9. [I] Referring to Fig. P9, determine the scale reading in the right arm of the suspension and the angle θ.

Figure P9

10. [I] The steel beam that hangs horizontally in Fig. P10 weighs 8.00 kN, and the lower cables hook on to it at angles of 70.0°. What are the tensions in the lower cables?

Figure P10

11. [I] In the static arrangement shown in Fig. P11, the pulleys and ropes are essentially weightless. If weight 1 is 15.0 N and weight 2 is 31.0 N, and the two angles are measured to be $\theta = 45.0°$ and $\phi = 20.0°$, determine the value of weight 3. Check that the system is in equilibrium.

Figure P11

12. [I] The hand in Fig. P12 exerts a downward force that holds the length of pipe at rest. (a) Compute that force. (b) What is the tension in the rope that is strung over the pulley? (c) What force supports the upper hook?

300 N

Figure P12

13. [I] Look at Fig. P13. If the hand pulls down so that the 100-N weight remains at rest, what is the tension in the long rope? What force is exerted down on the right-most ceiling bracket?

100 N

Figure P13

14. [I] What is the tension in each length of rope in Fig. P14, given that the load is at rest?

8000 N

Figure P14

15. [I] Referring to the pin-jointed truss in Fig. P15, compute the force in bars *BC, DE,* and *FG.*

40 000 lb

Figure P15

16. [I] What downward force on the rope in Fig. P16 will hold the 300-N block at rest assuming the pulleys are essentially weightless? Notice that there are two different ropes holding up the load.

17. [II] Determine the weight of the mass *m* in Fig. P17. Assume the pulleys and ropes are all essentially weightless.

18. [II] An essentially weightless rope is strung nearly horizontally over a light pulley, as shown in Fig. P18. The

Figure P16

Figure P17

Figure P18

spring balance is adjusted to read zero. A 100-N weight is then hung at the midpoint of the span, and the rope sags, descending 10 cm from the horizontal. Determine the scale reading.

19. [II] A 533.8-N tightrope walker dances out to the middle of a 20-m-long wire stretched parallel to the ground between two buildings. She is wearing a pink tutu, of negligible weight; the wire sags, making a 5.0° angle on both sides of her feet with the horizontal. Find the tension.

20. [II] If a force of 100 N in Fig. P20 holds the weight motionless, what is its mass? What is the tension in rope 1?

Figure P20

21. [II] Determine both the angle at which the pulley hangs and the tension in the hook supporting it in Fig. P21.

Figure P21

22. [II] The pin-connected symmetrical structure in Fig. P22 (which is used to support floodlights at a rock concert) is known as a Fink truss. If the lamps on top weigh 1.00 kN and the ones on the bottom weigh 2.00 kN each, compute the reaction forces exerted by the towers.

23. [II] The three bodies in Fig. P23 are at rest. If a fly lands on the 80.0-kg mass, and the system then begins to move, what is the static coefficient of friction for the box and surface? The pulleys are weightless and frictionless.

24. [III] Consider the arrangement depicted in Fig. P24. Now, compute the internal forces in each of the ropes and in the boom.

Figure P22

Figure P23

Figure P24

25. [III] Determine the reaction forces at the four contact points A, B, C, and D, in Fig. P25. The hollow balls, which are homogeneous and made of the same material, have masses of 102 kg and 204 kg, and the chamber they sit in is 55.98 cm wide. Everything is smooth, so that friction is negligible.

Figure P25

ROTATIONAL EQUILIBRIUM

26. [I] Determine the torques about the elbow and shoulder produced by the 20-N weight in the outstretched hand in Fig. P26.

Figure P26

27. [I] Compute the torque about the pivot O, generated by the 100-N force on the gearshift lever in Fig. P27.

Figure P27

28. [I] Figure P28 shows a tendon exerting an 80-N force on bones in the lower leg. Assume the tendon acts horizontally 0.06 m from the knee pivot. Draw a simplified version of the system and compute the tendon's torque about the pivot.

29. [I] While working with precision components, especially out in space, it is often necessary to use a torque wrench, a device that allows the user to exert only a pre-set amount

Figure P28

of torque. Having dialed in a value of 35.0 N·m, what maximum perpendicular force should be exerted on the handle of a wrench 25.0 cm from the bolt?

30. [I] Referring to the *xyz*-coordinate system, imagine a vector **A** lying on the positive *x*-axis and a vector **B** on the positive *y*-axis. Determine the direction of each of the following: **C**, where **C** = **A** × **B**; **D**, where **D** = **A** × **C**; **E**, where **E** = **C** × **B**; and **F**, where **F** = **E** × **D**.

31. [I] Two campers carry their gear (90.72 kg) on a light, rigid horizontal pole whose ends they support on their shoulders 1.829 m apart. If Selma experiences a compressive force of 533.8 N, where is the load hung on the pole, and what will Rocko feel?

32. [I] Harry, who weighs 320 N, and 200-N Gretchen are about to play on a 5.00-m-long seesaw. He sits at one end and she at the other. Where should the pivot be located, if they are to be balanced? Neglecting the weight of the seesaw beam, what is the reaction force exerted by the support on it?

33. [I] An 8000-N automobile is stalled one-quarter of the way across a bridge (see Fig. P33). Compute the additional reaction forces at supports *A* and *B* due to the presence of the car. Take the length of the bridge to be $\overline{AB}$.

8000 N

Figure P33

34. [I] The weightless ruler in Fig. P34 is in equilibrium. Determine both the unknown mass *M* and the left-hand scale reading.

35. [I] A 10.0-kg brass sphere rests in a 90.0° groove as shown in Fig. P35. Assuming that there is no friction and that the weight of the sphere acts at its center, determine the two reaction forces.

36. [I] The little bridge in Fig. P36 has a weight of 20.0 kN, which acts at its center. Calculate the reaction forces at points *A* and *B*.

37. [II] If the beam in Fig. P37 is of negligible mass, what value does the scale read (in kg)?

38. [II] Figure P38 shows a laboratory model of a roof truss constructed of two very light rods pinned at points *B* and *C*. A light cable tightened by a turnbuckle runs from *A* to

4.17 kg

10.0 cm

A   *B*

1.0 kg   2.0 kg   *M*

Figure P34

1   2

Figure P35

15 m — 35 m — 15 m — 20 m

A   *B*

8000 N   8000 N

20 000 N

Figure P36

10 cm

2.0 kg   1.0 kg

Figure P37

C to complete the truss. What do the two spring scales read in newtons? Determine the compression in rod *AB*. (Hint: Find out the reading of scale 1 and then draw a free-body diagram of *A*. Remember the forces exerted by the pinned rods act along their lengths.)

Figure P38

39. [II] The little temporary bridge in Fig. P39 consists of two very light planks. The 480-N person stands 3.00 m from the end of the upper board, which itself rests on rollers. What are the reaction forces at points *A*, *C*, and *D*?

Figure P39

40. [II] Find the force F_R in Fig. 6.28b by taking torques about both point *A* and the point where the lines-of-action of the two tensile forces on the ropes intersect.
41. [II] The hand in Fig. P41 exerts a force of 300 N on the hammer handle. Compute the force acting on the nail, just as the head begins to pivot about its front edge in contact with the board.

Figure P41

42. [II] A forceps is used to pinch off a piece of rubber tubing, as shown in Fig. P42. What is the force exerted on the rubber if each finger squeezes with 10.0 N? What is the force exerted on the pivot by each half of the forceps?
43. [III] Suppose a downward force of 80 N acts on the pedal of the bike shown in Fig. P43. The chain goes around the front chain wheel (of radius 10 cm) and the rear sprocket

Figure P42

Figure P43

wheel (of radius 3.0 cm). The pedal is attached to a 17-cm crank arm that turns around point *A*. What impelling force will the scale read (the bike, chain, and wheel are motionless at the moment depicted)? What provides that force?

44. [III] A sphere of mass 10.0 kg rests in a groove, as shown in Fig. P44. Assuming no friction and taking the weight of the sphere to act at its center, compute the reaction forces exerted by the two surfaces.

Figure P44

45. [III] Determine the internal forces acting at point *A* in Fig. P45 when a 1000-N weight is hung from the hook. All of the structural members can be taken as weightless.
46. [III] Figure P46 shows a smooth rod resting horizontally inside a bowl. Compute the force *F* that will maintain the rod in position, given negligible friction.

Figure P45

Figure P46

EQUILIBRIUM OF EXTENDED BODIES

47. [I] Two uniform blocks of wood 4 cm × 4 cm × 12 cm, each weighing 1.0 N, are glued together as shown in Fig. P47. Find the center-of-gravity of the system.

Figure P47

48. [I] A uniform piece of soft, thin copper wire 60 cm long and weighing 1.2 N is bent into three equal length segments to form two right angles, so that it has the shape of an angular letter "C." Locate the folded wire's center-of-gravity.

49. [I] A 0.50-m-long thin steel rod with a mass per unit

length of 0.50 kg/m is bent in half (making a right angle) into the shape of a letter "L." Locate its *c.g.*

50. [I] The glider depicted in Fig. P50 is descending at a constant speed. If the drag (F_D) is 600 N and the plane weighs 6000 N, determine both the angle of descent, θ, and the lift, F_L.

Figure P50

51. [I] Scale 1 in Fig. P51 reads 100 N. What does scale 2 read? How much does the suspended body weigh? Discuss the location of the vertical line passing through the *c.g.*

Figure P51

52. [I] Refer to the mobile in Fig. P52. If the star weighs 10.0 N, how much does the sphere weigh?

Figure P52

53. [I] Now suppose we displace the sphere in Problem 52 downward, whereupon the top rod tilts down at 30.0° with the horizontal. Will the mobile be in equilibrium?

54. [II] A pole vaulter carries a 6.0-m uniform pole horizontally. He holds one end with his right hand pushing downward, and 1.0 m away pushes upward with his left hand. If the pole weighs 40 N, compute both forces he exerts. What is the net force he applies? What would these forces be if the pole were vertical? What does this answer suggest about the carrying angle?

55. [II] We wish to stack five uniform wooden blocks so that they extend as far right as possible and still remain stable. How should each be positioned? Can the top block have its entire length beyond the edge of the bottom block (Fig. P55)?

Figure P55

56. [II] An essentially weightless beam 10.0 m long is supported at both ends, as in Fig. P56. A 300-N child stands 2.00 m from the left end, and a 6.00-m-long stack of newspapers weighing 100 N per linear meter is uniformly distributed at the other end. Determine the two reaction forces supporting the beam.

Figure P56

57. [II] Figure P57 shows two ropes supporting a body. Prove that the resultant of the two tensile forces, as indicated by the scale readings, equals the weight. Show that the plumb line passes through the body's c.g.

58. [II] A 55.0-kg woman is standing "rigid"—straight upright, hands at sides, feet together. At their widest, her two feet have a breadth of 20.0 cm and her c.g. is 95.0 cm above the ground. What minimum horizontal force in the

Figure P57

plane of her body (that is, perpendicular to the forward direction) will tilt her over if it acts along a shoulder-to-shoulder line 145 cm high?

59. [II] An 8.0-m-long, 30-kg uniform plank leans against a smooth wall, making an angle of 60.0° with the ground. Compute the reaction force, $\mathbf{F}_{RG}$, of the ground on the plank.

60. [II] In Problem 32, Harry and Gretchen were sitting on a weightless seesaw. Suppose now that the beam is uniform and weighs 200 N—find the new pivot point.

61. [II] A 65-kg woman is horizontal in a pushup position in Fig. P61. What are the forces acting on her hands and feet?

Figure P61

62. [II] A large box of breakfast cereal is 6.0 cm × 35 cm × 46 cm high and has a net mass of 1.0 kg. The contents are a uniformly distributed material. The box is next to an open window. Determine the speed of a uniform wind hitting the large surface if it just starts tilting the box over. The force exerted by a perpendicular wind in newtons on each square meter is given roughly by $0.6v^2$, wherein v is expressed in m/s. Assume the distributed force acts at the geometrical center of the face.

63. [III] One technique for measuring the c.g. of a person is illustrated in Fig. P63. The board is positioned according to the person's height (h), and the scales are then reset to zero. The participant lies down, and the scale readings are taken. Determine an expression for x_{cg} in terms of the measured quantities.

64. [III] A uniform ladder leans against a smooth wall, making an angle of θ with respect to the ground. The

Figure P63

Figure P65

ground's reaction force on the ladder is at an angle of ϕ with the horizontal. Determine the relationship between θ and ϕ. Which is larger?

63. [III] The uniform plank in Fig. P65 is 5.00 m long and

weighs 100 N. The cord that attaches to the plank 1.00 m from the bottom end holds it from sliding, there being no friction. Find the tension in the cord.

7

Gravity, According to Newton

AFTER 2000 YEARS OF thinking about gravity, we have a powerful, albeit still incomplete, understanding of the phenomenon. That understanding comes from several different theoretical perspectives. First, there is Newtonian Theory, which is reliable, practical, and simple. It was the formulation that guided us to a precise landing on the Moon (1969) and yet, on a grand scale, we know that it is inadequate. Then there is Einstein's General Theory of Relativity, which relates gravity to the fabric of space and time (p. 229). It subsumes Newton's Theory and goes far beyond it, allowing us to analyze events on a vast cosmic scale. But Einstein's formulation is

Nicolaus Copernicus (1473–1543), pictured here on a stamp as a young man in Torun, Poland.

On gravity. *For my part I think that gravity is nothing but a certain natural striving with which parts have been endowed . . . so that by assembling in the form of a sphere they may join together in their unity and wholeness.*

NICOLAUS COPERNICUS

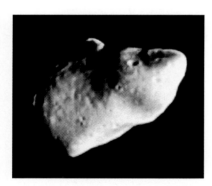

The asteroid Gaspra as photographed by the space probe *Galileo.* It's only 12 km by 20 km by 11 km, and is much too small to gravitationally crush itself into a sphere.

highly mathematical and quite unnecessary for most earthly considerations. Today, many theoreticians believe that all interactions arise from the continuous exchange of force-carrying particles (Chapter 33) and are therefore quantum mechanical in nature. They argue that General Relativity, which is not quantized, must be incomplete. This chapter focuses mainly on Newtonian theory, exploring how it applies on Earth and beyond.

THE LAW OF UNIVERSAL GRAVITATION

Newton's *Law of Universal Gravitation* states that the gravitational force F_G between any two bodies of mass m and M, separated by a distance r, is described by

$$F_G \propto \frac{mM}{r^2} \tag{7.1}$$

Gravity varies directly in proportion to the product of the masses of the interacting objects and inversely with their separation squared. Moreover, *it is a center-to-center attraction between all forms of matter.* Yet Newton's achievement was not so much in formulating the law as it was in applying it with supreme mathematical skill. He established that the formula was correct mainly by using it to explain the motions of the Moon and planets. As we will see, the pieces to the gravity puzzle were already there before Newton's great work of 1685.

7.1 Evolution of the Law

The word *gravity* derives from the Latin *gravitas,* meaning weight or heaviness. And since a force equal and opposite to a body's weight must be applied to keep it from falling, the tendency for things to drop toward the center of the Earth was recognized early on as part of the mystery of gravity.

Copernicus (1543) challenged the Earth-centered Aristotelian world view, maintaining instead that the Sun was at the center of the Universe. With that, the Earth became a spinning planet rushing through space. The dominant gravitational theory at the time was Aristotle's—things fall because they seek their "natural place" at the center of the Universe, the center of the Earth. But the new vision of planet Earth orbiting the Sun made that age-old wisdom implausible and, within two centuries, the Newtonian formulation replaced it.

The "universal" aspect of Newton's law alludes to the fact that *all* matter interacts via gravity. That insight, which still surprises some people today, developed gradually. Copernicus suggested that gravity was the tendency of *similar* matter to come together. He explained that the Sun, Moon, and Earth were all spherical because each kind of substance tends to pull itself inward into a symmetrical shape. That idea is in part correct, but there is no suggestion here that Earth-stuff might attract Moon-stuff. A century later, Galileo was still committed to the same parochial gravity, in which only like substances attract.

In 1600, William Gilbert published *De Magnete,* a book of remarkable insight, although he wrongly attributed the primary functions of gravity to magnetism. (Hold two powerful magnets close together and you immediately know the wonder of an invisible force reaching across space.) For a while, magnetism and gravity were a single mystery of matter joining matter.

Johannes Kepler, influenced by Gilbert, made the proper distinction between gravity and magnetism. Kepler argued correctly that both the Sun and Moon attract the "waters of the sea," causing the rising of the tides. The Earth, the Moon, the Sun, the waters—all apparently interact gravitationally. Extending this concept, Roberval (1644) hinted at an attraction between all "worldly matter." In 1670, Robert Hooke stated that gravity was an interaction between "All Celestial Bodies." And, finally, Newton proposed that the gravitational interaction exists between all material objects: "We must . . . universally allow that all bodies whatsoever are endowed with a principle of mutual gravitation."

The Earth pulls you downward with a gravitational force called *weight,* just as it pulls downward on apples, rocks, and frogs giving them weight as well. You, in turn, gravitationally attract the Earth with an equal and opposite force; that's the interaction pair. Matter somehow draws on matter, and you attract (weakly, to be sure) all the apples, rocks, and even the frogs—everything attracts everything else.

The realization that the force of gravity, Eq. (7.1), is proportional to the interacting masses comes from a suggestion by Gilbert, even though it was little more than a lucky guess. He observed that the force between two magnets depended on their sizes and weights—which just happened to be the case for the natural magnets available at the time. Carrying the idea over to gravity, Kepler (1609) maintained that "two stones . . . placed anywhere in space" would gravitationally attract and "come together . . . at an intermediate point [the center-of-gravity], each approaching the other in proportion to the other's mass."

The force of gravity decreases as the separation increases, Eq. (7.1), and we might also wonder how that insight came to be. In fact, it was Kepler who initiated it, albeit in a confused context. Kepler wrongly envisioned the planets propelled in orbit by a solar force that pushed them along. Since the farther a planet is from the Sun, the slower it moves, he proposed that this sweeping force decreased with distance. After all, the magnetic force between two bodies decreases as their separation increases. The concept of a force extending from the Sun to the planets would soon be more correctly applied to gravity itself. But how was the specific $1/r^2$ dependence arrived at?

When anything diffuses outward uniformly from a *point* source (Fig. 7.1), it spreads out, becoming less concentrated as it advances. We can imagine all of the flow passing successively through surrounding concentric spheres centered on the source, any one of which has a radius r and an area $4\pi r^2$. The same quantity of emanation that initially passes densely through a tiny spherical surface will later pass tenuously through a much larger one. The concentration of the emanation (the amount per unit area) decreases inversely with the area of the sphere—it *decreases inversely with the distance squared* $(1/r^2)$. Kepler certainly knew that this phenomenon was true for light; everyone knew you can only read by a candle flame if you stand near it. And he came close to embracing an inverse-square law for his fictitious solar force, once even writing that it "weakened through spreading from the Sun in the same manner as light." If gravity spreads out from a body uniformly in all directions, an inverse-square dependence would be reasonable to suppose.

In 1645, the astronomer Ismaël Bullialdus asserted that the Sun's attractive action on each planet *was along the center-to-center line joining the two and dropped off inversely with the distance squared.* Newton was aware of this work; in fact, he explicitly mentions the author by name. The idea that gravity acts along a line connecting the "centers" of the interacting bodies was not startling either. The Medieval scholar Albert of Saxony had defined weight as the tendency of the center-of-gravity of a body to approach the center-of-gravity of the Earth. Ever since Archimedes (Sect. 11.6), it had been common to think of weight acting at the *c.g.* Thus, the vision of a center-to-center gravitational interaction was quite natural.

Johannes Kepler (1571–1630).

The Moon causes the movement of the waters and the tides of the ocean.

The Sun (chief inciter of action in nature) . . . causes the planets to advance in their course.

The Moon alone of all planets directs its movements as a whole toward the Earth's centre, and is near of kin to Earth, and as it were held by ties to Earth.

The force which emanates from the Moon reaches to the Earth, and, in like manner, the magnetic virtue of the Earth pervades the region of the Moon.

WILLIAM GILBERT (1540–1603)

On gravity. *If the attractive force of the Moon reaches down to the Earth, it follows that the attractive force of the Earth, all the more, extends to the Moon and even farther.*

J. KEPLER

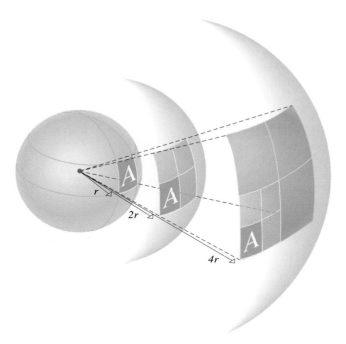

Figure 7.1 When the radius of a sphere r increases, the area of the sphere increases as r^2. Thus, if we double the radius, the little area A projects into an area of $4A$. Similarly, if the radius goes from r to $4r$, the area A projects out to an area of $16A$. The total area of the sphere, $4\pi r^2$, expands to $4\pi(4r)^2 = 16(4\pi r^2)$, or 16 times the original area.

7.2   Newton's Confirmation of $1/r^2$

Early in the 1660s, Hooke, already convinced of the variation of gravity with distance, attempted to arrive at its precise form experimentally. Being a practical man, he carried out a series of measurements to determine the minute differences in the weight of a body at various altitudes, within deep wells and high "on Paul's steeple and Westminster Abbey, but none that were fully satisfactory." Years later, Newton claimed to have made his own first test of the inverse-square dependence just around that same time, himself a young man of twenty-four. As he put it, I "compared the force requisite to keep the Moon in her orb with the force of gravity at the surface of the Earth; and found them answer pretty nearly."

Although its orbit is slightly elliptical we can envision the Moon circling the Earth at a center-to-center distance of 38.4×10^7 m under the influence of a centripetal force provided by gravity (Fig. 7.2). Its centripetal acceleration $a_C = v^2/r$, Eq. (5.10), can be easily determined numerically. The lunar orbital speed (v) is the distance traveled in one revolution ($2\pi r$) divided by the time it takes—the *period*. The orbit's circumference equals $2\pi(38.4 \times 10^7$ m), while the period is 27.32 days (or 2.360×10^6 s), which yields a speed of 1.02 km/s and an acceleration of

$$a_C = \frac{v^2}{r} = 0.002\,7 \text{ m/s}^2$$

Any object at that distance from Earth, whether orbiting or not, should experience this same downward acceleration via gravity.

Newton next compared this value with the acceleration at the planet's surface (6.371×10^6 m from its center). An object *on* the Earth experiences an acceleration due to gravity of 9.81 m/s². Assuming that the downward gravitational force trails off with distance from the planet as $1/r^2$, the resulting acceleration will do the same ($F = ma$). Let $g_\oplus$ be this distance-dependent acceleration associated with the Earth ($\oplus$ is the usual symbol for our planet). The Moon is 60.33 Earth-radii away, so it (and anything else at that distance) should experience a gravitational acceleration toward Earth that is about 60^2 times smaller than does an object at the planet's surface; thus,

$$g_\oplus = \frac{9.81 \text{ m/s}^2}{3600} = 0.002\,7 \text{ m/s}^2$$

Amazing! The Moon's centripetal acceleration equals the gravitational acceleration due to the Earth at that altitude: $a_C = g_\oplus$. The whole calculation, simple and profound, is wonderful.

Example 7.1   An astronaut out in space descends straight down toward the Earth in free-fall. Compute her acceleration at the moment she reaches an altitude of 1.911×10^4 km from the planet's surface. Neglect the presence of any other celestial object.

Solution: [Given: $h = 1.911 \times 10^4$ km, measured from the *surface*. Find: $g_\oplus$.] For the moment, we don't have an expression for $g_\oplus$ in terms of r (p. 216). We'll have to arrive at a value using the acceleration at the surface and the fact that $g_\oplus$ drops off as $1/r^2$. Remember that r is the

(continued)

(continued)

center-to-center distance. Letting $R_\oplus$ be the Earth's radius (6.371×10^3 km), $h = 3R_\oplus$, and the astronaut is at a distance $r = h + R_\oplus = 4R_\oplus$ from the planet's center. The gravitational acceleration at that altitude must be 4^2 or 16 times less than the surface value. Hence

$$g_\oplus = \frac{9.81 \text{ m/s}^2}{16} = \boxed{0.613 \text{ m/s}^2}$$

▶ **Quick Check:** $r = 25\,481$ km and $g_\oplus/(9.8 \text{ m/s}^2) = (6.4 \times 10^3 \text{ km})^2/(25 \times 10^3 \text{ km})^2$; hence, $g_\oplus = 0.6 \text{ m/s}^2$.

It would certainly seem that the force of gravity acting between two bodies (F_G) should be inversely proportional to their center-to-center (*c.g.*-to-*c.g.*) separation squared, or

$$F_G \propto \frac{1}{r^2}$$

7.3 The Product of the Masses

That all terrestrial objects at the surface of the planet accelerate downward at the same rate of $g = 9.81$ m/s^2 is both remarkable and suggestive. From Newton's Second Law ($F = ma$), we know that a body of mass m propelled by a force equal to its own weight F_W accelerates at a rate g such that $F_W = mg$. Consequently, $F_W/m = g$, but this equivalence is true for *all* masses, which can only be the case if F_W is proportional to m so that the ratio can be independent of m. Inasmuch as the weight of an object is the force of gravity pulling it down toward the planet ($F_W = F_G$), it follows that $F_G \propto m$ and

$$F_G \propto \frac{m}{r^2}$$

Arguing from Newton's Third Law, we know that, if the Earth pulls down on an object, the object must tug *up* on the Earth with an equal force. Imagine two interacting masses m and M. Since the force of gravity acting on any body depends on its mass, and since there is an equal and opposite force F_G on each of them, $F_G \propto m$ and $F_G \propto M$, implying that the gravitational interaction on either one of two objects is proportional to both m and M, represented mathematically as

$$F_G \propto \frac{mM}{r^2}$$

The force being proportional to the product mM agrees with our everyday experiences. We know that if one apple of mass m in a bag on a scale at the market has a weight of 1 N, then two identical apples will have a weight of 2 N. Doubling the mass in the bag (m) doubles the weight ($F_W = mg$) despite the immense mass of the Earth ($M_\oplus = 5.975 \times 10^{24}$ kg). This would not occur, for instance, if F_G were in proportion to the sum of the masses ($m + M_\oplus$) rather than their product ($mM_\oplus$).

To be rigorous, Newton's Law—Eq. (7.1)—ought to be applied only to idealized point masses. We would then, by mathematical means, determine the gravitational force due to a body of any shape by adding up the forces from all its constituent particles. Developing the ability to perform that calculation and then proving that the law as it stands applies between spherical masses (such as planets) was one of the main problems Newton had to overcome.

By all their influences, you may as well Forbid the sea for to obey the moon . . .

W. SHAKESPEARE (1564–1616)
The Winter's Tale

Universal Gravitation on a Celestial Scale. All Celestial Bodies whatsoever have an attraction or gravitating power towards their own Centres, whereby they attract not only their own parts . . . but that they do also attract all the other Celestial Bodies that are within the sphere of their activity.

ROBERT HOOKE (1670)

Figure 7.2 The acceleration due to gravity drops off inversely with distance squared. Since g at the Earth's surface is 9.8 m/s², it should be equal to 0.002 7 m/s² at the distance of the Moon. And that's precisely equal to the centripetal acceleration we know the Moon must have to maintain its orbit.

7.4 Gravitational Constant

The relationship between F_G, m, M, and r^2 can be recast as an equality (something Newton never had the need to do) by introducing a constant of proportionality G (see Appendix A-4). This constant also serves to make the resulting equation dimensionally correct—the units on both sides of the proportionality are not the same. Newton's Law of Universal Gravitation takes its final form as

$$F_G = G \frac{mM}{r^2} \tag{7.2}$$

Here, G has the units of N·m²/kg², and the force is properly in newtons. The numerical value of this *Universal Gravitational Constant* has to be determined experimentally.

That measurement was first done in 1798 by Lord Henry Cavendish, one of the richest men of his age and one of the most peculiar. Incredibly withdrawn, he rarely spoke to men and never, if he could help it, interacted with women. It's said that he would dismiss a female servant if she so much as entered a room he was in.

Eccentricities aside, determining G was a formidable technical accomplishment. To do it, Cavendish attached a small lead sphere about 2 in. in diameter to each end of a 6-ft-long, light, rigid rod, which was suspended horizontally from its middle on a thin vertical wire (Fig. 7.3). He then brought two lead balls, roughly 8 in. in diameter, into stationary positions, one in front of and one behind the small spheres. The resulting gravitational attraction subsequently caused the suspended rod-assembly to turn. The smaller balls swung very slowly toward the larger ones until the torque produced by the twisting wire brought them to rest. Having previously determined the force that would twist the wire through any angle, Cavendish promptly arrived at a value for F_G. Measuring m, M, r, and F_G, the constant G could then be computed. The presently accepted value is

$$G = 6.672\,59 \times 10^{-11} \text{ N·m}^2/\text{kg}^2$$

(Cavendish obtained the equivalent of 6.75×10^{-11} N·m²/kg²). It's interesting that of all the fundamental constants we will encounter, G is the least accurately known.

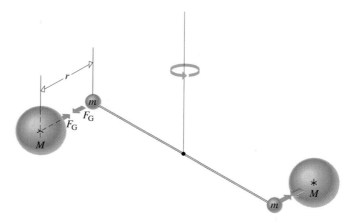

Figure 7.3 The Cavendish torsion pendulum for determining G. Drawn by their gravitational attraction, the two small balls move toward the large stationary balls, thereby twisting the wire.

Example 7.2 Compute the gravitational attraction between two 100-kg uniform spheres separated center-to-center by 1.00 m. (That is the rough equivalent of two 220-lb football players at arm's length.)

Solution: [Given: $m = 100$ kg, $M = 100$ kg, and $r = 1.00$ m. Find: F_G.]

$$F_G = G\frac{mM}{r^2}$$

$$F_G = (6.67 \times 10^{-11} \text{ N·m}^2/\text{kg}^2)\frac{(100 \text{ kg})(100 \text{ kg})}{(1.00 \text{ m})^2}$$

and

$$\boxed{F_G = 6.67 \times 10^{-7} \text{ N}}$$

which equals 0.1×10^{-6} lb, or just about the weight of a baby flea. No wonder we're not noticeably drawn to each other via gravity.

▶ **Quick Check:** The force between two 1-kg masses separated by 1 m is 6.7×10^{-11} N, and the force in our problem is 100×100 times greater.

Example 7.3 The mass of the Moon is 7.35×10^{22} kg and its distance from Earth is 384×10^3 km. Taking the planet's mass to be 5.98×10^{24} kg, compute what Newton called "the force requisite to keep the Moon in her orb."

Solution: [Given: $m_\mathbb{C} = 7.35 \times 10^{22}$ kg, $M_\oplus = 5.98 \times 10^{24}$ kg, and $r_{\oplus\mathbb{C}} = 384 \times 10^3$ km. Find: F_G.]

$$F_G = G\frac{m_\mathbb{C} M_\oplus}{r_{\oplus\mathbb{C}}^2}$$

$$F_G = (6.67 \times 10^{-11} \text{ N·m}^2/\text{kg}^2) \times$$

$$\frac{(7.35 \times 10^{22} \text{ kg})(5.98 \times 10^{24} \text{ kg})}{(384 \times 10^6 \text{ m})^2}$$

and

$$\boxed{F_G = 1.99 \times 10^{20} \text{ N}}$$

Incidentally, this value is approximately 4.5×10^{19} lb, or the equivalent of the thrust of 6 million million *Saturn 5* rockets.

▶ **Quick Check:** From p. 210, $g_\oplus = 0.0027$ m/s^2; hence, $F_G = ma = (7.4 \times 10^{22}$ kg$)(0.0027$ m/s$^2) = 2 \times 10^{20}$ N.

The quantity G is a measure of the strength of the gravitational force. Since it's an extremely small number, F_G will be appreciable only when at least one of the interacting masses is very large. Still, because it has indefinite range (dropping to zero when $r = \infty$), and because it is purely attractive (not being weakened by an opposing repulsive aspect), gravity rules the Universe on a grand scale. Consequently, gravity is central to all cosmological theories and it's from that background that have come, in recent times, some interesting suggestions that perhaps G is not a constant after all.

7.5 Terrestrial Gravity

Cavendish had an ulterior purpose when he performed his famous experiment—he wanted to determine the density of the Earth and ended up being the first person to effectively "weigh" the planet, as he put it. An apple of mass m, hanging on a scale at the Earth's surface a distance $R_\oplus = 6.37 \times 10^6$ m from its center, weighs $F_W = mg$. The scale and the apple are motionless on the Earth and therefore both are revolving with the planet. If, for a moment, we remove the small effect of that

A modern version of the Cavendish experiment. The two large lead balls are easily visible, one in front, one in back. One of the small balls can be seen just to the right of the red laser beam.

TABLE 7.1 Field events in the Solar System: long jump, high jump, and shot put on the Moon and nine planets*

Planet	Gravitational acceleration (m/s^2)	Long jump[1] distance (m)	High jump[2] height (m)	Shot put[3] distance (m)
Mercury	3.7	24	4.7	56
Venus	8.9	9.9	2.5	25
Earth	9.8	8.9	2.4	22
Moon	1.6	54	9.4	126
Mars	3.7	23	4.6	56
Jupiter	26	3.3	1.5	9.4
Saturn	12	7.5	2.2	19
Uranus	11	7.6	2.2	19
Neptune	12	7.3	2.1	19
Pluto	2	45	7.9	105

1. The athlete runs at 10 m/s and then jumps at 71° at 4.08 m/s.
2. The athlete jumps at 5.20 m/s.
3. The athlete throws the shot at 14.2 m/s at slightly less than 45°.

rotation, we must replace g by g_0, the nonspinning, *absolute acceleration due to gravity*, whereupon $F_W = mg_0$. This equation represents the "true" weight of an object due only to its gravitational interaction with the Earth and is therefore identical to $F_W = GmM_\oplus/R_\oplus^2$. Setting the two weight expressions equal and canceling m, we get *the absolute gravitational acceleration at the Earth's surface*

$$g_0 = \frac{GM_\oplus}{R_\oplus^2} \tag{7.3}$$

In the vicinity of the planet, the acceleration due to gravity is the same for all bodies independent of their masses. By appropriately substituting for M and R, the acceleration due to gravity can be determined for any celestial body (Table 7.1).

7.6 Earth's Density

Knowing both the value of g_0 (which differs from g by less than 0.4%) and $R_\oplus$, and having just determined G experimentally, Eq. (7.3) can be solved for the mass of the Earth

$$M_\oplus = \frac{R_\oplus^2 g_0}{G} = 6.0 \times 10^{24} \text{ kg}$$

and that's a splendid thing to be able to compute. Taking the planet to be a sphere of volume

$$V_\oplus = \tfrac{4}{3}\pi R_\oplus^3 = 1.1 \times 10^{21} \text{ m}^3$$

Cavendish found that the **average density**, or *mass per unit volume*, was

$$\frac{M_\oplus}{V_\oplus} = 5.5 \times 10^3 \text{ kg/m}^3$$

which is 5.5 times greater than that of water (whose density is 1.0×10^3 kg/m^3).

In vacuum, a feather and an apple fall at the same rate. The apple is more massive and must experience a greater force ($F_W = mg$) if it's to accelerate at a rate equal to that of the feather. But it has more mass and does weigh more ($F_W \propto mM_\oplus/R_\oplus^2$). Accordingly, the acceleration due to gravity is independent of the mass of the falling object.

We now believe that the Earth is a layered structure. The thin, cool surface *crust* composed of the lightest materials has an average density (Sect. 10.3) of only 2.5×10^3 kg/m³. It surrounds a *mantle* of hot rock that in turn encloses a dense *core* of iron and nickel.

Newton never had access to a measured value of G, but he could have guessed at the Earth's density and then just run the above calculation backwards to find G, had he so wished. We know that he made just such an extraordinary guess in the *Principia*, where he wrote: "Since . . . the common matter of our Earth on the surface thereof is about twice as heavy as water, and a little lower, in mines, is found about three or four, or even five times more heavy, it is probable that the quantity of the whole matter of the Earth may be five or six times greater than if it consisted all of water . . .".

7.7 Gravity of a Sphere

Let's look now at the one sticky problem sidestepped so far. An apple near the surface of the planet is apparently drawn downward by a gravity force that arises much as if the entire 6.0×10^{24} kg of Earth were located at a point 6.4×10^6 m away at the planet's center. And why that should be is certainly not obvious. Newton solved the problem in a complicated geometrical way, although it's suspected that he had already analyzed it using his own invention, the integral calculus. Sir Isaac showed that *a particle located outside of a very thin uniform spherical shell will be drawn toward the sphere's center as if all the shell's mass were concentrated at that point* (Fig. 7.4). The shell is imagined to be constructed of numerous circular strips that in turn are divided into minute segments of mass. The Law of Universal Gravitation holds exactly for each of these tiny mass segments interacting with the outside particle. If we pair up segments on diametrically opposite sides of any one strip, it's easy to see that for each pair of equal forces acting on the particle, only the two components along the center line will not cancel one another. By symmetry, the total force on the outside particle due to all mass segments—that is, due to the whole spherical shell—is directed inward along the center-to-center line. The rest of the analysis, which is too involved to do here, leads to the conclusion that the spherical shell behaves gravitationally like a central point-mass, behavior that is a consequence of the inverse-square dependence. No other force law would combine with the symmetry of the sphere to produce so simple a result.

The same combination of influences, of geometry and force, leads to another surprise when the particle is placed *anywhere inside* the spherical shell. In that case, the particle is pulled equally in all directions by the shell, so that it experiences no resultant force whatsoever. That much can been seen from Fig. 7.5, using an analysis very like Newton's original treatment. Shown are two narrow cones bounded by straight lines passing through the particle. The mass of the region of the shell corresponding to the base of each cone is proportional to the area thereof. Thus, each produces an oppositely directed gravitation force on the particle such that $F_{G1} = A_1/r_1^2$ and $F_{G2} = A_2/r_2^2$. Hence

$$\frac{F_{G1}}{F_{G2}} = \frac{A_1 r_2^2}{A_2 r_1^2}$$

The cones have the same apex angles and are similar; their corresponding sides are in the same ratios. Consequently,

Figure 7.4 The gravitational attraction of a hollow spherical shell on a point-mass *m*. Note that because of the symmetry, F_G is directly inward toward the center of the sphere (the transverse components of the forces toward the two area elements cancel).

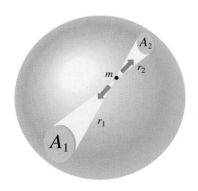

Figure 7.5 A point-mass, m, inside a spherical shell experiences a net gravitational force of zero. The little mass discs of area A_1 and A_2 produce equal and opposite forces on the point mass.

the base areas are in proportion to the squares of the sides

$$\frac{A_1}{A_2} = \frac{r_1^2}{r_2^2}$$

Substituting this expression into the preceding, it follows that $F_{G1}/F_{G2} = 1$: the forces are equal and exactly cancel. The small area nearby produces the same force as the large area far away—the areas go as r^2, and the forces as $1/r^2$. The whole surface can be divided into similar pairs of area elements with the overall effect that *no force acts on the internal particle*.

As a first approximation, we can imagine the Earth to be composed of a cluster of contiguous, uniform, spherical, thin shells (not necessarily with the same density) resembling a huge onion. On interacting with an external object such as an apple, each shell, no matter what its density, acts like a central point-mass. Accordingly, the whole cluster must behave as if the total mass $M_\oplus$ was at its center, pulling on the apple such that

$$g_\oplus = \frac{GM_\oplus}{r^2} \tag{7.4}$$

is the *absolute gravitational acceleration beyond the Earth*. Note that this expression becomes Eq. (7.3) at the planet's surface, where $r = R_\oplus$.

If the apple were to descend beneath the surface, into a deep vertical tunnel, we can imagine it traversing successive concentric shells. Once having passed through a shell, the latter's influence on the falling apple would vanish—the only thing pulling down on the apple at any moment is the spherical mass "beneath" it. When the apple is at a distance $r < R_\oplus$ from the planet's center, the volume V "beneath" it is $\frac{4}{3}\pi r^3$. The apple is outside of this spherical region, being drawn toward its center. If, for simplicity's sake, we take the density of the planet to be a constant ($\rho_\oplus$) then *the absolute gravitational acceleration beneath the Earth's surface is*

$$g_\oplus = \frac{G(V\rho_\oplus)}{r^2} = \frac{4}{3}\pi\rho_\oplus Gr \tag{7.5}$$

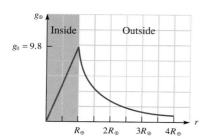

Figure 7.6 Absolute gravitational acceleration of the Earth, inside and outside, assuming the planet to have a uniform density.

The acceleration decreases linearly as the apple plunges downward, becoming zero at the Earth's center (Fig. 7.6) where it would truly be weightless. Notice that it's only at $r = 0$ and $r = \infty$ that the planet's gravity vanishes.

Example 7.4 By what percentage would your weight change if you ascended a height equivalent to that of the tallest building in the world, the Sears Tower in Chicago (1454 ft)? Take the Earth's diameter to be $1.274\,246 \times 10^7$ m.

Solution: [Given: $h = 4.432 \times 10^2$ m and $R_\oplus = 6.371\,23 \times 10^6$ m. Find: F_{wh}/F_w.] The ratio of the weights, from Eq. (7.2), is

$$\frac{F_{Wh}}{F_W} = \frac{1/(R_\oplus + h)^2}{1/R_\oplus^2} = \frac{(6.371\,23 \times 10^6 \text{ m})^2}{(6.371\,67 \times 10^6 \text{ m})^2}$$

which to four significant figures is $\boxed{99.99\%}$. The reduction in the weight, and therefore the reduction in $g_\oplus$, is a mere 0.01%.

▶ **Quick Check:** The ratio equals $[R_\oplus/(R_\oplus + h)]^2 = [1/(1 + h/R_\oplus)]^2 = 0.999\,9$.

7.8 An Imperfect Spinning Earth

The Earth is not a sphere and it is not uniform. Excluding its hills and valleys, the planet is slightly flattened at the poles, with a little bulge at the north giving the whole thing a pear shape ($R_\oplus$ at the equator is about 21.5 km greater than at the poles). At close range, the deviation in the force from an ideal $1/r^2$ dependence is small (about 0.1%), and yet it markedly affects the orbits of artificial Earth satellites. Farther out, at the distance of the Moon, the interaction with the Earth deviates from the inverse square by only about 0.00003%. In general, *any two mass configurations will interact in an approximately $1/r^2$ fashion when r is large in comparison to the sizes of the interacting objects.*

The planet's oblate shape, combined with the fact that it is rotating, produces a variation in the measured surface value of the gravitation acceleration (g) that is about five parts in a thousand (0.5%) as one moves from pole to equator (Table 3.2). To determine the effect of the Earth's rotation on g, consider an object of mass m supported by a scale situated at the *equator*. The two external forces acting on the object are its downward interaction with the planet (F_G) and the upward scale-force equal in magnitude to the object's measured weight (F_W). But since the object rotates as the planet turns, it must experience a centripetal acceleration. Hence, ($+\downarrow \Sigma F = ma$)

$$F_G - F_W = ma_C$$
$$\frac{GmM_\oplus}{R_\oplus^2} - mg = ma_C$$

and so
$$g = \frac{GM_\oplus}{R_\oplus^2} - a_C = g_0 - a_C$$

At the poles, $a_C = 0$, $g = g_0$, and the measured weight mg equals the *true weight* mg_0. As we saw in Example 5.9, $a_C = 0.0339$ m/s² at the equator. The absolute value of the gravitational acceleration gradually varies from $g_0 = 9.814$ m/s² at the equator to 9.832 m/s² at the poles, while $g = 9.780$ m/s² and 9.832 m/s², respectively. The planet's spin produces almost two-thirds of the observed change in g with latitude (Fig. 7.7).

Guidance systems for missiles and aircraft require a detailed knowledge of the Earth's gravity. Here U.S. Air Force technicians very precisely measure variations in F_G as the altitude increases.

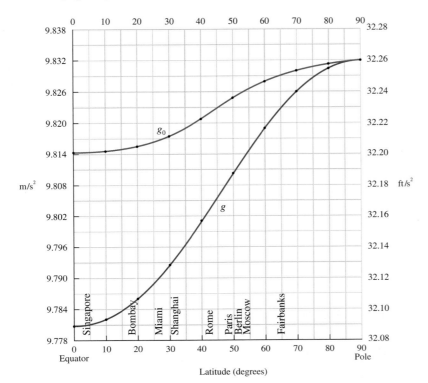

Figure 7.7 A plot of both the measured (g) and the absolute (g_0) values of the acceleration due to gravity on Earth. The former includes the rotational effects of the spinning planet, the latter does not.

Superimposed on the planetwide variation in gravity due to asphericity are innumerable small local alterations arising from geological details in the surface of the Earth. A modern instrument known as a gravimeter, usually consisting of a sensitive quartz spring balance, can measure minute changes in g of 1 part in 10^8 (or 0.000 001%), which is equivalent to a shift in elevation of only about 3 cm. Gravimeters are commonly used to survey the surface of the planet in order to explore geological structures and locate deposits of ore, gas, water, coal, and oil.

THE COSMIC FORCE

Brahe and Kepler met for the first time in Prague in 1600. Renowned as the leading astronomer of Europe, Tycho Brahe was 53. Self-conscious about his silver nose (the reminder of a college duel), he was aristocratic, suspicious, and overbearing. But he needed a theoretician to prove his own anti-Copernican model of the solar system. His opposite in every regard, Kepler, at twenty-nine, was sensitive, provincial, and an ardent Copernican. The young mystic, who came from a family of misfits and witches, was a little-known teacher of mathematics, a pauper, a caster of uncanny horoscopes, and a genius. Kepler went to work for Tycho to get his hands on the treasure of planetary data the latter had spent his life amassing. For 18 months they quarreled and reconciled and fought again, each trying to use the other to his own ends.

In 1601, Tycho, a dinner guest amongst illustrious companions, was drinking heavily and, being reluctant to leave the table, he (as Kepler put it) "held back his water beyond the demand of courtesy." Racked by pain, fever, and delirium, the greatest naked-eye astronomer of all time died, no doubt of uremia. Kepler quickly usurped the precious data and refused to hand it over to Tycho's rightful heirs who continued, with less than high motives, to harass him for years.

Finding patterns in the volumes of data, Kepler ultimately formulated his famous Three Laws of Planetary Motion. In time, Newton would come to rely on them as the touchstone of his theory of gravity.

7.9   The Laws of Planetary Motion

Unlike almost everyone before him* who insisted that the planetary paths must have the mystic perfection of the circle, *Kepler's First Law* maintains that *the planets move in elliptical orbits with the Sun at one focus* (Fig. 7.8). Granted, the paths are nearly circular—but the difference is crucial.

Kepler also noticed that each planet progressed at varying speeds over the course of its journey around the Sun. This would have been strange for a circular orbit; it wasn't for an elliptical one. Relying on the mathematics of his era—geometry—he succeeded in describing the motion in detail. *Kepler's Second Law* states that *as a planet orbits the Sun it moves in such a way that a line drawn from the Sun to the planet sweeps out equal areas in equal time intervals* (Fig. 7.9). For example, the Earth-Sun line will sweep out the same area in a week no matter where it is in orbit, whether it's a week in June or a week in January. Near the Sun the speed is great, the arc is long, and the slice of orbital pie is short and broad. Far from the Sun, the speed is low, the slice is narrow but tall, and still the area swept out is the same.

*Perhaps the one exception is the Muslim al-Zarqali (1081), "the blue-eyed one" who wrote about the "oval" orbit of Mercury.

Figure 7.8 The ellipse. Point *A* moves along an ellipse when the sum of the distances $\overline{F_1 A}$ and $\overline{F_2 A}$ is constant. Imagine a slack string tacked down at its ends, F_1 and F_2. A pencil holding the string taut will sweep out an ellipse, since $(\overline{F_1 A} + \overline{F_2 A})$ always equals the constant length of the string. As the two foci, F_1 and F_2, are set closer together, the figure more and more resembles a circle.

Figure 7.9 A line drawn from the Earth to the Sun sweeps out equal areas in equal times. Here each segment corresponds to a one-month interval and all have the same area. The Earth reaches its perihelion in January where it travels at maximum speed.

Armed with the physics of gravity, we can appreciate what's happening on another level. The planet, attracted to the Sun, swoops "down," accelerating as it rushes in and, as it does, the attraction increases even more. Reaching a maximum speed at its point of closest approach, the *perihelion,* it sails around and past. As it "climbs" against the Sun's gravity-pull, it decelerates. The force gradually weakens, and the planet continues to slow down until it reaches its minimum speed at the point farthest away, the *aphelion,* only to sweep around and begin the plunge once again.

As a rule, the more distant a body is from the Sun, the more slowly it moves in orbit—that much was already known (certainly Copernicus knew it). But unlike most of his predecessors, Kepler looked for precise law in such patterns. In 1618, after nearly 20 years of work he found the relationship between a planet's orbital period (T), the time it takes to travel once around the Sun, and its average distance from the Sun ($r_\odot$). *Kepler's Third Law* states that *the ratio of the average distance from the Sun cubed to the period squared is the same constant value for all planets.*

More succinctly

$$\frac{r_\odot^3}{T^2} = C_\odot \tag{7.6}$$

where $C_\odot$ is a to-be-determined constant that depends on the specific units, and is associated with the Sun as the central body. This equation with a different constant would describe the motion of all satellites about the Earth, and the same is true for any central body. The fact that all the planets orbiting the Sun share the same constant suggests that they are all ruled by the same underlying mechanism: gravity.

Example 7.5 Given that the mean solar distance for the Earth is 149.6×10^6 km, compute the corresponding value for Venus. The periods (in Earth days) of the Earth ($\oplus$) and Venus ($\female$) are 365.256 days and 224.701 days, respectively.

Solution: [Given: $r_{\odot\oplus} = 149.6 \times 10^6$ km, $T_\oplus = 365.256$ d, $T_\female = 224.701$ d. Find: $r_{\odot\female}$.] From Eq. (7.6)

$$\frac{r_{\odot\oplus}^3}{T_\oplus^2} = \frac{r_{\odot\female}^3}{T_\female^2} = C_\odot$$

Hence

$$r_{\odot\female} = \left(\frac{T_\female^2 r_{\odot\oplus}^3}{T_\oplus^2} \right)^{\frac{1}{3}} = (149.6 \times 10^6 \text{ km}) \left(\frac{224.701 \text{ d}}{365.256 \text{ d}} \right)^{\frac{2}{3}}$$

and $\boxed{r_{\odot\female} = 108.2 \times 10^6 \text{ km}}$. Notice that the division canceled the units of time, so no conversions were needed.

▶ **Quick Check:** We use the data for the Earth to determine $C_\odot = 2.509\,6 \times 10^{19}$ km³/d² and from that value and $T_\female$ find $r_{\odot\female} = \sqrt[3]{C_\odot T_\female^2} = 108.2 \times 10^6$ km, which at least shows that we did the math correctly.

Example 7.6 Knowing the period of the Moon ($27\frac{1}{3}$ days) and its distance from Earth, compute the Keplerian constant for our planet ($C_\oplus$) in units of m³/d². This same value is applicable to all artificial Earth satellites (see Problem 51).

Solution: [Given: $r_{\oplus\mathbb{C}} = 3.84 \times 10^8$ m, and $T_\mathbb{C} = 27\frac{1}{3}$ d. Find: $C_\oplus$.]

$$C_\oplus = \frac{r_{\oplus\mathbb{C}}^3}{T_\mathbb{C}^2} = \frac{5.66 \times 10^{25} \text{ m}^3}{746.9 \text{ d}^2}$$

and $\boxed{C_\oplus = 7.58 \times 10^{22} \text{ m}^3/\text{d}^2}$

Various numerical values of this constant result when different sets of units are used.

▶ **Quick Check:** $r_{\oplus\mathbb{C}} = \sqrt[3]{C_\oplus T_\mathbb{C}^2} = 0.384 \times 10^9$ m, so the math is okay. The best check would be to use Eq. (7.7), but we'll have to wait for that.

Kepler's solar constant $C_\odot$ has its simplest value, namely 1.00, when we measure distance in *astronomical units* (1 AU $= r_{\odot\oplus} \approx 1.5 \times 10^8$ km $\approx 93 \times 10^6$ mi) and time in *Earth-years*. Then, using Eq. (7.6), we can evaluate the constant for Earth, and that same value applies to all the planets orbiting the Sun; thus, $C_\odot = (1 \text{ AU})^3/(1 \text{ Earth-year})^2$. Table 7.2 gives the results for the planets known to Kepler.

7.10 Gravity and Kepler's Laws

Newton took the Sun as an immovable *center-of-force* toward which a planet is continuously impelled. He then showed that *an orbit will be elliptical if and only if the*

	Average distance to Sun (astonomical units) $r_\odot$	Period (Earth-years) T	Period squared T^2	Distance cubed $r_\odot^3$
Planet				
Mercury	0.39	0.24	0.058	0.059
Venus	0.72	0.62	0.38	0.37
Earth	1.00	1.00	1.00	1.00
Mars	1.53	1.88	3.53	3.58
Jupiter	5.21	11.9	142	141
Saturn	9.55	29.5	870	871

TABLE 7.2   Kepler's Third Law*

*Compare the last two columns.

centripetal force varies as the inverse square of the distance to the center-of-force, which is located at one of the two foci (Fig. 7.8). This statement is essentially Kepler's First Law. The needed **central force**, *a force always directed toward the same point,* is gravity and it must vary as $1/r^2$. Newton proved that a variety of orbits (closed and open) were possible under the influence of an inverse-square force. Thus, any object interacting gravitationally with some other stationary body, such as a star, moves through space along either a circular, elliptical, parabolic, or hyperbolic path.

As for Kepler's Second Law: Imagine an object moving at a constant speed in a straight line past some fixed point O, as in Fig. 7.10. In equal time intervals, Δt, a line from O to the object sweeps out areas A_1, A_2, A_3, and so on. Since each of these has the same base (Δt) and the same altitude, each area is equal. Suppose the object having reached point B in Fig. 7.11 receives a sudden momentary blow, a pulse of force directed toward O. That burst briefly accelerates the object, adding a component of velocity along the $\overline{OB}$ line to the original velocity that was along $\overline{BC}$. The result is that, in a time Δt, the object arrives at point D, having traversed the line $\overline{BD}$. The length $\overline{BE}$ is the distance the object would have traveled in Δt had it initially been at rest at B. Since $\overline{BE}$ is parallel to $\overline{CD}$, it follows that triangles BOD and BOC, which have the same base ($\overline{OB}$) and equal altitudes, are equal in area. The area A_2, which would have been swept out without the action of a centripetal force,

Halley's comet as it appeared in its most recent approach in 1985. The comet is held by gravity in an elongated elliptical orbit that causes it to swing near the Sun roughly every 75 years.

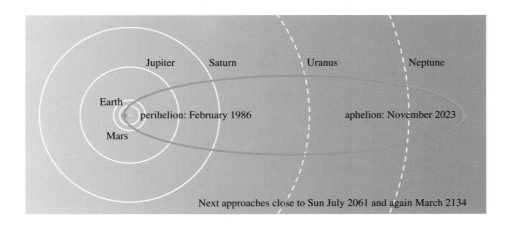

Earth

Mars

Jupiter   Saturn   Uranus   Neptune

perihelion: February 1986   aphelion: November 2023

Next approaches close to Sun July 2061 and again March 2134

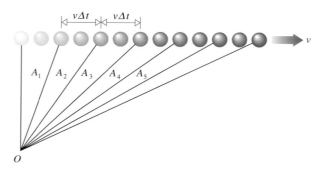

Area $\triangle BOC = A_2$
Area $\triangle BOD = A_2$

(a)

Figure 7.10 An object moving in a straight line past a point O sweeps out equal areas in equal intervals of time Δt.

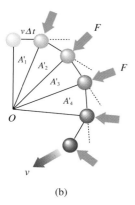

(b)

Figure 7.11 A planet moving in orbit under the influence of a central force sweeps out equal areas in equal times. (a) The altitude of a triangle BOD is obtained by dropping a perpendicular from D to $\overline{BO}$. Because $\overline{CD}$ is parallel to $\overline{BO}$, that altitude is equal to the altitude of triangle BOC, which is obtained by dropping a perpendicular from C to $\overline{BO}$. Thus, the area of triangle BOC equals the area of BOD, equals A_2. (b) A succession of pushes, all toward point O.

equals the area A_2', and so $A_1 = A_2$. Similarly, successive blows directed toward O and occurring regularly at time intervals of Δt will cause the object to sweep out areas of A_3', A_4', and so on, each of which will be equal. Note that these momentary forces need not be the same size for the swept areas to be constant. All that's required is that they be directed toward O.

And now for the final touch. Let the interval Δt be made vanishingly small as the number of blows becomes infinitely many, until the flood of distinct impulses blends into a sustained continuous center-seeking force and the segmented path dissolves into a smooth curve. *Under the influence of a centripetal force, an object moves about a center-of-force sweeping out equal areas in equal times.* The enigmatic machinery underlying Kepler's Second Law is revealed: gravity is the central force.

The last of Kepler's relationships is easy to understand, particularly in the special case of a circular orbit. A planet of mass m moving in a circle at a distance $r_\odot$ from the Sun must be under the influence of a centripetal force given by

$$F_C = ma_C = \frac{mv^2}{r_\odot} \qquad [5.11]$$

And if gravity is to provide such a force ($F_G = F_C$), then

$$F_G = G\frac{mM_\odot}{r_\odot^2} = \frac{mv^2}{r_\odot} = F_C$$

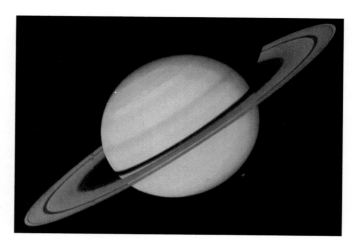

Saturn and its ring system. The rings are composed of billions of individual particles that typically range in size from pebbles to boulders. Gravity holds them in orbit and also, by its tidal action, keeps them from coalescing. Any moon orbiting closer than a certain distance will be ripped into pieces by its gravitation interaction with Saturn.

Figure 7.12 Just as a planet orbiting a star will displace the star, the Moon causes the Earth to snake along its orbit. It's the barycenter (the red dot) that moves in an elliptical orbit, as the Earth and Moon both weave in and out. The barycenter is 4672 km from the Earth's center (about 1000 miles beneath the planet's surface).

where $M_\odot$ is the mass of the central body. Since $2\pi r_\odot$ is the circumference, dividing this by the period ($2\pi r_\odot / T$) yields the orbital speed. Accordingly, canceling m and substituting for v, we obtain

$$G\frac{M_\odot}{r_\odot^2} = \frac{(2\pi r_\odot / T)^2}{r_\odot}$$

and

$$\frac{r_\odot^3}{T^2} = \frac{GM_\odot}{4\pi^2} \tag{7.7}$$

which is Kepler's Third Law [Eq. (7.6)], where the constants on the right side combine to equal $C_\odot$. Notice that *only the mass of the central body remains in the expression;* the mass of the orbiting object cancels out (see Problem 44 for a slightly more accurate version). Exchange $M_\oplus$ for $M_\odot$, and replace $r_\odot$ with $r_\oplus$ (the distance to our planet's center), and the same equation applies to the circling Moon or to any artificial satellite in Earth-orbit.

Newton soon realized that Kepler's Laws could only be approximately true. For an object to orbit the exact *c.g.* of a larger body, the latter's mass must be infinite. In real life, when a planet orbits a star, both revolve around a common center-of-gravity, the **barycenter** (see Problem 43). That means that a distant star orbited by an otherwise invisible planet should move through space with a slight wiggle (Fig. 7.12). Jupiter ($\mathfrak{2}$), the largest planet ($m_{\mathfrak{2}} = 318\, m_\oplus$) in the Solar System, displaces the Sun by almost a million miles. Half a dozen nearby stars show this telltale planetary wiggle.

When Newton saw an apple fall, he found
In that slight startle from his contemplation—
'Tis said (for I'll not answer above ground
For any sage's creed or calculation)—
A mode of proving that the earth turn'd round
In a most natural whirl, called "gravitation";
And this is the sole mortal who could grapple,
Since Adam, with a fall, or with an apple.

LORD BYRON (1788–1824)

Example 7.7 The Sun is one of an immense number of stars in a huge rotating pinwheel system known as the Milky Way galaxy. The Sun is located about 3×10^{20} m (or 3×10^4 light-years) from the center of the spherical hub about which it orbits once every 250 million years. Make a *very crude estimate* of the equivalent mass of galactic material acting on the Sun and compare that with its own mass of $M_\odot \approx 2 \times 10^{30}$ kg. Assume all the mass within the confines of the Sun's orbit can be taken to act as if it were at the center of the galaxy.

Solution: [Given: $r = 3 \times 10^{20}$ m and $T_\odot = 250 \times 10^6$ years $= 7.88 \times 10^{15}$ s. Find: M, the mass of the galaxy.] From Eq. (7.7), where now the central body corresponds to the galactic mass, we have

$$\frac{r^3}{T_\odot^2} = \frac{GM}{4\pi^2}$$

Solving this equation, we get

$$M = \frac{4\pi^2 r^3}{GT_\odot^2} = \frac{4\pi^2(3 \times 10^{20}\text{ m})^3}{(6.7 \times 10^{-11}\text{ N·m}^2/\text{kg}^2)(7.9 \times 10^{15}\text{ s})^2}$$

(continued)

(continued)

and $M \approx 3 \times 10^{41}$ kg $\approx \boxed{10^{11} M_\odot}$

This result approximately agrees with estimates that our galaxy contains about 10^{11} stars. Incidentally, there are likely some 10^{11} galaxies in the Universe.

▶ **Quick Check:** There are $\approx \pi \times 10^7$ seconds per year, so the period is right. As given above, the units of M are m³/(N·m²/kg²)s² = m/(N·s²/kg²); since m/s² is acceleration and N is force, 1 N/(m/s²) = 1 kg; hence, M has the units of kg²/kg = kg.

7.11 Satellite Orbits

Long before *Sputnik* and its thousands of progeny, Newton had observed that any projectile launched horizontally is, in a sense, an Earth *satellite* (a word coined by Kepler). A rock thrown from a tall building sails in a modest orbit that soon intersects the Earth not far from its point of launch (Fig. 7.13). Galileo maintained that this trajectory should be parabolic, but that assumed a flat world. On a spherical Earth, a horizontally hurled baseball (neglecting friction) theoretically arcs along an elliptical orbit, with its far focus at the planet's center. All the planet's mass acts as if it were at its *c.g.*, and the ball attempts to orbit that point. If the Earth were squashed down small enough, the baseball would become a moonlet. In real life, the Earth looms up in the way, and the motion is abruptly interrupted when the projectile crashes into the ground.

If the ball were fired more swiftly to start with, it would travel further; the ellipse would flatten out, becoming less elongated (Fig. 7.14). Further increasing the speed would result in ever larger, rounder elliptical paths and more distant impact points. Finally, at one particular launch speed, the ball would glide out just above

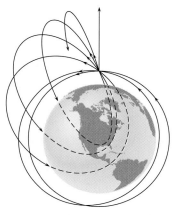

Figure 7.13 Various orbits for projectiles with the same speeds launched in different directions. In each case, the major or long axis is unchanged.

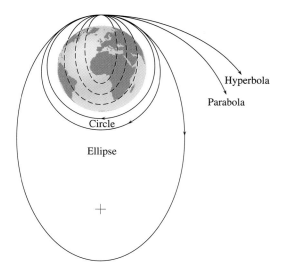

Figure 7.14 The orbits of projectiles fired horizontally close to the Earth's surface at various speeds (based on a drawing by Newton). The circular orbit occurs at a launch speed of about 8 km/s. From 8 to 11.2 km/s, the orbit is elliptical. At 11.2 km/s, it's parabolic and, beyond that, it's hyperbolic.

Y^e suns conversion doth turn the planet out of this line framing its motion into a circular, but the former desire of ye planet to move in a streight line hinders the full conquest of ye Sun, and forces it into an Ellipticke figure.

JEREMIAH HORROCKS (1618–1641)

the planet's surface all the way around to the other side without ever striking the ground. Like a little leather moon, it would wheel around the globe in a circular orbit until something (air friction, collision with birds or buildings) brought it down. At successively greater launch speeds, the ball would revolve in ever-increasing elliptical orbits until it moved so fast initially that it sailed off in an open parabolic or (if even faster) into a still flatter, hyperbolic orbit, never to come back to its starting point.

A low, circular flight path means that the satellite follows the planet's curvature. In effect, it continually drops toward the center-of-force, but since the ground "falls away" at the same rate, it never gets any closer to the surface. Figure 7.15 shows how the curvature of the Earth's surface drops a distance h beneath a tangent of length L such that the Pythagorean theorem yields

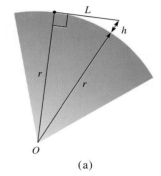

$$(r + h)^2 = r^2 + L^2$$
$$r^2 + 2rh + h^2 = r^2 + L^2$$
and
$$2rh + h^2 = L^2$$

Here we are looking at a very small segment of a large curve, and so h is tiny and h^2 by comparison with $2rh$ (since r is big) is quite negligible. Hence

$$2rh \approx L^2$$
and
$$L \approx \sqrt{2rh}$$

Now, suppose we are dealing with an orbit just above the planet's surface and the small segment corresponds in time to 1 s of motion. Then, L is the straight-line distance the object would travel if there were no gravity and h is the height it would fall due to gravity alone (see Fig. 4.2). From Eq. (3.9), $h = \frac{1}{2}g(1 \text{ s})^2 = 4.9$ m and, with $r \approx R_\oplus$, we have

$$L \approx \sqrt{2R_\oplus h} \approx 7.9 \text{ km}$$

This result means that if an object were initially fired horizontally with a speed such that it traveled 7.9 km in 1 s (that is, 18×10^3 mi/h), it would subsequently orbit the Earth at the *lowest possible altitude*, effectively dropping 4.9 m each second.

An alternative way of seeing this, which is less graphic but more general, comes from equating F_G and F_C. An object of mass m at a distance r from the center of a large body of mass M will travel a circular path with a *tangential orbital speed* v_o when the gravitational force exactly equals the centripetal force, whereupon

$$G\frac{mM}{r^2} = \frac{mv_o^2}{r}$$

and so
$$v_o = \sqrt{\frac{GM}{r}} \qquad (7.8)$$

Any satellite *regardless of its mass* (provided $M \gg m$) will move in a circular orbit of radius r, so long as its speed obeys Eq. (7.8), and it does so *independently of any rotation of the central body* (Fig. 7.16).

Figure 7.15   (a) The Earth drops away by an amount h below the tangent line of length L. (b) If a projectile is fired with a speed such that it travels a distance L in the time it falls a height h, it will remain in orbit.

Example 7.8   Use Eq. (7.8) to compute the necessary launch speed for an impractical treetop-height, circular Earth-orbit.

Solution: [Given: $r = R_\oplus = 6.4 \times 10^6$ m and $M = M_\oplus = 6.0 \times 10^{24}$ kg. Find: v_o.] If the object is to orbit just above ground level, it must have a horizontal speed of

(continued)

(continued)

$$v_o = \left(\frac{GM_\oplus}{R_\oplus}\right)^{\frac{1}{2}}$$

and

$$\boxed{v_o = 7.9 \times 10^3 \text{ m/s}}$$

$$v_o = \left[\frac{(6.67 \times 10^{-11} \text{ N·m}^2/\text{kg}^2)(6.0 \times 10^{24} \text{ kg})}{(6.4 \times 10^6 \text{ m})}\right]^{\frac{1}{2}}$$

▶ **Quick Check:** This speed is the same one we obtained from the analysis where we used Fig. 7.15 and the Earth's curvature.

Astronauts orbiting Earth are effectively weightless in free-fall.

The independence of v_o on m makes the point that an astronaut inside an orbiting spaceship is also orbiting at that same speed. And so, too, is all the loose paraphernalia in the vehicle—all of which is in orbit within the confines of the craft independent of it. The pilot will float around inside the ship just as she would float around outside of it, moving along at speed v_o in orbit. And since the only force acting is gravity, the crew will be *effectively weightless*. Indeed, orbiting astronauts are in *free-fall* toward the Earth at a rate $g_\oplus = a_C$. The concept of "falling" is generally thought of from a flat-Earth perspective, where the thing that falls gets closer to the surface in the process, which need not be the case on a spherical planet.

A launch vehicle initially rises vertically, gradually rolling over. If on separation the payload is moving tangentially at a speed v_o, satisfying Eq. (7.8), it will be injected into a circular orbit. If the speed is less than v_o, the craft will descend in an elliptical orbit (Fig. 7.14). If the speed is greater than v_o but less than $\sqrt{2}v_o$, it will ascend into a large elliptical orbit. At $\sqrt{2}v_o$, it will escape into an open parabolic path that carries it out and away forever (p. 301). At speeds in excess of $\sqrt{2}v_o$, the escape trajectory becomes hyperbolic and increasingly flattened (Fig. 7.17).

One of the most desirable orbital arrangements, especially for communications purposes, is the *geostationary* or *geosynchronous* orbit. While parked in one of these, a satellite, having been lofted directly above the equator, will circle once around in a day, revolving in synchronization with the rotating Earth. As seen from anywhere on the globe, the craft will remain in a fixed location in the sky and therefore in continuous line-of-sight communication with a ground station. Well over 100 military and civilian spacecraft are crowded into the geosynchronous band with more being added every year.

Figure 7.16 Because the Sun represents 99.9% of the mass of the Solar System, each planet essentially "sees" only the Sun as the central gravitating body. Consequently, for all planets $v_o \propto 1/\sqrt{r}$; the larger the orbit, the slower the orbital speed.

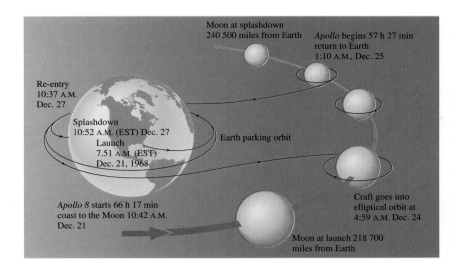

Re-entry
10:37 A.M.
Dec. 27

Splashdown
10:52 A.M. (EST) Dec. 27
Launch
7.51 A.M. (EST)
Dec. 21, 1968

Earth parking orbit

Apollo 8 starts 66 h 17 min
coast to the Moon 10:42 A.M.
Dec. 21

Moon at splashdown
240 500 miles from Earth

Apollo begins 57 h 27 min
return to Earth
1:10 A.M., Dec. 25

Craft goes into
elliptical orbit at
4:59 A.M. Dec. 24

Moon at launch 218 700
miles from Earth

Figure 7.17 This drawing summarizes the *Apollo 8* journey to the Moon in 1968. The spacecraft first went into a parking orbit around the Earth. After the crew briefly fired its rockets and sped it up, the ship coasted off to the Moon.

Example 7.9 Compute the altitude above the planet's surface and the necessary injection speed for a geostationary orbit.

Solution: [Given: $M_\oplus = 5.98 \times 10^{24}$ kg and the most important fact (though not stated explicitly) $T = 24$ h. Find: the altitude h.] Remember that $h = r - R_\oplus$, the difference between the distance from the center of the Earth to the satellite and the Earth's radius. Setting $F_G = F_C$, canceling m, and using $v = 2\pi r/T$, we get, once again

$$\frac{GM_\oplus}{r^2} = \frac{(2\pi r/T)^2}{r}$$

and so

$$r = \left(\frac{M_\oplus T^2 G}{4\pi^2}\right)^{\frac{1}{3}}$$

hence $r = (7.544 \times 10^{22} \text{ m}^3)^{\frac{1}{3}} = 42.25 \times 10^6$ m

We want $h = r - R_\oplus = \boxed{3.59 \times 10^7 \text{ m}} \approx 5.6R_\oplus \approx 22 \times 10^3$ mi. Do you suppose this is why the nation of Colombia, which straddles the equator, claims air space above it up to 23 000 miles? As for the speed—in one transit of the orbit, the satellite travels a distance of $2\pi r_\oplus$ in a time T. Hence

$$v_0 = \frac{2\pi r}{T} = \frac{2\pi(42.25 \times 10^6 \text{ m})}{(24 \text{ h})(60 \text{ min/h})(60 \text{ s/min})}$$

and

$$\boxed{v_0 = 3.1 \times 10^3 \text{ m/s}}$$

or 6.9×10^3 mi/h. By the way, *Syncom II* was the first of the species successfully launched in July 1963.

▶ **Quick Check:** Using Eq. (7.8), $v_0 = [(6.67 \times 10^{-11} \text{ N·m}^2/\text{kg}^2)(5.975 \times 10^{24} \text{ kg})/(42.25 \times 10^6 \text{ m})]^{\frac{1}{2}} = 3.1 \times 10^3$ m/s.

7.12   The Gravitational Field

An apple loosed from a branch drops earthward under the influence of gravity. But how does the apple get the message? How does it know which way is down and how hard it's supposedly being pulled?

We are content with the idea of *direct contact*—two things pressed against one another, force transmitted by touching. But matter is composed of separate, spaced atoms and the idea of *direct* contact becomes meaningless. Atoms certainly interact, yet they never touch in the familiar way we think of macroscopic objects touching each other. Even here, in the very essence of contact, there is *action-at-a-distance*, the idea of objects exerting influences across an intervening void. When your finger

All things by immortal power,
Near or far,
Hiddenly
To each other linked are,
That thou canst not stir a flower
Without troubling of a star.
FRANCIS THOMPSON (1859–1907)

"touches" this book, it approaches close enough for the electromagnetic repulsion between the electron clouds of finger atoms and book atoms to become palpable.

Each particle of matter has associated with it a surrounding field of influence, and it is that field that carries the interaction between separated bits of matter. Euler, in the mid-eighteenth century, devised a mathematical description of fluid flow that assigned a velocity to every point in a moving liquid, thereby creating a field of motion. Euler's concept was the origin of Field Theory. In much the same way, we could associate a thermometer reading with every point in a room and speak about the corresponding scalar temperature field. The idea became indispensable to physics in the late nineteenth century with the conceptualization of the electric and magnetic force fields (Chs. 17 and 21).

A **field of force** exists in a region of space when an appropriate object placed at any point therein experiences a force. Apparently, we can envision a gravitational force field surrounding any object of mass M. Suppose that we hang a small test mass m on a scale and take measurements moving from one point to another throughout the space surrounding the object in question. At each point, we could read off the force F_G (that is, the weight of the test mass), record its magnitude and direction, and thereby map the vector force field of the object.

One disadvantage to this scheme is that all the field values depend on the size of the test mass. It would be far more useful to conceive a new field quantity varying only with the mass of the object that was creating it, and not with the probe. Thus, we define the **gravitational field strength $\mathfrak{g}$** as

$$\mathfrak{g} = \frac{\mathbf{F}_G}{m} \tag{7.9}$$

This is the force on the test mass, divided by the test mass, and it's called the *gravitational field intensity,* or just the *gravitational field* of the mass distribution

Figure 7.18 This summary of theoretical considerations gives us a picture of where we have been and where we are going. Thus far, we have considered the relationship between symmetry and conservation as put forth in Noether's Principle, and we have learned about Conservation of Linear Momentum. Now, we've encountered the concept of the force field in the form of gravity.

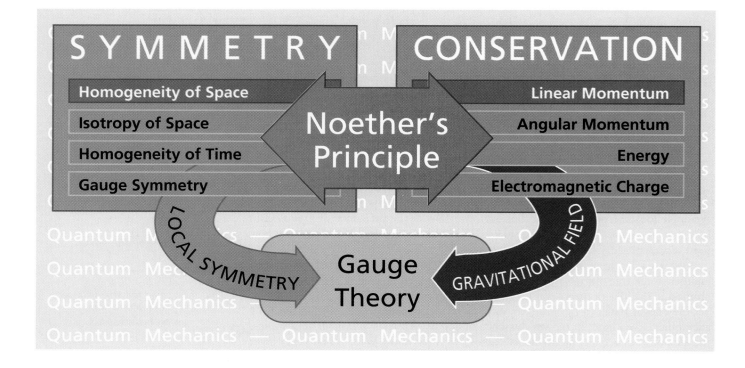

M. The gravitational field at any point is equal to the force exerted on a unit test mass located at that point (Fig. 7.18). Inasmuch as the "true" weight of an object is proportional to the absolute gravitational acceleration ($F_W = mg_0$), we can equate the field with the free-fall acceleration of the test mass.

From the definition of field strength, the appropriate units are N/kg, or, equivalently, m/s^2. In the workaday world of measuring gravitational fields, the principal unit used is the *gal*, named after Galileo. Defined as 1 cm/s^2 or 10^{-2} m/s^2, 1 gal is about one-thousandth of the Earth's surface field strength (9.8 m/s^2). Determinations of g can be made accurate to 1×10^{-6} gal by measuring the free-fall time of a body over a distance fixed via laser interferometery.

Example 7.10 Derive an expression for the magnitude of the gravitational field strength, g, at any distance *r* from a point mass *M*.

Solution: [Given: *M*, the mass of a particle. Find: g at any *r*.] We start with the definition of g [Eq. (7.9)] in terms of the force F_G on a test mass *m*. But we already know that for a point mass *M* interacting with a point mass *m*

$$F_G = G\frac{mM}{r^2} \qquad [7.2]$$

Taking *m* as our test mass and substituting it into Eq. (7.9), the field strength anywhere in the surrounding space beyond *M* becomes

$$g = \frac{GM}{r^2} \qquad (7.10)$$

▶ **Quick Check:** According to Eq. (7.10), g has the units of GM/r^2 or $(\text{N·m}^2/\text{kg}^2)\,(\text{kg})/\text{m}^2 = \text{N/kg}$, which is what it should be.

To get a picture of the field of an object, we could draw a vector representing **g** at a number of points in the surrounding region, as in Fig. 7.19. *All such vectors are directed toward the object,* since each one points in the direction of the force on a test mass and that force is always attractive. Alternatively, it's far more convenient to construct *lines-of-force,* called *field lines*. At each point in space, the direction of the field is tangent to the line-of-force (Fig. 7.20). Visualized as a three-dimensional distribution, the number of lines per unit area, the density of lines, is proportional to the field strength.

We think of every object that has mass permeating the surrounding space with a gravitational field that extends out indefinitely, its strength dropping off with distance. The field is a physical entity, and any other object immersed in it will interact directly with it, experiencing an attractive force. This explanation of action-at-a-distance raises fascinating questions that still await definitive answers. Is the field separate from *M* or is it the nonlocalized aspect of a single thing called mass? Does a particle of mass fill the surrounding space and exist everywhere its field exists? What, if anything, does the particle continue to give off as it sustains the expanding field?

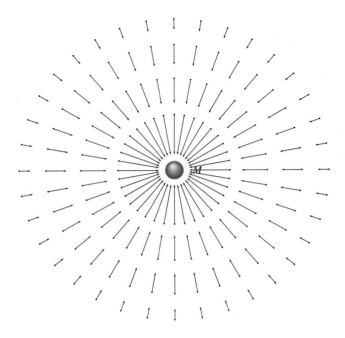

Figure 7.19 The gravitational field of mass *M*. At every point in space surrounding *M* we draw a vector corresponding to the force on a test mass *m* divided by *m*.

7.13 New Directions

In early conceptions, space was a vessel and the field was strung across it like a web. But Einstein's General Relativ-

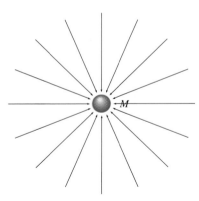

Figure 7.20 The gravitational field of a mass *M*. Rather than drawing a vector field, it's more convenient to draw the lines to which the vectors are tangent. These are the lines-of-force.

Variations in the strength of the gravitational field across North America. The units are milligals. The field is strongest (red) in the Atlantic basins off the continental shelf and weakest in the high reaches of the Rocky Mountains (purple).

ity has space-time itself as the medium of gravity. Matter is viewed as a localized deformation of the space, distortions of which propagate out into the surroundings as the field. A material object simply moves within and across the warped space created by other objects. There is no attractive force per se; there is only *free* motion along contours in space-time. Since the introduction of Relativity Theory, it has become widely accepted that the speed of light in vacuum (c) represents the upper bound to motion. This suggests that changes in the gravitational field of an object move out into space at that speed, too.

That finite propagation rate leads to an apparent conflict with Newton's Third Law whenever there's a change in the relationship between any two interacting objects. For example, suppose you jump into the air, moving in the Sun's gravity field in such a way that the solar tug on you changes. Yet your field, in which the Sun is immersed and with which *it* interacts, will remain unaltered in the vicinity of that star for about eight minutes. For all that time, the interaction pair between you and the Sun will be out of tune.

It is from the study of the microworld that we have learned to expect that *all* forces are mediated by streams of special particles. Contemporary Quantum Field Theory maintains that the forces between objects arise from the exchange of these field particles. The carrier of the gravitational interaction is the still hypothetical massless **graviton**. These particles, streaming at the speed of light from Earth to apple and vice versa, carry the interaction from one to the other. And with the reasonable assumption that gravitons possess momentum (the field transmits momentum), the law of Conservation of Momentum (p. 112) holds exactly at all times.

By comparison to the other kinds of forces to be encountered in later chapters (namely, electromagnetic, weak, and strong), gravity is by far the faintest. Not surprisingly, then, the graviton remains elusive, its existence neither confirmed nor denied. For the time being, the graviton is a theoretical entity in a marvelous scheme that has replaced invisible tentacles with invisible particles. On the frontier of modern research are theories that attempt to unite the force particles of Quantum Field Theory with the geometrical space-time perspective of General Relativity.

We now believe that the Universe was created some 17 thousand million years ago in a violent burst known as the Big Bang. That primeval space-time fireball grew and evolved into the cosmos as we know it, still expanding today. It has been suggested (by A. H. Guth, 1979) that within the very first moments of time ($\approx 10^{-35}$ s to $\approx 10^{-33}$ s), the Universe experienced an *inflationary era* during which it underwent an immensely rapid expansion. Under bizarre circumstances General Relativity allows the possibility of negative gravity. And so, driven by a momentary spasm of repulsive gravitation, the Universe may have expanded, perhaps by a factor of as much as 10^{90}. Reaching a kind of saturation, a dramatic transition occurred and the motherlode of material particles condensed into existence. The simplest of all substances, hydrogen, soon formed as particles captured one another. And $\approx 17 \times 10^9$ years later, the very same primordial atoms, dating back almost to the beginning of time, are there in every cell of your body.

Core Material

The primary insight of this chapter is Newton's **Law of Universal Gravitation**, that is,

$$F_G = G\frac{mM}{r^2} \qquad [7.2]$$

where $G = 6.672\,59 \times 10^{-11}\ \text{N·m}^2/\text{kg}^2$. This equation represents the attractive interaction of gravity between two masses M and m separated by a center-to-center distance of r. The expression holds exactly for point masses or uniform spheres. In general, any two mass configurations will interact in an approximately $1/r^2$ fashion when r is large in comparison to the sizes of the interacting objects.

On Earth's surface, the *absolute acceleration due to gravity* is

$$g_0 = \frac{GM_\oplus}{R_\oplus^2} \qquad [7.3]$$

Kepler's *Three Laws of Planetary Motion* are

1. The planets move in elliptical orbits with the Sun at one focus (and nothing at the other).
2. Any planet moves in such a way that a line drawn from the Sun to its center sweeps out equal areas in equal time intervals.
3. The ratio of the average distance from the Sun cubed to the period squared is the same constant value for all planets. That is

$$\frac{r_\odot^3}{T^2} = C_\odot \qquad [7.6]$$

Newton's formulation immediately leads to an expression for this constant, namely

$$\frac{r_\odot^3}{T^2} = \frac{GM_\odot}{4\pi^2} \qquad [7.7]$$

A satellite in a circular orbit about a large mass M, at a distance r from its center, moves with a speed

$$v_0 = \sqrt{\frac{GM}{r}} \qquad [7.8]$$

It's useful to imagine that any mass is accompanied by a *gravitational field* that extends out into the surrounding space. The *field intensity,* or *strength,* is defined as

$$\mathbf{g} = \frac{\mathbf{F}_G}{m} \qquad [7.9]$$

and is represented pictorially by a series of lines converging down to the source mass.

Suggestions on Problem Solving

1. Make sure that your answers are reasonable. Compare your results with known values to see if they fit within recognized extremes. Accordingly, a value of $g_\oplus$ computed anywhere inside or outside the Earth should be between 0 and roughly 9.8 m/s². A value of 98 m/s² must be wrong!

2. As ever, check your answers whenever possible by recomputing using a different approach.

3. Don't confuse G, g, g_0, $g_\oplus$ and g—they are all different.

4. Consider units—don't enter a number in *kilometers* when you want *meters,* or use 32 m/s² when you should use 9.8 m/s².

5. Remember that F_G varies inversely with the square of the distance. An object moved from the surface of the Earth to a point 100 Earth-radii from the planet's *center* drops in weight to 1/10 000th its original value, *not* 1/100th.

6. Many of the equations in the chapter call for the radial distance. When given the numerical value of a diameter remember to divide by 2 before substituting. Keep in mind that the altitude above the surface of a planet is not the radial distance to its center.

Discussion Questions

1. The following quote is from *Bioastronautics Data Book,* 2nd ed., NASA SP-3006 (1973), p. 149. Why is it unmitigated nonsense?

In its third form, acceleration occurs as a component of the attraction between masses. The resulting force is directly proportional to the product of the masses and inversely proportional to the square of the distance between them. The proportionality constant is the gravitational constant g which represents an acceleration of 32.24 feet per second (fps) within the terrestrial field of reference. This is the accepted unit of measurement of acceleration.

2. Why do you lose weight when you enter a tunnel passing horizontally through a mountain, or walk inside a skyscraper?

3. A comic book space colony occupies a huge central cavity symmetrically hollowed out of a spherical planetoid. The idea is to provide a home for any space traveler not wishing to return to a traditional gravity environment on the surface of some ordinary planet. Is such a weightless heaven possible? Explain. Can

the lighthearted inhabitants of the colony detect the arrival of a spacecraft landing on the planet's surface via their gravitational interactions, or are they *shielded* inside the sphere?

4. Jupiter has a mass 318 times that of the Earth and yet its "surface" acceleration due to gravity is roughly only 26 m/s². Explain how this could be.

5. It's convenient, particularly with weather and spy satellites, to have them always point radially downward toward the planet. One imaginative *passive* system for accomplishing that is to extend from the bottom of the craft a long boom with a small mass at the end and let the Earth's gravity do the rest. Explain qualitatively what effects might be at work here. (Hint: Think of the boom swinging slightly about its *c.g.* away from verticality.) The operative mechanism is called *differential gravity*. The satellite is also stable with the boom pointing up and, believe it or not, that has actually happened—somewhere in space a spy satellite is upside down and useless.

6. Suppose that an astronaut in space is in a high circular Earth orbit moving eastward (that is, counterclockwise looking down on the North Pole). Imagine that she releases an apple into space. What will be the flight path of the apple? What if the apple is hurled directly forward (eastward)? What if it's thrown in the backward direction? What do you think would happen to it if it were thrown radially toward the planet? (This last question goes beyond the material we have studied so far, and is only meant as a test of your intuition.)

7. Why is it easier to launch Earth satellites in an easterly direction than in a westerly direction? The John F. Kennedy Space Center is on which coast of Florida? Why? Why is it in Florida?

8. The Sun seems to move across the star field more rapidly in the winter than it does in the summer. What does this phenomenon imply about the Earth-Sun distances at these two times?

9. A number of spacecraft, including the several *Apollo* Command Modules, have been put into low lunar orbits. Why are such low paths possible around the Moon but not around the Earth?

10. Is it possible to have one object orbit another in a closed path if the mutual interaction is repulsive?

11. Figure Q11 shows a spaceship changing from a low to a high circular orbit. Explain how it's being done. Compare the initial and final orbital speeds. Assume an ordinary chemical rocket that can be fired in limited duration bursts.

12. Figure Q12 shows a maneuver that a rocket can use to escape from a circular orbit about a planet into a hyperbolic unpowered flight path. What must happen at point *A*? If the process is run backward and the craft free-falls toward the planet, what must happen at *A* to effect capture?

13. Figure Q13 depicts a double-thrust hyperbolic departure of a rocket from a circular orbit. Explain how it happens. Incidentally, for high speeds at great distances, this maneuver is more fuel-efficient than the one considered in the previous question.

14. The Sun gravitationally attracts the Moon. Does the Moon orbit the Sun? Explain.

15. It has become commonplace to see large (≈9-ft-diameter) parabolic TV antennas on the roofs of buildings, especially bars and hotels. In what direction do they point, and at what are they pointed?

Figure Q11

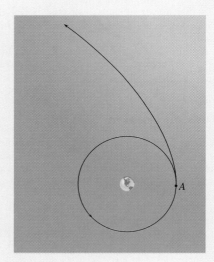

Figure Q12

16. Imagine that all the mass of the Earth were somehow compressed into a sphere < 200 m in diameter. It can be shown that a *solid steel* satellite 1.0 m in diameter descending toward the planet would be pulled apart as soon as it approached to within 100 m of the center of this miniworld. Similarly, if the Moon were to orbit Earth at a center-to-center distance of less than $2.9R_\oplus$, it, too, would be torn apart. Explain.

17. The third outermost moon of Saturn is Hyperion, an irregular object 400 km long by 220 km wide. It is one of the largest lopsided bodies in the Solar System. Why do you think Hyperion isn't spherical like most other moons?

18. Figure Q18 shows the Earth and Moon and a number of lines forming some sort of pattern. Identify the pattern and discuss its important features. (See Problem 18.)

Figure Q13

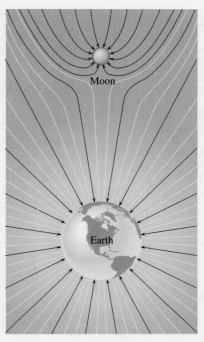

Figure Q18

19. In 1978, the *International Sun-Earth Explorer 3* satellite (*ISEE-3*) was launched into a remarkable orbit. What's so extraordinary is that it orbits a *libration point* in space where no central mass exists. This point lies on the Earth-Sun center-to-center line and is the place where the gravitational forces of these two bodies exactly counterbalance each other. In other words, the gravitational field there is zero. The plane of the orbit is perpendicular to the center line and contains the libration point. Explain, qualitatively, how *ISEE-3* could be held in orbit this way.

20. Make a sketch of the gravitational field both inside and outside a hollow spherical shell of mass m, assuming no other masses are present. Why is it necessary to add that last requirement?

Multiple Choice Questions

1. If the dimensions of length, time, and mass are L, T, and M, respectively, then the dimensions of G are   (a) $[L^2/MT^2]$ (b) $[L^3/MT]$ (c) $[L^3/MT^2]$ (d) $[L^3/M^2T]$ (e) $[L^2/M^3T]$.

2. A 1.00-kg chicken weighs 9.8 N on the surface of the Earth. At a distance of one Earth-radius above the planet's surface   (a) its weight is 4.9 N   (b) its mass is 0.50 kg   (c) its weight is 19.6 N   (d) its mass is 2.00 kg   (e) none of these.

3. Call the gravitational attraction between you and the planet, your *true* weight. Because of the rotation of Earth, the reading that you get when you stand on a scale   (a) will be more than your *true* weight   (b) will everywhere be less than your *true* weight   (c) will always be your *true* weight   (d) will only be your *true* weight at the two poles   (e) will only be your *true* weight at the equator.

4. Suppose you were transported to the mythical planet Mongo, which is four times as massive as Earth and has twice the diameter. Your Mongoian weight as compared to your present weight would be   (a) 4 times larger   (b) the same   (c) 2 times smaller   (d) 4 times smaller   (e) none of these.

5. Mars has a mass of 0.1074 $M_{\oplus}$ and is at a mean distance from the Sun which is 1.52 times larger than that of

Earth. By comparison to the gravitational force exerted on Mars by our world, the force exerted on Earth by Mars is   (a) 0.1074 times smaller   (b) 0.1074 times larger   (c) the same   (d) 1.52 times less   (e) none of these.

6. The acceleration of a meteor at a height above the Earth of $1R_{\oplus}$ is   (a) about 2.5 m/s^2 (b) 9.8 m/s^2 (c) not enough information to say   (d) can be anything from 9.8 m/s^2 to 0   (e) none of these.

7. The asteroid Geographos (one of the Apollo group, each of which crosses the Earth's orbit on the way around the Sun) has a radius of $2.4 \times 10^{-4}R_{\oplus}$ and a mass of $8.4 \times 10^{-12}M_{\oplus}$. How does the gravitational acceleration on its surface compare to the corresponding value g on the Earth? It equals   (a) $2.4 \times 10^{-4}g$ (b) $8.4 \times 10^{-12}g$ (c) $1.5 \times 10^{-4}g$ (d) $3.5 \times 10^{-8}g$ (e) none of these.

8. Suppose a deep vertical shaft is drilled into the Moon and a rock is slowly lowered down. It will be found that the rock's   (a) mass and weight both decrease   (b) mass decreases but the weight increases   (c) mass and weight both remain unchanged   (d) mass is constant but its weight decreases gradually   (e) none of these.

9. The Earth's value of the acceleration due to gravity   (a) is constant inside   (b) varies by about 0.5% over the surface

(c) is constant over the surface   (d) is constant beyond the surface   (e) none of these.

10. An astronaut on the Moon has a mass that by comparison to his mass on Earth is   (a) unchanged   (b) six times greater   (c) six times less   (d) not enough information to say   (e) none of these.

11. The acceleration due to gravity, as measured by a spring balance determination of the weight of an object ($F_W = mg$), varies from place to place on Earth because   (a) the mass changes   (b) g is affected by the rotation of the planet only   (c) g depends on the shape of the planet only   (d) g depends on both the rotation and shape of the planet   (e) none of these.

12. If *Martian Orbiter 1* is sailing about the planet in a circle with an orbital radius nine times that of *Orbiter 2*, whose speed is v_2, what is the speed of *Orbiter 1*?   (a) $\frac{1}{3}v_2$ (b) $3v_2$ (c) v_2 (d) $81v_2$ (e) none of these.

13. Figure MC13 shows a spaceship in orbit about a star. If its speeds at the four points shown are v_A, v_B, v_C, and v_D, respectively, then   (a) $v_A < v_B < v_C < v_D$ (b) $v_A > v_B > v_C > v_D$ (c) $v_A > v_B = v_D > v_C$ (d) $v_A < v_B = v_D < v_C$ (e) none of these.

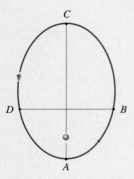

Figure MC13

14. A spacecraft is in a circular orbit about a planet located at point O in Fig. MC14. When it reaches point A on its orbit, the First Mate throws a canister out the front port straight ahead; the canister goes into a new orbit shown in   (a) part a   (b) part b   (c) part c   (d) part d   (e) none of these.

(a)   (b)   (c)   (d)

Figure MC14

15. It is desired that a spacecraft initially in a circular orbit drop straight down to the planet below. Figure MC15 shows the planned double-thrust maneuver, which supposedly will save fuel. The craft   (a) slows down at B and brakes to zero speed at D (b) speeds up at A and brakes to zero speed at E (c) brakes slightly at A and then again at E (d) the maneuver is impossible   (e) none of these.

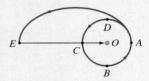

Figure MC15

16. The gravitational field intensity at a point 1.00 m away from the center of a small 1.00-kg spherical mass (assuming no other sources exist) is   (a) 1.00 N/kg   (b) 1.00 gal   (c) 6.67×10^{-11} m/s^2 (d) 9.8 m/s^2 (e) none of these.

Problems

Miscellaneous data: $M_\oplus = 5.975 \times 10^{24}$ kg, $M_\odot = 1.987 \times 10^{30}$ kg, $M_\mathbb{C} = 7.35 \times 10^{22}$ kg, $R_\odot = 6.97 \times 10^8$ m, $R_\oplus = 6371.23$ km, $R_\mathbb{C} = 1.74 \times 10^6$ m, $r_{\odot\oplus} = 1.495 \times 10^{11}$ m, $r_{\oplus\mathbb{C}} = 3.844 \times 10^8$ m.

THE UNIVERSAL LAW OF GRAVITATION

1. [I] What would happen to the weight of an object if its mass was doubled while its distance from the center of the Earth was also doubled?

2. [I] The gravitational attraction between a 20-kg cannonball and a marble separated center-to-center by 30 cm is 1.48×10^{-10} N. Compute the mass of the marble.

3. [I] Suppose that two identical spheres, separated center-to-center by 1.00 m, experience a mutual gravitational force of 1.00 N. Compute the mass of each sphere.

4. [I] At what center-to-center distance from the Earth would a 1.0-kg mass weigh 1.0 N?

5. [I] Suppose the Earth were compressed to half its diameter. What would happen to the acceleration due to the gravity at its surface?

6. [I] Compare the gravitational force of the Earth on the Moon to that of the Sun on the Moon.

7. [I] If the average distance between Uranus ($\hat{\delta}$) and Neptune (Ψ) is 4.9×10^9 km, and $M_{\hat{\delta}} = 14.6M_\oplus$ while $M_\Psi = 17.3M_\oplus$, compute their average gravitational interaction.

8. [I] Imagine two uniform spheres of radius R and density ρ in contact with each other. Write an expression for their mutual gravitational interaction as a function of R, ρ, and G.

9. [I] If you can jump 1.00 m high on Earth, how high can you jump on Venus, where $g_\varphi = 0.88g_\oplus$? Assume the same takeoff speed.

10. [I] What fraction of what you weigh on Earth would you weigh in a rocket ship firing its rockets so that it was stationary with respect to the planet $4R_\oplus$ from its surface?

11. [I] Consider two subatomic particles, an electron and a proton, which have masses of 9.1×10^{-31} kg and 1.7×10^{-27} kg, respectively. When separated by a distance of 5.3×10^{-11} m, as they are in a hydrogen atom, the electrical attraction (F_E) between them is 8.2×10^{-8} N. Compare this with the corresponding gravitational interaction. How many times larger is F_E than F_G?

12. [I] Show that Eq. (7.3) is equivalent to Eq. (7.4) when $r = R_\oplus$.

13. [I] The acceleration due to gravity on the surface of Mars is 3.7 m/s². If the planet's diameter is 6.8×10^6 m, determine the mass of the planet and compare it to Earth.

14. [II] Imagine an astronaut having a mass of 70 kg floating in space 10.0 m away from the *c.g.* of an *Apollo* Command Module whose mass is 6.00×10^3 kg. Determine the gravitational force acting on, and the resulting accelerations (at that instant) of, both the ship and the person.

15. [II] Venus has a diameter of 12.1×10^3 km and a mean density of 5.2×10^3 kg/m³. How far would an apple fall in one second at its surface?

16. [II] Given that $M_{\mathbb{C}}/M_\oplus = 0.012\,30$ and $R_{\mathbb{C}}/R_\oplus = 0.2731$, compute the ratio of an astronaut's Moon-weight ($F_{W\mathbb{C}}$) to Earth-weight ($F_{W\oplus}$).

17. [II] Taking the surface value of $g_\oplus$ to be g_0, show that

$$g_\oplus = g_0(R_\oplus/r)^2 \quad \text{for } r \geq R_\oplus$$

18. [II] Locate the position of a spaceship on the Earth-Moon center line such that, at that point, the tug of each celestial body exerted on it would cancel and the craft would literally be weightless. (See Fig. Q18.)

19. [II] Mars has a mass of $M_\delta = 0.108M_\oplus$ and a mean radius $R_\delta = 0.534R_\oplus$. Find the acceleration of gravity at its surface in terms of $g_0 = 9.8$ m/s².

20. [II] Three very small spheres of mass 2.50 kg, 5.00 kg and 6.00 kg are located on a straight line in space away from everything else. The first one is at a point between the other two, 10.0 cm to the right of the second and 20.0 cm to the left of the third. Compute the net gravitational force it experiences.

21. [II] It is believed that during the gravitational collapse of certain stars, such great densities and pressures will be reached that the atoms themselves will be crushed, leaving only a residual core of neutrons. Such a *neutron star* is, in some respects, very much like a giant atomic nucleus with a tremendous density of roughly about 3×10^{17} kg/m³. Compute the surface acceleration due to gravity for a one-solar-mass neutron star.

22. [II] Fig. P22 shows two concentric, thin, uniform spherical shells of mass m_1 and m_2, at the center of which is a small ball of lead of mass m_3. Write an expression for the gravitational force exerted on a particle of mass m at each point A, B, and C located at distances r_A, r_B, and r_C from the very center.

23. [III] Write an expression for the gravitational force on a small mass m imbedded in a uniform spherical cloud of mass M and radius R. Take the particle to be at $r < R$.

24. [III] Two 2.0-kg crystal balls are 1.0 m apart. Compute the magnitude and direction of the gravitational force they

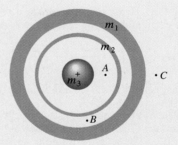

Figure P22

exert on a 10-g marble located 0.25 m from the center-to-center line as shown in Fig. P24.

2.0 kg

1.0 m   0.25 m

2.0 kg

Figure P24

25. [III] A neutron star (see Problem 21) can be envisioned as an immense nucleus held together by its self-gravitation. What is the shortest period with which such a star could rotate, if it's not to lose mass flying off at the equator? Take $\rho = 1 \times 10^{17}$ kg/m³. It is widely believed that *pulsars*, strange celestial emitters of precisely pulsating radiation, are rapidly rotating neutron stars (see p. 260, Chapter 8).

26. [III] Draw a graph of the weight of an object of mass m due to the Earth versus the height above the surface, out to roughly 700 km. What can you say about the curve (so long as $R \gg h$)?

27. [III] Let M be the mass of a uniform spherical planet of radius R. If h is the height above its surface, show that the absolute gravitational acceleration g_p varies with h as

$$g_p = \frac{GM}{R^2}\left(1 + \frac{h}{R}\right)^{-2}$$

This expression can be approximated using the binomial expansion. (See Appendix A-5.)

$$(a + x)^n = a^n + na^{n-1}x + \tfrac{1}{2}n(n-1)a^{n-2}x^2 + \cdots$$

where $x^2 < a^2$. Here, $a = 1$, $x = h/R$, $n = -2$, and we limit the calculation to the case where $h \ll R$. Show that

$$g_p \approx \frac{GM}{R^2}\left(1 - \frac{2h}{R}\right)$$

Notice that GM/R^2 is the surface value of g_p occurring when $h = 0$, as in Eq. (7.3).

28. [III] Calculate the acceleration due to gravity 10 000 m above the Earth's surface in the following two ways: (1) using Eq. (7.3); and (2) using the approximation of Problem 27. Compare your results.

29. [III] In light of Problem 27, determine the acceleration due to gravity experienced by the Lunar Module when it was 100 m above the Moon's surface. Is it appreciably different from the surface value?

THE COSMIC FORCE

30. [I] Using the data for the Earth's orbit, compute the mass of the Sun.

31. [I] Determine the approximate speed of a Lunar Orbiter revolving in a circular orbit at a height of 62 km. Take the Moon's radius as 1738 km.

32. [I] Each of the *Apollo* Lunar Modules was in a very low orbit around the Moon. Given a typical mass of 14.7×10^3 kg, assume an altitude of 60.0 km and determine the orbital period.

33. [I] For any Earth satellite in a circular orbit, show that both its period (in seconds) and its distance from the center of the planet (in meters) are related by way of

$$T = 3.15 \times 10^{-7} (r_\oplus)^{\frac{3}{2}}$$

34. [I] *Sputnik I,* the first artificial satellite to circle the planet (October 1957) had a mean orbital radius of 6950 km. Compute its period.

35. [I] A satellite is to be raised from one circular orbit to another twice as large. What will happen to its period?

36. [I] Referring to Problem 35, compare the two orbital speeds v_1 and v_2.

37. [I] What is the acceleration due to the gravity of the Moon at the Earth?

38. [I] Find the gravitational field strength 1.00 m from a small sphere whose mass is 1.00 kg.

39. [I] What is the gravitational field strength of the Sun at the Earth?

40. [II] By definition, the Earth is a distance of 1.000 AU from the Sun. Using the fact that Jupiter is, on average, 5.202 8 AU from the Sun, compute its period in Earth-years.

41. [II] Determine the period in Earth-years of a satellite placed in a circular solar orbit with a radius of 371.6 million miles.

42. [II] Imagine a central body of mass M_B (be it a star, planet, or moon) about which another object is orbiting such that its Keplerian constant is C_B. Show that C_B/M_B is a universal constant, the same for all bodies.

43. [II] Imagine the two comparable masses m_1 and m_2 of Fig. P43 orbiting their barycenter O at distances r_1 and r_2, respectively, with a common period T. Since their mutual gravitational interaction provides their individual centripetal forces F_{C1} and F_{C2}, these must be equal. Show that

$$\frac{r_1}{r_2} = \frac{m_2}{m_1}$$

and compare this with the definition of the *c.g.*

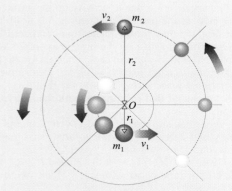

Figure P43

44. [II] Referring to Problem 43 and Fig. P43, write explicit expressions for $F_G = F_{C1}$ and $F_G = F_{C2}$, cancel out whatever you can, and then combine them to get Kepler's Third Law in the more accurate Newtonian version, namely

$$\frac{(r_1 + r_2)^3}{T^2} = \frac{G(m_1 + m_2)}{4\pi^2}$$

Unlike Eq. (7.7), which applies when the central body (here the Sun) has essentially infinite mass, this expression is more realistic. Notice that the equations are identical when $m_1 \gg m_2$, so that m_2 is negligible; of course, $r = r_1 + r_2$.

45. [II] For an elliptical orbit, the *mean or average distance* from the central body to the orbiter is just half the main symmetry axis. This *semi-major axis* is also equal to the average of the aphelion and perihelion distances. The tiny asteroid Icarus (which has a mass of 5.0×10^{12} kg and a radius of just 0.7 km) orbits the Sun in an elongated elliptical path. It crosses the Earth's orbit as it sweeps in to a perihelion of only 0.186 AU and out to an aphelion at 1.97 AU. Compute its mean distance from the Sun and its orbital period in Earth-years.

46. [II] The Andromeda galaxy, known as M31, is a great spiral star-island 2.2×10^6 light-years away. Measurements show that a star out at its extremities—5×10^9 AU from the center—orbits the nucleus at a speed of 200 km/s. Approximate the mass of M31.

47. [II] What is the value of the gravitational field intensity halfway down inside of a solid uniform sphere of mass M and radius R? (See Problem 23.)

48. [II] Determine the gravitational field at the center of a uniform ring of mass M and radius R.

49. [II] Galileo discovered the four major moons of the planet Jupiter in 1610. The nearest one, Io, has a period of 1.769 9 d and is 5.578 Jovian radii ($R_{\unicode{x2643}}$) from the center of the planet. Use this information to calculate the mean density of Jupiter.

50. [III] The binary stars Sirius A and Sirius B each orbit their barycenter with a period of 50 years. Their separation is measured to be 20.0 AU (or 2.99×10^{12} m) with the much fainter star, Sirius B, being twice as far from the barycenter as is Sirius A. Compute both their masses. (Hint: Go back and look at Problems 43 and 44.)

51. [III] We wish to put an artificial Earth satellite in a circular orbit halfway out to the Moon. Compute its period and the necessary orbital speed.

52. [III] Two identical masses of value m are each located on the y-axis a distance d above and below the origin. Show that the gravitational field intensity at any point P on the x-axis at a distance from the origin of x is given by

$$g = \frac{2Gmx}{(x^2 + d^2)^{\frac{3}{2}}}$$

Notice that the distance from either mass to P is $r = (x^2 + d^2)^{\frac{1}{2}}$.

8

Rotational Motion

MOTION APPEARS IN TWO basic types: **translational** and **rotational** (Fig. 8.1). If a body is moving such that a line drawn between any two of its internal points remains parallel to itself, the body is *translating*—it may be traveling either in a straight line (*rectilinear translation*) or along an arc (*curvilinear translation*). If, on the other hand, that line between any two points does not remain parallel to itself, the body is *rotating*.

At this very instant, the Earth is rotating and revolving around the Sun, which is itself spinning and revolving around the center of the galaxy, which is itself spinning through space. At the other extreme, the atoms that form

Rectilinear translation
(a)

Curvilinear translation
(b)

Rotation (about a
point outside of the body)
(d)

Rotation (about a
point within the body)
(c)

Rotation and translation
(e)

Figure 8.1   The movement of all rigid bodies in space can be described in terms of rotational and translational motion.

all matter are composed of far smaller parts that are in perpetual rotation.

On a more pragmatic level, many of our machines, from washing machines, tape recorders, CD players, jet engines, and pencil sharpeners to clocks and carousels, involve rotation. After all, the most basic means of mechanically transmitting power is by way of a rotating drive shaft.

This chapter focuses on the kinematics and dynamics of rotational motion. Happily, the conceptual framework is very similar to that of translational motion. There will be a few fresh definitions, a new circularization of imagery, but you have already mastered the essence of the treatment (see Chapters 2–6). Just as linear momentum is the measure of translational motion, *angular momentum* is the measure of rotational motion.

At the center of this treatment is a grand insight: Conservation of Angular Momentum. This is another of those rare principles that reveals the inner symmetry of the Universe (p. 267)—empty space is the same in all directions, and so the laws of physics do not depend on direction.

THE KINEMATICS OF ROTATION

Imagine an object revolving about some point; it might, for instance, be a wheel, a line of hand-holding skaters, or a string of beads as in Fig. 8.2. Whirled in a circle, each bead sails along a different arc, but each sweeps through the same angle θ. Up to now, we have measured angles in *degrees;* 360° to a circle. Yet, there is nothing special about "degrees"—they are a remnant of the ancient Babylonian number system, whose base was 60 rather than 10.

The angle θ can be specified in a less arbitrary way by relating it back to an essential aspect of the circle, its circumference equal to $2\pi r$. Each bead in Fig. 8.2 moves an arc-length (l) at some radial distance from the center (r). The larger is r, the larger is l. In each case, θ is the same, suggesting that we might formulate θ in terms of the ratio of l to r. Thus, let

$$\theta = \frac{l}{r} \quad \text{or} \quad l = r\theta \tag{8.1}$$

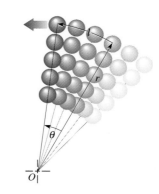

Figure 8.2 A revolving string of beads all moving around their respective circles in step with each other.

The angle θ is now framed as a distance over a distance and is *unitless*—it's a pure number. The degree is not a physical unit in the strict sense either—it is not referenced back to meters, kilograms, or seconds. When $l = r$, $\theta = 1$ and as a reminder that we are measuring angles via Eq. (8.1), we call this 1 *radian* or 1 rad. If we ever meet up with aliens from some other world, they will probably use the equivalent of radians, too.

Draw a circle of radius r (Fig. 8.3a). Now lay off arc-lengths equal to r around the circle. You will end up with six such segments plus an additional fraction of one. That's because the circumference equals $2\pi r$, which is $(6.283185 \ldots)r$. Figure 8.3b shows the relationship between an arc-length l and the angle θ subtended by it at point O. When $l = r$, $\theta = 1$ rad, but it also equals $1/2\pi$ of the angle corresponding to a whole turn (360°), which means that

Figure 8.3 The relationship between arc-length, radius, and angle. (a) The circumference of a circle equals $2\pi r$. (b) As r moves through the angle θ, its end point sweeps out a length l. (c) When $l = r$, $\theta = 1$ rad. This shape is close to that of an equilateral triangle where $\theta = 60°$ (d).

$$1 \text{ rad} = \frac{360°}{2\pi} = \frac{360°}{(6.283\ldots)} = 57.3° \tag{8.2}$$

as in Fig. 8.3c. Compare this angle to that formed in an equilateral triangle, which has to be a little wider (Fig. 8.3d). Clearly, 2π rad = 360°. Any angle θ in radians can be transformed into degrees by simply taking proportions, such that

$$\frac{\theta(\text{radians})}{2\pi(\text{radians})} = \frac{\theta(\text{degrees})}{360(\text{degrees})} \tag{8.3}$$

Thus, π rad = 180°, $\pi/2$ rad = 90°, and so forth (see Table 8.1).

The Moon has a diameter $D_{\mathbb{C}}$ equal to about 3.4×10^6 m, and it is a distance $r = 3.8 \times 10^8$ m from the Earth's surface. If we approximate its straight-line diameter as an arc-length, then the angle θ subtended at the Earth (Fig. 8.4) by the Moon is

$$\theta = \frac{l}{r} = \frac{D_{\mathbb{C}}}{r} = \frac{3.4 \times 10^6 \text{ m}}{3.8 \times 10^8 \text{ m}} = 0.009 \text{ rad}$$

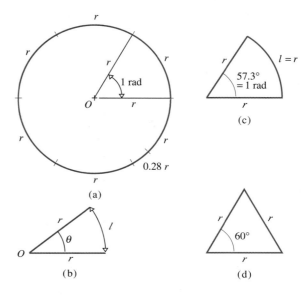

TABLE 8.1 Angles in degrees and radians

Degrees	Radians in terms of π	Radians in decimals
0	0	0
15	$\pi/12$	0.261
30	$\pi/6$	0.524
45	$\pi/4$	0.785
57.3	1.0	1.0
60	$\pi/3$	1.047
90	$\pi/2$	1.571
120	$2\pi/3$	2.094
180	π	3.142
270	$3\pi/2$	4.712
360	2π	6.283

or 0.5°. By comparison, the diameter of the Sun (1.4×10^9 m) is immensely larger, but its distance to the Earth (1.5×10^{11} m) is also much greater. Coincidentally, the Sun also subtends an angle of 0.009 rad. That's why the Sun and Moon seem to be the same size.

Example 8.1 The Andromeda galaxy is a giant spiral star-island whose mass is that of 300 thousand million suns. You can see it with the naked eye as a faint elongated cloud in the night sky. Inasmuch as it subtends an angle of 4.1° and is known to be larger than our own galaxy [163×10^3 light-years (ly) in diameter for Andromeda as compared to 100×10^3 ly for our galaxy], how far away is it?

Solution: [Given: the diameter of Andromeda $D_A = 163 \times 10^3$ ly and its subtended angle $\theta = 4.1°$. Find: r_A.] First, convert θ into radians, which yields 0.072 rad. Then, approximate the arc-length by the diameter. Hence, $\theta \approx D_A/r_A$. Consequently

$$r_A = \frac{D_A}{\theta} = \frac{163 \times 10^3 \text{ ly}}{0.072} = \boxed{2.3 \times 10^6 \text{ ly}}$$

▶ **Quick Check:** When the arm r_A-long sweeps through θ, its end point traces an arc of length $(4.1°\pi/180°)(2.3 \times 10^6 \text{ ly}) = 1.6 \times 10^5$ ly.

Example 8.2 A decent spy-satellite camera can distinguish objects that subtend angles of as little as 0.5×10^{-6} rad. How small an object could such a device detect from 200 mi away?

Solution: [Given: $\theta = 0.5 \times 10^{-6}$ rad and $r = 200$ mi. Find: l.] This analysis involves a straightforward substitution into Eq. (8.1). Using

$$r = (200 \text{ mi})(1.609 \text{ km/mi}) = 321.8 \text{ km}$$
$$l = r\theta = (321.8 \times 10^3 \text{ m})(0.5 \times 10^{-6} \text{ rad}) = 0.160\,9 \text{ m}$$

or, to one significant figure, $\boxed{l = 0.2 \text{ m}}$.

▶ **Quick Check:** $l = (200 \text{ mi})(0.5 \times 10^{-6} \text{ rad}) = 100 \times 10^{-6}$ mi, or $l = 0.5$ ft.

Incidentally, the Muslim Alhazen (A.D. 1000) correctly suggested that visual perspective arises because the angle an object subtends gets smaller as it recedes from us—a nearby tree appears large, whereas a distant one, the same size, seems minute.

8.1 Angular Speed

Imagine the last bead in Fig. 8.2 moving in a circle of radius r. As it travels along the circular segment l, the radius sweeps through angle θ, the **angular displacement**. If the bead revolves through six complete counterclockwise turns and comes back to its starting point, we will take $\theta = 6 \times (2\pi)$ rad rather than $\theta = 0$. Define θ swept out counterclockwise as positive and clockwise as negative, and the same for l. Consequently, four turns in one direction followed by four turns in the other brings the bead back to where it started, $\theta = 0$ ($l = 0$); in that sense, we are dealing with angular displacements.

If the bead travels the distance $l = r\theta$ in a total time t, divide both sides to get

$$\frac{l}{t} = r\frac{\theta}{t} \tag{8.4}$$

The ratio on the left is evidently the average speed (that is, $v_{av} = l/t$). The ratio on the right is the **average angular speed**, denoted by the lowercase Greek letter *omega* (ω) such that

$$\omega_{av} = \frac{\theta}{t} \tag{8.5}$$

The units of angular speed are rad/s although *degrees per second, revolutions per second,* and *rotations per minute* (rpm) are commonly used. Notice that each revolution swings through 2π rad, so that 1 rev/s $= 2\pi$ rad/s. Only when the angular speed is specified in rad/s will

$$v_{av} = r\omega_{av} \tag{8.6}$$

Example 8.3 On August 24, 1968, a very fast horse named Dr. Fager finished a 1.00-mi race in 1.00 min 32.2 s. Assuming he ran once around a circular track, what was his average speed and average angular speed?

Solution: [Given: $l = 1.00$ mi on a circular path and $t = 1.00$ min 32.2 s. Find: v_{av} and ω_{av}.] By Eq. (2.1)

$$v_{av} = \frac{l}{t} = \frac{1.00 \text{ mi}}{1.537 \text{ min}} = 0.651 \text{ mi/min}$$

or $\boxed{v_{av} = 62.8 \text{ km/h}}$. To find the radius, we use the fact

that $2\pi r = 1.00$ mi; hence, $r = 0.159$ mi $= 0.256$ km. From Eq. (8.6)

$$\omega_{av} = \frac{v_{av}}{r} = \frac{0.017\,4 \text{ km/s}}{0.256 \text{ km}} = \boxed{0.068\,2 \text{ rad/s}}$$

▶ **Quick Check:** You could use Eq. (8.5) instead. Since $\theta = 2\pi$ rad

$$\omega_{av} = \frac{\theta}{t} = \frac{2\pi \text{ rad}}{1.537 \text{ min}} = \frac{6.28 \text{ rad}}{1.537 \text{ min}} = 4.09 \text{ rad/min}$$

Figure 8.5 depicts the slightly different circumstance in which both θ and l are measured from some reference line; in this case, the x-axis. Here, the finite intervals $\Delta l = l_f - l_i$ and $\Delta\theta = \theta_f - \theta_i$, as before, are related by

$$\Delta l = r\Delta\theta \tag{8.7}$$

If these are traversed in a time duration Δt, then

$$\frac{\Delta l}{\Delta t} = r\frac{\Delta\theta}{\Delta t} \tag{8.8}$$

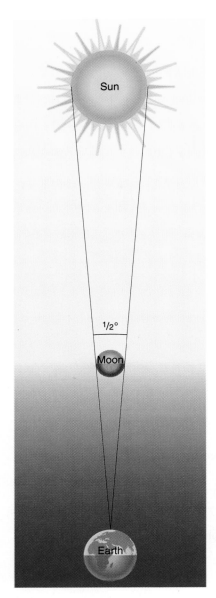

Figure 8.4 The Moon is closer to us than the Sun. Therefore, even though it is much smaller than the Sun, it happens to subtend very nearly the same angle when viewed from Earth.

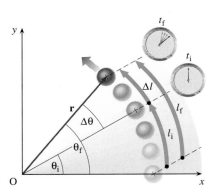

Figure 8.5 As the ball moves from angular position θ_i at l_i to angular position θ_f at l_f on the arc, the angle changes by $\Delta\theta$ and the arc-length measured up from the x-axis changes by Δl.

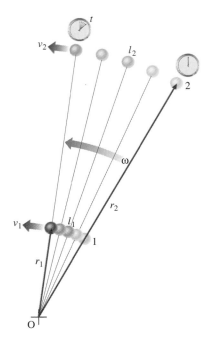

Figure 8.6 Here, beads 1 and 2 move in step, swinging around their respective circles together. Bead 2 travels a larger path and must therefore be moving faster ($v_2 > v_1$), even though both beads have the same ω.

which provides an alternative statement for the average speeds

$$v_{av} = \frac{\Delta l}{\Delta t} \quad \text{and} \quad \omega_{av} = \frac{\Delta\theta}{\Delta t} \tag{8.9}$$

These reduce to Eqs. (2.1) and (8.5) when $\theta_i = 0$, $\theta_f = \theta$, $t_i = 0$, and $t_f = t$. The advantage to this notation is that we can define the **instantaneous angular speed** (or just the angular speed ω) by finding the limit of the average angular speed as the averaging time interval becomes smaller and smaller. In other words

$$\omega = \lim_{\Delta t \to 0} \left[\frac{\Delta\theta}{\Delta t}\right] \tag{8.10}$$

Thus, ω *is the time-rate-of-change of* θ. Taking the limit as $\Delta t \to 0$ of Eq. (8.8) yields

$$v = r\omega \tag{8.11}$$

where ω is in rad/s. Keep in mind that the velocity is always tangent to the curved path. At any instant, if we know ω, we can find v, and vice versa.

It's evident from Fig. 8.6 that the more distant a bead is from the center of rotation, the more rapidly it must be moving. Once around for bead 1 is a small circle, while bead 2 travels an appreciably larger path in the same time. Thus, if they stay abreast, $v_2 > v_1$ even though ω is the same for each of them. That's why the outermost skater in a long rotating line has to be the fastest, and why the inner rail position in a race is most advantageous.

Example 8.4 A phonograph record of Beethoven's Ninth Symphony is whirling around at $33\frac{1}{3}$ rpm. Two 1.0×10^{-5}-kg lint balls are at rest on the disk, one on the end of the paper label 5.0 cm from the central spin axis and the other 15 cm out on the very edge of the record. Determine their respective linear speeds and the corresponding friction forces acting on them.

Solution: [Given: ω equivalent to $33\frac{1}{3}$ rpm, $m_1 = m_2 = 1.0 \times 10^{-5}$ kg, $r_1 = 0.050$ m and $r_2 = 0.15$ m. Find: v_1, v_2, F_{f1}, and F_{f2}.] Convert $33\frac{1}{3}$ rpm to rad/s: $\omega = (33\frac{1}{3} \text{ rpm})(2\pi \text{ rad/rot})/(60 \text{ s/min}) = 3.491$ rad/s. Hence, from Eq. (8.11), the lint balls are moving tangentially at speeds of

$$v_1 = r_1\omega = (0.050 \text{ m})(3.491 \text{ rad/s}) = \boxed{0.17 \text{ m/s}}$$

and $\quad v_2 = r_2\omega = (0.15 \text{ m})(3.491 \text{ rad/s}) = \boxed{0.52 \text{ m/s}}$

The rad unit is dropped when dealing with linear quantities.

The balls are held in place by friction, which must supply the needed centripetal force. It follows from Eq. (5.11) that

$$F_{f1} = F_{C1} = \frac{m_1 v_1^2}{r_1} = \frac{(1.0 \times 10^{-5} \text{ kg})(0.17 \text{ m/s})^2}{0.050 \text{ m}}$$

and $\boxed{F_{f1} = 5.8 \times 10^{-6} \text{ N}}$

Moreover

$$F_{f2} = F_{C2} = \frac{m_2 v_2^2}{r_2} = \frac{(1.0 \times 10^{-5} \text{ kg})(0.52 \text{ m/s})^2}{0.15 \text{ m}}$$

and $\boxed{F_{f2} = 1.8 \times 10^{-5} \text{ N}}$

▶ **Quick Check:** $F_C = mr\omega^2$; hence, $F_{C1} = (1 \times 10^{-5} \text{ kg})(0.050 \text{ m})(3.5 \text{ rad/s})^2 \approx 6 \ \mu\text{N}$ and $F_{C2} = (1 \times 10^{-5} \text{ kg})(0.15 \text{ m})(3.5 \text{ rad/s})^2 \approx 20 \ \mu\text{N}$.

Example 8.5 A horizontal cylinder (Fig. 8.7) with a radius of 50 cm has a rope wrapped around it that supports a weight at its end. At what constant angular speed must the shaft revolve if the weight is to descend at 1.0 m/s?

Solution: [Given: $R = 0.50$ m and $v = 1.0$ m/s. Find: ω.] As the shaft turns through an angle θ, a length of rope l unwinds such that $l = R\theta$, which *is* the distance the weight drops. Hence, $v = R\omega$ and

$$\omega = \frac{v}{R} = \frac{1.0 \text{ m/s}}{0.50 \text{ m}} = \boxed{2.0 \text{ rad/s}}$$

▶ **Quick Check:** $v = R\omega = (0.5 \text{ m})(2 \text{ rad/s}) = 1$ m/s.

Figure 8.7 As the cylinder revolves, the rope wrapped around it unwinds, and the weight descends.

When a rigid body rotates about an axis at some angular speed ω, *that rate characterizes the motion and is the same for the entire object.* By contrast, each constituent point of the body moves with a linear speed determined by its perpendicular distance from the spin axis.

8.2 Angular Acceleration

Variations in ω are as commonplace as are variations in v. Just imagine a rock whirling faster and faster on the end of a string, or a carousel slowing down as it comes to rest. If a rotating body changes its angular speed by an amount $\Delta\omega$ in a time interval Δt, its **average angular acceleration** (denoted by the lowercase Greek letter *alpha*) is

$$\alpha_{av} = \frac{\Delta\omega}{\Delta t} = \frac{\omega_f - \omega_i}{t_f - t_i} \tag{8.12}$$

The unit of angular acceleration is radians per second-per second or rad/s^2. Once again, if we take the limit of this expression as $\Delta t \to 0$, the average angular acceleration approaches the **instantaneous angular acceleration** and

$$\alpha = \lim_{\Delta t \to 0}\left[\frac{\Delta\omega}{\Delta t}\right] \tag{8.13}$$

For simplicity, we restrict our study to situations where α is constant.

From Eq. (8.11), it follows that a change in angular speed will be accompanied by a change in linear speed, $\Delta v = r\Delta\omega$. And Eq. (8.12) can be rewritten as

$$\alpha_{av} = \frac{\Delta v}{r\Delta t} = \frac{1}{r}a_{av} \tag{8.14}$$

Taking the limit as $\Delta t \to 0$ of this equation provides the link between the linear and angular acceleration, namely

$$\alpha = \frac{a_T}{r} \tag{8.15}$$

as long as α is in rad/s^2 (Table 8.2). Here, a_T is a *tangential acceleration arising only from a change in the tangential speed*—a_T doesn't exist if v is constant; that is, if $\Delta v = 0$. It is the scalar value of an acceleration vector drawn from the object tangent to the path of motion at any instant (Fig. 8.8):

$$a_T = r\alpha \tag{8.16}$$

Centripetal acceleration a_C arises from a change in the direction of the motion (not a change in speed): v may or may not be constant. So long as the direction of **v** changes, there will be an a_C, and that always occurs for circular motion. When both the direction and magnitude of **v** vary in time, a_T will exist acting perpendicular to a_C. This is true for a runner who speeds up around a turn, or a monkey who slows while swinging on a vine.

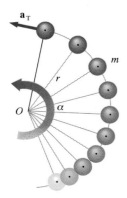

Figure 8.8 When an object moving in a circle goes faster (or slower), it accelerates tangentially. The angle that marks its position (θ) changes at a rate (ω) that is not constant. The object has an angular acceleration (α) such that $a_T = r\alpha$.

TABLE 8.2	Linear and angular kinematic parameters	
Equation	**Units**	
$l = r\theta$	l (m)	θ (rad)
$v = r\omega$	v (m/s)	ω (rad/s)
$a_T = r\alpha$	a_T (m/s^2)	α (rad/s^2)

Example 8.6 At a particular moment in a race, a car roaring around a turn with a radius of 50 m had an angular speed of 0.60 rad/s and an angular acceleration of 0.20 rad/s². Compute its linear speed, its centripetal acceleration, its tangential acceleration, and its *total* linear acceleration (Fig. 8.9).

Solution: [Given: $r = 50$ m, $\omega = 0.60$ rad/s, and $\alpha = 0.20$ rad/s². Find: v, a_C, a_T, and the total acceleration a.] The linear speed is

$$v = r\omega = (50 \text{ m})(0.60 \text{ rad/s}) = \boxed{30 \text{ m/s}}$$

From which it follows that

$$a_C = \frac{v^2}{r} = \frac{(30 \text{ m/s})^2}{50 \text{ m}} = \boxed{18 \text{ m/s}^2}$$

The tangential acceleration is

$$a_T = r\alpha = (50 \text{ m})(0.20 \text{ rad/s}^2) = \boxed{10 \text{ m/s}^2}$$

To find the total linear acceleration a, refer to the figure, which shows the relationship between $\mathbf{a}_C$, $\mathbf{a}_T$, and $\mathbf{a} = \mathbf{a}_C + \mathbf{a}_T$. Accordingly,

$$a = \sqrt{a_C^2 + a_T^2} = \sqrt{(18 \text{ m/s}^2)^2 + (10 \text{ m/s}^2)^2}$$

and

$$a = 21 \text{ m/s}^2$$

at an angle of

Figure 8.9 A car moving faster and faster around a curve has both a tangential ($\mathbf{a}_T$) and a centripetal ($\mathbf{a}_C$) acceleration. Its total acceleration is the vector $\mathbf{a} = \mathbf{a}_T + \mathbf{a}_C$.

$$\phi = \tan^{-1}\frac{a_C}{a_T} = \boxed{61°}$$

▶ **Quick Check:** It's always best to avoid making a calculation that depends on a previous one. Since we computed v, we really should have used

$$a_C = r\omega^2 = (50 \text{ m})(0.60 \text{ rad/s})^2 = 18 \text{ m/s}^2$$

Moreover, $a_C = a \sin \phi = (21 \text{ m/s}^2)(\sin 61°) = 18 \text{ m/s}^2$.

8.3 Equations of Constant Acceleration

In Chapter 3, we considered linear motion at constant acceleration and arrived at five important equations. With the proper interpretation, these same expressions also apply to curvilinear motion. Hence, if we take l to be the arc-length traveled and require a *constant* tangential acceleration a_T, then

$$v_f = v_i + a_T t \tag{8.17}$$
$$v_{av} = \tfrac{1}{2}(v_i + v_f) \tag{8.18}$$
$$l = v_{av}t = \tfrac{1}{2}(v_i + v_f)t \tag{8.19}$$
$$l = v_i t + \tfrac{1}{2}a_T t^2 \tag{8.20}$$
and
$$v_f^2 = v_i^2 + 2a_T l \tag{8.21}$$

As usual, the positive direction for l determines the positive directions of both v and a_T.

Example 8.7 A space station in the shape of a 100-m-diameter wheel is spinning so as to impart a linear tan-gential speed of 22.1 m/s to a point on its outermost wall, thereby providing an artificial gravity of 1g. A

(continued)

(continued)

pair of thrusters is to be fired to decelerate that point at a rate of 1.00 m/s². How long should the burn last to bring the station to rest? How large an arc-length will a reference point on the rim traverse during the process? What will happen if the rockets burn too long?

Solution: [Given: Take the initial direction of rotation as positive; thus, $v_i = +22.1$ m/s, $R = 50$ m and, because we are dealing with a deceleration here, $a_T = -1.00$ m/s². Find: t and l for $v_f = 0$.] Let the origin ($l = 0$) correspond to the location of the reference point at the moment the thrusters fire ($t = 0$). From Eq. (8.17), we obtain t, as

$$t = \frac{(v_f - v_i)}{a_T} = \frac{(0 - 22.1 \text{ m/s})}{(-1.00 \text{ m/s}^2)} = \boxed{22.1 \text{ s}}$$

Knowing t, we can use either one of Eqs. (8.19), (8.20), or (8.21) to find l, thus

$$l = \frac{(v_f^2 - v_i^2)}{2a_T} = -\frac{-(22.1 \text{ m/s})^2}{2(-1.00 \text{ m/s}^2)} = \boxed{244 \text{ m}}$$

If the rockets burn longer than 22.1 s, the station will begin to rotate in the opposite direction, picking up speed such that v_f becomes negative and l decreases, ultimately becoming negative too.

▶ **Quick Check:** $l = v_{av}t = \frac{1}{2}(22.1 \text{ m/s} + 0) \times (22.1 \text{ s}) = 244$ m.

These equations of curvilinear motion can most easily be transformed into an equivalent angular description when the path is circular. We know that

$$l = r\theta \tag{8.1}$$

$$v = r\omega \tag{8.11}$$

$$a_T = r\alpha \tag{8.16}$$

It is then a simple matter of substitution into Eqs. (8.17) through (8.21) to arrive at

$$\omega_f = \omega_i + \alpha t \tag{8.22}$$

$$\omega_{av} = \frac{1}{2}(\omega_i + \omega_f) \tag{8.23}$$

$$\theta = \omega_{av}t = \frac{1}{2}(\omega_i + \omega_f)t \tag{8.24}$$

$$\theta = \omega_i t + \frac{1}{2}\alpha t^2 \tag{8.25}$$

and

$$\omega_f^2 = \omega_i^2 + 2\alpha\theta \tag{8.26}$$

where the direction in which θ is positive is the direction in which ω and α are both positive.

Example 8.8 Mounted in a bus is a 2.0-m-diameter flywheel, a massive disk often used to store rotational energy. If it is accelerated from rest at a constant rate of 2.0 rpm per second, what will be the angular speed of a point on the rim of the flywheel after 5.0 s? Through what angle will that point have rotated?

(continued)

(continued)

Solution: [Given: $r = 1.0$ m, $\alpha = +2.0$ rpm/s and $\omega_i = 0$. Find: ω_f and θ at $t = 5.0$ s.] Let's first get α in rad/s²; that is,

$$\alpha = \frac{(2.0 \text{ rpm/s})(2\pi \text{ rad/rot})}{(60 \text{ s/min})} = 0.209 \text{ rad/s}^2$$

To find the final angular speed, use Eq. (8.22)

$$\omega_f = \omega_i + \alpha t = 0 + (0.209 \text{ rad/s}^2)(5.0 \text{ s})$$

and $\qquad \omega_f = 1.045$ rad/s

or, to two figures, $\boxed{\omega_f = 1.0 \text{ rad/s}}$. Equation (8.25) pro-

vides θ:

$$\theta = \tfrac{1}{2}\alpha t^2 = \tfrac{1}{2}(0.209 \text{ rad/s}^2)(5.0 \text{ s})^2 = \boxed{2.6 \text{ rad}}$$

▶ **Quick Check:** Notice that at $\alpha = 2.0$ rpm/s, the flywheel's angular speed should pick up 10 rpm in 5 s; that is, $(2\pi 10 \text{ rad/min})/(60 \text{ s/min})$, or $\omega_f = 1.05$ rad/s, which gives the flywheel an ω_{av} of $(1.05/2)$ rad/s, so it will revolve through $\theta = \omega_{av}t = (1.05/2)(5 \text{ rad}) = 2.6$ rad.

The situation of a freely rolling disk, cylinder, or sphere (Fig. 8.10) merits a closer study. The term *freely rolling* means there is no slipping at the point of contact with the ground—no skidding and no spinning in place. The wheel in the figure rolls to the right and point O on the axis moves to O' as A moves to A', and B moves to B'. The arc-length from B to A equals l just as length $\overline{BA'} = \overline{OO'} = l$. Thus, the rectilinear distance traveled by the centerpoint is $l = R\theta$, and so $v = R\omega$ and $a = R\alpha$. *The center of the wheel has a linear speed and acceleration equal to that of any point on its rim.* Notice that the velocity of point B, while in contact with the ground, measured with respect to O, namely $\mathbf{v}_{BO}$, is to the left. The velocity of O with respect to the ground, $\mathbf{v}_{OG}$, is to the right. Furthermore, $v_{BO} = v_{OG} = R\omega$; hence

$$\mathbf{v}_{BG} = \mathbf{v}_{BO} + \mathbf{v}_{OG} = 0$$

The velocity of point B with respect to the ground at the moment of touching is zero—the contact point doesn't slip across the surface. (See Fig. 5.28.)

Figure 8.10 As a body rolls, revolving about its center at a rate ω, with an angular acceleration α, point O moves at a speed $v = R\omega$, with an acceleration at that instant of $a = R\alpha$.

Example 8.9 A cyclist traveling at 5.0 m/s uniformly accelerates up to 10.0 m/s in 2.0 s. Each tire of the bike has a 35.0-cm radius, and a small pebble is caught in the tread of one of them. (a) What is the angular acceleration of the pebble during those two seconds? (b) Through what angle does the pebble revolve? (c) How far does the pebble travel during that accelerating interval?

Solution: [Given: $v_i = 5.0$ m/s, $v_f = 10.0$ m/s, $t = 2.0$ s and $R = 0.35$ m. Find: (a) α, (b) θ, and (c) l.] The bike moves along with the center of each wheel, accelerating at a linear rate of

$$a = \frac{(v_f - v_i)}{t} = \frac{(10 \text{ m/s}) - (5.0 \text{ m/s})}{2.0 \text{ s}} = 2.5 \text{ m/s}^2$$

This, in turn, equals $R\alpha$ and therefore

$$\alpha = \frac{a}{R} = \frac{2.5 \text{ m/s}^2}{0.35 \text{ m}} = \boxed{7.1 \text{ rad/s}^2}$$

which is the angular acceleration of each wheel and, hence, of the pebble. Inasmuch as the linear speed of the pebble equals the linear speed of the bike, $v_i = r\omega_i = 5.0$ m/s, $\omega_i = 14.29$ rad/s and so, from Eq. (8.25)

$$\theta = \omega_i t + \tfrac{1}{2}\alpha t^2$$

(continued)

(continued)

Therefore

$$\theta = (14.29 \text{ rad/s})(2.0 \text{ s}) + \tfrac{1}{2}(7.1 \text{ rad/s}^2)(2.0 \text{ s})^2$$

and $\boxed{\theta = 43 \text{ rad}}$. Both bike and pebble travel distances

of $l = R\theta = \boxed{15 \text{ m}}$.

▶ **Quick Check:** At an average speed of 7.5 m/s, moving for 2.0 s, the cyclist traverses 15 m, and everything checks.

THE DYNAMICS OF ROTATION

Newton's Three Laws are the foundation of our understanding of dynamics, both linear and rotational. A body revolves as a single entity when its various parts are constrained to stay together by internal forces. That being the case, we may describe the overall rotation in terms of our new kinematic concepts θ, ω, and α and introduce new dynamical quantities, as well.

Newton's First Law has its rotational equivalent in the concept that *a body at rest tends to stay at rest while a body in uniform rotational motion tends to stay in such motion, except insofar as it is acted upon by a torque.* Remember the discussion in the section "Rotational Equilibrium" (p. 173), where we considered the two determining conditions of nonaccelerating systems: $\Sigma \mathbf{F} = 0$ and $\Sigma \tau = 0$. Force is the changer of all motion. Specifically, the *moment of the force* is the *torque,* and *torque is the changer of rotational motion.*

The resistance to the change in motion of an object is called *inertia,* and that resistance is physically embodied in the *inertial mass* (or just *mass,* for short). This same property of matter can appear as a resistance to the change in rotational motion, where it is called *rotational inertia.* This rotational resistance is associated with both the amount of mass and its distribution with respect to the axis of rotation. Euler called that composite concept the *moment of inertia.*

8.4 Rotational Inertia

In swinging closed a heavy door or turning a massive wheel, it's evident that if the thing is to accelerate rotationally, a force must be applied, and it must be applied with some moment-arm. There must be a net torque.

Envision a very small mass m constrained to move on a circle of radius r, about an axis passing through O, under the influence of an applied tangential force F (Fig. 8.8). The particle will experience a tangential acceleration via the Second Law

$$F = ma_T$$

or, using Eq. (8.16) $F = mr\alpha$

The torque about O arising from F is rF, and multiplying both sides of the above equation by r yields

$$\tau_0 = rF = mr^2\alpha \tag{8.27}$$

The equation suggests that the rotational equivalent of mass (of *inertia*) is the quantity mr^2. This is the **moment of inertia of a point-mass** about a given axis, and it is designated as $I. = mr^2$. Hence, $\tau_0 = I.\alpha$ just as $F = ma$.

A rigid body consists of a great many interacting particles. When such a body is set into rotation, Eq. (8.27) describes any one of those component particles. The sum of all the torques acting on all the constituent point-masses results in an overall angular acceleration of the body such that

$$\Sigma\tau_0 = (\Sigma mr^2)\alpha \tag{8.28}$$

The summation on the right is the moment of inertia of the body about the axis of rotation passing through O, hence

$$I = \Sigma mr^2 \tag{8.29}$$

Each particle (1, 2, 3, . . .) with its own mass and perpendicular distance from O has its own moment of inertia. All these can be added so that $I = m_1 r_1^2 + m_2 r_2^2 + m_3 r_3^2 + \ldots$

Calculus is usually used to compute moments of inertia, but we'll rely on Table 8.3, which lists values of I for some uniform symmetrical bodies about various axes. *The more mass and the farther it is from the axis, the greater will be I, and the greater will be the resistance to the change in the rotational motion* (Fig. 8.11). That's why the thin walled ring in Table 8.3 has twice the moment of inertia of the

TABLE 8.3   Moments of inertia

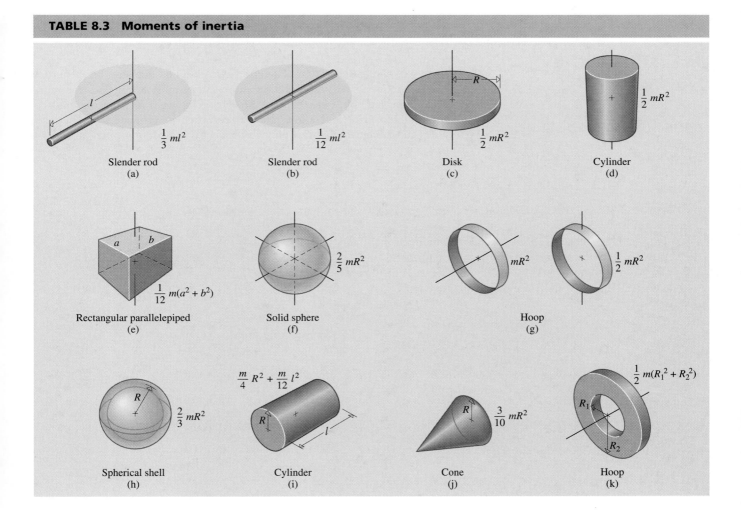

$\frac{1}{3}ml^2$	$\frac{1}{12}ml^2$	$\frac{1}{2}mR^2$	$\frac{1}{2}mR^2$
Slender rod (a)	Slender rod (b)	Disk (c)	Cylinder (d)

$\frac{1}{12}m(a^2 + b^2)$ $\frac{2}{5}mR^2$ mR^2 $\frac{1}{2}mR^2$

Rectangular parallelepiped (e)   Solid sphere (f)   Hoop (g)

$\frac{2}{3}mR^2$ $\frac{m}{4}R^2 + \frac{m}{12}l^2$ $\frac{3}{10}mR^2$ $\frac{1}{2}m(R_1^2 + R_2^2)$

Spherical shell (h)   Cylinder (i)   Cone (j)   Hoop (k)

Figure 8.11 With the limbs drawn in near the spin axis as in (a), the moment of inertia is comparatively small. With bent knees I increases, as in (b), and reaches a maximum with arms and legs extended perpendicular to the axis (c).

(a)   (b)   (c)

solid disk with the same mass, and why the moment of inertia of the hollow sphere is $\frac{5}{3}$ that of the solid sphere. The expression

$$\Sigma \tau_0 = I\alpha \qquad (8.30)$$

is the primary dynamical equation of rotating systems. The farther a bit of mass is from the axis, the larger will be I, and the more torque must be applied to attain a given α. "Choking up" on a baseball bat (that is, gripping it in from the end) decreases I about the pivot and allows the same torque to produce a larger α. If a youngster can't swing the bat fast enough to meet the incoming pitch, a larger α will help.

Example 8.10 The evil Dr. Doom, involved in an extortion threat to de-spin the Earth, plans to mount a series of surplus rockets tangentially all along the Equator. Taking the planet to be a uniform sphere of radius 6.37×10^6 m and mass 5.98×10^{24} kg, how much continuous total thrust would the rockets need to apply to accomplish the heinous deed in the course of 12 hours?

Solution: [Given: $R_\oplus = 6.37 \times 10^6$ m and $M_\oplus = 5.98 \times 10^{24}$ kg. Find: F such that $\omega_f = 0$ when $\Delta t = 24$ h.] The necessary torque is given by Eq. (8.30), so we must first determine I and α. According to Table 8.3, the moment of inertia of a solid sphere with the mass of the Earth is

$$I = \tfrac{2}{5}mR_\oplus^2 = \tfrac{2}{5}(5.98 \times 10^{24} \text{ kg})(6.37 \times 10^6 \text{ m})^2$$

or $I = 9.71 \times 10^{37}$ kg·m². It must decelerate from its present angular rate to zero in 12 h. The former (ω_i) can

be computed accurately enough assuming the planet now spins through 2π rad in 24 h; that is, $\omega_i = 0.727 \times 10^{-4}$ rad/s, and

$$\alpha = \frac{\omega_f - \omega_i}{\Delta t} = \frac{0 - 0.727 \times 10^{-4} \text{ rad/s}}{12 \text{ h} \times 60 \text{ min/h} \times 60 \text{ s/min}}$$

and

$$\alpha = -1.68 \times 10^{-9} \text{ rad/s}^2$$

The total required torque is then

$$\Sigma \tau_0 = I\alpha$$
$$\Sigma \tau_0 = (9.71 \times 10^{37} \text{ kg·m}^2)(-1.68 \times 10^{-9} \text{ rad/s}^2)$$

and

$$\Sigma \tau_0 = -1.63 \times 10^{29} \text{ N·m}$$

Since each rocket acts about the spin axis with a moment arm of $R_\oplus$, it follows that

$$\Sigma \tau_0 = R_\oplus \Sigma F$$

(continued)

(continued)

and the total thrust must be

$$\Sigma F = \frac{\Sigma \tau_0}{R_\oplus} = \frac{-1.63 \times 10^{29} \text{ N·m}}{6.37 \times 10^6 \text{ m}} = \boxed{-2.6 \times 10^{22} \text{ N}}$$

The minus sign shows that the thrust must oppose the positive ω_i. The value 2.6×10^{22} N is a realistically large force. Compare it to the 3.4×10^7 N thrust of a *Saturn* Moon rocket. Dr. Doom would have to fire over

700 million million such rockets continuously over the whole 12 hours! So, not to worry.

▶ **Quick Check:** $I \approx 0.4(6 \times 10^{24} \text{ kg})(36 \times 10^{12} \text{ m}^2) \approx 9 \times 10^{37}$ kg·m²; $\omega_i(24 \text{ h} \times 60 \text{ min/h} \times 60 \text{ s/min}) = 6.28 \text{ rad} = 2\pi$, so $\omega_i = 0.727 \times 10^{-4}$ rad/s is okay. Using Eq. (8.24), $\theta = (3.6 \times 10^{-5} \text{ rad/s}) \times (4.3 \times 10^4 \text{ s}) = 1.6 \text{ rad}$. From Eq. (8.26) $\alpha = -\omega_i^2/2\theta = -1.7 \times 10^{-9}$ rad/s².

Equation (8.30) governs all sorts of angular motion including, for example, that of a pole-carrying tightrope walker. The long balance pole has a large moment of inertia about its center. Since the acrobat cannot topple over sideways off the wire without rotating the pole, and since such angular acceleration is resisted by the large rotational inertia of the pole, it can be quite handy. Any kid walking on a narrow curb knows to stick both arms out for "balance."

Legs swinging about hip joints also have rotational inertia. Long, heavy legs require large torques to accelerate them. That's why animals with four lightweight legs can usually run faster than erect creatures with two necessarily big feet and relatively massive legs. One thing to do to reduce the rotational inertia I about the hip joint when high accelerations are needed is to bend the legs. Most people must know that, since few run stiff-legged.

Example 8.11   A 10.0-kg mass hangs on a rope wrapped around a freely rotating 2.00-kg cylinder of radius 10.0 cm as shown in Fig. 8.12. Determine the tension in the rope and the accelerations of both the cylinder and the mass. Use $g = 9.81$ m/s².

Solution: [Given: $m = 10.0$ kg, $M_c = 2.00$ kg, and $R = 10.0$ cm $= 0.100$ m. Find: F_T, α, and a.] There are three unknowns, and we'll need three equations (the sum-of-the-torques, the sum-of-the-forces, and the relation between a and α). The tension of the rope produces a torque on the cylinder such that

$$\curvearrowright^+ \Sigma \tau_0 = F_T R = I\alpha$$

and so $F_T = I\alpha/R$. Furthermore, for the descending mass traveling in the vertical direction

$$\downarrow^+ \Sigma F_V = mg - F_T = ma$$

As we saw earlier, the rope accelerates at a rate equal to that of a point on the cylinder, $a = R\alpha$. All three equations yield

$$mg - F_T = ma = mR\alpha$$

$$mg - \frac{I\alpha}{R} = mR\alpha$$

Figure 8.12   Up to this point, we neglected the motion of the pulley in this sort of problem. Now, however, we can take into account that it really isn't massless. The pulley has a moment of inertia and will reduce the acceleration of m.

(continued)

(continued)

and so
$$\alpha = \frac{mg}{mR + \dfrac{I}{R}}$$

Since

$$I = \tfrac{1}{2}M_c R^2 = \tfrac{1}{2}(2.00 \text{ kg})(0.100 \text{ m})^2 = 0.010 \text{ kg·m}^2$$

$$\alpha = \frac{98.1 \text{ kg·m/s}^2}{(1.00 \text{ kg·m} + 0.100 \text{ kg·m})} = \boxed{89.2 \text{ rad/s}^2}$$

This result corresponds to $a = R\alpha = \boxed{8.92 \text{ m/s}^2}$ and

a tension of $F_T = m(g - a) = \boxed{8.90 \text{ N}}$.

▶ **Quick Check:** Let's examine the expression for α at its extremes; namely, when $m \gg M_c$ and $m \ll M_c$. In the first case ($m \gg M_c$), I/R would be negligible compared to mR. Thereupon $\alpha \approx g/R$; that is, $a \approx g$; $F_T = 0$, and we would have free-fall. When $m \ll M_c$, $I/R \gg mR$, and $\alpha \approx mgR/I$, which is vanishingly small—hardly any motion occurs. Both conclusions are reasonable.

8.5 The Center-of-Mass

Given an extended object, it is possible to find a point, the **center-of-mass** (*c.m.*), *where all the mass* (*M*) *of the object can be thought of as concentrated,* as the object translates in compliance with Newton's Second Law: $\Sigma \mathbf{F} = M\mathbf{a}_{cm}$.

To locate the *c.m.*, envision an object of mass M composed of particles bound together via internal interactions (Fig. 8.13). Suppose the body, initially at rest, is acted upon by several externally applied forces that *do not cause it to rotate*—$\Sigma\tau = 0$. If the sum of those forces is nonzero, the body must translate in the direction of the resultant. Assume that the net force is in the y-direction. An acceleration (*a*) of the body in that direction will then occur. What we have to find is the point, the *c.m.*, such that if the net force vector acted through that point it would produce the same translational motion of the object as did the group of forces. The body would then behave as if its entire mass was at the *c.m.*

The answer is evident for the simple case of two identical spheres of mass m rigidly tied to one another by a massless rod, as in Fig. 8.14. If the net force $\Sigma\mathbf{F}$ is applied directly to one of the spheres causing it to accelerate, the other will momentarily lag behind and swing around. If the net force is now applied near to one or the other of the spheres, the system will again rotate. Only when $\Sigma\mathbf{F}$ passes through the midpoint—the *c.m.*—will both spheres undergo the same acceleration. The resulting torque about the *c.m.* will then be zero, and no rotation will occur.

The more complex object in Fig. 8.14, uniformly accelerating at a upward in the y-direction, behaves exactly as if it were at rest in a *uniform* downward gravitational field (wherein $g = a$). We established that behavior much earlier in problems dealing with elevators. In these two equivalent pictures, the upward net accelerating force ($\Sigma F = ma$) is equal in magnitude and opposite in direction to the effective weight (mg), which acts through the *c.g.* If $\Sigma\mathbf{F}$ is to be applied through the *c.m.* so that there is no rotation, the net torque about the *c.m.* must equal zero. We know from Section 6.5 that in a gravitational field the torque due to the weight of each particle will be zero about the *c.g.*, where

$$x_{cg} = \frac{\Sigma F_W x}{\Sigma F_W} = \frac{\Sigma mgx}{\Sigma mg} \qquad [6.7]$$

In the present case, the acceleration due to gravity is equated with a, so that

$$x_{cm} = \frac{\Sigma max}{\Sigma ma}$$

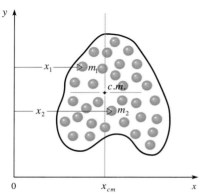

Figure 8.13 An object composed of a number of point-masses $m_1, m_2, \ldots$

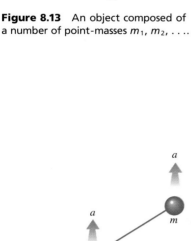

Figure 8.14 The center-of-mass of two identical spherical masses.

Since the body accelerates as a whole at a single rate, a is the same on the top and bottom of this equation, and cancels, yielding

$$x_{cm} = \frac{\Sigma mx}{\Sigma m} \tag{8.31}$$

Two unequal-mass balls fastened together with a string are free to slide along the central horizontal rod. The system can be spun about the vertical shaft. Only when that vertical axis passes through the *c.m.* of the two balls will they remain where they are as the system rotates. A slight displacement to either side will unbalance the centripetal forces and cause the balls to slide to that side.

Similar expressions hold in three dimensions for y_{cm} and z_{cm}.

In the event that the gravitational field is precisely uniform over the body, g can be canceled in Eq. (6.7), and the *c.g.* is then identical to the *c.m.* For a symmetric homogeneous body (for example, a sphere, cylinder, cube, disk, hoop, doughnut, or rectangular parallelepiped), the *c.m.* is at the geometric center. Usually, the field is adequately uniform for the *c.g.* and *c.m.* to be at essentially the same point. A rigid body has a fixed *c.m.* independent of the gravitational environment; the location of the *c.g.*, however, depends on the gravitational field. Indeed, several artificial satellites have maintained their orientations in space using the $1/r^2$ drop in the Earth's gravity. A long tube extending down from a satellite experiences a nonuniform field—the *c.g.* is displaced downward from the *c.m.*, allowing torques to develop that maintain a stable alignment (Discussion Question 7.5, p. 232).

We started this discussion by assuming that the applied torque was zero and the motion was purely translational. If the opposite is true—namely, $\Sigma \mathbf{F} = 0$ and $\Sigma \tau \neq 0$—the body will rotate accelerating about the *c.m.*, which cannot itself accelerate—the motion will be purely rotational. Adding in an additional net force will superimpose a translational acceleration of the *c.m.* onto the rotation about the *c.m.* A wrench or a baseball bat hurled through the air may tumble around (if $\Sigma \mathbf{F}$ does not pass through its *c.m.* at launch), but the center-of-mass will sail along a parabola as if it marked the motion of a point-mass. Only if you hit a cue ball along a line through its *c.m.* will it move away spinless, and only if you push this book along a line through its *c.m.* will it slide without rotating.

A body in motion does not have to be rigid for its mass to behave as if it were at the *c.m.* A diver in the air (Fig. 8.15) can tumble and twist, changing the location of her *c.m.* with respect to her body from one instant to the next via *internal forces,* and the *c.m.* will still free-fall along a parabola. Her *c.m.* will move in a ballistic arc as she rotates around it.

A tumbling wrench shown every 1/30 of a second. The center-of-mass (indicated by a spot of light paint) moves along a straight path, while the wrench revolves around it.

8.6 Rolling Down an Incline

The solid sphere (of radius R) in Fig. 8.16 is rolling freely, *without slipping* down an inclined plane. The point (A) on the sphere in contact, for an instant, with the incline would tend to slide down the plane were it not for the friction force. Thus, F_f points up the incline, resisting the downward motion of A and thereby generating a torque about the *c.m.*, which turns the sphere. *It is F_f that causes the rotation of the ball;* without friction, it would slide downward with point A in constant contact with the ramp.

The motion can be analyzed as a translation of the *c.m.* coupled with a rotation about the *c.m.* As ever, $\Sigma \mathbf{F} = m\mathbf{a}$, where m is the mass of the sphere and $\mathbf{a}$ is the linear acceleration of its center-of-mass. Consequently

$$\searrow + \Sigma F_\parallel = F_W \sin \theta - F_f = ma \tag{8.32}$$

Because the sphere is rotating, we also have $\Sigma \tau = I\alpha$. Taking the torque with reference to the *c.m.*, we get

$$\circlearrowleft + \Sigma \tau_{cm} = F_f R = I_{cm} \alpha$$

for the rotation about the *c.m.* Now, from Table 8.3, $I_{cm} = \frac{2}{5}mR^2$. Moreover, we saw earlier that $a = R\alpha$; the two motions are interrelated. Substituting into the torque equation, we get

$$F_f R = \tfrac{2}{5}mR^2(a/R) \tag{8.33}$$

Using the fact that $F_W = mg$, Eqs. (8.32) and (8.33) can be solved simultaneously to yield

$$a = \tfrac{5}{7}g \sin \theta \tag{8.34}$$

and

$$F_f = \tfrac{2}{7}mg \sin \theta \tag{8.35}$$

The acceleration is independent of both m and R. It's *not* $g \sin \theta$ as it would have been if the sphere had *slid* down the incline; instead, it's appreciably smaller because of the rotational inertia.

The requirement that no slipping occurs demands that the frictional force be *static* friction, but that doesn't mean that the sphere is just about to slip. It cannot be assumed that $F_f = \mu_s F_N$. In Problem 63, you will find that the minimum coefficient of friction needed to prevent slipping is $\mu_s = \frac{2}{7} \tan \theta$. Tilt the plane beyond that, and the sphere slides.

Notice that, had the sphere been hollow, its moment of inertia ($I_{cm} = \frac{2}{3}mR^2$) would have been larger and the resulting acceleration smaller. But again a would be

Figure 8.15 Since the motion of the *c.m.* can only change with the application of an external force, and no new force is applied, the *c.m.* moves in a ballistic arc regardless of the diver's twisting.

Figure 8.16 A solid sphere rolling down an inclined plane. The sphere will accelerate at a rate of $a = \frac{5}{7}g \sin \theta$, as compared to its sliding acceleration of $a = g \sin \theta$. Interestingly, Galileo overlooked this relationship in his experiments (p. 72), but it didn't affect his conclusion since the acceleration is still constant.

independent of both R and m. In a no-slip-rolling-down-an-incline race, the solid sphere will always beat the hollow one, while on a *frictionless* surface, it will be a tie every time (Problem 54). The acceleration depends only on the shape of the rolling body.

8.7 Angular Momentum

The rotational equivalent of force is torque, the *moment of the force*. Likewise, the rotational equivalent of linear momentum ($m\mathbf{v}$) is **angular momentum** (**L**), the *moment of momentum*.

Figure 8.17 depicts a particle of mass m and speed v moving along a straight line past a point O. We define the magnitude of the angular momentum with respect to O as

$$L_0 = r_\perp p = r_\perp mv \tag{8.36}$$

with units of kg·m²/s. Note that as long as $\mathbf{v}$ is constant, $r_\perp$ is constant and L_0 is constant. It might seem surprising that a particle moving in a straight line has *angular* momentum. Yet experience tells us that a straight-blowing wind can impart angular motion to a windmill and that a door can be swung closed (imparting angular momentum to it) by bouncing a ball straight off it.

Just as linear momentum is inherently a vector quantity, angular momentum is also. Because angular momentum is a measure of the "quantity-of-angular-motion" of a body, it is reasonable that oppositely acting contributions must tend to cancel. For instance, a system of two identical particles heading toward each other at the same speed along a straight line has no net linear momentum and should similarly have no net angular momentum about any reference point.

Equation (8.36) is reminiscent of the relationship between force and torque, Eq. (6.3). If we recognize that $r_\perp = r \sin \phi$, then

$$L_0 = rp \sin \phi$$

where ϕ is the smallest angle between the lines-of-action of $\mathbf{r}$ and $\mathbf{p}$. And, as with the torque (the moment of the force, $\boldsymbol{\tau}_0 = \mathbf{r} \times \mathbf{F}$), this suggests a vector cross product via Eq. (6.5). In other words, if we define the angular momentum vector (the moment of the momentum) as

$$\mathbf{L}_0 = \mathbf{r} \times \mathbf{p} \tag{8.37}$$

we specify both the required magnitude and a meaningful direction.

The direction of $\mathbf{L}_0$ is found by first drawing $\mathbf{r}$ from the axial point of rotation O to the mass point, as in Fig. 8.17b. Next, slide $\mathbf{r}$ along its line-of-action so that it has a common origin at m with $\mathbf{p}$. Then cross $\mathbf{r}$ into $\mathbf{p}$ through the smallest angle (via the method of p. 178) to get $\mathbf{L}_0 = \mathbf{r} \times \mathbf{p}$.

Now, examine the flat extended object made up of point masses in Fig. 8.18. It's revolving about a perpendicular axis through point O. Each mass point has a distinctive angular momentum even though they all have the same angular speed ω. In this case of an essentially two-dimensional body, all the constituent angular momenta are directed perpendicular to the plane of the body, all are parallel to each other, and their magnitudes can be added algebraically. Since, for every particle, $\mathbf{r}$ is perpendicular to $\mathbf{v}$, the magnitude of the body's total angular momentum L_0 about O is

$$L_0 = \Sigma rmv$$

Because $v = r\omega$,

$$L_0 = (\Sigma mr^2)\omega$$

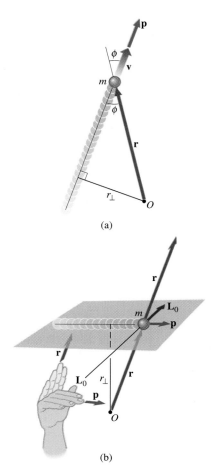

Figure 8.17 (a) A particle of mass m has an angular momentum L_0 with respect to point O equal to $r_\perp mv$. (b) The direction of $\mathbf{L}_0$ is perpendicular to the plane of $\mathbf{r}$ and $\mathbf{p}$ such that $\mathbf{L}_0 = \mathbf{r} \times \mathbf{p}$. Slide $\mathbf{r}$ along itself until it's tail-to-tail with $\mathbf{p}$ and then form $\mathbf{r} \times \mathbf{p}$. Determine the smallest angle between $\mathbf{r}$ and $\mathbf{p}$ by placing your fingers along $\mathbf{r}$ so they close naturally into $\mathbf{p}$. Your thumb will point in the direction of $\mathbf{L}_0$.

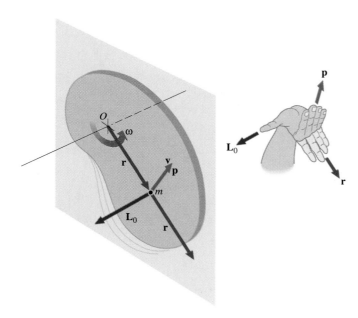

Figure 8.18 To find L_0 for point mass m, slide r so it's tail-to-tail with p, and then construct $r \times p$.

The term in parentheses is the body's total moment of inertia about O [Eq. (8.29)], and so

$$L_0 = I_0 \omega$$

Here point O, which locates the axis of rotation, happened to lie on the body, but it need not have. The O subscript tells us that L is referenced to that point; it's really not crucial and is often omitted. The last equation shows angular momentum to be analogous to linear momentum: $p = mv$. In transforming to the rotational regime, m becomes I, v becomes ω, and p becomes L.

The similarities between the linear and angular formalisms can be made complete by assigning a direction to the angular velocity, recognizing its vector nature. To that end, we take ω to lie *on the axis of rotation* and define its direction via the right-hand-rule; namely, the thumb on the right hand points in the direction of the vector ω when the fingers curl around in the direction of rotation of the body. Thus, in Fig. 8.2, ω is out of the page as it is in Fig. 8.9, whereas in Figs. 8.7, 8.10, 8.12, and 8.16, it is into the page.

For the flat object illustrated in Fig. 8.19, ω is parallel to L_0, and so we can extend the last equation, writing it as

$$\mathbf{L}_0 = I_0 \boldsymbol{\omega} \qquad (8.38)$$

This important result is a little misleading and commonly misunderstood. In general, for a three-dimensional body (Fig. 8.20), the angular momentum of each point-mass is along a direction that need not correspond to ω. Thus, *the net angular momentum vector of an arbitrary revolving body need not be parallel to ω* and Eq. (8.38) must be modified; I_0 cannot be a scalar. If the body is symmetrical about the axis of rotation, then for every point-mass on one side there will be one diametrically opposite it, the same distance away from the axis. For every such pair of point-masses, the two components of $\mathbf{L}_0$ perpendicular to the axis will cancel, leaving a resultant angular momentum parallel to ω. So Eq. (8.38) is true, as is, for the important case of symmetrical objects revolving about their symmetry axes, and we will limit our treatment to these.

Example 8.12 Taking the Earth to be a uniform sphere of radius 6.37×10^6 m and mass 5.98×10^{24} kg, compute its angular momentum about its spin axis. What are the directions of **L** and $\boldsymbol{\omega}$?

Solution: [Given: $R_\oplus = 6.37 \times 10^6$ m and $M_\oplus = 5.98 \times 10^{24}$ kg. Find: **L**.] As we saw earlier, $I = \frac{2}{5} \times mR_\oplus^2 = 9.71 \times 10^{37}$ kg·m², while $\omega = 0.727 \times 10^{-4}$ rad/s; hence

$$L = I\omega = (9.71 \times 10^{37} \text{ kg·m}^2)(7.27 \times 10^{-5} \text{ rad/s})$$

and $\boxed{L = 7.06 \times 10^{33} \text{ kg·m}^2/\text{s}}$

Looking down onto the North Pole, the Earth rotates counterclockwise. Hence, $\boldsymbol{\omega}$ points up northward along the spin axis and **L** and $\boldsymbol{\omega}$ are parallel.

▶ **Quick Check:** Let's see if the units are right: kg·m²/s = m(kg·m/s), which are the units of $L = rmv \sin \phi$. $L \approx (10^{38} \text{ kg·m}^2)(7 \times 10^{-5} \text{ rad/s}) \approx 7 \times 10^{33} \text{ kg·m}^2/\text{s}$.

On an Atomic Level

A variety of experiments performed during this century have revealed a surprising fact: *angular momentum is an intrinsic property of matter on the atomic and subatomic level.* Each of the elementary particles that constitutes ordinary matter (the leptons and quarks—see Chapter 33) has a characteristic angular momentum, just as it has a characteristic mass and electric charge. The building blocks of the atom—electrons, protons, and neutrons—each have an intrinsic angular momentum. The discrete measure of angular momentum associated with these particles turns out to be the smallest observed amount—namely, $h/4\pi = 5.27 \times 10^{-35}$ kg·m²/s. Because it comes in specific doses, we say that angular momentum is *quantized*—it exists only in whole-number multiples of $h/4\pi$. Even the particles of light—photons—each have an angular momentum of $h/2\pi$. The quantity h is the fundamental constant of Quantum Mechanics known as **Planck's Constant**. The angular momentum of macroscopic objects is so large compared to $h/4\pi$ that we never have to worry about its inherently quantized nature.

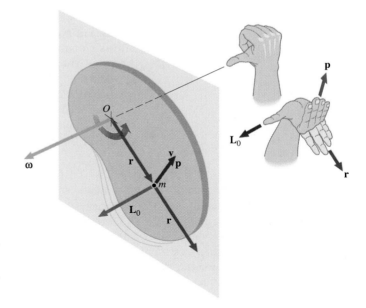

Figure 8.19 For this flat object, $\mathbf{L}_0$ is in the same direction as $\boldsymbol{\omega}$ for all point-masses.

8.8 Conservation of Angular Momentum

Suppose that a constant net torque τ acts on a body for a time interval Δt. There will then be a constant angular acceleration such that

$$\tau = I\alpha = I\frac{\Delta\omega}{\Delta t}$$

Forasmuch as I too is constant, any change in the angular momentum is a result of a change in ω; that is, $\Delta L = I\Delta\omega$, and therefore

$$\tau = \frac{\Delta L}{\Delta t}$$

The torque equals the time-rate-of-change of the angular momentum. To be more inclusive, we can write this expression as a vector equation and so

$$\boldsymbol{\tau} = \frac{\Delta \mathbf{L}}{\Delta t} \qquad (8.39)$$

where the change in $\mathbf{L}$ may be in magnitude, direction, or both.

Remember that we are limiting the discussion to symmetrical objects revolving around symmetry axes. And so, when the torque is parallel to the angular momentum (when $\boldsymbol{\tau}$ is in the same direction as $\boldsymbol{\omega}$), $\Delta\mathbf{L}$ is parallel to $\mathbf{L}$, and the body speeds up. In the same way, when $\boldsymbol{\tau}$ and $\mathbf{L}$ are antiparallel, $\Delta\mathbf{L}$ is opposite to $\mathbf{L}$, and the body slows down. This behavior occurs in both instances without a change in the direction of $\mathbf{L}$; the body does not change its orientation. By contrast, if $\boldsymbol{\tau}$ is not co-linear with $\mathbf{L}$, then the direction of $\mathbf{L}$, and therefore of $\boldsymbol{\omega}$, must change.

Equation (8.39) is the rotational equivalent of Newton's Second Law and, as before, it points to a grand underlying insight. If the net torque applied to a body is

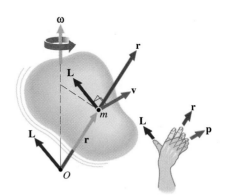

Figure 8.20 An asymmetrical body may revolve such that $\mathbf{L}$ and $\boldsymbol{\omega}$ are not parallel. Here, we see how that can happen for a single point-mass m. In a symmetrical body revolving about a symmetry axis, there will always be a point-mass on the opposite side of the axis so that the net $\mathbf{L}$ will be parallel to $\boldsymbol{\omega}$.

zero, there will be no change in its angular momentum: $\tau = 0$ and $\Delta L = 0$. **In the absence of a net external torque acting on a system, its angular momentum remains constant in magnitude and direction.** This is the law of **Conservation of Angular Momentum.** And it applies when ΔL is due to a change in ω, or I, or both.

The tendency for matter to sustain rotational motion is utilized in a number of practical devices. The modern automobile engine is fitted with a flywheel that is set in high-speed rotation by the engine. Mounted on the end of the crankshaft, it stores angular momentum and keeps the shaft turning smoothly between successive strokes of the pistons.*

Example 8.13 By pushing on the ground, a skater with arms extended (as in Fig. 8.21a) manages to whirl around at a maximum angular speed of 1.0 rev/s. In that configuration, her moment of inertia about the spin axis is 3.5 kg·m². The point-masses making up her arms and outstretched leg are, for the most part, fairly far from the vertical spin axis, and therefore her moment of inertia is relatively large. What will happen to her spin rate as she draws her arms in and stands upright in a pirouette (Fig. 8.21b), whereupon her moment of inertia is only 1.0 kg·m²? Assume the friction forces acting on her are negligible.

Solution: [Given: $\omega_i = 1.0$ rev/s, $I_i = 3.5$ kg·m², and $I_f = 1.0$ kg·m². Find: ω_f.] Since $\Sigma \tau = 0$, $L =$ constant; hence, from Eq. (8.38)

$$L_i = L_f$$

$$I_i \omega_i = I_f \omega_f$$

and so

$$\omega_f = \frac{I_f}{I_i} \omega_i = \frac{3.5 \text{ kg·m}^2}{1.0 \text{ kg·m}^2} (1.0 \text{ rev/s})$$

yielding $\boxed{\omega_f = 3.5 \text{ rev/s}}$.

(a)   (b)

Figure 8.21 (a) The skater revolves as rapidly as possible, maximizing ω_i while maintaining a large I about the spin axis, which produces a substantial $L_i = I_i\omega_i$. (b) By pulling in her arms and legs, her moment of inertia decreases ($I_f < I_i$). Since there are no appreciable external torques, $L_i = L_f$ and $\omega_f > \omega_i$; her spin rate increases.

▶ **Quick Check:** The skater speeds up as her moment of inertia decreases, all the while conserving her angular momentum of 3.5 kg·m²/s.

The same mechanism that governs a pirouetting skater governs a whirling hurricane or a spinning star. A fuel-spent star will collapse as it dies, gravitationally crushing itself to immense densities. The tiny remnant of this cosmic compression is a neutron star. If the thing is spinning at the outset, as it collapses, I decreases tremendously and ω increases in proportion. The end result is a rapidly rotating source of radiant energy—a *pulsar*. One such object, perhaps only 25 km (15 mi) in diameter and having about the same mass as the Sun, is at the center of the Crab Nebula. Whirling around 30 times each second it emits regular pulses of light and X-rays that sweep the sky every 33 milliseconds.

*The flywheel is geared via a disengaging mechanism to the starting motor, and if you've ever heard a horrible grinding sound when turning the ignition key for the second or third time, it's because the flywheel was still spinning.

An upwelling of air, here above warm water, causes a drop in pressure that draws in air from all sides. If there is any circulation of the surrounding air, it spirals inward, moving faster and faster as its distance from the center decreases. Thus, a great vortex forms, conserving angular momentum all the while.

The Crab Nebula is the remnant of a star that exploded (a supernova) in A.D. 1054. At its center are the crushed remains now in the form of a spinning neutron star. It's called a pulsar because it emits radiant energy like a lighthouse beam that sweeps past, blinking on and off. You can see it appear and disappear in the center of this sequence of closeup photos.

Once the jumper is airborne, no external torques are applied and angular momentum is conserved. Since he wants to land legs extended, he jumps with arms up. Then, while in midair, he pulls his arms down (producing angular momentum into the page) as he draws his legs up (producing an equal amount of angular momentum out of the page). The net change in **L** is zero, and the jumper lands feet first.

In this old-fashioned long jump, the athlete rotates his legs forward (with an angular momentum vector out of the page) while pulling his arms backward (with an angular momentum vector into the page). Once airborne, his net angular momentum cannot change because there are no external torques on him.

On a far less grand scale, a high diver leaping into a pool can exert a torque about his *c.m.* while still in contact with the board and so begin to rotate. Once in the air and "torqueless," **L** is constant no matter how he tumbles, and tumble he must. Just as a translating body will continue to move with a constant **v** unless acted upon by a force, a rotating body will continue to rotate forever except insofar as it is acted upon by an *external* torque.

Example 8.14 A very small sphere of mass 0.20 kg is being whirled in a horizontal circle at the end of a 2.0-m-long string at a constant speed of 1.0 m/s. (a) Determine its orbital angular momentum about the axis of rotation. Suppose the string is quickly reeled in leaving a radius of only 1.0 m. (b) How fast will the sphere be moving then? (c) Talk about the tension in the rope.

Solution: [Given: $m = 0.20$ kg, $v_i = 1.0$ m/s, $r_i = 2.0$ m, and $r_f = 1.0$ m. Find: L_0 taking O at the axis of rotation, and v_f.] The angular momentum can be obtained by approximating the sphere as a point located at its *c.m.* whereupon, via Eq. (8.36), we have

Figure 8.22 A body spiraling inward due to the tension in an attached string.

(a) $\qquad L_0 = rmv = (2.0 \text{ m})(0.20 \text{ kg})(1.0 \text{ m/s})$

and $\qquad \boxed{L_0 = 0.40 \text{ kg·m}^2\text{/s}}$

The string exerts a centripetal force on the ball that passes through O and, therefore, no torque about the axis is produced. No net torque means no change in the ball's angular momentum—the initial value of the angular momentum (mr_iv_i) before the string is reeled in must equal the final value (mr_fv_f) afterwards, consequently

$$mr_iv_i = mr_fv_f$$

or $\qquad I_i\omega_i = I_f\omega_f$

In either case

(b) $\qquad v_f = \dfrac{v_ir_i}{r_f} = \dfrac{(1.0 \text{ m/s})(2.0 \text{ m})}{(1.0 \text{ m})} = \boxed{2.0 \text{ m/s}}$

(continued)

(continued)

(c) One might wonder which force causes the sphere to accelerate from v_i to v_f. Figure 8.22 shows that as the body is pulled out of its circular orbit by an increased tensile force $\mathbf{F}_T$, the latter can be resolved into two components, one parallel ($\mathbf{F}_{T\parallel}$) and one perpendicular ($\mathbf{F}_{T\perp}$) to the motion. While $\mathbf{F}_{T\perp}$ changes the direction, it is $\mathbf{F}_{T\parallel}$ that tangentially accelerates the body. The *net* torque about O is always zero, and $\mathbf{L}$ is constant throughout.

▶ **Quick Check:** The moment of inertia decreases and as a result the angular speed increases, as does the linear speed. The radius is halved; the speed must double.

Planetary Motion

The string in the above example behaves like the gravitational field acting on an orbiting planet. Each—the string and the field—is the instrument of a centripetal force that passes through O and therefore produces no net torque τ_0. Thus, *a planet moving under the influence of a gravitational force directed toward the center of the Sun will maintain a constant orbital angular momentum: r may change and v may change, but* $\mathbf{L}$ *will be constant*. Since the force on the planet is always pointing toward the Sun, the planet cannot acquire any velocity out of the plane in which it was originally moving. Hence, *each planetary orbit lies within a plane*. That is, since $\mathbf{L}$ is constant, its direction is constant and therefore perpendicular to a plane, the plane formed by $\mathbf{r}$ and $\mathbf{v}$, the plane containing the motion.

Figure 8.23 depicts a planet in orbit moving a distance Δl in an interval of time Δt. Take Δt to be so small that Δl is essentially a straight line. The area of the little triangular section (equal to one-half the base Δl times the altitude $r_\perp$) swept out by the radius is

$$\Delta A \approx \tfrac{1}{2}\Delta l\, r_\perp$$

an approximation that gets better as Δt gets smaller. Dividing by Δt and taking the limit of both sides as $\Delta t \rightarrow 0$ leads to

$$\lim_{\Delta t \to 0}\left[\frac{\Delta A}{\Delta t}\right] = \frac{1}{2}vr_\perp = \frac{1}{2m}(r_\perp mv) = \frac{1}{2m}L$$

The left side of the equation is the time-rate-of-change of the area swept out by the radius vector drawn from the Sun to the planet. We have already seen, as a result of Conservation of Angular Momentum, that L is constant for central forces. Thus, the right side of the equation is constant, and so the left must be—the radius sweeps out equal areas in equal times. Of course, this is Kepler's Second Law.

Example 8.15 The Earth's nearest (perihelion) and farthest (aphelion) distances from the Sun are 1.47×10^8 km and 1.52×10^8 km, respectively. If its orbital speed at its closest approach is 30.3 km/s, how fast is the planet moving when most distant from the Sun?

Solution: [Given: $r_a = 1.52 \times 10^8$ km, $r_p = 1.47 \times 10^8$ km and $v_p = 30.3$ km/s. Find: v_a.] Since the angular momentum of orbiting planets is constant, the values of L at aphelion and perihelion are equal; accordingly

$$M_\oplus r_a v_a = M_\oplus r_p v_p$$

where, at both these locations, $\mathbf{r}$ is perpendicular to $\mathbf{v}$. Hence

$$v_a = \frac{r_p}{r_a}v_p = \frac{1.47 \times 10^8 \text{ km}}{1.52 \times 10^8 \text{ km}}30.3 \text{ km/s}$$

and $\boxed{v_a = 29.3 \text{ km/s}}$.

▶ **Quick Check:** The Earth, reasonably enough, slows down as it pulls away from the Sun. The radius increases by a factor of 1.034, so the speed must decrease by the same factor: $v_a = (30.3 \text{ km/s})/1.034 = 29.3 \text{ km/s}$.

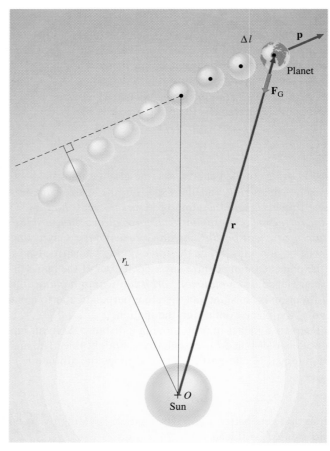

Figure 8.23   A planet orbiting the Sun. The direction of L_0 is into the page parallel to $r \times p$ and perpendicular to the plane of the orbit. Since F_G passes through O, there is no torque about O, and L_0 cannot change.

The Earth and Moon form a single system bound together by gravity. This photograph was taken by the *Galileo* spacecraft from a distance of 6.2×10^6 km.

The Moon gravitationally draws the Earth's crust, waters, and atmosphere, generating tidal motions in each. These, in turn, create frictional forces that gradually slow the planet's spin rate. The effect is quite small, an increase in period of roughly 25×10^{-9} s each day. Still, one thousand million years from now, a day may be approximately 3 h longer as a result (see Problem 41). What makes these predictions a bit unreliable is that the rate of de-spin is varying. Nonetheless, it's been suggested that 1.5×10^9 years ago a day may have been only 9 or 10 hours long, with roughly 900 such days per year.

The source of the torque causing the de-spin is the gravitational interaction of the Moon, which is *external* to the Earth; hence, the planet's angular momentum must change. Alternatively, if this same torque is envisioned as *internal* to the Earth-Moon system, the angular momentum of that larger system must then be conserved. This means that as the planet loses angular momentum, its moon, to which it is coupled gravitationally, must gain angular momentum, and it does. The Moon is receding from the Earth and thereby increasing its distance, r, at a rate of 1 foot in about 30 years and, consequently, increasing its orbital angular momentum ($F_C = F_G$, therefore $mv^2/r \propto m/r^2$ and $v^2 \propto 1/r$, so $mvr = L \propto m\sqrt{r}$).

Gyroscopic Stability

If there is no net torque acting on a revolving body, then its angular momentum vector will be fixed in space, in both magnitude and orientation. There is no more important example of this phenomenon than the spinning Earth itself, whose tilted axis of rotation points very nearly in the same direction all the while the planet orbits the Sun. It's this effect that gives rise to the seasons and that keeps the star Polaris due north all year long.

The tendency of the spin axis to remain fixed has important implications, even when a torque is present. We know, from Eq. (8.39) wherein I is constant, that

$$\boldsymbol{\tau}\Delta t = \Delta \mathbf{L} = I\boldsymbol{\omega}_f - I\boldsymbol{\omega}_i$$

and hence
$$I\boldsymbol{\omega}_f = I\boldsymbol{\omega}_i + \boldsymbol{\tau}\Delta t \tag{8.40}$$

Imagine that we have a massive body spinning with a large initial angular speed $\boldsymbol{\omega}_i$. A torque subsequently applied produces a term, $\boldsymbol{\tau}\Delta t$, in Eq. (8.40). Suppose this contribution is small, either because the torque itself or the time of application is small. Adding it vectorially to the large value of $I\boldsymbol{\omega}_i$ can only result in a final $I\boldsymbol{\omega}_f$ that is not very different from its initial value. By comparison, if the body isn't initially rotating ($\boldsymbol{\omega}_i = 0$), the entire final rotation ($I\boldsymbol{\omega}_f$) will be due to the action of the torque.

Thus, a spinning body has a kind of **gyroscopic stability**. A casually tossed football might easily be made to tumble about its *c.m.* from the slightest torque, but one first set spinning around its long axis maintains its orientation throughout the flight (so does a Frisbee and a discus). A modern rifle or pistol fires spinning bullets to resist tumbling that would otherwise drastically increase air resistance and diminish accuracy. The *Space Shuttle* has a turntable in its cargo bay that is used to spin up satellites just prior to releasing them into space. Gyroscopic stability also accounts for the wonderful way a spinning top stays upright on its pointed end, just as it accounts for the way a twirling quarter, balanced on edge, seems to defy gravity. The fact that a freely spinning mass keeps its axis fixed in space allows us to use that mechanical reference line as the heart of guidance systems for missiles, airliners, submarines, and spaceships.

Figure 8.24 shows a top spinning about its symmetry axis with an angular velocity $\boldsymbol{\omega}_s$. The spin axis can simultaneously swing around, or **precess** (ω_p), and it can also nod up and down, or *nutate* (ω_n). Still, by making $\boldsymbol{\omega}_s$ much larger than either of the other two angular speeds, the analysis becomes tremendously simplified. In that case, $\mathbf{L}$ is very nearly fixed in the body along $\boldsymbol{\omega}_s$, and Eq. (8.38) again applies.

The tilted top in Fig. 8.24 (which we require to be rapidly spinning) experiences a constant torque due to its own weight acting downward at its *c.g.* As shown in Fig. 8.24b, the torque $\boldsymbol{\tau} = \mathbf{r}_{cg} \times \mathbf{F}_W$ is perpendicular to the spin axis, along which $\mathbf{L}$ lies. In a small time Δt, it will produce an angular momentum change parallel to itself, $\Delta\mathbf{L} = \boldsymbol{\tau}\Delta t$, as in Fig. 8.24d. The angular momentum will go from some

This photo, made at a location well below the equator, was taken by holding the camera's shutter open for several hours. As the Earth revolved, the stars left circular streaks about the spin axis. A similar photo, taken in the Northern Hemisphere, would show Polaris, the North Star, as a bright, tiny arc very near the center of the star traces.

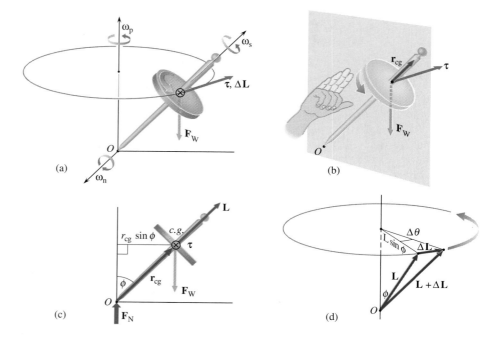

Figure 8.24 (a) A top spinning about a tilted axis. (b) The weight acts down at the *c.g.* producing a torque $\tau = r_{cg} \times F_W$. (c) The torque $\tau = r_{cg}F_W \sin \phi$ produces a change in the angular momentum. (d) The ΔL that results is parallel to τ.

initial value $\mathbf{L}$ to some final value $\mathbf{L} + \Delta \mathbf{L}$, only to continuously change again as the torque continues to act. There is no component of torque in the direction of $\mathbf{L}$, and so L is constant. Without falling, the spin axis of the top, tilted at some constant angle ϕ, will sweep around in a uniform conical wobble, or *precession*. Only when the spin rate becomes small will the top flop over.

The actual precessional angular velocity ω_p can be determined with the help of Fig. 8.24d and the realization that $\omega_p = \Delta\theta/\Delta t$ in the limit as $\Delta t \to 0$. From the geometry and Eq. (8.1), $\Delta\theta = \Delta L/(L \sin \phi)$, but since $\Delta L = \tau \Delta t$ and $\tau = mgr_{cg} \sin \phi$, then $\Delta L = (mgr_{cg} \sin \phi)\Delta t$ and so $\Delta\theta = (mgr_{cg} \sin \phi)\Delta t/(L \sin \phi)$. Hence, $\Delta\theta/\Delta t = mgr_{cg}/L$ and in the limit as $\Delta t \to 0$, using $L = I\omega_s$ and assuming $\omega_s \gg \omega_p$, we get

$$\omega_p = \frac{mgr_{cg}}{I\omega_s} \tag{8.41}$$

Of course, a nonspinning top would have to be supported by an appropriate, additional vertical force in order to be held at an angle ϕ. That is to say (when $\omega_s = 0$), for nonzero ϕ, $F_W > F_N$, and without an additional force $\Sigma F_V \neq 0$, and down goes the *c.g.* The net torque produces a parallel $\Delta \mathbf{L}$ and a parallel rotation ($\boldsymbol{\omega}$) about O. By contrast, for a spinning top $F_N = F_W$, $\Sigma F_V = 0$ and the *c.g.* cannot descend. There is still a torque, but it produces a $\Delta \mathbf{L}$ that now is not parallel to the existing $\mathbf{L}$. In general, $\mathbf{L}$ tends to move toward $\boldsymbol{\tau}$ and the top stays at ϕ while it precesses. Clearly, a torque does not have the same effect on both a spinning and a nonspinning object, and that's what makes the gyroscope seem so magical.

It is common to demonstrate the effect with a wheel whose spin axis is initially horizontal (Fig. 8.25). When let loose, the wheel precesses as expected, with its axis seemingly floating in air pivoting about its supported end. But something is being overlooked—where does the precessional angular momentum come from? There are no torques about the vertical axis. What happens is this: the spin axis actually tilts down a little below the horizontal. This gives the spin angular momentum a

A spinning gyroscope, balanced and precessing in the manner depicted in Fig. 8.24.

downward vertical component that cancels the upward precessional angular momentum, thereby conserving vertical angular momentum at a value of zero.

The phenomenon of precession isn't limited to tops: we steer bicycles and motorcycles by tilting them so they precess. For a guidance gyro we want to minimize ω_p, which means making I and ω_s large and r_{cg} small. The best guidance gyros are military secrets that probably have precession rates around 0.02° per day (the Earth as a gyro is over a thousand times better than that).

Because of its bulging middle, the Earth experiences an unbalanced attraction to both the Sun and Moon and a corresponding torque. That, in turn, gives rise to a precession of the spin axis that takes 25.78×10^3 years to sweep once around. Thus, the pole of the heavens slowly shifts, and when the ancient Chinese wrote about the "celestial pivot," their pole star was not Polaris. Approximately 4000 years ago, Thuban in the constellation Draco shone as the North Star, and in A.D. 14,000 bright Vega will have that small honor.

Figure 8.25 A spinning wheel, initially horizontal when let loose, drops slightly and precesses.

Angular Momentum and Symmetry

Noether's Principle (p. 6) maintains that every conservation law is linked to a symmetry of some kind (Fig. 8.26). We change an aspect of the physical system and some characteristic remains unchanged or conserved—that's what a symmetry is. Rotate the disk of Fig. 1.3 and its appearance does not change. But the same logic also applies to subtle conceptual symmetries. Conservation of Linear Momentum arises from the spatial homogeneity of the fabric of the Universe: the physical nature of empty space is the same everywhere (p. 116). We can move our laboratory to any place in the Universe and the laws of physics would apply unaltered—they are invariant under translation through a homogeneous space. That symmetry is equivalent to Conservation of Linear Momentum.

Figure 8.26 Conservation of Angular Momentum arises from the isotropy of free space. The laws and constants of physics are independent of orientation in space. The four fundamental forces of nature are isotropic, and so the net angular momentum of an isolated system cannot change.

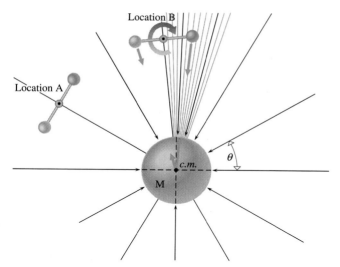

Location B

Location A

c.m.

M

θ

Figure 8.27 A dumbbell at two locations in the anisotropic gravity field of a uniform sphere in an imaginary anisotropic space. At location B, the dumbbell will have an angular acceleration and a nonzero **L**. The torque on the dumbbell is now not equal to the torque on the sphere, and, as a result, Conservation of Angular Momentum will be violated.

Similarly, we find that Conservation of Angular Momentum is a consequence of the isotropy of empty space. If space is *isotropic* (that is, the same in all directions), the constants and the laws of physics must be independent of orientation. Imagine a laboratory in a spaceship far from any other objects. If the ship is rotated through some arbitrary angle, none of the experiments performed in the lab will change. A magnet will attract a piece of iron oriented one way or another. The gravitational field of a uniform sphere is the same in every direction, and the speed of light is independent of the way the source is pointing. Natural law is invariant under rotation within an isotropic space. That this invariance demands Conservation of Angular Momentum is not easy to prove, but we can at least make it reasonable.

Suppose space were anisotropic and therefore the laws of physics were dependent on spatial orientation. For example, imagine a uniform spherical mass M having a hypothetical gravitational field that varies with angle (that is, θ in Fig. 8.27). A dumbbell with two identical balls, one red and one blue, is pivoted at its center so that it can turn freely. Holding it at rest at location A in a region of uniform field, the gravitational attraction on both small masses is equal, and the dumbbell remains motionless. But holding it at location B is like putting a paddlewheel in a waterfall. Although initially $L = 0$, as soon as the dumbbell is allowed to, it will start to revolve clockwise. The red ball in the strong-field region will descend, swing past the vertical and enter the weak-field region, whereupon the blue ball will fall through the strong field, and the dumbbell will accelerate around. On the other hand, the forces acting on the large spherical mass pass through its center. They do produce a torque about the *c.m.* of the system, but because the *c.m.* is so close to the sphere's center, that torque is negligible. Thus, the dumbbell experiences a large torque and a large $\Delta \mathbf{L}$, while the sphere experiences a small oppositely directed torque and a small, oppositely directed $\Delta \mathbf{L}$. The system has a net $\Delta \mathbf{L}$ and Conservation of Angular Momentum is violated.

We assume that in our Universe the converse is true: in all cases, the interactions between two bodies are such that the torque on body 1 due to body 2 is equal in magnitude and opposite in direction to the torque on body 2 due to body 1. **Empty space is isotropic (physics is indifferent to orientation) and therefore angular momentum is conserved.**

Core Material

The basic measure of **angular displacement** θ is defined by

$$\theta = \frac{l}{r} \quad \text{or} \quad l = r\theta \quad [8.1]$$

where l is the *arc-length*. The units of θ are *radians* such that 1 rad = 57.3° and π rad = 180°. The *average angular speed* ω_{av} equals $\Delta\theta/\Delta t$ and in the limit as $\Delta t \to 0$ that ratio approaches ω, the **instantaneous angular speed**, where

$$v = r\omega \quad [8.11]$$

and ω has units of rad/s. The **instantaneous angular acceleration**, α, equals the limit of $\Delta\omega/\Delta t$ as $\Delta t \to 0$, and

$$a_T = r\alpha \quad [8.16]$$

The units of a_T are m/s², whereas those of α are rad/s² (p. 246). When α *is constant*,

$$\omega_f = \omega_i + \alpha t \quad [8.22]$$

$$\omega_{av} = \tfrac{1}{2}(\omega_i + \omega_f) \quad [8.23]$$

$$\theta = \omega_{\mathrm{av}}t = \tfrac{1}{2}(\omega_{\mathrm{i}} + \omega_{\mathrm{f}})t \qquad [8.24]$$

$$\theta = \omega_{\mathrm{i}}t + \tfrac{1}{2}\alpha t^2 \qquad [8.25]$$

$$\omega_{\mathrm{f}}^2 = \omega_{\mathrm{i}}^2 + 2\alpha\theta \qquad [8.26]$$

These equations are identical to those of translational motion in Chapter 3, where $s \rightarrow \theta$, $v \rightarrow \omega$, $a \rightarrow \alpha$, and $t \rightarrow t$.

For a body composed of a number of particles of mass m, the moment of inertia about an axis passing through some point O is

$$I = \Sigma mr^2 \qquad [8.29]$$

Table 8.3 lists many common shapes and the associated moments of inertia for homogeneous bodies. The rotational equivalent of $F = ma$ is

$$\Sigma \tau_0 = I\alpha \qquad [8.30]$$

It is possible to find a point, the **center-of-mass**, where, as far as its motion is concerned, all of the mass of a body can be thought of as concentrated, thus

$$x_{\mathrm{cm}} = \frac{\Sigma mx}{\Sigma m} \quad \text{and} \quad y_{\mathrm{cm}} = \frac{\Sigma my}{\Sigma m} \qquad [8.31]$$

Angular momentum, the moment of momentum, is defined as

$$L_0 = r_\perp p = r_\perp mv \qquad [8.36]$$

and $\qquad \mathbf{L}_0 = \mathbf{r} \times \mathbf{p} \qquad [8.37]$

For symmetrical bodies spinning around symmetry axes

$$\mathbf{L} = I_0\boldsymbol{\omega} \qquad [8.38]$$

The rotational equivalent of the Second Law is

$$\tau = \frac{\Delta \mathbf{L}}{\Delta t} \qquad [8.39]$$

In the absence of a net torque, the angular momentum of the system remains constant; this is the law of **Conservation of Angular Momentum** (p. 259), and it has its origins in the isotropy of free space.

Suggestions on Problem Solving

1. Make a quick sketch of what is happening and always use free-body diagrams if forces are involved.

2. Watch out for those unit conversions! Do not confuse rpm with rev/s or rad/s. Remember the factor of 2π.

3. Keep your left hand in your pocket—the vector cross products are done only with the right hand.

4. Remember that m, I, $\mathbf{F}$, $\boldsymbol{\tau}$, $\mathbf{p}$, and $\mathbf{L}$ are dynamical variables; seeing any of these in a problem should call to mind certain specific equations from the second half of the chapter. For example, it's not very likely that you will ever come across a numerical problem that talks about angular momentum and does not require either Eq. (8.38) or Eq. (8.39). Similarly, the mere mention of the word *torque* suggests angular acceleration and vice versa, although a time-varying angular momentum should come to mind as well. It's certainly possible to have a torque problem without ever seeing the word explicitly. Such "disguised" torque problems work by giving $\mathbf{F}$ and r instead—when you see these, think torque. The crucial step is to relate the "Given" and "Find" via one or more familiar equations once you recognize the general type of problem. The simple problems require one application of one of the basic equations. More difficult problems may require solving for things that weren't asked for.

Discussion Questions

1. When a car turns, do the wheels on its right side have the same angular velocity as those on its left side? What does the answer suggest about mounting the wheels rigidly on a common axle?

2. People have been predicting that a rise in the temperature of the planet would result in the melting of the polar ice caps. Would such massive melting, in turn, have any effect on the length of the day? Explain.

3. Figure Q3 shows a gymnast in three postures, spinning about the frontal axis passing through her *c.m.* Which has the greatest moment of inertia about that axis and which the least? Explain. Draw a rectangular parallelepiped in the shape of a deck of playing cards. Fix the *c.m.* at its geometric center. Draw, passing through the *c.m.,* the three axes perpendicular to the three pairs of faces, numbering them 1, 2, and 3 as the area of each increases. About which axis is the moment of inertia a maximum? A minimum?

4. A diver shown in Fig. Q4 leaves the springboard an instant after being projected by it into the air. Resolve the normal force on his feet into two components, one along the line through his *c.m.* and one perpendicular to it. Discuss the motion that results from each component.

Figure Q4

(a)

(b)

(c)

Figure Q3

Figure Q9

5. Can a small force acting on a body produce a larger angular acceleration than that produced by a bigger force? Explain. If the angular velocity of a body is nonzero, must the sum of the torques acting on it also be nonzero? Explain.

6. Imagine that you spin two eggs, one raw and one hardboiled, on a smooth flat surface at the same rate. Which would come to a stop first and why? Again, spin the raw egg but now stop it just for an instant with a fingertip and then let it go. Why does it start spinning again?

7. Consider the extraordinary situation of a polished sphere rolling without slipping on a flat horizontal surface with a constant angular speed. What can you say about the force of friction on the sphere?

8. Suppose you were playing pool with balls floating in space. How would you hit a motionless cue ball to propel it forward in a straight line, making it rotate as if it were in contact with a table? (By the way, on Earth where the game is only played *on* a table, this is known as a "follow shot" because the cue ball follows after the ball it hits.)

9. The vaulting gymnast executing a push-off at the moment shown in Fig. Q9 experiences a vertical normal force acting at her fingertips and somewhat exceeding her weight. Determine the resulting direction of **L**, **ω**, and **τ** about her *c.m.* Describe her motion soon after.

10. Explain why most helicopters have a small additional propeller at the tail end that revolves in a vertical plane.

11. Imagine Superman in a black leather jacket sitting on a motorcycle floating out in space. Describe what would happen as he gunned the engine (presuming it was modified to run out there).

12. An astronaut floating in space has zero angular momentum initially. He then proceeds to rotate both his extended arms around an axis passing across his chest through his two shoulders. While looking at his right side, you see his arms revolving counterclockwise. What motion, if any, will be executed by the rest of his body? What will happen when he stops?

13. If a person throws a ball, its angular momentum is increased, but the person doesn't rotate. Is angular momentum conserved in the ball-person system? What, if any, external torque acts on the person? Is the angular momentum of the ball-person-Earth system conserved? What, if anything, happens to the motion of the planet? What would happen to someone floating in space who threw a ball in an overhand pitch?

14. One of the best ways of transporting the magnetic tape in a recorder is by a direct-drive system where a motor turns the capstan that moves the tape (see Problem 28). Such motors are often fitted with heavy flywheel housings that revolve with the shaft. Why do you think this might be done? (See Fig. Q14.)

Capstan

Base

Motor

Housing

Flywheel

Shaft

Figure Q14

15. A motorboat generally noses up out of the water, and the more it does so, the faster it travels. Make a sketch locating the *c.m.* of the motorboat and the level of its propeller. Draw in all the forces acting on the boat while it is traveling at a constant speed—don't forget the upward force of the water on the boat and the friction on the hull bottom. Explain what's happening in terms of $\Sigma\mathbf{F}$ and $\Sigma\tau_{\text{cm}}$.

16. When a rear-wheel drive car or motorcycle rapidly accelerates, it tends to rotate about its *c.m.* so that the front end noses upward. What forces are acting *on* the vehicle? Make a sketch and explain what is happening. What would happen if the front brakes engaged first and locked during a quick stop?

17. Imagine a gradual, inwardly spiraling groove cut just into the surface of a flat tabletop. Now, suppose we propel a small sphere into the channel with a speed *v* so that it rolls toward the center point *O* (with no appreciable losses due to friction). What, if anything, will happen to its speed and why?

18. In the past, many satellites have had to be put into orbit spinning more rapidly than was operationally desirable. A cute de-spin system consists of two masses each (known as a yo-yo) tied to a long cable. These are wound around the craft and the weights are held in place on opposite sides. On command, the weights are released and they unwind as the satellite spins until the ends of the cables unhook and each weight, trailing its cable, flies away into space. Explain how this system works to reduce the satellite's spin rate—usually by a factor of about 10.

19. A physics student, with arms stretched out to either side, holds a weight in each hand. While sitting on a bar stool, he is swung around by a helpful friend. What will happen to his angular speed, if anything, when he lets the weights fall to the floor? Think this one through carefully.

Multiple Choice Questions

1. A flying saucer in a TV show banks and just blocks out the disk of the Moon (which subtends an angle of 0.009 rad) when a nearby radar device determines its range to be 10 km away. How big was the craft?   (a) 9 m   (b) 0.09 km   (c) 0.9 km   (d) 0.9 m   (e) none of these.

2. Every point-mass on a rigid body that is rotating has (a) the same angular speed and angular acceleration (b) a different angular speed but the same angular acceleration   (c) the same linear and angular speed (d) the same angular speed and the same linear acceleration   (e) none of these.

3. At what latitude on Earth would you have the greatest angular velocity?   (a) 0°   (b) 45° N   (c) 45° S (d) 60° N   (e) none of these.

4. A body free of support and experiencing a zero net applied force   (a) must be accelerating   (b) must only be translating at a constant speed   (c) must only be rotating about its *c.m.*   (d) may be at rest or moving at a constant speed and may be rotating about its *c.m.* as well   (e) none of these.

5. Given that object 1 exerts a torque on object 2 about some axis. Then with respect to that same axis, object 2 (a) exerts an identical torque on object 1   (b) exerts an equal but oppositely directed torque on object 1 (c) exerts no torque on object 1   (d) may or may not exert a torque on object 1   (e) none of these.

6. Two slender uniform rods have the same cross section and the same mass. If one is light plastic and the other lead, what can you say regarding their respective moments of inertia about an axis through the very center perpendicular to each rod's long axis?   (a) Nothing.   (b) The moments of inertia of the rods are equal.   (c) The lead rod must have a greater moment of inertia.   (d) The plastic rod must have a greater moment of inertia.   (e) None of these.

7. The moment of inertia of a spinning body about its spin axis depends on its   (a) angular speed, shape, and mass   (b) angular acceleration, mass, and spin-axis position   (c) mass, shape, spin-axis position, and size (d) mass, size, shape, and speed   (e) none of these.

8. When a single force is applied to a free body floating in space, it is possible that there will be a resulting change in (a) only its translational motion   (b) both its translational and rotational motions   (c) both (a) and (b) are possible (d) only its rotational motion   (e) none of these.

9. For the net force acting on a body to result in purely linear motion   (a) it must be zero   (b) it must pass through the *c.m.*   (c) it must be less than the weight of the body (d) it must not pass through the *c.m.*   (e) none of these.

10. Two straight inclined grooved tracks are lined up next to one another. Number 1 track is coated with a slippery film and number 2 is not; otherwise, they are identical. Two steel balls are placed one on each track and released so that ball 1 slides down and ball 2 rolls all the way. (a) Both reach the bottom at the same time.   (b) What happens depends on the exact tilt of the tracks and so we cannot say who wins.   (c) Ball 1 wins always. (d) Ball 2 wins always.   (e) Ball 1 wins as often as ball 2.

11. A solid cylinder, a solid sphere, and a hoop all of the same mass but different radii, roll without sliding down an inclined plane. The body that gets to the bottom first will invariably be   (a) the hoop   (b) the sphere   (c) the

cylinder   (d) they all arrive together   (e) not enough information.

12. The turntable platter, the large circular revolving disk of a record player, is designed to resist small variations in speed that would occur with slight fluctuations in the drive motor's output. This design is accomplished by (a) making the platter as small as possible   (b) using a platter as heavy as possible   (c) using a very light platter (d) making the platter's moment of inertia as large as possible   (e) none of these.

13. A body isolated from all other objects can alter, by internal interactions among its various parts, its (a) angular momentum   (b) linear momentum   (c) both (a) and (b)   (d) neither (a) nor (b)   (e) none of these.

14. The dimensions of angular momentum are   (a) $[ML^2T]$ (b) $[M^{-1}L^2T]$ (c) $[ML^{-2}T]$ (d) $[ML^2T^{-1}]$ (e) $[ML^{-2}T^{-1}]$.

15. Low-altitude artificial Earth satellites traversing wisps of atmosphere are initially slowed down by air-drag so that they cannot remain in their original orbits and inevitably fall back to the planet. Spiraling down ever closer to the Earth, they usually break up and burn up before hitting ground. The speed of such a device   (a) increases as it descends and its angular momentum is not conserved (b) decreases as it descends and its angular momentum is conserved   (c) increases and then decreases, conserving picture is its angular momentum   (d) must remain constant, conserving its angular momentum   (e) none of these.

16. At a given moment, the net angular momentum of a system of two particles about a certain point is zero. Accordingly, we know that the particles   (a) must be at rest   (b) must be moving on the same circle in opposite directions (c) must be moving on straight perpendicular paths (d) must have the same linear momentum   (e) none of these.

17. A particle of mass m and speed v has zero angular momentum with respect to a point A in space. We know that the particle   (a) has already passed through A (b) will never pass through A (c) is moving along a line that is perpendicular to the plane containing it and A (d) is moving along a line that passes through A (e) will go somewhere, but we don't have enough information to comment.

18. Imagine a particle traveling with a linear momentum **p** past a point O with respect to which it has an angular momentum of **L**. If **r** drawn from O to the particle makes an angle of 45° with **v**, the angle between **L** and **p** is (a) 0°   (b) 45°   (c) 90°   (d) 180°   (e) none of these.

19. When a rigid body rotates about a symmetry axis with no external torque being applied   (a) the angular acceleration is a nonzero constant   (b) the angular velocity is a nonzero constant   (c) the angular speed changes uniformly   (d) the angular acceleration changes uniformly   (e) none of these.

Problems

THE KINEMATICS OF ROTATION
1. [I] What does 1.000 0° equal in radians?
2. [I] How many radians are swept out by the minute hand of a clock as it goes from 12:00 noon to each of the following times: 1:45 P.M., 2:30 P.M., and 3:17 P.M.?
3. [I] Determine, via geometry, the angle in radians subtended by a 2.00-m-wide car seen head-on from a distance of 1.00 m away. What would that angle be as seen standing 100 m away? Compute this last part exactly (using the straight-line width) and approximately (taking the width as an arc-length). Now, go back to the first part and do it with the same approximation. Compare your results.
4. [I] A 25-cent coin is roughly 2.5 cm in diameter. How far away should it be held if it is to subtend the same angle as does the Moon?
5. [I] A well-adjusted laser beam is emitted as an exceedingly narrow diverging cone of light with a typical spread angle of about 8×10^{-4} rad. How large a circle will be illuminated on the surface of the Moon by such a device $(r_{\mathbb{C}} = 3.8 \times 10^8$ m)?
6. [I] The 10-point center-circle bull's eye on the official slow fire pistol target is $\frac{7}{8}$ in. in diameter. What angle does it subtend as seen by a shooter 50 ft away? How does this compare with the apparent size of the Moon?
7. [I] Assume the human eye can resolve details separated by about 1.0 minute of arc (where there are 60 minutes per degree). How far away must you sit from a TV set (whose

picture is 30 cm high) in order to no longer distinguish the horizontal scan lines? (Typically, 525 horizontal lines make up a modest TV picture.)
8. [I] An ant positioned on the very edge of a Beatles record that is 13 cm in radius revolves through an angle of 100° as the disk turns. How far does he travel?
9. [I] A ball 30 cm in diameter rolls 65 m without slipping. How many revolutions did it make in the process?
10. [I] What is the equivalent of 1.00 rev/min in rad/s?
11. [I] A well-trained runner can swing a leg forward in the vertical plane around a horizontal axis through the knee at a rate of about 700° per second. How much is that in rad/s?
12. [I] What is the angular speed of the 10-cm-long second hand of a clock? What is ω for the minute hand?
13. [I] A videodisc revolves at 1800 rpm beneath a laser read-out head. If the beam is 12 cm from the center of the disc, how many meters of data pass beneath it in 0.10 s?
14. [I] A variable speed electric drill motor turning at 100 rev/s is uniformly accelerated at 50.0 rev/s² up to 200 rev/s. How many turns does it make in the process?
15. [I] Thirty seconds after the start button is pressed on a big electric motor, its shaft is whirling around at 500 rev/s. Determine its average acceleration.
16. [I] A large motor-driven grindstone is spinning at 4 rad/s when the power is turned off. If it rotates through 100 rad as it uniformly comes to rest, what is its angular acceleration?
17. [I] A steam engine is running at 200 rpm when the

engineer shuts it off. The friction of its various parts produces torques that combine to decelerate the machine at 5.0 rad/s/s. How long will it take to come to rest?

18. [I] What angular acceleration is required to stop a turbine blade spinning at 20 000 rpm in 50.0 s?

19. [II] Take the Earth to be moving in a circular orbit ($r_{\oplus\odot} = 1.495 \times 10^{11}$ m) about the Sun. Compute both its linear and angular speeds.

20. [II] A bicycle with 24-in.-diameter wheels is traveling at 10 mi/h. At what angular speed do the wheels turn? How long do they take to turn once around?

21. [II] A 1.0-m-diameter nontranslating disk is made to accelerate from rest up to 20 rpm at a rate of 5 rad/s². Through how many turns will it revolve in the process? How far will a point on its rim travel while all this is happening?

22. [II] The motor in Fig. P22 is mounted with a 20-cm-diameter pulley and revolves at 100 rpm. We would like to drive the attic fan so that no point on each of its four 1.0-m-long blades exceeds a speed of 7.0 m/s. What size pulley should it have?

Figure P27

Figure P22

23. [II] A train rounds a turn 1000 ft in radius while traveling at a speed of 20 mi/h. Determine its angular speed and centripetal acceleration. Give the answer in SI units.

24. [II] A bicycle has its gears set so that the front sprocket, which is driven by the pedals, has 52 teeth whereas the rear one, to which it is attached via the chain, has 16 teeth. If it's pedaled at a most efficient 50 rpm and has 24-in. wheels, how fast is the bicycle moving? Give the answer in SI units.

25. [II] A chimp sitting on a yellow unicycle with a wheel diameter of 20 in. is pedaling away at 100 rpm. How fast does it travel? Give the answer in SI units.

26. [II] Imagine two ordinary gears of different diameters meshed together, with the larger being the driver. If the larger gear has 100 teeth around its circumference and rotates at 5.0 rad/s, the smaller gear, which has only 25 teeth, will rotate at what speed?

27. [II] The pulley on the left of Fig. P27 is 0.6 m in diameter and rotates at 1.0 rpm. It is attached by a twisted belt to the 0.2-m-diameter hub of a compound pulley whose outer diameter is 0.8 m. If the driver turns clockwise, find the velocity of the suspended body.

28. [II] Figure P28 shows the structure of a magnetic tape cassette. The tape is transported across the heads at a

Figure P28

constant speed of $1\frac{7}{8}$ inches per second. Contrary to popular belief, this is not done by driving the feed and take-up hubs. Why not? Actually, the tape is moved by being squeezed between a precisely rotating shaft—the capstan—and a soft pinch roller. If the capstan has a diameter of 3/32 in., how fast must it turn? Give the answer in SI units.

29. [II] An electric circular saw reaches an operating speed of 1500 rpm in the process of revolving through 200 turns. Assuming the angular acceleration is constant, determine its value. How long does it take to get up to speed? Now redo the problem using the average angular speed.

30. [II] A wheel of radius R is freely rolling along a flat surface at an angular rate of ω. Determine, analytically, the linear speed with respect to the ground at any instant of (a) the *c.m.*, (b) the topmost point, and (c) the point in contact with the ground.

31. [II] New York City is traveling around at a tangential speed of about 790 mi/h as the Earth spins. Assuming the planet is a sphere of radius 6371 km, how long is the perpendicular from the city to the spin axis? Compute the city's latitude.

32. [II] Two spacecraft are freely coasting in circular orbits, one 2 km beneath the other. Referring to the conclusions of Chapter 7, compare their angular speeds and show that the lower ship will pull ahead of the upper one.

33. [II] A gear train consists of five meshed gears arranged so that the first (20 teeth) drives the second (80 teeth), which

drives the third (40 teeth), and so on. If the first is mounted on the shaft of a 1500-rpm motor rotating clockwise, at what speed and in what direction will the fourth (25 teeth) and fifth (75 teeth) gears rotate? Figure out a general rule for the speed and direction of the last gear in such a train.

34. [II] The pegged wheel in Fig. P34 was a precursor of the modern cam. Here it's shown operating a trip hammer the same way it was used to crush ore and forge metal 1000 years ago in Europe. At what speed should the central drive shaft be turned, presumably by a waterwheel, for the hammer to pound away at one smash every 3 s?

Figure P34

35. [III] A cylindrical wheel is rotating with a constant angular acceleration of 4.0 rad/s². At the instant it reaches an angular speed of 2.0 rad/s, a point on its rim experiences a total acceleration of 8.0 m/s². What is the radius of the wheel?

36. [III] A cord is wrapped around a 0.24-m-diameter pulley and the free end is tied to a weight. That weight is allowed to descend at a constant acceleration. A stopwatch is started at the instant when the weight reaches a speed of 0.02 m/s. At a time of $t = 2.0$ s, the weight has dropped an additional 0.10 m. Take $y = 0$ to occur at $t = 0$ and show that

$$y = (0.02t + 15 \times 10^{-3}t^2) \text{ m}$$

Prove that for any point on the circumference

$$a_C = (33 \times 10^{-4} + 10 \times 10^{-3}t + 75 \times 10^{-4}t^2) \text{ m/s}^2$$

and
$$a_T = 0.03 \text{ m/s}^2$$

37. [III] If the hand in Fig. P37 pulls down on the rope moving it at a speed of 2.0 m/s, what will be the resulting velocity of the hanging mass?

38. [III] If the hand in Problem 37 uniformly accelerates the rope downward from rest at $1\frac{1}{3}$ m/s², how fast will the hanging mass be moving 1.0 s later?

39. [III] A 30-cm-diameter turntable platter of a record player is attached, by a belt (to reduce vibrations), to a motor-driven pulley 2.0 cm in diameter (Fig. P39). Determine the motor's angular acceleration if the platter is to reach $33\frac{1}{3}$ rpm in 6.0 s. How many rotations will it make before reaching operating speed?

40. [III] An inexpensive rim-drive record player has a 6.0-cm-diameter rubber wheel rotating in contact with the

Figure P37

Figure P39

Figure P40

inner surface of the 30-cm-diameter platter (Fig. P40). If the turntable is to go from $33\frac{1}{3}$ rpm to 78 rpm in 3.0 s, what must be the angular acceleration of the drive wheel?

41. [III] Because of tidal friction, the Earth is de-spinning. Its period is increasing at a rate of roughly 25×10^{-9} s per day. Approximate the corresponding angular acceleration in rad/s². Compute the decrease in the Earth's angular speed that would exist one thousand million years from now, assuming α is constant. What will its new period be?

THE DYNAMICS OF ROTATION

42. [I] Figure P42 depicts a tone arm, turntable, and record showing the friction force exerted by the record *on the stylus* (the needle). Determine the corresponding frictional torque on the arm when $F_f = 0.008$ N, $\theta = 4.5°$, and $l = 20$ cm. This torque tends to make the arm "skate" across the record whenever the stylus happens to jump out of the groove. Which way will it skate?

43. [I] Consider a thin ring or hoop of mass M and radius R of negligible thickness. Compute the moment of inertia about its *c.m.* by imagining it to be made up of a large number of small segments.

44. [I] Two solid 3.0-kg cones, each with a base radius of 0.30 m, are glued with their flat faces together. Determine the moment of inertia of this device about the symmetry

Figure P42

axis passing through both vertices.

45. [I] A gymnast in the process of performing a forward somersault in midair increases his angular velocity by 450% while going from an arms-overhead layout position to a tight tuck. What can you say about the change, if any, in the moment of inertia about the frontal axis (parallel to the shoulder-to-shoulder line) passing through his *c.m.* (in one side and out the other)?

46. [I] An astronaut of mass 70 kg rotates his arms backward about a shoulder-to-shoulder axis at a rate of 1 rev/s. *Both of his arms together* have a mass of 12.5% of his body mass and a total moment of inertia about the shoulders of 1 kg·m². Compute the angular momentum imparted to the rest of his body in the process.

47. [I] The **Parallel-Axis Theorem** states that if I_{cm} is the moment of inertia of a body about an axis through the *c.m.*, then I, the moment of inertia about any axis parallel to that first one, is given by

$$I = I_{cm} + md^2$$

Here, m is the object's mass and d is the perpendicular distance through which the axis is displaced. Accordingly, use the theorem for a thin rod of length l; $I_{cm} = \frac{1}{12}ml^2$ to compute I about one end.

48. [I] Show that a small object (such as a sphere) that orbits a distant axis can be approximated by a point mass when you want to compute its moment of inertia. Begin with the theorem of the previous problem.

49. [I] A tiny hummingbird with a mass of only 2×10^{-3} kg is circling around a flower in a 1.0-m-diameter orbit. If it travels once around in 1.0 s, approximate its angular momentum with respect to the blossom (look at Problem 48).

50. [I] Envision a solid cylindrical wheel of radius r and mass m resting upright on a flat plane. With Problem 47 in mind, compute the wheel's moment of inertia about the line of contact with the surface.

51. [I] Two small rockets are mounted tangentially on diametrically opposite sides of a cylindrically shaped artificial satellite. The spacecraft has a 1.0-m diameter and a moment of inertia about its central symmetry axis of 25 kg·m². The rockets each develop a thrust of 5.0 N, are oppositely directed, and are aligned to produce a maximum spin-up of the craft. What's the resulting angular acceleration when they are both fired?

52. [II] Compute the orbital angular momentum of Jupiter ($M_{2\!\!\downarrow} = 1.9 \times 10^{27}$ kg, $r_{2\!\!\downarrow} = 7.8 \times 10^{11}$ m, and $v_{av} = 13.1 \times 10^3$ m/s) and then compare it to the spin angular momentum of the Sun ($M_{\odot} = 1.99 \times 10^{30}$ kg, $R_{\odot} = 6.96 \times 10^8$ m). Assume the Sun, whose equator rotates once in about 26 days, is a rigid sphere of uniform density. It would seem that most of the angular momentum of the Solar System is out there with the giant planets that also rotate quite rapidly.

53. [II] A 26.0 in.-diameter bicycle wheel supported vertically at its center is spun about a horizontal axis by a torque of 50.0 ft·lb. If the rim and tire together weigh 3.22 lb, determine the approximate angular acceleration of the wheel. Ignore the contribution of the spokes. Of course, these ugly units should cancel out.

54. [II] Derive an expression for the acceleration of a hollow sphere of mass m and radius R rolling, without slipping, down an incline of angle θ. Compare this result with Eq. (8.34) and discuss the implications of the absence of m and R in both those formulas.

55. [II] A block of mass m is tied to a light cord that is wrapped around a vertical pulley of radius R and moment of inertia I. Assuming the bearings are essentially frictionless, write an expression for the rate at which the block will accelerate when released.

56. [II] A 10-kg solid steel cylinder with a 10-cm radius is mounted on bearings so that it rotates freely about a horizontal axis. Around the cylinder is wound a number of turns of a fine gold thread. A 1.0-kg monkey named Fred holds on to the loose end and descends on the unwinding thread as the cylinder turns. Compute Fred's acceleration and the tension in the thread.

57. [II] How would Fred manage in Problem 56 if the bearings exerted a frictional torque of 0.060 N·m?

58. [II] A small mass m is tied to a string and swung in a horizontal plane. The string winds around a vertical rod as the mass revolves, like a length of jewelry chain wrapping around an outstretched finger. Given that the initial speed and length are v_i and r_i, compute v_f when $r_f = r_i/10$.

59. [II] A very thin 1.0-kg disk with a diameter of 80 cm is mounted horizontally to rotate freely about a central vertical axis. On the edge of the disk, sticking out a little, is a small, essentially massless, tab or "catcher." A 1.0-g wad of clay is fired at a speed of 10.0 m/s directly at the tab perpendicular to it and tangent to the disk. The clay sticks to the tab, which is initially at rest, at a distance of 40 cm from the axis. What is the moment of inertia (a) of the clay about the axis?   (b) of the disk about the axis?   (c) of both clay and disk about the axis?   (d) What is the linear momentum of the clay before impact?   (e) What is the angular momentum of the clay with respect to the axis just before impact?   (f) What is the angular speed of the disk after impact?

60. [II] A water turbine (having a moment of inertia of 1000 kg·m²) is essentially a wheel with a lot of blades attached to it much like its predecessor, the waterwheel. When a high-speed blast of water hits the blades, it drives the turbine into rotation. Suppose the input valve is closed and the turbine winds down from its operating speed of 200 rpm, coming to rest in 30 minutes. Determine the frictional torque acting.

61. [II] A hoop of mass m and radius R rolls without slipping down an incline at an angle of θ. Write an expression for the acceleration of its *c.m.* and find its numerical value when $m = 1.0$ kg, $R = 1.0$ m, and $\theta = 30°$.

62. [II] Write an expression for the orbital angular momentum of a small artificial satellite of mass m in a circular flight-path of radius r about the Earth. Check the units. What happens to angular momentum as the orbit increases?

63. [II] A solid sphere of mass m and radius r rolls without sliding down an inclined plane tilted at an angle θ. Show that the minimum coefficient of static friction needed to prevent slipping is $\mu_s = (2/7) \tan \theta$.

64. [III] Figure P64 shows two masses strung over a 0.50-m-diameter pulley whose moment of inertia is 0.035 kg·m². If there is a constant frictional torque of 0.25 N·m at the bearing, compute the acceleration of either mass. (The tension in the rope is not the same on either side of the pulley.)

Figure P64

65. [III] An astronaut is working out at the far reach of a 100-m tether, the opposite end of which is attached to a pivoting ring on the nose of the space station. She does not notice that an air hose on her backpack has developed a leak. The little blast of gas produces a tangential thrust and a corresponding acceleration of $a_T = 1.0 \times 10^{-3}\, g$. After two minutes, she realizes what has happened and shuts off the leak. At that point, she and her life support system (total mass of 150 kg) are sailing along at a tangential speed of _____ with an angular momentum of _____ . Annoyed, she decides to return to the craft by pulling hand-over-hand on the tether. If she manages to get 5.0 m from the ring on the ship, her tangential speed would be _____ . Considering that the centripetal force she would have to hold against is then _____ , it's unlikely she would ever get that close.

66. [III] A solid cylindrical pulley of mass m and radius R is held upright and a number of circular turns of light cord are wrapped around it within the groove. The loose end of the cord is held vertically and the pulley is released so it falls straight downward, unwinding like a yo-yo. Please compute its linear acceleration, in terms of g, and the tension in the cord, in terms of m and g, before it reaches the end of its rope.

67. [III] A solid cylinder is placed at rest on an inclined plane at a vertical height h and allowed to roll down without slipping. Show that its speed at the bottom is given by $v = 2\sqrt{\frac{1}{3}gh}$.

68. [III] A cylinder of radius R and mass m mounted with an axle is placed on a horizontal surface. It's pulled forward via a mechanical arrangement so that the applied horizontal force F passes through the *c.m.* Assuming the cylinder rolls without slipping, determine both its acceleration and the friction force at the surface in terms of F.

69. [III] Consider a long slender rod of mass M and length l pivoted at its far end at point O. Now compute I_0 directly by dividing the rod into 100 equal-mass parts and summing up the moments of inertia for all of these. You will need to know that

$$1^2 + 3^2 + 5^2 + \cdots + (2n - 1)^2 = \frac{n}{3}(2n + 1)(2n - 1)$$

9

Energy

THUS FAR WE HAVE studied Newtonian Mechanics, the description of motion in terms of momentum. This was the unrivaled theoretical formalism into the 1800s. But by the beginning of that century, a powerful alternative, predicated on the concept of *energy*, was already taking shape. One of the great achievements of that era was the formulation of the Law of Conservation of Energy (p. 297). In modern times, the Special Theory of Relativity and the work of Amalie Noether have made it clear that momentum and energy are complementary ideas rooted in the nature of space and time.

We now begin a treatment of energy, first looking at how it is trans-

It is important to realize that in physics today, we have no knowledge of what energy is.

R. P. FEYNMAN
Nobel laureate

ferred through *work* and then at how it is manifest in its basic forms—*kinetic* and *potential*. The emphasis is on mechanics, although the concept of energy is all-embracing. In one way or another, it influences our thinking about every branch of physics.

THE TRANSFER OF ENERGY

Over the centuries, the word *energy* has had different meanings. The nontechnical usage derives from the Greek *en* (which means *in*) and *ergon* (which means *work*). Hence, energy is the capacity to do work, an inherent vigor. The word has been used in this way at least since the late 1500s. Galileo (1638) made use of the term *l'energia,* though he never defined it. Only in the last 200 years has the idea taken on a scientific meaning, albeit one not altogether satisfactorily defined.

In very general terms, *energy describes the state of a system in relation to the action of the Four Forces.* It is a property of all matter and is observed indirectly through changes in speed, mass, position, and so forth. There is no universal energy meter that measures energy directly. The change in the energy of a system, which is all we can ever determine experimentally, is a measure of the physical change in that system. **Force is the agent of change; energy is a measure of change**. Because a system can change through the action of different forces in different ways, there are several distinct manifestations of energy.

We buy *electrical energy* "packaged" from the hardware store and "on tap" from the power company. *Chemical energy* can be gotten from a slice of pizza or a tank of gasoline. Rubber bands, ligaments, and girdles all store *elastic energy.* Snowflakes fall because they have *gravitational energy.* We bake apple pies with *thermal energy* and defend them with *nuclear energy.* You don't play on railroad tracks because trains have lots of *kinetic energy. Radiant energy* floods in from the Sun to warm us and from TV stations to entertain us.

The most important characteristic of energy is that it is transferred from one configuration of interactions to another, such that the total amount of energy remains unchanged. Thermal energy can be converted into electrical energy, and some of that turned into light, and back again into thermal energy, but the net amount of energy is always the same—energy is conserved.

Energy is a scalar quantity associated in various amounts with all the "things" that exist, from minute massless particles to immense whirling galaxies. By observing the changing behavior of matter, we infer the presence of one form or another of energy. Like linear and angular momentum, energy is not an entity in and of itself—there is no such thing as pure energy.

9.1 Work

The ancient Greeks seem to have had a vague concept of work; it appears just beneath the surface in their explanations of how a large weight could be lifted by exerting a small force on a lever. In the early 1600s, Galileo was only beginning to grope toward the essential idea. He considered the behavior of a pile driver and recognized that the combination of the weight of the hammer and the distance through which it fell determined its effectiveness. Distance and force are related in some crucial way. Newton, in his *Principia,* doesn't mention energy, but it is clear that he had a strong intuitive sense of the concept.

We learned earlier that the product of *force* and the *time* over which it acts is the *change in momentum.* Now, we will see that *the product of force and the distance over which it acts is a measure of the change in energy.* When that idea was finally formalized by Gaspard Coriolis (1829), he called the product of force and distance **work**. That word is still used, although *mechanical change in energy,* or *energy mechanically transferred,* would be more to the point. **Work is the change in the energy of a system resulting from the application of a force acting over a distance.**

At the heart of the concept of work is the notion of *movement against resistance,* be it the resistance produced by gravity, or friction, or inertia, or whatever. Work can be done on a body to get it moving, to keep it moving, or to change the way it is moving. Work is done to overcome, to move against the resistance of, some force.

Work Along a Straight Line: **F** ∥ **s**

Let's begin with the simplest case of a point-mass, or equivalently, a finite object that is perfectly rigid so it acts as though it were a point-mass located at its center-of-mass. The idea is to avoid the complexities that exist with deformable bodies. Figure 9.1 shows a constant external force (F) exerted on a rigid object, moving it through a horizontal displacement (s). The applied force is parallel to the straight-line motion and, in that case, a preliminary definition of work (W) is

$$W = \pm Fs \tag{9.1}$$

As long as the object moves along the line-of-action of the force while the force acts on it, work is being done. Here **F** is parallel to the displacement **s**; it sustains the motion and does positive work on the object. When **F** is antiparallel to **s**, it opposes the motion and does negative work on the object.

The SI unit of work is the *newton-meter.* To be more concise and to honor J. P. Joule, the work done by a 1-newton force moving a body through 1 meter is defined as 1 *joule* (J), or

$$1 \text{ J} = 1 \text{ N·m}$$

Thus, if the force in Fig. 9.1 is 20 N and the displacement 2.0 m, the work done on the body is 40 N·m, or 40 J. In U.S. Customary units, still used in engineering, work is expressed in *foot-pounds,* where

$$1 \text{ J} = 0.7376 \text{ ft·lb} \qquad \text{and} \qquad 1 \text{ ft·lb} = 1.356 \text{ J}$$

Positive work is done on an object when the point of application of the force moves in the direction of the force. If the object experiencing a force does not move,

> *As we cannot give a general definition of energy, the principle of the conservation of energy simply signifies that there is* something *which remains constant. Well, whatever new notions of the world future experiments may give us, we know beforehand that there will be something which remains constant and which we shall be able to call energy.*
>
> HENRI POINCARÉ (1854–1912)

Figure 9.1 (a) A man dragging a crate uses a force *F* to pull it along the ground. (b) When *F* acts over a distance *s*, an amount of work (*W*) is done by the man.

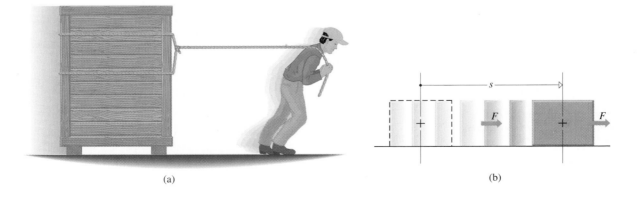

(a)

(b)

there is no work done. Raise a 100-kg barbell into the air against its downward weight and you do work. Hold it above your head, and you do none.

Example 9.1   A locomotive exerts a constant forwardly directed force of 400 kN on a train that it pulls for 500 m. (a) How much work does the engine do on the train (W_{et}) during that period? Coming toward a station, the locomotive slows the train, applying a constant force of 100 kN in the opposite direction. (b) How much work does it then do on the train over a distance of 1000 m?

Solution: [Given: (a) $F = +400$ kN, $s = 500$ m, (b) $F = -100$ kN, $s = 1000$ m. Find: W_{et} in both cases.] (a) **F** and **s** are parallel and so $F > 0$; hence

$$W_{et} = +Fs = +(400 \text{ kN})(500 \text{ m}) = \boxed{+200 \times 10^6 \text{ J}}$$

The locomotive does 200 MJ of positive work *on* the train. (b) Now **F** and **s** are antiparallel and so

$$W_{et} = -Fs = -(100 \text{ kN})(1000 \text{ m}) = \boxed{-100 \times 10^6 \text{ J}}$$

The engine opposes the motion and does negative work on the train. Equivalently, the train does work ($W_{te} = -W_{et}$) on the engine.

▶ **Quick Check:** (a) $W_{et} = +4 \times 5 \times 10^7$ J. (b) $W_{et} = -1 \times 1 \times 10^8$ J.

Work in General

Imagine a rigid crate being pulled along by a person tugging on a rope that makes an angle θ, as in Fig. 9.2a. Part of the force **F** acts upward (tending to lift the body and lessen the load), and part acts in the direction of the motion and does work. The force is a vector; it is equivalent to two perpendicular component vectors, one vertical ($F \sin \theta$) and the other horizontal ($F \cos \theta$). The situation is identical to two people pulling on the crate at the same time via two ropes, one up and one forward. Assuming it doesn't lift off the floor, the crate only moves horizontally and only the horizontal puller does work.

Equation (9.1) can be generalized by requiring that *the work done on a body by an applied force is the product of the component of the force in the direction of the displacement multiplied by the magnitude of the displacement*:

$$W = (F \cos \theta)s = Fs \cos \theta \qquad (9.2)$$

A working force must to some extent be in, or opposite to, the direction of the motion. While walking across a room at a constant speed with a barbell above your head, you do no work on it because the supporting force, counteracting gravity, is vertical and the displacement is horizontal ($\theta = 90°$).

Figure 9.2   Often the force propelling an object is not parallel to the displacement—a stick pushing on a hockey puck or a foot pressing on the pedal of a bike are familiar examples. Only the component of the force parallel to the displacement does work. In both (a) and (b), that component is $F \cos \theta$.

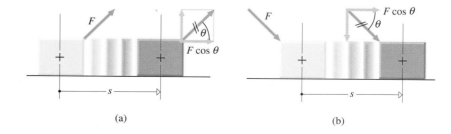

(a)   (b)

Example 9.2 The truck in Fig. 9.3 is dragging a stalled car up a 20° incline. The tensile force on the tow-line is constant, and the two vehicles accelerate at a constant rate. If the chain makes an angle of 30° with the road and the tension is 1600 N, how much work was done by the truck on the car in pulling it 0.50 km up the incline?

Solution: [Given: $F_T = 1600$ N, $\theta = 30°$, and $s = 500$ m; the angle of the incline is irrelevant here. Find: W_{tc}.] The motion is along the incline, and the component of the force acting in the direction of the displacement is therefore $F_T \cos 30°$, no matter what the tilt of the hill. Hence

$$W_{tc} = (F_T \cos 30°)s = (1600 \text{ N})(0.866)(500 \text{ m})$$

and
$$\boxed{W_{tc} = 6.9 \times 10^5 \text{ J}}$$

▶ **Quick Check:** The working force is 1386 N acting over 500 m; hence, $W_{tc} \approx (1.4 \times 10^3 \text{ N})(5 \times 10^2 \text{ m}) \approx 7 \times 10^5$ N·m.

Figure 9.3 (a) A tow truck pulling a car up an inclined plane. Only the component of the force parallel to the displacement does work. (b) That working component is $F_T \cos 30°$.

Example 9.3 Figure 9.4 depicts someone very slowly raising a rigid 1.0-kg block at a constant speed through a vertical distance of 2.0 m at the Earth's surface. The center-of-mass rises from point A to point B. Neglecting the tiny additional force needed to begin the motion and assuming g to be constant, determine the work done on the block by the hand.

Solution: [Given: $m = 1.0$ kg, $s = 2.0$ m, and $g = 9.8$ m/s². Find: W_{hb}.] To raise the block at a constant speed ($a = 0$), the sum-of-the-forces acting vertically must be zero ($\Sigma F = ma = 0$), meaning that the applied force $\mathbf{F}$ must be equal in magnitude and opposite in direction to the weight of the block $\mathbf{F_W}$. The force is straight up, as is the displacement, so that in Eq. (9.2) $\theta = 0$, $\cos \theta = 1$, and

$$W_{hb} = Fs = F_W s = mgs = (1.0 \text{ kg})(9.8 \text{ m/s}^2)(2.0 \text{ m})$$

and
$$\boxed{W_{hb} = 20 \text{ J}}$$

This is the work done in overcoming gravity.

▶ **Quick Check:** $W_{hb}/F = s = 20$ N·m/9.8 N ≈ 2 m.

In Example 9.3, the hand pushes up on the block and the block pushes down on the hand. The latter force comes from the block-Earth gravitational interaction. The work done by the hand ($W_{hb} = +20$ J) on the block is positive because the force is in the direction of the displacement. By contrast, while the block is moved upward, the force it exerts on the hand acts downward; $\theta = 180°$, $\cos \theta = -1$, and so the work done on the hand is $W_{bh} = -20$ J. It is negative because the weight of the block resists the upward motion. When the work done *on* system 1 (the block) *by* system 2 (the hand) is positive, system 1 receives energy from system 2. Thus, while one of the *interacting* systems gains energy, the other loses an equal amount of energy; while one has positive work done on it, the other does the positive work, and has negative work done on it.

Example 9.4 A 30.0-cm-diameter horizontal drum capable of rotating at 10.0 rpm has a rope wound around it. A 200-N load is hung from the rope. How much work is done on the load by the rope in raising it at a constant speed for 2.0 min?

Solution: [Given: $D = 30.0$ cm, 10.0 rpm, $F_W = 200$ N, and $t = 2.0$ min. Find: W_{rl}.] Since the load is not accelerating, the tension F_T acting up equals the weight F_W, down. Since there are πD meters per rotation, the load is raised a distance

$$s = (\pi D \text{ m/rot})(10 \text{ rpm})(2.0 \text{ min})$$

and $$s = \pi(0.30 \text{ m})20 = 18.85 \text{ m}$$

It is the tension in the rope pulling up that does the work in the amount of

$$W_{rl} = F_T s = (200 \text{ N})(18.85 \text{ m}) = \boxed{3.8 \times 10^3 \text{ J}}$$

▶ **Quick Check:** 20 revolutions of the drum (circumference = 0.94 m) lift the 200-N load 18.8 m, doing 38×10^2 N·m of work.

Despite the beads of sweat on your brow after hours of trying to lift a barbell, if it hasn't been raised, you have done no work *on it*. Still, you will certainly have generated a good deal of "heat"; your metabolic rate and oxygen consumption will both have gone up. This seeming contradiction can be understood by examining how muscles function. When a muscle contracts, exerting a force over a distance, it does work, but a muscle maintaining a constant tension is doing something as well. The skeletal muscles are composed of bundles of elongated cells known as *fibers*. In response to nerve signals, these cells individually contract (generating a brief pulse of tension) and then relax, all in a matter of a few milliseconds. Each time a fiber contracts it does work on the muscle. The net result of all of these tiny, short-lived contributions is the apparently constant muscle force that may or may not be doing work on some other body.

Deformable Objects and Friction

Until now, s was the displacement of every point-mass constituting a rigid object, and it was the same for all of them. Thus, s was also the displacement of the center-of-mass, s_{cm}. But not all statements about work that are true for a point-mass are necessarily true for a deformable object. Portions of a real extended object can move in ways different from that of its *c.m.* The concept of work has to be applied to extended bodies more cautiously than was sometimes done in the past. We will come back to this later when we examine the energetics of walking, swimming, and so forth. For the moment, let's consider the example of friction.

Kinetic friction arises from the electromagnetic interaction at the interface between two objects in relative motion. Imagine a block being pulled at a constant speed along the floor by an applied force F equal to the force of kinetic friction F_f (Fig. 9.5). On a macroscopic level, we deal with the friction force as if it were continuous, and it's tempting to assume that the work done *on the block* by F_f as it

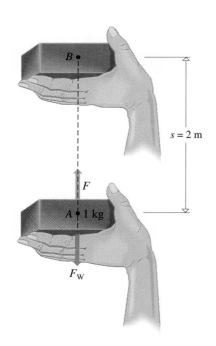

Figure 9.4 The hand slowly raises the block by exerting an upward force F negligibly greater than its weight F_W. Work is done to overcome gravity—that is, to move against the block-Earth gravitational interaction.

moves a distance s_{cm} is simply $W_f = -F_f s_{cm}$ *but it is not!* The friction force is an average effect arising from a complex of occurrences on the atomic level. Neither the body nor the floor is rigid. Regions in contact abrade each other, microscopic welds are formed and pulled apart, and work is done. In the process, atoms are jostled and energy is transferred to *both* the block and the floor. *There is no way to calculate the work done on just the block by the friction force:* F_f is not the force on each weld and it does not act over a distance s_{cm}. However, the net work done on the block-floor system is Fs_{cm}. This is the total mechanical energy delivered to the system and, since $F = F_f$, it does equal $F_f s_{cm}$. As we will soon see, the energy transferred into the system by the action of F appears primarily as thermal energy in *both* the block and the floor.

Figure 9.5 A force F is applied to a block that also experiences a friction force F_f. An amount of work $W_f = F_f s_{cm}$ goes into overcoming friction in the block-floor system, and that energy ultimately appears as thermal energy.

Example 9.5 A youngster weighing 250 N is sitting on the grass holding on to a large dog via a leash stretched horizontally. The dog pulls on the leash with a force of 100 N and drags the kid at a slow, constant speed 20 m straight across the yard and then stops. How much work did the dog do on the child and the ground in overcoming kinetic friction?

Solution: [Given: $F = F_f = 100$ N and $s = 20$ m.

Find: W_f.] The force F exerted by the dog was constant and in the direction of the displacement; hence, from Eq. (9.1)

$$W_f = F_f s = (100 \text{ N})(20 \text{ m}) = \boxed{2.0 \times 10^3 \text{ N·m}}$$

▶ **Quick Check:** 10^2 N $\times$ 2 $\times$ 10^1 m = 2 $\times$ 10^3 N·m.

When there is a displacement and positive work is done on one system by another, an amount of energy equal to W is transferred to the first from the second. In Example 9.5, the dog was the source of 2.0 kJ that ended up warming the child and the ground. If it takes $+2.0$ kJ to drag the youngster 20 m to the right, it will take another $+2.0$ kJ to slide him back. The friction force changes direction as the motion changes, and the work done against friction is always positive. *The net work done equals the force applied to overcome friction times the total path-length traveled.* Thus, when a body returns to its starting point so that its displacement is zero, the work done in overcoming friction is not zero!

Example 9.6 The force needed to overcome air friction for a person walking at a constant normal rate is only about 1 ounce (0.3 N). How much work does someone do against air friction in walking twice around a circular track having a radius of $\frac{1}{4}$ km?

Solution: [Given: $F = 0.3$ N, $R = \frac{1}{4}$ km, $s =$ twice around $= 2 \times 2\pi R$. Find: W_f.] Since F is in the direction of the motion

$$W_f = Fs = (0.3 \text{ N})(2 \times 2\pi R) = \boxed{9 \times 10^2 \text{ J}}$$

▶ **Quick Check:** The path is $2 \times 2\pi R = 2 \times 2(3.1416)(250 \text{ m})$, or $10^3 \pi$ m $\approx 3 \times 10^3$ m, and $W \approx (0.3 \text{ N})(3 \times 10^3 \text{ m}) \approx 0.9$ kJ.

Changing Forces

When we defined work ($W = Fs \cos \theta$), it was assumed that the driving force F was constant over the displacement. Although there are many important situations where this *is* true, there are many more where it is not. For instance, the force of gravity

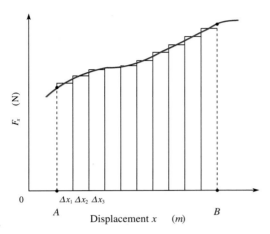

Figure 9.6   A force that varies from point to point in space plotted against distance. The area under the curve is the work done in moving from A to B.

actually varies as $1/r^2$; the force exerted by a hammer on a nail or a fist on a nose will vary with contact distance.

Imagine an object being displaced along the *x*-axis by a varying force **F** that has a component F_x that does the work. Figure 9.6 is a plot of F_x versus *x*. (Recall the treatment that led to Fig. 4.13, where the momentum change associated with a time-varying force was determined.) We cannot simply plug into Eq. (9.2) to find the work done because the force varies from point to point, but we can use Fig. 9.6, which is a plot of exactly how F_x changes. The area under the curve (F_x times *x*) is the desired work. Therefore, divide that curve into a number of very narrow intervals: Δx_1, Δx_2, Δx_3, and so forth. Assume that the force is constant at some average value over each of these segments. We have a definition for work done by a constant force; hence, for any segment $\Delta W = F_x \Delta x$. The net work done in going from $x = A$ to $x = B$ is approximately the sum of all these contributions. And that approximation improves as the Δx intervals get narrower and narrower because the average force in each case then more closely approaches the instantaneous force; that is, the corresponding point on the F_x-curve. As these intervals approach zero in width and the number of them increases toward infinity, *the sum approaches the area under the curve between the two points, which is the total work done by the variable force in moving the object from A to B.* The work done is the area under the force-displacement curve.

If we can plot the curve, the *work diagram,* we can determine the area and therefore the work. The area can be found directly, by counting boxes on graph paper; mechanically, by tracing the curve with a gadget called a planimeter; or electronically, using a microcomputer. It's common practice when evaluating all kinds of devices, from mechanical hearts to steam engines, to automatically produce work diagrams.

Example 9.7   Figure 9.7a shows a mass attached to a spring in a zero-gravity environment. The mass is initially at a height of 0.3 m. It is subsequently pulled down to 0.0 m and released. The force-displacement curve for the spring is shown in Fig. 9.7b. Describe the varying force and compute the work done by the spring on the mass in moving from 0.0 m to 0.3 m. How much work is done by the spring as the mass goes from 0.0 m to 0.6 m?

Solution: [Given: the curve. Find: the work done.] The spring-force acts along a vertical line such that it is parallel to the displacement ($\theta = 0$). It follows from the curve that the force varies linearly with the distance from the unstretched equilibrium position (0.3 m). The force is up and maximum (3.0 N) at the very bottom (0.0 m); zero at equilibrium (0.3 m); and down and maximum (-3.0 N) at the top (0.6 m). From 0.0 m to

Figure 9.7   (a) A mass in a zero-gravity environment attached to a vertical spring. It is pulled down to point 0.0 and released. (b) The plot is a graph of the variable force exerted *by the spring* versus displacement. We'll study the spring-force in detail later; for the moment, notice that it's linear.

(continued)

(continued)

0.3 m, the spring-force causes the motion and, from 0.3 m to 0.6 m, it opposes it. The triangular area under the curve from 0.0 m to 0.3 m equals $\frac{1}{2}$(base)(altitude); hence, the work done by the spring on the mass is

$$W_{sm} = \frac{1}{2}(0.3 \text{ m})(3.0 \text{ N}) = 0.45 \text{ N·m}$$

or

$$\boxed{W_{sm} = +0.5 \text{ J}}$$

Once the mass passes the equilibrium point on the way up, the spring-force opposes the motion. The work done by the spring is negative; the mass does work on the spring. In this case, the area between 0.3 m and 0.6 m is negative and $W_{sm} = -0.5$ J. Thus, the total amount of work done by the spring in the complete upward journey is zero.

▶ **Quick Check:** The total area under the curve is zero.

Meandering Paths

Imagine a person holding a bowling ball with a constant upward force F. Suppose that the ball is to be slowly brought up a hill along a smooth path (Fig. 9.8a) so that it is at rest at both A and B. After each step, the ball will have been raised a small amount vertically (Δs_V) and translated a small amount horizontally (Δs_H). The ball is displaced at each step by $\Delta s = \Delta s_V + \Delta s_H$. Since the weight always acts downward and F, accordingly, always acts upward, positive work is done ($\Delta W = F\Delta s_V$) only during the rising part of the motion—no work is done on the ball in moving it horizontally. Although the force doing the lifting is constant from point to point along the route, the angle between F and the tangent to the path changes. Hence, we divide the route into tiny intervals, Δs, that are so small that θ can be taken as constant over each one (Fig. 9.8b). In any one such interval, $\Delta W = (F \cos \theta) \Delta s$. The total work done in going from A to B is

$$W(A{\rightarrow}B) = \Sigma \Delta W = \Sigma F(\Delta s \cos \theta)$$

Because F is constant, it can be factored out of the summation; thus

$$W(A{\rightarrow}B) = F\Sigma \Delta s \cos \theta = F\Sigma \Delta s_V$$

The sum of all the little vertical displacements is the total vertical displacement ($s_V = \Sigma \Delta s_V$) and

$$W(A{\rightarrow}B) = Fs_V \tag{9.3}$$

(a)

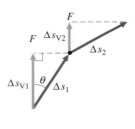

(b)

Figure 9.8 The curved path from A to B is divided into tiny straight segments Δs_1, Δs_2, Δs_3, $\cdots$. The small amount of work done during each minute step has the form $\Delta W = F\Delta s \cos \theta$; hence, $W(A{\rightarrow}B) = F\Delta s_{V1} + F\Delta s_{V2} + F\Delta s_{V3} + \cdots = Fs_V$, where $\Delta s_{V1} + \Delta s_{V2} + \Delta s_{V3} + \cdots = s_V$.

The work depends only on the net vertical distance traversed. **In going from A to B in a gravitational field, the work done is independent of the path taken—it is determined solely by the weight of the body and the vertical displacement**. The ball could be carried from A up, down, and around the hill in any zigzag path to B; negative work will subtract from positive work and, in the end, the net work will always be the same.

9.2 Power

When you want a day's labor and are concerned about getting every minute's worth, you buy a certain amount of work per hour. That's the practical language of industry, and it was in the beginning of the Industrial Revolution that the idea of *power* became quantified. **Power** is *the rate of doing work,* that is

$$\text{Power} = \frac{\text{work done}}{\text{time interval}}$$

More generally, power is the rate at which energy is transferred into or out of a system. An engine that will do the same amount of work as another in half the time is twice as powerful. If the time interval Δt is finite, we are talking about the *average power,* or

$$P_{av} = \frac{\Delta W}{\Delta t} \tag{9.4}$$

To express the moment-by-moment power, the **instantaneous power**, take the limit of this ratio as the time interval goes toward zero, whereupon

$$P = \lim_{\Delta t \to 0} \left(\frac{\Delta W}{\Delta t} \right) \tag{9.5}$$

In many cases, the power will be constant and Eq. (9.4) will be P.

The idea of standardizing the measure of power using the output of a horse had been around for a while before James Watt (1783) formalized it. Watt found that a large draft horse could exert a pull of about 150 lb while walking at about $2\frac{1}{2}$ mi/h for a considerable amount of time. That rate of doing work is equivalent to 33 000 foot-pounds per minute or *550 ft·lb/s,* which he called 1 **horsepower** (hp). Even though *horsepower* is now an outdated measure, its use persists in the United States.

In the SI system, the unit of power is the **watt** (W), equal to 1 joule of work done per 1 second: 1 W = 1 J/s. This unit is familiar because almost all electrical devices are rated in watts to indicate how much power they require. Since a watt is a fairly small amount of power, the more convenient kilowatt (1 kW = 1000 W) is often used. Thus,

$$1 \text{ hp} = 550 \text{ ft·lb/s} = 746 \text{ W} = 0.746 \text{ kW}$$

Presumably, someday all forms of power will be specified in SI units so that automobile engines will be rated in kilowatts along with toasters (Table 9.1).

Example 9.8 The express elevator in the Sears Tower in Chicago averages a speed of 1800 ft/min in its climb to the 103rd floor, 1340 ft above ground. Assuming a total load of 1.0×10^3 kg, what average power must the lifting motor supply? Always use SI units.

(continued)

(continued)
Solution: [Given: $v_{av} = 30.00$ ft/s $= 9.144$ m/s, $\Delta s =$ 1340 ft $= 408.4$ m, and $m = 1.0 \times 10^3$ kg. Find: P_{av}.] From Eq. (9.4) it's clear that we need ΔW, which, in turn, equals the weight (F_W) times the displacement (Δs). Hence

$$\Delta W = F_W \Delta s = mg\Delta s$$
$$\Delta W = (1.0 \times 10^3 \text{ kg})(9.8 \text{ m/s}^2)(408.4 \text{ m})$$

and

$$\Delta W = 4.0 \times 10^6 \text{ J}$$

To find Δt, remember that $v_{av} = \Delta s/\Delta t$ and so

$$\Delta t = \frac{408.4 \text{ m}}{9.144 \text{ m/s}} = 44.66 \text{ s}$$

Finally

$$P_{av} = \frac{\Delta W}{\Delta t} = \frac{4.0 \times 10^6 \text{ J}}{44.66 \text{ s}} = 89.57 \text{ kW} = \boxed{90 \text{ kW}}$$

▶ **Quick Check:** 90 kW will lift the 9.8×10^6-N load a distance of 408 m in a time of about $(9.8 \times 10^3 \text{ N}) \times (408 \text{ m})/(90 \text{ kW}) = 44$ s.

A person in good physical condition can deliver work at a fairly continuous rate of roughly $\frac{1}{10}$ hp, or about 75 W. Working at a steady pace, the human body develops power in proportion to the amount of oxygen it consumes. A rate of consumption of about 1 liter (10^3 cm^3) per minute is what's required for an output of 75 W. An athlete can raise that to about 5.5 liters per minute and develop continuous power at close to 400 W. A sustained power level of about twice this amount is the limit that a human can maintain for a period in excess of a minute or so. Over short time intervals (for example, the time it takes to leap into the air or throw a ball), much higher power levels can be attained. When muscles are rested, they have an additional short-term supply of oxygen that may also be called upon. This limited reserve can supply bursts of power of, say, 450 W for a minute or so, or as much as several kilowatts for a fraction of a second. Once exhausted, that reserve supply takes some time to replenish, and any athlete is aware of its one-shot nature.

TABLE 9.1   Power equivalencies

Quantity	Equivalent
1 Btu per hour	0.293 W
1 joule per second	1 W
1 foot-pound per second	1.356 W
1 kilocalorie per hour	1.16 W
1 kilowatt-hour per day	41.7 W
1 kilocalorie per minute	69.77 W
1 horsepower	745.7 W
1 kilowatt	1000 W
1 Btu per second	1054 W
1 gallon of gasoline per hour	39 kW
1 gallon of oil per minute	2.5 MW
1 million barrels of oil per day	73 GW

Power and Motion

When a constant force F acts on a body which, in the process, moves through a displacement Δs, we know from Eq. (9.2) that an amount of work is done, $\Delta W = (F \cos \theta)\Delta s$. If this occurs in a time interval Δt

$$P_{av} = \frac{\Delta W}{\Delta t} = (F \cos \theta)\frac{\Delta s}{\Delta t} = (F \cos \theta)v_{av}$$

or, in the limit, as $\Delta t \rightarrow 0$

$$P = Fv \cos \theta \qquad (9.6)$$

Power equals the product of the component of the force in the direction of motion and the speed. At any instant, it depends on the instantaneous speed at which the point of application of the force is moving. In many cases, **F** and **v** are

In 1979, the human-powered Gossamer Albatross flew across the English Channel. The pedaling pilot averaged 190 W, or about 0.25 horsepower.

parallel, $\theta = 0$, and the expression becomes

$$[\theta = 0] \qquad\qquad P = Fv \qquad\qquad (9.7)$$

Example 9.9   In 1935, R. H. Goddard, the American rocket pioneer, launched several A-Series sounding rockets. Given that the engine had a constant thrust of 200 lb, how much power did it transfer to the rocket traveling at a maximum speed of 1130 km/h?

Solution: [Given: $F = 200$ lb and $v = 1130$ km/h. Find: P.] Putting things in SI units, we obtain $F = (200$ lb$)(4.448$ N/lb$) = 889.6$ N and 1130 km/h =

313.9 m/s. Using Eq. (9.7)

$$P = Fv = (889.6 \text{ N})(313.9 \text{ m/s}) = \boxed{279 \text{ kW}}$$

▶ **Quick Check:** Let's recalculate using U.S. Customary units; (1130 km/h)(0.9113 ft·h/km·s) = 1029 ft/s; hence, P = (200 lb)(1029 ft/s) = 20.6 × 10⁴ ft·lb/s. Using Table 9.1, we multiply this value by 1.356 W/ft·lb/s, yielding 279 kW.

MECHANICAL ENERGY

There are four basic forces in nature as it exists today, and when work is done by or against any of them—and that's the only way work can be done—change occurs and energy is transferred.*

All matter interacts gravitationally, resulting in a web of force that binds every entity in the Universe together. It's been suggested that inertia arises from that invisible web (Mach's Principle, 1872). If that's true (and the idea is quite speculative), we can anticipate that a change in the distribution of matter effected by a moving body will correspond to an energy of motion.

9.3 Kinetic Energy

Huygens never cared for the idea that the basic measure of motion was momentum. Nonetheless, that view was widely held in the seventeenth century by the followers of Descartes. To be meaningful, momentum had to be directional. A bomb at rest could explode into two pieces, rapidly flying apart in opposite directions, and yet the total momentum of the system would remain zero—that's Conservation of Momentum. But what kind of essential quality of motion equals zero when the bodies being described are hurtling through the air? That thought did not sit well with Huygens, who searched to find a different measure independent of direction, one that would vanish only when motion ceased. His studies of rigid colliding balls led him to conclude that there was something special about the product of mass and speed-squared. Remarkably, adding the values of mv^2 for each ball prior to collision yielded a total that was essentially the same after collision, even though all the velocities had changed. Squaring the speed removed any dependence on the sign of v (that is, on the direction of **v**); mv^2 is always positive and only vanishes when v vanishes.

Leibniz picked up on the idea, calling mv^2 the *vis viva*. He showed that, for a falling body, mv^2 was proportional to Galileo's product of weight and height. It was not until 1807 that Thomas Young, an English physicist and physician, shifted the

Christian Huygens (1629–1695), Dutch mathematician, physicist, and astronomer.

*The common dictionary definition, *energy is the ability to do work,* is at best misleading (see Discussion Question 9). We will learn that the First Law of Thermodynamics maintains that energy is conserved, while the Second Law tells us that the availability of energy to do work tends to decrease as processes occur.

imagery and spoke of mv^2 for the first time as *energy*. He concluded that "labour expended in producing any motion, is proportional . . . to the energy which is obtained." In other words, work that causes motion equals the resulting change in energy.

Work and the Change in KE

Today, we call the energy associated with motion **kinetic energy** (KE), a term introduced in 1849 by one of the great figures of the nineteenth century, Lord Kelvin. Imagine a *rigid* body moving under the influence of a constant net force, such as a bullet decelerating as it penetrates soft clay. In each case, $W = Fs$ and, since $F = ma$, $W = mas$. From Eq. (3.10), $v_f^2 = v_i^2 + 2as$; hence, $as = \frac{1}{2}v_f^2 - \frac{1}{2}v_i^2$, and multiplying both sides by m yields

$$W = \tfrac{1}{2}mv_f^2 - \tfrac{1}{2}mv_i^2 \tag{9.8}$$

Here, s is the displacement of the *c.m.* Actually, the net force need not be constant. We would get the same result using the average force, producing an average acceleration. Under the influence of a net force ($\theta = 0$), a body accelerates as positive work is done on it to overcome inertia. It increases its speed and gains kinetic energy (KE) in the process. We define the kinetic energy of any object of mass m traveling at speed v as

$$KE = \tfrac{1}{2}mv^2 \tag{9.9}$$

such that Eq. (9.8) is the difference between the final and the initial KE:

$$W = KE_f - KE_i = \Delta KE \tag{9.10}$$

The net work done accelerating a rigid object equals its change in kinetic energy (Fig. 9.9). From the derivation, the units of KE are the units of W; namely, *joules*. That can be confirmed by the following: $F = ma$, $1\ N = 1\ kg\cdot m/s^2$, and $W = Fs$; hence, $1\ J = 1\ N\cdot m = 1\ (kg\cdot m/s^2)m = 1\ kg(m/s)^2$. All forms of energy will be measured in joules (Table 9.2).

Equation (9.10) cannot always be applied to deformable bodies. For example, throw a wad of soft clay against a stone wall. The clay comes to a stop, ΔKE can be quite large, but W_{wc} is, for all practical purposes, zero. The wall exerts forces on the clay across the essentially stationary clay-wall interface. The wall is not perfectly rigid, and there is a tiny motion of the points of application of force as the wall distorts, but $W_{wc} \approx 0$. This sort of **nonworking force** will become important to us later in the discussion.

Force

Distance

$\Delta\ KE$

Figure 9.9 A force acting over a distance results in a change in kinetic energy.

TABLE 9.2   Rough values of the energies of various occurrences

Occurrence	Energy (J)	Occurrence	Energy (J)
Creation of the Universe	10^{68}	Burning a cord of wood	10^{10}
Emission from a radio galaxy	10^{55}	One gallon of gasoline	10^{8}
Supernova explosion	10^{44}	One day of heavy manual labor	10^{7}
Yearly solar emission	10^{34}	Explosion of 1 kg of TNT	10^{6}
Earth moving in orbit	10^{33}	Woman running for 1 h	10^{6}
Earth spinning	10^{29}	Burning match	10^{3}
Earth's annual sunshine	10^{25}	Hard-hit baseball	10^{3}
Yearly U.S. sunshine	10^{22}	Apple falling 1 m	1
Annual tidal friction	10^{20}	Human heartbeat	0.5
Exploding volcano (Krakatoa)	10^{19}	Depressing typewriter key	10^{-2}
Severe earthquake (Richter 8)	10^{18}	Cricket chirrup	10^{-3}
100-megaton H-bomb	10^{17}	Hopping flea	10^{-7}
Burning a million tons of coal	10^{16}	Proton in supercollider	10^{-7}
Hurricane	10^{15}	Fission of 1 uranium nucleus	10^{-11}
Thunderstorm	10^{15}	Electron mass-energy	10^{-13}
Atomic bomb (Hiroshima)	10^{14}	Electron in atom	10^{-18}
Saturn V rocket	10^{11}	Photon of light	10^{-19}
Lightning bolt	10^{10}	KE of room-temperature air molecule	10^{-21}

Example 9.10   A Boeing *747* airliner weighing 2.2×10^6 N (5.0×10^5 lb) at takeoff is cruising at a ground speed of 268 m/s (600 mi/h). Compute its kinetic energy. If 1 kg of TNT yields 4.6×10^6 J, how much TNT is the KE equivalent to?

Solution: [Given: $F_W = 2.2 \times 10^6$ N and $v = 268$ m/s. Find: KE.] We need the mass to compute KE; hence, $m = F_W/g = (2.2 \times 10^6 \text{ N})/(9.8 \text{ m/s}^2) = 2.24 \times 10^5$ kg. Consequently

$$KE = \tfrac{1}{2}mv^2 = \tfrac{1}{2}(2.24 \times 10^5 \text{ kg})(268 \text{ m/s})^2$$

and

$$\boxed{KE = 8.0 \times 10^9 \text{ J}}$$

This is equivalent to the amount of energy given off when $\boxed{1.7 \times 10^3 \text{ kg}}$ of TNT explodes.

▶ **Quick Check:** 8.0×10^9 J $= \tfrac{1}{2}mv^2$; $v^2 = 2(8.04 \times 10^9 \text{ J})/(2.24 \times 10^5 \text{ kg})$; $v = 268$ m/s.

Example 9.10 suggests a fundamental point: KE *is a relative quantity;* we effectively choose the zero of KE by selecting the coordinate system with respect to which v is measured. Here the Earth was the motionless reference frame, and ground speed was used in the calculation. If you were in that jet sitting at "rest" next to your 20-kg suitcase, it would have no KE with respect to you and yet, to someone at "rest" on the ground watching, your bag would flash past with almost 3×10^3 J. **As a rule, we are not interested in the absolute energy of a system, but in the energy added to or taken from it**.

Work → KE → Work

Your arm can do work on a bowling ball and thereby increase the ball's kinetic energy ($W = \Delta KE$). The ball, going from rest to some launch speed, travels down the lane, crashes into the pins, does work on them, and loses a corresponding amount of KE. A moving object can do a maximum amount of work equal to $\tfrac{1}{2}mv^2$. Work →

$KE \rightarrow$ work, the transformation goes either way. When work is done on the system, its KE can increase; when work is done by the system, its KE can decrease. Every system that moves transports energy in the form of KE from one place to another.

Example 9.11 A man of average size having a mass of 70 kg can run at a maximum speed of about 10 m/s. Compute his maximum KE.

$$KE_{max} = \tfrac{1}{2} m v_{max}^2 = \tfrac{1}{2}(70 \text{ kg})(10 \text{ m/s})^2$$

and

$$\boxed{KE_{max} = 3.5 \times 10^3 \text{ J}}$$

Solution: [Given: $m = 70$ kg and $v_{max} = 10$ m/s. Find: KE_{max}.] His maximum kinetic energy is

▶ **Quick Check:** $v_{max}^2 = 2KE_{max}/m = 100$ m²/s²; $v = 10$ m/s.

Example 9.12 According to the record books, Aleksandr Zass (known to his admirers as "Samson") would, when he wasn't bending iron bars, catch a 104-lb (463-N) woman fired from a cannon at around 45 mi/h. Assuming that Samson brought her to rest uniformly in a distance of 1.00 m, compute the force he exerted on our heroine. As a guess, take her "landing speed" to be 8.94 m/s (20 mi/h).

Solution: [Given: $F_W = 463$ N, $v = 8.94$ m/s, and $s = 1.00$ m. Find: F.] To find the force, we need the work done in stopping her, and that work equals her KE at impact. To find her KE, we need her mass: $m =$

$F_W/g = (463 \text{ N})/(9.81 \text{ m/s}^2) = 47.2$ kg. Hence

$$KE = \tfrac{1}{2}(47.2 \text{ kg})(8.94 \text{ m/s})^2 = 1.89 \text{ kJ}$$

Samson had to do 1.89 kJ of work on her to bring her to rest; hence, $W = Fs = \Delta KE$ and

$$F = \frac{\Delta KE}{s} = \frac{1.89 \times 10^3 \text{ J}}{1.00 \text{ m}} = \boxed{1.89 \text{ kN}}$$

▶ **Quick Check:** It is sometimes useful to check a problem by reworking it in different units. KE = $\tfrac{1}{2}(3.3$ lb/ft/s²$)(29$ ft/s$)^2 = 1.4 \times 10^3$ ft·lb, $(1.4 \times 10^3$ ft·lb$)(1.356$ J/ft·lb$) = 1.9$ kJ.

Note that doubling the speed of an object (the young lady fired from the cannon or perhaps the family car on the highway) quadruples the kinetic energy and also quadruples the work needed either to get it up to speed or to stop it. Insofar as the braking force on a car is fairly constant, once the brakes are engaged, it will take four times the distance to stop an auto traveling at 40 km/h than it takes for one moving at 20 km/h.

9.4 Potential Energy

Imagine a force that is *continuously* exerted on a body; for example, the gravitational interaction producing the body's weight. To move upward against that downward pull requires the application of a counterforce (provided by leg muscles or elevators) and the doing of work. The crucial point is that the force will continue to act after the displacement. A body raised in a gravitational field still experiences a downward force while held up there at rest. When let loose, it will fall; it will be driven back down toward where it came from and kinetic energy will be imparted to it in the process.

All of that is fine; work done on the system is ultimately converted into KE. But what is happening while the body is held motionless, high in the gravitational field? KE is not "liberated" until the body is released. Apparently, it is possible to do

Wind blows against the sails, exerting a force F and driving the ship forward. Air and water resistance produce a net drag F_f on the craft. When $F > F_f$, $\Sigma F = F - F_f = ma$, the ship accelerates and its KE increases.

The dogs pull on the sled and do work on it, W_{ds}, over some displacement. This energy goes into overcoming drag due to both the air and snow, W_f. If the sled accelerates over the run, $W_{ds} = W_f + \Delta KE$.

work on a system and not have it immediately appear as KE. Yet the potential for generating that energy is there because of the ever-present force. *Energy is stored in the system of interacting objects,* waiting to be let loose. This *retrievable stored energy, energy by virtue of position or configuration in relation to a force,* is known as **potential energy** (PE), a name suggested by W. Rankine in the mid-1800s. The concept of PE gives continuity to the idea of energy; W done on a body against a force such as gravity goes into changing its PE, which may later go into changing its KE.

The change in the potential energy of a body incurred in moving from one point to another equals the work done in overcoming the interaction that stores the energy. PE exists only in relation to systems of interacting objects.

Gravitational-PE

When a painter climbs a flagpole, she does work to overcome the downward pull of gravity. Work corresponds to a change in energy, and we associate this change with the *gravitational*-PE; inasmuch as she is motionless, high on her perch, $W = \Delta PE_G$. That work equals the product of the force she exerts (that is, a force insignificantly greater than her weight F_W) and the vertical height through which she ascends, Δh. Thus

$$\Delta PE_G = F_W \Delta h \qquad (9.11)$$

Given that the painter has a mass m, we can write Eq. (9.11) in terms of her initial (h_i) and final (h_f) heights independent of the path taken (p. 285); accordingly

$$\Delta PE_G = mg\,\Delta h = mg(h_f - h_i) \qquad (9.12)$$

For the time being, we limit our activities to the Earth and assume g is constant. Potential energy, like KE, is a relative quantity; even the idea of "height" is relative. How high is your own nose? Above what—your lip, the ground, or perhaps sea level? Any of these would do; the zero-reference level of PE_G is arbitrary and can be taken wherever it's convenient. We simply say that the height at some level is zero and take the PE_G with respect to that level. Usually, we are concerned only with changes in PE_G due to changes in altitude, which then removes the need to deal with the zero-reference level. In fact, **the only thing we are able to measure is the change in PE_G, the change in energy**. That tells us something very profound about the nature of the Universe (p. 308).

It is useful, nonetheless, to have an expression for the PE_G of a body at any height h above the chosen zero-reference, especially if that reference is a commonly used one, such as the surface of the planet. Here, what is meant by "the height of a body" is the height of its center-of-gravity. Accordingly, Eq. (9.12) becomes

$$PE_{Gf} - PE_{Gi} = mg(h_f - h_i)$$

which suggests that

$$PE_{Gf} = mgh_f \qquad \text{and} \qquad PE_{Gi} = mgh_i$$

If h_i is the reference level (that is, $h_i = 0$), $PE_{Gi} = 0$, and we set $PE_G = PE_{Gf}$ (at any elevation $h = h_f$ measured from $h = 0$). Hence

$$PE_G = mgh \qquad (9.13)$$

The work done raising the mass a height h (namely, $F_W h$) equals the increase in potential energy. Our planet is spinning; consequently, the g here should represent the absolute gravitational acceleration. It's common practice, however, to overlook the small discrepancy and just take the acceleration as 9.81 m/s².

Example 9.13   The flagpole painter in Fig. 9.10 is carrying a 2.00-kg can of white paint. She has climbed (raising her can's *c.g.* 10.0 m) up the pole from its base, where she just finished having lunch on the roof of a 30.0-m-tall tower. (a) What is the increase in the potential energy of the can as a result of that last ascent? (b) What is the total increase in the can's potential energy above the value it had while it sat on the ground? (c) If we fix $PE_G = 0$ at the top of the roof, what was the PE_G of the paint can while it sat on the ground? Assume the can's *c.g.* is 10.0 cm above its bottom.

Solution: [Given: heights of 30.0 m and 10.0 m, $m = $ 2.00 kg, and the *c.g.* 10.0 cm above the can's bottom. Find: appropriate PEs.] (a) Using Eq. (9.12)

$$\Delta PE_G = mg\Delta h = (2.00 \text{ kg})(9.81 \text{ m/s}^2)(10.0 \text{ m})$$

and so $\boxed{\Delta PE_G = 196 \text{ J}}$

(b) Its *c.g.* rose 40.0 m; hence

$$\Delta PE_G = mg\Delta h = (2.00 \text{ kg})(9.81 \text{ m/s}^2)(40.0 \text{ m})$$

therefore $\boxed{\Delta PE_G = 785 \text{ J}}$

(c) The *c.g.* is 0.100 m above the ground when the can is standing upright. At such a time, the *c.g.* would be (30.0 − 0.100) m *below* the zero level, and so from Eq. (9.13)

$$PE_G = mgh = (2.00 \text{ kg})(9.81 \text{ m/s}^2)(-29.9 \text{ m})$$

and $\boxed{PE_G = -586 \text{ J}}$

▶ **Quick Check:** If we take the roof as the zero of PE_G, raising the can from the ground to the top of the pole will change the PE_G by 196 J − (−586 J) = 785 J.

Figure 9.10   A flagpole painter has *gravitational*-PE given by $PE_G = mgh$. Here, h is measured with respect to any level taken as $h = 0$. Only the change in PE_G is measurable, so the location of the $h = 0$ level is arbitrary.

A descending roller-coaster car converts *gravitational*-PE into KE. An ascending car does just the reverse.

Depending on where the zero-reference is chosen to be, PE_G can be positive or negative. Figure 9.11 shows a car on a roller coaster. It starts at the top and drops to the first low point. If we take the ground level as the zero of *gravitational*-PE, $\Delta PE_G = (mgh_f - mgh_i) = -mg\Delta h$, where Δh is positive. PE_G decreases, ΔPE_G is negative, but each value of PE_G is positive. If, however, we take the very top to be the zero-PE_G level, then $PE_{Gi} = 0$ and $\Delta PE_G = PE_{Gf} = -mg\Delta h$. The PE_G is negative everywhere on the ride, but the change in potential energy is the same as before, when the ground was the zero level.

Forces Varying with Distance

The prior discussion is true provided that g is constant, but as seen in Eq. (7.4), it isn't! So $PE_G = mgh$ is an approximation, a fine one when h is small compared to the radius of the Earth, but an approximation all the same. This circumstance arises because an object's weight is a function of distance ($F_G = GmM/r^2$). That means that the work done in moving against gravity must be determined by finding the area under an appropriate portion of the force-displacement curve.

Figure 9.12a is a plot of the external, inverse-square gravity-force for a planet of radius R. When an object of mass m is raised from the surface R to some distance r, the work done on it will equal the area beneath the curve from R to r. And so we divide the distance into short segments Δr, over each of which we can assume F_G is constant. Each little rectangular region is topped off at a value equal to the average force for that region. The smaller is Δr, the greater the number of little regions, and the better the average force approximates the actual force at every point on the curve. The first interval is bounded by the forces GmM/R^2 and GmM/r_1^2, and we need the average force somewhere between. On the left is $1/RR$ and on the right $1/r_1r_1$. That suggests using $1/Rr_1$ as the average. The average force is then GmM/Rr_1. As $\Delta r \to 0$, $r_1 \to R$ and $1/R^2 \to 1/r_1^2 \to 1/Rr_1$, which is fine.

The work done in traversing the first interval is the average force times $\Delta r = (r_1 - R)$, namely

$$\Delta W_1 = G\frac{mM}{Rr_1}(r_1 - R) = GmM\left(\frac{1}{R} - \frac{1}{r_1}\right)$$

In the same way

$$\Delta W_2 = GmM\left(\frac{1}{r_1} - \frac{1}{r_2}\right)$$

and so on, until

$$\Delta W_n = GmM\left(\frac{1}{r_{n-1}} - \frac{1}{r_n}\right)$$

where the number of little segments is n, and $r = r_n$ finishes the trip. In the end, we will cause $\Delta r \to 0$, while $n \to \infty$. Now add all of these work contributions, whereupon $W = \Delta W_1 + \Delta W_2 + \cdots + \Delta W_n$. Realize that each ΔW expression contains a $1/r$ term that cancels a corresponding identical term in the next expression. Thus, the $1/r_1, 1/r_2, 1/r_3, \cdots, 1/r_{n-1}$ terms all cancel, leaving only

$$W = GmM\left(\frac{1}{R} - \frac{1}{r}\right)$$

no matter how many segments there are (even as $n \to \infty$). In general then, *the change in the* PE *experienced by a body of mass* m *relocating from* r_i *out to* r_f *in the gravita-*

Figure 9.11 A roller coaster showing a car at its highest point, where the potential energy is PE_{Gi}. The zero-PE_G level can be located anywhere. When the car drops to the lowest point, it has a potential energy of PE_{Gf}. If PE_{Gi} is set equal to zero, the PE_G anywhere on the slope is negative; thus, $PE_{Gf} = -mg\Delta h$.

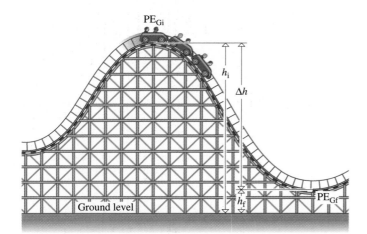

PE_{Gi}

h_i

Δh

Ground level

h_f

PE_{Gf}

tional field of a body of mass M is

$$\Delta PE_G = GmM\left(\frac{1}{r_i} - \frac{1}{r_f}\right) \qquad (9.14)$$

Since $\Delta PE_G = PE_{Gf} - PE_{Gi} = (-GmM/r_f) - (-GmM/r_i)$, if at $r_f = \infty$ we choose PE_{Gf} to be zero, the $1/r_f$ term goes to zero and the i subscript can be dropped. Then $PE_{Gi} = PE_G$ and

$$PE_G = -\frac{GmM}{r} \qquad (9.15)$$

is the *gravitational*-PE of a mass m at any distance r from the center of a mass M. The farther m is from M, the larger is r and the larger is the PE_G, inasmuch as it gets to be a less negative number. At $r = \infty$, the PE_G has its greatest value: zero.

We can apply Eq. (9.14) to the surface of the Earth (Fig. 9.13) to show that it agrees with the more limited statement of Eq. (9.13). Neglecting spin, Eq. (7.3) tells us that $gR_\oplus^2 = GM_\oplus$. Using that, we can find the change in potential energy of a mass m raised to a low altitude ($h = r - R_\oplus$) for which $r \approx R_\oplus$. Equation (9.14) becomes

$$\Delta PE_G = mgR_\oplus^2\left(\frac{1}{R_\oplus} - \frac{1}{r}\right) = mgR_\oplus^2\left(\frac{r - R_\oplus}{rR_\oplus}\right) = mgR_\oplus\frac{h}{r} \approx mgh$$

Raising his *c.m.*, this climber increases the *gravitational*-PE stored via the interaction between him and the Earth. Considering the gravitational field as a mapping of the interaction, the PE can be thought of as stored in the field.

(a)

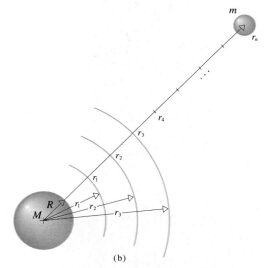

(b)

Figure 9.12 (a) A spherical mass of radius R is surrounded by a gravitational force field that drops off inversely with the square of the distance r. (b) The distance from the surface out to a mass m is divided into a large number n of segments.

Example 9.14 What is the change in potential energy of a 100-kg meteorite if it free-falls from an altitude of 1000 km down to the surface of the Earth? Take the mass of the planet to be 6.0×10^{24} kg and the diameter to be 1.28×10^7 m.

Solution: [Given: $m = 100$ kg, $h = 1000$ km, $M_\oplus = 6.0 \times 10^{24}$ kg, $R_\oplus = 0.64 \times 10^7$ m, and $G = 6.67 \times 10^{-11}$ N·m²/kg². Find: ΔPE_G.] Falling from r to $R_\oplus$, the PE_G decreases and so ΔPE_G must be negative:

(continued)

(continued)

$$\Delta PE_G = GmM_\oplus \left(\frac{1}{r} - \frac{1}{R_\oplus} \right)$$

With $r = R_\oplus + h = 7.4 \times 10^6$ m

$$\Delta PE_G = (4.0 \times 10^{16} \text{ N·m}^2)$$
$$\times (1.35 \times 10^{-7} \text{ m}^{-1} - 1.56 \times 10^{-7} \text{ m}^{-1})$$

Hence

$$\boxed{\Delta PE_G = -8.5 \times 10^8 \text{ J}}$$

▶ **Quick Check:** $\Delta PE_G \approx mg\Delta h = (100 \text{ kg}) \times (9.81 \text{ m/s}^2)(-10^6 \text{ m}) = -9.8 \times 10^8$ J.

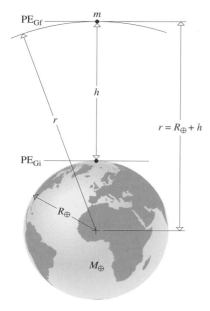

Figure 9.13 A mass m at a distance r from the center of the Earth. Given that the altitude of m is h, it follows that $r = R_\oplus + h$. When h is small, $r \approx R_\oplus$ and $\Delta PE_G = PE_{Gf} - PE_{Gi} \approx mgh$.

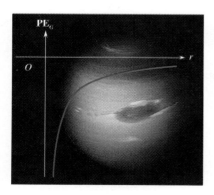

Potential energy is often formulated so that it is negative when the interaction is attractive.

9.5 Internal Energy

It is convenient to distinguish two basic forms of energy, KE (energy of motion) and PE (energy of position). For objects that have structure (that is, interacting parts), it's helpful to further recognize two manifestations of each of these forms called *internal* and *external*. A bowling ball at rest has all its atoms randomly vibrating about their equilibrium positions at fairly high speeds (typically ≈ 0.4 km/s). We refer to that disorganized internal kinetic energy as *thermal energy*. A bowling ball flying through the air has an organized motion superimposed on all its jiggling atoms, and we have called this external energy *kinetic energy*. Similarly, a bowling ball in the air has external potential energy by virtue of its interaction with the Earth. When the ball is distorted (for example, during a collision) it has *elastic energy*, one of several kinds of internal PE.

Imagine two structureless particles interacting gravitationally. As we have seen, the work done in bringing a mass from any point A to any point B is independent of the path taken. It follows that moving a particle from A to B and back to A results in zero work done. That feature, namely, that *zero work is done in moving a particle around a closed path*, defines a **conservative force**. To be conservative, the force on a particle at any given location in space must be constant, no matter how or *when* the particle gets there. Whatever energy goes into such an interaction on moving from A to B, comes out on moving from B to A; the process is reversible. The conservative interaction stores PE.

As we continue our study, we will encounter a number of ways in which the Four Forces act to store PE both internally and externally. A stretched garter retains *elastic*-PE; a bright red stick of dynamite is gift-wrapped *chemical*-PE; a battery supplies *electrical*-PE; two magnets pulled apart wait to reunite with *magnetic*-PE; and all those neatly painted warheads hiding in their silos threaten with *nuclear*-PE.

Internal Energy Sources and Work

The work done on passive objects (crates, blocks, balls, and sailboats) by externally applied forces has been our primary concern thus far. Now, consider self-propelled objects (cars, people, helicopters, frogs, etc.) that have their own internal energy sources. Each can be accelerated by a net external force ($F = ma$) arising from its interaction with the environment. *As a rule, such a force does no net positive work on the active nonrigid body, and $W \neq \Delta KE$.* No energy is transferred to the body from the environment via that reaction force, even though that force accelerates the body.

A swimmer's hand (Fig. 9.14a) pushes back on the water and the water pushes forward on the hand, accelerating the person. But the hand is moving backward all the while. The hand does positive work on the water; the force it exerts is in the same direction as the displacement. By contrast, the water pushes in the forward direction on the hand. It does negative work on the swimmer: $W_{ws} < 0$ even though $\Delta KE > 0$. The water gets energy from the swimmer—it gets warmer. The swimmer is self-powered and uses the water to generate a reaction force so she can accelerate. Just imagine a motorboat suspended above the water with its engine running at full speed, going nowhere. All the energy needed is provided by the fuel, but without the water to push on, the boat cannot accelerate.

The energy to walk, climb (Fig. 9.14b), skate, or jump comes from internal energy stored in the person (see Discussion Questions 2 and 3). When you jump (Fig. 5.12), the upward reaction force that accelerates you acts at the stationary foot-floor interface. If the floor is rigid, *there is no motion of the point of application of the force, and no work is done by the floor on you:* $W_{fu} = 0$ even though $\Delta KE > 0$. Actually, the floor will sag slightly while it exerts a normal force on you and W_{fu} is positive though very small, since s is very small.

When a car accelerates, the tires on the drive wheels push back on the ground; the ground pushes forward on the tires and the car accelerates forward. But the region of contact between tire and road is motionless, and no work is done on the car by the ground. Of course, the energy equivalent to ΔKE comes from the fuel via the engine—a car doesn't drain energy out of the ground, it just pushes off it.

CONSERVATION OF MECHANICAL ENERGY

It was Descartes who proclaimed the Law of Conservation of Momentum (1644). And so when Huygens sought to replace mv as the essential measure of motion, it was natural for him to look for a quantity that was likewise conserved (that is, constant in time). The realization that there were various interchangeable forms of an otherwise unchanging total energy had already been proposed by several people before Rankine (1853) provided us with the ringing phrase "Conservation of Energy." That era gave rise to the study of Thermodynamics with its premier insight that "heat" was another form of energy. Yet it was not until 1905 that Einstein added the final piece to the construct. He showed that mass can be thought of as equivalent to energy; that by virtue of its very existence, a particle of mass possesses a store of energy.

The grandest generalization in all of physics is the fully developed statement of the Law of Conservation of Energy:

The total energy of any system that is isolated from the rest of the Universe remains constant, even though energy may be transformed from one kind to another within the system.

Assuming that *the energy of the Universe is constant,* the energy of any portion of the Universe isolated from the rest must also be constant. If there is no flow of energy into or out of a system (and that's what is meant by *isolated*), then there can be no change in energy either there or in the remainder of the Universe.

9.6 Mechanical Energy

Let's begin with the case of a system acted on by both externally applied contact forces and gravity. To keep things simple, we eliminate any possible changes in internal energy—no friction, no deformation. The effect of the contact forces is re-

(a)

(b)

Figure 9.14 (a) A swimmer pulls on the water with a force F_{sw}, drawing it back a distance x and doing positive work on it. The forward reaction of the water on the swimmer's hand (F_{ws}) is opposite to the displacement of the hand, and the work done on the swimmer is negative. Energy, supplied by her food, is transferred to the water. (b) A person climbing a ladder exerts forces down on the rungs, but the point of application does not move and no work is done on the ladder or on the climber. The increase in PE_G comes from internal energy.

flected in the net work they perform on the system: W_{net}. The *gravitational*-PE of the system delineates the influence of the gravitational field. And the energy associated with inertia is accounted for via ΔKE. Thus, energy delivered to or removed from the system can go into changing the *gravitational*-PE and/or the KE, hence

$$W_{net} = \Delta PE_G + \Delta KE \qquad (9.16)$$

There are other conservative forces besides gravity, and we can add appropriate PE terms to Eq. (9.16) later.

We define the **mechanical energy** (E) of a system as *the sum of the KE and gravitational-PE of all its parts.* Therefore

$$W_{net} = \Delta E = E_f - E_i$$

and

$$E_f = E_i + W_{net}$$

If no contact forces are applied W_{net} is zero, no energy is transferred into or out of the system, $E_f = E_i$, and *mechanical energy is conserved.* This is an early, limited formulation of what would later develop into the concept of conservation of all forms of energy.

Assuming that g is constant, Eq. (9.16) becomes

$$W_{net} = (\tfrac{1}{2}mv_f^2 - \tfrac{1}{2}mv_i^2) + (mgh_f - mgh_i)$$

When the only force acting is gravity, $W_{net} = 0$ and

$$\tfrac{1}{2}mv_f^2 + mgh_f = \tfrac{1}{2}mv_i^2 + mgh_i \qquad (9.17)$$

At every instant in the motion, the total mechanical energy is constant; if the body's KE increases, its PE_G must decrease, and vice versa. Fire a ball straight up into the

Figure 9.15 (a) A cannonball fired upward loses KE as it rises and gains PE_G. Fired with a $KE_i = 100$ J, it will always have a total energy of 100 J. Of course, in real life, some energy will be lost to air friction. (b) When the ball drops from rest, it loses PE_G and gains KE.

(a)   (b)

air and in the process do work on it, giving it a kinetic energy of 100 J (Fig. 9.15a). Let's focus on the launched projectile. Once fired (ignore air friction), no forces other than gravity act and $W_{net} = 0$. At $h_i = 0$, $PE_G = 0$, while the KE, equal to 100 J, is a maximum (equal to E). As the ball rises, increasing in PE_G, it slows down, decreasing in KE such that at any moment $KE + PE_G = E = 100$ J. At peak altitude, where the ball momentarily comes to rest, $KE = 0$, and $PE_G = 100$ J is a maximum (equal to E). As it falls back (Fig. 9.15b), PE_G decreases to zero, its speed increases, and the KE attains a maximum value of 100 J (equal to E) as it returns to the point from which it was launched. The similar situation of a swinging pendulum is depicted in Fig. 9.16.

During a running jump, a person can only get as high as his vertical velocity will carry him. However, using a pole to vault into the air provides a means of transforming nearly all of a runner's KE into PE_G. At the start, the person runs as fast as possible to gain as much KE as possible. Once the athlete jumps off the ground and is pole-borne, his run-up KE is gradually converted into both *gravitational*-PE and *elastic*-PE (p. 329) as the pole continues to bend while he rises. It soon straightens out, gives up its *elastic*-PE, and hurls him even higher.

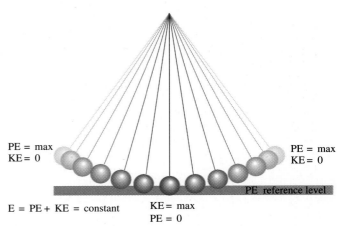

$$E = PE + KE = constant$$

Figure 9.16 The transfer of energy from one form to another in the swinging pendulum.

Example 9.15 A vaulter carrying a graphite-fiberglass pole (a thin-walled tube weighing around 4 lb) is about to make a jump. At what speed must he run in order to clear the 20.0-ft (6.096-m) mark? Neglect all possible energy losses, take his *c.g.* to be 1.00 m above the floor while standing, and presume that it just clears the bar.

Solution: [Given: $h_i = 1.00$ m, $\Delta h_{cg} + h_i = 6.10$ m. Find: v_i when $v_f = 0$.] Neglect friction. It follows from Eq. (9.17) that the final PE_G equals the initial KE. Take the zero-PE level at the initial height of the *c.g.*; thus, if the *c.g.* rises an additional $\Delta h_{cg} = 6.096$ m − 1.00 m, then

$$0 + mg\Delta h_{cg} = \tfrac{1}{2}mv_i^2 + 0$$

The mass cancels, yielding

$$v_i^2 = 2g\Delta h_{cg} = 2(9.81 \text{ m/s}^2)(5.096 \text{ m}) = 99.949 \text{ m}^2/\text{s}^2$$

and $\boxed{v_i = 10.0 \text{ m/s}}$. This is a high speed for someone running with a pole. Actually, at the last moment, high in the air, as he approaches the bar, the athlete will straighten his arms, lifting his *c.g.* above his hand position and even above the end of the pole.

▶ **Quick Check:** With a launch speed v_i, the peak altitude (p. 80) reached by a projectile is $v_i^2/2g \approx 10^2/(<20)$ m $\approx$ >5 m above the 1-m lift-off height.

9.7 The Energy Perspective

Unlike Newton's Laws, which require us to know the details of the motion in order to carry out the analysis, the energy formulation provides an overview. With it we can determine the final state of a system from its initial configuration, without looking at the intermediate stages it passes through. For instance, knowing that the identical cannonballs in Fig. 9.17 were launched with the same initial speed, and therefore with the same KE, we can be sure that at any one height (where their potential energies are the same), they will all be moving with equal speed regardless of the path taken. Similarly, knowing the height of a snowy slope, we can compute a skier's speed at the bottom of a run without any knowledge of the route she took, or even the shape of the mountain, so long as friction is negligible.

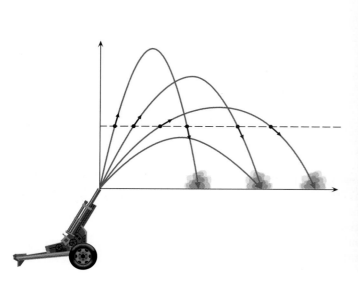

Figure 9.17   Cannonballs fired with the same muzzle speed will all be traveling at the same speed at the same altitude.

A vaulter transfers his run-up KE into *elastic*-PE as the pole bends. When the pole straightens out, it imparts energy to the vaulter. At maximum altitude, he's motionless and subsequently drops straight down.

Example 9.16   A 60-kg skier starts from rest at the top of a 60-m-high slope. She descends without using her poles. (a) What is her initial *gravitational*-PE with respect to the level ground at the bottom? (b) Assume friction is negligible and compute at what speed she will ideally arrive at the bottom. (c) She reaches the bottom of the run, traveling at 25 m/s. What is the net energy transferred via friction to her, the skis, the slope, and the air?

Solution: [Given: $m = 60$ kg, $h_i = 60$ m, $v_i = 0$, $v_f = 25$ m/s, and $s = 10$ m. Find: (a) PE_G, (b) v_f, no friction, and (c) W_F.] (a) Taking zero PE at the bottom, at the top of the hill

$$PE_{Gi} = mgh_i = (60 \text{ kg})(9.8 \text{ m/s}^2)(60 \text{ m})$$

and

$$\boxed{PE_{Gi} = 3.5 \times 10^4 \text{ J}}$$

(b) With no work done against friction, $E_f = E_i$ and from Eq. (9.17), with $v_i = 0$ and $h_f = 0$, $KE_f = PE_{Gi}$, and

$$\tfrac{1}{2} mv_f^2 + 0 = 0 + mgh_i = 3.5 \times 10^4 \text{ J}$$

Hence

$$v_f = \sqrt{\frac{2(3.5 \times 10^4 \text{ J})}{60 \text{ kg}}} = \boxed{34 \text{ m/s}} = 76 \text{ mi/h}$$

(c) With friction acting on the system, $E_f = E_i + W_F$; hence

$$\tfrac{1}{2} mv_f^2 + 0 = 0 + mgh_i + W_F$$

and

$$W_F = \tfrac{1}{2} mv_f^2 - mgh_i = \tfrac{1}{2}(60 \text{ kg})(25 \text{ m/s})^2$$
$$- (60 \text{ kg})(9.8 \text{ m/s}^2)(60 \text{ m})$$

Hence, $W_F = 18\,750 \text{ J} - 35\,280 \text{ J} = \boxed{-1.7 \times 10^4 \text{ J}}$. The work done on the system by friction is negative because it opposes the motion.

▶ **Quick Check:** Working backwards, $\tfrac{1}{2} mv_f^2 = \tfrac{1}{2}(60 \text{ kg})(34 \text{ m/s})^2 = 35 \text{ kJ} = PE_{Gi}$.

Satellites and Spaceships

As a further example of the power of the energy formulation, consider the following: Why does a planet in orbit speed up as it approaches the Sun? We know how to find the answer from Newton's force picture, but now we have an alternative. A planet's speed and KE both increase because its *gravitational*-PE decreases as it approaches the Sun. Any energy losses are negligible and the total mechanical energy is everywhere constant. At the point of closest approach and smallest PE, the planet must have its maximum KE and be traveling with maximum speed (Fig. 9.18).

Reconsider Eq. (9.15), $PE_G = -GmM/r$. At $r = \infty$, PE_G is zero, while everywhere else it's negative. A negative PE is characteristic of attractive forces. Had the interaction been repulsive, as between two magnets with like poles, the PE would have been positive at finite distances. When the force is repulsive, work must be done on one body to bring it from infinity "down" to the vicinity of the other body, and so the change in PE (from zero) must be positive. By contrast, if the force is attractive, the bodies will come together on their own. The force field will do the work, and the change in PE (from zero) will be negative. *If a system can return to its zero-PE configuration by doing positive work, then the PE stored in the system is positive.* This is the case with two magnets with like poles repelling and separating out to infinity. *If positive work must be done on the system to bring it back to the zero-*PE *state, then the PE stored is negative.* This is the case with two attracting masses being pulled apart.

It's been suggested that since *gravitational*-PE is negative, if the total of all other forms of energy is equal in magnitude to the total PE_G, then the net energy of the entire Universe is zero. It follows that such a system could have spontaneously appeared out of "nothingness" without violating energy conservation.

Figure 9.18 A planet in an elliptical orbit around the Sun. As it gets closer to the Sun, it speeds up. The speeds are the same for equal distances, either moving toward or away from the Sun, since there the PEs must be the same and therefore the KEs must also be the same.

Example 9.17 How much work must be done on a projectile of mass m to slowly raise it from the surface of a motionless planet (of radius R and mass M) out to an altitude that is essentially infinitely far away? Neglect all frictional losses.

Solution: [Given: $v \approx 0$, $r_i = R$ and $r_f = \infty$. Find: W, the work done on the rocket.] We know that

$$E_f = E_i + W$$

hence $$KE_f + PE_{Gf} = KE_i + PE_{Gi} + W$$

Using Eq. (9.15),

$$W = PE_{Gf} - PE_{Gi} = 0 - \left(-\frac{GmM}{R}\right)$$

And so the work done, the energy that must be supplied, is

$$\boxed{W = \frac{GmM}{R}}$$

▶ **Quick Check:** The units of W are those of force (GmM/R^2) times distance (R), as it should be. As $R \to \infty$, $W \to 0$, which is appropriate. The smaller R is, the larger W gets, which is also reasonable.

With the results of the Example 9.17, suppose we wish to fire a craft into deep space with a single short-lived burst of energy. The minimum blast-off speed, the lowest speed with which we could fire a projectile straight up and never have it fall back to the planet from which it was launched, is the **escape velocity**, $\mathbf{v}_{esc}$. When given that speed measured with respect to the planet's *c.m.*, the object will have enough KE to overcome the planet's gravity. It will sail on forever, gradually slow-

TABLE 9.3 Escape speeds from the surfaces of celestial bodies

Body	Escape speed (km/h)
Eros asteroid	55
Moon	8×10^3
Mercury	15×10^3
Mars	18×10^3
Venus	37×10^3
Earth	40×10^3
Neptune	83×10^3
Saturn	130×10^3
Jupiter	216×10^3
Neutron star*	$\approx 0.8 \times 10^9$

*That's about 80% c.

ing down, but nonetheless arriving at infinity (whatever that means) as the final speed becomes zero. At launch, at the planet's surface (a distance R from its center)

$$E_i = KE_i + PE_{Gi} = \tfrac{1}{2}mv_{esc}^2 - \frac{GmM}{R}$$

and no matter what this equals, it must be forever constant, provided nothing external acts on the craft. Ideally, the object will reach infinity, where $PE_{Gf} = 0$, traveling at zero speed such that $KE_f = 0$. Hence, $E_f = 0$ at $r = \infty$, and since E is everywhere constant, it must be that $E_i = 0$. Thus

$$\tfrac{1}{2}mv_{esc}^2 - \frac{GmM}{R} = 0$$

and

$$v_{esc} = \sqrt{\frac{2GM}{R}} \tag{9.18}$$

At the surface of the Earth, where, from Eq. (7.4), $mg_\oplus = GmM_\oplus/R_\oplus^2$

$$v_{esc} = \sqrt{2g_\oplus R_\oplus} \tag{9.19}$$

Although the escape speed is independent of the mass of the projectile, the needed KE certainly is not. That's why it takes big rockets to raise big payloads. Notice that this is also the speed with which an object would collide with the planet, having fallen from a very great distance. Recall Eq. (7.8) for the speed of a body in a circular orbit; namely, $v_0 = \sqrt{GM/r}$. The escape speed from *any altitude* (where $r = R$) is equal to $\sqrt{2}$ times the orbital speed at that altitude. The higher the orbit, the slower the speed, and the easier the escape (Table 9.3).

Example 9.18 In 1865, Jules Verne published a story called *From the Earth to the Moon*. In it, a gigantic cannon fires a capsule carrying three men to the Moon. The cannon was located at "27°7'N. lat. 5°7'W. long. of the meridian of Washington." (That, incidentally, turns out to be in Florida about 100 mi south of the Kennedy Spaceflight Center.) Compute the muzzle speed needed to escape the Earth. Use $R_\oplus = 6.4 \times 10^6$ m.

Solution: [Given: $R_\oplus = 6.4 \times 10^6$ m and $g_\oplus = 9.8$ m/s^2. Find: v_{esc}.] Using

$$v_{esc} = \sqrt{2g_\oplus R_\oplus} \tag{9.19}$$

$$v_{esc} = [2(9.8 \text{ m/s}^2)(6.4 \times 10^6 \text{ m})]^{\frac{1}{2}} = \boxed{1.1 \times 10^4 \text{ m/s}}$$

This is equal to 25×10^3 mi/h, which is formidable. In fact, enough friction would be encountered in the dense lower atmosphere at that speed to burn up the ship. This is one reason why we use rockets that build up speed gradually, only reaching v_{esc} in the thin upper atmosphere.

▶ **Quick Check:** From Table 9.3, $v_{esc} = 40 \times 10^3$ km/h $= (40 \times 10^6/3600)$ m/s $= 1.1 \times 10^4$ m/s.

Because gravity drops off inversely with r^2, most of the energy required to travel away from a large body like the Earth is needed to get beyond the first few hundred thousand miles. While the engine-cutoff speed to escape to deep space from Earth is 25×10^3 mi/h, it takes a launch at about 99% of that speed just to get as far as the Moon. The Earth orbits the Sun, and attempting to leave our star requires a still greater speed. By taking advantage of the planet's orbital motion and sailing off in a judicious direction, this speed can be reduced to 16.4 km/s or 37×10^3 mi/h. The *Pioneer 10* space probe passed beyond the orbit of Pluto in

1983 (traveling at 30 613 mi/h) to become the first humanmade object to leave the Solar System.

Example 9.19 In his *Exposition of the System of the World* (1798) the great French mathematician Laplace pointed out that an object of the same density as the Earth with a diameter 250 times that of the Sun would have so large a gravitational attraction that its escape speed would exceed the speed of light. He concluded that such a body, which could prevent light from leaving its surface, would be invisible to us. (a) Show that for a body of a certain density ρ the escape speed is proportional to its radius R. (b) Verify Laplace's assertion.

Solution: [Given: $R = 250\ R_\odot$. Find: v_{esc}.] (a) Mass equals density times volume, $M = V\rho = \frac{4}{3}\pi R^3 \rho$. It follows from Eq. (9.18) that

$$v_{esc} = \left(\frac{2G\frac{4}{3}\pi R^3 \rho}{R}\right)^{\frac{1}{2}} = \boxed{R(\tfrac{8}{3}G\pi\rho)^{\frac{1}{2}}}$$

(b) The ratio of the diameter of the Sun to the Earth is $(1.393 \times 10^9\ \text{m})/(1.274 \times 10^7\ \text{m}) = 109$, which means that an object 250 times the size of the Sun is $(109)(250)$, or 2.7×10^4 times the size of the Earth. Hence, its escape speed is 2.7×10^4 times 11.2 km/s, or $\boxed{3.1 \times 10^8\ \text{m/s}}$ and that slightly exceeds c. Collapsed stars that are so dense and so powerful gravitationally that light cannot escape from them seem to exist. Nowadays, they are known as **black holes**. Although Laplace had the wrong theory for both gravity and light, he ended up predicting what we think is a real entity.

▶ **Quick Check:** The units of $R(\frac{8}{3}G\pi\rho)^{\frac{1}{2}}$ are $m[(\text{N}\cdot\text{m}^2/\text{kg}^2)(\text{kg/m}^3)]^{\frac{1}{2}} = m[(\text{kg}\cdot\text{m/s}^2)(1/\text{kg}\cdot\text{m})]^{\frac{1}{2}} = \text{m/s}$.

The Universe is an explosion that began $\approx 17 \times 10^9$ years ago. Its galactic star-islands are still rushing outward to this day; the Universe—all there is—is expanding. If it has enough KE to override its own gravitational field, if the galaxies, and whatever else is out there, have the needed escape speed, the Universe will go on expanding forever. If not, it will ultimately stop and fall back upon itself thousands of millions of years from now.

9.8 Collisions

Much of what we know about the atomic and subatomic domains has been learned from the observation of collisions. Beams of particles are hurled against targets, and the motion of the scattered fragments is analyzed. Similar relationships govern both these exotic events and the more mundane ones, such as clubs hitting golf balls. *A collision is marked by the transfer of momentum between objects in relative motion resulting from their interaction via at least one of the Four Forces.* **In all cases where there are no external forces, the total momentum of the colliding objects is conserved.**

Inelastic Collisions

An **inelastic collision** *is one where the final KE of the system is different from the initial KE.* A tennis ball dropped against a concrete floor distorts, momentarily comes to rest, and then springs back, popping into the air. But the squashing produces some internal heating as the molecules shift position (and even the crack of sound carries off a little energy) and so the ball only returns about two-thirds of the way back up. One-third of the initial KE has been converted into other energy forms. If, after an impact between two macroscopic objects, either one increases in temperature or remains distorted, the collision is inelastic.

Most of the KE of the moving car is transformed into thermal energy during the crash. The yellow plastic bumper extends the collision time and therefore decreases the force exerted on the car, as indicated in Eq. (4.3), p. 104.

Macroscopic bodies always collide more or less inelastically. What usually happens is that the system either changes its configuration, or transforms some of the impact KE into thermal energy, or both. Even the most rigid macroscopic objects convert some small amount of KE into thermal energy. To be precise, only subatomic particles collide elastically.

The *completely inelastic* collision is at one extreme—where the impacting objects stick together—and the maximum amount of KE is transformed (Fig. 9.19). A wad of oatmeal landing on a table, a bug hitting the windshield of a moving bus, or a neutron absorbed by a uranium nucleus are all representative of inelastic collisions. Less drastic is the case where the colliding objects bounce apart and yet lose KE in the process. A lead ball rebounding from a steel anvil will not even retain 1% of its original KE, whereas a soft cork sphere might retain around 35%. The harder the colliders, the less KE is transformed, which is why billiard balls are not soft. A glass marble will bounce off the anvil with about 95% of its initial KE. Our purposes will be served by examining only the two extremes of elastic and completely inelastic collisions.

Early in its existence (1663), the British Royal Society sponsored a scientific study of collisions. Among the leading scholars who participated were Huygens and the English mathematician John Wallis. Wallis, in a paper of 1668, was probably the first to apply the Law of Conservation of Momentum correctly. When two objects of mass m_1 and m_2, initially moving with speeds v_{1i} and v_{2i}, collide head-on and fly off with speeds v_{1f} and v_{2f}, initial momentum equals final momentum, and

$$m_1 v_{1i} + m_2 v_{2i} = m_1 v_{1f} + m_2 v_{2f} \tag{9.20}$$

The head-on restriction requires that the motion of both bodies at the moment of impact be along a common straight line connecting the two centers-of-mass. As a result, the final motion is also along that line. Hitting off-center or with noncolinear velocities results in the objects moving away at angles to the original direction. Any pool player knows how to strike a ball on the side so it sails away at an angle. Momentum is still conserved, but now the components in the x and y directions will require separate applications of Eq. (9.20).

INITIALLY   IMPACT   FINALLY

Figure 9.19 Inelastic collisions. (a) Here the skaters flying toward each other have equal masses and speeds. The total momentum is initially zero, and it remains zero after the collision, as they simply come to rest and stay that way. (b) Here number 9 initially has all of the system's momentum. After impact, since the total momentum is constant and they stay together, the moving mass is twice what it was, and therefore the speed must be halved. The two sail off at half the speed number 9 came in with. (c) Taking revenge, number 2 comes flying at number 9, at twice 9's speed. The total momentum is $m(2v) + mv$. After the collision, the two skaters sail off together ($2m$) with a speed ($3v/2$) that conserves the initial momentum ($3mv$).

Example 9.20 During a rainy day football game, a 192-lb quarterback is standing holding the ball looking for a receiver when he's unkindly hit by a 288-lb tackle charging in at 6.1 m/s (20 ft/s). (a) At what speed do the two men, tangled together, initially sail off onto the wet field? Assume friction is negligible and that the impact

(continued)

(continued)

is head-on. (b) How much mechanical energy is lost to friction? As always, work the problem through in SI units only.

Solution: [Given: $F_{Wq} = 192$ lb, $F_{Wt} = 288$ lb, $v_{qi} = 0$ and $v_{ti} = 20$ ft/s. Find: the final speed of the two, v_f and ΔKE.] First change everything to SI: $F_{Wq} = 854$ N, $F_{Wt} = 1281$ N, $v_{qi} = 0$, and $v_{ti} = 6.10$ m/s. Since the two slide away together, this collision is completely inelastic—momentum is conserved, but neither E nor KE is conserved. (a) Consequently, $p_i = p_f$ and

$$m_q v_{qi} + m_t v_{ti} = (m_q + m_t)v_f$$

Using $m = F_W/g$, and taking the direction of the initial motion as positive

$$0 + \frac{1281 \text{ N}}{9.81 \text{ m/s}^2}6.10 \text{ m/s} = \left(\frac{854 \text{ N}}{9.81 \text{ m/s}^2} + \frac{1281 \text{ N}}{9.81 \text{ m/s}^2}\right)v_f$$

$$796.8 \text{ N/s} = (217.7 \text{ N·s}^2/\text{m}) \, v_f$$

and $\boxed{v_f = 3.7 \text{ m/s}}$

The two players move off in the positive direction. Without ever coming to rest, the tackle captures the quarterback and carries him off. This, of course, is why tackles are designed with so much mass. (b) No change in potential energy occurs provided no one falls, but there is a change in KE, since

$$\text{KE}_i = \tfrac{1}{2}m_t v_{ti}^2 = 2.4 \text{ kJ}$$

while $\quad \text{KE}_f = \tfrac{1}{2}(m_q + m_t)v_f^2 = 1.5 \text{ kJ}$

Hence, $\boxed{0.9 \text{ kJ}}$ was transformed, mostly to thermal energy. This problem has one unknown and can be solved using only Conservation of Momentum.

▶ **Quick Check:**

$$0 + \frac{288 \text{ lb}}{32 \text{ ft/s}^2}20 \text{ ft/s} = \left(\frac{192 \text{ lb}}{32 \text{ ft/s}^2} + \frac{288 \text{ lb}}{32 \text{ ft/s}^2}\right)v_f$$

and $v_f = 12$ ft/s. Similarly, $\text{KE}_i = 1.8 \times 10^3$ ft·lb and $\text{KE}_f = 1.1 \times 10^3$ ft·lb; $\Delta\text{KE} = 0.7 \times 10^3$ ft·lb = 0.9 kJ.

For completely inelastic collisions between two bodies, only one of which (m_1) is initially moving, the ratio of the total final KE to the total initial KE is

$$\frac{\text{KE}_f}{\text{KE}_i} = \frac{\tfrac{1}{2}(m_1 + m_2)v_f^2}{\tfrac{1}{2}m_1 v_{1i}^2}$$

Using momentum conservation, $m_1 v_{1i} = (m_1 + m_2)v_f$; hence, $v_f = m_1 v_{1i}/(m_1 + m_2)$, and so squaring and substituting yields

$$\text{KE}_f = \left(\frac{m_1}{m_1 + m_2}\right)\text{KE}_i$$

The loss of energy generally runs a system down—in this case, a bouncing golf ball. With each impact, a certain amount of energy goes into random thermal motion, warming the ball and floor; less is returned as organized KE to the ball, so it rises more slowly and doesn't bounce as high. The closer the images of the ball, the slower it is moving.

which suggests that $\text{KE}_f \approx \text{KE}_i$ when m_1, the moving mass, is much greater than m_2. Then $(m_1 + m_2) \approx m_1$ and the term in parentheses becomes 1. If a sports car locks bumpers with the back of a stopped garbage truck, the two will roll slowly ahead—KE_i will be greater than KE_f, and the difference will go into mangling the participants. On the other hand, if the truck did the crashing with the same KE, the two would roll away at nearly the same speed as the truck came in at, and the collision would be less destructive. A small, high-speed projectile in an inelastic collision with a movable target will do more damage than a massive, slow one with the same KE.

Elastic Collisions

A collision is said to be **elastic** *when KE is conserved.* Particles on the submicroscopic level collide elastically, presumably because they don't suffer any permanent rearrangement of their component parts (if there are any). Thus, suppose a head-on collision takes place between mass 1 and mass 2 on a horizontal plane with m_2 at

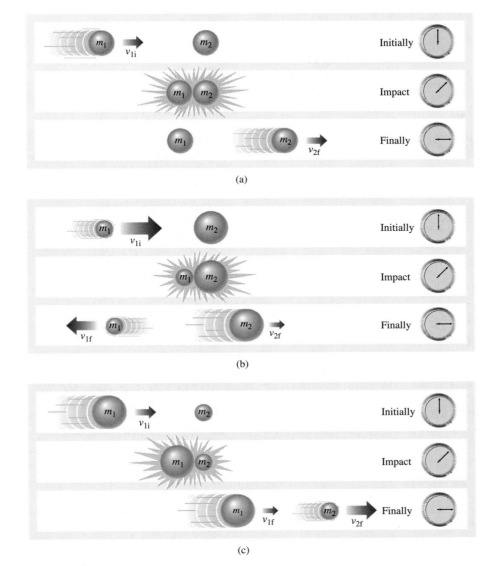

Figure 9.20 Three elastic collisions in which a moving mass m_1 crashes into a stationary mass m_2. In (a) $m_1 = m_2$ and they essentially exchange speeds. In (b) $m_1 < m_2$, and the balls move off in opposite directions. In (c) $m_1 > m_2$, and both balls move off in the direction in which m_1 was originally traveling.

rest, where this time the impact is completely elastic (Fig. 9.20). Conservation of Momentum provides

$$m_1 v_{1i} + 0 = m_1 v_{1f} + m_2 v_{2f}$$

but now the participants move off independently. Since $\Delta PE_G = 0$, Conservation of Energy provides

$$KE_{1i} + KE_{2i} = KE_{1f} + KE_{2f}$$
$$\tfrac{1}{2} m_1 v_{1i}^2 + 0 = \tfrac{1}{2} m_1 v_{1f}^2 + \tfrac{1}{2} m_2 v_{2f}^2$$

Now there can be two unknowns, usually the two final speeds, because there are two independent equations. Accordingly, we can solve for v_{1f} and v_{2f} by first re-arranging these two equations as

$$m_1(v_{1i} - v_{1f}) = m_2 v_{2f} \tag{9.21}$$

and

$$m_1(v_{1i}^2 - v_{1f}^2) = m_2 v_{2f}^2$$

Two identical pucks on an air table. The blue puck has a positive y-component of momentum mv_{By}. The yellow puck has a negative y-component of momentum $-mv_{Yy}$, where $mv_{By} > mv_{Yy}$. After colliding, the blue puck has no y-component and the yellow puck moves off with a positive y-component equal to the initial net y-component ($mv_{By} - mv_{Yy}$). Both the x- and y-components of momentum are conserved independently.

Factoring this last expression yields

$$m_1(v_{1i} + v_{1f})(v_{1i} - v_{1f}) = m_2 v_{2f}^2 \qquad (9.22)$$

Dividing the right side of Eq. (9.22) by the right side of Eq. (9.21) and the left by the left gets rid of the masses, and

$$v_{1i} + v_{1f} = v_{2f} \qquad (9.23)$$

Hence

$$v_{1i} = v_{2f} - v_{1f}$$

The "relative speeds" of the two bodies before and after impact are equal: they initially approach one another at the same speed with which they finally separate. Substituting v_{2f} from Eq. (9.23) into Eq. (9.21) leads to the sought-after expression for the final speed of mass 1, which is

$$v_{1f} = \frac{m_1 - m_2}{m_1 + m_2} v_{1i}$$

Substituting this expression into Eq. (9.23), we get $v_{2f} = [v_{1i}(m_1 - m_2)/(m_1 + m_2)] + v_{1i}$, and the expression for the final speed of mass 2 becomes

$$v_{2f} = \frac{2m_1}{m_1 + m_2} v_{1i}$$

If we take the direction of $\mathbf{v}_{1i}$ as positive, the last two equations describe the resulting motion for the three possible cases.

(1) When $m_1 = m_2$, as with two billiard balls, $v_{1f} = 0$ and $v_{2f} = v_{1i}$. The moving ball slams to a stop and the struck ball slides (no rolling, please) away with the initial speed. The balls, in effect, exchange velocities (Fig. 9.20a). (2) When $m_1 < m_2$, v_{1f} is negative, and the incoming object rebounds in the opposite direction while the initially stationary object moves off in the positive direction, as in Fig. 9.20b. Indeed, when $m_1 \ll m_2$, $v_{1f} \approx -v_{1i}$ and $v_{2f} \approx 0$, which is like bouncing a ball (m_1) off the back of a stationary bus (m_2). (3) Finally, when $m_1 > m_2$, v_{1f} is positive as is v_{2f}; both move off in the initial direction of motion (Fig. 9.20c). When a tennis racket slams into an essentially motionless ball during a serve, it continues to move in the forward direction but at a diminished speed. In the extreme, when $m_1 \gg m_2$, $v_{1f} \approx v_{1i}$, and since m_2 is negligible in the denominator of the last equation for v_{2f}, it follows that $v_{2f} \approx 2v_{1i}$. Hit a Ping Pong ball with a massive paddle and it will fly off with twice the paddle's speed (*and still the relative speeds will be unchanged*).

9.9 Energy Conservation and Symmetry

Linear momentum is a vector quantity and Conservation of Linear Momentum arises from the homogeneity of free space (p. 116). The laws and physical constants of nature are the same everywhere—they are invariant with respect to place. Position is relative in the sense that no experiment performed in an isolated laboratory can produce results that depend on its location in empty space. Thus, there is spatial displacement symmetry. Similarly, angular momentum is a vector quantity and Conservation of Angular Momentum arises from the isotropy of free space. There is no absolute zero angle of orientation, no preferred direction. The laws and constants of nature are the same at every orientation.

Noether showed that Conservation of Energy is a logical consequence of the temporal homogeneity of the Universe (Fig. 9.21). Energy is a scalar quantity; in an isolated laboratory, it is constant in time, independent of the lab's position and orientation in free space. Time is relative in the sense that no experiment performed

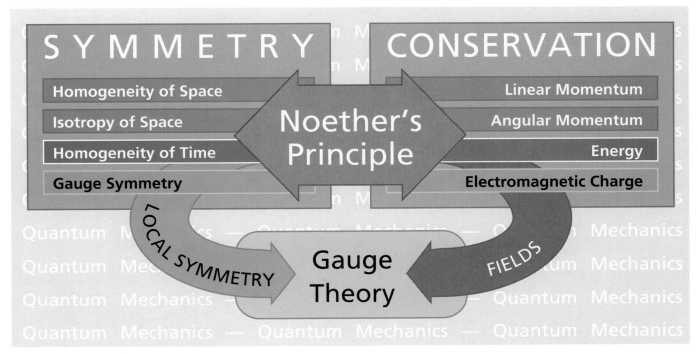

Figure 9.21 Thus far, we have examined the ideas of symmetry, conservation, and Noether's Principle. We have seen that linear and angular momentum are conserved. Now we know that energy is also conserved. Our Universe exhibits temporal displacement symmetry, which requires Conservation of Energy.

in the laboratory can produce results that depend on its location in time. All experiments repeated exactly will behave in precisely the same way today as they did yesterday. If, as we believe is the case, physical law is timeless throughout the Universe, then energy must be conserved.

Though the rigorous proof of this assertion is beyond the level of our treatment, we can make the conclusion plausible. To that end, imagine an isolated lab in which you separate two gravitationally interacting masses, thereby doing work and storing a certain amount of PE_G in that system. Assume, for the sake of argument, that the laws of physics change in time. If you hold the objects motionless and simply wait for gravity to change, a time will come when the PE_G is either larger or smaller than it was at the start. At that moment, you let the objects fall back to where they were originally. The lab is returned to its original configuration, but there is now an energy imbalance—Conservation of Energy is violated.

The nature of the fundamental interactions (Newton's Third Law) ensures that the transfer of energy everywhere, while the laws of physics are constant, will be such as to maintain a fixed total energy. When any two systems interact, both will change; one will lose the energy the other gains. Insofar as our Universe exhibits temporal displacement symmetry, energy is and will be conserved.

Core Material

The measure of the amount of energy transferred into or out of a system via the action of an applied force is known as **work**, where

$$W = Fs \cos \theta \qquad [9.2]$$

The unit of work is the *newton-meter* or *joule*.

A force is **conservative** if *the work done in moving an object from one point to another against the force depends only on the locations of the points and not on the path taken* (p. 296).

Power is *the rate of doing work,* the rate at which energy is transferred. *Average power* is

$$P_{av} = \frac{\Delta W}{\Delta t} \qquad [9.4]$$

The SI unit of power is the **watt**: 1 W = 1 J/s. When an object is in motion at a speed v under the influence of a force

$$P = Fv \cos \theta \qquad [9.6]$$

When work is done on a system to overcome inertia, it goes into **kinetic energy** (KE), where

$$KE = \tfrac{1}{2}mv^2 \qquad [9.9]$$

The energy *stored in a system of interacting objects* is known as **potential energy** (PE). Assuming g is constant

$$PE_G = mgh \qquad [9.13]$$

The change in the potential energy experienced by a body in moving from the surface of a planet out to some distance r from the center is

$$\Delta PE_G = GmM\left(\frac{1}{R} - \frac{1}{r}\right) \qquad [9.14]$$

The Law of *Conservation of Energy* (p. 297) maintains that

The total energy of any system that is isolated from the rest of the Universe remains constant, even though energy may be transformed from one kind to another within that system.

In the absence of friction and any other force but gravity, mechanical energy is conserved; thus

$$\tfrac{1}{2}mv_f^2 + mgh_f = \tfrac{1}{2}mv_i^2 + mgh_i \qquad [9.17]$$

The lowest speed at which we could fire a projectile so that it never falls back to the place from which it was launched is the **escape speed**

$$v_{esc} = \sqrt{\frac{2GM}{R}} \qquad [9.18]$$

For *inelastic* collisions between two bodies, only one of which (m_1) is initially moving

$$KE_f = \left(\frac{m_1}{m_1 + m_2}\right)KE_i$$

When the collision is completely *elastic*, the relative speeds of the two bodies before and after impact are equal, and

$$v_{1f} = \frac{m_1 - m_2}{m_1 + m_2}v_{1i}$$

and

$$v_{2f} = \frac{2m_1}{m_1 + m_2}v_{1i}$$

Suggestions on Problem Solving

1. A common error in *work* problems involving forces that are not parallel to the displacement is to forget to use the component of F, namely ($F \cos \theta$), in computing W. If the force is not parallel to the displacement, check whether or not the perpendicular component ($F \sin \theta$) is affecting the system. For example, ($F \sin \theta$) can increase or decrease the normal force and thus affect the friction and W.

2. When a body moves under the influence of a constant force, the power developed is given by $P = Fv$. If you know the power provided by a rocket motor traveling at a given speed, you can determine its thrust. Or if you know the maximum power of a locomotive and the force exerted on it, you can calculate its maximum speed.

3. Remember to *square* the speed when doing a KE calculation. When considering the change in KE, bear in mind that $v_f^2 - v_i^2$ is *not* equal to $(v_f - v_i)^2$.

4. Momentum is a vector quantity. If the velocity vectors of two bodies lie along a line connecting their centers, a collision problem can be treated via a single scalar momentum expression. This is the case provided that you first assign a sign to the appropriate directions. Then colinear vectors will properly add (or subtract) as scalars. If the initial velocity vectors are not colinear, their separate component momenta in, say, the x- and y-directions will be conserved individually.

5. We are restricted to the two extreme classes of collision problems, and that in itself is helpful to remember. Thus, if the colliding bodies stick together, the collision involves only Conservation of Momentum, and there will be one equation and only one unknown. When the two bodies move off together, there will be one final velocity for both. By contrast, if the collision is elastic, we can also use Conservation of Energy, and there will be two equations and there can be two unknowns. The final velocities of the bodies will be different.

6. Proceed cautiously when confronted with a collision problem where it's not explicitly stated that mechanical energy is conserved. A bullet that squashes against a steel plate or imbeds in a phone book or passes right through a wooden board will "lose" an appreciable amount of energy. Nonetheless, provided no external forces are acting on the bullet-board system, momentum will be conserved. Watch out for situations where any of the objects taking part suffer deformation or generate "heat" via friction.

Discussion Questions

1. A rocket is held in place as its engines are fired up. Is any work done on the rocket? Once released, the rocket rises into the air. What force accelerates it? Considering both the vehicle and the exhaust, talk about the KE of each and the nature of the forces doing work. Where does the energy that ends up as KE come from? How is this different from swimming? Even though the thrust is constant, the longer the engine fires, the faster the rocket goes. What does that say about the power transferred to the craft as a function of time?

2. In the process of rowing, what external force propels the boat? Describe the work done on the boat and on the water. Where does the change in KE of the boat come from? How is this similar to the operation of a propeller on an airplane?

3. A person walks at a constant speed by pushing backward on the ground with a force *F*. If he travels a distance *s* in a straight line on level ground, (a) what force is he overcoming? (b) Suppose his feet slide a little as he pushes back—does the reaction force of the ground do work on him? (c) What if his feet don't slip as he walks; is work done on him by the floor? (d) Where does the energy expended come from?

4. A lump of soft clay of mass *m* falls from a height *h* above the floor, where it crashes and comes to rest. What force brings the clay to a stop? How much work is done by the floor on the clay? What is the net amount of work done on the clay during the fall? Where does that energy end up?

5. A youngster on a bicycle starts at rest and pedals a straight path on level ground at a constant speed, coming to a stop some distance away. (a) Is it reasonable to assume that he is the energy source for the journey? (b) What *external* force propels the bicycle and rider? (c) Does the ground do positive work on the bike?

6. Work is done against resistance. In each of the following cases, describe the forces overcome: (a) hammering a nail into a piece of wood; (b) stretching a piece of rubber; (c) cutting a loaf of bread; (d) putting a book on a high shelf; (e) opening the door on a modern refrigerator; (f) pulling a piece of masking tape off a roll.

7. Two identical balls travel along frictionless grooves in a flat surface. One groove is straight and runs horizontally along the surface, while the other dips down below the surface and up again smoothly in a vertical plane. The two runways begin and end parallel and adjacent to each other. Assume that neither ball leaves its track and that both are given the same initial speed. Describe their motions in a race where the starting and finish lines are across horizontal portions of the grooves.

8. A rock at the bottom of a well has a negative amount of *gravitational*-PE with respect to the ground level. What does that mean? What will happen if it's launched upward with an equal positive amount of KE? Can a body have a negative KE? Discuss the significance of the situation in which a body has a positive KE and a negative PE, where $|PE| > KE$. What happens when $|PE| < KE$?

9. The late physicist and Nobel laureate Richard P. Feynman once wrote, "It is important to realize that in physics today, we have no knowledge of what energy *is*." What do you think he meant by that? The usual definition of energy is *the ability to do work*. Discuss the circular nature of that definition. Consider the alternative definition, energy is *the ability to impart vis viva* (mv^2); is this definition any better?

10. Potential energy belongs to the entire system of interacting objects and not to any one of them. Explain this statement and discuss how it applies to the *gravitational*-PE of a frog sitting on a rock on the planet Earth.

11. A ball on a string whirls around in a circle at a constant speed. Is there a net force on the ball? Does it do work? Is there a corresponding change in either PE or KE? How can the ball be made to accelerate tangentially? Will there then be work done on it and a change in either PE or KE?

12. The number One billard ball moving with a speed v_{1i} strikes the Two ball, which is at rest, hitting it off-center and sailing away in an elastic collision. With respect to the initial line of motion, the two balls move off at angles θ and ϕ, respectively. Draw a diagram showing that $\mathbf{p}_{1i}$, $\mathbf{p}_{1f}$, and $\mathbf{p}_{2f}$ form a closed triangle. Explain your reasoning.

13. The cars of a roller coaster are usually tugged up to the top of the first hill by a chain drive and then released to coast the rest of the way around. Consider that friction is acting all along the journey. (a) Where is the *gravitational*-PE maximum? (b) How should the ride be configured to produce the greatest value of the maximum speed? (c) Having once attained its maximum *gravitational*-PE, can it reach that value again? (d) Having once attained its maximum KE (which may well not be the maximum possible KE for a structure of that height), can it reach that value again? (e) Must each successive hill be lower than the previous one? (f) Must the second hill be lower than the first?

14. It is possible to hold a large rock or a massive lump of soft clay and slam it with a hammer and yet not hurt the hand holding the object. Explain what is happening in terms of energy and momentum.

15. A glass marble is dropped from some height onto wet sand and penetrates to a depth *d*. The experiment is repeated, but this time from twice the original height. How deep will the sphere penetrate and why? What assumptions are made?

16. You wish to hit a nail into a board with a hammer. It is your intention to slam the nail so the hammer remains in contact with the nail without rebounding. Would you do better with a light or a heavy hammer? Explain.

17. A crate of mass *m* rests on a flatbed truck. The truck accelerates from rest to a speed *v*, during which time the crate does not slide on the bed. How much work is done on the crate by friction? Is that work positive or negative? What is the work done against? In what final form is the energy that was provided to the crate? Now suppose the truck comes to a gradual stop. Given this situation, answer the same questions again.

18. During the first Moon-landing flight (*Apollo 11*, July 1969), the spacecraft was put into a circular parking orbit at an altitude of 119 mi above the Earth. The engines were later fired, sending the ship off to the Moon at 24 245 mi/h. Compare this speed to the escape speed and explain the difference. When we calculated the formula for the escape speed for a planet, we neglected the fact that our particular dirtball is spinning. Discuss the effect, if any, of the spin if launch is from some parking orbit about the Earth. What if lift-off is from the surface?

Multiple Choice Questions

1. A 10-kg mass is held 1.0 m above a table for 25 s. How much work is done during that period? (a) none (b) 10 J (c) 250 J (d) 9.8 J (e) none of these.

2. The net work done on an object moving along a closed path in a force field is zero when it returns to the origin. The force is (a) conservative (b) nonconservative (c) impossible (d) liberal (e) none of these.

3. A typical adult male's heart pumps about 160 milliliters of blood per beat. It beats around 70 times per minute and does roughly 1 J of work per beat. How much work does it do in a day? (a) 10^5 J (b) 10^6 J (c) 70 J (d) 70×10^5 J (e) none of these.

4. Which of the following is not a measure of the same quantity as the others? (a) newton-meter per second (b) kilogram-meter^2 per second^3 (c) joule per second (d) watt (e) none of these.

5. If a 20-kW engine can raise a load 50 m in 10 s, how long will it take for it to raise that same load 100 m? (a) 20 s (b) 40 s (c) 5.0 s (d) not enough information (e) none of these.

6. If a 25-hp motor can raise an elevator 10 floors in 20 s, how long will it take a 50-hp motor to do the same? (a) 40 s (b) 10 s (c) 20 s (d) 5.0 s (e) not enough information.

7. Which of the following is not a measure of the same quantity as the others? (a) foot-pound (b) newton-meter (c) watt (d) joule (e) none of these.

8. If this book is placed on an ordinary table and slid along a path that brings it back to where it started, (a) no net power will have been required (b) work will certainly have been done (c) assuming a conservative gravitational field, no net work will be done (d) not enough information is given to say anything about the work done (e) none of these.

9. A fairly small asteroid (1000 kg) out in deep space is to be accelerated from rest up to 10 m/s. Inasmuch as it is weightless, will work have to be done on it during the acceleration and, if so, how much? (a) no (b) yes, 10 000 J (c) yes, 50×10^3 J (d) yes, 10 000 N (e) yes, 50×10^3 N

10. Work is done on an object far out in space where it has negligible *gravitational*-PE. If in the process there is no net change in its KE, we can conclude (a) that friction *may* have been operative (b) that this situation is impossible (c) that the energy of the object has decreased (d) that the object's speed decreased (e) none of these.

11. A rocket coasting along in space at some speed v fires its engines thereupon doubling its speed, but at the same time it jettisons some cargo, reducing its mass to half its previous value. In the process, its KE is (a) doubled (b) tripled (c) quadrupled (d) unchanged (e) none of these.

12. A kid in a wagon rolls from rest down a hill reaching the bottom at 12 m/s. On the next run, she gets a push and starts down at 5.0 m/s. At what speed does she now arrive at the bottom? (a) 12 m/s (b) 17 m/s (c) 7 m/s (d) 13 m/s (e) none of these.

13. An arrow is fired, via a bow, straight up. It rises for a while and then drops back to the ground. The process, taking the arrow from loading to just prior to touch down, can best be described by a series of energy transformations corresponding to (a) work, *elastic*-PE, KE, *gravitational*-PE, KE (b) work, KE, *elastic*-PE, KE (c) KE, *gravitational*-PE, work (d) *elastic*-PE, *gravitational*-PE, KE (e) none of these.

14. How fast must a 1.0-kg mass be moving if its kinetic energy is 1.0 J? (a) 1.0 m/s (b) 9.8 m/s (c) 1.414 m/s (d) 10 m/s (e) 0.102 m/s.

15. An open wagon is rolling along frictionlessly on a level road when it begins to snow. After a load of snow has fallen vertically into the wagon, its speed, kinetic energy, and momentum will each, respectively (a) increase, increase, stay the same (b) decrease, stay the same, decrease (c) decrease, decrease, stay the same (d) stay the same, increase, increase (e) decrease, stay the same, increase.

16. How high above the Earth's surface must a 1.0-kg mass be for it to have a *gravitational*-PE of 1.0 J with respect to that surface? (a) 9.8 m (b) 1.0 m (c) 0.10 m (d) 0.01 m (e) 32 m.

17. While a ball rolls down the circularly curved track shown in Fig. MC17, its speed, acceleration, and kinetic energy, respectively (a) increase, increase, increase (b) decrease, decrease, decrease (c) increase, decrease, decrease (d) decrease, increase, increase (e) none of these.

Figure MC17

18. A 458 Winchester magnum cartridge has a 500-grain (1 gr = 0.064 8 g) bullet that attains a muzzle speed of 644.9 m/s (2120 ft/s). After traveling 456 m (500 yd), it would still be moving at 365.6 m/s (1202 ft/s). What fraction of its original KE does it have at that distance? (a) 23% (b) 50% (c) 32% (d) 68% (e) 77%.

19. Taking infinity as the zero of *gravitational*-PE, through what height at the Earth's surface must a 1.0-N body be raised if its PE_G is to be increased by 1.0 J? (a) ∞ (b) 10 m (c) 1.0 m (d) −1.0 m (e) none of these.

20. If Superman really is "more powerful than a locomotive," then with regard to freight trains (a) he can pull more cars at the same speed (b) he can pull the same number of cars faster (c) he can pull the same number of cars up steeper hills (d) all of these (e) none of these.

21. A 9.5-g wad of chewing gum traveling at 1.0 m/s horizontally crashes into and sticks to a 9.5-g puck floating frictionlessly on an air table. They stick together and sail away at (a) 1.0 m/s (b) 2.0 m/s (c) 0.0 m/s (d) 0.50 m/s (e) none of these.

Problems

THE TRANSFER OF ENERGY

1. [I] A book is slowly slid at a constant speed 1.5 m across a level table by a 15-N horizontal force. How much work is done on the book and what is it done against?

2. [I] A 1.0-kg mass is raised 10 m into the air by a 10-N force exerted vertically. Determine the amount of work done by the force. Now suppose a 10-N force pulling horizontally moves a 1.0-kg mass through 10 m over a frictionless floor. How much work is done here? What is the work done against in both instances?

3. [I] A delicatessen owner holds a knife horizontally above a 10-cm-thick cheese. With one smooth motion, exerting a downward force of 20 N, the cheese is cut in two. How much work was done by the owner?

4. [I] A 2224-N piano is rolled on little wheels 3.1 m across a horizontal floor by a 890-N piano mover. If, while lifting upward on one end with 111 N, he pushes horizontally with 445 N, how much work will he do?

5. [I] When a solid-fuel rocket burns, the flame front advances into the material in a direction perpendicular to the ignited surface. By configuring the fuel core in various ways and lighting it along the entire length, one can obtain all sorts of performance characteristics. Figure P5 shows a starred-cavity motor. How much work was done by the device in the first 90 s of flight if the rocket rose vertically to a height of 2.0 km in that time?

Figure P5

6. [I] A newspaper delivery boy pulls a rubber-wheeled cart that has a coefficient of rolling friction of 0.02 and a mass of 25 kg. It travels 10 km on level streets. How much work does he do in overcoming friction between the cart and the ground?

7. [I] A nurse pushes someone in a wheelchair 100 m doing 400 J of work in the process. What average force did the nurse exert in the direction of motion?

8. [I] Prove that 1 hp = 746 W.

9. [I] A runner traverses a 50-m-long stretch on a horizontal track in 10 s at a fairly constant speed. All the while she experiences a retarding force of 1.0 N due to air friction. What power was developed in overcoming that friction?

10. [I] A small hoist can raise 100 kg of bricks to the top of a construction project 30 m above the street in half a minute. Determine the power provided.

11. [I] The per capita power consumption in the United States is around 10 kJ each and every second. At what speed would you have to push a car exerting a force of 1.0 kN on it all year, day in and day out, to be equivalent to your share?

12. [I] A rigid steel crowbar is rested on an upright brick standing near the back of a car. The end of the bar touches the bottom of the bumper 30 cm from its pivot point on the brick. Someone pushes straight down on the other end of the bar, 270 cm from the pivot point, and raises the car's chassis 5.0 cm. Given ideally that the work-in equals the work-out, if it takes 3200 N to raise the car, how much force did the person exert on the bar?

13. [I] Assuming no friction and weightless pulleys, how much rope will have to be drawn off if the weight in Fig. P13 is to be raised 1.0 m? Remember: ideally, work-in equals work-out.

Figure P13

14. [II] A pickup truck is hauling a barge along a canal at a constant speed. The truck, driving parallel to the waterway, is attached to the barge by a cable tied to the bow making a 30° angle with the forward direction. If the truck exerts a force of 1000 N on the cable, how much work is done in overcoming friction as the barge is moved 10 km?

15. [II] A 100-N box filled with books is slid along a 13-m-long ramp up to a platform 5.0 m above the ground. How much work is done if friction is negligible? How much work would have been done if the box were lifted straight up to the platform?

16. [II] A 50-kg keg of beer slides upright down a 3.0-m-long plank leading from the back of a truck 1.5 m high to the ground. Determine the amount of work done on the keg by gravity.

17. [II] The newspaper boy in Problem 6 pushes his cart 25 m along a road inclined at 10°. How much work does he do to overcome road friction?

18. [II] Five fat dictionaries, each 10 cm thick and each 2.5 kg, are resting side by side flat on a table 1.0 m high. How much work would it take to stack them one atop the other? How high would that raise the center-of-mass of all the books?

19. [II] A constant force of 100 N is applied to an object over a distance of 2.0 m parallel to the displacement. Draw a work diagram (*F* vs. *s*) with a scale of 1.0 cm = 10 N and 1.0 cm = 0.25 m. (a) What was the total work done? (b) How much work does an area of 1.0 cm² equal? (c) What is the area under the curve?

20. [II] The Zambesi River in Africa rushes over Victoria Falls at a rate of 25×10^6 gallons per minute. The falls are 108 meters high, 1 gallon equals 0.133 7 cubic foot, and each cubic foot weighs 62.4 pounds. Determine the power developed by the water in SI units.

21. [II] A 2.5-hp motor drives a hoist that can raise a load of 50 kg to a height of 20 m. At full power, how long will the hoist take to do it?

22. [II] The belt connecting an auto engine to an air conditioner is moving at 40 m/min and has an effective tension of 20 N. How much work does it do in an hour? How much power does it transmit?

23. [II] An animal's power level is known as its **metabolic rate**. If a man typically uses 10 MJ in the course of a day, what is his metabolic rate? While resting but awake, an average person has a *basal* power level of about 90 W. This figure, however, depends on the size of the person, being larger for larger people, who have greater surface areas and thus lose more energy. And so to standardize the notion, making it independent of body size, it is customary to divide by the skin surface area (typically 1.5 m²)—the number thus arrived at is the *basal metabolic rate* or BMR. What is the BMR for our average person?

24. [II] The oxygen taken in by the body reacts with fats, carbohydrates, and protein, liberating energy *internally* at a rate of about 2.0×10^4 J per liter. If a 70-kg man requires 77 W of power even while sleeping, what is his rate of oxygen consumption?

25. [II] Refer to Problems 23 and 24. What is the BMR of a person who (according to the standard charts) has a surface area of 1.8 m² and is consuming oxygen at a rate of 0.40 liters per minute?

26. [II] Efficiency is defined as work-out/work-in. Muscles operate with an efficiency of about 20% in converting energy internally into work externally. Accordingly, how much energy will be expended by a 60-kg person in the process of ascending several flights of stairs to a height of 25 m?

27. [II] A variable force, depicted in Fig. P27, acts on a 10-kg body. How much work does it do in moving the body from $x = 2$ to $x = 6$? What is the total amount of work done in going from $x = 0$ to $x = 10$? Over which 1-m interval was the smallest amount of work done? How much work is done in going from $x = 2$ to $x = 12$?

MECHANICAL ENERGY

28. [I] A major concern in the design of spacecraft is the presence in space of tiny, high-speed meteoroids. Micrometeoroids, as they're called, have been detected traveling as fast as 70 km/s. Compute the kinetic energy of a 1.0-g mass moving at that rate.

Figure P27

29. [I] It has been suggested that a controlled fusion reaction (a mini-H-bomb) could be achieved by causing a small (on the order of 1 mm), extremely high-speed projectile to hit a stationary target, both made of the appropriate materials. Determine the kinetic energy of a 0.5-g mass traveling at 200 km/s. If the collision takes 10 ns, how much power would be provided to the target?

30. [I] One ton of uranium-235 can provide about 7.4×10^{16} J of nuclear energy. If that much energy went into accelerating a 3.5×10^6-kg spaceship (that's the size of a fully loaded *Saturn V* Moon rocket) from rest, what would its final speed be?

31. [I] A 6.5-g bullet is fired from a 2.0-kg rifle with a speed of 300 m/s. What is the recoil speed of the gun? Compare the kinetic energies of gun and bullet.

32. [I] What is the kinetic and potential energy of a Boeing *747* airliner weighing 2.22×10^6 N, flying at 600 mi/h at an altitude of 20×10^3 ft? Work the problem in SI, please.

33. [I] A single barrel of oil contains the equivalent *chemical*-PE of about 6×10^9 J. How high in the air could that much energy raise a million kilogram load, assuming it is all converted to *gravitational*-PE?

34. [I] A uniform rod 6.0 m long, weighing 60 N, is pivoted about a horizontal axis 1.0 m up from its *c.m.* so that it hangs vertically. Neglecting friction, (a) what's the net work done on the rod by gravity in the process of raising its end, making it horizontal? (See if you can come up with two ways to do this.) (b) What change in potential energy, if any, has the rod experienced?

35. [I] A car with a mass of 1000 kg is at rest at the base of a hill. It accelerates up the incline reaching a speed of 20 m/s at a height of 100 m. What is the total increase in mechanical energy at that point?

36. [I] A 17.8-kN car is to be accelerated from 0 to 96.5 km/h (60 mi/h) in 10 s. Neglecting all frictional losses, how much power must be supplied in the process (independent of the exact nature of the acceleration)? Incidentally, air friction increases rapidly with speed, becoming a major concern above 64 km/h (40 mi/h).

37. [II] (a) What is the KE of a 17.8-kN car traveling at 64 km/h? (b) Assuming the car gets 20 miles to the gallon at that speed, and each gallon of gasoline provides

1.3×10^8 J of chemical energy, what's its power consumption? Incidentally, of this total, only about 20% gets into the mechanical system (pumps, transmission, etc.). About 10^4 J/s goes into propelling the car, of which half is dissipated in overcoming tire and road friction and half in air friction. Older cars typically lose about 10^3 J/s in unburned fuel that evaporates from the carborator and never even gets into the engine.

38. [II] A 60-kg person is slowly doing chin-ups, during each one of which her *c.g.* rises just about $\frac{1}{2}$ m. If the biceps contract roughly 4.0 cm in the process of each lift, how much tension is there, on average, in the muscles of each arm?

39. [II] A 2×10^3-kg spacecraft has an ion engine that delivers 150 kW of propulsive power via its six thrusters. Assuming it to be far from any massive object, how much time would it take to accelerate from "rest" to 268 m/s (600 mi/h)?

40. [II] A 1000-kg car at rest at the top of a hill accelerates down the road, reaching a speed of 20 m/s after descending a height of 100 m. What was its total change in mechanical energy as of that moment?

41. [II] A 100-N weight sitting on the floor is tied to a light rope that passes over a very light frictionless pulley 20 m above it. The other end of the rope hangs down to the floor where it is being held by a 10-N monkey named George. Suppose George now climbs 10 m above the floor. (a) How much work does he do? (b) How much rope ends up on the floor? (c) What is the total change in *gravitational*-PE of the system?

42. [II] During a vertical jump, a person crouches down to lower his or her *c.m.* and then leaps straight up. The leg muscles essentially do all the work, accelerating the body over a push-off distance of about $\frac{1}{3}$ m. Typically, people can support an additional load using the legs equal to their own weight, but only with considerable effort. Let's suppose then that our person can push off with a muscle-force equal to 1.5 times body-weight. Neglecting losses due to friction, how high can our friend jump; that is, how much above the approximately 1.0-m standing position can we expect the *c.m.* to rise?

43. [II] (a) Determine the value of the gravitational acceleration at the surface of the Moon. (b) What is the change in the *gravitational*-PE of a 1000-kg spacecraft when it has risen 100 m above the surface? Take the radius of the Moon to be $\frac{1}{4}$ that of the Earth and the mass to be about 100 times less.

44. [II] An object explodes into two pieces as, for example, when an atom emits an α-particle. Show that the ratio of their energies equals the inverse ratio of their masses—the smaller particle carries off the larger share of KE.

45. [III] A locomotive exerts a maximum force of 22.7×10^4 N while pulling a freight train up a $\frac{1}{2}$% grade (that is, the rise in the roadbed is $\frac{1}{2}$ m per 100 m). The train is 45 cars long, and each one loaded weighs 4.0×10^5 N. The total friction varies with the load and speed, but in this case, 35 N per 10^4 N of weight is a reasonable number. Using energy considerations, how far will the train travel while accelerating from a speed of 6.7 m/s to twice that?

46. [III] A runner accelerates to a top speed of 9.9 m/s taking a number of strides (N). If the average horizontal force exerted on the ground is 1.5 times body-weight, and if it acts during each stride over a length of $\frac{1}{3}$ m, determine the value of N.

47. [III] The second half of a smooth horizontal track is bent upward to form an incline at 20°. A solid homogeneous sphere 20 cm in diameter with a mass of 2.0 kg is rolled along the track so that it reaches the sloped portion at a speed of 5.0 m/s. Part of the sphere's energy of motion is associated with its rotation about its *c.m.*, and that quantity is given by the expression $\frac{1}{2}I\omega^2$. In addition, there is the usual KE $= \frac{1}{2}mv^2$ term. Neglecting friction, how far along the incline will the sphere progress before momentarily coming to rest?

CONSERVATION OF MECHANICAL ENERGY

48. [I] A 60-kg stuntperson runs off a cliff at 5.0 m/s and lands safely in the river 10.0 m below. What was the splashdown speed?

49. [I] Two automobiles of weight 7.12 kN and 14.24 kN are traveling along horizontally at 96 km/h when they both run out of gas. Luckily, there is a town in a valley not far off, but it's just beyond a 33.5-m-high hill. Assuming that friction can be neglected, which of the cars will make it to town?

50. [I] A ball having a mass of 0.50 kg is thrown straight up at a speed of 25.0 m/s. (a) How high will it go if there is no friction? (b) If it rises 22 m, what was the average force due to air friction?

51. [I] An athlete whose mass is 55.0 kg steps off a 10.0-m-high platform and drops onto a trampoline, which, while stretching, brings her to a stop 1.00 m above the ground. Assuming no losses, how much energy must have momentarily been stored in the trampoline as she came to rest? How high will she rise?

52. [I] While traveling along at 96 km/h, a 14.2-kN auto runs out of gas 16 km from a service station. Neglecting friction, if the station is on a level 15.2 m above the elevation where the car stalled, how fast will the car be going when it rolls into the station, if in fact it gets there?

53. [I] A 90-kg signal relay floating in space is struck by a 1000-g meteoroid. The latter imbeds itself in the craft and the two sail away at 5.0 m/s. What was the initial speed of the meteoroid?

54. [I] What is the potential energy of a 1.0-kg mass sitting on the surface of the Earth if we take the zero of PE_G at infinity?

55. [II] Referring to Problem 54, compute the *gravitational*-PE of a 1.0-kg object at distances of 1, 2, 3, 4, 5, and 10 Earth-radii. Draw a plot of PE_G against r measured from the center of the planet in units of Earth-radii. Now suppose the object is fired upward with a KE of 5.2×10^7 J. What is its initial total mechanical energy? Draw a horizontal line on your diagram representing E. Where does it cross your curve and what is the significance of that point?

56. [II] A 1000-kg car racing up a mountain road runs out of gas at a height of 35 m while traveling at 22 m/s. Cleverly, the driver shifts into neutral and coasts onward. Neglecting all friction losses, will he clear the 65-m peak? Would it help to throw out any extra weight or even jump out and run alongside the car? Not having any brakes, at what speed will he reach the bottom of the mountain?

57. [II] If the mass of the Moon is 7.4×10^{22} kg and its radius is 1.74×10^6 m, compute the speed with which an object

would have to be fired in order to sail away from it, completely overcoming the Moon's gravity pull.

58. [II] A pendulum consists of a spherical mass attached to a rope so that its *c.g.* hangs down a distance L beneath the suspension point. The bob is displaced so that the taut string makes an angle of θ_i with the vertical, whereupon it is let loose and swings downward. Show that its maximum speed is given by

$$v_{max}^2 = 2gL(1 - \cos \theta_i)$$

59. [II] It's been suggested that we mine either the Moon or some of the asteroids in order to get raw material from which to build space stations. The idea is that removing material from the Moon "would consume only five percent of the energy needed to lift the same payload off Earth." Show that this conclusion is roughly true.

60. [II] At what speed should a space probe be fired from the Earth if it is required to still be traveling at a speed of 5.00 km/s, even after coasting to an exceedingly great distance from the planet (a distance that is essentially infinite)?

61. [II] A satellite is in a circular orbit about the Earth moving at a speed of 1500 m/s. It is desired that by firing its rocket, the craft attain a speed that will allow it to escape the planet. What must that speed be?

62. [II] In an arrangement for measuring the muzzle velocity of a rifle or pistol, the bullet is fired up at a wooden mass, into which it imbeds. The wood is blasted straight up into the air to a measured height h. Assuming negligible losses to friction, write an expression for the velocity in terms of the known masses and height. If a 100-grain (6.48-g) 25-06 Remington rifle bullet is fired into a 5.00-kg block that then rises 4.0 cm into the air, what was the muzzle speed of that bullet?

63. [II] An 8.0-kg puck floating on an air table is traveling east at 15 cm/s. Coming the other way at 25 cm/s is a 2.0-kg puck on which is affixed a wad of bubblegum. The two slam head-on into each other and stick together. Find their velocity after the impact.

64. [II] As seen from the window of a space station, a 100-kg satellite sailing along at 10.0 m/s collides head-on with a small 300-kg asteroid, which was initially at rest. Taking the collision to be elastic, and neglecting their mutual gravitational interaction, what are the final velocities of the two bodies? What were their "relative speeds" before and after the collision?

65. [II] Two billiard balls, one heading north at 15.0 m/s and one heading south at 10 m/s, collide head-on. Take the collision to be perfectly elastic. What is the post-impact speed of each ball? (Hint: It will help if you remember that for completely elastic collisions, the "relative speeds" before and after are equal.)

66. [II] A skater with a mass of 75 kg is traveling east at 5.0 m/s when he collides with another skater of mass 45 kg heading 60° south of west at 15 m/s. If they stay tangled together, what is their final velocity?

67. [III] A light rope is passed over a weightless, frictionless pulley and masses m_1 and m_2 are attached to its ends. The arrangement is called Atwood's Machine. The starting configuration corresponds to both masses held at rest at the same height. The two are then released. During some arbitrary interval of time, the heavier one falls a distance y, while the lighter one rises a distance y. Derive an

expression for the speed of either body in terms of g, y, and the masses.

68. [III] The coefficient of restitution of two colliding bodies is defined as the ratio of their "relative speeds" (after the impact to before the impact). Imagine that we drop a sphere made of some material of interest from an initial height onto a test anvil and measure the final height to which the ball bounces. Derive an expression for the coefficient of restitution in terms of these two heights. If the coefficient for glass on steel is 0.96, to what height will a glass marble bounce off a steel plate when dropped from 1.0 m?

69. [III] A soft clay block is suspended so as to form a so-called ballistic pendulum, as shown in Fig. P69. A bullet is fired point-blank into the block, imbedding itself therein and raising the latter to a height h. Write an expression for the muzzle speed of the bullet in terms of the masses g and h. Remember that although friction on the pendulum is negligible, clay-bullet friction losses are not, and *mechanical energy is not conserved* in the collision.

Figure P69

70. [III] Referring to Problem 69, derive an expression for the percentage of the kinetic energy converted into internal energy during the bullet-clay impact.

71. [III] Two identical hard spheres, one at rest and the other moving, slam into one another in a *non-head-on* elastic collision, as shown in Fig. P71. Prove that they will always fly apart at 90° to each other.

Figure P71

10

Solids

ONE OF OUR CONTINUING concerns in studying the physical Universe is *matter:* What is it? How does it behave, and why? This chapter is the beginning of an exploration into the mechanical, thermal, electrical, magnetic, and optical properties of matter. It develops a sketch of the atomic landscape, introducing ideas that will be elaborated throughout the remainder of the text.

The behavior of materials is of great practical interest. After all, few of us go through life without ever breaking a bone, chipping a tooth, or tearing a tendon. And although collapsing buildings, falling bridges, and toppling tow-

The fuselage of this Aloha Airlines jet ripped off during a flight in 1988. Experts attribute the failure to metal fatigue.

It is probably as meaningless to discuss how much room an electron takes up as to discuss how much room a fear, an anxiety, or an uncertainty takes up.

SIR JAMES JEANS (1877–1946)

ers are a rarity, these things happen too, especially with some provocation like an earthquake. Boats still snap in half on the high seas and airplanes occasionally come apart in awkward ways. Our world is filled with material structures (trees, cars, people, skyscrapers, heart valves, and even the Earth itself), all of which we rely on to function in predictable ways. To understand why falling off a ladder might shatter a leg bone, we have to know a little about the materials. This chapter is a practical one, dealing as it does with the mechanical properties of solids: how they stretch and compress, fatigue, break, and shear. It examines an idea—Hooke's Law—that will have important ramifications throughout our study. That law describes elastic systems, and the atom itself is the most fundamental of all such systems. Atoms vibrate much as springs vibrate, and so as we study the common spring, we are laying the groundwork for far grander things to come.

ATOMS AND MATTER

It might seem easy to hold a thing in hand and know what matter is—it's stuff, substance, the material aspect of the Universe. But language fails us if we try to go beyond that intuitive apprehension. Matter cannot satisfactorily be defined, certainly not in terms of more fundamental concepts.

The observable Universe is composed of material particles and electromagnetic radiation, but these are the same, as much as they are different. Radiant energy may be thought of as "the most refined form of matter." A century ago, one could have responded that matter is that which possesses mass. But we now know that there are things called *neutrinos* (Sect. 33.5)—tiny, chargeless particles that carry momentum and energy, but which are nonetheless apparently massless.

The old maxim that "matter is that which is impenetrable and occupies space" is also not very useful. A steel cannonball may appear solid enough, but it's mostly empty space. Bulk matter is composed of separated atoms and does not fill space in any ordinary sense. The particles that make up the atoms, that make up the cannonball, and everything else, are fluttery, elusive things that recoil from localization quite unlike the imagery of tiny billiard balls. The idea of atoms as ultimate impenetrable specks of solid matter is an ancient mirage.

Roughly 110 basic materials are known, and they are called **elements** (oxygen, iron, hydrogen, and so on). *These cannot be further simplified or broken down into another substance by any chemical or physical process* short of nuclear meddling. A sample of a given element is composed of identical submicroscopic entities called **atoms**, each of which is the smallest representative sample of the element. Chemical **compounds** are formed when elements combine. The basic unit of a compound, the smallest piece that behaves chemically like the whole, is a **molecule**, itself made up of atoms.

An adult human being is a collection of over 10^{27} atoms bound together as $\approx 10^{14}$ cells. We are a constantly changing aggregate of atoms, themselves created thousands of millions of years ago. The living cells are recent enough, though almost without exception the atoms that form us are at least as old as the Solar System and often much older than that. They have circulated around the Earth for the last four-and-a-half thousand-million years, through air and water, through fish and fowl and trees and burgers and dung, and back to the soil and then, for the moment,

This regular pattern of bumps is formed by iodine atoms absorbed on a platinum surface.

to you. Formed in the thermonuclear fires of stars long gone, the atoms you ate for breakfast extend your lineage back in time to the dawn of creation. We *are* stardust, borrowers in the ancient ritual of "ashes to ashes."

10.1 Atomism

The concept of the *atom* is an old idea that dates back to the Greek Leucippus (ca. 450 B.C.). If a substance, a piece of gold, say, is divided over and over again into ever smaller portions, could the process go on forever, or would some last indivisible fragment—the *atomos (that which cannot be further cut)*—be reached? The most influential early opponent of atomism was Aristotle, and not until his hold on Western thought started to collapse in the seventeenth century was atomism again taken seriously. Robert Boyle embraced the concept of the atom as an explanation of the "springiness of air," (in a balloon, for example). In 1665, Robert Hooke proposed that certain crystalline structures resulted from the close packing of tiny spherical particles—atoms stacked like cannonballs. Still others were convinced of the reality of atoms by the nature of the mixing process. That salt dissolves into water suggests that the liquid, instead of being continuous, is rife with empty spaces. A cup of alcohol and a cup of water will combine to form less than two cups of "good cheer"—the atoms seem to be sliding in between one another, filling holes. A droplet of ink released into a beaker of water will spread out uniformly, just as the smell of perfume can fill a room in minutes, suggesting a mingling of colliding particles.

Despite the long list of believers—Galileo, Bacon, Boyle, Hooke, Newton, and Leibniz, to name a few—the quantitative evidence in favor of the atom was far from impressive. The breakthrough came in the early 1800s at the hands of an unknown country schoolmaster, a gruff, awkward man named John Dalton. It had already been shown that *elements combine to form compounds in certain definite proportions.* One gram of hydrogen burned with eight grams of oxygen combine to produce nine grams of water (H_2O). Dalton recognized that these definite proportions were just the observed manifestations of an invisible atomic structure. He argued that if water is formed of equal numbers of hydrogen and oxygen atoms (and here he was wrong) in a mass ratio of 1 to 8, then each O atom must be 8 times as massive as an H atom. Suspecting, and rightly so, that hydrogen was the lightest of the atoms, he set its mass equal to 1. The mass of an oxygen atom became 8, and so on down the list of elements involved in the known chemical reactions.

The table of *relative atomic masses* that Dalton drew up had many errors, but they were soon corrected. Water was electrically broken down and its true formula established: one oxygen atom has 8 times the mass of *two* hydrogens—that is, 8×2, or 16. Nowadays, we take carbon (^{12}C) as the standard instead of hydrogen, set its mass equal to exactly 12.000 00 *unified atomic mass units* (u), and measure all the other atoms relative to it (Table 32.1). Thus, a hydrogen atom has a mass of 1.007 825 u.

10.2 Avogadro's Number

In 1811, Amedeo Avogadro, Count of Quaregna, set forth two brilliant proposals (coining the word *molecule* in the process): (1) *The gaseous elements can exist in molecular form.* (2) *Equal volumes of molecular gases (under the same conditions of temperature and pressure) contain the same number of molecules.*

Molecular hydrogen (H_2) has a relative mass of 2. Suppose we admit 2 g of it into an expandable chamber that can be maintained at **standard temperature** (0°C)

A few droplets of ink in a beaker of water. The warmer the water, the faster the ink diffuses.

A gram-mole of mercury, water, glass, salt, and iron. The glass marble can be used for establishing the scale.

and pressure (1 atm)—STP. We would find that the gas occupied a volume of 22.4 liters (that is, 22.4×10^3 cm^3). According to Avogadro, 22.4 liters of oxygen, or of any gas, would contain the same number of molecules. Since each O$_2$ molecule has a mass of 2×16 u, or 32 u, we expect any volume of oxygen to be 16 times more massive than the same volume of hydrogen. Consequently, 22.4 liters of oxygen should have a mass of 32 g. Indeed, *22.4 liters of any gas at STP has a mass in grams numerically equal to its molecular mass.* This highly useful measure is called a *gram-molecular mass,* and it's often spoken of as a **mole.** For instance, one molecule of carbon dioxide (CO$_2$) has a relative mass of 12 u + 2 × 16 u or 44 u; therefore, 1 mole of it has a mass of 44 grams and, as a gas at STP, occupies 22.4 liters.

Similarly, *a **kilomole** is the mass of any substance in kilograms that is numerically equal to the molecular mass of that substance.* Thus, a kilomole (abbreviated kmol) of any gas at STP occupies 22.4 m^3. Many elements exist as collections of separate atoms rather than molecules, and the word *kilomole* then stands for a *kilogram-atomic mass.*

We could start out with 22.4 m^3 of a gas and cool it down to a liquid or freeze it into a solid and still have a kilomole. Thus, a kilomole of uranium, a shiny 238-kg block, has the same number of atoms as there are molecules in 44 kg of carbon dioxide, whether it is gaseous or solid (dry ice). To honor the good Count, the number of molecules in a mole is known as **Avogadro's number** (N_A), though he himself had no idea of its numerical value. The best current measure of that quantity is

$$N_A = (6.022\ 136\ 7 \pm 0.000\ 003\ 6) \times 10^{23} \text{ molecules/mole}$$

Example 10.1   Determine (a) the mass of a carbon atom in kilograms and (b) the equivalent of one atomic mass unit in kilograms. Use both sets of units.

Solution: [Given: the mass of a carbon atom is 12.00 u, and a mole or 12.00 g of carbon contains N_A atoms. Find: the mass of a carbon atom in kg, and the equivalent of 1.000 u in kg.] The mass of a carbon atom equals the mass of a mole of carbon divided by the number of atoms in a mole; thus

$$m_C = \frac{12.00 \text{ g}}{6.022 \times 10^{23}} = 1.993 \times 10^{-23} \text{ g}$$

and

$$\boxed{m_C = 1.993 \times 10^{-26} \text{ kg}}$$

Since $12.00 \text{ u} = m_C = 1.993 \times 10^{-26}$ kg

$$1.000 \text{ u} = \frac{1.993 \times 10^{-26} \text{ kg}}{12} = \boxed{1.661 \times 10^{-27} \text{ kg}}$$

▶ **Quick Check:** In SI units, $N_A = 6.022 \times 10^{26}$ molecules/kilomole and

$$m_C = \frac{12.00 \text{ kg}}{6.022 \times 10^{26}} = 1.993 \times 10^{-26} \text{ kg}$$

10.3 Density

One of the more evident features of bulk matter is that different substances are not packed with equal density—a kilogram of lead has a much smaller volume than a kilogram of feathers. When dealing with the behavior of materials rather than individual objects, it's often more convenient to use quantities that are *independent of volume.* Thus, we define the **density** of any material *as the mass per unit volume:*

$$\text{Density} = \frac{\text{mass}}{\text{volume}}$$

TABLE 10.1 Densities of some materials

Substance	Density (kg/m³)	Substance	Density (kg/m³)
Interstellar space	10^{-18} to 10^{-21}	Chloroform	1.53×10^3
Hydrogen*	0.090	Sugar	1.6×10^3
Oxygen	1.43	Magnesium	1.7×10^3
Helium	0.178	Bone	$(1.5–2.0) \times 10^3$
Air, dry (30°C)	1.16	Clay	$(1.8–2.6) \times 10^3$
Air, dry (0°C)	1.29	Ivory	$(1.8–1.9) \times 10^3$
Styrofoam	0.03×10^3	Glass	$(2.4–2.8) \times 10^3$
Balsa wood	0.12×10^3	Cement	$(2.7–3.0) \times 10^3$
Cork	$(0.2–0.3) \times 10^3$	Aluminum	2.7×10^3
Pine wood	$(0.4–0.6) \times 10^3$	Marble	2.7×10^3
Oak wood	$(0.6–0.9) \times 10^3$	Diamond	$(3.0–3.5) \times 10^3$
Ether	0.74×10^3	Moon	3.34×10^3
Ethyl alcohol	0.79×10^3	Planet Earth,	
Acetone	0.79×10^3	average	5.25×10^3
Turpentine	0.87×10^3	Iron	7.9×10^3
Benzene	0.88×10^3	Nickel	8.8×10^3
Butter	0.9×10^3	Copper	8.9×10^3
Olive oil	0.92×10^3	Silver	10.5×10^3
Ice	0.92×10^3	Lead	11.3×10^3
Water (0°C)	$0.999\,87 \times 10^3$	Mercury	13.6×10^3
Water (3.98°C)	$1.000\,00 \times 10^3$	Uranium	18.7×10^3
Water (20°C)	$1.001\,80 \times 10^3$	Gold	19.3×10^3
Tar	1.02×10^3	Tungsten	19.3×10^3
Seawater	1.025×10^3	Platinum	21.5×10^3
Blood plasma	1.03×10^3	Osmium	22.5×10^3
Blood, whole	1.05×10^3	Pulsar	$10^8–10^{11}$
Ebony wood	$(1.1–1.3) \times 10^3$	Nuclear matter	$\approx 10^{17}$
Rubber, hard	1.2×10^3	Neutron star, core	$\approx 10^{18}$
Brick	$(1.4–2.2) \times 10^3$	Black hole	
Sun, average	1.41×10^3	(1 solar mass)	$\approx 10^{19}$

*Gases are at 0°C and 1 atm unless otherwise indicated.

The symbol for density is the lowercase Greek letter ρ (rho), and so

$$\rho = \frac{m}{V} \qquad (10.1)$$

Density is a characteristic of a substance; mass is a characteristic of an object (Table 10.1).

At one time, the centimeter-gram-second (cgs) unit of g/cm³ was widely used; it still persists where small quantities of substances are dealt with, as in medicine. Notice that since

$$1 \text{ g/cm}^3 = \frac{1 \times 10^{-3} \text{ kg}}{(1 \times 10^{-2} \text{ m})(1 \times 10^{-2} \text{ m})(1 \times 10^{-2} \text{ m})} = 10^3 \text{ kg/m}^3$$

the numerical values only differ by a factor of 10^3. Because the gram was defined as the mass of a 1-cm³ volume of water, the density of water is 1 g/cm³, or 10^3 kg/m³. The pungent-smelling metal osmium, whose atoms are massive at 190 u (though not the most massive), is so closely packed that it is the densest substance on Earth.

The pale blue, almost transparent solid resting on a mound of shaving cream is silica aerogel. Its density is only about 3 times that of air and yet it can support a load $\approx$1600 times its own weight.

Example 10.2   Determine the mass of a gold sphere 0.10 m in diameter. How many atoms are in it? The atomic mass of gold is 197 u.

Solution: [Given: $D = 0.10$ m, from Table 10.1 $\rho = 19.3 \times 10^3$ kg/m^3. Find: m and the number of atoms N.] The volume of the sphere

$$V = \tfrac{4}{3}\pi R^3 = \tfrac{4}{3}3.141\,6\,(0.050\text{ m})^3 = 0.524 \times 10^{-3}\text{ m}^3$$

together with the density provides the mass, which is

$$m = \rho V = (19.3 \times 10^3\text{ kg/m}^3)(0.524 \times 10^{-3}\text{ m}^3)$$

and

$$\boxed{m = 10\text{ kg}}$$

A kilomole of gold is 197 kg, so this sphere is the equivalent of 0.051 27 kmol and therefore

$$N = N_A\,0.051\,27 = (6.022 \times 10^{26}\text{ atoms/kmol})$$
$$\times\,(0.051\,27\text{ kmol}) = \boxed{3.1 \times 10^{25}}$$

▶ **Quick Check:** $(3.1 \times 10^{25}$ atoms$)(197$ kg·kmol$^{-1})/$ $(6.022 \times 10^{26}$ atoms/kmol$) = 10$ kg.

10.4 The Size of Atoms

By the 1880s, a number of experiments had established a rough measure of the size of atoms and molecules. One of the easiest to appreciate takes a known volume of oil and allows it to spread out on the surface of water. There are good reasons to assume that the resulting film would thin out until it was only one molecule thick. Measuring the area of the slick gives us the thickness of the film, and therefore the diameter of the oil molecule. Nowadays, we know that *atoms of every kind are nearly the same size, varying in diameter from roughly 1×10^{-10} m to about 3×10^{-10} m.*

We can get a rough estimate of the dimensions of atoms and molecules in a solid or liquid by ignoring the space between them, assuming that they "fill" the volume of the material, however tenuously. They do not, but there cannot be a great deal of interatomic space since neither solids nor liquids can be compressed very much. If we take each atom as a little cube of volume L^3, the volume V of a mole would then be $N_A L^3$. If the mass of that mole is M_m

$$\rho = \frac{M_m}{V} = \frac{M_m}{(N_A L^3)}$$

and

$$L = \left(\frac{M_m}{N_A \rho}\right)^{\frac{1}{3}} \tag{10.2}$$

This means that a water molecule should have a size of very roughly

$$L = \left[\frac{18\text{ g}}{(6.022 \times 10^{23}\text{ atoms/mole})(1.0\text{ g/cm}^3)}\right]^{\frac{1}{3}}$$

which is 3×10^{-10} m, or 0.3 nm. Interestingly, if we do this same calculation for most simple solids and liquids, they generally turn out to have atomic dimensions within a factor of 2 or so of water (a result confirmed by X-ray scattering). For example, $L = 0.5$ nm for ethyl alcohol.

Example 10.3   Assuming that the atoms in a solid are packed shoulder to shoulder, and taking a typical diameter to be 0.2 nm, (a) how big a cube corresponds to a mole of almost any solid? (b) What is a typical value for the molar volume (that is, the volume of a mole) of a solid?

(continued)

(continued)
Solution: [Given: $D = 0.2$ nm, and N_A. Find: the volume of a 1-mole cube.] (a) The cube root of 6.022×10^{23} is 8.445×10^7, and this must be the number of atoms along each edge of the cubic volume. Each atom is 0.2 nm in diameter, so the edges must be $(8.445 \times 10^7)(0.2 \times 10^{-9}$ m$) = 16$ mm ≈ 2 cm long. (b) And the molar volume is approximately 8 cm^3.

▶ **Quick Check:** 1 mole of gold has a mass of 197 g and a volume of

$$V = \frac{m}{\rho} = \frac{(197 \text{ g})}{(19.3 \text{ g/cm}^3)} = 10 \text{ cm}^3$$

which is close to our approximation.

The numbers of atoms in even modest amounts of material are incredibly vast. If we took all the molecules in a tablespoon of water, about 18 g or 1 mole, and put them end to end, the column would be $(6.022 \times 10^{23})(3 \times 10^{-10}$ m$) = 1.8 \times 10^{14}$ m long and would stretch from here to the Sun (1.495×10^{11} m) and back over 600 times! There are more molecules in a tablespoon of water than there are stars ($\approx 10^{22}$) in all the Universe.

The atom is actually a complex structure, a thing of interacting, whirling parts, many of whose intricacies evade us still. Each atom consists of a dense, minute core—the *nucleus*—surrounded by a tenuous mist of electrons. The nucleus, whose diameter is only about one ten-thousandth the diameter of the atom, is a ball of rapidly moving, positively charged protons and neutral neutrons bound together by the strong force (Sect. 32.3). The electrons, which tumble around at tremendous speeds in the outer reaches of the atom, are retained by the attractive electromagnetic force (Fig. 10.1).

Since each proton and neutron is roughly 2000 times more massive than an electron, a typical nucleus is many thousands of times more massive than the accompanying electrons. The tiny atomic core where practically all the mass resides is surrounded by a far-distant, relatively massless, negative wisp of charge. Thus, the space occupied by the atom as a whole is mostly devoid of matter.

We no longer envision neat little well-defined electron orbits. Instead, the atomic landscape is calculated in terms of the patterns in which the electrons are likely to reside, patterns known as electron clouds. The diameter of an atom is not perfectly defined because its electrons are in constant shifting motion. Indeed, an atom's size varies slightly with the different kinds of experiments used to measure it.

10.5 The States of Matter

Atoms in solids and liquids can be imagined as tightly packed (that is, cloud to cloud), whereas in gases they are quite a bit farther apart—gases under ordinary conditions are about a million times more compressible than solids. The regions between the nuclei of adjacent atoms in all forms of bulk matter are comparatively immense, and comparatively empty; that's true even within solids. If you could drink Alice in Wonderland's growth potion and swell up until the core of an atom in your body were the size of a cherry pit, you would stand about a thousand million kilometers tall! And most of your huge volume would be void, with hundreds of meters of emptiness separating the pit-sized concentrations of matter. Atoms combine under the influence of the electromagnetic force to produce all the

Figure 10.1 The atoms in a solid are roughly 0.2 nm apart, center-to-center, and have almost all their mass concentrated in the tiny 10^{-15} m to 10^{-14} m nuclei.

10^{-15} m - 10^{-14} m

$\sim 2 \times 10^{-10}$ m

TABLE 10.2   Molecular number-densities

Substance	State	Number-density (molecules/m³)
Gold	solid	5.9×10^{28}
Ice	solid	3.1×10^{28}
Water	liquid	3.3×10^{28}
Mercury	liquid	4.2×10^{28}
Oxygen	gas	27×10^{25}
Air	gas	27×10^{25}

things of our world. The kinds of atoms grouping together determine the strengths of their electrical interactions, which define the ultimate structure of the substance. Under ordinary planet-Earth conditions, matter exists as solid, liquid, gas, or plasma. If the interatomic forces are strong enough, the collection of particles independently *maintains its shape and volume.* Along with its elastic properties, that ability to maintain shape is a distinguishing feature of a **solid**. A **liquid** is characterized by weaker binding, so that *it flows, assuming the shape of its container, although it maintains a constant volume regardless of that shape.* If the forces are still weaker, the material exists as a **gas**; the atoms or molecules tend to disperse, and *the substance flows, assuming both the shape and volume of its container.* If some of the atoms of a gas are *ionized* (that is, electrons are removed or added) *the mix of atoms, ions, and electrons is a conducting fluid* called a **plasma**.

Some materials can exist as solid, liquid, or gas and readily shift from one form, or **phase**, to another and back, depending on the prevailing conditions of temperature and pressure. Both internal motion and disorder progressively increase as a sample transforms from the solid to the liquid and then to the gaseous state. Most solids become less dense by a few percent on melting (ice is a notable exception, being less dense than water). The space between molecules thereby increases, but not by very much. On the other hand, a tremendous decrease in density results when a liquid is transformed to a gas. Table 10.2 shows the effect in terms of *number-density,* the number of molecules per cubic meter, which provides a sense of the intermolecular spacing.

10.6   Solids

Many more materials in our moderate-temperature environment are in the solid state than in either the liquid or gaseous states. One way to classify these solids is as either *crystalline, quasi-crystalline, amorphous,* or *composite.* As with liquids and gases, the atoms of a solid are in motion, but it's a small jiggling, vibratory motion about fixed, closely spaced positions. When the atoms arrange themselves in *an orderly, three-dimensional pattern that repeats itself over and over again,* the solid is **crystalline**. Most minerals and all metals and salts exist in at least one crystalline configuration.

If a substance is composed of atoms that are *not arranged in any orderly and repetitive array,* the solid is said to be **amorphous**. Rubber, pitch, resin, plastics, and various glassy materials are examples. The atoms of glasses are strung out in an irregular mesh that lacks any *long-range* order (Fig. 10.2). What order there is, is patchy, localized, and nothing like the neat repetitive array of row upon row of atoms in a crystal. In contrast, the long-chain molecules of rubber and many plastics resemble the entwined, but still not total, chaos of a bowl of spaghetti.

Between these two long-recognized traditional configurations—crystalline and amorphous—there is a new form of solid matter that was only discovered in 1983. When the atoms arrange themselves in *an orderly three-dimensional pattern that nonetheless does not repeat itself with a regular periodicity,* the solid is **quasi-crystalline**.

And lastly, **composite** solids such as wood, concrete, fiberglass, bone, and blood vessel walls are composed of several different materials bonded together.

Glassy

Crystalline

Figure 10.2 Glassy and crystalline solids—short- and long-range order.

Crystals

Many materials are made up of tiny crystal grains, each no more than a fraction of a millimeter in size. Indeed, ordinary metal objects (knives, forks, and hubcaps) are all **polycrystalline** (many-crystalled).

A distinguishing property of most crystalline solids is their sharp transition to the liquid state at a well-defined constant temperature. Crystals melt suddenly. Because the distances between atoms are well defined, the interatomic bonds that hold the crystal together are all nearly equal in strength, and all rupture almost simultaneously. This behavior is unlike that of a noncrystalline material (for example, glass or tar) that melts gradually over a broad range of increasing temperature.

The smooth, flat, angular faces of crystals reflect their internal order, an order that can be appreciated by imagining the atoms arranged in minute, identical building blocks. A particular basic grouping of atoms forms a type of building block. Countless such blocks are then stacked one upon another, filling the space of the sample and forming its overall geometrical design. There are only seven basic structures (Fig. 10.3) from which all crystals can be constructed. The key point here is that crystals are filled, their atoms are packed in, leaving no large empty regions. These imagined blocks must also completely fill the space when they are stacked, and not every nice geometrical figure does that. The idea is easy to see in two dimensions by just considering a tiled floor. Without leaving empty spaces between

Zinc crystals on a galvanized parking lot rail guard. Crystals like these can be seen on streetlamps and garbage cans. Old brass doorknobs etched by sweat and years of handling also show a fine-grain polycrystalline structure.

Type	Geometry	Example
Cubic	all 90° angles all edges equal	gold, lead, silver, diamond, salt, pyrite, garnet, alum, iron
Tetragonal	all 90° angles two of three edges at a corner are equal	zircon, tin, rutile, urea, indium, chlorine
Orthorhombic	all 90° angles no two edges at a corner are equal	sulphur, iodine, codeine, morphine, dextrose, topaz, epsom salt
Monoclinic	no two edges at a corner are equal eight angles not 90°	cane sugar, gypsum, mica, borox, cocaine, saccharin, baking soda, meerschaum, talc
Triclinic	no two edges at a corner are equal no 90° angles	boric acid, copper sulphate, feldspar
Trigonal	all edges are equal no 90° angles	ruby, bismuth, sapphire, hematite, quartz, tourmaline
Hexagonal	hexagonal ends 90° sides	zinc, nickel, magnesium, calcite, mercury, hydrogen, osmium, ice, graphite

Figure 10.3 Crystal building blocks.

Figure 10.4 (a) The packing of regular objects with 5, 6, 7, and 8 sides in two dimensions. Triangles, rectangles, and hexagons (shown here formed out of circles) fill the space leaving no gaps between their edges as do the other figures. (b) Regular hexagons fill the flat two-dimensional space of this floor.

(a)

(b)

Simple cubic (SC)

Body-centered cubic (BCC)

Face-centered cubic (FCC)

Figure 10.5 The three basic variations of the cubic crystal form.

adjacent tiles, we can cover the floor with triangles, squares, rectangles, parallelograms, or hexagons. But it cannot be done with regular pentagons or any other shape (Fig. 10.4). There isn't a washroom in the Universe with a flat floor tiled solid in white pentagons.

The simplest building block is cubic, and there are three variations on the theme (Fig. 10.5). The points at the corners, faces, and center of these cells are *lattice points.* They locate the positions of the atoms or molecules of the crystal. In Fig. 10.5 a single atom is seen at each lattice point, although in complex materials a cluster of hundreds of atoms may occupy each lattice point. The simple cubic (SC) structure is not uncommon with solids composed of ions—cesium chloride (CsCl), for instance—and metallic alloys. Of the elements, only polonium is SC, and then only in a particular temperature range. If we imagine the atoms as "touching" spheres, the SC structure can fill only about 52% of the space available. More effi-

This polycrystalline mass of calcium carbonate ($CaCO_3$) grew in a human kidney until it rather painfully found its way down and out. Fortunately this photo is about six times life-size.

A single crystal of table salt (NaCl). It's a face-centered cubic crystal. The atomic structure (a square roughly 0.56 nm on a side) has two ions, a sodium and a chlorine, at each lattice point. Most brands of table salt come as tiny cubes.

cient, at 68% packing, is the body-centered cubic (BCC) structure utilized by several metals, including iron and sodium. The densest packing, 74%, is achieved in the face-centered cubic (FCC) arrangement assumed by many elemental metals including silver, gold, nickel, copper, aluminum, and lead.

There are two ways to stack spheres (or atoms) so as to minimize the unfilled, or interstitial, volume. One arrangement is the face-centered cubic, and the other, which also fills 74% of the volume (Fig. 10.6), is the hexagonal close-packed (HCP). There is nothing very exotic about this arrangement; it's the way many things are usually stacked—from plums to cannonballs. We can expect that atoms that are attracted uniformly in all directions will have the spherical symmetry of cannonballs and so pack relatively densely. Those that do not combine into asymmetrical molecules usually form densely packed crystals; atoms of the metals and rare gases (helium, neon, and argon) are typical. Thus, argon is face-centered cubic, while zinc is hexagonal close-packed.

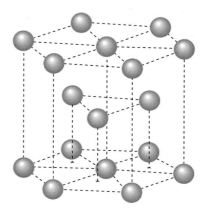

Figure 10.6 The hexagonal close-packed (HCP) arrangement of atoms.

ELASTICITY

Galileo began his book, *Two New Sciences* (1638), with a discussion "treating of the resistance which solid bodies offer to fracture." By the middle of the 1600s, there were some other tentative studies of the strengths of wires and rods, but it remained for that remarkably homely, lascivious genius, Robert Hooke, to really get things going in what they called "*ye* science of *elasticity*"—that is, the analysis of the behavior of materials and structures under the influence of applied forces.

10.7 Hooke's Law

In 1676, wishing to establish priority for some new discoveries, Hooke published two cryptic anagrams. Two years later he decoded them, announcing the invention of the spring balance and the discovery of what has since come to be known as **Hooke's Law**. When an object is acted on by a force, it may be compressed, stretched, or bent. If, after the force is removed, the object rapidly returns to its original configuration, we say that it is *elastic*. If the force is not too large, materials such as steel, bone, rock, tendon, glass, flesh, and rubber are elastic. The metal hair-spring in a pocket watch coils and uncoils elastically over 170 000 times a day. By contrast, solids or quasi-solids, such as chewing gum, lead, moist clay, and putty, do not return to their original configurations once distorted. These materials are categorized as *plastic*.

What Hooke found was that, besides being elastic, **many materials deform in proportion to the load they support**. That's Hooke's Law, and it applies to *linearly elastic* or *Hookean* objects—to steel bars, rods, and wires, to springs, diving boards, and more or less to rubber bands, among other things. Thus, if an ordinary coiled (helical) spring in equilibrium ($\Sigma F = 0$) is either stretched or compressed by an applied force F, the spring will be deformed (extended or contracted) by a distance s that is proportional to the force: $F \propto s$ (Fig. 10.7). Doubling the stretching force doubles the extension of the spring.

When one parameter is proportional to another, the relationship can be transformed into an equality by introducing a constant-of-proportionality. Here, this constant, k, takes care of the differences between units on both sides of the equation. It introduces the physical characteristics of that particular spring into the formula. Presumably, if we knew more physics, we could express the constant in terms

An HCP stack of plums. Notice how each plum is surrounded by six nearest neighbors.

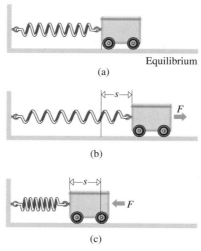

(a)   Equilibrium

(b)

(c)

Figure 10.7   A force F applied to a spring (a) can either (b) stretch or (c) compress it by a distance s. Provided s is not large, $F = ks$.

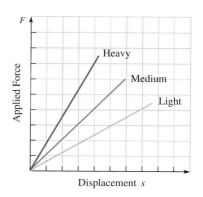

Figure 10.8   The force F applied to a spring stretches it a distance s. This figure is a plot of F versus s for three different springs. The slope of each curve is the spring's elastic constant, k. The less rigid the spring (heavy, medium, and light), the smaller the value of k.

Hooke's Law. The Power [that is, force] of any Spring is in the same proportion with the Tension [meaning, at the time, extension] thereof; That is, if one power [force] stretch or bend it one space, two will bend it two, three will bend it three and so forward.

ROBERT HOOKE

of the atomic parameters that give rise to it, but that's still beyond our means. Hooke's Law is then

$$F = ks \qquad (10.3)$$

This is the expression for the force that must be exerted *on* the spring to produce a corresponding displacement: when F is positive, s is positive. From Newton's Third Law, just the opposite would be true for the force exerted *by* the spring (Fig. 9.7, p. 284). Because this force always tends to return the system to its equilibrium configuration, it's called a *restoring force*.

The **spring constant** or **elastic force constant** k is defined by Eq. (10.3). It has the dimensions of force over distance and the units of N/m. *For a given force, the larger the value of k, the smaller the extension of the spring—that is, the more rigid the spring* (Fig. 10.8). The greater the slope (k) of the force-displacement curve, the stiffer the spring, rod, or diving board. The elastic constant varies with the size and shape of the object as well as with the material: the element, compound or alloy, its impurities, heat treatment, crystal structure, and so on.

Example 10.4   A spring balance (Fig. 10.9) has a 10-cm graduated scale that can read mass from zero to 500 g. How does the device work? Determine the spring constant. What displacement would the pointer experience if you pulled down on the support hook with 10 N? Would this balance be useful on the Moon?

(continued)

(continued)

Solution: [Given: mass range 0–500 g, displacement 0–10 cm, and a load of 10 N. Find: k and then s for 10 N.] The spring scale indicates s, which is proportional to F, or the weight of the load. On Earth, that force is more or less constant and proportional to the constant mass of the load and so, if properly calibrated on this planet, such a scale can be used to measure mass. If the scale was set to read newtons directly, it would function perfectly on the Moon. However, when set to read grams, it is assumed the device is on Earth, where $F_W/(9.8 \text{ m/s}^2) = m$; hence, it would have to be recalibrated at 1/6th scale.

The weight of a 500-g mass is

$$F_W = mg = (0.500 \text{ kg})(9.8 \text{ m/s}^2) = 4.9 \text{ N}$$

producing a displacement of 10 cm, or 0.10 m; hence

$$k = \frac{F}{s} = \frac{4.9 \text{ N}}{0.10 \text{ m}} = \boxed{49 \text{ N/m}}$$

If you pulled down on the balance with a force of 10 N, a displacement

$$s = \frac{F}{k} = \frac{10 \text{ N}}{49 \text{ N/m}} = \boxed{0.20 \text{ m}}$$

would be necessary, which for this device is physically impossible because it's only 10 cm long.

Figure 10.9 A spring scale much like the one invented by Robert Hooke. A load either stretches or compresses a spring, changing its length proportionately.

▶ **Quick Check:** Each 1 cm corresponds to 500 g/10 = 0.050 kg; hence, $k = F/s = mg/s = (0.050 \text{ kg}) \times (9.8 \text{ m/s}^2)/(0.010 \text{ m}) = 49 \text{ N/m}$.

You may have seen a Slinky mercilessly stretched until portions of it became distorted, unable to return to its original unstretched condition. Any rod, girder, spring, or wire can be bent or stretched beyond its **elastic limit** so that it becomes permanently deformed, losing its springiness. In such a sorry state, it will no longer obey Eq. (10.3); it will no longer be *linearly* elastic. Accordingly, those nice straight-line curves in Fig. 10.8 don't go on forever.

Elastic Potential Energy

One of the practical functions served by springs is to store energy (Fig. 10.10). The mainspring in a clock (a device invented by Hooke) replaces the *gravitational*-PE of a falling weight or pendulum with the *elastic*-PE of a spring. As we saw in Chapter 9, *the work done on a system against a nonconstant force, such as the force of a spring, is the area under the force-displacement curve which, if the force is conservative, equals the change in potential energy of the system.* Thus, for any of the plots in Fig. 10.8, the *elastic*-PE (PE_e) stored in stretching the spring a distance s is the area of the triangle under the line—namely, $\frac{1}{2}$ the base (s), times the corresponding altitude ($F = ks$):

$$\Delta PE_e = \tfrac{1}{2} s(ks)$$

Figure 10.10 The suspension system of an automobile. When the wheel of a car is jolted upward by a bump in the road, a coil spring is compressed, slowing and smoothing out the vertical motion. The shock absorbers convert the stored mechanical energy to thermal energy, keeping the car from bouncing on the springs.

Heavy trucks use leaf springs and shock absorbers to smooth out the ride. The leaf spring is a stack of thick steel bars of decreasing length strapped together. They are slightly bent so that an impulsive vertical force tends to straighten them out.

and so

$$\Delta PE_e = \tfrac{1}{2} k s^2 \qquad (10.4)$$

This is the same as saying that the work done by the displacement-dependent spring-force is its average value, $F_{av} = \tfrac{1}{2}(F_f + F_i) = \tfrac{1}{2}(F_f + 0) = \tfrac{1}{2} k s$ times the displacement s: $W = F_{av} s = \tfrac{1}{2} k s^2 = \Delta PE_e$.

Example 10.5 Four people each with a mass of 70 kg enter a car of mass 1100 kg—two in front and two in back. The vehicle settles down in a little while on its four identical springs, compressing them a distance of 2.5 cm. Determine the elastic constant of each spring and the amount of energy stored in each. What is the equivalent elastic constant of the suspension taken as a single unit? Where does the energy come from?

Solution: [Given: mass of each person, 70 kg; mass of car, 1100 kg; and $s = 2.5$ cm. Find: each k, each ΔPE_e, and the net spring constant k'.] The spring system can be taken to be in equilibrium when the car is empty, so its mass is irrelevant. The total mass of people supported by the four springs is 4×70 kg and this weighs

$$F_W = (280 \text{ kg})(9.8 \text{ m/s}^2) = 2744 \text{ N}$$

Each spring compresses the same amount (0.025 m) and supports the same load; hence, the spring-force exerted upward by each is $\tfrac{1}{4}$ the total load

$$F = \tfrac{1}{4}(2744 \text{ N}) = ks = k(0.025 \text{ m})$$

and

$$\boxed{k = 2.7 \times 10^4 \text{ N/m}}$$

Taking the springs together as one, exerting 2744 N upward, the equivalent force constant is

$$k' = \frac{2744 \text{ N}}{0.025 \text{ m}} = \boxed{1.1 \times 10^5 \text{ N/m}}$$

which is 4 times the individual spring constant.

The *elastic*-PE stored in each spring is

$$\Delta PE_e = \tfrac{1}{2} k s^2 = \tfrac{1}{2}(2.7 \times 10^4 \text{ N/m})(0.025 \text{ m})^2 = \boxed{8.4 \text{ J}}$$

while the whole system stores 4 times that. The energy stored elastically is a fraction of the decrease in the *gravitational*-PE—the remainder is lost as thermal energy.

▶ **Quick Check:** The total stored PE$_e$ is $\tfrac{1}{2} k' s^2 = 34$ J, which equals 4(8.4 J).

The elastic energy of a spring arises from pulling against the interatomic bonds holding the material together—it's really *electromagnetic*-PE. In fact, whenever a contact force acts on any kind of body, the interaction is accompanied by a distortion of the body and the resulting electrical restoring forces among the many millions of atoms involved. There is no such thing as an absolutely rigid body.

Elastic Materials

Suppose we take a sample of some material, apply an increasing force F to it, and plot the resulting change in length ΔL, as in Fig. 10.11. If the sample is animal tissue, the graph will usually be a curve like that of Fig. 10.11b, whereas if it's a synthetic stretchy polymer, it will look more like Fig. 10.11c. For moderate extensions, the data goes up and back along the two curves, and both represent elastic behavior. By contrast, some elastic materials manifest a linear relationship between F and ΔL that corresponds to Hooke's Law

$$F = k \Delta L \qquad (10.5)$$

Typically, however, a linear response only occurs over a limited range of applied force and that's determined by the nature of the interatomic bonding being stretched. Removing a small load from a sample allows the atoms, which have only been slightly shifted, to return to their equilibrium configuration, and the material comes back to its original length.

Consider what happens to a sample as the load on it gets very large. Depicted in Fig. 10.12 is the behavior of several different solid rods under increasing tension. The longest curve with the greatest value of ΔL corresponds to a material like pure iron or aluminum. Such a sample will elongate a little linearly until its **elastic limit**, or **yield point**, is reached and the bonds between atoms begin to be overcome. Rows of atoms then slide past one another in a process known as *slip,* and the material plastically deforms in an unrecoverable way. Once the plastic mode is initiated, the elongation occurs more dramatically until ultimately the specimen ruptures at the **breaking point**. You have probably never stretched a steel rod to the breaking point, but the same thing happens with a piece of taffy or a Tootsie Roll.

As a rule, removing the load anywhere between the elastic limit and the breaking point will leave the sample permanently longer than it was originally. A material that can be stretched thin or drawn without breaking even while cold (like chewing gum) is *ductile*. Metals such as copper, gold, and silver, are quite ductile at STP and elongate considerably via slipping. Such substances are *tough* and resist cracking. The more carbon is fused into iron, the less ductile is the resulting alloy. Steel is iron with less than about 1% carbon. Pure iron has a low elastic limit, low-carbon steel a moderate one, and high-carbon steel a still greater one (Fig. Q3, p. 347). If we built a bridge out of pure iron, it would permanently sag the first time it was heavily loaded.

Carbon embedded among iron atoms serves as a kind of grit that impedes slipping. Brittle substances such as glass, stone, ceramics, and cast iron (which contains about as much carbon as it can, ≈4%) break close to their elastic limits, without appreciable deformation. Even though glass is stronger than steel in both compression and tension, you are not likely to see glass houses because glass does not yield rapidly at STP—it's just too brittle to be a major structural material. If you want to shatter an old cast iron household radiator, just hit it with a sledge hammer.

The degree to which a material will display ductile behavior depends on several factors: temperature, the type of loading, and the rate of loading. A substance that is

(a)

(b)

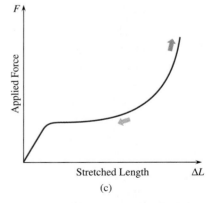

(c)

Figure 10.11 (a) A length (L_0) of material stretches an amount ΔL under a load F. (b) Soft animal tissue stretches elastically, forming a J-shaped curve and contracting back along the same arc as the force is reduced. (c) Rubbery synthetic solids stretch elastically along an S-shaped curve. They also return to $\Delta L = 0$ along the curve as F goes to zero.

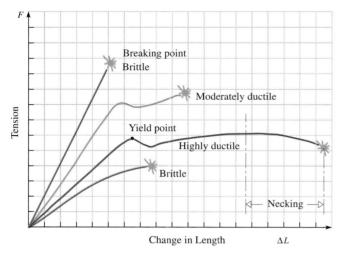

Figure 10.12 Load or tension versus elongation for several different engineering materials. Brittle substances rupture suddenly without much elongating. Ductile substances elongate via plastic flow. A ductile metal might be Hookean up to 0.25%–0.5% elongation and fracture at around 50%. Some nonmetallic materials such as wood and fiberglass fracture after elongations of 1%–3%.

(a) When a sample is put under tension, it elongates linearly by less than ≈0.5%. If it's ductile, it continues to elongate several hundred times as much under slightly increased loads. (b) After a maximum load is reached, the sample begins to neck down until (c) it fails. (d) If the material is brittle, it fractures without deforming plastically.

brittle at −200°F may well be ductile at room temperature, and red-hot glass stretches like chewing gum. All across the northern United States, cheap plastic outdoor garbage pails bounce in July and shatter in January. The strongest structural material used today is steel, but because it becomes plastic at 1200°F, steel columns and beams in buildings have to be fireproofed. At the other extreme, steel becomes brittle at around −30°F, so much so that a railroad bridge in northern Canada broke up when a heavy train rattled over it on an especially cold winter evening.

For the most part, *brittle materials tend to rupture sooner in tension than compression* because of the presence of flaws (for example, microscopic cavities and cracks) that weaken the specimen far more in tension than in compression.

Bone is a composite structure made of crystals of hydroxyapatite (a calcium compound good in compression) and collagen fibers (good in tension). Bone, which is varyingly brittle, remains elastic almost to the point of rupture and cannot be either permanently compressed or elongated unless the loads are applied for a very long time. Infants' bones are high in collagen and very flexible. By young adulthood, hydroxyapatite has built up to the point (≈67%) where bones are both flexible and strong. As adults age, collagen becomes less flexible and calcium tends to be reabsorbed from the hydroxyapatite, making bones brittle and more breakable.

10.8 Stress

Remarkably, about a century and a half elapsed before Hooke's ideas were further developed by the French mathematician Augustin Louis Cauchy. It is likely that this long period of idleness was in no small part due to Newton, who detested Hooke and, outliving him by 25 years, had ample opportunity to denigrate the latter's practical science. Sir Isaac was a loner—chaste, sensitive, and vulnerable—while Hooke was argumentative, vain, and a womanizer, a testy, earthy sort who took pleasure in controversy as much as Newton shrank from it. It can be argued that Hooke contributed to Newton's nervous breakdown, and it can also be argued that Sir Isaac had his revenge, at least on the theory of elasticity.

In any event, Cauchy (who himself was widely held to be selfish and narrow-minded) introduced the central notions of *stress* and *strain* (1822). What severely limits Hooke's Law is that it deals with the specimen or spring or structure in its entirety. The elastic constant depends on the details of the object under test and not just on the material of which it's made. When we bend a tree and compare the applied force and resulting deflection, we arrive at the elastic constant of that particular tree and not some clever insight about the behavior of oak or pine. Recall Example 10.5. The elastic constant of the four springs acting as one was 4 times as great as the constant for any one of the springs, even though nothing different was happening inside them in either case. What we need is a reformulation of Hooke's Law that focuses not on the forces and deformations of the thing as a whole, but on the internal processes at work that are independent of the external dimensions.

Cauchy introduced the notion of *stress* to describe the distribution of force within a solid acted upon by a load. Ordinary interactions between bodies are always distributed over some contact area; the idea of a force acting at a point is an idealization. **The stress experienced within a solid is defined as the magnitude of the force acting, divided by the area over which it acts:**

$$\text{Stress} = \frac{\text{force}}{\text{area}}$$

Using the Greek letter sigma (σ) for stress, this equation becomes

$$\sigma = \frac{F}{A} \qquad (10.6)$$

The idea is very much like the concept of pressure, and like pressure it has the SI units of N/m^2, or the U.S. Customary units of $lb/in.^2$ (psi). To make matters a bit more concise, $1 \ N/m^2$ is called a *pascal* (Pa) honoring the French physicist by that name. Since this quantity is actually very small, stress is more often given in megapascals (MPa = 10^6 Pa) or even gigapascals (GPa = 10^9 Pa).

The hanging bar in Fig. 10.13b experiences both a downward force and the upward supporting force provided by the ceiling. The line-of-action of these two equal-magnitude forces is along the rod's central axis, and so the loading is said to be *axial*. Imagine a plane cutting perpendicularly across the bar. The corresponding stress in that cross section is given by Eq. (10.6) and it's known as the *normal stress* and, in particular, since the bar is in tension, it's called the **normal tensile stress** (σ_t). Similarly, under an axial compressive load, Eq. (10.6) would provide the **normal compressive stress** (σ_c). Notice in Fig. 10.13 that while a load F is supported by a single rod (which stretches a distance ΔL in the process), two such rods will support twice that load ($2F$), each elongating by that same ΔL, a situation that corresponds to Hooke's Law. Now if each rod has a cross-sectional area A, the tensile stress (F/A) in the single rod exactly equals the stress in the double-rod suspension ($2F/2A$), even though the elastic constants are not the same. That's the great virtue of the idea of stress—it reflects how strongly the atoms are being forced to interact.

When a force acts over a finite area, it's called a *distributed force*. Pressure then becomes a very useful concept. Here, the pressure exerted on the platform is the net load divided by the total contact area. By contrast, the stress anywhere within either leg is the magnitude of the force on that leg divided by the cross-sectional area at that point. Where the leg is narrow, the stress is large. And that's one reason elephants have stout legs.

Figure 10.13 (a) A sample of an elastic material. (b) Under a load *F*, it elongates by an amount Δ*L*. (c) Two identical samples loaded with 2*F* also elongate by the same Δ*L*.

Figure 10.14 A pair of equal-magnitude, parallel forces shearing a sample. Think of how you tear a piece of cardboard or snip a twig or wire.

Human skin is under tension like a rubber glove. As a person ages, the skin wrinkles because it loses its elasticity.

In addition to these normal stresses, there is also something known as **shear stress** (σ_s) that exists when the force-pair acts parallel to the area A (Fig. 10.14). A scissors cuts paper and a pruning clipper cuts twigs by applying a pair of equal-magnitude, oppositely directed forces parallel to the cross section to be severed. These instruments produce a shearing stress that is great enough to sever the material. Thus, the shear stress, again given by Eq. (10.6), is a measure of the effort needed to cause the atoms in one part of a solid to slide past an adjacent part—like a pushed deck of cards that edges over at a slant or one that is fanned out. The latter reflects the fact that shear stress also occurs when the sample (a wrench, wrist, or screwdriver) is twisted.

Example 10.6 Figure 10.15 depicts a punch used to make circular holes in a stack of paper or a sheet of soft metal. The cylindrical punch has a cross-sectional diameter of 1.0 cm and is driven downward with a force of 100 kN. The punch and the die beneath produce a ring of vertical force on the paper that will ultimately shear it, cutting out a disk-shaped slug. Compute the shear stress that exists in the 2.0-mm-thick stack along a vertical cylindrical region beneath the edge of the punch.

Solution: [Given: $D = 1.0$ cm, $F = 100$ kN, and $d = 2.0$ mm. Find: σ_s.] The area of shear is that of the vertical cylindrical surface (corresponding to the edge of the die) of circumference $2\pi R$ and height d; $A = 2\pi \times (D/2)d = \pi(1.0 \times 10^{-2}$ m$)(2.0 \times 10^{-3}$ m$) = 6.283 \times 10^{-5}$ m^2. Hence

$$\sigma_s = \frac{F}{A} = \frac{10 \times 10^4 \text{ N}}{6.283 \times 10^{-5} \text{ m}^2} = \boxed{1.6 \text{ GPa}}$$

Figure 10.15 A paper punch. The cylindrical punch is forced down through the stack of paper above the die hole. It shears the paper along the circular outline of the punch.

which is about 2.3×10^5 psi.

▶ **Quick Check:** $F = \sigma_s A = (1.6 \times 10^9 \text{ Pa})2\pi \times (0.50 \times 10^{-2} \text{ m})(2.0 \times 10^{-3} \text{ m}) = 10 \times 10^4 \text{ N}$.

Several decades ago, women's high-heeled shoes with a bare steel shank in the heel were fashionable. Ordinarily, a shoe might have roughly 20 in.² of contact area with the ground and a 140-lb woman, her weight shifted to one foot, would produce a meager stress of about 48 kN/m² (7 psi). But when that weight was concentrated on the spiked heel of one shoe (with an area of about 0.12 in.²), the stress would soar to 8 MN/m² (1.1 × 10³ psi). The resulting dents, holes, and chipped tiles prompted some shopkeepers to ban the menace.

10.9 Strain

Until now, we have talked about rods and the way they elongate or shorten by some amount ΔL under a tensile or compressive load. But that change in length depends on the original dimensions of the rod. Imagine that a tensile force pulling on the atoms stretches their bonds and increases the interatomic separation by, say, 0.01%. Then, with all its atoms 0.01% farther apart axially, the rod must be 0.01% longer, no matter how long it was originally. Different length rods will deform by different amounts, but the percentage change will be the same, and it's the change in the interatomic spacing that is crucial here. It's that change that generates the counter-force to the load. Hence, we define the **normal strain under an axial load as the change in the length over the original length**:

$$\text{Strain} = \frac{\text{change in length}}{\text{original length}}$$

which can be compressive or tensile. Commonly, the Greek letter epsilon (ε) represents the strain, which symbolically becomes

$$\varepsilon = \frac{\Delta L}{L_0} \tag{10.7}$$

where it is assumed that the sample has a constant cross-sectional area. Strain is a measure of by how much the atoms are shifted and, therefore, by how much the bonds are being stretched or compressed. It is a *dimensionless quantity*. In engineering applications, strain is generally kept within ±1.0% and rarely even exceeds ±0.1%. The columns holding up a high-rise building 300 m tall, whether made of concrete or steel, will be designed to compress by little more than about 3 cm. A 100-m elevator cable that stretches 10 cm under a load has a strain of 10 × 10^{-2} m/100 m = 1.0×10^{-3} = 0.1%.

Clearly, there must be a strain as a solid—any solid—yields under the influence of a stress, however small. That's true for normal stress, as well as shear stress, and there must be a **shear strain** (ε_s), a corresponding change in shape. The solid block shown in Fig. 10.16 is being twisted

Floor tiles sheared by spiked heels. The downward force increases as a person shifts back onto the heels just before sitting.

The threadlike structures of collagen magnified tens of thousands of times.

Figure 10.16 Two equal-magnitude, oppositely directed forces twist the solid block through an angle γ that corresponds to the shear strain.

by a force-couple. The larger the side l_0, the more the edge is displaced, Δl, but the shifting on an atomic level will be the same no matter what l_0 is, provided the angle γ is constant. Hence, *we define the shear strain as the angle of distortion,* γ (Greek lowercase gamma), measured in radians, which makes it dimensionless. Since the angle of distortion is generally small, $\tan \gamma \approx \gamma$ and

$$\varepsilon_s = \gamma \approx \tan \gamma \approx \frac{\Delta l}{l_0} \qquad (10.8)$$

An elastic material is one that returns to its original shape after an initial deformation, and that applies to shear strain as well. Metals, bone, stone, and concrete display *elastic* shear strain of less than about 1°, which is typical of many hard solids. Still, stone bends more than one might imagine; indeed, every building bends in the wind. The towers of the World Trade Center heave over as much as 3 ft in a strong wind (you can even hear them creak) and will bend away from the vertical by 6 or 7 ft in a hurricane. By comparison, stretchy things such as soft biological tissue can recover from shear strains of upwards of 40°. Take hold of the skin on your belly and pull; the way it returns depends on such things as dehydration and, of course, age.

Figure 10.17 is a *stress-strain diagram* for several materials. It's similar to the force-displacement graph of Fig. 10.12. Here, provided the strains are never very large, the test pieces behave elastically, returning to their original shapes in a Hookean fashion. In Fig. 10.17b, the curve is the idealized S-shape associated with

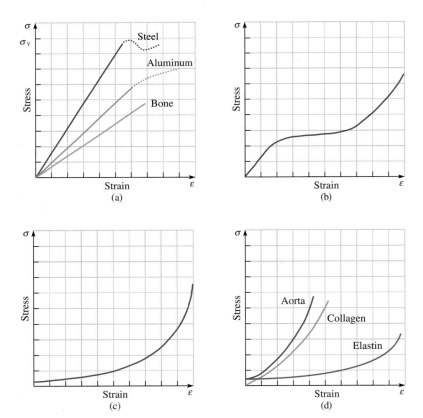

Figure 10.17 (a) The stress-strain curves for several Hookean materials. The ductile samples, such as soft steel, stop behaving linearly at their yield points (σ_Y). (b) When stressed, rubbery polymers first elongate by straightening out their molecules and thereafter by tugging on the chemical bonds. (c) Most biological materials are under tension even when not strained. Your skin is like a rubber glove stretched over your body. (d) Elastin is usually reinforced with collagen in biological systems such as arteries. Tendon is made principally of collagen.

elastomers, stretchy polymers such as Spandex. Their giant molecules are intertwined, coiled strands that straighten out when stretched and spring back when released. They are elastic, even out to strains of several hundred percent, but not Hookean.

Figure 10.17c depicts the behavior of a typical biological tissue sample—pull on your lip or ear lobe. Soft connective animal tissue often contains a mix of two different protein materials: elastin and collagen. The elastin fibers, abundant in the vessels of the human circulatory system, are easily stretched, requiring little stress to produce considerable strain. These are frequently reinforced with a network of quasi-Hookean collagen fibers, as in a human artery. Collagen (p. 335) is the strongest of the connective tissues. It's the collagen that gives flesh its toughness so it does not tear easily. Collagen makes meat tough, although at high temperatures ($\approx$100°C) it breaks down to soft gelatin, leaving the other more tender constituents like elastin behind—that's one reason why we cook meat.

Some biological tissue can stretch elastically well beyond the 800% value that's typical of soft materials such as rubber.

10.10 Strength

In broad terms, *strength* means the ability to withstand a load without failure, but there is a difference in the definition of this term when referring to structures as opposed to materials. When applied to structures, the strength of a truss, building, bridge, or tower equals the load, in newtons, that will initiate collapse. On the other hand, the **strength** *of a material is the stress, in newtons per meter-squared, that will cause failure in a sample of that substance.* Strength is inherently linked to the atomic behavior of the material and tells us how any sample of the stuff should behave, all else being equal (Table 10.3).

In the case of ductile materials, several different strengths are of interest. Figure 10.18 is the stress-strain curve for a sample of an aluminum alloy. *The stress at which slip is first observed, at which the curve stops being linear, is the* **yield strength**. Any mechanical part that will have to support a load without plastically deforming will have to have an appropriately high yield strength—we can't make the girders supporting a building out of soft steel. Yet, if a sheet of aluminum is to be squashed into a mold so it comes out looking like a beer can, its yield strength needs to be lower than the applied stress.

The **tensile strength**, *or* **ultimate strength**, *is the stress corresponding to the greatest applied force*—it's the high point on the stress-strain curve. Usually a ductile sample will deform nonuniformly over its cross section, tending to narrow or **neck** down. The start of that process also marks the ultimate strength (see photo on p. 332). This concept can be misleading, since by the time a piece of material reaches its point of ultimate strength, it has already been irreversibly elongated. We don't often want a ladder that will permanently sag when the first person climbs it.

The stress corresponding to the rupture of the sample is the **breaking strength**, and that applies whether the sample is being stretched, compressed, or sheared. Brittle materials show little or no necking, and their ultimate strengths equal their breaking strengths (see photo on p. 332). Notice in Fig. 10.19 that the stress-strain curve can be quite different in tension than in compression, and the ultimate strengths can be appreciably different too.

TABLE 10.3 Approximate ultimate strengths of various materials

Material	Tensile (MPa)	Compressive (MPa)	Shear (MPa)
Natural rubber	0.03		
Muscle	0.1		
Bladder wall	0.2		
Intestine	0.5		
Cartilage	3		
Concrete	4	30–40	
Brick	5.5	10–21	
Skin	10		
Marble	10	110	22
Lead	12		12
Granite	20	240	35
Ceramics	35–350		
Glass	40–175	50	
Leather	41		
Polystyrene	48	90	55
Fir, Douglas	50	50	8
Tendon	65–82		
Pine	100	27–50	9
Bronze	100–600		
Bone, compact	110	150	
Brass	120–400		
Femur, horse	121	145	
Femur, human	124	170	
Iron, cast-gray	170	650	240
Hair	190		
Iron, wrought	200		
Spider's silk	240		
Magnesium alloy	280		150
Catgut	350		
Cotton	350		
Aluminum	300–570		200–330
Silk	350		
Steel, structural	400		
Nylon thread	1100		
Kevlar	2760		
Steel piano wire	3100		

Example 10.7 The human thigh bone or femur at its narrowest point resembles a hollow cylinder with an outer radius of roughly 1.1 cm and an inner radius of just about half of that. Taking the compressive strength of the bone to be 170 MPa, how much force will be required to rupture it?

Solution: [Given: $R_o = 1.1$ cm, $R_i = 0.5R_o$, and $\sigma_c = 170$ MPa. Find: F.] The cross-sectional area of bone matter is

$$A = \pi(R_o^2 - R_i^2) = \pi R_o^2(1 - 0.5^2) = \tfrac{3}{4}\pi(1.1 \times 10^{-2}\text{ m})^2$$

therefore $A = 2.85 \times 10^{-4}\text{ m}^2$

Hence, since $\sigma_c = F/A$

$$F = \sigma_c A = (170 \times 10^6\text{ Pa})(2.85 \times 10^{-4}\text{ m}^2)$$

and so $\boxed{F = 4.8 \times 10^4\text{ N}}$

(continued)

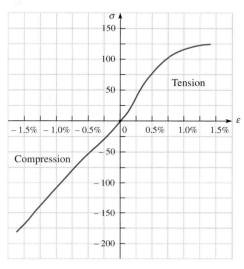

(continued)
or about 11×10^3 lb. An adult who steps off a kitchen table and lands on the heel of one bare foot with the knee locked will experience a force large enough to shatter the femur.

▶ **Quick Check:** $A = 3.8 \times 10^{-4}$ m$^2 - 0.79 \times 10^{-4}$ m$^2 = 3 \times 10^{-4}$ m^2; $F/A = \sigma_c \approx (50$ kN$)/(3 \times 10^{-4}$ m$^2) \approx 0.17$ GPa.

Figure 10.18 The stress-strain curve for an aluminum alloy. There are several different strengths that are of interest to someone using the material: (1) The yield strength tells us when it will cease being elastic and begin to permanently deform. (2) The ultimate strength is the maximum stress it can experience. (3) The breaking strength is the stress at which it fails.

Figure 10.19 Stress-strain for compact (dense) bone.

Because strength depends on many factors (the presence of impurities, holes, surface scratches, cracks, and so on), engineers tend to be cautious in their designs. They usually incorporate a "safety factor" of 3 or 4, or if the situation is potentially dangerous, perhaps 10 or more. Thus, while a large structure might typically experience a stress of 80 MN/m^2 to 95 MN/m^2, mild steel (over 95% of all steel manufactured is of this type) has a strength of around 450 MN/m^2, affording a safety factor of about 4 to 5. In the early 1800s, one out of every four bridges built collapsed. We've come a long way from that sorry record, but earthquakes still play havoc with our cities.

Our forearms and calves house massive concentrations of muscle. Muscle tissue has a very low strength (0.1 MPa), and there has to be a good deal of it to maintain the needed cross-sectional area. Hands and feet, which must move around quickly, need to be light, agile, and still strong. The solution is to bunch our muscles fairly far away and link them to our fingers and toes by strong (82 MPa), and therefore affordably thin, light tendons. Tendon, though its strength is less than that of mild steel (≈ 450 MPa), is about 7 times less dense and so is actually stronger pound for pound.

Some modern plastics such as Kevlar (1972) are considerably stronger, per unit weight, and stiffer than steel. This strength derives from the highly ordered arrangement of its long rodlike molecules. With a tensile strength of 2.8 GPa (4 ×

The groundwork for a modern building. Reinforced concrete (concrete with rods of steel embedded in it) has added tensile strength. This is important because concrete is strong in compression and weak in tension.

10^5 psi), Kevlar has been used for such things as reinforcing tires, making special mooring cables, and for bulletproof vests.

10.11 Fatigue

Many ordinary situations exist in which a metal component can experience a cyclical application of stress. Even though the applied stress is well below the yield strength of the metal, it is still possible for the component to rupture after repeated applications of stress. This failure mode, induced by a relatively small, cyclic stress, is known as **fatigue**. It is characterized by brittle behavior even in metals that are ordinarily ductile, and so the effect must always be considered in dealing with varying loads.

The traffic on a bridge creates fluctuating stress levels, but the complete reversal of the loading is an even more severe condition. Picture the axle of a railroad car supporting the downward weight of the vehicle and cargo. Every time the axle turns through half a rotation, the portion that was on top in compression is now on the bottom in tension, and vice versa—the bending of the axle completely reverses as it rotates. That's why the axles on trains are inspected regularly, and it's also why cracks appear in the structural members of buses bouncing along potholed city streets. Most of the United States' B-1B bomber fleet was grounded in 1992 because of cracks in the landing gear assembly. Similarly, a paper clip, which certainly would be impossible to pull apart by hand, ruptures after about ten 180°-bending cycles.

Figure 10.20 is a plot of stress (σ) versus number-of-cycles-to-failure (N) for two typical metals, tool steel and aluminum. Accordingly, a machine part designed to last through 10^5 cycles will rupture at a stress of 620 MPa (90×10^3 psi) if made of tool steel, and at about half that if made of aluminum. At higher stress levels, the part will survive fewer operating cycles before failing. Steel generally shows a behavior that levels off at what is called the *endurance,* or *fatigue, limit.* That's where the curve becomes horizontal. Presumably, a cyclic stress at a level less than the endurance limit will "never" result in fatigue failure regardless of the number of stress cycles. Incidentally, N is ordinarily likely to be quite large. Even a slow-moving crane might be loaded and unloaded a few hundred times a day and, at 5 days a week

Figure 10.20 Fatigue curves for tool steel and aluminum alloy. At any particular value of stress (σ), there is a corresponding number-of-cycles-to-failure (N). Steel below a certain stress never fails. The curve for aluminum continues down and can fail at any level.

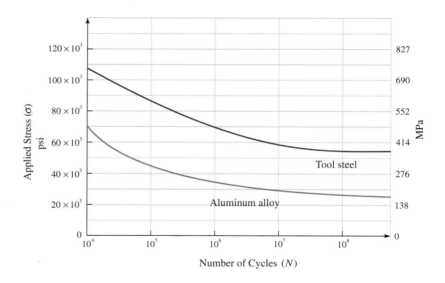

for 30 years, that is over a million and a half cycles. By comparison, parts in the propulsion system of an automobile can move through 500 times as many cycles in their much shorter lifetimes.

In contrast to steel, nonferrous metals and alloys, aluminum and copper among them, have σ-versus-N curves that give no indication of ever leveling off, and so no fatigue limit exists for these materials. This means that an aluminum part can be expected to fail at any applied stress sooner or later—the idea is to make the "later" long after the designed lifetime.

Figure 10.21a is an X-ray of a broken ulna in the forearm of a man showing the steel brace used to hold the two segments together. Unfortunately, the bone did not knit and so whenever the patient used his arm, especially during physical therapy, the brace carried the load and experienced cyclic stresses, often reversing completely. Of course it should have been expected that the fairly thin brace would fail sooner or later due to fatigue, but no one knew the bone had not fused. Finally, the brace ruptured at the place where the steel was narrowest and the stress the greatest, right across a screw hole.

MODULI AND ENERGY

In 1807, Dr. Thomas Young, one of the most brilliant minds of the era, introduced a notion he called "the modulus of the elasticity" (*modulus* is Latin for "a small measure"). But it was the French engineer Claude Navier who in 1826 provided the modern formulation of what has come to be known as **Young's Modulus** (Y), or the **Modulus of Elasticity**. What Young had found was a measure of the stiffness of materials.

10.12 Young's Modulus

The normal stress-strain graphs for many materials begin with a linear region where the substance, at least at low strain, is Hookean (Fig. 10.17a). This means that the slope of the curve is a constant, which reflects how the material elastically strains under an increasing normal stress. This **elastic stress per unit strain** defines Young's Modulus:

$$\text{Young's Modulus} = \frac{\text{stress}}{\text{strain}}$$

or symbolically

$$Y = \frac{\sigma}{\varepsilon} \tag{10.9}$$

which corresponds to either tension or compression (Table 10.4). The units of Y are the same as those for stress—namely, N/m^2 or Pa. The steeper the slope, the less distortion the sample experiences for a given stress, the stiffer it is, and the greater is Y. A stiff material more nearly maintains its size and shape under a load in the elastic regime.

Referring to the loaded rod in Fig. 10.13b, since $\sigma = F/A$ and $\varepsilon = \Delta L/L_0$, Eq. (10.9) rearranged, $\sigma = Y\varepsilon$, leads to $F/A = Y\varepsilon$, and so

$$F = \frac{YA}{L_0}\Delta L \tag{10.10}$$

which is the equivalent of Hooke's Law, $F = k\Delta L$; hence, $k = YA/L_0$.

Figure 10.21 An X-ray of a steel plate used to bridge a broken bone. Months later, the plate fatigued and broke. The break occurred at a hole, right where you would expect.

Example 10.8 Return to Example 10.7, which deals with a broken femur, and calculate the fractional change in length of the bone, assuming Y is constant right up to failure. Use Table 10.4.

Solution: [Given: $\sigma_c = 170$ MPa. Find: ε_c.] From

Eq. (10.9) and Table 10.4

$$\varepsilon_c = \frac{\sigma_c}{Y} = \frac{170 \text{ MPa}}{10 \text{ GPa}} = 17 \times 10^{-3} = \boxed{1.7\%}$$

▶ **Quick Check:** Figure 10.19 agrees with 1.7%.

TABLE 10.4	Approximate values of Young's Modulus for various solids		
Material	**Young's Modulus (GPa)**	**Material**	**Young's Modulus (GPa)**
Biological tissue,		Compact bone,	
soft	0.000 2	compression	10
Rubber	0.007	tension	22
Egg membrane	0.008	Concrete	25–30
Cartilage, human	0.024	Magnesium	42
Collagen	0.6	Marble	50
Tendon, human	0.6	Aluminum	56–77
Nylon	2	Glass	65
Polystyrene	3	Iron, cast, gray	70–145
Spider thread	3	Granite	70
Catgut	3	Gold	79
Nylon fiber	5.5	Brass, yellow	105
Plywood	7	Copper	120
Hair	10	Bronze	120
Pine, southern	11	Iron, wrought	180–210
Fir, Douglas	13	Steel, stainless	190
Oak	14	Steel, structural	200
Brick	14	Tungsten	360
Lead	16	Sapphire	420
		Osmium	560
		Diamond	1200

The stronger the interatomic bonding, the greater the force that must be applied to separate the atoms and, in turn, to stretch the material. Thus, we should expect that for materials with larger and larger bonding forces, as indicated by increasing melting points, Young's Modulus would increase as well (Table 10.5).

Notice that when a substance elongates to twice its original length $\Delta L = L_0$, whereupon the strain is 100%. Young's Modulus then equals the corresponding stress (presuming the thing does not break before then). *We can estimate the stress needed to double the separation of the atoms in a molecule, and that should provide something resembling Y.*

The energy needed to pull a molecule apart, its **binding energy**, is determined by the specific bonding mechanisms at work. Table 10.6 provides a range of values for the energies associated with the various types of atomic bonding. Thus, while it takes about 9×10^{-19} J to separate the sodium from the chlorine in a molecule of table salt, just about half as much energy will break up a molecule of water into hy-

TABLE 10.5	Young's Modulus and the melting point for various solids	
Material	**Young's Modulus (GPa)**	**Melting point (°C)**
Lead	16	327
Magnesium	42	650
Aluminum	56–77	660
Copper	120	1083
Tungsten	360	3410

TABLE 10.6	Binding energies
Bond mechanism	**Energy (eV/particle)***
Weak	<0.4
Metallic	1–4.5
Covalent	0.6–8.4
Ionic	6–11

*1 eV = 1.6×10^{-19} J.

drogen and oxygen. In fact, the majority of atoms are bound to their respective molecules and neighboring atoms in a solid with an energy, within a factor of 10 or so, of 1.6×10^{-19} J. This particular quantity is called an **electron volt** (eV) and it's especially convenient in the microworld: 1.6×10^{-19} J = 1 eV.

To approximate Young's Modulus, let's take the interatomic separation of the material as 0.2 nm and the binding energy as 2 eV = 3.2×10^{-19} J. If we assume that pulling the atoms completely apart is not going to require appreciably more energy than separating them by an additional 0.2 nm, then the work done in so doing is approximately the binding energy. Since $\Delta W = F_{av}\Delta r$, for each atom

$$F_{av} = \frac{\Delta W}{\Delta r}$$

and

$$F_{av} = \frac{3.2 \times 10^{-19} \text{ J}}{0.2 \times 10^{-9} \text{ m}} = 16 \times 10^{-10} \text{ N}$$

Because the atoms are originally separated by 0.2 nm, there are $(0.2 \text{ nm})^{-1}$ atoms per meter, and about $(0.2 \text{ nm})^{-2}$, or roughly 2.5×10^{19}, atoms per square meter. Thus, the average stress is $F_{av}/A = (16 \times 10^{-10} \text{ N/atom})(2.5 \times 10^{19} \text{ atoms/m}^2)$ = 40 GPa, which is a fine value of Young's Modulus for a metal.

As something of a review, keep in mind that nylon fishing line and catgut are both *flexible* (as exemplified by low Y values), and yet both are quite *strong*. Cartilage, cooked spaghetti, and Jello are also all *flexible*, but each is relatively *weak* (they have low breaking strengths). On the other hand, a hard pretzel or a dried noodle is *stiff* (large Y) but *weak*, breaking easily, whereas a steel crowbar is *stiff* and *strong*.

10.13 Shear and Bulk Moduli

The stress-strain curves for shear are also often linear initially and therefore an elastic **Shear Modulus** (S), or **Modulus of Rigidity**, can be introduced:

$$\text{Shear Modulus} = \frac{\text{shear stress}}{\text{shear strain}}$$

that is

$$S = \frac{\sigma_s}{\varepsilon_s} = \frac{F/A}{\gamma} \qquad (10.11)$$

where the force acts parallel to the area, and we have made use of Eq. (10.8). The Shear Modulus is typically 2 or 3 times less than the Young's Modulus for the same material (Table 10.7).

Thomas Young (1773–1829), physicist and physician, has a long list of accomplishments to his credit, including his introduction of the Modulus of Elasticity. He is best known for his work in optics.

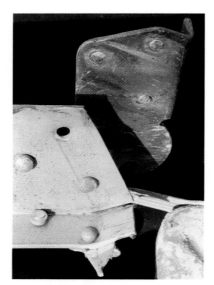

The top of a failed extension ladder. Each side had a steel holding plate fastened to the body with three aluminum rivets. The steel plates are much harder than the rivets, and when there was enough torque, they sheared the rivets in half. The plate twisted upward, still holding its half of the three severed rivets. One disembodied head has dropped out of its hole in the body and rests on the flange. The ladder collapsed and the man on it luckily broke only his wrist.

TABLE 10.7 Shear and bulk moduli		
Material	Shear Modulus (GPa)	Bulk Modulus (GPa)
Rubber	0.000 8–0.001 6	
Ethyl alcohol		0.9
Water		2
Seawater		2.1
Pine	0.6	3
Glycerin		4.5
Lead	5	8
Mercury		25
Quartz	8–30	35
Compact bone	3.5	12
Marble		70
Aluminum	30	70
Glass	8–30	40
Iron, cast	50	60
Granite		50
Brass	40	70
Copper	50	140
Iron, wrought	85	60
Steel, stainless	85	70
Steel, structural	80	160
Tungsten	150	200
Diamond		620

When an ordinary helical spring is "stretched," the coiled wire itself does not elongate. Rather than being in tension or compression, the wire is actually twisted by a load that tends to pull it out into a straight length. Thus, the spring constant is actually dependent on the Shear Modulus.

Figure 10.22 shows the geometry for a tube, resembling a hollow human bone, being twisted by a torque τ. The resulting angle $\gamma \approx (R_0 \phi)/l$ is proportional to τ/S. If some structure or component is twisted (a bone, for example) through a large enough angle ϕ, it may fracture, which is what often happens to the tibia when a skier's leg is twisted in a fall (Fig. 10.23). Table 10.8 lists a few values of both the breaking torque and the corresponding value of ϕ for some human bones.

Similarly, we can define a **Bulk Modulus** (B) as the ratio of the volumetric stress (the pressure P) to the volumetric strain:

$$\text{Bulk Modulus} = -\frac{\text{volumetric stress}}{\text{volumetric strain}}$$

that is

$$B = -\frac{\sigma_v}{\varepsilon_v} = -\frac{F/A}{\Delta V/V_0} \tag{10.12}$$

The negative sign makes the modulus positive since an increase in stress causes a decrease in volume and therefore a negative strain (Table 10.7). Notice that the larger the value of B, the smaller the change in the volume ΔV, the less compressible is the material.

(a) (b) (c)

Figure 10.22 As an example of the theoretical Shear Modulus, consider a hollow thin-walled tube: $S = (F/A)/\gamma$. (a) A torque is applied and the rod is twisted through an angle γ. A small region in the untwisted state (b) experiences shear (c) in the twisted state.

(a) (b) (c)

Figure 10.23 Common bone breaks. (a) A spiral fracture results from a torque that twists the bone. This type of break is often associated with skiing accidents. (b) Under compression, the bone may fail and the fractured ends break into pieces. (c) An applied transverse force bends the bone. The side away from the force is stretched, and since bone is weak in tension, it shatters. The near side is in compression, and if the bone is flexible, it might not break there at all.

	τ	ϕ
TABLE 10.8 Torques and breaking angles for human bones		
Bone	(N·m)	(degrees)
Ulna or radius (forearm)	20	15
Humerus (upper arm)	60	6
Femur (thigh)	140	1.5
Tibia (lower leg)	100	3

Example 10.9 Determine the change in the volume of a 1.00-m³ block of granite when it's submerged about 3 km in the ocean, where the pressure on all its surfaces is about 300 times ordinary atmospheric pressure (1.013×10^5 Pa).

Solution: [Given: $P = F/A = 300(1.013 \times 10^5$ Pa), $V_0 = 1.00$ m³. Find: ΔV.] From Table 10.7, $B = 50$ GPa; hence from Eq. (10.12)

$$\Delta V = -\frac{V_0 F/A}{B}$$

$$\Delta V = -\frac{(1.00 \text{ m}^3)(3.039 \times 10^7 \text{ Pa})}{50 \times 10^9 \text{ Pa}} = \boxed{-6.1 \times 10^{-4} \text{ m}^3}$$

▶ **Quick Check:** $B = -(F/A)/(\Delta V/V_0) \approx (-3 \times 10^7)/(-6 \times 10^{-4})$ Pa $\approx 0.5 \times 10^{11}$ Pa.

Clearly, solids do not compress very much. By comparison, water is about 25 times more compressible than granite, but that still isn't a great deal. Nonetheless, estimates indicate that if water were truly incompressible, sea level would be roughly 30 m higher than it is.

Core Material

An **element** is a homogeneous material that cannot be broken down into another substance by any chemical process. Carbon (^{12}C) is the mass standard, equal to 12.000 00 *unified atomic mass units* (u). A volume of 22.4 liters of any gas at STP has a mass in grams numerically equal to its molecular mass and that constitutes a **mole** (p. 320). The number of molecules per mole is Avogadro's number: $N_A \approx 6.022 \times 10^{23}$ molecules/mole.

Density is defined as

$$\text{Density} = \frac{\text{mass}}{\text{volume}}$$

$$\rho = \frac{m}{V} \qquad [10.1]$$

Solids can be classified as either *crystalline, quasi-crystalline, amorphous,* or *composite* (p. 324).

Hooke's Law for a spring is

$$F = ks \qquad [10.3]$$

where s is either the distance it is elongated or compressed and k is the elastic constant with units of N/m. The elastic potential energy stored in such a spring is

$$\Delta PE_e = \tfrac{1}{2}ks^2 \qquad [10.4]$$

Many materials manifest a linear relationship between the load F they support and the resulting change in length of the sample (ΔL):

$$F = k\Delta L \qquad [10.5]$$

Stress is defined as the force acting, divided by the area over which it acts:

$$\sigma = \frac{F}{A} \qquad [10.6]$$

Strain is the fractional change in some geometrical measure of a sample under stress. Thus, for a specimen experiencing an axial load, the *normal strain,* compressive or tensile, is

$$\varepsilon = \frac{\Delta L}{L_0} \qquad [10.7]$$

Similarly, the *shear strain* equals the angle of distortion γ:

$$\varepsilon_s = \gamma \approx \tan \gamma \approx \frac{\Delta l}{l_0} \qquad [10.8]$$

The **strength** of a material is the stress, in N/m², that will cause failure of some sort in a sample.

The *elastic stress per unit strain* defines **Young's Modulus**:

$$Y = \frac{\sigma}{\varepsilon} \qquad [10.9]$$

in either tension or compression. The units of Y are the same as those for stress—namely, Pa. Alternatively

$$F = \frac{YA}{L_0}\Delta L \qquad [10.10]$$

which is the equivalent of Hooke's Law, $F = k\Delta L$.

The elastic **Shear Modulus**, or **Modulus of Rigidity**, is the shear stress over the shear strain:

$$S = \frac{\sigma_s}{\varepsilon_s} = \frac{F/A}{\gamma} \qquad [10.11]$$

Similarly, the **Bulk Modulus** is the ratio of the volumetric stress (the pressure P) to the volumetric strain:

$$B = -\frac{\sigma_v}{\varepsilon_v} = -\frac{F/A}{\Delta V/V_0} \qquad [10.12]$$

Suggestions on Problem Solving

1. Area conversions (for example, cm^2 to m^2) are common in this chapter, so be careful with them. Don't mix up diameters and radii when calculating areas.

2. We deal extensively with the abbreviated notation of mega- and giga-, as for example MPa and GPa, which is often the way data is given in the literature. Get in the habit of entering the numbers in your calculations in scientific notation right away. It's easier to substitute the numbers in MPa, but it's also easier to lose the factor of 10^6 that way.

3. Young's Modulus provides an insight into the meaning of the spring constant by way of Eq. (10.10). Hence, when confronted with a problem dealing with the physical parameters that describe the object being loaded (length, cross section, etc.) think in terms of Y and that equation.

Discussion Questions

1. Where in your home might you find each of the following?

aluminum	tungsten	carbon	oxygen
mercury	zinc	chlorine	chromium
copper	sodium	iron	nickel
lead	magnesium	cobalt	

2. Prestressed concrete structural forms are made by setting up steel rods within a mold and then pouring in concrete around the rods, which are kept stretched until the concrete sets. Why is this manner of construction advantageous?

3. Figure Q3 shows the stress-strain curves for several kinds of steel. (a) Which is the most ductile? (b) Which is the most brittle? (c) What can be said about their Young's Moduli? (d) Which has the greatest ultimate strength? (e) Which has the greatest rupture strength? (f) Which has the least yield strength? (g) Which has the greatest stiffness in the elastic region? (h) Which has the greatest strain at rupture?

Figure Q3

4. Refer to Fig. Q3. If a high-strength steel is substituted for a low-strength steel and all else is kept constant, what will happen to the load-carrying capacity and stiffness of the structure?

5. The area under the force-displacement curve is called the **strain energy**. (a) Discuss the physical meaning of this quantity. Similarly, the area under a stress-strain curve is called the **strain-energy density**. (b) Discuss its physical meaning. What are the units of strain-energy density? (c) Figure Q5 shows a σ-ε curve for a material strained to a value of ε_F beyond the yield point. It is unloaded and returns to zero stress, but not zero strain. There is a permanent strain ε_p. What is the significance of the colored area?

Figure Q5

6. Why are "steel-belted" tires belted with steel? Or, alternatively, why is it advisable to cover a recurring crack in a ceiling with cloth mesh before replastering it?

7. Imagine two vertical hanging ropes that are identical except that one is considerably longer than the other. A gradually increasing load is applied to each of their bottom ends until one or both break. (a) What can you say about the two breaking loads? (b) Compare the amounts of strain energy stored—that is, the amount of energy pumped in—in order to break the ropes. (See Question 5.)

8. Human bones in the embryo stage are mostly collagen and only later are they gradually reinforced with calcium and phosphorus. Compare the corresponding values of Young's Modulus and discuss the implications in regard to the mechanical properties of children.

9. Mountains are typically shaped like a cone. Why is that configuration more able to reach great heights than a straight vertical columnlike structure? Explain why a column of marble about 4.2 km high would collapse under its own weight. Given that granite has about the same density as marble, how tall a column could be made of it before it crushed itself?

10. There is a general principle (attributed to Galileo) that deals with scaling structures—that is, increasing or decreasing their sizes proportionately. For example, King Kong was a scaled-up gorilla. The *square-cube rule* points out that while the volume of a thing being scaled-up increases as the cube of the dimensions, the cross-sectional area of the load-supporting members only increases as the square of the dimensions. How does the weight of the structure increase? What happens to the stress in the load-bearing members? Assuming King Kong was

made out of normal gorilla-stuff, why would he have to be very careful when standing up?

11. The strengths and rigidities of animal bones do not vary all that much from one type of mammal to the next. Human bone (femur), for instance, has a relatively high compressive strength of 170 MPa; at the other extreme, pig bone has a low compressive strength of 100 MPa—not that much different. With Question 10 in mind, what does this suggest about the relative safety of hopping rabbits as compared to jumping elephants? What advice would you give to a horse about prancing around?

12. Rubber at a working elastic strain of 300% has a stress of 7 MPa, and the area under its nonlinear stress-strain curve is 10 MJ/m³. What is the significance of this area (which, by the way, is about 10 times the corresponding value for spring steel)? Why do we make the heels of shoes out of rubber? (See Question 5.)

13. The stress-strain diagram for vulcanized rubber is shown in Fig. Q13. The sample returns to its original length when the load is removed, but not along the same curve—this phenomenon is called **elastic hysteresis**. Discuss the work done *on* the sample to deform it as compared to the work done *by* the rubber on returning to zero strain. What happens to the difference? Why do we put rubber legs under vibrating pieces of equipment or rubber bumpers on automobiles and tugboats? Stretch a heavy rubber band quickly a few times—what will happen to its temperature?

14. The inverse of the Bulk Modulus is known as the *compressibility*. Is this name appropriate? Explain.

Figure Q13

15. For a Hookean material, the area under the linear portion of the stress-strain curve is called the **resilience**. It's a measure of how much energy the material or structure can store without permanently distorting. The area under the curve all the way out to rupture is known as the **toughness**. It's a measure of the ability of a material or structure to withstand a blow without rupturing. In other words, it's a measure of the energy it takes to break the object. What can you say about the resilience of a human tendon or a bridge or a steel spring? Is a glass bowl or a china plate tough? Explain. Suppose you were designing a bow for shooting arrows or a pole for vaulting. Do you want these to be tough or resilient?

Multiple Choice Questions

1. The dimensions of density are   (a) $[L/M]$ (b) $[L^3/M]$
   (c) $[M^3/L]$ (d) $[M/L^3]$ (e) $[M/L]$.
2. If we totally decompose 11 g of carbon dioxide, how many grams of oxygen would be liberated?   (a) 11   (b) 8
   (c) 3   (d) 44   (e) none of these.
3. Figure MC3 shows a nail that was left in a concentrated solution of sugar for several days. What has grown on the nail can best be described as   (a) a single crystal   (b) an amorphous mass   (c) a glassy solid   (d) a polycrystalline mass   (e) none of these.
4. Methane melts at −184°C, whereas table salt melts at 801°C. From this we can conclude that the crystal's cohesive forces, the bonding forces,   (a) are much larger in sodium chloride than methane   (b) have nothing to do with melting   (c) are both ionic   (d) are both van der Waals   (e) none of these.
5. Figure MC5 shows a stack of cotton swabs. In two dimensions, it suggests a crystal system known as
   (a) FCC   (b) HCP   (c) SC   (d) LSD   (e) none of these.
6. A simple molecule such as water has a size of about
   (a) 0.3 nm   (b) 3.0 nm   (c) 3×10^{-6} m   (d) 0.3 mm
   (e) none of these.
7. The dimensions of stress are   (a) $[L/M]$ (b) $[L^3T/M]$
   (c) $[M^3T/L]$ (d) $[M/T^2L]$ (e) $[M/T^3L]$.
8. A metal part fails via fatigue   (a) when it gets tired
   (b) as a result of the repeated application of a cyclic stress
   (c) when the stress exceeds the breaking strength
   (d) when the metal is very old and the load very large
   (e) none of these.

Figure MC3

9. An ordinary piece of blackboard chalk is   (a) ductile
   (b) quite tough   (c) brittle   (d) highly resilient
   (e) none of these.

Figure MC5

10. A rod of tungsten is quite stiff, and so one can say that tungsten   (a) is soft   (b) is tough   (c) has a large conductivity   (d) has a large Young's Modulus   (e) none of these.
11. A spring with a large elastic force constant can be said to be fairly   (a) soft   (b) rigid   (c) long   (d) short   (e) none of these.
12. A spring balance with a linear scale can be used to measure mass because   (a) springs are strong   (b) metal springs are nearly weightless   (c) a spring-force is linear because it has a large Young's Modulus   (d) the spring-force is linear in displacement via Hooke's Law   (e) none of these.
13. A piece of wire is stretched by a certain amount and then allowed to return to its original configuration. It is then stretched twice that initial amount, without exceeding its elastic limit. Compared to the first stretching, the second elongation stored   (a) twice as much energy   (b) four times as much energy   (c) half as much energy   (d) no additional energy   (e) none of these.
14. A concrete block supported horizontally at its two ends can be broken with a downward karate chop to the middle because   (a) the bottom of the block will be in tension and concrete is weak in tension so it will crack on the bottom face   (b) the top will be in compression and concrete is weak in compression so it will crack at the top   (c) the bottom will be in compression so it will crack on the bottom   (d) the top will be in tension so it will crack on top   (e) none of these.
15. It is possible, while tightening the nuts on the wheel of your car, to use too much torque and break off one of the bolts. This happens when the   (a) shear stress exceeds the breaking strength   (b) compressive stress becomes too large   (c) screw has too low an elastic constant   (d) volumetric stress on the bolt is too great   (e) none of these.
16. The units of strain are   (a) m/s   (b) N/m   (c) m^2 (d) N/m^2 (e) none of these.
17. The breaking strength of a ductile material is generally   (a) lower than the yield strength   (b) very low   (c) higher than the yield strength   (d) higher than the ultimate strength   (e) none of these.
18. A steel wire stretches by 10 cm when a load is hung from it. If a new wire made of the same material with twice the cross-sectional area supports the same load, it will stretch   (a) the same amount   (b) not at all   (c) half as much   (d) twice as much   (e) not enough information given.
19. A dress made of cloth will hang nicely, conforming to the shape of the body beneath, because cloth has both   (a) a large Young's Modulus and a low Bulk Modulus   (b) a low Shear Modulus and a low Modulus of Elasticity   (c) a high Young's Modulus and a low Modulus of Toughness   (d) a low Bulk Modulus and a high Shear Modulus   (e) none of these.
20. Of the several moduli, the only one applicable to liquids and even to gases is   (a) Young's Modulus   (b) the Modulus of Toughness   (c) the Shear Modulus   (c) the Elastic Modulus   (d) the Bulk Modulus   (e) none of these.

Problems

ATOMS AND MATTER
1. [I] How many molecules of trichloromonofluoromethane are there in 22.4×10^3 cm^3 of the gas at STP?
2. [I] What quantity of water will be produced when 5.0 g of hydrogen are burned in the presence of 40 g of oxygen? What if 6.0 g of hydrogen are burned in the presence of the same amount of oxygen?
3. [I] White dwarf stars are very small and very massive. A 1-in.^{3} chunk of such star-stuff brought to Earth would weigh about 1 ton (2000 lb). Determine its density in SI units.
4. [I] The mass of the Earth is 5.975×10^{24} kg, and its mean diameter is $1.274\,246 \times 10^7$ m. Determine its average density. Given that the density of nuclear matter is about 2×10^{17} kg/m^3, if the Earth were squashed into a small sphere, how big would it have to be to have a comparable density?
5. [I] A typical human being contains roughly 13 gallons of water, where 1.00 gallon = 3.785 liters. What is the corresponding mass of water?
6. [I] How many grams of salt are in a gram-mole of NaCl?
7. [I] Imagine that a kilomole of a substance undergoes some chemical change that liberates 2.0 eV per reaction, per molecule. How much energy will be released by the entire kilomole?
8. [I] A cubic centimeter of water has a mass of 1.00 g. How many *molecules* does it contain? How many atoms are in a 1.00-g ice cube?
9. [I] How many molecules of pure sugar ($C_{12}H_{22}O_{11}$) are in 684 g of the stuff?
10. [I] A ring of pure gold has a mass of 19.7 g. How many atoms are in it?
11. [I] Benzoylmethylecgonine (cocaine) is $C_8H_{13}N(OOCC_6H_5)$ $(COOCH_3)$. What is its atomic weight?

12. [I] Assuming that you weigh 150 lb and that your body is roughly 70% water, how many H_2O molecules are you?

13. [I] Make a rough estimate of the number of atoms in all the stars in the Universe. (Hint: Assume an average atomic mass of, say, 10 u; a star-mass of 10^{33} g; 10^{11} stars in a galaxy; and 10^{11} galaxies in the Universe.)

14. [II] Derive a general expression for the number of molecules there are in a cubic meter (the number-density) of any substance at STP.

15. [II] Using the results of Problem 14, determine the number-density for water—that is, the number of molecules there are per cubic meter. (Check your answer with Table 10.2.)

16. [II] The mass of a proton is 1.673×10^{-27} kg, and it can be considered to be a sphere of roughly 1.35×10^{-15}-m radius. Determine its density and compare it with that of most solids. How much is that in tons per cubic inch (1 ton = 2000 lb)?

17. [II] A cubic foot of water weighs 62.4 lb. How many kilograms of hydrogen are in that much water?

18. [II] What is the mass of a single water molecule, and how many of them are in 1.000 g of pure water? The atomic masses of hydrogen and oxygen are 1.008 u and 16.00 u, respectively.

19. [II] Ethyl alcohol, C_2H_5OH, has a density of 0.789×10^3 kg/m³. Estimate the size of the molecule.

20. [II] Assume that a water molecule is a sphere of diameter 0.3 nm. Approximately how many of these spherical molecules will be in 18 cm³ of water? How does this result compare with the correct answer? Why is it in error?

21. [III] What is the approximate separation between gas molecules at STP? How does this compare to the separation between molecules in a solid?

ELASTICITY MODULI AND ENERGY

22. [I] Given a spring with an elastic constant of 500 N/m, how much will it contract when pushed on by an axial force of 10 N?

23. [I] A helical spring 20 cm long extends to a length of 25 cm when it supports a load of 50 N. Determine the spring constant.

24. [I] How much energy is stored in a helical spring with an elastic constant of 50 N/m when compressed 0.05 m?

25. [I] An axial force of 200 N is applied to a rod with a cross-sectional area of 10^{-4} m². What is the normal stress within the rod?

26. [I] A man leans down on a cane with a vertical axial force of 100 N. The narrowest part of the cane has a cross-sectional area of 1.0 cm². What is the maximum compressive stress in the cane?

27. [I] A 10-m-long wire strung between two trees has a "Curb Your Dog" sign weighing 10 N hung from it. If the wire stretches 1.5 cm, what strain is it under?

28. [I] A rubber band is stretched to twice its original length by a kid who is going to shoot it across a room. What strain is it experiencing?

29. [I] If the normal strain experienced by a rod in compression is 5.000×10^{-4} and its original length is 40.00 cm, what's its final length?

30. [I] Given that collagen has an ultimate tensile strength of 60 MPa, what maximum load could be supported by a fiber with a cross-sectional area of 10^{-6} m²?

31. [I] A length of tendon L_0 mm long stretches an amount ΔL under a load F. A second piece twice as long, with twice the cross-sectional area, experiences the same load. By how much will it elongate in comparison?

32. [I] Pure natural rubber has a tensile strength of 21 MPa. When carbon particles are added to reinforce it, the tensile strength rises to 31 MPa (which is one reason why you see so much black rubber around). If a rod of natural rubber can support a maximum load of 1000 N, how much will an identical rod of carbon-reinforced rubber support?

33. [I] A rod made of structural steel, 4 m long and 2 mm in diameter, supports a 1.0-lb microphone hanging 3 m above the floor of a gym. A 181.4-kg (400-lb) Sumo wrestler reaches up, grabs the mike, and hangs from it. How much will the rod stretch, assuming it doesn't break?

34. [I] A piece of fresh human cartilage, with a 1.0 cm² cross-sectional area, is being studied in a laboratory. It is found that when loaded with 100 N, its length increases by 4.2%. Determine its Elastic Modulus.

35. [I] A dictionary weighing 10 N rests face down on a table. The book's cover is 20 cm by 30 cm. What is the pressure it exerts on the table?

36. [I] A cube of marble 1.00 m on each edge is submerged in the sea so that there is a distributed force F acting perpendicularly inward on each of its faces. What must that force be if the block is to decrease in volume by 1.00%?

37. [II] A helical spring is 55 cm long when a load of 100 N is hung from it and 57 cm long when the load is 110 N. Find its spring constant.

38. [II] Two identical helical springs are attached to one another, end-to-end, making one long spring. If $k = 500$ N/m is the elastic constant of each separate (essentially weightless) spring, what is the constant for the combination?

39. [II] Estimate the value of the spring constant for an atom bound in a molecule. Take the potential energy stored when the displacement from equilibrium is 0.2 nm to be 2 eV = 3.2×10^{-19} J.

40. [II] Two identical wires each of length L_0 are attached to one another producing a single long wire. If $k = 500$ N/m was the elastic constant of each separate (essentially weightless) wire, what is the constant for the combination?

41. [II] A brass wire 1.5 m long is to support a 200-N "Eat At Joe's" sign without permanently elongating. Given that the yield strength of this variety of yellow cold-rolled brass is 435 MPa, what minimum diameter wire is called for?

42. [II] A thin rod of structural steel 5.0 m long is to carry a tensile load such that it's just about to begin to deform plastically. Given that the yield strength in tension of steel is 250 MPa, how much will the rod stretch?

43. [II] A rubber motor belt with a circular cross section of radius 1.25 cm breaks at a tensile load of 7363 N. If a new belt was made of the same material with twice the cross-sectional radius, at what load would it break?

44. [II] What is the largest working strain you should design for when dealing with a structural steel member that is not to be permanently deformed by the tensile load? The yield

strength of the steel in tension is 250 MPa as compared to its tensile strength of 400 MPa.

45. [II] Marble crushes under a compressive stress of 110 MPa. What maximum strain will it experience?

46. [II] The compressive strength and maximum strain of the crown of a human tooth are 150 MPa and 2.3%, respectively. Assuming the crown to behave elastically until it ruptures, determine its Young's Modulus and (remembering Discussion Question 15) its resilience.

47. [II] The cable on a crane is to be made of high-strength, low-alloy steel with a yield strength in tension of 345 MPa. If the maximum load is to be 25 000 lb, how big in diameter should the cable be? What maximum strain can it experience given that its Modulus of Elasticity is 200 GPa?

48. [II] A certain kind of brick has a density of 2×10^3 kg/m^3 and a compressive strength of 40 MPa. What's the highest vertical parallel-walled tower that can be built that will not crush the bottom bricks?

49. [II] The problem of bending a rod of length L and radius r is a fairly complicated one to analyze. Nonetheless we can approach it, if only crudely, using Eq. (10.11). Accordingly, suppose a young pine tree experiences a horizontal force of 200 N at its top and bends through an angle of roughly 0.001 rad. Approximate its diameter. Actually, $S \approx (\sigma_s/\gamma)\frac{1}{2}(L/r)^2$ is a much better approximation than Eq. (10.11)—the details of the geometry are important.

50. [II] A structural steel bar 4.0 cm in diameter juts horizontally 50 cm straight out of a wall. A 90-kg (200-lb) person reaches up, grabs the bar at its end, and hangs from it. By approximately how much will the bar deflect? (Hint: See Problem 49.)

51. [II] Two 1.00-cm-thick steel plates on a bridge are riveted to one another with five 4.00-cm-diameter steel rivets. Given that the yield strength in shear of structural steel is 145 MPa, what maximum force parallel to the plates can they be expected to handle without permanent deformation?

52. [II] Show that the *elastic*-PE stored in a normally strained material of uniform cross-sectional area A and original length L_0 is given by

$$PE_e = \frac{F^2 L_0}{2AY}$$

53. [II] The bar in Fig. P53 is made from a single piece of material. It consists of two segments of equal length $\frac{1}{2}L_0$

Figure P53

and cross-sectional areas of A and $2A$. Show that the *elastic*-PE stored under the action of an axial force F is

$$PE_e = \frac{3F^2 L_0}{8AY}$$

(Hint: Look at Problem 52.)

54. [III] A solid lead sphere is placed in a pressure chamber so that the force per unit area acting on the sphere is hundreds of times atmospheric pressure (1 atm = 1.013×10^5 Pa). What pressure will produce a 1% increase in the density of the lead?

55. [III] If a uniform rod of cross-sectional area A and length L_0 can sustain a maximum stress of σ_R without rupture, show that it stores an amount of elastic energy given by

$$PE_e = \frac{\frac{1}{2}AL_0\sigma_R^2}{Y}$$

in getting to that stress level.

56. [III] A robot's main body is supported on two legs, each containing two vertical rods one above the other jointed at the knee in human fashion. Each rod has a cross-sectional area of 5 cm^2 and a length of 50 cm and is made of a bonelike material that can sustain a maximum stress of 150 MPa without rupturing and has a Young's Modulus (in compression) of 10 GPa. How much elastic energy can be stored in these two leg supports (as a result of jumping in a gravity field) without having to go back to the shop for repair? If the robot has a mass of 70 kg, how high a drop can it sustain on Earth?

11

Fluids

LIQUIDS AND GASES FLOW, so we call both *fluids*. Their atoms and/or molecules can move around fairly freely, and that contributes to a range of shared properties. Still, there are difficulties in defining the term "fluid" precisely. Even diamond can be made to flow like soft wax at high enough pressures ($>1.7 \times 10^{11}$ Pa). It might be more practical to say that a fluid is an aggregate of atoms that does not significantly resist shear.

The mobility inherent in fluids makes them crucial to all known life forms. The human body itself is a fluid-dynamical system; we breathe, drink, bleed, and excrete fluids. This chapter is about the physics of gases and liquids. It first treats characteristics that are independent of any bulk motion: hydrostatic pressure, hydraulics, and surface tension, which are all examples of *fluid statics*. The chapter then deals with behavior that derives from bulk motion of the medium: fluid flow, Bernoulli's Equation, and vortices, all of which are examples of *fluid dynamics*.

FLUID STATICS

When most solids (such as ice or steel) are heated, their atoms and/or molecules pick up KE in the form of random vibrational motion: they oscillate more vigorously about their equilibrium locations. Once enough energy is obtained to overcome the intermolecular forces, a solid melts. Small groupings of associated molecules still persist in the *liquid*, but as the patterns shift, they link and unlink. There is order, but it's local and changing. A web of cohesive, fairly long-ranged, intermolecular force remains, though the molecules are now energetic enough to easily move against it. They stay relatively near to each other, still interacting appreciably, but the locked, powerful bonding of the solid is gone.

Raising the temperature further brings the liquid to the boiling point, where bonds between molecules are overcome. For some molecules, their random thermal KE exceeds the cohesive potential energy, and they escape the liquid altogether. The short-range groupings disperse and the liquid vaporizes—it turns into a *gas*.

11.1   Liquids

Very few liquids occur naturally in any great quantity. Apart from water and petroleum, most others, such as gasoline, are manufactured. Not surprisingly, living organisms are typically composed of between 65 and 95% water. In fact, many of the familiar liquids—from blood and beer to berry juice—are primarily water.

The balance between bonding energy and thermal energy determines the state of an atomic system. The liquid is a transition state, existing between the random violence of the gas and the relative calm of the solid. At ordinary temperature and pressure, the liquid state is most effectively formed of either highly polarized molecules with electrically positive and negative ends (like water) or large heavy molecules (like petroleum). As with Goldilocks's porridge, the temperature of the liquid has to be "just right" with respect to the intermolecular forces—too hot and it's a gas, too cold and it's a solid.

The ability to flow, the hallmark of the fluid, varies with the cohesive force from one substance to the next—from acetone to water to motor oil to molasses to tar. This notion was first treated analytically by Newton (ca. 1687). *Viscosity* is the internal resistance, or friction, offered to an object moving through a fluid. Usually small molecules, such as water and benzene, move around easily and manifest little viscosity compared to large complex molecules, such as tar. Much of our discussion will be simplified by treating *ideal liquids* that are incompressible and nonviscous—characteristics well approximated by water and many other real liquids under ordinary conditions.

11.2   Gases

Despite their basic mechanical similarity, gases display a diversity of characteristics. The majority of them (nitrogen, oxygen, helium, carbon dioxide, and so forth) are

No physicist would have predicted the existence of a liquid state from our present knowledge of atomic properties.

VICTOR WEISSKOPF (1972)
Physicist

transparent. A few are colorful: chlorine is pale yellow, bromine is orange-brown, and iodine vapor is violet. Radon is radioactive and nasty, oxygen is the breath of life, carbon monoxide is deadly, and nitrous oxide is amusing. Natural gas burns, hydrogen explodes, and carbon dioxide extinguishes. Ozone has a sweet metallic odor, hydrogen sulfide smells rotten-egg vile, and water vapor has no scent at all.

The alchemist van Helmont (ca. 1620) transmuted the Greek word for *chaos* into Flemish and came up with the term *gas*—and a raging chaos it is. At this very instant, each square inch of your face, and every other exposed surface at sea level, is being bombarded by 2×10^{24} air molecules every second! They pelt, ricochet, hit other molecules, and rebound, over and over in a tireless fury (powered, for the most part, by energy from the Sun). At its greatest, the density of the atmosphere at our planet's surface is only 1/800 that of water. Still, at STP there are roughly 3×10^{19} molecules careening around in every cubic centimeter. Though the distance between them is constantly changing, a typical separation is about 10 times the size of the molecules themselves or around 3 nm, which is relatively large. Except for a very slight electromagnetic cohesion and an even slighter attraction due to gravity, these widely spaced molecules behave independently.

Aside from breezes, an average air molecule (Table 11.1) travels at a remarkable 450 m/s, or about 1000 mi/h! Usually a molecule will get no farther than 8×10^{-6} cm before crashing into some other one, only to recoil and collide again. This aimless bombardment goes on for each molecule in a gas at a rate of roughly 6×10^{9} collisions per second. Such a jostling mass of molecules will spread out in a container until it encounters and rebounds from the walls. Swelling of its own internal violence, a gas more or less uniformly fills whatever vessel it's in (provided gravity does not play a significant role). The room you are in at this moment is happily "filled" with some 10^{26} to 10^{30} molecules of air, so you needn't worry about walking into an empty region.

Gravity can function on a large scale to restrain a gas much as a container does. It is the internal gravity that causes an interstellar gas cloud to contract into a star, and it is gravity that holds down the Earth's atmosphere. A balance of molecular speed against strength of attraction keeps that thin layer of air (99.999% lies below 90 km) from escaping and leaving us all breathless.

Nitrogen, the main constituent of air, becomes a liquid at a very low temperature ($-196°C$). It looks and pours like water but evaporates much more quickly.

11.3 Hydrostatic Pressure

Rather than forces that act at a few points on a body, we now treat *distributed forces* that act over an extended area. Accordingly, **pressure** (P) is defined as *the magnitude of the force acting perpendicular to, and distributed over, a surface divided by the area of that surface:*

$$P = \frac{F_\perp}{A} \qquad (11.1)$$

Pressure is a scalar quantity; at any point, it has magnitude but no direction (Table 11.2). The units of P are N/m² or pascals (Pa).

TABLE 11.1 The composition of the Earth's atmosphere

Gas	Percent by volume
Dry air	
Nitrogen	78.08
Oxygen	20.95
Argon	0.9
Carbon dioxide	0.03
Neon	0.002
Helium	0.000 5
Methane	0.000 1
Krypton	0.000 1
Nitrous oxide	0.000 05
Hydrogen	0.000 05
Ozone	0.000 007
Xenon	0.000 009
Typical air	
Nitrogen	76.9
Oxygen	20.7
Water vapor	1.4
All others	1.0

TABLE 11.2   A sampling of pressures

Source	Pressure (Pa)
Sun, at the center	2×10^{16}
Earth, at the center	4×10^{11}
Laboratory, highest sustained	1.5×10^{10}
Ocean, deepest point	1.1×10^{8}
Spiked heels	$\approx 10^{7}$
Atmospheric, Venus	90×10^{5}
Earth	1.0×10^{5}
Mars	≈ 700
Air, Earth 1-km altitude	90×10^{3}
10-km altitude	26×10^{3}
100-km altitude	≈ 0.1
Air, top of Sears Tower, Chicago	95×10^{3}
Air, top of Mt. Everest	30×10^{3}
Air, commercial jet altitude (11 km)	23×10^{3}
Solar radiation, at Earth	5×10^{-6}
Laboratory vacuum (very good)	10^{-12}

Imagine a container filled with a liquid—a cup of tea, say, or a barrel of wine. Gravity acts on the substance and very slightly compresses the fluid, pushing molecule against molecule and ultimately transmitting the force to the bottom of the container, which pushes upward on the liquid. Like a mound of dry sand, the liquid would run off sideways were it not for the walls of the tank. A fluid exerts an outward push on the walls of the container, which restrains it with a counterforce—a counterpressure. This outward pushing of the liquid is observed when water spurts from a hole in the side of a container.

The force exerted by a fluid at rest acting on any rigid surface is always everywhere perpendicular to that surface. It cannot be otherwise since the fluid has no rigidity and will not sustain shear stress. A wooden stick certainly could be used to push on a wall in a direction that is not perpendicular, and the same thing could also be done with a stream of water from a hose, but a fluid *at rest* pushes only perpendicularly. If it somehow managed to exert a tangential force on the wall, the wall would react with a counter tangential force (via Newton's Third Law), and the fluid, unable to resist shear stress, would flow.

A fluid at rest within a container is typically in static equilibrium under the perpendicular compressive forces exerted by the walls. It is important to realize, however, that a liquid, by virtue of its internal (electromagnetic) cohesive force, can also sustain tensile stress (Fig. 11.1). Although liquids usually push outward, they can also pull inward. *The pressure exerted by a fluid will be taken as positive when the fluid is under compression,* as is most often the case (Discussion Question 11).

The mobility of the molecules is the mechanism that transmits pressure *independent of direction.* The value of P measured at any point in a liquid is independent of the orientation of the measuring gauge. Underwater or in a jet plane you can turn your head any way you like and your ears will still feel stuffy.

Gravity and Hydrostatic Pressure

Gravity is the ultimate cause of hydrostatic pressure. To see this, suppose there is a tank of liquid with a postage stamp in it having an area A parallel to and below the surface at a depth h (Fig. 11.2). The downward normal force on the top face of the stamp due to the liquid must be equal to the weight of the column of fluid above the stamp. Assuming the density ρ is constant (because the fluid is essentially incompressible), the mass of the column is its volume $V = Ah$ times its density; that is, $m = \rho A h$. Hence, the weight ($F_{\mathrm{w}} = mg$) of the column is $\rho A h g$, and the average pressure on the stamp due to the liquid *only* is

$$P_{\mathrm{l}} = \frac{F_{\mathrm{w}}}{A} = \frac{\rho A h g}{A}$$

By supporting a load at a number of "points," the normal force at each is small. The pressure at each "point" is then also relatively small.

or

$$P_{\mathrm{l}} = \rho g h \qquad (11.2)$$

Figure 11.1   A liquid in a cylinder being stretched as the tight-fitting piston is pulled down. This is resisted by the liquid's intermolecular (electromagnetic) cohesive forces pulling inward. Pure water has a tensile strength of about 30 MN/m².

Figure 11.2   The weight of the column of liquid above the stamp, of height h, determines the pressure at that level.

Example 11.1   What pressure (due to only the water) will a swimmer 20 m below the surface of the ocean experience?

Solution: [Given: $h = 20$ m and (from Table 10.1) for seawater $\rho = 1.025 \times 10^3$ kg/m³. Find: P_1.] From Eq. (11.2)

$$P_1 = \rho g h = (1.025 \times 10^3 \text{ kg/m}^3)(9.8 \text{ m/s}^2)(20 \text{ m})$$

and $\boxed{P_1 = 2.0 \times 10^5 \text{ Pa}}$

or, since 1 Pa $= 1.45 \times 10^{-4}$ lb/in.², this is 29 lb/in.², or 29 psi, which is about the pressure in an automobile tire.

▶ **Quick Check:** The pressure increase per 1 m of depth is roughly $(1 \times 10^3$ kg/m³$)(10$ m/s²$)(1$ m$) = 10^4$ Pa; at a depth of 20 m, the pressure must be $\approx 20 \times 10^4$ Pa.

For a "weightless" fluid, one far out in space or, more practically, one in free-fall, no internal pressure exists (other than the trivial amount caused by the surface tension). That's true provided the amount of fluid is modest, which means its self-gravity is negligible. This is not the case for something huge like a star. Generally, the force on the postage stamp in any fluid, and therefore the pressure at that level, is determined by the weight of the column above, which is the case whether or not the density varies, as it does in the atmosphere. If the fluid is not homogeneous, Eq. (11.2) will not apply. Still, it can be used successively if there are several homogeneous layers of fluid floating one above the other. When a uniform pressure P_s exists on the surface of the liquid (Fig. 11.3), the total pressure at any point within that liquid is the sum of the contributing pressures:

$$P = P_s + P_1 = P_s + \rho g h \qquad (11.3)$$

The Dutch boy who held back the North Sea by sticking his finger in the dike could really have accomplished the feat. The pressure that he had to overcome de-

Figure 11.3   In a single liquid, the total pressure at a point a distance h beneath the surface is the sum of the pressure on the surface P_s and $\rho g h$.

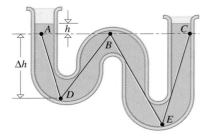

Figure 11.4 The pressure at point A is $P_s + \rho g h$. It increases in going down to D by $\rho g \Delta h$, but decreases by the same amount rising back up to B. Again, it increases even more in descending to E, but drops back that same amount in rising the same vertical distance to point C. Hence, the pressure at A, B, and C is the same.

The pressure at the surface of each liquid column must be the same. Therefore, the level of each surface must be independent of the shape of the column and all must stand at the same height.

Figure 11.5 The pressure at a given depth is independent of the shape of the vessel. It depends only on the density of the liquid and the depth h.

pended on h, the depth in the liquid. The force on the end of his legendary finger acted perpendicularly, tending to pop it out of the hole like a cork in a champagne bottle. But it didn't matter whether he was holding back a leak from an ocean or a bathtub; there is nothing in Eq. (11.3) about the total amount of water present.

Pressure in a Container

It follows from successive applications of Eq. (11.3) that **the pressure at every point at a given horizontal level in a single body of fluid at rest is the same** (Fig. 11.4). Consequently, the pressure at any point on the flat horizontal bottom of a container of fluid is the same, regardless of how weirdly shaped the container may be, since P only depends on the depth beneath the open surface (Fig. 11.5). The validity of this conclusion was proven experimentally by Simon Stevin in 1586 (Fig. 11.6).

What's happening in Fig. 11.6 is that the walls of the container are interacting with the liquid and participating in the process. Examine the long, narrow-necked vessel with the wide shoulders shown in Fig. 11.7. The pressure at point A differs from that at point B by an amount determined by the weight of the column of liquid between A and B, namely $\rho g \Delta h$. The pressure at B equals that at D and so the pressure at C, also a height Δh above D, must equal that at A, even though C is under the glass shoulder. Anywhere at the level of point A, the pressure in the fluid is the same whether there is liquid or glass directly above the point. The pressure generates an upward normal force on each unit area of the horizontal shoulder, and the glass, in turn, pushes down on the liquid with the same force per unit area. The pressure is precisely equal to that produced by the weight of the column of fluid above A in the neck of the vessel. A guppy swimming around could not tell from the pressure whether it was under the shoulder or the neck.

Atmospheric Pressure and the Barometer

A formidable problem plaguing the European mining community in the seventeenth century was the seepage of groundwater into the deep shafts of iron, silver, alum, and salt mines. The real frustration arose from the mysterious inability of even the best suction pumps to raise the water any more than 34 ft above the flood level. In Italy, Galileo was consulted about the strange phenomenon. The maestro knew that air had weight. He had determined that much experimentally by weighing a glass bulb sealed under room conditions and then reweighing it after forcing in air under pressure. Yet it was only after Galileo died that his assistant, Evangelista Torricelli, was able to recognize that the two apparently unrelated ideas were cause and effect (1643).

Being 13.6 times denser than water, mercury would presumably rise only about 30 in. instead of 34 ft and so would be far more convenient to work with in the laboratory. Cleverly, Torricelli sealed one end of a glass tube about 2 m long, filled it with mercury, and covering the opening with a finger, upended the tube into a bowl of mercury (Fig. 11.8). On removing his finger from the submerged end, some of the liquid poured out of the tube into the bowl, and the shiny column dropped to a height of about 76 cm (30 in.) above the open mercury level. Since no air had entered the tube, the space above the mercury column was apparently empty—the vacuum, forbidden by Aristotelian dogma, was there in the tube, at last.

"On the surface of the liquid which is in the bowl," wrote Torricelli, "there rests the weight of a height of fifty miles of air." Air pressure produces a normal force that pushes down on the open liquid surface, driving the mercury up the tube. The column settles with a height such that the downward force it exerts, at the level of the air-mercury interface, equals the upward force beneath and supporting the column. In other words, *the pressure is equal everywhere at the level of the open surface, both inside the liquid in the tube and out* (Fig. 11.9). The column of mercury effectively takes the place of a much taller, though equally weighty, column of air, producing the same pressure inside as the air produces outside. Clearly, the height of the mercury is a direct measure of the **atmospheric pressure** acting on the device. Torricelli had constructed the first mercury *barometer.*

"We live immersed at the bottom of a sea of . . . air," he observed—a fluid sea wherein pressures are generated just as they are in a liquid. Approximately 5×10^{18} kg of air press down on the planet, giving rise to an average sea-level atmospheric pressure (P_A) of 1.013×10^5 N/m^2 or, in proper SI units, 1.013×10^5 Pa. The study of pressure is particularly rife with anti-quated units that are still in common usage (Table 11.3). Thus, an *atmosphere* (abbreviated atm) is defined as 1 atm = $1.013\,25 \times 10^5$ Pa, which equals 14.7 lb/in.^{2}. The *torr* (for Torricelli) was formerly the "millimeter of mercury" (or mm Hg), a pressure corresponding to a barometric mercury height of 1 mm and equal to 133.3 Pa. The mm Hg still survives, however, especially in medical literature.

Figure 11.6 Stevin's experiment to determine the pressure on the bottoms of differently shaped containers. The balance holds up the bottom of each test vessel with a force equal to the load on the right-hand pan. When the weight of fluid poured into the left exceeds the load on the right, the bottom drops out on the left.

Figure 11.7 The walls of the container contribute to the process that leads to the pressure being dependent only on the depth in the liquid.

Figure 11.8 The atmosphere pressing down on the open mercury surface can support a 30-in. column. The tube is first filled and then upended in the bowl of mercury. The column drops a little, leaving a partial vacuum in the space above it.

Figure 11.9 Two equivalent versions of the barometer. The pressure at A is essentially zero. The pressure at B equals that at C, which is atmospheric. Hence $\rho g h$ must equal atmospheric pressure.

TABLE 11.3 Pressure conversions

Quantity	Equivalent in pascals
1 atm	1.013×10^5
1 lb/in.^{2} (psi)	6.895×10^3
1 mm Hg (torr)	1.333×10^2
1 in. Hg	3.386×10^3
1 in. H$_2$O	2.491×10^2
1 bar	1.000×10^5

Equivalent of 1 atmosphere
1.013×10^5 Pa
14.70 lb/in.^{2}
7.60×10^2 mm Hg (0°C)
29.92 in. Hg (0°C)
1.013 bar
2.117×10^3 lb/ft^2
4.068×10^2 in. H$_2$O

One standard *atmosphere* corresponds to a barometric reading of 29.9 in. of mercury (76 cm), although air pressure ordinarily varies from 29 to 30 in. of mercury. Incidentally, water vapor is less dense than air (a liter of moist air has a mass of 1.18 g, whereas a liter of dry air has a mass of 1.29 g). As the air becomes drier, it becomes heavier, and a rising barometer usually portends fair weather, whereas falling pressure heralds foul weather.

In 1644, Torricelli wrote a letter to a friend describing the barometer, explaining the effect of the atmosphere, and discussing the vacuum created above the mercury. In the next few years, that letter was circulated around Europe. A copy went to a young man named Blaise Pascal who subsequently (1648) proved that the level of a barometer dropped as it went higher in the atmosphere. At the top of a 3200-ft mountain, the mercury column was about 3 in. lower than before the ascent (p. 394).

Example 11.2 One *atmosphere* is defined as the pressure equivalent to that produced at 0°C by exactly 76 cm of mercury of density 13.5950×10^3 kg/m^3 under standard gravitational conditions, where $g = 9.80665$ m/s^2. Show that a mercury barometer column 76.00 cm (29.921 in.) high corresponds to an air pressure of 1.013×10^5 Pa.

Solution: [Given: Hg column 76.00 cm high. Find: pressure beneath it.] Consider the barometer in Fig. 11.9. The height of the column above the open-surface level is $h = 0.7600$ m. The pressure at point B is

$$P = P_s + P_l = 0 + \rho g h$$

Since there is vacuum above the mercury in the tube ($P_s = 0$), the pressure at B, at the open-surface level, is

$$P = \rho g h = (13.5950 \times 10^3 \text{ kg/m}^3)(9.80665 \text{ m/s}^2)$$
$$\times (0.7600 \text{ m}) = \boxed{1.013 \times 10^5 \text{ Pa}}$$

▶ **Quick Check:** This is also the pressure outside at the open surface of the mercury—namely, at point C where $P_l = 0$. Since $h = 0$, $P = P_s = P_A = 1.013 \times 10^5$ Pa.

The medicine dropper and its larger culinary cousin, the meat baster, are devices on the family tree of the barometer. The low-pressure region created by squeezing the rubber bulb and allowing it to expand back to size corresponds to the near vacuum in the top of the barometer. Air acting on the open surface of the liquid forces it up the medicine dropper tube. A good bit of residual air left in the bulb pushes down on the top of the liquid column, effectively adding to its weight. The result is that a much smaller amount of liquid is forced up the tube, which is fine, since one rarely wants a 34-ft-long eye dropper. The pipette and the drinking straw are the same machine, with the rubber bulb replaced by a mouth. A drop in mouth pressure of a few percent will operate the thing nicely, while an excess of mouth pressure will force the liquid in the straw below the open-surface level, allowing the operator to "make bubbles."

Gauge Pressure

In the practical world of pumps, automobile tires, and compressed gas tanks, pressure is measured with a gauge that is set to read zero in the open air and thus neglects the effect of the atmosphere. This pressure above or below atmospheric is the **gauge pressure** (we will distinguish it with a subscript G). The pressure referenced to the perfect vacuum as zero is the **absolute pressure** (P). Until now we have only considered absolute pressure. It's given as the sum of the gauge and atmospheric pressures:

$$P = P_A + P_G \qquad (11.4)$$

There is nothing profound about this relationship. If we agreed to limit ourselves to absolute pressure, P_G would be superfluous. But people measure the pressure in veins, chests, and scuba tanks using gauge pressure, which makes Eq. (11.4) a practical necessity.

For the fluid in Fig. 11.10 to flow from the collapsible pouch into the vein, the gauge pressure at the needle ($P_G = \rho gh$) must exceed the gauge blood pressure in the arm. The gauge pressure in a vein (Table 11.4) is less than about 2 kPa (15 mm Hg), and so the pouch must be raised at least 20 cm above the arm.

Figure 11.10 The liquid's gauge pressure at the needle is $P_G = \rho gh$, where ρ is the density of the fluid being injected.

TABLE 11.4 Some fluid pressures in the human body		
	Gauge pressure	
Location	mm Hg	kPa
Brain, surrounding fluid	5–12	0.7–1.6
Eye, aqueous humor	12–24	1.6–3.2
Gastrointestinal	10–20	1.3–2.7
Lungs		
inhalation	−2	−0.3
exhalation	3	0.4
Chest (intrathoracic)		
inhalation	−6	−0.8
exhalation	−2.5	−0.3
Venous blood		
venules	8–15	1–2
veins	4–8	0.5–1
major veins	4	0.5
Arterial blood at heart level		
maximum (systolic)	100–140	13–19
minimum (diastolic)	60–90	8–12
Bladder		
average	0–25	0–3
during urination	15–30	2–4

Example 11.3 A straw is to be used to raise water from a glass to a mouth 15.0 cm above the liquid's surface. Determine the absolute pressure that must be maintained in the mouth. What is the corresponding gauge pressure?

Solution: [Given: $h = 15.0$ cm and $\rho = 1.00 \times 10^3$ kg/m^3. Find: P_s in the mouth at the surface of the water at the top of the straw.] The column is to rise 0.150 m above the open-surface level where the pressure is P_A. Then in the liquid within the straw at that open-surface level

$$P = P_1 + P_s = P_A$$

Hence

$$P_s = P_A - \rho gh = 1.013 \times 10^5 \text{ Pa}$$
$$- (1.00 \times 10^3 \text{ kg/m}^3)(9.81 \text{ m/s}^2)(0.150 \text{ m})$$

and $\boxed{P_s = 0.998 \times 10^5 \text{ Pa}}$. That's 98.5% of atmospheric pressure. To find the gauge pressure just subtract atmospheric pressure—that is

$$P_G = P_s - P_A = -1.53 \times 10^3 \text{ Pa}$$

▶ **Quick Check:** The weight of the water column of cross-sectional area A is $\rho ghA \approx (10^3 \text{ kg/m}^3) \times (10 \text{ m/s}^2)(0.15 \text{ m})A$ and the pressure is ≈ 1.5 kPa: air pressure must exceed mouth pressure by this amount.

For several centuries the only accurate practical pressure gauge was the mercury manometer (Fig. 11.11), a U-shaped open tube filled with mercury. The gauge pressure at point A equals the pressure at point B equals the chamber pressure. When h is above or below point B, the chamber pressure is above or below atmospheric: $P_G = \pm \rho gh$. Because of the manometer, blood pressure is measured in mm Hg to this day.

Vacuum

After centuries of philosophical insistence on Aristotle's idea that an empty void cannot exist, the "Torricellian vacuum" was seen by many as miraculous. The great Descartes, who debated the existence of vacuum with Pascal, never accepted the notion. In fact, just before his death (1650), he suggested that Pascal had a void in his head.

The next advance in vacuum technology was made by the German engineer Otto von Guericke. In 1650, he succeeded in constructing the first air-extraction pump, which was a modified version of a commonplace fire extinguisher—"a brass squirt-pump" that operated with a movable piston like a large hypodermic syringe. Later, von Guericke constructed a pair of bronze hemispheres, 22 in. in diameter, between which was placed a soft ring gasket. These were then evacuated with his pump until the outside air pressure forced the hemispheres together (Fig. 11.12a). The normal force is everywhere radially inward (Fig. 11.12b), but the symmetry is such that for every component acting downward above the central z-axis, there is an equal upward component below that axis. The end result is that all force components in the x- and y-directions must cancel on each half-sphere. Only a net z-axis force remains inward on each, driving the hemispheres together.

If each half-sphere was independently sealed with a flat plate of radius R and separately re-evacuated, each would again be in equilibrium (Fig. 11.12c). There would be the same net axial force F acting on the curved surface as before, but now it would be balanced by an atmospheric-pressure force, $(\pi R^2)P_A$, acting in opposition on the flat face. It follows, assuming the pressure inside is zero, that the net atmospheric force on each hemispherical surface is

$$F = \pi R^2 P_A$$

With $R = 11$ in. and $P_A = 14.7$ lb/in.^{2}, it would take a force of 5.6×10^3 lb pulling on each half to yank them apart. On May 18, 1654, with Emperor Ferdinand III

Figure 11.11 A manometer. It is usually filled with mercury, but water is also used in medical applications, such as measuring lung pressure, because mercury vapor is poisonous. Here, the gas pressure in the chamber is being measured. It is assumed that the weight of the column of gas above B is negligible (that is, the pressure in the chamber is uniform). As shown, the chamber pressure exceeds atmospheric by ρgh.

Magnified about 1.5×10^5 times, this staphylococcus bacterium explodes after being exposed to a low-level dose of antibiotic. With an internal pressure of between 25 and 30 times atmospheric pressure, the cell wall ruptured at its weakest point.

Put a little water in an open empty gallon can, and gently heat it until the escaping steam displaces most of the air that was inside. Cap it tightly, cool it, and as the vapor condenses and the internal pressure drops, it will gradually collapse under the weight of the atmosphere. Notice how the can is pushed in on all sides and not just flattened top down. The pressure is in all directions—it's a scalar.

looking on, von Guericke orchestrated a bit of scientific theater, a gigantic tug-of-war between horses and hemispheres. Sixteen powerful animals strained to separate the two halves.

We are not aware of this potentially crushing force produced by the atmospheric sea because we are filled with, and surrounded by, air. Were the air to be drawn out

Figure 11.12 (a) The two hemispheres are evacuated and the force due to air pressure (b) acting radially inward squeezes them together. (c) Here, the hemispheres are separated, and each is sealed with a flat plate.

Von Guericke's Magdeburg sphere demonstration from his *Experimenta Nova*, published in 1672.

This house exploded during a hurricane. The pressure outside dropped below the inside pressure, which raised the roof. Had the occupants left several windows open, the explosion would not have happened.

of a vessel (as with a vacuum-sealed jar), that container had better be able to withstand 1.013×10^5 N/m², or it will be crushed. Youngsters often squash paper cups on their faces by sucking the air out of them, wearing the crumpled remains as a muzzle. An ordinary window (1 m by 1.5 m) experiences an inward atmospheric force of 1.5×10^5 N (34×10^3 lb), which is usually counterbalanced by an equal outward force since the building is filled with air at the same pressure. Sometimes, for example, during a tornado, the pressure outside can suddenly drop and, if the windows are closed, a house can explode.

The Honourable Robert Boyle, the seventh son of the Earl of Cork, read about von Guericke's work in 1657 and immediately set out to better it. By 1659, Boyle and his assistant Hooke had built an extraction air-pump (a so-called vacuum pump, though it certainly doesn't pump vacuum) far superior to any in existence at the time. They soon established that air was necessary to both combustion and respiration (at least for flies, bees, birds, and mice). They even put a barometer in a chamber and confirmed that the mercury column descended as the pressure dropped. Today, vacuum systems are commonplace in the production of a variety of familiar items, from TV picture tubes and light bulbs to computer circuits, freeze-dried food, and coated sunglasses.

11.4 Boyle's Law

Boyle's early work with vacuum pumps led him to the study of gases, a research that culminated in the recognition of a simple interdependence between pressure and volume. Although R. Towneley and H. Power actually published the relationship a year earlier, it is still most often referred to as Boyle's Law.

Figure 11.13 The apparatus used by Boyle to determine the law bearing his name. The gas captured in the space on the left in (b) is at atmospheric pressure. By adding 30 in. of mercury, the pressure is doubled and the volume is halved, as in (c).

Boyle's approach was simple and elegant. He bent a glass tube about 2 m long into the shape of a J and sealed off the short arm (Fig. 11.13a). Then some mercury was poured in, usually trapping a quantity of air under pressure in the short segment. By tipping the device "so that air might freely pass from one leg into the other," the pressure was equalized at atmospheric and the mercury level made the same on both sides (Fig. 11.13b). The device, turned back upright, was ready to use. Adding more mercury caused the liquid level to rise unequally in both sections until the volume of gas in the smaller arm was halved and the open column had risen another 30 in. (Fig. 11.13c). The pressure in the trapped gas was then twice atmospheric—the volume of the gas had been halved and the pressure doubled. Similarly, at 3 atm of pressure the volume was further reduced to $\frac{1}{3}$, and so forth. **Keeping the temperature constant, the volume of a gas varies inversely with the pressure**, which is equivalent to saying that the **pressure times the volume is constant. Boyle's Law** is then

$$PV = \text{constant} \qquad (11.5)$$

Consider a balloon floating upward in the air or a bubble rising in a liquid. As the external pressure drops, the internal counterpressure must also decrease. If PV is to be constant, the volume must increase; the balloon grows larger as the pressure decreases. Similarly, Fig. 11.14 shows how Boyle's Law applies to the way we breathe.

The imagery of a gas as a swarm of tiny bouncing, colliding atoms or molecules is called the Kinetic Theory. It was first applied analytically by Daniel Bernoulli (1738), who succeeded in deriving Boyle's Law using the concept. A molecule impacting on, and ricocheting off, a chamber wall changes its momentum in the process. That, in turn, corresponds to a force on the molecule and an equal-magnitude oppositely directed force on the wall. The ongoing bombardment of millions of molecules blurs into a seemingly continuous force and pressure.

Suppose we have a gas in an upright cylindrical chamber fitted with a vertical descending piston (Fig. 11.15). Compressing the gas to half its original volume doubles the density, giving the same number of molecules half the room to fly around in. As the piston slowly drops, halving the top-to-bottom distance, the average time to traverse the chamber vertically is halved, since the speed of the particles is unchanged. Therefore, the number of impacts per second is doubled, and so is the pressure. Similarly, lowering the piston halves the exposed side-wall area without altering the average time to traverse the chamber horizontally. That, in turn, results in the same number of impacts per second happening on half as much area; that is, the pressure on the walls also doubles—halving the volume doubles the pressure.

Boyle's Law is an idealization that works over a wide range of temperature and pressure. It applies especially well when the intermolecular forces are negligible, as they are when the molecular separations are large. It is most in error at very high pressures and at temperatures near liquefaction, where the molecules are relatively close and do attract one another. Carbon dioxide (CO_2), a gas that is fairly easily liquefied, deviates appreciably from Boyle's Law, whereas helium and hydrogen comply with it rather well under ordinary conditions. The make-believe stuff for which Boyle's Law holds exactly would be a *noninteracting* collection of independent molecules known as an **ideal gas**. Insofar as all real gases resemble the ideal gas to some extent, Boyle's Law is quite useful.

11.5 Pascal's Principle

Sometime around 1651, Pascal wrote a treatise entitled *On the Equilibrium of Liquids*. It contained the first precise statement* of what has come to be known as **Pascal's Principle: An external pressure applied to a fluid confined within a closed container is transmitted undiminished throughout the entire fluid**. When pressure is put on some region of a confined liquid (as, for example, when a piston pushes down on the liquid in a cylinder), the fluid compresses slightly and distributes the pressure uniformly everywhere therein. This process is quite different from the internal pressure generated by gravity, and would exist even in a

*Archimedes had come close to it 2000 years before.

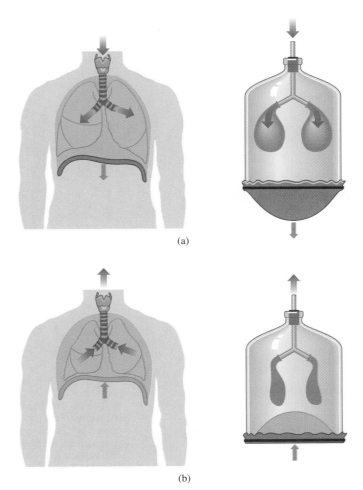

(a)

(b)

Figure 11.14 A bottomless jar surrounding two balloons and closed off with a rubber membrane can make a simple model of the lung system. (a) To inhale, the diaphragm pulls downward, enlarging the chest volume and thereby dropping the pressure (PV = constant). The chest cavity is usually at a negative gauge pressure (to keep the lungs open against their surface tension), but now the lung pressure is also negative and air flows in. (b) When the chest compresses and the diaphragm moves up, chest pressure rises, the gauge pressure in the lungs becomes positive, and air is expelled.

Figure 11.15 When the pressure exerted on a gas is doubled, the volume is halved; when it is tripled, the volume is reduced to one-third.

Figure 11.16 Pascal's syringe is a bottle punctured with holes and fitted with a tight piston. Pushing down on the piston increases the pressure on the fluid, which blows out the holes uniformly in all directions. This phenomenon suggests that the applied pressure is distributed equally throughout the liquid.

Valve closed

Valve open

Figure 11.17 An aerosol spray can contains a gas under pressure called the propellant. It pushes down on the surface of the liquid that is to be sprayed. When the valve is opened, the top end of the long tube is at atmospheric pressure, and the bottom end is at a pressure well above that. The difference propels the liquid up and out.

On work—long before the notion was formalized. And it is truly admirable that there is encountered in this new [hydraulic] machine the constant rule which appears in all the older machines, such as the lever, the wheel and axle, the endless screw, etc., which is, that the path is increased in the same proportion as the force.

BLAISE PASCAL
Traité de l'Equilibre des Liqueurs

weightless liquid. Pascal's syringe (Fig. 11.16) illustrates the point nicely, as does the common seltzer bottle and the aerosol spray can (Fig. 11.17).

One reason you don't let people sit on your belly when your bladder is full is Pascal's Principle. In the same way, pressure on the abdomen of a pregnant woman is transmitted to the fetus via the amniotic fluid. This mechanism is also responsible for the motion of soft-bodied animals, like the earthworm, that have hydrostatic skeletons. Using a mesh of perpendicular muscles, a worm squeezes itself into shape, becoming long and thin or short and fat as needed.

Hydraulic Machines

The same amount of pressure can be produced within a liquid by pistons of different sizes acting with proportionately different forces (Fig. 11.18)—the larger the cross-sectional area of the piston, the larger the force needed to create a given pressure. It was at this juncture that Pascal recognized the tremendous practical significance of his principle. For the first time since antiquity, a new class of force multipliers known as *hydraulic* machines (from the Greek for *water* and *pipe*) was possible (al-

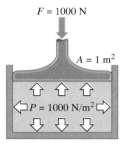

$F = 1000$ N

$A = 1$ m^2

$P = 1000$ N/m^2

$F = 10$ N

$A = 0.01$ m^2

$P = 1000$ N/m^2

Figure 11.18 The force divided by the area of the piston determines the pressure. Different piston areas and forces here produce the same pressure.

F_i

A_i

F_o

A_o

$P_i = P_o$

Figure 11.19 A hydraulic lift. The input force F_i creates a pressure F_i/A_i that is transmitted to the output cylinder, where that same pressure equals F_o/A_o.

though a practical device was not built until Bramah devised a functioning pressure-seal in 1796).

If two chambers fitted with different-sized pistons are connected so they share a common working fluid, the pressure generated by one will be transmitted undiminished to the other (Fig. 11.19). Years ago, gas stations dug service pits in which the mechanics stood while working on cars parked overhead. Nowadays, automobiles are unceremoniously lofted into the air on hydraulic lifts. Most are activated by compressed air pressing on oil, but the simplest arrangement is a U-tube, narrow on one side, wide on the other, with sealed movable pistons at both ends. A downward input-force F_i acting over the small input-area A_i of the narrow piston generates an input-pressure $P_i = F_i/A_i$. But this pressure is distributed uniformly and so equals the output-pressure ($P_i = P_o$), which, in turn, is given by $P_o = F_o/A_o$. Hence, $F_i/A_i = F_o/A_o$, and so

$$\frac{F_o}{F_i} = \frac{A_o}{A_i} \tag{11.6}$$

Although the pressures in and out are equal, the forces are certainly not. When the output piston under the car has 100 times the area of the input piston, it will experience an upward force 100 times greater than the input-force. We exert a force of 200 N on a hydraulic jack, and it exerts a force of 20 000 N on the car. Like all machines, this is not a *work* multiplier: at best, when energy losses (for example, those due to friction) are negligible, work-in equals work-out. In this example, the

Today, hydraulic devices that push and pull are commonplace.

small piston will have to descend a distance $y_i = 100$ cm for every 1 cm the large one rises (y_o). In other words, the volume of liquid displaced at the input side is $A_i y_i$, which is equal to the volume displaced at the output $A_o y_o$. Accordingly,

$$\frac{A_o}{A_i} = \frac{y_i}{y_o} \tag{11.7}$$

One person doing a lot of pumping over a lot of distance (y_i) can jack a car up (y_o) using a relatively small force, but *work-in* $\geq$ *work-out*.

Example 11.4   A barber's chair rests on a hydraulic piston 10 cm in diameter. The input side has a piston with a cross-sectional area of 10 cm², which is pumped on using a foot pedal. If the chair and the client together have a mass of 160 kg, what force must be applied to the input piston?

Solution: [Given: $A_i = 10$ cm² $= 0.0010$ m², $r_o = 5.0$ cm $= 0.050$ m, and $m_o = 160$ kg. Find: F_i.] Using

Eq. (11.6)

$$F_i = \frac{F_o A_i}{A_o} = \frac{m_o g (0.0010 \text{ m}^2)}{\pi (0.050 \text{ m})^2} = (1568 \text{ N})(0.1273)$$

and

$$\boxed{F_i = 2.0 \times 10^2 \text{ N}}$$

▶ **Quick Check:** $A_i / A_o = \pi (5 \text{ cm})^2 / (10 \text{ cm}^2) = 7.85 = F_o / F_i = 1.6 \text{ kN} / F_i$ and $F_i = 0.2$ kN.

Keep an eye out for hydraulic machines. They can be found on garbage compacters, forklifts, robots, cherry pickers, plows, tractors, airplane landing gear, elevators, and even in the brake system of the family car.

11.6   Buoyant Force

It can be argued that the study of hydrostatics was begun by Archimedes in the third century B.C. The greatest physicist of ancient time, Archimedes was apparently a kinsman of Hieron II, Tyrant of Syracuse. Legend has it that the king ordered a solid gold crown to be made. But when the piece was delivered, although its weight was right, Hieron suspected that his jeweler had substituted silver for gold in the hidden interior. Archimedes was given the challenge of determining the truth without damaging the royal treasure. After pondering the problem for some time, its solution came to him while he was musing in a warm tub at the public baths. The distracted philosopher leaped from the water and ran home naked, shouting through the streets, *"Heureka! Heureka!"* I have found it! I have found it! What he found, as we shall soon see, was far more valuable than Hieron's crown.

A completely submerged body displaces a volume of liquid equal to its own volume. Experience also tells us that when an object is submerged, it appears lighter in weight; the water buoys it up, pushes upward, partially supporting it somehow. That much would be obvious to anyone who ever tried to submerge an inflated tire tube or a beach ball. Archimedes quantified the phenomenon. His **Buoyancy Principle** asserts that *an object immersed in a fluid will be lighter (that is, it will be buoyed up) by an amount equal to the weight of the fluid it displaces.* The upward force exerted by the fluid is known as the **buoyant force**. A 10-N body that displaces 2 N of water will "weigh" only 8 N while submerged.

Buoyant force is caused by gravity acting on the fluid. It has its origin in the pressure difference occurring between the top and bottom of the immersed object, a difference that always exists when pressure varies with depth (as it does for a fluid in the Earth's gravitational field). Imagine a solid cube, each face of which has an area A, somewhere below the surface of a fluid of density ρ_f, as depicted in Fig. 11.20. The gauge pressure on the bottom, $P_b = \rho_f g h_b$, is greater than the gauge pressure on the top, $P_t = \rho_f g h_t$, and that difference $\Delta P = \rho_f g (h_b - h_t) = \rho_f g h$ gives rise to the *buoyant force* F_B. The force pushing up on the bottom is greater than the force pushing down on the top by an amount

$$F_B = A\,\Delta P = \rho_f g A h$$

but $Ah = V$, the volume of the body or equivalently the volume of the fluid displaced. It follows, since the mass of fluid displaced is $m_f = \rho_f V$, that

$$F_B = g\rho_f V = m_f g \qquad (11.8)$$

The buoyant force equals the weight of the fluid displaced.

Figure 11.20 The buoyant force on a cube is the difference between the downward force on its top face and the larger upward force on its bottom face.

Example 11.5 An empty spherical weather balloon with a mass of 5.00 kg has a radius of 2.879 m when fully inflated with helium. It is supposed to carry a small load of instruments having a mass of 10.0 kg. Taking air and helium to have densities of 1.16 kg/m³ and 0.160 kg/m³, respectively, will the balloon get off the ground?

Solution: [Given: The bag plus the load has a mass of 15.0 kg, $\rho_{He} = 0.160$ kg/m³, $\rho_f = 1.16$ kg/m³, and $R = 2.879$ m. Find: the buoyant force compared to the weight.] To find the weight of fluid displaced, we need the volume of the spherical body: $V = \frac{4}{3}\pi R^3 = 100$ m³. The buoyant force is the weight of 100 m³ of air; hence

$$F_B = \rho_f V g = (1.16 \text{ kg/m}^3)(100 \text{ m}^3)(9.81 \text{ m/s}^2)$$

and

$$\boxed{F_B = 1.14 \times 10^3 \text{ N}}$$

By comparison, the weight of the helium is

$$\rho_{He} V g = (0.160 \text{ kg/m}^3)(100 \text{ m}^3)(9.81 \text{ m/s}^2) = 156.9 \text{ N}$$

Thus, the total weight of the helium, balloon, and load is

$$F_W = (15.0 \text{ kg})(9.81 \text{ m/s}^2) + 156.9 \text{ N} = \boxed{304 \text{ N}}$$

Clearly, the buoyant force far exceeds the weight, and the balloon will accelerate upward rapidly.

▶ **Quick Check:** $F_B \approx (100 \text{ m}^3)(1 \text{ kg/m}^3)(10 \text{ m/s}^2) \approx 10^3$ N.

To see how Archimedes might have solved Hieron's problem, suppose the crown had a mass, $m = 0.982$ kg. By suspending the crown immersed in water from a balance (Fig. 11.21), he could measure $F_W - F_B = g(m - m_w)$, where m_w is the mass of water displaced. Suppose this *reduced weight* turned out to be $g(0.922 \text{ kg})$. This means that $g(0.922 \text{ kg}) = g(m - m_w)$ and so $m_w = 0.982 \text{ kg} - 0.922 \text{ kg} = 0.060$ kg. The mass of water displaced is 0.060 kg and its volume must be 60 cm³, which is the volume of the crown. Is it gold? If it was, it would have a mass given by the density of gold (19.3×10^3 kg/m³) times the volume: $(19.3 \times 10^3 \text{ kg/m}^3) \times (60 \times 10^{-6} \text{ m}^3) = 1.158$ kg, not 0.982 kg. It is *not* pure gold. There is a shortage in the mass of $(1.158 \text{ kg} - 0.982 \text{ kg}) = 0.176$ kg, which is assumed to have arisen from the inclusion of silver. Alas, most versions of this legendary tale end with the untimely demise of the anonymous goldsmith.

11.7 Specific Gravity

The notion of density can be made more convenient by forming the ratio of the density of any material to that of water. The result is the idea of *relative density,* and it's especially useful when we float things in water. Thus, suppose an object of mass m

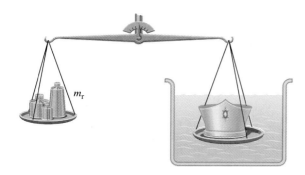

Figure 11.21 A crown of mass m, immersed in water, has a reduced mass of $m_r = m - m_w$, where m_w is the mass of water displaced. We can measure m and m_r directly and then calculate $m/(m - m_r)$, which equals the specific gravity.

is immersed in water such that the mass of fluid displaced is m_w. Using a balance, we could then directly measure the object's *reduced mass* in the water: $m_r = m - m_w$, which provides a convenient measure of the *relative density* ρ/ρ_w:

$$\frac{m}{m_w} = \frac{m}{m - m_r} = \frac{\rho}{\rho_w} \qquad (11.9)$$

Equation (11.9) is equivalent to saying that *the ratio of the weight of a body to the weight of an identical volume of water is equal to the relative density*, or, as it has inappropriately been called for centuries, the **specific gravity**. Alternatively, the weight of an object in air divided by the difference between its weight in air and its weight in water is its specific gravity. The idea was introduced by the Persian al-Biruni sometime around A.D. 1025 in order to use unitless quantities and so avoid the muddle of different units existing throughout the Muslim Empire. The relative density tells us how many times more or less dense a material is than water. Equation (11.9) provides a convenient method for measuring the average density of a human body by submerging a person, sitting on a scale, in what is fondly called a "fat tank."

Example 11.6 Hieron's crown having a mass of 0.982 kg displaces 0.060 kg of water when totally submerged. Is it solid gold?

Solution: [Given: $m_g = 0.982$ kg and $m_w = 0.060$ kg. Find: the relative density of the crown.] From Eq. (11.9)

$$\frac{m_g}{m_w} = \frac{0.982 \text{ kg}}{0.060 \text{ kg}} = \frac{\rho_g}{\rho_w} = 16$$

whereas the relative density of gold (Table 10.1) is 19.3—this crown isn't solid gold!

▶ **Quick Check:** For the water $m_w/\rho_w = V$, for the gold $V\rho = m$; because the volumes are equal, $m_w\rho_g/\rho_w = m_g = (0.06 \text{ kg})(19.3 \times 10^3 \text{ kg/m}^3)/(10^3 \text{ kg/m}^3) = 1.2$ kg and not 0.982 kg.

Floating

When an object weighs more than the total volume of fluid it can displace, it sinks. If the fluid is water, this is equivalent to saying that an object whose relative density exceeds 1.0 will sink. A solid cubic foot of gold, copper, concrete, or glass weighs more than a cubic foot of water and will sink in spite of the (278-N, or 62.4-lb) buoyant force.

When an object weighs less than the total volume of fluid it can displace, it will settle down until the buoyant force equals the weight and it floats partially submerged. In water, an object whose relative density is less than 1.0 will float partially submerged. A steel ship can encompass a great deal of empty space and so have a large volume and a relatively small average density. Provided it weighs less than the maximum amount of water it can displace (that is, provided it weighs less than the maximum buoyant force), a ship—cargo and all—will float. *The weight of the water displaced by the submerged portion of a floating object equals the weight of the object* (Fig. 11.22).

People are slightly less dense than water, especially when the lungs are filled with air, and so float partially submerged. The average specific gravity of the human body varies from person to person and breath to breath. Body fat, which is $\approx 18\%$

Volume displaced

Figure 11.22 A floating object displaces its own weight of liquid.

for a male and $\approx 28\%$ for a female, has a specific gravity of ≈ 0.8. Muscle has a specific gravity of ≈ 1.0 and bone about 1.5–2.0. A lean, muscular body will tend to sink. In general, young people and women have lower average specific gravities, but 0.98 is a typical lungs-filled value. Thus, a person can float with at most 2% of the body out of the water. In seawater, the relative density $(0.98 \times 10^3 \text{ kg/m}^3)/(1.025 \times 10^3 \text{ kg/m}^3) = 0.956$, and so about 4% of the body will be above the surface. The problem in swimming for humans is keeping the dense head above the water in such a way as to be able to breathe.

Figure 11.23 A way to displace more liquid than is actually present.

The greater the density of the fluid, the more the buoyant force. That's why it is easier to swim in the ocean than in a salt-free pool and why a penny will float on mercury. A fresh egg, one that has developed little or no gas, will be relatively dense and will sink in a glass of tap water. Dissolving a few tablespoons of salt into the water will float the egg up to the surface.

Surprisingly, *it is possible to displace more fluid than might be present,* and it is the *displaced* fluid that counts with regard to floating. However paradoxical that sounds, it really isn't, and Fig. 11.23 should make the business clear. Again, the walls of the container play a crucial role as the fluid transmits the load to them. It doesn't matter how much water is surrounding the inner beaker.

When an object's weight equals the weight of the total amount of fluid it can displace—that is, equals the maximum buoyant force—it neither sinks nor rises but hovers beneath the surface in static equilibrium. Some fish accomplish this balance by storing gas in a bladder, thereby being able to remain at a fixed depth without the need to move about. By contrast, sharks, which are rather primitive in design, lack such systems and must continually swim or they will sink. To rise, or dive, or hover even without operating its engines, a submarine takes on and discharges seawater into ballast tanks.

A fat tank provides a measure of a person's average density.

Example 11.7 An object of volume V floats in water with a volume V_u above the surface. (a) Write an expression for its average density. (b) Apply this solution to find the average density of a loaded barge with 20% of its volume above the waterline.

Solution: [Given: The total volume is V, the volume above water is V_u. Find: ρ.] (a) The volume of water displaced is $(V - V_u)$ and the buoyant force is

$$F_B = (V - V_u)\rho_w g$$

But this equals the weight of the object, which is also

$\rho g V$, therefore

$$\rho g V = (V - V_u)\rho_w g$$

and

$$\boxed{\rho = \frac{V - V_u}{V}\rho_w}$$

(b) Thus $\rho = (1 - 0.20)\rho_w = \boxed{0.80\rho_w}$

▶ **Quick Check:** It is always wise to check the extremes. To hover below the surface, $V_u = 0$ and $\rho = \rho_w$, which makes sense. For the entire object to be above water, $V_u \to V$ and $\rho \to 0$, which also makes sense.

11.8 Surface Tension

In a liquid under the influence of gravity, each molecule descends until it can drop no more, ultimately settling into a zero net-force equilibrium. The system then has the lowest possible potential energy, which is why a liquid assumes the shape of its containers. But in the absence of gravity, or in an effectively weightless environ-

The density of the salt-rich water of Great Salt Lake is appreciably greater than that of pure water. Consequently, the swimmer can float with much of her body out of the water.

Astronauts in Earth orbit in the *Space Shuttle Columbia* examine a ball of water. In the effectively weightless environment, surface tension pulls the liquid into a minimum energy sphere.

ment, a glob of water would draw itself into a sphere. Free-falling raindrops are roughly spherical—though only roughly because of the air friction. In the same way, if we shake some mercury in an Earth-bound jar, thousands of tiny, almost spherical beads will be formed on the glass, beads quite unwilling to conform to the shape of the container's walls. These familiar liquid "tricks" are the outward manifestations of internal interactions that give rise to what we call *surface tension*.

Any molecule within the body of a liquid is surrounded by other molecules. These pull on it more or less uniformly in all directions via the cohesive interactions that bind the substance together. By contrast, a molecule on a liquid's *free surface* (that is, one bounded by gas or vacuum), will have none of its comrades above it. It will be drawn more strongly to its neighbors on the side and below. Consequently, the surface sheet of molecules reacts as if it were an elastic membrane confining the liquid, being drawn in, assuming the least area it can. About 200 years ago, Dr. Thomas Young suggested this stretchy membranous imagery—it is helpful, *but should not be taken literally*.

The force exerted by a stretched elastic membrane is proportional to the amount of stretch via Hooke's Law: no displacement from equilibrium, no reaction force. By contrast, a liquid surface film is always under tension and that tension is independent of any displacement. The surface "membrane" *is* able to stretch in the usual sense, but only slightly. When the surface film is enlarged, it is accomplished by bringing up more molecules from the interior of the liquid. This increases the amount of surface, creating new exposed regions, rather than increasing the intermolecular distances within a permanent surface film, as would be the case with a true elastic membrane.

To quantify the surface tension, let's measure the force exerted by such a surface directly. The simplest practical scheme, if not the most accurate, is to attach a U-shaped wire frame (Fig. 11.24) to a balance and measure its weight with the two ends just submerged. That done, the frame is lowered straight down completely into the liquid and then raised back out, forming a film across the inverted U. The additional upward force F (beyond the original weight of the frame and the negligible weight of the film) required to hold the wire in equilibrium at its starting level is a direct measure of the downward tug F_t of the liquid. That force is parallel to the plane of the film, points in the direction of expansion, and is distributed along a

Most ships have markings on their hulls showing how deep they are resting in the water and therefore how heavy their cargos are.

TABLE 11.5   Surface tensions of various liquids

Substances	Temperature °C	Surface tension N/m
Helium–vapor	−269	0.12×10^{-3}
Hydrogen–vapor	−258	2.9×10^{-3}
Oxygen–vapor	−183	13.2×10^{-3}
Ether–vapor	20	17.0×10^{-3}
Olive oil–water	20	18.2×10^{-3}
Ethanol–vapor	20	22.3×10^{-3}
Benzene–air	20	29×10^{-3}
Soap solution–air	20	$\approx 25 \times 10^{-3}$
Whole blood–air	37	58×10^{-3}
Water–air	100	58.9×10^{-3}
Glycerin–air	20	63×10^{-3}
Water–air	50	67.9×10^{-3}
Water–air	20	72.8×10^{-3}
Blood plasma–air	37	73×10^{-3}
Water–air	15	73.6×10^{-3}
Water–air	0	75.6×10^{-3}
Mercury–water	20	0.375
Mercury–air	20	0.465
Copper–air	1130	1.1
Tungsten–vapor	3410	2.5

Figure 11.24   (a) An arrangement for determining surface tension. (b) The wire frame pulls up a two-sided film of length L.

narrow interface of length L. The **surface tension** γ is defined as *the force per unit length exerted by the surface*. Gamma, which has the same value everywhere on the surface, is measured in N/m. Here, where the single liquid film contained within the frame has *two* surfaces

$$F_t = 2\gamma L \qquad (11.10)$$

Surface tension depends on both substances forming the interface (Table 11.5). Furthermore, as you might expect, increasing the temperature generally (there are a few exceptions) decreases the surface tension until it vanishes at a point where the substance can no longer exist in the liquid form.

Work must be done in order to bring a molecule up from the inside of a liquid, to push apart surface molecules and force this new one into the boundary layer. In the process of creating the film, in raising the U-shaped frame a distance Δy, an amount of work ΔW was done on the film, thus

$$\Delta W = F_t \Delta y = 2\gamma L \Delta y = \gamma \Delta A \qquad (11.11)$$

where ΔA is the net increase in surface area (on both sides of the film). This, in turn, corresponds to an increase in potential energy stored in the surface—an idea introduced by the great astronomer and mathematician Friedrich Gauss. Alternatively, γ can be understood as the increase in stored surface potential energy per unit area. In that case, it has the units of J/m^2, which is equivalent to N/m.

Generally, a system under the influence of forces moves toward an equilibrium configuration that corresponds to a minimum PE. The sphere contains the most volume for the least surface area and therefore minimizes surface energy. Thus, *surface*-PE diminishes as a droplet approaches sphericity. That's why there are no

A water strider resting on the surface of a pond. Its "feet," or tarsi, work something like snowshoes. They are covered with fine hairs, which increase the area in contact with the liquid.

cubic raindrops. Lead shot used to be manufactured by dropping molten metal within tall temperature-controlled towers. One NASA project may lead to the commercial production of perfect ballbearings in facilities orbiting the planet.

Example 11.8   An atomizer, used for misting plants, is pumped several times, transforming 100 cm³ of water into a fog of droplets with an average diameter of 50 μm. Determine the total amount of energy that went into creating the droplets. Ignore the surface area of the water initially.

Solution: [Given: $D = 50 \times 10^{-6}$ m, $V = 100$ cm³ $= 100 \times 10^{-6}$ m³, and $\gamma = 0.072\,8$ J/m². Find: ΔW.] The energy required is the work done in producing the new area. From Eq. (11.11) that work done is $\Delta W = \gamma \Delta A$, and so we need the increase in area. Each new sphere

has an area of $4\pi R^2$; hence, we need to find the number of spheres created, N. Since the volume of each droplet is $\frac{4}{3}\pi R^3$, it follows that $N = V / \frac{4}{3}\pi R^3$. The total surface area of the spheres is therefore $\Delta A = 4\pi R^2 N = 3V/R$ and

$$\Delta W = \gamma \frac{3V}{R} = \frac{(0.072\,8 \text{ J/m}^2)3(10^{-4} \text{ m}^3)}{(25 \times 10^{-6} \text{ m})} = \boxed{0.87 \text{ J}}$$

▶ **Quick Check:** For each sphere $\Delta W_s = (0.072\,8$ J/m²$)4\pi(25 \times 10^{-6}$ m$)^2 = 0.57$ nJ. $N = 3(10^{-4}$ m³$)/4\pi(25 \times 10^{-6}$ m$)^3 = 1.5 \times 10^9$ and $\Delta W_s N = 0.87$ J.

Plastic-coated paper clips supported by the surface tension of water. Try this demonstration yourself, but make sure the clips are flat.

Surface tension is responsible for a range of familiar occurrences from soap bubbles to teardrops. The bristles of a fine paintbrush tend to stand apart in a bushy tuft until they are wet with paint, and then the surface tension pulls the bristles together. That's why your hair, wet after a shower or swim, is a matted clump. An old-fashioned double-edged razor or a plastic-coated paper clip can be supported on top of water by surface tension even though either one is much denser and should otherwise sink.

FLUID DYNAMICS

A moment does not go by without each of us somehow interacting with fluids in motion. We walk, drive, and fly through the air, all the while breathing at least 6 quarts of it per minute. With blood pumping in our veins, we ourselves are *hydrodynamic* systems. On a larger scale, every thriving city on Earth has a lifeline of water pouring through its veins; New York City alone consumes 1.5×10^9 gallons per day. How do liquids flow through pipes, pumps, and arteries? What happens to the pressure as air blows over the wing of a plane or up the face of a skyscraper? How can we quantify fluid flow?

11.9   Fluid Flow

Experiments by O. Reynolds (1883) on the motion of fluids in pipes showed that there are two distinct flow regimes: *laminar* and *turbulent*. Gently blow air through your lips and the well-defined stream resembles the extreme of idealized smooth flow. Cough, and the burst of air is a complex swirling turbulence that represents the other extreme.

Laminar Flow

When a fluid moves such that the velocity within it at any point in space is fixed, we have steady-state motion. In the real world where fluids have internal friction, steady-state motion usually means slow flow. The velocity may be different at different points, but any speck of fluid arriving at a given spot in space will always have a fixed velocity at that location. We can inject a trickle of dye at several points within a liquid (or smoke in a gas) and trace the paths taken by the particles of fluid as they move in orderly procession, each following the route of the particle that arrived just before it. These unchanging lines of flow (Fig. 11.25) are called **streamlines**. At every point on any one of them, the tangent to the curve is in the direction of the velocity of the fluid at that point. There is no flow perpendicular to the streamlines. No two streamlines ever cross, since that would mean that at the point of intersection a particle of fluid would have two different velocities. This kind of streamline flow is also called **laminar flow** because successive layers of fluid molecules move smoothly past one another. For a normal person, about half the volume of blood is made up of cells, and so blood flow, which is smooth in most vessels, cannot be perfectly laminar, though it usually comes close.

   At every point in space within the stream there is a velocity, and if this *velocity field* (Fig. 11.25) reminds you of the gravitational field, it is because the latter was derived conceptually from it. *The streamlines crowd together as the speed gets higher just as the gravity-field lines pack closer as that field gets stronger.* The boundary of a bundle of streamlines defines a *tube of flow,* wherein the fluid travels as if it were contained within an invisible pipe (Fig. 11.26).

(a)

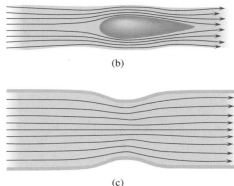

(b)

(c)

Figure 11.25   (a) The laminar flow pattern revealed by drawing the velocity vector at each point in the stream. (b) Streamlines in the laminar flow around an obstacle and (c) in a pipe. Note that the closer the lines, the higher the speed.

Turbulent Flow

Turbulent flow corresponds to nonsteady, chaotic, changing motion (Fig. 11.27). With increasing speed, the molecules of a real fluid meeting discontinuities or obstructions swirl into little shifting whirlpools, energy-carrying vortices that are spun off as curls in the flow lines (Fig. 11.28). At high enough speeds the fluid has sufficient momentum to sail past obstructions, so the flow no longer simply conforms to the shape of those obstructions. The situation is like a stream of motorcy-

Figure 11.26   The fluid in a tube of flow travels as if in an invisible pipe, neither leaving the confines nor mixing with the contents of adjacent tubes.

The smoke from a cigarette first rises in a laminar column and then breaks into turbulent flow.

(a)

(b)

Figure 11.27 (a) An ideal fluid flows smoothly around an obstruction. (b) A real fluid cannot always follow the solid surface and forms a jumble of whirlpools that corresponds to turbulence.

Figure 11.28 As the flow speed of a real fluid increases, its ability to follow the contours of a solid obstacle decreases. It tears away from the surface and forms a wake of turbulence that carries away energy.

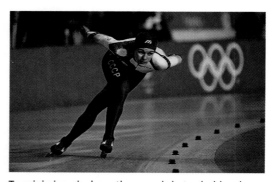

To minimize air drag, the speed skater holds a low profile and keeps one arm behind the back. Smooth clothing and a hood also lessen drag.

Figure 11.29 The fluid in contact with the solid surface is at rest. Its speed increases, depending on the viscosity, until it reaches the unencumbered free-stream speed, after which it's constant. The transition region is the boundary layer.

clists riding on hilly terrain—the faster they go, the more they tend to sail over the mounds and ruts rather than riding down them.

A real fluid streaming through pipes and around objects is influenced by its viscosity. Moreover, the molecules of a fluid easily come into intimate contact with those of a solid, and the short range (electromagnetic) adhesive force produces appreciable effects. *The layer of fluid in contact with a solid will adhere to that surface and remain at rest with respect to it.* The speed of flow increases from zero up to the free-stream value as the distance from the surface increases (Fig. 11.29). That takes place across a relatively thin region, depending on the medium, called the **boundary layer**.

Although we might expect fan blades running at high speeds to stay clean, they certainly don't. The dust particles they pick up low to the surface come into the static-air layer and are dragged along with the blades. Instead of the wind blowing the whirling blades clean, the dust remains at rest in the film of calm air adhering to the solid surface. Similarly, one might imagine that driving a car at 80 km/h would keep it clean, but it does not.

When the flow over an object is very rapid, the fluid may not be able to make a smooth transition from zero speed at the surface layer to some high value nearby. The result is instability and turbulence. This state is characterized by a diminished flow rate (there is a lot of backward swirling), substantial mixing (indeed, turbulence is required if efficient mixing is to occur), and noise (it is the turbulence at the heart valves that makes that familiar thumping sound). A healthy person breathes quietly, but obstructions in the air passages cause turbulence and noise that can be heard with a stethoscope.

For an object moving through a liquid or gas, the resistance offered by the medium increases with the turbulence. While the viscous drag force varies directly with speed, the turbulent drag force is proportional to the speed-squared. Since the power expended in overcoming drag is force times speed, that power varies as the speed-cubed! That is why auto manufacturers are finally streamlining their cars and why big square-front trucks are being retrofitted with wind deflectors, or fairings. The price of fuel demands that less energy be wasted making vortices. A fairing can reduce drag by 20% and save thousands of dollars a year. One wonders why it took the industry so long to figure out what kids, crouching on bicycles, have known all along.

The sloped fairing above the driver's compartment channels air up and over the body of the truck, cutting drag by about 20%.

11.10 The Continuity Equation

The constancy of the density of a flowing liquid is the basis of a fundamental relationship that allows us to understand how liquids progress in pipes and veins. Envision a tube of flow in a liquid (Fig. 11.30). There are no *sources* within the volume of the tube (no nozzles squirting in fluid), and there are no *sinks* (no drains tapping off fluid). The fluid enters at boundary 1, where area A_1 is perpendicular to the flow lines, and emerges at boundary 2, where A_2 is perpendicular to the flow. Let v_1 and v_2 be the average speeds of the fluid over A_1 and A_2. In a tiny time Δt, during which

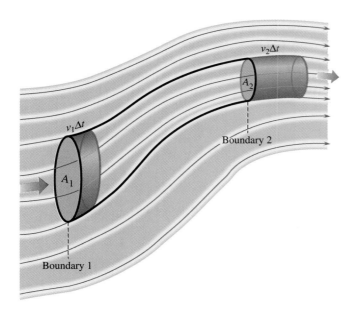

Figure 11.30 A fluid in laminar flow passes from left to right. Assuming incompressibility, the volume entering the flow tube per second at boundary 1 must equal the volume per second leaving at boundary 2, or anywhere else along its length. The flow tube, which is part of a broader stream, could also be envisioned as a pipe with a changing cross section.

American sprint cyclists in the 1992 Olympics wore suits with ribbing on the back and shoulders. The ridges reduce turbulence and cut drag by 3%.

Figure 11.31 As the water falls, its speed increases and so its cross-sectional area decreases as mandated by the Continuity Equation.

the particles of fluid entering the tube travel a distance $v_1\Delta t$, the particles leaving it travel a distance $v_2\Delta t$. Since the volume entering equals the volume leaving, we have

$$(A_1 v_1 \Delta t) = (A_2 v_2 \Delta t)$$

and

$$A_1 v_1 = A_2 v_2 \qquad (11.12)$$

This is the **Continuity Equation**, and it points out that, as the cross-sectional area increases, the speed decreases and vice versa (Fig. 11.31). A river flowing leisurely along a broad bed speeds up as the bed narrows and all the water rushes through a smaller cross-sectional area. If instead, the river branches into a number of identical tributaries so that the *total* cross-sectional area is increased, the speed in each narrow branch will be slower than in the river. Thus, in the human body, as the blood-carrying tubing branches from the single aorta (net area $\approx 2.5 \times 10^2$ mm^2) into the many arteries (net area $\approx 2 \times 10^3$ mm^2) into the far more numerous arterioles (net area $\approx 4 \times 10^3$ mm^2) into the billions of capillaries (net area $\approx 2.5 \times 10^5$ mm^2), the speed of flow decreases proportionately even though the individual blood vessels are getting narrower.

The quantity Av, which is constant within any tube, is denoted by the symbol J and is called the *discharge rate* or **volume flux**—the latter name applies because at both boundaries $v = \Delta s/\Delta t$ and $\Delta V = A\,\Delta s$; hence

$$J = Av = \frac{\Delta V}{\Delta t} \qquad (11.13)$$

This quantity is the volume of fluid flowing past a point in the tube per interval of time Δt, and it has SI units of m^3/s, though liters/min and gal/min are also common in the literature. Multiplying J by the density of the fluid provides the *mass flux,* or rate of mass flow in kg/s. Envision a system pumping water into a hose at a fixed rate. If a nozzle at its end necks the area of the opening down by a factor of 2, the exit speed must go up by a factor of 2—if the hose carries 0.000 5 m^3/s, the nozzle must discharge 0.000 5 m^3/s. Similarly, in the human circulatory system, the amount of blood pumped out of the heart's ventricles equals the amount returning to the atria.

Example 11.9 Blood is pumped out of the heart via the thick-walled (2-mm) tube known as the aorta (inside diameter about 18 mm) at an average speed, for an adult at rest, of 0.33 m/s. (a) Compute the discharge rate. The aorta branches into about 32 major arteries that are each roughly the same size—namely, 4-mm inside diameter. (b) Determine the speed of the blood through these arteries. The smallest branches of the system are the capillaries, about 8×10^{-6} m in inner diameter. (c) Given that the net cross-sectional area of capillaries is 2.5×10^5 mm^2, what is the speed of flow in a capillary?

Solution: [Given: $v_A = 0.33$ m/s, $D_A = 18$ mm, 32 arteries with $D_a = 4$ mm, $A_c = 2.5 \times 10^5$ mm^2. Find: (a) J_A, (b) v_a, (c) v_c.] (a) For the aorta, which we subscript with an A

$$J_A = Av = \pi\left(\frac{D_A}{2}\right)^2 v = \pi(9.0 \times 10^{-3} \text{ m})^2(0.33 \text{ m/s})$$

and

$$\boxed{J_A = 8.4 \times 10^{-5} \text{ m}^3/\text{s}}$$

(b) The aorta branches into 32 arteries. To find the speed of the blood, we need the *net cross-sectional area of all of these arteries:* $A_a = 32\pi(D_a/2)^2 = 4.02 \times 10^{-4}$ m^2. Thus, since the flow rate from the aorta equals that in all the arteries taken together

$$J_A = J_a = A_a v_a$$

$$8.4 \times 10^{-5} \text{ m}^3/\text{s} = (4.02 \times 10^{-4} \text{ m}^2)v_a$$

and

$$\boxed{v_a = 0.21 \text{ m/s}}$$

(continued)

(continued)
(c) Similarly

and

$$J_A = J_c = A_c v_c$$

$$8.4 \times 10^{-5} \text{ m}^3/\text{s} = (2.5 \times 10^{-1} \text{ m}^2)v_c$$

$$\boxed{v_c = 3.4 \times 10^{-4} \text{ m/s}}$$

▶ **Quick Check:** $A_a/A_c = (4 \times 10^{-4} \text{ m}^2)/(0.25 \text{ m}^2) = 16 \times 10^{-4}$, which should equal $v_c/v_a = 16 \times 10^{-4}$.

11.11 Bernoulli's Equation

The first modern treatment of hydrodynamics was a work completed in 1734 by the Swiss mathematician, physicist, and physician Daniel Bernoulli. He was a disciple of Leibniz and an intellectual adversary of the Newtonian school. Shunning Newton's Laws, Bernoulli used the idea of conservation of *vis viva* (p. 288) to derive the central formula of fluid dynamics. That derivation, though it came long before the concept of *energy* was formalized, turned out to be equivalent to the principle of Conservation of Energy.

The first point to consider is that a pressurized fluid must contain energy by virtue of the work done on it to establish that pressure. Pop the cap off a bottle of beer that has been well shaken, and fluid will come blasting out with plenty of KE. Clearly, potential energy is stored in the pressurized system and *a fluid that undergoes a pressure change undergoes an energy change.*

Now, imagine an ideal incompressible fluid for which there are no losses due to the conversion of mechanical energy into rotational, thermal, or any other form of energy. The pressure acting on a moving glob of such a fluid does work on it that appears as a net change in the kinetic and/or potential energy of the system; that is

$$\Delta W = \Delta KE + \Delta PE_G$$

Let's first determine ΔW for a sample of fluid.

Figure 11.32 depicts a narrow tube of flow where the specified values of pressure, speed, displacement, and height are to be thought of as averages. The pressure-force $F_1 = P_1 A_1$ acting on the glob of liquid, pushing it in the direction of motion, does an amount of work *on* it of $F_1 \Delta s_1$. In the process, molecules throughout the tube are shifted to the right with the net effect that an equal-volume glob of length Δs_2 is displaced into region 2. This time, the fluid external to the tube pushes to the left with a force $F_2 = P_2 A_2$, which is opposite in direction to the displacement. Here, the liquid in the tube is doing work pushing on the surrounding fluid, and so the work done *on* it, $F_2 \Delta s_2$, is negative. Hence, the net work done *on* the fluid is

$$\Delta W = F_1 \Delta s_1 - F_2 \Delta s_2 = P_1 A_1 \Delta s_1 - P_2 A_2 \Delta s_2$$

which can be put in a more convenient form. Recall that $\Delta s_1 = v_1 \Delta t$ and $\Delta s_2 = v_2 \Delta t$; thus

$$\Delta W = P_1 A_1 v_1 \Delta t - P_2 A_2 v_2 \Delta t$$

and from the Continuity Equation, $A_1 v_1 = A_2 v_2 = Av$; hence

$$\Delta W = vA \Delta t (P_1 - P_2)$$

Since the mass of the glob of liquid is $\Delta m = \rho \Delta V = \rho(vA \Delta t)$,

$$\Delta W = \frac{\Delta m}{\rho}(P_1 - P_2) \tag{11.14}$$

The glob of fluid, in moving from a higher to a lower pressure, has work done on it.

Original System

Region 2

C *D*

$F_1 = P_1 A_1$ Region 1   Region 3

y_1

Δs_1

(a)

A *B*

Low pressure $F_2 = P_2 A_2$

C *D*

Δs_2

High pressure

y_2

A *B*

Final System

(b)

Figure 11.32 The energy analysis for an incompressible fluid in laminar flow. In a time interval Δt, the tube of liquid essentially shifts from where it was in (a) to where it is in (b). A mass of fluid Δm has been raised in the gravitational field; there's been a change in pressure and a change in speed.

Now for the accompanying change in KE. The system of interest originally is the tube of liquid (Fig. 11.32a) extending from position *A* to *C*. The glob of fluid Δm is moving at v_1 in region 1 (*A* to *B*). After a time, Δt (Fig. 11.32b), that glob has moved into region 3 and an equivalent mass of fluid that was in region 3 has moved into region 2. The system now extends from *B* to *D*. In effect, Δm has gone from region 1 at speed v_1 to region 2 at v_2. The change in kinetic energy of the fluid in the tube is

$$\Delta KE = \tfrac{1}{2}\Delta m(v_2^2 - v_1^2) \tag{11.15}$$

The change in the *gravitational*-PE arises from the fact that the shifting of all the molecules in the tube has effectively brought a mass Δm from region 1 to region 2. Measuring the *y* values up from some arbitrary reference level, we can write

$$\Delta PE_G = \Delta mg(y_2 - y_1) \tag{11.16}$$

Using the last three equations, we obtain

$$\Delta W = \Delta KE + \Delta PE_G$$

$$\frac{\Delta m}{\rho}(P_1 - P_2) = \tfrac{1}{2}\Delta m(v_2^2 - v_1^2) + \Delta mg(y_2 - y_1)$$

and $(P_1 - P_2) = \tfrac{1}{2}\rho(v_2^2 - v_1^2) + \rho g(y_2 - y_1)$ (11.17)

Rearranging terms, we get **Bernoulli's Equation**

$$P_1 + \tfrac{1}{2}\rho v_1^2 + \rho g y_1 = P_2 + \tfrac{1}{2}\rho v_2^2 + \rho g y_2 \tag{11.18}$$

In the steady flow of an ideal liquid, *following along a streamline*

$$P + \tfrac{1}{2}\rho v^2 + \rho g y = \text{constant} \tag{11.19}$$

which is often referred to as *Bernoulli's Theorem*.

Each term has the dimensions of energy per unit volume, or energy density. There is a *kinetic-energy density* ($\tfrac{1}{2}\rho v^2$) associated with the bulk motion of the fluid, a *potential-energy density* ($\rho g y$) associated with changes in location within the gravitational field, and a *pressure-energy density* (P) arising from internal forces on the moving fluid. This latter energy is similar to that stored in a compressed spring in the sense that it's a manifestation of the interatomic electromagnetic interaction. In effect, Eq. (11.19) says that, as long as no energy enters or leaves, *the net energy density contained in the fluid is constant all along a given flow tube*. Though it was derived for an incompressible fluid, Bernoulli's Equation can be applied to gases, provided the pressure changes are only a few percent.

Example 11.10   Figure 11.33 depicts an uncovered vat of brewing beer and the narrow pipe used to take samples from it. The cross-sectional area of the vat is 1.50 m². At a given instant, the liquid level is falling at

(continued)

(continued)
1.0 cm/s while the beer is traveling at 50 cm/s past the gauge. Determine the absolute pressure reading at that point within the pipe, at that moment. Take the density of beer to be 1.0×10^3 kg/m³.

Solution: [Given: gauge 2.0 m below surface, area of tank 1.50 m², speed of surface 1.0 cm/s, speed in pipe 0.50 m/s. Find: pressure in pipe.] Take the influx end of a tube of flow as region 1 located at the open surface of the liquid in the vessel. Take region 2 in the pipe near the pressure gauge. Since the reference level is arbitrary when dealing with potential energy, let $y_2 = 0$. The height of the liquid surface (above the gauge) is $y_1 = h = 2.0$ m. Bernoulli's Equation

$$P_1 + \tfrac{1}{2}\rho v_1^2 + \rho g y_1 = P_2 + \tfrac{1}{2}\rho v_2^2 + \rho g y_2$$

becomes

$$P_A + \tfrac{1}{2}\rho v_1^2 + \rho g h = P_2 + \tfrac{1}{2}\rho v_2^2 + 0$$

and so

$$P_2 = P_A + \tfrac{1}{2}\rho(v_1^2 - v_2^2) + \rho g h$$
$$P_2 = 1.013 \times 10^5 \text{ Pa} + \tfrac{1}{2}(1.0 \times 10^3 \text{ kg/m}^3)$$
$$\times [(1.0 \times 10^{-2} \text{ m/s})^2 - (0.50 \text{ m/s})^2]$$
$$+ (1.0 \times 10^3 \text{ kg/m}^3)(9.8 \text{ m/s}^2)(2.0 \text{ m})$$

thus

$$P_2 = 1.013 \times 10^5 \text{ Pa} - 1.25 \times 10^2 \text{ Pa} + 1.96 \times 10^4 \text{ Pa}$$

Figure 11.33 A vat of beer is sampled via a narrow tube with a pressure gauge mounted 2.0 m below the surface.

and
$$\boxed{P_2 = 1.2 \times 10^5 \text{ Pa}}$$

▶ **Quick Check:** The pressure at a point that is 2 m down must be $P_A \approx 100$ kPa, plus the contribution due to the weight of the column of fluid $\rho g h \approx (10^3 \text{ kg/m}^3) \times (10 \text{ m/s}^2)2 \text{ m} \approx 20$ kPa, plus a drop in pressure (≈ 0.1 kPa) due to an increase in speed or ≈ 120 kPa.

Notice that when the fluid is at rest, $v_1 = v_2 = 0$, and Eq. (11.18) becomes

$$P_1 - P_2 = \rho g(y_2 - y_1)$$

which is equivalent to the result (p. 356) for the pressure difference between any two points in a liquid.

Torricelli's Result

Consider the closed tank of liquid shown in Fig. 11.34 that has an orifice of area A_2, out of which is pouring the fluid. The rate at which the fluid leaves, the *speed of efflux* (v_2), is the quantity we want to determine. The vessel is taken to be very large so that neither the liquid level nor the pressure above it (P) changes significantly. Since the flow tube goes from region 1 out to region 2, which is open in the air, let $P_1 = P$. The emerging stream in region 2 is surrounded by air at atmospheric pressure, and nothing is there to sustain a pressure difference. *A free fluid jet must be at the same pressure as the enveloping air,* $P_2 = P_A$. Since $y_1 - y_2 = h$, Bernoulli's Equation is then

$$P + \tfrac{1}{2}\rho v_1^2 + \rho g h = P_A + \tfrac{1}{2}\rho v_2^2$$

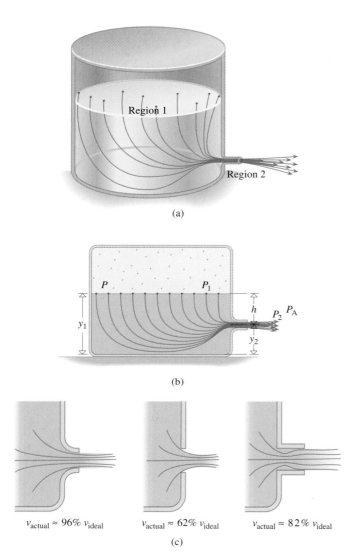

(a)

(b)

$v_{actual} \approx 96\% \, v_{ideal}$ $v_{actual} \approx 62\% \, v_{ideal}$ $v_{actual} \approx 82\% \, v_{ideal}$

(c)

Figure 11.34 (a) The streamlines for a fluid flowing out of a closed vessel, where the pressure on the liquid's surface is P. (b) To apply Bernoulli's Equation, we take the top of the liquid as region 1, the outside of the nozzle as region 2, and follow a streamline from 1 to 2. (c) In actuality, the shape of the orifice affects the shape of the jet and its speed.

$PE_G = mgh$

$v = \sqrt{2gh}$

$KE = \frac{1}{2}mv^2$

Figure 11.35 The emerging jets reach about the same heights, independent of the diameters of the nozzles. They don't have the same discharge rates, although the efflux speeds are equal. This is a different situation from that of a pump forcing *a fixed discharge rate* through two different orifices. It's also different from the garden hose fed by miles of pipe in which friction affects the flow. Note that some lines are behind others—streamlines never cross.

and

$$v_2^2 = v_1^2 + \frac{2(P - P_A)}{\rho} + 2gh \qquad (11.20)$$

From the Continuity Equation ($v_1 A_1 = v_2 A_2$), it follows that, when $A_1 \gg A_2$ and $v_2 \gg v_1$, then v_1^2 is negligible compared to v_2^2. In the special, though common, case where the tank is open to the air ($P = P_A$), the pressure-energy density vanishes—the fluid runs out via gravity without also being pushed out by a pressure difference, and

$$v_2 = \sqrt{2gh} \qquad (11.21)$$

This relationship is *Torricelli's Result*.

Being frictionless, in theory at least, the liquid spurts from the hole with a speed equal to that which it would have gained in free-falling through the height h. Torricelli observed that the emerging jet, if directed straight up, would almost reach the original level (Fig. 11.35). The reason the jet doesn't quite attain the initial height is that some energy is invariably converted to thermal energy via friction.

Example 11.11 A boat on a lake crashes into a submerged rock, ripping a hole 40 cm^2 in its hull 1.00 m below the waterline. If the craft can take on 10.00 m^3 of water before getting its cargo drenched, roughly how much time does the crew have to do something clever?

Assume the water flows straight in the hole and neglect any frictional effects.

Solution: [Given: area of hole 40×10^{-4} m^2, $h = 1.00$ m. Find: time to take on 10.00 m^3 of water.] Let

(continued)

(continued)

region 1, the input end of a flow tube, be at the surface of the lake and region 2, the output end, be at the hole—water flows down and into the boat. Assuming there is atmospheric pressure in the boat at the hole, Bernoulli's Equation is

$$P_1 + \tfrac{1}{2}\rho v_1^2 + \rho g y_1 = P_2 + \tfrac{1}{2}\rho v_2^2 + \rho g y_2$$

and

$$P_1 + \tfrac{1}{2}\rho v_1^2 + \rho g(y_1 - y_2) = P_2 + \tfrac{1}{2}\rho v_2^2$$

Since the level of the lake is not going to fall very rapidly

$$P_A + 0 + \rho g h = P_A + \tfrac{1}{2}\rho v_2^2$$

which gives us Torricelli's Result

$$v_2 = \sqrt{2gh} \qquad [11.21]$$

We should have anticipated this result to begin with—the container with the leak is the whole lake pouring

water into the boat, so

$$v_2 = \sqrt{2(9.81 \text{ m/s}^2)(1.00 \text{ m})} = 4.43 \text{ m/s}$$

Now

$$J = v_2 A_2 = (4.43 \text{ m/s})(40 \times 10^{-4} \text{ m}^2)$$

or

$$J = 1.77 \times 10^{-2} \text{ m}^3/\text{s}$$

and so it will take on 10.00 m³ of water in

$$\frac{(10.00 \text{ m}^3)}{(1.77 \times 10^{-2} \text{ m}^3/\text{s})} = 565 \text{ s} = \boxed{9.4 \text{ min}}$$

This solution assumes the boat doesn't sink appreciably lower in the water in the process, thereby changing h and so J.

▶ **Quick Check:** The water effectively falls 1 m; hence, $y = \tfrac{1}{2}gt^2$ and $t = 0.45$ s; thus, $v = gt = 4.43$ m/s. $J \approx 160 \times 10^{-4}$ m³/s; 10 m³/$J \approx$ 10 min.

The Venturi Effect

Some of the most practical applications of fluid dynamics arise from the interdependence of pressure and speed. There is a class of situations in which the change in *gravitational*-PE is ignorably small and Bernoulli's Equation then relates differences in pressure to differences in kinetic energy, and therefore speed. For example, envision a length of pipe (with a transverse area A_1) that necks down to a smaller area (A_2) and then opens out again (Fig. 11.36). We know that

$$A_1 v_1 = A_2 v_2 \qquad [11.12]$$

and, as we saw earlier, decreasing the area traversed by the flow of a gas or liquid results in an increase in its speed.

In any situation where the heights of the incoming stream and the constricted region are about the same ($y_1 \approx y_2$), Bernoulli's Equation can be used to express the driving pressure difference:

$$P_1 + \tfrac{1}{2}\rho v_1^2 + \rho g y_1 = P_2 + \tfrac{1}{2}\rho v_2^2 + \rho g y_2$$

becomes

$$P_1 + \tfrac{1}{2}\rho v_1^2 = P_2 + \tfrac{1}{2}\rho v_2^2$$

Using the Continuity Equation to eliminate v_1, we get

$$P_1 - P_2 = \tfrac{1}{2}\rho v_2^2 \frac{(A_1^2 - A_2^2)}{A_1^2} \qquad (11.22)$$

Inasmuch as $A_1 > A_2$, the right side of the equation is positive and $P_1 > P_2$—there is a pressure drop in the narrow region. As the fluid enters region 3, the higher pressure slows it down, and the speed returns to its original value. This decrease in pressure accompanying an increase in speed is variously called Bernoulli's Effect, or the *Venturi Effect,* after the Italian researcher who first studied it (1791).

Figure 11.36 In region 1, the cross-sectional area is large, the speed is low, and the pressure is high. In region 2, the area is small, the speed is high, and the pressure is low. In region 3, the area is again large, the speed is small, and the pressure is high.

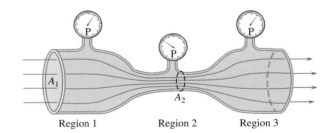

Region 1   Region 2   Region 3

Figure 11.37 Two ships traveling side by side or moored in moving water. The deflection of the flow between the two vessels causes a Venturi drop in pressure, and the ships experience a force pushing them together.

Figure 11.38 The flow of a fluid through a constricted channel.

Figure 11.39 Flow past a horizontal asymmetrical airfoil. Note the high-speed, low-pressure regions where the flowlines are closer together. Compare this flow pattern with that of Fig. 11.38.

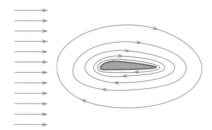

Figure 11.40 Circulation around an airfoil, in cross section. The sum of the circulation and the free-air flow produces the pattern in Fig. 11.39, where the speed is higher above the airfoil.

It is possible to mechanically increase the speed of a fluid, as for example, with a pump. The KE thereby added comes from the work done on the fluid. Thus, the pressure in a high-speed free jet produced by a pump is still atmospheric. That's not to be confused with the Venturi Effect, where no additional energy enters the system and an increase in KE must come from a decrease in pressure energy, and vice versa. In that case, **a portion of a fluid made to move rapidly sustains a lower pressure than does the portion of the fluid moving more slowly.** As a large truck or bus passes alongside a car, the pressure between the two will often drop from the rushing air, giving the driver of the auto the disturbing feeling of being pulled toward the larger vehicle (Fig. 11.37).

Arteriosclerosis arises when plaque builds up on the inner walls of the arteries, restricting the flow of blood. Any such obstruction results in a Venturi pressure drop (Fig. 11.38). In an advanced stage of the disease, the pressure difference, inside and out, can be great enough to cause the artery to momentarily collapse. Backed-up blood pressure will soon force it open, only to collapse again and so on, producing a vascular flutter that can be heard with a stethoscope. The Venturi Effect in airstreams is responsible for other flutterings, such as the sound of snoring and the more melodious tones of reeded woodwinds.

11.12 Airfoils

An **airfoil** is an object whose shape is such that it generates a desired reaction force from the fluid through which it moves. For instance, the wing-shaped object shown in Fig. 11.39 is an asymmetrical airfoil, and so, unlike Fig. 11.38, the streamline pattern is asymmetrical. Air moves with respect to the long, curved upper surface, at a more rapid speed than the airstream beneath the wing. With an appropriately shaped wing, the corresponding drop in pressure above, as compared to below, creates a net upward force known as **lift**. Lift makes possible the level flight of powered airplanes, holds gliders in the air, and supports soaring birds.

Still, it's not clear at this point why the air moves at a higher speed above the wing than below. Most popularized versions suggest that the two streams must take the same time to arrive at the rear edge and so travel at different speeds, but that explanation doesn't account for how the air finds out that it has a scheduled arrival time to meet, nor, in fact, is it even true—the traversal times need not be the same. What actually happens is affected by the viscosity of the air and its adhesion to the wing. The airfoil induces a circulation of air (Fig. 11.40) around the wing. This circulation combines with the free-air flow to produce a higher speed stream above than below the wing.

The amount of circulation generated depends on the shape of the airfoil, its speed, and its orientation with regard to the air flow. For example, a plane can fly horizontally, nose up, with its wings tilted at the so-called *attack angle*. The pressure above drops and the pressure below rises even more, thereby increasing the lift (Fig. 11.41). There is a practical limit to the process (typically around 15°): too much tilt and the boundary-layer air tears away from the top of the wing, becoming turbulent—the wing *stalls,* lift vanishes, and the plane falls.

There's a popular engineering adage that if a plane's engines have enough power, they can drag almost anything into the air. That being the case, the curvature of a plane's wings becomes less important. Wings that are symmetrical airfoils get their lift primarily by being pushed through the air at a nonzero attack angle—in effect, the air moving at the underside is deflected downward and transfers an upward momentum to the wing. That's how planes manage to fly upside down—if the only thing occurring was a reduced pressure due to wing curvature, the craft would be driven into the ground by the inverted "lift," but that doesn't happen.

A simple argument can be made, via Bernoulli's Equation (Problem 69), showing that *lift is proportional to the product of airspeed-squared and wing area.* This relationship turns out to be true, and it explains why big planes and big flying birds have relatively big wings. It's also why big birds have a higher minimum air speed that they must reach to take off and must maintain to stay airborne.

11.13 Vortices

A vortex is a whirling mass of fluid bounded by a region that is irrotational. Tornadoes and hurricanes are large-scale powerful vortices; the whirling spiral of water draining from a bathtub and the great mushroom cloud of an atomic bomb are also vortex structures.

Whenever an object moves with respect to a fluid (a plane through the air or an oar through the water), it tends to spawn vortices. Small invisible whirlwinds are everywhere—behind moving ships and cars and planes and baseballs. Stirring a cup of coffee creates vortices, which is why it's done—to churn up turbulence that will disperse the ingredients.

Vortices are formed when there is a discontinuity in the flow pattern, often due to the presence of an obstruction. When two currents at different speeds come together, a transition region exists wherein the rapidly moving layer progresses beyond and around the slower layer, which essentially drags the former down. The process is like a forward moving row of skaters linked arm-in-arm right to left. If the person on the far right skates onto a bed of sand, the right side will slow dramatically while the left side of the row, still moving rapidly, swings around clockwise.

Imagine a cylindrical body immersed in a moving fluid (Fig. 11.42). Generally, there will be a more or less irregular flow pattern behind the cylinder. But if conditions are right (the speed of the fluid is high enough, its viscosity low enough), two parallel rows of equally spaced alternating eddies will be created—known as a *von Kármán vortex street.* Rows of vortices are easy to generate by moving a cylindrical object such as a pen or a finger at various speeds across the surface of a deep pan of water. The periodic shedding of vortices by an obstruction results in an oscillating pressure on the body that often tends to set it vibrating in a direction *perpendicular* to the flow. During World War II, the vibration of machine-gun barrels jutting from airplane turrets was a frequent problem; and skyscrapers are driven into oscillation by the same vortex mechanism. Similarly, vibrating (so-called singing) phone and power lines have often led to mechanical problems, sometimes ending in catastrophic failure due to fatigue.

Surfaces that develop lift, such as airfoils, create vortices at their two ends, due to the difference in pressure at top and bottom. Air flows outward away from the fuselage, up and around the wing tips from the high-pressure region beneath to the low-pressure region above the wings (see the photo on p. 387). Figure 11.43 shows how

The skier becomes an airfoil.

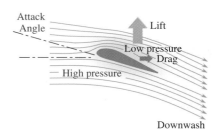

Figure 11.41 The downwash tells us that there is an upward reaction force exerted on the airfoil by the deflected air stream. The time-rate-of-change of the momentum of the air must equal the magnitude of the upward force.

This photo of the Great Red Spot on the surface of Jupiter was taken by *Voyager 2.* The spot is a giant stable vortex that has been sustained in the deep atmosphere of that rotating planet for at least 300 years.

A satellite photo of vortices in the cloud cover formed in the wake behind an island in the ocean.

A vortex street.

different city situations can also spawn troublesome vortices. These vary from the street-level variety that steals hats to the high-altitude kind that blows out windows.

11.14 Blood Pressure

The human heart is actually two twin-chambered pumps (Fig. 11.44) mounted in a single unit: a low-pressure device (the right atrium and ventricle) and a high-pressure device (the left atrium and ventricle). The right half delivers oxygen-depleted blood to the nearby lungs. With each *systole* or rhythmic contraction of both ventricles, the right half discharges 70 to 80 cm^3 of blood at a mean gauge pressure of 2 kPa (15 mm Hg) into the pulmonary artery on the way to the lungs. An equal amount of oxygen-rich blood is returned from the lungs to the left half of the heart (at $\approx$ 8 mm Hg) for delivery to the rest of the body. The left ventricle pumps the blood out, maintaining a pulsating pressure that rises and falls 70 to 80 times per minute. The peak, or *systolic pressure*, developed during pumping (Fig. 11.45) is around 16 kPa (120 mm Hg) and the minimum or *diastolic pressure*, which occurs when the heart is essentially relaxed and refilling, is around 11 kPa (80 mm Hg).

Figure 11.42 A fluid moving past a cylindrical obstruction. If conditions are right, an alternating series of vortices will be formed in the wake.

Example 11.12 Compute the average output power of the human heart at rest. Take the average output pressure to be 1.33×10^4 Pa (100 mm Hg) and the flow rate as 5.0 liters per minute. Compare your result with the fact that the heart consumes energy at a rate of about 10 W.

Solution: [Given: $P_{av} = 1.33 \times 10^4$ Pa and $J =$ 5.0 liter/min. Find: output power.] The force doing the work equals $P_{av} \times A$, and the power developed is $P = Fv = P_{av}Av = P_{av}J$. First change the units of the flow rate; since 1 liter = 1000 cm^3, then $J =$ 5.0 liter/min = (5.0 liter/min)(0.001 m^3/liter)(1/60 s/min) and $J = 8.33 \times 10^{-5}$ m^3/s. The heart takes that much blood

every second from a gauge pressure of near zero to an average of about 1.33×10^4 Pa. Consequently,

$$P = P_{av}J = (1.33 \times 10^4 \text{ Pa})(8.33 \times 10^{-5} \text{ m}^3/\text{s})$$

and

$$\boxed{P = 1.1 \text{ W}}$$

which makes it roughly 10% efficient.

▶ **Quick Check:** 1 liter = 10^3 cm^3 = $10^3 \times 10^{-6}$ m^3; 1 liter/min = 10^{-3} m^3/60 s = 1.66×10^{-5} m^3/s. Power per 1 liter/min is P = 13 kPa(1.66×10^{-5} m^3/s) = 0.22 W.

Blood is a viscous fluid, and there will be some loss in energy and a corresponding drop in pressurization along the flow route—a real liquid drops in pressure as it negotiates even a smooth channel. The effects of viscous drag become less

Figure 11.43 City winds and vortices. Poorly designed and positioned buildings can create dangerous windstorms at street level. The average person can be blown off balance by a 65-km/h (40-mi/h) wind, and turbulent winds of 16 km/h (10 mi/h) impede walking. A gust of wind in Boston's Copley Square once blew over a half-ton mail truck. High winds can also be created by building-lined streets via the Venturi Effect.

Tip vortices made visible during crop spraying. The presence of powerful vortices behind large aircraft, such as the Boeing *747*, determines the maximum rate at which planes can follow one another down an airport runway.

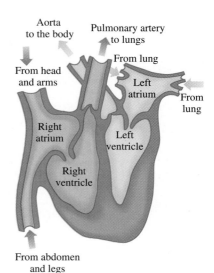

Figure 11.44 Both atria contract at the same time, forcing blood into their corresponding ventricles. Thereafter, the ventricles contract, forcing blood into the arteries and out to the body and lungs. The bluish regions carry oxygen-depleted blood, the reddish regions oxygen-rich blood.

troublesome as the pipe gets larger and more of the fluid is moving farther from the walls. Arteries have relatively large cross sections and blood progresses through them with little frictional loss and little diminution of pressure energy. That's why the pulses of blood pressure are most easily detected with arteries, especially ones close to the surface at the neck and wrist.

Insofar as viscous effects in the main arteries are not significant, Bernoulli's Equation applies:

$$P_1 + \tfrac{1}{2}\rho v_1^2 + \rho g y_1 = P_2 + \tfrac{1}{2}\rho v_2^2 + \rho g y_2 \quad [11.18]$$

Since the flow rates and cross-sectional areas in the main arteries are fairly uniform, the speeds of flow are approximately equal as well. We want to compare the pressure at the heart (P_H) and feet (P_F), so take these to be regions 1 and 2, respectively. For someone standing, $y_1 = y_H$ and $y_2 = 0$; hence

$$P_H + \rho g y_H = P_F$$

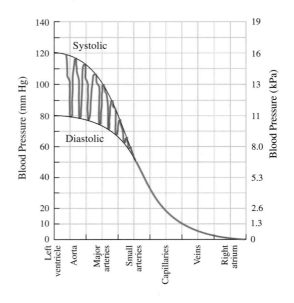

Figure 11.45 Blood pressure in the adult human male body. As the heart goes through its pumping cycle, the pressure in the main vessels fluctuates. The systolic and diastolic pressures are the maximum and minimum values developed in the arteries. Typical female pressures are slightly lower (110 mm Hg/70 mm Hg).

Figure 11.46 The mercury manometer reads the gauge pressure in the cuff wrapped around the arm. The small rubber bulb pumps up the cuff until it cuts off the blood flow in the brachial artery in the upper arm. The pressure is released gradually and, with the use of a stethoscope, the blood can be heard to start flowing.

Blood pressure varies over the body, in part due to gravity. A person who jumps up from a prone position can feel momentarily dizzy because of the sudden drop in blood pressure. Similarly, standing still can cause blood to pool in the legs, and dizziness can again occur. Fainting usually lowers the head.

In effect, the weight of the blood causes an increase in pressure at the feet, just as it would in the static case. This is why it's so civilized to have a nap after lunch—the heart doesn't have to work quite as hard pumping against gravity.

The most common way to measure arterial blood pressure, though it's not very accurate, is with an inflatable bag or cuff attached to a manometer (Fig. 11.46). The cuff is tightly wrapped around the upper arm *at the level of the heart* (so there is no $\rho g y$ contribution to P) and then inflated with a rubber bulb pump. Squeezing the arm, it cuts off blood flow in the brachial artery. The overpressure in the cuff is then allowed to gradually drop while the observer listens with a stethoscope placed downstream on the artery. When the pressure in the cuff (as indicated by the manometer) equals the peak or systolic gauge pressure, blood begins to spurt past the constriction, and the turbulent, noisy flow is easily heard in the stethoscope. Further reduction of the cuff pressure is accompanied by a changing sound until at the diastolic pressure the flow is again laminar and the characteristic noise of turbulence vanishes (Table 11.6).

TABLE 11.6 **The human cardiovascular system**	
Blood density (37°C)	1.0595×10^3 kg/m^3
Number of red cells per mm^3 of blood	5×10^6
Number of white cells per mm^3 of blood	8×10^3
Flow rate at heart	
resting	8×10^{-5} m^3/s (5 liters/min)
intense activity	40×10^{-5} m^3/s (25 liters/min)
Time for tracer to circulate	18–24 s
Time for complete circulation	54 s
Blood volume (70-kg male)	5.2×10^{-3} m^3 (5.2 liters)
Gauge pressure, right ventricle, peak	3 kPa (25 mm Hg)
Gauge pressure, left ventricle, peak	16 kPa (120 mm Hg)
Mean gauge pressure, large veins	1.07 kPa
Mean gauge pressure, large arteries	12.8 kPa
Mean gauge pressure, at the heart	13.3 kPa
Pressure drop per cm of aorta	4.3 Pa (3.2×10^{-2} mm Hg)
Power required by heart at rest	≈ 10 W

Core Material

Pressure (P) is *the magnitude of the total force acting perpendicular to a surface divided by the area of that surface:*

$$P = \frac{F_\perp}{A} \qquad [11.1]$$

The units of pressure are N/m^2 or *pascals*. The pressure a distance h beneath the surface of a homogeneous fluid is

$$P_l = \rho g h \qquad [11.2]$$

and if there is a pressure on its surface P_s, then

$$P = P_s + P_l = P_s + \rho g h \qquad [11.3]$$

The pressure at every point at a given horizontal level in a single fluid at rest is the same.

An *atmosphere* is defined as 1 atm $= 1.013 \times 10^5$ Pa, which equals 14.7 lb/in.^{2}. The absolute pressure (P) is the sum of the gauge and atmospheric pressures, or

$$P = P_A + P_G \qquad [11.4]$$

Keeping the temperature constant, the volume of any gas varies inversely with the pressure. This is **Boyle's Law**:

$$PV = \text{constant} \qquad [11.5]$$

Pascal's Principle states that an external pressure applied to a fluid that is confined within a closed container is transmitted undiminished throughout the entire fluid. Hence, for two connected pistons

$$\frac{F_o}{F_i} = \frac{A_o}{A_i} \qquad [11.6]$$

The **Buoyancy Principle** of Archimedes states that *an object immersed in a fluid will be lighter (that is, it will be buoyed up) by an amount equal to the weight of the fluid that it displaces, and the corresponding upward force exerted by the fluid is known as the buoyant force.* The *relative density*, ρ/ρ_w, is called the *specific gravity*.

Surface tension, γ, is *the force per unit length exerted by the surface.* Where the single liquid film has *two* surfaces

$$F_t = 2\gamma L \qquad [11.10]$$

The **Continuity Equation** is

$$A_1 v_1 = A_2 v_2 \qquad [11.12]$$

The quantity Av is given the symbol J and called the *discharge rate*, or *volume flux*. Applying Conservation of Energy to the ideal fluid, we get **Bernoulli's Equation**

$$P_1 + \tfrac{1}{2}\rho v_1^2 + \rho g y_1 = P_2 + \tfrac{1}{2}\rho v_2^2 + \rho g y_2 \qquad [11.18]$$

or

$$P + \tfrac{1}{2}\rho v^2 + \rho g y = \text{constant} \qquad [11.19]$$

The speed of efflux from a hole in an open container a distance h below the surface is

$$v_2 = \sqrt{2gh} \qquad [11.21]$$

which is known as **Torricelli's Result**.

Suggestions on Problem Solving

1. When doing buoyancy problems with balloons and the like, don't forget to add in, along with the load, the weight of the gas doing the lifting—it is part of the weight of the craft.

2. Some people are accustomed to working in the cgs system when it comes to density and since $\rho_w = 1.00$ g/cm^3 and, moreover, since the specific gravity of water is 1.00, there's a tendency to lose the factor of 10^3 when substituting $\rho_w = 1.00 \times 10^3$ kg/m^3. Remember, 1.00 g/cm$^3 = 1.00 \times 10^3$ kg/m^3. Also remember that 1 cm$^2 = (1$ cm$)(1$ cm$) = (1 \times 10^{-2}$ m$) \times (1 \times 10^{-2}$ m$) = 10^{-4}$ m^2. Many problems deal with pressure beneath the water in an open container—*don't forget atmospheric pressure acts as well.*

3. There are a number of situations in this chapter where we have the same quantity on both sides of an equation—for example, $P_i = P_o$. In such cases, there are the same constants on both sides, and we can take some short cuts with the units. If both pressures in the expression $P_i V_i = P_f V_f$ are given, say, in mm Hg, they need not be converted to Pa before substitution because the units, whatever they are, cancel. This can save a lot of effort, but it should go without saying that you must be careful. *For the time being, use such short cuts only as checks.*

4. There are only two basic equations in this treatment of fluid dynamics, the Continuity Equation and Bernoulli's Equation. To these are added the derived expressions for Torricelli's Result and the equation for the Venturi Effect. Read the problem carefully and determine which of these concepts is involved. If the problem deals with fluid flowing through a hole or a constricted channel, one of the last two equations might be called for rather than going back to Bernoulli's Equation to rederive those results each time.

5. If you are considering a pipeline of constant diameter, the Continuity Equation requires that the speed of an ideal fluid be constant regardless of twists and turns, rises and falls.

6. When using Bernoulli's Equation, draw a diagram. If possible, locate region 1, where the fluid pressure, elevation, and speed are known. At a place where the section area is much larger than elsewhere in the flow, the fluid speed there may be negligible. The pressure in a fluid jet is the same as that of its surroundings. Locate region 2, where the unknown quantity is to be determined. Pick a zero y-level. The values of y_1 and y_2 may turn out positive or negative. If there are two unknowns, use the Continuity Equation as well. You can use either gauge or absolute pressure, but don't mix them. Bernoulli's Equation is applicable to the flow from one region to another. Accordingly, Eq. (11.18) is applicable *along a tube of flow;* in general, you cannot use the equation for *any* two points in the fluid and get exact results.

Discussion Questions

1. Imagine that you are in an elevator with a tank of water and three goldfish. A device in the gravel measures gauge pressure at the bottom of the tank. (a) Suppose it reads 4000 N/m^2 with the elevator car at rest. Will it increase, decrease, or stay the same when the car accelerates upward? Explain. (b) What will it read if the car drops at 9.8 m/s^2?

2. Consider the mercury barometer (a word coined by Boyle). What, if anything, would happen to the reading (that is, the difference in height between the two surface levels) if someone floated a coin on the open mercury? Explain.

3. Imagine a closed bottle filled with water into which is sealed a drinking straw. Could you "suck" the liquid up the straw? Why is the idea of sucking up fluids misleading? What would happen if we now punched a hole in the lid? How does the rubber suction cup on a dart stick to a wall? Why is it a common practice to wet the suction cups first? NASA is developing suction-cup shoes for use in space stations. Will they be of any use outside, for walking on the hull?

4. Figure Q4 shows a spherical (the distortion is optical, not actual) glob of olive oil (0.9 g/cm^3) suspended in a solution of roughly 50% alcohol (0.8 g/cm^3) and water. Explain what you see. Incidentally, by using a mix of bromobenzene (1.497 g/cm^3) and xylene (0.867 g/cm^3), we can determine the specific gravity of blood in this same way.

5. When a helium balloon soars into the sky, it gains KE and *gravitational*-PE. Where does that energy come from?

6. In Fig. Q6, the tube is filled, pinched off, and then its ends are positioned as shown and released. The liquid is *siphoned* until the vessel on the left is emptied or the two levels

Figure Q4

become equal. What makes it function? (Hint: A siphon can work in vacuum.)

7. Picture yourself on a raft floating in a large tank of water. On board with you is a tin cup, a pine bust of Napoleon, a banana, and a chicken. What happens to the level of the water in the tank if: (a) you reach out, fill your cup with water and drink

Figure Q6

it? (b) you toss the bust overboard? (c) you eat the banana? (d) the chicken flies away?

8. Imagine a metal vessel completely closed except for a small threaded hole on top. It is partially filled with water and placed on a scale that then reads 400 N. Holding on to a solid rod, you lower it down into the water. (a) Will the scale reading change? (b) If the rod were weightless, would that alter your answer? (c) What would happen to the scale reading if the rod is fixed in place by a very light collar that screws into the hole and holds the rod so that you can let go?

9. The photo in Fig. Q9 shows a glass tube open on top and bottom standing in a quantity of trichloroethylene (cleaning fluid) of density 1.44×10^3 kg/m^3. Water (with a little black ink in it for contrast) was subsequently poured into the tube. Explain what the photo shows.

Figure Q9

10. The underwater living chamber depicted in Fig. Q10 is wide open on the bottom for easy entrance. What keeps the water from rushing in? What's the pressure inside? What would happen if it were punctured at the top? Describe what happens to the volume of gas inside as the thing is hauled up to the surface.

Figure Q10

11. Figure Q11 shows an arrangement for producing negative pressures (P_T) at the top of the liquid column. Write an expression for P_T in terms of the pressure at the level of the liquid surface, the density, and the height of the column. Explain what is happening. Pressures of up to −300 atm have been achieved using water.

Figure Q11

12. Pascal performed a rather startling demonstration: breaking a barrel using water pressure. He fit the cask with a very tall narrow tube sealed tightly into the lid (Fig. Q12). The barrel and then the tube were filled with water. Explain what happened.

Figure Q12

13. Sprinkle some pepper on the surface of a cup of water and then touch the center of the distribution with a freshly washed and dried finger (no skin oil should be left). Now rub a little soap onto your fingertip and do it again. What happens and why? Put a little oil on your finger and do it again.

14. Rubbing alcohol and acetone both reduce the surface tension of water. With this fact in mind, float two wooden matchsticks in a bowl of water parallel to one another about 2 cm apart. What do you think will happen if you very gently introduce a drop of water between the sticks? Now do that again, this time with a drop of acetone or vodka—what happens?

15. A Ping-Pong ball can be supported in the airstream

Figure Q15

from a hair dryer even when it is tilted at an appreciable angle (Fig. Q15). Explain.

16. Concoct a bubble-making solution of 1 part water to 1 part dishwashing liquid detergent. If you have some glycerin (available at drugstores), add it in (3 parts detergent, 2 parts water, 1 part glycerin). Construct a wire ring about 2 or 3 in. in diameter and tie a small loop of thread to it. Now dip the ring in the solution so that a film stretches across the whole ring and the loop "floats" around within it. Puncture the film within the sagging thread-loop (use a hot pin if necessary). What happens? See Fig. Q16.

Figure Q16

17. The spinning baseball in Fig. Q17a carries along a boundary layer of air that revolves with it. If the spin axis is horizontal and the ball is thrown as shown in the figure, what will its flight path look like? Figure Q17b shows the flow pattern around a rotating cylinder. An experimental ship built in the

(a)   (b)

Figure Q17

1920s crossed the Atlantic using two large rotating vertical columns and the wind as its source of power. Explain.

18. Figure Q18 shows two sets of spheres in streams of ideal fluid. Describe the forces, if any, that appear to be acting between them. An ideal fluid flows parallel to an essentially infinite smooth wall that serves as a boundary. Immersed in the fluid is a solid sphere. Describe the force, if any, that appears to exist between the sphere and the wall.

(a)   (b)

Figure Q18

19. Figure Q19 shows a wing that has recently begun to move and around which has just developed a clockwise circulation of air giving rise to lift. Behind the wing we see a so-called *starting vortex,* which always accompanies the initiation of air circulation around a wing. Once formed, the starting vortex washes away downstream. Discuss why you might expect such an eddy to exist solely from momentum considerations. What do you think would happen if the wing is accelerated from rest and then rapidly stopped?

Figure Q19

20. If you cut your finger badly, why might it be wise to hold it high above your head until the bleeding stops?

Multiple Choice Questions

1. Suppose we fit a mercury barometer with a snug washer-shaped weightless disk with a hole in the middle that sits on the open surface like a piston. What would happen to the pressure reading if a bird landed on the disk?

(a) nothing   (b) the pressure would exceed atmospheric and the column would rise with respect to the open surface   (c) the pressure would fall slightly, and so the central column of mercury would drop a little   (d) the weight of the bird would press on the mercury, increasing the pressure and decreasing the height of the column   (e) none of these.

2. A sample of gas in a chamber sealed by a movable piston is kept at a constant temperature. The volume is slowly changed until it is twice its original value and the pressure therefore   (a) is halved   (b) is doubled   (c) is unchanged   (d) decreases by some unknown amount   (e) none of these.

3. Boyle's Law is PV = constant, where the units of the constant are the same as those of   (a) momentum   (b) acceleration   (c) impulse   (d) work   (e) none of these.

4. Is there a buoyant force acting on you at this very moment?   (a) no   (b) not enough information to say   (c) only if you are floating around in water   (d) yes, the atmosphere exerts a buoyant force   (e) none of these.

5. Float an ice cube in a glass of water filled to the very brim. Will it overflow as the ice, jutting above the rim, melts?   (a) no, it simply fills the displaced volume and the level is unchanged   (b) yes, but only when it melts rapidly   (c) no, but that is only provided not much ice is in the glass   (d) yes, no matter how fast it melts   (e) none of these.

6. Figure MC6 shows glass tubing, a rubber bulb, and two bottles. What has taken place in order to produce the situation depicted?   (a) nothing, the arrangement is impossible   (b) the tubing was filled with a single liquid, turned over, and inserted into the bottles   (c) the bulb was squeezed while the tubing was in place as shown, and the two different-density liquids rose   (d) the bulb was squeezed more tightly on one side than the other, so the single liquid rose as shown   (e) none of these.

Figure MC6

7. Suppose you stick a drinking straw into a deep cup of liquid, tightly cover the upper end with a finger, and raise the straw up and out.   (a) some liquid stays in the straw but its level initially drops, the gas pressure above the liquid in the straw drops, and the straw works like a poor barometer   (b) the liquid simply runs out of the straw, there being nothing to keep it up   (c) the liquid stays in the straw by surface tension alone   (d) the liquid stays in the straw primarily because of friction with the long walls, but surface tension contributes as well   (e) none of these.

8. Consider a helium-filled balloon floating around in a car. The driver suddenly stops the car. With respect to the car,   (a) the driver, the air in the car, and the balloon are pushed backward   (b) the driver is pushed backward, but the balloon moves forward   (c) only the driver moves— and that is forward   (d) the driver is pushed forward as is the air in the car, so the balloon moves backward   (e) none of these.

9. A device that reads gauge pressure is suspended in a fixed position above the bottom in a large swimming pool. A helicopter with pontoons lands on the surface. Does the pressure reading change, and if so, how?   (a) the water level rises and therefore the pressure rises   (b) the pressure drops because the water level drops   (c) the pressure remains unchanged   (d) the water level remains unchanged, but the pressure drops   (e) none of these.

10. The water level in a large tank supplying a small town is 37 m high. Water is pumped up for storage in the evenings and allowed to free-fall down as needed during the day. The street-level water pressure is   (a) undetermined   (b) 37 Pa   (c) 36×10^4 Pa   (d) 53×10^4 Pa   (e) none of these.

11. Figure MC11 shows a hollow, narrow, open cylinder immersed in a beaker of water. Before being pushed into the fluid, a light plastic disc with a string glued to it was positioned on the bottom end of the cylinder and held in place by pulling up on the thread. Once immersed, the string could be let loose and the disc held in place by the water. That done, a liquid can now be poured into the cylinder from above until the bottom drops out. What is the main point of this experiment?   (a) that it's fun to play with water   (b) that liquids are more dense than air   (c) that the pressure in a fluid is greater than the pressure outside   (d) that the pressure in a liquid is due to the weight of the column of fluid above   (e) none of these.

Figure MC11

12. Consider Fig. MC12, which shows a filled fish tank in the atmosphere. At which of the labeled points must the pressure always be equal?   (a) A and E (b) B and C (c) A and B (d) C and D (e) none of these.

13. In the fish tank of Fig. MC12, it is possible that the pressure be equal at points   (a) A and B, and B and E (b) B and C and E (c) A and C and E (d) A and C and B (e) none of these.

Figure MC12

14. The long glass tube of a barometer weighs 1.0 N, and the quantity of mercury almost filling it weighs 10 N (Fig. MC14). If the tube (which is very thin so the mercury-buoyant force can be ignored) is supported by a spring scale, it will read   (a) 11 N   (b) 9 N   (c) zero   (d) 1.0 N   (e) none of these.

Figure MC14

15. The elderly lady in the park was running a once-in-a-lifetime sale, so you bought the 1.5-kg "solid" gold paperweight. When lowered into a fish tank back at home, it displaced 100 cm^3 of water. The known density of gold is 1.93×10^4 kg/m^3.   (a) you made a great purchase   (b) it may not have been a bargain, but at least it's got the right mass   (c) bad luck—The density is wrong, so it's definitely not solid gold   (d) the density is too high, so something must be wrong   (e) none of these.

16. A stone tied to a string is supported from a spring scale that reads 10 N. A pail of water rests on a platform scale that reads 80 N. The stone is lowered into the water, and the spring scale then reads 5 N. It follows that the platform scale   (a) still reads 80 N   (b) now reads 70 N   (c) now reads 90 N   (d) now reads 75 N   (e) none of these.

17. A block of wood floats in a pot of water on the floor of an elevator. Ignoring surface tension, how is the depth at which it floats changed when the elevator accelerates downward at a rate less than g?   (a) the block rises because the water sinks   (b) the block lowers because the water rises   (c) nothing happens to it   (d) the depth decreases because the water gets effectively lighter   (e) none of these.

18. Water traveling at a speed of 0.1 m/s enters a pipe with a 1.0-cm radius. It emerges from a section of pipe with a radius of $\frac{1}{2}$ cm at a speed of   (a) 0.4 m/s   (b) 0.2 m/s   (c) 1.0 m/s   (d) 0.04 m/s   (e) none of these.

19. A filled milk container has two holes punched in its side, one at point 1, 7.5 cm down from the liquid level, and the other at point 2, 15 cm down from that level. The two speeds of efflux are related by   (a) $v_1 = v_2$ (b) $v_2 = 2v_1$ (c) $v_1 = 2v_2$ (d) $v_2 = \sqrt{2}\,v_1$ (e) none of these.

20. Two tall glass containers are connected by a narrow tube closed off with a stopcock. One container is filled with water, and when the stopcock is opened, the water passes into the other container until both levels are equal. At the end of this process, in which the "fluid seeks its own level,"   (a) the *gravitational*-PE has increased to a maximum   (b) the KE has increased   (c) the *gravitational*-PE has become a minimum   (d) the *gravitational*-PE is unchanged   (e) none of these.

Problems

1. [I] A swimming pool 5 m wide by 10 m long is filled to a depth of 3 m. What is the pressure on the bottom due only to the water?

2. [I] A rectangular tank 2.0 m by 2.0 m by 3.5 m high contains gasoline, with a density of 0.68×10^3 kg/m^3, to a depth of 2.5 m. What is the gauge pressure anywhere 2.0 m below the surface of the gasoline?

3. [I] An oxygen tank sitting in the corner of a laboratory has an internal gauge pressure 5.00 times atmospheric pressure. What outward force is exerted per square centimeter on the inner wall of the tank?

4. [I] At about 10 miles up in the atmosphere, air pressure drops to roughly 2 lb/in.^{2} (down from 14.7 lb/in.^{2} at sea level). How much is that in *atmospheres* and *pascals*?

5. [I] Prove that 1.000 lb/in.^{2} = 6.895×10^3 N/m^2, using the fact that 1.000 lb = 4.448 23 N.

6. [I] A typical automobile tire has a gauge pressure of around 30 lb/in.^{2}. How much is that in pascals? If a car weighs 8897 N (2000 lb), how much area (in SI units) on each tire is in contact with the road?

7. [I] When Pascal's brother-in-law climbed to the top of the 3200-ft-high Mont Puy-de-Dôme, the barometer he carried showed a drop in mercury of about 3.0 in. What was the corresponding pressure drop?

8. [I] Suppose von Guericke had used two half-cubes instead of hemispheres to demonstrate atmospheric pressure. Compute the force each team of horses would have had to apply to a 1-meter-on-a-side cube. Assume zero pressure inside. How could von Guericke have been less theatrical and saved the cost of one team of horses?

9. [I] How deep must you dive in fresh water before the gauge pressure equals 1.00 atmosphere?

10. [I] A swimming pool 5.0 m wide by 10 m long is filled to a depth of 3.0 m. What is the absolute pressure on the bottom?

11. [I] Determine the gauge pressure (in both Pa and psi) at the bottom of the deep ocean trenches that reach depths of about 11 km. Assume the density of the sea is constant.

12. [I] A swimming pool 5.0 m wide by 10 m long is filled to a depth of 3.0 m. What is the total force exerted on the bottom due to the water?

13. [I] The initial pressure of the air inside a hypodermic syringe extended to 10 cm³ is 1.013×10^5 Pa. If the cap is kept on so that the needle end is sealed, what pressure would exist when the gas is slowly compressed to 2.5 cm³ with no change in temperature?

14. [I] A hydraulic lift consists of two interconnected pistons filled with a common working liquid. If the areas of the piston faces are 64.0 cm² and 3200 cm², and a 900-kg car rests on the latter, how much force must be exerted to raise the vehicle very slowly? If the car is to be raised 2.00 m, how far must the input piston be depressed?

15. [I] An ancient coin, which X-rays show is solid and homogeneous, has a mass of 0.010 0 kg. When submerged, it displaces 0.952 g of water. What is its specific gravity and what is it probably made of?

16. [I] Envision a thin-walled jar having a mass of 10.0 g containing 0.100 g of hydrogen at atmospheric pressure. Estimate the difference we can expect to observe if we weigh the filled jar in dry air at 0°C, as opposed to weighing it in vacuum. The density of hydrogen is 90×10^{-6} g/cm³.

17. [I] Figure P17 shows a U-shaped wire closed with a movable section of length $L = 0.10$ m. The device is then covered with a glycerin film. How much force is needed to pull the wire at a constant rate, thereby increasing the area of the film?

Figure P17

18. [I] A needle of length L and mass m rests on the surface of a liquid having a surface tension of γ. Show that the force needed to raise the needle is

$$F = 2L\gamma + mg$$

19. [II] The medical literature uses yet another pressure unit: the cm of water (water columns are convenient for measuring modest gauge pressures). Thus, one reads that a newborn baby is capable of developing a "momentary intrathoracic negative pressure of the order of 40 cm of water." How much is that in pascals?

20. [II] To remove unwanted fluids from the body, an aspirator such as that shown in Fig. P20 is used. A small electric pump provides the suction (typically at gauge pressure levels of −90 to −120 mm Hg), while the bottle containing water provides the control. Explain how the depth beneath the water surface of the control tube establishes the pressure in the drain tube. What happens to the suction pressure as the drain tube in the bottle becomes submerged in liquid aspirated from the patient? When will the system shut itself off? Suppose the procedure starts with the control tube 10 cm into the water—what is the suction pressure to the patient?

Figure P20

21. [II] Given that most people cannot "suck" water up a straw any higher than about 1.1 m, what's the lowest gauge pressure they can create in the lungs?

22. [II] The gauge pressure in a basement water pipe supplying a tall building is 3.00×10^5 Pa. In an apartment several floors up, the pressure is half this value. How high up is the apartment?

23. [II] A rubber pipe is attached to one end of an open U-tube water-filled manometer (such as the one in Fig. 11.11). A patient exhales strongly into the rubber pipe and a difference in height between the two columns of 61 cm results. What absolute pressure was developed by the lungs?

24. [II] A 227-kg block of cement, of density 2.8×10^3 kg/m³, rests on a pedestal in front of the Al Capone Memorial Library. How much did it weigh submerged while being hauled out of the river (freshwater)?

25. [II] A ring weighs 6.327×10^{-3} N when measured in air and 6.033×10^{-3} N when submerged in water. What is its volume (to three significant figures)? What is it likely made of? What is its relative density (to three significant figures)?

26. [II] A woman weighing 500 N jumps into a swimming pool 10 m × 10 m by 5 m deep and floats around. By how much does the water level change as a result of her arrival?

27. [II] Icebergs float in the ocean with much of their huge volumes hidden below the surface. What fraction is visible

above the water? What portion of an ice cube floats above the surface of a glass of tap water?

28. [II] Consider a catfish on the bottom of a freshwater lake. It opens its mouth to let loose a bubble of gas with a 1.00-mm³ volume. Now suppose that the bubble's volume just before it bursts at the surface is 12.0 mm³. How deep was the lake? (Assume a uniform temperature and no reabsorption.)

29. [II] A bottle of oxygen having a volume of 100 cm³ is at a pressure of 228 cm Hg. The gas is fed into an expandable chamber maintained at atmospheric pressure and the same constant temperature. What volume will the gas occupy?

30. [II] The piston in equilibrium, shown in Fig. P30, has a face area of 0.10 m² and experiences a downward force of 1.00 kN. The mercury in the vessel is 10.0 cm deep. What is the gauge pressure at the very bottom of the chamber?

Figure P30

31. [II] A submarine rests in 20.0 m of water. How much force must a diver exert against the pressure of the sea in order to pull open a hatch 1.0 m × 0.50 m, assuming the internal pressure in the boat is 90% of atmospheric?

32. [II] A person floats on the Great Salt Lake, which has a density of 1.15×10^3 kg/m³. Approximately how much of the swimmer's body is above the water?

33. [II] Determine the mass of helium needed to provide enough buoyancy (in dry air at 0°C) to lift a balloon and its load having a net mass of 454 kg.

34. [II] A 226.7-kg polar bear standing 6 ft tall walks onto a floating sheet of ice 0.305 m thick. How big is the ice sheet if it sinks just below the surface while supporting the bear?

35. [II] Given a pine raft, of density 0.50×10^3 kg/m³ and dimensions 3.05 m by 6.10 m by 0.305 m, how much of a load can this raft take on for each 2.54 cm it settles into freshwater? How deep does it sink into the water when unloaded?

36. [II] The open U-tube in Fig. P36 contained some water before a less dense liquid was poured in on the right side. If the density of the unknown liquid is ρ_x, show that

$$\rho_x = \frac{\rho_w h_w}{h_x}$$

37. [II] A platinum wire ring, 6.0 cm in diameter, is suspended horizontally from a delicate balance. The ring is lowered into a liquid, and the balance is reset to zero. The

Figure P36

ring is raised again, and the force needed to pull it free of the surface is then measured. What force would be required if the liquid was water at 20°C?

38. [III] A glass plate 0.12 m × 0.20 m is mounted on the top of a submerged camera box to make an angle of 30° with the horizontal. The uppermost 0.12-m edge is 0.20 m below the surface of the water. The box is sealed, and there is air inside at atmospheric pressure. Please compute the net force acting on the plate due to the water.

39. [III] Figure P39 deals with the problem of showing that the pressure within a fluid is independent of orientation. Shown is the end of a wedge-shaped volume of fluid where the average pressure on each of its faces is P_a, P_b, and P_c. The lengths of the sides of the top face are both L, and it makes an angle of ϕ with the bottom. Find the net force on each face, and show that the fluid wedge is in equilibrium when

$$P_a L^2 \sin \phi = P_c L^2 \sin \phi$$

and $$P_b L^2 \cos \phi = P_c L^2 \cos \phi + \tfrac{1}{2}\rho g L^3 \sin \phi \cos \phi$$

Now, let the wedge shrink and show that $P_a = P_b = P_c$ in the limit as $L \to 0$.

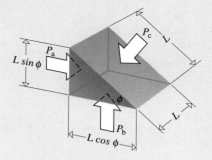

Figure P39

40. [III] The diving bell shown in Fig. P40 allows swimmers to comfortably remain under water for long intervals. Show that

$$d = \frac{P_A h}{(H - h)\rho g} + h$$

Figure P40

where P_A is the original pressure in the bell (usually atmospheric). (Hint: Don't forget atmospheric pressure on the water.)

41. [III] A swimming pool 5 m wide by 10 m long is filled to a depth of 3 m. What is the total force on either narrow wall due only to the water?

42. [III] A bronze wire, with a Young's Modulus of 120 GPa, supports a 0.000 45-m³ block of metal. When the block is lowered completely into water, the wire decreases in length by 0.000 80 m. If the diameter of the wire is 0.10 mm, how long is it?

43. [III] An object has a weight measured in air of F_{Wa} and when immersed in water, its effective weight is F_{Ww}. It is then submerged in an unknown liquid, and its effective weight is found to be F_{Wl}. Determine the density of the unknown liquid in terms of these measured quantities and the density of water.

44. [III] Figure Q16 shows a loop of thread that will be pulled out into a circle of radius R by surface tension. Prove that the tension in the thread is given by

$$F_T = 2R\gamma$$

FLUID DYNAMICS

45. [I] Petroleum flows along a 10-cm diameter pipeline at a speed of 2.0 m/s. Determine the flow rate, assuming the liquid is ideal.

46. [I] A pipe 5.0 cm in diameter carries gasoline with a density of 0.68×10^3 kg/m³ at a speed of 2.5 m/s. Calculate the mass-flow rate, assuming the liquid is ideal.

47. [I] Gasoline is flowing in a pipeline having a 0.50-m diameter. Taking the fluid to be ideal, what pressure change results when the pipe descends 4.0 m down an embankment?

48. [I] Determine the blood pressure at the brain, taking its height above the heart to be 40 cm.

49. [I] Someone cuts a finger and, to control the bleeding, raises the wound 85 cm above the level where the hand was being bandaged. What is the change in the pressure of the blood? Compare that to the pressure at the heart.

50. [I] A horizontal water main with a cross-sectional area of 200 cm² necks down to a pipe of area 50 cm². Meters mounted in the flow on each side of the transition coupling show a change in gauge pressure of 80 kPa. Determine the flow rate through the system, taking the fluid to be ideal.

51. [I] Derive an expression for the speed of efflux of an ideal liquid from a hole in a large tank filled with fluid in terms of the gauge pressure at the level of the orifice and the density of the liquid.

52. [I] A length of tubing is used as a siphon to empty a large upright cask of wine. At a given moment, the lower opening of the tube from which the wine emerges is 20 cm below the liquid level in the cask. Compute the speed at which the wine then emerges.

53. [I] A large vat of clear chicken soup is about to spring a leak from a small hole a distance h below the surface of the broth. Knowing that the hole is a height (y) above the floor, at what distance (x) away from the vat should you place a bowl on the floor to catch the flow, assuming it to be ideal?

54. [II] A cylindrical water tank 2.0 m in diameter has a spigot with a hole area of 5.0 cm² located 4.0 m below the surface. Show that for a point at the surface the $\frac{1}{2}\rho v^2$ term in Bernoulli's Equation is properly ignorable compared to the corresponding term outside of the spigot in the escaping water jet. Assume no contraction of the stream as it leaves the tank.

55. [II] A Texas pipeline carrying natural gas ($\rho = 0.90$ kg/m³) with a mass-rate of flow of 1.0 kg/s is 35 cm in diameter. Determine the average speed at which the gas is moving along.

56. [II] Refer to the photo on p. 000. Suppose a person standing erect or sitting upright is accelerated upward at a rate a. Write an expression for the resulting blood pressure in the brain in terms of the pressure at the heart and discuss your results.

57. [II] A Pitot tube, shown in Fig. P57, is a fluid-speed measuring device that has applications in many fields, including physiology. Often used on airplanes to measure the airspeed of the craft, it can be seen mounted on the wings or fuselage. At the front opening, the streamlines part and there is a **stagnation point** there. The fluid rushes away from that point in all directions, and *its speed there is zero.* Accordingly, the pressure there and in the inner tube (P)—the stagnation pressure—is relatively high. On the other hand, the speed of the fluid over any of the small holes in the outer cylinder is the undisturbed free-stream speed (v). Derive an expression for v in terms of the pressure difference indicated by the attached manometer. By the way, the device was first used by Henri Pitot to measure water speeds in the Seine River (1730).

58. [II] A small plane, having a mass of 3000 kg, is flying in air of density 1.0 kg/m³. Air moves over the top and bottom surfaces of the wings at 160 m/s and 130 m/s,

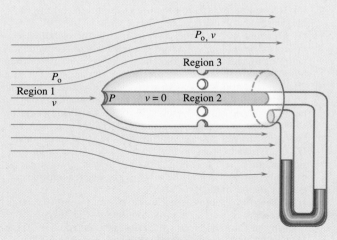

Figure P57

respectively. What is the net minimum wing area needed? Use the results of Problem 69.

59. [II] The instrument depicted in Fig. P59 is inserted in a pipeline to determine flow rates and speed. Known as a Venturi meter, it consists of a throat and two manometer tubes. Derive an expression for the speed of flow in the pipe in terms of Δy, the difference in the column heights, and the known cross-sectional areas of pipe and throat.

Figure P59

60. [II] The flowmeter in Fig. P59 shows a difference in height of 5.0 cm on an oil pipeline 200 cm^2 in cross section. If the throat of the device has a diameter of 10 cm, what is the pipeline flow rate? (Assume the liquid to be ideal.)

61. [II] A vaccination gun forces vaccine through a small aperture a few thousandths of an inch in diameter, at high pressures (550 psi), and so does away with the need for hypodermic needles. Compute the speed at which the fluid leaves the gun. Take the flow speed of the vaccine ($\rho = 1.1 \times 10^3$ kg/m^3) in the reservoir within the body of the gun to be negligible.

62. [II] A cardboard milk container has two holes, one above the other, punched in its side. At a given moment, the two escaping streams of milk strike the table the container is standing on, at the same point. If the heights of the upper and lower holes are, respectively, y_u and y_l, write an expression for y, the level of milk at that instant, in terms of these heights.

63. [II] A large-diameter open cylindrical storage tank stands on a high platform. It has a small horizontal spigot at its very bottom, a height Y above the ground and a depth h below the surface of the water. Use Bernoulli's Equation to follow a tube of flow from point 1 at the surface, to point 2 just outside the spigot, to point 3 at the point of impact with the ground. Write expressions for v_2 and for v_3 (the speeds at the spigot and at the point of impact with the ground) in terms of g, h, and Y. Did you expect these results?

64. [II] A large, open storage tank is being filled with water from a pipeline at a rate J. Unfortunately, it springs a leak via a hole of area A, at its base. Write an expression for y, the equilibrium height to which the water rises in the tank. Assume the liquid emerges without any contraction of the jet.

65. [II] Figure P65 shows a submerged orifice discharging a liquid from a large tank. Determine an expression for the ideal efflux speed at point 2.

Figure P65

66. [III] During a storm, a 50-m/s (112-mi/h) wind blows horizontally across the flat level roof of a supermarket. If the roof has an area of 220 m^2, what is the most force that's likely to be exerted on it? Assume the place is locked up tight for the night. (Take the density of air to be 1.1 kg/m^3.)

67. [III] Figure P67 shows a stream of liquid emerging from a tube in the base of an open tank. Use Bernoulli's Equation applied between points 2 and 3 to get an expression for y in terms of θ and h. Can y exceed h?

Figure P67

68. [III] A highly pressurized liquid escapes from a large, closed tank through a small nozzle with an aperture of

area A. Show that a thrust results, given by

$$F = \rho A v^2$$

which is essentially only dependent on the orifice area and the gauge pressure in the tank. There are toy rockets that you fill with water and then pressurize with compressed air via a handpump—they blast out the water and lift off quite effectively.

69. [III] Show that the lift on the wing of an airplane of area

A is given by

$$F_L = \tfrac{1}{2}\rho(v_\alpha^2 - v_\beta^2)A$$

where v_α is the speed of the air above, v_β the speed below, and ρ is the density of air. Although it's not clear how to predict these two speeds, it is reasonable to assume that each is proportional to the airspeed of the craft, and hence $(v_\alpha^2 - v_\beta^2) \propto v^2$.

12

Oscillations
and
Waves

A MOTION THAT REPEATS itself over and over again, in successive constant time intervals, is said to be *periodic*. Such behavior is remarkably commonplace and seems fundamental to the physical nature of the Universe. A minute electron displays periodic characteristics that are describable in the same language we apply to analyzing a spinning neutron star. The revolving of the Earth around the Sun, the sway of a gently rocked cradle, or the rhythmic pumping of a heart are all more or less periodic (Fig. 12.1). We can distinguish two forms of repetitive motion: movement along a closed path in one angular direction (like the Earth around the Sun), and back-and-forth

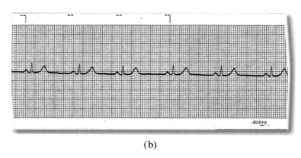

(a)

(b)

Figure 12.1 (a) The giant star Delta Cephei swells and then shrinks every 5.4 days. The plot of its brightness is therefore a regularly rising and falling curve.

(b) This graph is periodic, too, but it's just an electro-cardiogram of a physicist's heart. (See Sect. 19.5.)

oscillation along the same path (like the cradle). The former was introduced in Chapter 8, and the latter, *vibrational* motion, will be considered here. This analysis leads naturally to a discussion of waves produced in vibrating mechanical systems and then, later, to sound waves (Chapter 13).

HARMONIC MOTION

The world abounds with all sorts of vibrating systems: the balance wheel in a watch oscillates back and forth; puckered lips blowing a trumpet, or a kiss, vibrate; a walker's swinging arms oscillate; so does a singing vocal cord. Vibratory periodic motion is referred to as *harmonic motion,* because of its relationship to sound and the harmony therein.

12.1 Simple Harmonic Motion

A single sequence of moves that constitutes the repeated unit in a periodic motion is called a *cycle.* During one cycle, the system progresses in some way, returning on completion to its initial physical configuration and motion. The child on the swing leaves her father's raised hands, swoops down, rises high into the air, stops, descends backward, and comes up again to meet his hands and finish the cycle. *The time it takes for a system to complete a cycle is a* **period** *(T).* The period of the Earth's rotation about its spin-axis is roughly 23 h 56 min. The period is *the number of units of time per cycle;* the reciprocal of that—*the number of cycles per unit of time*—is known as the **frequency** (*f*), after Galileo who called it the *frequenza:*

$$f = \frac{1}{T} \tag{12.1}$$

The SI unit of frequency is the *hertz* (Hz), so named to honor Heinrich Hertz, where 1 Hz = 1 cycle/s = 1 s^{-1}. The appropriate units of *cycles per second* got distorted in common usage during the first half of this century. Radio people, especially, talked about kilocycles and megacycles, forgetting the "per second," so today, to

The balance wheel of a mechanical clock oscillates about its central axis.

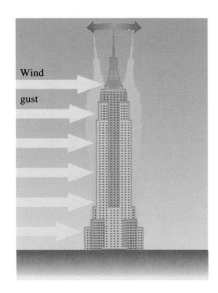

Figure 12.2 The 1250-ft-tall Empire State building set oscillating by a strong wind. Even though the displacement as drawn is exaggerated, tall buildings do sway up to several meters. In fact, on a windy day, the Twin Towers of the World Trade Center in Manhattan shift 6 or 7 ft, and the elevators have to be slowed down.

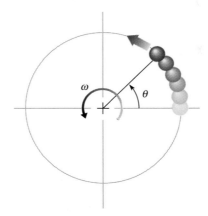

Figure 12.3 An object moving in a circle with a constant angular speed ω. Its angular position $\theta = \omega t$ changes at a constant rate.

(a) $t = 0$

(b) $t = \frac{1}{4}T$

(c) $t = \frac{1}{2}T$

(d) $t = \frac{3}{4}T$

Figure 12.4 This Scotch yoke produces a vertical oscillation that is SHM as the wheel revolves uniformly.

avoid that difficulty, we have the hertz. For example, once set swaying by a wind, the Empire State building oscillates through one cycle in about 8 s, that is $T = 8$ s and $f = 1/T = 0.1$ s^{-1} = 0.1 Hz (Fig. 12.2).

Imagine an object moving in a circular orbit (Fig. 12.3) at a constant rate: its angular speed ω is constant. Each time it makes one complete orbit, one cycle, the object sweeps through an angle of 2π rad. Since the number of cycles it makes per second is f, *the number of radians it moves through per second is $2\pi f$*, and that's exactly what angular speed ω is. Hence

$$\omega = 2\pi f = \frac{2\pi}{T} \tag{12.2}$$

and it is common practice to refer to ω as the **angular frequency**.

Most real oscillatory phenomena wiggle about in a complicated fashion with lots of different frequencies occurring at different strengths, all at once. Even so, many important systems vibrate with a single dominant frequency that far outweighs all the others. We can approximate this behavior using a single harmonic function; that is, *the periodicity has a single frequency and the motion is sinusoidal*—described by a sine or cosine function. This idealization is known as **simple harmonic motion**, or SHM for short.

Uniform rotational motion, simple harmonic motion, and periodic wave motion are all intimately related and can be analyzed with the same mathematics. We can get a sense of that interrelationship from Fig. 12.4, which depicts an arrangement

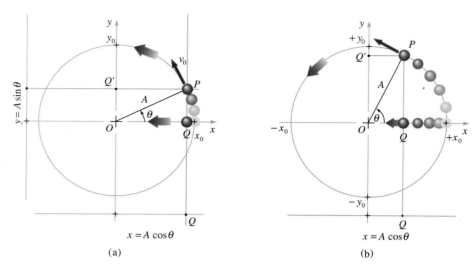

Figure 12.5 A particle P revolving around a circle at a speed v_0. Its projection on the x-axis is point Q at a distance $x = A \cos \theta$ from the nondisplaced origin O. As P revolves, Q oscillates between $+x_0$ and $-x_0$ in SHM. (a) and (b) show two locations as the motion progresses.

called a Scotch yoke. As the disk turns uniformly, the rod oscillates linearly in SHM. The frequencies and periods of the two motions are the same.

We next derive the general expressions that describe SHM for an oscillating point. These results will later be applied to specific systems, such as vibrating springs and pendulums.

Displacement in SHM

An object oscillating in SHM can be located at any instant with respect to a fixed point O called the *origin*. To do that, imagine a particle P moving at a uniform speed v_0 counterclockwise along a circle of radius A (Fig. 12.5) centered on O. At any moment, the position of the particle is given by θ. Since the linear speed is constant, the corresponding angular speed ω is also constant, $v = r\omega$. Moreover, after a time t, it follows from Eq. (8.5) that $\theta = \omega t$. The projection of P perpendicularly down onto the x-axis locates the point Q, and as P circles around with a constant speed, Q oscillates from $+x_0$ to $-x_0$ and back at a single constant frequency in SHM. The projection onto the y-axis locates Q' and it will also display SHM. The displacement of Q from 0 is given by

$$x = A \cos \theta = x_0 \cos \omega t = x_0 \cos 2\pi f t \qquad (12.3)$$

where $\omega = 2\pi f$. The displacement of Q' from 0 is given by

$$y = A \sin \theta = y_0 \sin \omega t = y_0 \sin 2\pi f t \qquad (12.4)$$

Because $2\pi f t = 2\pi t/T$, this factor varies from 0 when $t = 0$, to π when $t = \frac{1}{2}T$, to 2π when $t = T$, as shown in Fig. 12.6. The location of Q provided by Eq. (12.3) can be thought of as describing the top of the Empire State building wavering in the wind, or anything else oscillating in SHM about its undeflected ($x = 0$) position.

The maximum displacement of Q from 0 is the length $A = x_0$, the radius of the **reference circle**, which is called the **amplitude** of the oscillation. At any instant, the value of x is the size of the displacement from 0. The value of x varies as Eq. (12.3) and may be positive or negative; the amplitude is constant and positive.

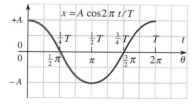

Figure 12.6 A plot of the cosine function. As t varies from 0 to T, $\cos 2\pi t/T$ varies from $\cos 0 = 1$ to $\cos 2\pi = 1$.

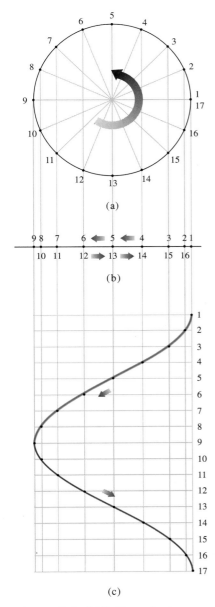

Figure 12.7 (a) Uniform circular motion projected onto a line (b), resulting in linear SHM projected on a "moving" axis (c), which results in a sinusoidal oscillation.

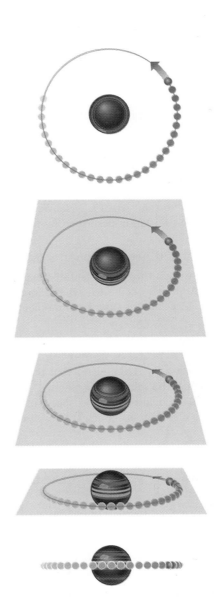

Figure 12.8 A schematic representation of a moon circling around Jupiter. As the viewing angle gets narrower, the orbit flattens, and the moon is seen to oscillate in SHM along a line in the plane of the motion.

Figure 12.9 The pencil oscillates in SHM along a vertical line. Pulling the paper to the right at a constant speed results in a sinusoidal curve.

The oscillator starts out ($t = 0$) at its maximum displacement ($\cos 0 = 1$); that is, $x = A$. We now know how to describe the position mathematically as it varies in time for an object in SHM. The relationship between uniform circular motion, SHM, and the sinusoidal function is explored in Figs. 12.7, 12.8, and 12.9.

(a)

(b)

(c)

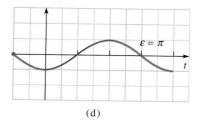

(d)

Figure 12.10 The addition of a phase term ε shifts the curve. When ε is negative, any given value occurs later in time; when ε is positive, it occurs earlier.

Notice in Eq. (12.3) that at $t = 0$, $x = +x_0$. That's fine, but what if we want our particle somewhere else at the start of the game ($t = 0$)? The argument of the sinusoidal function is the **phase**, which is often signified by the letter ϕ. In Eq. (12.3) $\phi = \theta = \omega t$, but more generally the phase will have a nonzero value at $t = 0$, a value ε, which is known as the **initial phase**. With $\phi = \omega t + \varepsilon$, we have

$$x = A \cos (\omega t + \varepsilon) = x_0 \cos (2\pi f t + \varepsilon) \tag{12.5}$$

This is the most inclusive way to express the displacement of an object undergoing SHM.

The oscillations depicted by Eqs. (12.3) and (12.5) are *out-of-phase* with respect to each other by an amount ε (Fig. 12.10). If ε is positive, the function in Eq. (12.5) will reach any desired value of x at an earlier time than the function in Eq. (12.3). In other words, $\phi = (\omega t + \varepsilon)$ will reach a given value at a smaller t when $\varepsilon > 0$ than when $\varepsilon = 0$. We say that the oscillation described by Eq. (12.5) *leads* the oscillation described by Eq. (12.3).

Velocity and Acceleration in SHM

We have found that the displacement of a system in SHM varies sinusoidally with time. As that's occurring, what happens to the system's velocity and acceleration? These, too, must change from moment to moment.

Velocity. Point P moves around its circular path ($r = A$) with a constant linear speed $v_0 = A\omega = 2\pi fA$, while its projection, point Q, always remains just below it on the x-axis. This means that the x-component of the velocity vector ($\mathbf{v}_0$) for P must equal the velocity of Q, namely, $\mathbf{v}_x$. Figure 12.11 shows the geometry. To write the scalar equation for the speed of Q, we must be careful about the signs. As shown, $\mathbf{v}_x$ points in the negative x-direction ($0 < \theta < \pi$) and so v_x is negative:

$$v_x = -v_0 \sin \theta = -A\omega \sin \theta \tag{12.6}$$

Therefore

$$v_x = -A\omega \sin \omega t = -2\pi f x_0 \sin 2\pi f t \tag{12.7}$$

For a given x_0, the maximum value of the speed ($v_0 = 2\pi f x_0$) must depend on the frequency—the point cannot oscillate over a range of $2x_0$ at a high frequency without having a high speed.

The oscillator is momentarily at rest at its two maximum displacements ($x = \pm x_0$), where $t = 0$ or $t = \frac{1}{2}T = 1/2f$ and, correspondingly, $\theta = 0$ or π, and $\sin \theta = 0$. As it begins a cycle ($t = 0$), the particle in Fig. 12.11b heads left, picking up speed that rises to a maximum at $x = 0$ when $t = \frac{1}{4}T = 1/4f$ and $\theta = \frac{1}{2}\pi$, whereupon $v_x = -v_0$. Sailing past $x = 0$, the speed of Q diminishes until it is again zero, when $t = \frac{1}{2}T$ at $x = -x_0$. Like a ball thrown straight up in Earth's gravity field, it stops for an instant at its maximum displacement.

We can also derive an expression for the speed of Q in terms of x. From Fig. 12.5 and the Pythagorean Theorem, $\overline{PQ} = y = \sqrt{A^2 - x^2}$ and so $\sin \theta = \sqrt{(A^2 - x^2)}/A$. Using $v_0 = A\omega$, we obtain

$$v_x = -A\omega \sin \omega t = \mp\omega\sqrt{A^2 - x^2} = \mp v_0\sqrt{1 - (x/A)^2} \tag{12.8}$$

which is the speed of a harmonic oscillator at any position x in terms of its maximum speed.

The trigonometric identity $\cos (\alpha \pm \beta) = \cos \alpha \cos \beta \mp \sin \alpha \sin \beta$ tells us that $\cos (\omega t + \pi/2) = -\sin \omega t$, which, in turn, means that Eq. (12.7) could be written as $v_x = v_0 \cos (\omega t + \pi/2)$. Therefore, as can be seen in Fig. 12.12, the speed leads the displacement by 90°. The speed, which is a cosine function shifted one-quarter period ($t = \frac{1}{4}T$ or $\omega t = \frac{1}{2}\pi$) to the left, results in a curve that is an inverted sine function—that is, minus sine.

Acceleration. Particle P experiences a constant centripetal acceleration, $a_C = v^2/r = r\omega^2 = A\omega^2$, directed toward the origin (Fig. 12.13). Because Q always follows along, the horizontal component of the centripetal acceleration has to equal Q's acceleration. As before, to write $\mathbf{a}_x$ as a scalar we must include a negative sign, because it points in the negative x-direction:

$$a_x = -a_C \cos \theta = -A\omega^2 \cos \theta$$

and
$$a_x = -A\omega^2 \cos \omega t = -A(2\pi f)^2 \cos 2\pi ft \qquad (12.9)$$

The oscillator has its maximum acceleration at the limits of its motion ($x = \pm A$) where the speed is zero. Prior to this chapter, we studied systems where the acceleration was either zero or constant, but that's not the case here. Moreover, because the acceleration of P is centripetal, the acceleration of Q is also center-seeking, pointing toward the equilibrium position, $x = 0$.

By comparing Eqs. (12.3) and (12.9), it is evident that

$$a_x = -\omega^2 x \qquad (12.10)$$

independently of time. **The acceleration of a simple harmonic oscillator is proportional to the negative of its displacement from the midpoint of the motion.** That's the hallmark of SHM.

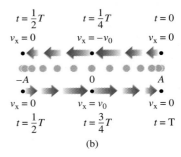

Figure 12.11 The point Q oscillates horizontally from $x = +A$ to $x = -A$, moving from a speed $v_x = 0$, to a maximum of $v_x = -v_0$ at $x = 0$, to $v_x = 0$ back at $x = -A$. (a) This reference circle shows the vector $\mathbf{v}_0$ and its component v_x. (b) Here we see the time variation of the speed of a harmonic oscillator.

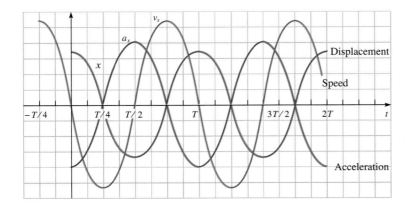

Figure 12.12 Plots of the speed, displacement, and acceleration for a simple harmonic oscillator. Note that the speed leads the displacement by $\pi/2$, while the acceleration lags the displacement by $\pi/2$.

Example 12.1 A spot of light on the screen of a computer is oscillating to and fro along a horizontal straight line in SHM with a frequency of 1.5 Hz. The total length of the line traversed is 20 cm, and the spot begins the process at the far right. Determine (a) its angular frequency, (b) its period, (c) its maximum speed,

(continued)

(continued)

and (d) its maximum acceleration. (e) Write an expression for x and find the location of the spot at $t = 0.40$ s.

Solution: [Given: $f = 1.5$ Hz and $A = 10$ cm. Find: (a) ω, (b) T, (c) v_x(max), (d) a_x(max), (e) x in general, and x at $t = 0.40$ s.]

(a) $\omega = 2\pi f = 2\pi(1.5 \text{ Hz}) = 9.4$ rad/s $= \boxed{3.0\pi \text{ rad/s}}$

(b) $T = 1/f = 1/1.5$ Hz $= \boxed{0.67 \text{ s}}$

(c) v_x(max) $= v_0 = A\omega = 2\pi f A = 2\pi(1.5 \text{ Hz})(0.10 \text{ m})$

and $\boxed{v_x(\text{max}) = 0.94 \text{ m/s}}$

(d) From Eq. (12.9), a_x(max) $= A\omega^2 = A(2\pi f)^2$ and

a_x(max) $= (0.10 \text{ m})(2\pi 1.5 \text{ Hz})^2 = \boxed{8.9 \text{ m/s}^2}$

(e) $x = A \cos \omega t = \boxed{(0.10 \text{ m}) \cos (9.4 \text{ rad/s})t}$

At $t = 0.40$ s

$x = (0.10 \text{ m}) \cos (3.76 \text{ rad}) = (0.10 \text{ m})(-0.81)$

and $\boxed{x = -8.1 \text{ cm}}$

▶ **Quick Check:** From (e) at $t = 0$, $x = +A$, as it should be. The value of x at $t = 0.40$ s; namely, -8.1 cm is reasonable since the period is 0.67 s, and so after $\frac{1}{2}T = 0.34$ s, the spot is on the negative side at $-A = -10$ cm heading right. At 0.06 s later, it's not surprising that it be at $x = -8.1$ cm.

12.2 Elastic Restoring Force

When a system oscillates naturally (without being driven by some external source of energy), it does so by moving against a force that tends to return it to its undisturbed equilibrium condition. That *restoring force* is usually gravitational or elastic (that is, electromagnetic), but other phenomena, such as surface tension, will sustain vibrations as well—raindrops oscillate as they fall. The system in equilibrium is first distorted. An additional quantity of potential energy is thereby stored in it, and then it's let loose to return toward equilibrium. Invariably, it overshoots (there being no tendency to stop at its undisplaced configuration since the system carries an excess of energy and still possesses momentum when it reaches that equilibrium point). It sails past its undistorted configuration only to be displaced in the opposite direction, to again become distorted and thereby begin the process anew. As the system vibrates, PE goes into KE and back to PE, and so on (with no losses) indefinitely. That "lossless" single-frequency ideal vibrator is known as a **simple harmonic oscillator**,

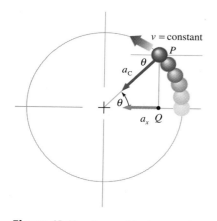

Figure 12.13 As an object moves in a circle, there is a centripetal acceleration a_C that has a horizontal component a_x. The latter is the acceleration of the oscillator.

An Oscillating Spring

If a spring with a mass attached to it is slightly stretched or compressed and then let loose, it will oscillate in a way very closely resembling SHM. Many elastic systems (buildings, flagpoles, airplane wings, etc.) behave in a similar fashion, and so the spring merits a detailed examination.

Reconsider Hooke's Law [see Eq. (10.3)], $F = ks$: the force that must be exerted on an object—a spring or a wire or a tree—to make it distort *slightly* is proportional to the displacement from equilibrium that the object undergoes. Provided the distortion (stretching, compressing, twisting, bending, whatever) is not too great, $F \propto s$, and the object behaves elastically. From Newton's Third Law, the force exerted *by* an elastically stretched spring or a bent pole (the *restoring force* pulling back against that which is doing the stretching) is $F = -ks$.

If a body of mass m is attached to a helical spring of negligible mass (as in Fig. 12.14), displaced a small distance $s = x$ from equilibrium and then let go, the body will thereafter experience a force $F = -kx$, exerted on it by the spring. Yet since $F = ma$, the resulting acceleration of the body is given by $a_x = -(k/m)x$. *The acceleration of the body is proportional to the negative of its displacement*—the mass oscillates about $x = 0$ in SHM. This description assumes no energy is lost to friction (either in the spring, on the table, or through the air); otherwise the motion is *damped harmonic,* not SHM, and it will die out.

Example 12.2 Use energy considerations to derive an expression for the maximum speed of the mass oscillating in Fig. 12.14 in terms of k, m, and A. That done, confirm Eq. (12.8) for the speed as a function of displacement of a harmonic oscillator.

Solution: [Given: k, m, and A. Find: an expression for v_0.] From Eq. (10.4), we know that *elastic*-PE is given by $\Delta PE_e = \frac{1}{2}kx^2$. Let's look at the extremes: At $x = 0$, the KE is maximum and the PE is zero; hence, the total mechanical energy is $E = \frac{1}{2}mv_0^2$. At $x = \pm A$, the KE is zero ($v_x = 0$), the PE is maximum, and $E = \frac{1}{2}kA^2$. Since E is conserved

$$\tfrac{1}{2}mv_0^2 = \tfrac{1}{2}kA^2$$

and the maximum speed is

$$\boxed{v_0 = A\sqrt{\frac{k}{m}}}$$

In general

$$E = \tfrac{1}{2}mv_x^2 + \tfrac{1}{2}kx^2 = \tfrac{1}{2}kA^2$$

Solving for v_x^2, we obtain

$$v_x^2 = \frac{k}{m}(A^2 - x^2) = \frac{k}{m}A^2\left(1 - \frac{x^2}{A^2}\right) = v_0^2\left(1 - \frac{x^2}{A^2}\right)$$

which is equivalent to Eq. (12.8).

▶ **Quick Check:** We have already seen that $v_0 = A\omega$, and we will show presently that $\omega = \sqrt{k/m}$. Now let's just check that the units are right. Using $F = ks$, it follows that k has units of N/m. Thus, k/m has units of $(N/m)/(N \cdot s^2/m) = 1/s^2$, and $\sqrt{k/m}$ correctly has the units of 1/s.

The restriction that the spring, diving board, or suspension bridge (whatever it is that's being stretched) must behave in a Hookean fashion is equivalent to saying that F is linear in x, which is equivalent to saying a is linear in x, which is the hallmark of SHM. Still, precise linear behavior is more an idealization than a reality. To the degree that a system behaves in a way that is approximately linear, it will oscillate in a way that is approximately SHM. To that end, the spring must only be stretched slightly and the string plucked gently—nature may love simplicity, but twist it too far and it gets very complicated.

Frequency and period. When the mass on the spring is initially displaced some distance $\pm x_0$ (that is, via compression or elongation) and *released from rest,* that displacement will remain the maximum value attainable, and the mass will oscillate between $+x_0$ and $-x_0$. Of course, this is the *amplitude* of the oscillation: $x_0 = A$. Now, compare the body's acceleration $a_x = -(k/m)x$ with Eq. (12.10) for the acceleration of an object in SHM; namely, $a_x = -\omega^2 x$. The angular frequency equals the square root of k/m, but here we will use the symbol ω_0 instead of ω to remind us that it is the *natural* angular frequency, the specific frequency at which a physical system oscillates all by itself once set in motion (as we will see, the spring can be made to vibrate at other frequencies by driving it). Thus

$$\omega_0 = \sqrt{\frac{k}{m}} \tag{12.11}$$

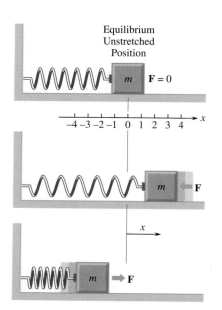

Figure 12.14 A mass on a spring vibrating horizontally in simple harmonic motion. Here, **F** is the force exerted by the spring, and there is no friction.

Figure 12.15 A simple harmonic oscillator shown every quarter cycle for two cycles. Indicated are the corresponding speeds and displacements.

and with $\omega_0 = 2\pi f_0$, we have

$$f_0 = \frac{1}{2\pi}\sqrt{\frac{k}{m}} \tag{12.12}$$

Since $T = 1/f_0$

$$T = 2\pi\sqrt{\frac{m}{k}} \tag{12.13}$$

Figure 12.15 shows the relationship among x, v_x, t, and T.

The stiffer the coil spring or wire or rubber band, the larger the elastic force constant k and therefore (with a given mass), the higher the vibrational frequency and the shorter the period. The springs on a car must have a large k to keep the up-and-down excursions small, but because the mass of the car is large, the frequency will nonetheless be low. For a given spring (that is, a given k) if the oscillating mass is increased, the inertia of the system will be increased, causing it to accelerate more slowly and therefore increasing the period.

A skyscraper can be made as stiff as the builder is willing to pay for, up to a point at least. It is the financial as much as the structural concerns that dictate how rigid a tall building will be. As a rule, such structures are designed so that the top never displaces more than about 1/400 of the height when acted upon by a sustained pressure of 1.4 kPa (30 lb/ft^2), which is equivalent to a 150-km/h (95-mi/h) wind. The period with which the building will rock is then determined by its mass and stiffness (k). The World Trade Center towers sway with a period of 10 s whereas the smaller Citicorp building, also in Manhattan, has a period of only 7 s.

Example 12.3 The cart in Fig. 12.16 has a mass of 1.00 kg, and someone reaches down and displaces it 5.00 cm to the right with an axial horizontal force of 10.0 N. (a) Assuming no friction, what is the period of the resulting oscillation when the cart is released? (b) Where will it be 0.200 s after release? (c) What is the elastic force constant for the system if one of the two identical springs is removed? (d) Determine the new frequency.

Solution: [Given: $m = 1.00$ kg, $A = 0.0500$ m, and $F = 10.0$ N. Find: T and x at $t = 0.200$ s, k with one spring, and f.] (a) The period can be gotten from Eq. (12.13) once we have the elastic force constant for the system. Inasmuch as the applied force produces a displacement x such that $F = kx$

$$k = \frac{F}{x} = \frac{10.0 \text{ N}}{0.0500 \text{ m}} = 200 \text{ N/m}$$

This is the elastic force constant of the system—both springs acting at once, and both undergoing the displacement x as if they were attached side by side to the same end of the cart. Consequently

$$T = 2\pi\sqrt{\frac{m}{k}} = 2\pi\sqrt{\frac{1.00 \text{ kg}}{200 \text{ N/m}}} = \boxed{0.444 \text{ s}}$$

(b) To find where the cart will be at $t = 0.200$ s, we go back to the equation of x as a function of time, thus

$$x = A \cos \theta = A \cos \omega_0 t = A \cos 2\pi f_0 t \quad [12.3]$$

where the motion began at $x = A$, at $t = 0$, so $\varepsilon = 0$. We need either ω_0 or f_0, and since we have T

$$f_0 = \frac{1}{T} = 2.25 \text{ Hz}$$

Figure 12.16 To displace the cart horizontally, energy must be provided to the system. That energy is conserved and ideally transforms back and forth from PE$_e$ to KE as the cart oscillates in SHM. In reality, the oscillation will continue to diminish until all the energy ends up as thermal energy via friction and internal spring losses.

Therefore

$$x_{0.2} = A \cos 2\pi f_0 t$$
$$x_{0.2} = (0.0500 \text{ m}) \cos 2\pi (2.25 \text{ Hz})(0.200 \text{ s})$$

and
$$\boxed{x_{0.2} = -0.0476 \text{ m}}$$

(c) If a spring is removed, displacing the body will require only half the force it did before, when one spring was being stretched as the other was compressed. It follows that now $\boxed{k = 0.100 \text{ kN/m}}$, half its previous value.

(d) The resulting oscillatory frequency, Eq. (12.12), will be $1/\sqrt{2}$ of what it was—namely, $\boxed{1.59 \text{ Hz}}$.

▶ **Quick Check:** The answer to (b) is reasonable since 0.200 s is very nearly $\frac{1}{2}T$, at which time $x = -A = -0.050$ m. (c) The new period is $T = 0.444$ s × (1.414) = 0.63 s; $f = 1/T = 1/0.63$ s = 1.6 Hz.

Gravity and oscillations.   An elastic band hanging freely from a hook, as in Fig. 12.17, stretches a length ΔL when it is made to support a load of mass m. At that point, the restoring force up equals the weight down: $k\,\Delta L = mg$ and the system is in equilibrium. Here, k is the net elastic force constant of the two supporting lengths of material acting together. Now, imagine that we pull down on the mass, displacing it from equilibrium an arbitrary distance $+y$ (*taking down to be positive* so that the restoring force will be negative). Newton's Second Law states that at any instant the sum-of-the-forces equals ma_y. The upward rubber band force $-k(\Delta L + y)$ plus the downward weight mg equals ma_y:

$$-k(\Delta L + y) + mg = ma_y$$

but $k\,\Delta L = mg$; hence, for any value of y

$$a_y = -\left(\frac{k}{m}\right)y$$

This is the same relationship we obtained for the horizontally oscillating mass on a spring. Gravity acting on the load establishes a new equilibrium position about which the mass will then oscillate.

Example 12.4   A 2.0-kg bag of candy is hung on a vertical, helical, steel spring that elongates 50.0 cm under the load, suspending the bag 1.00 m above the head of an expectant youngster. The candy is pulled down an additional 25.0 cm and released. How long will it take for the bag to return to a height of 1.00 m above the child?

Solution: [Given: $m = 2.0$ kg, $\Delta L = 50.0$ cm, and $A = 25.0$ cm. Find: t when $y = 0$.] The loaded system is in equilibrium when the mass is 1.00 m above the small person. It returns there from its lowest point, the starting position ($x = A$), after $\frac{1}{4}$ cycle—that is, at $t = \frac{1}{4}T$. To find T, we must first find k.

Initially

$$-k(\Delta L) + mg = 0$$

and so

$$k = \frac{mg}{\Delta L} = \frac{(2.0\text{ kg})(9.8\text{ m/s}^2)}{0.50\text{ m}} = 39.2\text{ N/m}$$

We know from Eq. (12.13) that

$$T = 2\pi\sqrt{\frac{2.0\text{ kg}}{39.2\text{ N/m}}} = 1.4\text{ s}$$

and so

$$\boxed{\tfrac{1}{4}T = 0.35\text{ s}}$$

▶ **Quick Check:** $F = k\,\Delta y = mg$, $m/k = \Delta y/g$, $T = 2\pi\sqrt{\Delta y/g} = 2\pi\sqrt{0.5\text{ m}/g} = 1.4$ s.

The mass of the spring.   Continuing from Eq. (12.13) for the period, it follows that $T^2 \propto m$. A plot of T^2 versus m, for a number of different masses hung on a vertical coil spring, should be a straight line with a slope of $4\pi^2/k$, passing through the origin. In practice, for small values of A, the graph is straight—the motion is SHM—but it *does not* pass through the origin (Fig. 12.18). That's because the spring itself is not massless. The complete analysis is complicated by the fact that the spring's mass oscillates along with the load, but each segment of it has a different amplitude. It turns out that the effective mass of the system is actually the mass of the load plus one-third the mass of the spring.

12.3   Gravitational Restoring Force: Pendulums

Legend has it that Galileo, while an undergraduate at Pisa, was in the cathedral (1581) when an attendant pulled a candelabrum hanging from the ceiling off to one side to light it. After it was let loose, it swayed to and fro, and Galileo sensed a rhythm in the smooth, repetitive motion. At first the lamp rushed quickly along large arcs, and then as the swing diminished, the speed slowed with it so that the rhythm remained strangely constant. Being a student of medicine, Galileo naturally used the beat of his own pulse to time the swing (there were no watches then). To his astonishment, the duration of each cycle (what he named the *periodo*) was constant.

Galileo subsequently performed a series of experiments using balls of different weight hung on various lengths of string. What he soon discovered was remarkable: *the period of a pendulum is constant and determined only by the square root of its length.* The implication was clear: the string deflected the ball along a curved path, but it was otherwise free to fall, and since all objects fall at the same rate, two pendulums of the same length should, indeed, swing in step.

The Simple Pendulum

The pendulum in Fig. 12.19 is depicted in flight at some angle θ. The bob is displaced a positive distance l measured to the right along the arc of the path from the vertical zero-θ axis. Taking θ in radians, $l = L\theta$. At that moment, the weight of the bob (mg) is, as always, straight down, but it has a component tangent to the arc. It is this component, this unbalanced force

$$F = -mg \sin \theta$$

that drives the pendulum back to equilibrium ($\theta = 0$). The minus sign is needed because **F** points in the negative l-direction, toward decreasing θ. Since $F = ma_{\mathrm{T}}$, the tangential acceleration is

$$a_{\mathrm{T}} = -g \sin \theta \qquad (12.14)$$

which is proportional to the *sine* of θ and not to l, and so it is *not* the condition for SHM. Still, for small angles, the value of θ in radians is very nearly the value of $\sin \theta$; they differ by less than 2% out to about 20°. Hence, $\sin \theta \approx \theta = l/L$ and

$$a_{\mathrm{T}} \approx -\left(\frac{g}{L}\right)l \qquad (12.15)$$

which is SHM, provided θ is small. It follows from $a_x = -\omega_0^2 x$ and Eq. (12.15) that $\omega_0 = \sqrt{g/L}$ and so

$$f_0 \approx \frac{1}{2\pi} \sqrt{\frac{g}{L}} \qquad (12.16)$$

and

$$T \approx 2\pi \sqrt{\frac{L}{g}} \qquad (12.17)$$

At any location on Earth, the period of a pendulum is dependent only on the square root of its length. For small initial angular displacements, less than 23°, the actual periods vary by less than 1% from the predictions of Eq. (12.17), and the motion is approximately SHM. The agreement improves tremendously for still smaller starting angles.

To get Eq. (12.14), we set ma_{T} equal to $-mg \sin \theta$ and then canceled the mass. Well, that really was more significant than it might seem. The mass in $F = ma$ is

Figure 12.17 Hanging a mass m on an elastic band stretches it a length ΔL, to a new equilibrium position. Pulling down further extends it a distance y. When it is let loose, it will oscillate in SHM, provided the displacements are small and the restoring force is linear.

Figure 12.18 Since $T = 2\pi\sqrt{m/k}$, it follows that $T^2 = 4\pi^2 m/k$, and a plot of T^2 versus m must be a straight line with a slope of $T^2/m = 4\pi^2/k$. The reason it doesn't pass through the origin is that the spring itself has mass, which adds inertia to the system and increases the period.

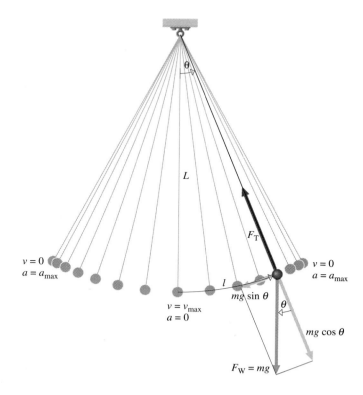

Figure 12.19 A pendulum is driven into oscillation by the nearly horizontal component of its weight $mg \sin \theta$. Provided θ is small, the mass will move in SHM.

the **inertial mass**, the mass associated with the object's tendency to resist changes in its motion. It seems to have nothing to do with gravity and perhaps might even be labeled m_i to underscore the difference in its conceptual origins. On the other hand, the weight of a body is determined by a physical property it possesses called **gravitational mass**, m_g. That property is proportional to the gravitational interaction between objects and seemingly has nothing to do with inertia. Thus, assuming these masses to be different (since they certainly seem to define different characteristics), $T \approx 2\pi \sqrt{m_i L/m_g g}$. But experiments going back to Newton himself all confirm (nowadays to within 1 part in 10^{12}) that the period is independent of the mass of the bob, which implies that $m_i = m_g$. The assertion of the equality of inertial and gravitational mass is known as the *Equivalence Principle* and it is central to the General Theory of Relativity. Perhaps gravity—the Universe acting gravitationally on each mass—*is* the cause of inertia.

It is believed that Galileo described the design of the pendulum clock to his assistant Viviani a year or so before he died.

Example 12.5 How long should a pendulum be if it is to have a period of 1.00 s at a place on Earth where the acceleration due to gravity is 9.81 m/s²?

Solution: [Given: $T = 1.00$ s and $g = 9.81$ m/s². Find: L.] From Eq. (12.17), $T^2 = 4\pi^2(L/g)$; hence

$$L = \frac{T^2 g}{4\pi^2} = \frac{(1.00 \text{ s})^2(9.81 \text{ m/s}^2)}{4\pi^2} = \boxed{0.248 \text{ m}}$$

▶ **Quick Check:** A pendulum 1 m long is sometimes called a "seconds pendulum" because one swing takes 1 s: $T = 2\pi\sqrt{L/g} = 2.0$ s. The pendulum in this problem has half that period and therefore should have $\frac{1}{4}$ the length.

Huygens recognized that Eq. (12.17) provides a simple means of measuring g. Geologists have been using the pendulum for a long time to determine local variations in g. A fairly accurate value of g can be determined by just swinging a mass on a string and measuring T averaged over 10 or 15 cycles. Try it!

12.4 Damping, Forcing, and Resonance

Thus far we have talked about freely vibrating systems, but that's an idealization. There are usually external forces acting on an oscillator in addition to the restoring force. These forces may either impede (that is, damp) the motion, or they may drive it so that its amplitude is sustained, or even grows. Sometimes this latter effect can cause catastrophic failures, as when large buildings are toppled during earthquakes.

Damping

In practice, oscillating mechanical systems lose energy in a variety of ways via friction, and the sinusoidal behavior of SHM is invariably replaced by a damped *anharmonic* motion that gradually dies away (Fig. 12.20). For example, a common friction mechanism is **viscous damping** (we are, after all, surrounded by air). At low speeds, viscous damping is proportional to the speed of the body in the damping fluid (Fig. 12.21). Any undriven system oscillating in the air must inevitably come to rest.

With weak damping, a system can continue to oscillate a relatively long time before stopping at its zero displacement position. This is what happens to an ordinary pendulum losing energy via friction. Its motion slowly decays, just as the vibrations of a tuning fork or a gong slowly decay. With still more friction, the amplitude decreases rapidly, and the mass undergoes fewer oscillations before coming to rest. Such a system is **underdamped** (Fig. 12.22). It's still periodic, although the frequency is lower than it would be without damping.

If friction is further increased, a displaced system can return to equilibrium without overshooting, in which case there will be no oscillation at all. The shock absorbers on a car should damp out any oscillations in less than a cycle. You don't want your car bobbing up and down long after it hits a bump in the road. When this nonvibratory motion occurs in the shortest amount of time, such that once let loose the system sweeps right back to its undisturbed configuration, it is said to be **critically damped**. It would be nice to have the swinging door at the kitchen of a restaurant or the entrance of a supermarket close swiftly without overshooting and slamming back to hit you from behind.

With even more damping, the system again does not oscillate, but it takes much longer to return to equilibrium. Heavy doors in public buildings almost all have hydraulic devices at their tops to keep them from slamming, and if you have ever waited interminably for one to close, you have experienced the results of **overdamping**.

A variety of damping devices are used to suppress vibrations of all kinds in physical structures and machines. Most are as simple and straightforward as a rubber collar on the shaft of a fan or inflated tires on a car, but some are more exotic. Olympic gold medals have been won using skis equipped with vibration dampers—

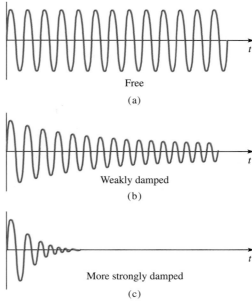

Free
(a)

Weakly damped
(b)

More strongly damped
(c)

Figure 12.20 (a) An ideal harmonic oscillator, free of all forms of energy loss, will vibrate harmonically forever with no decrease in amplitude. (b) When friction is present, the oscillator will be damped, and its amplitude will decrease in time. (c) The more the damping, the more repressed is the oscillation.

Figure 12.21 The shock absorber (or dashpot) uses viscous damping, which is proportional to the speed of the piston. A rapid jerk of the piston, as might be produced when a car hits a pothole, is met with a large damping force. On the other hand, a gradual depression of the piston is hardly resisted.

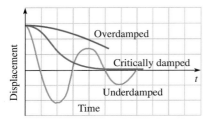

Figure 12.22 When a damped oscillating system experiences still more friction, it will stop vibrating and the motion will gradually decay. The system is then overdamped. When the motion decays as quickly as possible, the system is critically damped.

Figure 12.23 By moving up and down, the hand pumps energy into the system, forcing it to oscillate at the driving frequency.

sandwiches of lead plates and spongy absorbent foam. There are space satellites that contain little carts that ride back and forth along tracks, set into motion by any inadvertent wobbling of the vehicle. Eventually the wobble is damped out and the carts come to rest—the kinetic energy of the satellite having been converted into thermal energy via friction in the carts. The World Trade Center in New York contains visco-elastic dampers: large plates coated with a sticky polymer that drag across one another whenever the towers sway in the wind. They convert troublesome shear energy into harmless thermal energy.

Forced Oscillation and Resonance

One way to sustain the oscillations of a system suffering damping is to periodically pump in energy via a force that does positive work on the system. Pushing a playground swing can keep it going indefinitely, despite friction. When mom pushes, it's best done after junior has reached the peak of the swing and is moving away from her so she does positive work on the system (**F** and **Δs** are parallel). Energy is efficiently transferred to the system when it is pumped in in step with the natural oscillation. Accordingly, mom can push at the natural frequency, or miss a beat and push at one-half the natural frequency, or miss two and push at one-third the natural frequency, and so on.

The application of an *external alternating* driving force to a system capable of vibrating produces **forced oscillation**. The specific physical causes are many. Pulsations of pressure in lines carrying fluids will cause vibrations—some houses have "singing" plumbing, and refrigerators and air conditioners often buzz.

Forced oscillation is easily seen with either a Slinky or a mass on a long elastic band (Fig. 12.23). Displace either and let it loose to oscillate at its natural frequency (f_0), which for the Slinky will be around 1 Hz. That done, stop the oscillation and now move your supporting hand up and down with a small amplitude ($\approx$2 cm) at a very low frequency ($f \approx 0.3$ Hz, $T \approx 3.0$ s), thereby driving the elastic system into motion. The Slinky will follow your hand (the driver), moving upward when your hand moves up, and downward when it moves down; that is, the Slinky will oscillate in-phase with the driver. Moreover, it will oscillate at the same frequency (f) as the driver but with a relatively small amplitude, not much different from that of your hand. Little energy is transferred from the driver to the Slinky as the process continues.

Similar behavior occurs when the driving frequency is much greater than the natural frequency ($f \gg f_0$) except now the two will be nearly 180° out-of-phase—when your hand goes down, the Slinky will move up, and vice versa. One thing will become obvious when you further vary the driving frequency—*as f approaches the natural frequency of the system, the resulting oscillation will dramatically increase in amplitude* (Fig. 12.24). The vibrational amplitude will reach a maximum when $f = f_0$, a condition known as **resonance**. At that special driving frequency, which is also known as the **resonant frequency**, energy is most efficiently transferred to the system, whether it's a hand-pumped Slinky, a windblown skyscraper, or a human internal organ set into vibration by the overamplified tumult of a rock band. It seems that every material object can oscillate somehow. As a rule, a body can bend, twist, and elongate in several different ways, displaying one or more natural frequencies at which it can resonate. Similarly, a composite object, such as a bus, in addition to vibrating as a whole, will have lots of resonant modes for its various parts—windows, doors, seats, and so on.

At resonance, most of the small amount of energy available during each cycle is stored in both the elastic medium and the moving mass; none is returned to the

An earthquake caused the collapse of the double-decked Nimitz Freeway in Oakland, California, in 1989. This portion of the highway rests on fine-grained sediments and mud that transmitted frequencies of around 2 Hz. This value is close to the natural frequency of the highway, which experienced violent resonant vibrations. Fifty-three cars were crushed when the supports for the top deck gave way.

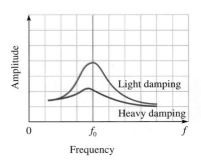

Figure 12.24 The less damping there is, the sharper and taller is the amplitude versus frequency curve. The system responds more selectively to being driven at its resonant frequency.

driver, though a bit is invariably lost to friction. The less damping there is, the sharper and taller is the curve (Fig. 12.24), the greater the resulting amplitude of the vibration and the nearer that peak is to the resonant frequency. In a lightly damped system such as a bell, gong, or turbine blade, the stresses occurring at resonance can be upwards of 300 times larger than those resulting from the same applied force acting at a much different frequency.

The phenomenon of resonance is one of the most important in all of physics. It is at the heart of a tremendous diversity of occurrences from the atomic absorption and emission of light to the tuning of a TV set, from the swaying of bridges to the rattling of china cabinets when a low-flying plane passes overhead.

Example 12.6 Some bridges have exposed steel gridwork embedded in the road, and certain highway turns have been roughened with rows of ridges to improve traction on wet days. Given that a car has 60-cm wheels, if the road ridges are 8.0 cm apart, at what speed will a vehicle traveling on such a surface go into resonance if the natural frequency of its "front end" (steering and suspension system) is 15.0 Hz?

Solution: [Given: $D = 60$ cm, $d = 8.0$ cm, and $f_0 = 15.0$ Hz. Find: v for resonance.] There are $1/0.080 = $ 12.5 ridges per meter and so at speed v (in m/s), the car is moving over $12.5v$ ridges per second, which is the driving frequency. Hence

$$12.5v = 15.0 \text{ Hz}$$

and $\boxed{v = 1.2 \text{ m/s}}$ or 3 mi/h, regardless of wheel size. That's why your car shakes so much at low speeds on that kind of surface.

▶ **Quick Check:** $v = x/T = x/(1/f_0) = (0.080 \text{ m}) \times (15.0 \text{ Hz}) = 1.2 \text{ m/s}$.

Tap a good quality wineglass and it will ring, undriven, at its natural frequency, somewhere around 900 Hz. If it's now forced at that frequency, for example, by a blast of vibrating air (that is, sound) up near the rim, the glass will efficiently ab-

A man playing a tune on a street in New Orleans by resonating (via stick-slip friction) different sized and pitched glasses.

A wineglass oscillating in two of its several possible modes. The bright region is the stationary portion of the glass. The photos were made from interferometric holograms.

sorb energy and begin to vibrate. When the note is sustained and carries enough energy to overcompensate for internal damping, the rim of the glass will oscillate with increasing amplitude and ultimately will shatter because glass is brittle. The same resonance mechanism causes the ashtray in your car to clatter annoyingly to the beat of the radio, and it's the reason why troops marching over bridges are usually ordered to break step. At least two large bridges collapsed in the nineteenth century, one near Manchester, England (1831), the other over the Maine in Anjou, France (1849), both having been set into resonance by the pounding, low-frequency cadence.

Example 12.7 A room at the top of the Citicorp building houses a 400-ton block of concrete that slides "floating" on a layer of oil pumped under pressure. One end of the block is attached to pneumatic springs fixed to one side of the building while the other end is attached to a hydraulic piston that acts like a large shock absorber, mounted to the other end of the building. The device is called a "tuned-mass damper" because it's tuned to oscillate at the natural frequency of the building. As the tower begins to sway, the block goes into resonant oscillation out-of-phase with the building. The wind's energy is transferred to the tower, and then to the block, and then, by moving against the shock absorber, it's finally dissipated as thermal energy. Take the period of oscillation of the building to be 7.0 s and calculate the elastic force constant of the pneumatic spring that will produce resonance with the damping removed. (See the photo on the opposite page.)

Solution: [Given: $T = 7.0$ s and $F_W = 400$ tons $= 800 \times 10^3$ lb. Find: k.] The natural period of the spring-mass system must match that of the building—namely, 7.0 s. The period is given by Eq. (12.13), from which it follows that

$$k = \frac{4\pi^2 m}{T^2} = 4\pi^2 \frac{3.63 \times 10^5 \text{ kg}}{49 \text{ s}^2} = \boxed{2.9 \times 10^5 \text{ N/m}}$$

▶ **Quick Check:** Compute T using k; $\omega_0 = \sqrt{k/m} = \sqrt{(2.925 \times 10^5 \text{ N/m})/(3.63 \times 10^5 \text{ kg})} = 0.9$ rad/s; $f_0 = \omega_0/2\pi = 0.14$ Hz; $T = 1/f_0 = 7$ s.

12.5 Self-Excited Vibrations

Remarkably, it is possible to initiate and sustain the vibration of a particular kind of system even though the energy source is nonoscillatory. In such situations, the response of the vibrating system itself produces alternations in applied force. These **self-excited vibrations** are commonplace and often quite dangerous, though a kiss (sucking in a stream of air so that the lips vibrate) can be harmless enough.

Take a bow and a violin and attempt the business of making music—or at least sound. We wish to set the strings vibrating; they will set the air vibrating, and that's what a concerto is, more or less. Moving in one direction, say, to the right, the bow is made to brush over a string. Initially at rest with respect to the bow, the string is drawn to the right, and bent. The displacement continues until the growing restoring force of the string overcomes the static friction force, and it breaks loose from the bow. The string vibrates at a natural frequency, oscillating almost freely for a little while (losing energy in the form of sound). It encounters a markedly increased friction force when its relative speed with respect to the bow is low. Thus, the moving bow will recapture the string, again displacing it to the right, pumping energy in, and beginning the process over once more. The energy source is the constantly moving bow, but the motion of the system—the string—is vibratory.

The tuned-mass damper at the top of the Citicorp tower. (See Ex. 12.7.)

This so-called **stick-slip friction** is the process responsible for the screech of a fingernail drawn across a blackboard. The high-pitched creak of an unoiled door is due to a torsional vibration of the short, stiff hinge pin. As Galileo pointed out, a wineglass can be made to ring at its resonant frequency in much the same way. Just run a moist, clean finger around the upper edge of the lip, gently pressing downward—stick-slip friction will do the rest.

There are many *self-excited vibrational* phenomena that are driven by constant streams of fluid, singing being one of them. Blowing across the mouth hole of a flute causes vortices to peel off periodically, creating a fluctuating pressure and setting the instrument vibrating. Similarly, a constant wind streaming over a banner stretched across a street can set it vibrating. The effect is easily duplicated by blowing over the broad faces of a wide rubber band.

The first bridge across the Tacoma Narrows at Puget Sound, Washington, had a main span 2800 ft long and 39 ft wide with 8-ft-tall steel stiffening girders. Opened for traffic on July 1, 1940, it soon became infamous for oscillating wildly whenever the wind was blowing. Thrill-seeking motorists were lured from miles around for the ride of a lifetime. On the morning of November 7, 1940, a wind of 40 to 45 mph set the span rippling as usual, at a frequency of 36 vibrations per minute. When the amplitude became too large, the bridge was cleared. At around 10:00 A.M., the north cable became loose in its collar, allowing the main deck to abruptly begin to vibrate in a twisting resonant mode ($f_0 = 0.2$ Hz) around the yellow centerline of the roadway.

Although there are still questions to be answered, the basic mechanism that caused the catastrophe seems to have been vortex-induced self-excited vibrations. Air rushing toward the tall windward stiffening girder broke into two streams, shedding vortices alternately above and below the deck. Once the structure began to oscillate, that motion led to the formation of other motion-induced vortices (Fig. 12.25). These were formed as the deck rose and fell, and so matched the natural vibrational frequency. The rate at which energy was absorbed from the wind soon exceeded the frictional loss, and a state existed in which the oscillations quickly grew in amplitude. Shortly after 11:00 A.M., the center span tore into shreds like so much cotton ribbon.

(a)

(b)

Figure 12.25 (a) Cross-sectional view of the roadway deck of the Tacoma Narrows Bridge, initially set into oscillation by alternating vortices. (b) As the deck oscillated, it caused other vortices to be induced that drove the bridge at its resonant frequency. These so-called motion-induced vortices were created in step with the twisting deck and furthered the motion.

MECHANICAL WAVES

There are only two fundamental mechanisms for the transport of energy and momentum: a streaming of particles and a flowing of waves. And even these two seemingly opposite conceptions are subtly intertwined—there are no waves without particles and no particles without waves, but we will come back to that later.

A wave is a disturbance of a medium, and that medium can be either a field (such as the gravitational field) or a material substance (a solid or fluid). Here, the focus is on waves in material media, and these are known as **mechanical waves**. Sound is such a wave, but because of its special relationship to us, it will be treated on its own in Chapter 13.

12.6 Wave Characteristics

Consider an object—a bell, a rope, or even the Earth. Each is a vast collection of atoms forming an essentially continuous elastic medium. If the atoms or molecules in a material are pushed together, they repel and, if separated, they attract. Electrically interacting, the atoms behave as if they are connected to one another via springs. As a result of this restoring force, the medium returns to an equilibrium configuration that is its normal stable state. Now imagine that a driving force acting on the medium displaces some of its atoms; for example, the bell is struck with a hammer. Energy is imparted to these impacted atoms, which are forced off their equilibrium positions, thereby interacting more strongly with neighboring atoms, doing work on them and displacing them. These newly shifted atoms interact with still others, and so the displacement process continues, with energy transferred from atom to atom (Fig. 12.26). *The state of being displaced moves through the medium as a wave.* The disturbance of a medium under the influence of a restoring force is common to all mechanical waves.

A **progressive** or **traveling** wave *is a self-sustaining disturbance of a medium that propagates from one region to another, carrying energy and momentum.* There are a great variety of progressive mechanical waves, among which are waves on strings, surface waves on liquids, sound waves in the air, and compression waves in both solids and fluids. In all cases, although the energy-carrying disturbance advances through the medium, the individual participating atoms remain in the vicinity of their equilibrium positions: *the disturbance advances, not the material*

The Tacoma Narrows Bridge oscillating and ultimately collapsing. The stalled car belonged to a reporter who crawled off the bridge, leaving the car behind. The only fatality was a dog initially trapped in the car. Professor Farquharson, who was studying the bridge, risked his own life to free the dog, who then chose not to leave the scene (the dog is the small spot left of center).

medium. That's the crucial aspect of a wave that distinguishes it from a stream of particles. The wind blowing across a field sets up "waves of grain" that sweep by, even though each stalk only sways in place. Da Vinci seems to have been the first person to recognize that a wave does not transport the medium through which it travels, and it is precisely for this reason that waves are capable of propagating very rapidly.

Longitudinal and Transverse Waves

There are two basic forms of waves: **longitudinal** and **transverse**. In addition to these, there are also torsion waves, which are a variation of the latter, and water waves, which are a combination of the two. Envision a long loose-coil spring, like a Slinky, held in such a way that it is straight and horizontal (Fig. 12.27). The spring itself is to be the medium that will sustain the disturbance. Now, compress several of its coils by pulling to the left and release them all at once. The region of compression, the disturbance from equilibrium, will rapidly advance to the right along the whole length of the spring. *When the sustaining medium is displaced parallel to the direction of propagation, the wave is longitudinal* (meaning "lengthwise"). The compression wave in a bell is of this variety, as are both sound waves and certain seismic waves.

If the end of the spring is displaced up and down, a hump will be produced (Fig. 12.28). A momentary disturbance will move, effectively as a pulse of energy, along the length of the spring. *When the sustaining medium is displaced perpendicu-*

Figure 12.26 Energy is pumped into the horizontal rod by the impact of the ball on the left. A compression wave travels down the rod and slams into the ball on the right. The wave of displaced atoms transports energy and momentum from one ball to the other.

Figure 12.27 A longitudinal wave in a spring.

Figure 12.28 A transverse wave in a spring.

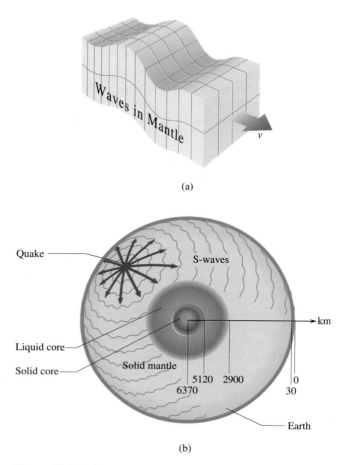

(a)

(b)

Figure 12.29 (a) Transverse seismic waves like this one are called shear waves, or S-waves. (b) Because of the Earth's molten core, which cannot support transverse waves, the S-waves from a quake on one side of the planet are picked up everywhere except diagonally across on the other side. Shear waves in the mantle travel at about 8 km/s.

lar to the direction of propagation, *the wave is transverse* (meaning "lying across"). Plucked guitar strings oscillate as transverse waves. All forms of electromagnetic radiation from radio waves to X-rays and light are envisioned as transverse waves.

Transverse mechanical waves can be supported provided the medium resists shear, which is something fluids don't do very well. Accordingly, we can expect these waves will not be sustained within the body of a liquid or gas (Fig. 12.29).

Waveforms

Waves in strings, wires, and ropes are of considerable practical concern to piano tuners and people who worry about vibrating powerlines, but our study of them is more general. For us, they provide an especially simple visual model that can be applied to all forms of waves. Figure 12.30 depicts a transverse **wavepulse** set up in a taut rope driven by a hand that smoothly moves once up and back down again. The outline, *profile* or shape of the pulse, is determined by the motion of the driver (the hand), while the physical aspects of the medium (the tension and inertia of the rope)

Figure 12.30 A wavepulse riding along a taut rope.

determine the speed of the wave. A single wavepulse has its equivalent in many familiar effects, such as the sound of a shotgun blast or an ocean tidal wave.

If the hand now oscillates up and down in a regular way, it can generate a disturbance that alternately appears above and below the equilibrium zero-line like the **wave-train** depicted in Fig. 12.31. A disturbance of this sort can be viewed as a steady sinusoidal oscillation called the **carrier**, which gradually varies in amplitude. A wavetrain, whatever the details of its shape, has a beginning and an end. A burst on a whistle will produce an acoustical wavetrain.

It is possible to sustain a vibration for a long time and thus generate a continuing wave. Figure 12.32 shows the waveform produced via a microphone by a sustained note played on a saxophone. It could equally well have been created on a taut rope through some fancy jiggling. Overlooking the very slight variations, the wave can be visualized as made up of a single repeated profile-element (Fig. 12.32b). However long the overall wavetrain is, it is finite—there was a time before which the note was blown and there will be a time when it ends. That's the nature of real waves as compared with mathematical ones. In many situations, the disturbance may be extremely long with an enormous number of repetitions of the same profile-element. It then becomes much simpler mathematically to assume that the wave is infinitely long. Such an idealized disturbance composed of endless repetitions of the same profile-element is a **periodic wave**.

To Find the Velocity of Waves

The **temporal period** (T) of a periodic progressive wave is the time it takes for one profile-element to pass a given point in space. The inverse of that ($1/T$) is the **frequency** f, the number of profile-elements passing per second. The distance in space over which the wave executes one cycle of its basic repeated form (one profile-element) is the **spatial period** or **wavelength**, λ (Greek lowercase lambda).

Imagine that you are at rest and a periodic wave on a string is progressing past you. The number of profile-elements that sweep by per second is f, and the length of each is λ. In 1 s, the overall length of the disturbance that passes you is the product $f\lambda$. If, for example, each is 2 m long and they come at a rate of 5 per second, then in 1 s, 10 m of wave flies by. This is just what we mean by the speed of the wave (v)—the rate in m/s at which it advances. Said slightly differently, since a length of wave λ passes by in a time T, its speed must equal $\lambda/T = f\lambda$. We now have an expression for the speed of any progressive *periodic* wave, be it sound, water ripple, or light:

$$v = f\lambda \qquad (12.18)$$

Incidentally, Newton derived this relationship in the *Principia* (1687) in a section called "To find the velocity of waves."

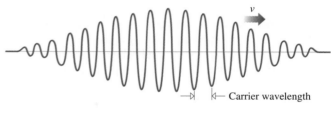

— Carrier wavelength

— Pulse length —

Figure 12.31 This wavetrain one can be thought of as a constant-frequency carrier whose amplitude is modulated— that is, made to vary in time.

Figure 12.32 (a) The waveform generated by a saxophone. We can imagine any number of profile-elements (b) that, when repeated, create the waveform (c). The length over which the wave repeats itself is called the wavelength, λ.

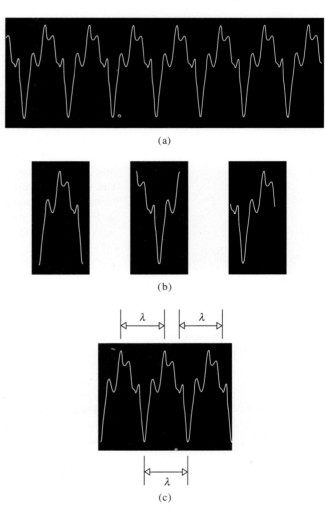

(a)

(b)

(c)

Example 12.8 A youngster in a boat watches waves on a lake that seem to be an endless succession of identical crests passing with a half-second pause between each. If one takes 1.5 s to sweep straight down the length of her 4.5-m-long boat, what is the frequency, period, and wavelength of the waves?

Solution: [Given: The waves are periodic; there is $\frac{1}{2}$ s between crests; $t = 1.5$ s, $L = 4.5$ m. Find: f, T, and λ.] The time between crests is the period, so $\boxed{T = \frac{1}{2}\text{ s}}$; hence, $f = 1/T = \boxed{2.0\text{ Hz}}$. As for the speed

$$v = \frac{L}{t} = \frac{4.5\text{ m}}{1.5\text{ m}} = 3.0\text{ m/s}$$

We now know T, f, and v and must determine λ. Thus, from Eq. (12.18)

$$\lambda = \frac{v}{f} = \frac{3.0\text{ m/s}}{2.0\text{ Hz}} = \boxed{1.5\text{ m}}$$

▶ **Quick Check:** $Tv = \lambda = (\frac{1}{2}\text{ s})(3.0\text{ m/s}) = 1.5$ m. $v = f\lambda = (2.0\text{ Hz})(1.5\text{ m}) = 3.0$ m/s.

Harmonic Waves

The easiest of repeating disturbances to treat theoretically is the idealized **harmonic wave**, which rises and falls sinusoidally without end. This is the fundamental waveform because, as we will see later, all real waves can be synthesized from overlapping harmonic waves.

Remembering the treatment of SHM, examine Fig. 12.33, which is a plot of

$$y = A \sin \frac{2\pi x}{\lambda} \tag{12.19}$$

The argument of the sine function $(2\pi x/\lambda)$ is the **phase** and it's unitless, as it must always be. Equation (12.19) describes the profile of a harmonic wave frozen at $t = 0$. At any location x, the ratio x/λ is the number of wavelengths from the origin out to that point and, since there are 2π rad per wavelength, $2\pi x/\lambda$ is the number of radians out to the point at x. The profile repeats itself with a wavelenth λ so that $y = 0$ when $x = 0, \lambda, 2\lambda, 3\lambda, \ldots$. The *amplitude* is A and it is always positive even though y may be negative. The wave builds up and falls off between values of $+A$ and $-A$. **The energy associated with a wave is proportional to the amplitude of the wave squared**, and that's true for all waves. The amplitude of a sound wave determines how loud it is, and the amplitude of a light wave determines how bright it is. Figure 12.34 shows a harmonic wave advancing 1 wavelength during a time interval of 1 period (T).

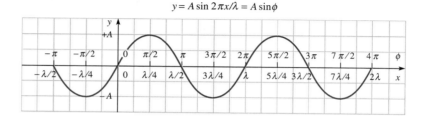

$$y = A \sin 2\pi x/\lambda = A \sin\phi$$

Figure 12.33 A harmonic function, which is to serve as the profile of a harmonic wave. One wavelength corresponds to a change in phase ϕ of 2π rad.

Example 12.9 The profile of a harmonic wave, traveling at 1.2 m/s on a string, is given by

$$y = (0.02\text{ m}) \sin (157\text{ m}^{-1})x$$

Determine its amplitude, wavelength, frequency, and period.

Solution: [Given: the profile, and $v = 1.2$ m/s. Find: A, λ, f, and T.] Comparing y with Eq. (12.19) tells us that $\boxed{A = 0.02\text{ m}}$. Moreover,

$$\frac{2\pi}{\lambda} = 157\text{ m}^{-1}$$

(continued)

(continued)

and so $\lambda = 2\pi/(157 \text{ m}^{-1}) = \boxed{0.040\,0 \text{ m}}$. The relationship between frequency and wavelength is fixed by Eq. (12.18), $v = f\lambda$, and so

$$f = \frac{v}{\lambda} = \frac{1.2 \text{ m/s}}{0.040\,0 \text{ m}} = \boxed{30 \text{ Hz}}$$

The period is the inverse of the frequency, and therefore $T = 1/f = \boxed{0.033 \text{ s}}$.

▶ **Quick Check:** $f/v = 1/\lambda$; hence the phase $(2\pi/\lambda)x = (2\pi f/v)x = [2\pi(30 \text{ Hz})/(1.2 \text{ m/s})]x = (157 \text{ m}^{-1})x$.

Imagine a point on the wave in Fig. 12.34 with a value of y of, say, $A/2$. Since the amplitude is constant, the point we are watching must have a particular fixed value of the phase; in this instance, the phase is $\pi/6 = 30°$. As time goes on, the wave progresses, but for the point we are watching, $y = A/2$ must always correspond to a phase of $\pi/6$. Thus, when we talk about the speed of a harmonic wave, we are referring to the rate at which a point with constant phase moves, and that's called the **phase velocity**, or phase speed.

12.7 Transverse Waves: Strings

When a mechanical disturbance is created, the speed of the wave is determined by the inertial and elastic properties of the medium and not in any way by the relative motion of the source. The central concerns are how much mass is being accelerated and with how much force does the medium resist deformation. This insight can be shown rigorously for waves on strings using the calculus, but for our purposes a slightly limited though much simpler analysis (formulated by P. G. Tait in the 1800s), will do nicely.

Consider a smooth pulse traveling with a speed v to the right along a lightweight, flexible string (Fig. 12.35). The string is kept taut by the application of a constant external force that maintains a tension F_T in the system. Imagine that the wave is viewed from a reference frame moving with the pulse so it is seen to be at rest. Although the disturbance is motionless in this frame, the string is observed to move to the left at a constant speed v. The situation appears as if the string were being pulled to the left through a stationary invisible tube bent in the shape of the pulse. A tiny segment of the string that has a radius of curvature r, a length Δl, and mass Δm is essentially revolving around point C (the center of curvature of that segment) with a speed v. Since it is moving at a constant speed, the net force acting on the segment must be radially inward toward C and equal to the required centripetal force $F_C = \Delta m(v^2/r)$. From the diagram, it follows that each end of the segment experiences a force equal to F_T acting tangent to the rope. The sum of these two tensile force vectors (Fig. 12.35b) is a single radial vector of magnitude $2F_T \sin(\frac{1}{2}\Delta\theta)$: the tangential components cancel since there is no tangential acceleration.

Setting that sum equal to the centripetal force yields

$$\Delta m \left(\frac{v^2}{r}\right) = 2F_T \sin\left(\tfrac{1}{2}\Delta\theta\right)$$

but the segment is tiny compared to r; hence, $\sin(\frac{1}{2}\Delta\theta) \approx \frac{1}{2}\Delta\theta$ and

$$\Delta m \left(\frac{v^2}{r}\right) \approx F_T \Delta\theta \approx \frac{F_T \Delta l}{r}$$

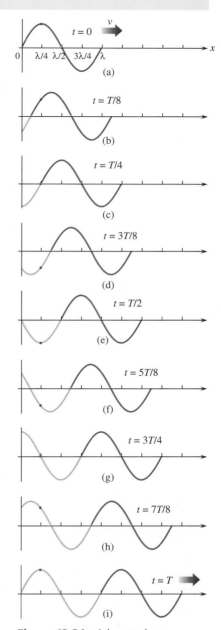

Figure 12.34 A harmonic wave moving along the x-axis during a time of one period. Note that any one point on the rope only moves vertically.

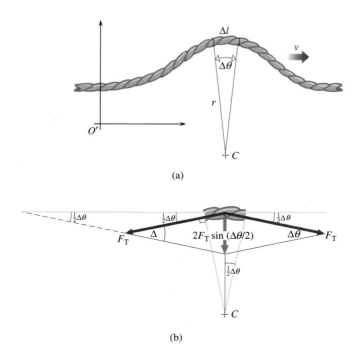

Figure 12.35 (a) A pulse on a stretched rope traveling at a speed *v*. (b) The tension in the rope, F_T, determines *v*.

where $\Delta\theta = \Delta l/r$. If we define the *linear mass-density* μ to be the mass per unit length of the string, $\mu = \Delta m/\Delta l$, and

$$v^2 \approx \frac{F_T}{\mu}$$

Provided that the pulse is not very steep, this relationship holds for all segments and

$$v = \sqrt{\frac{F_T}{\mu}} \qquad\qquad (12.20)$$

The speed of the wave only depends on the force exerted and the inertial characteristics of the medium. When μ is large, there is a lot of inertia and the speed is low. When F_T is large, the string tends to spring back rapidly, and the speed is high.

Example 12.10 A 2.0-m-long horizontal string having a mass of 40 g is slung over a light frictionless pulley, and its end is attached to a hanging 2.0-kg mass. Compute the speed of the wavepulse on the string. Ignore the weight of the overhanging length of rope.

Solution: [Given: a string of $l = 2.0$ m, $m = 40$ g supporting a 2.0-kg load. Find: v.] The speed of the wave is gotten from Eq. (12.20) and so we need F_T and μ. The tension is just the load in newtons; namely, (2.0 kg) $\times$

(9.81 m/s^2) = 19.62 N. While the linear mass-density is the mass of the string divided by its total length, (0.040 kg)/(2.0 m) = 0.020 kg/m. Hence

$$v = \sqrt{\frac{F_T}{\mu}} = \sqrt{\frac{19.62 \text{ N}}{0.020 \text{ kg/m}}} = \boxed{31 \text{ m/s}}$$

which is equivalent to 70 mi/h.

▶ **Quick Check:** If the mass was 1 kg, $F_T = 9.8$ N and $v = 22$ m/s, and this quantity times 1.414 is 31 m/s.

Reflection, Absorption, and Transmission

Every variety of wave from sound to light can undergo reflection, absorption, and transmission on interacting with material media. Interesting things happen to waves at discontinuities in the media in which they travel. For the moment, we focus on waves on ropes, but the conclusions are widely applicable to other types.

In Fig. 12.36, one end of the rope is held stationary while energy is pumped in at the other end. The **reflected** wave ideally carries away all the original energy. Because it is inverted, it's said to be 180° out-of-phase with the incident wave. Alternatively, if the far end of the rope is free (Fig. 12.37), it will rise up as the pulse arrives until all the energy is stored elastically. The rope then snaps back down, producing a reflected wavepulse that is right side up. A similar effect occurs when an ocean wave slams into a breaker wall and climbs high above the crests behind it.

If we draw off energy from the oscillating end point of a rope via friction, the reflected pulse has a proportionately diminished amplitude and we call the process **absorption**. It occurs most effectively when the rope encounters friction that is speed-dependent. To understand why this is the case, realize that any small segment of the rope only moves vertically. Moreover, the vertical force exerted on any such segment due to the rope just before it is proportional to that segment's vertical speed. When a wave travels from left to right, at any point the segment on the left does work on the segment immediately to the right of it, and energy flows along with the disturbance.

Consider what would happen were we to mount a little dashpot damper to the end of the rope (Fig. 12.38). A dashpot, you recall, is a fluid-filled cylinder sealed with a movable piston. The dashpot opposes motion, via viscous damping, with a force that varies as the speed of the point to which it is fixed (that is, the speed of the piston). If the dashpot exerts just the right amount of drag on the rope, the segment immediately to the left of it experiences the same motion as if the rope continued beyond it indefinitely—the wave's energy flows via friction into the dashpot. In this ideal case, the wave will be totally absorbed. A similar absorption process happens with other waves, and that makes it of considerable interest.

When a wave passes from one medium to another having different physical characteristics, there will be a redistribution of energy. Figure 12.39 shows a wavepulse initially traveling in a low-density rope impinging on the interface with a high-density rope. Like the fixed point of Fig. 12.36, the large inertia of the second medium at the junction retards the easy motion of the boundary. There is again an oppositely directed reaction force and the reflected wave is phase-shifted by 180°. But the second medium is also displaced, and a portion of the incident energy will appear as a **transmitted** wave. The time it takes to generate the reflected and transmitted waves must be identical since they share a common source. Yet the speeds of the pulses must be different because they have the same tension and different densities. It follows that the pulses have different lengths.

When the first medium is denser than the second, the situation resembles that of Fig. 12.37 with a free end point: no phase shift results and the transmitted wave has a greater length. The process of reflection and transmission at an interface between media occurs for all waves regardless of whether they are longitudinal or transverse, though the details, such as phase shifts, differ.

Evidently, if the incident wave is periodic, the transmitted wave has the same frequency but a different speed and therefore a different wavelength: *the larger the density of the transmitting medium, the smaller the length of the wave.* The fact that the frequencies of the incident, reflected, and transmitted waves are normally the same is true for every sort of disturbance from sound to light.

Figure 12.36 The reflection of a pulse on a rope with a fixed end point. As the pulse arrives, it exerts a vertical force on the fixed anchor point, which in turn exerts an equal and opposite force on the string. When the string tugs up, the anchor point tugs down. This downward force on the rope generates an upside-down reflected pulse traveling in the opposite direction.

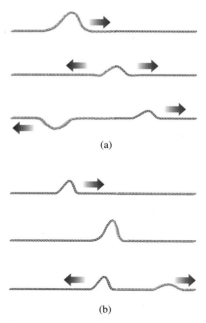

Figure 12.37 The reflection of a pulse on a rope with a free end point. That free end rises until all the energy of the end segment is stored elastically. It comes to rest at a maximum vertical displacement of twice the height of the crest. Carried up by its inertia, the end segment pulls upward on the rope, generating a reflected wavepulse that travels back toward the source, right side up and simply reversed.

Figure 12.38 A wave in a string terminated on a dashpot. The vertical speed of the rope is proportional to the slope of the wave. The drag exerted by the dashpot is proportional to the vertical speed of the piston.

Figure 12.39 Reflection and transmission of a pulse at the junction of two media. The darker rope has a greater linear mass-density.

Example 12.11 An incident periodic disturbance of wavelength λ_i traveling at a speed v_i in one medium impinges on an interface with another medium of different density. Derive an expression relating the speeds and wavelengths of the incident and transmitted waves.

Solution: The speeds and wavelengths within the incident and transmitting media are related via $v = \lambda f$; hence

$$v_i = \lambda_i f_i \quad \text{and} \quad v_t = \lambda_t f_t$$

But we know as a rule that $f_i = f_t$ and so dividing the first equation by the second yields the desired relationship

$$\boxed{\frac{\lambda_i}{\lambda_t} = \frac{v_i}{v_t}}$$

▶ **Quick Check:** Both sides are unitless, which is fine. Recall Eq. (12.20), which tells us that when μ increases there's more inertia, and v decreases. Thus, if the speed of the wave decreases in the transmitting medium, the wavelength decreases and that makes sense.

12.8 Compression Waves

Atoms of an elastic substance can be shifted by an applied force, and the resulting state of being compressed will travel through the medium as a *compression wave*. In solids, such a disturbance is called a *longitudinal elastic wave*. In fluids, which cannot support transverse disturbances, it's called an *acoustic wave*. An earthquake or a large truck rumbling down the street can shake a building, rattling everything in it with streams of compression waves. When the frequency lies in the range from about 20 Hz to 20 kHz, compression waves are called *sound*. There is nothing special about sound waves beyond the fact that we happen to be able to *hear* them.

Let's model a succession of atoms or molecules in an elastic medium (either solid or fluid of sufficient density) using a row of spheres separated by springs (Fig. 12.40). The particle at the far left is made to vibrate in SHM along the line by some source of energy. Wherever the springs are elongated, the density of particles is diminished, and that is a *rarefaction*. Wherever the springs are compressed, the density is increased, and that is a *condensation*. **The propagation of a compression wave takes place in the direction along which the particles of the medium oscillate, and it is marked by a series of alternate condensations and rarefactions**. There are tremendous numbers of atoms in an ordinary medium, and they behave as though the distribution were continuous; rarefactions gradually blend into condensations. Figure 12.41 shows that at any instant the particles within the condensations move in the propagation direction, while those in the rarefactions move in the opposite direction.

The Speed of Compression Waves

We now derive an expression for the speed of a compression wave. It can be anticipated that any such formula involves both the restoring force generated by the sustaining medium, via some measure of its elastic properties, and its density.

Consider a compression pulse of uniform high pressure traveling to the right through a liquid at rest in a tube (Fig. 12.42, p. 431). The liquid is at a pressure P, and the pulse pressure exceeds that by ΔP. When viewed from a coordinate system moving to the right with the pulse at a speed v, the normally stationary fluid is seen to be moving left at v while the pulse is motionless. In a time Δt, a small cylinder of fluid of length $v \Delta t$ enters the high-pressure region of the pulse. The forces acting on this cylinder, the pressure changes it undergoes, and the alterations in its volume will allow us to determine the change in its speed as it becomes part of the pulse.

Time

(a)

(b)

(c)

(d)

(e)

(f)

(g)

(h)

(i)

(j)

λ

Figure 12.40 A sequence of views of a line of masses attached by springs. In (a) the system is undisturbed. The first mass on the left oscillates in SHM. The resulting disturbance propagates to the right [(b) through (j)] as a series of compressions and elongations that repeats after a distance λ. Each mass oscillates sinusoidally in SHM.

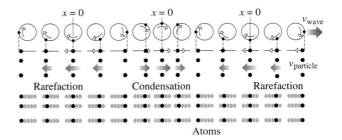

$x = 0$ $x = 0$ $x = 0$

v_{wave}

v_{particle}

Rarefaction Condensation Rarefaction

Atoms

Figure 12.41 A compression wave passing to the right through a distribution of elastically interacting atoms. The peaks of the condensations and rarefactions occur at zero atomic displacements ($x = 0$). Note that the atoms in the condensations move in the direction of propagation of the wave. Start at the right and examine each reference circle in turn. Remember that each atom oscillates around its equilibrium position ($x = 0$) in SHM.

The volume of this cylinder with its cross-sectional area A is $Av\Delta t$ and its mass is therefore $\rho Av\Delta t$, where ρ is the density of the liquid. The pressure difference it experiences between its two ends is ΔP, and so it has acting on it a net force $A\,\Delta P$ to the right. By Newton's Second Law, this force produces an acceleration of $-\Delta v/\Delta t$ (the minus sign is here because a is opposite to v). In other words, $F = ma$ and

$$A\,\Delta P = (\rho Av\Delta t)\left(-\frac{\Delta v}{\Delta t}\right) \tag{12.21}$$

The elastic properties of the medium are embodied in its ability to resist compression under pressure. Recall that the Bulk Modulus B (p. 343) is defined as

$$B = -\frac{\Delta P}{\Delta V/V_0}$$

When the liquid cylinder enters the high-pressure region, it experiences a decrease in speed Δv; hence, while it originally had a volume $V_0 = Av\Delta t$, it decreases in volume by $\Delta V = A\,\Delta v\,\Delta t$ and

$$\frac{\Delta V}{V_0} = \frac{A\,\Delta v\,\Delta t}{Av\,\Delta t} = \frac{\Delta v}{v}$$

Equation (12.21) becomes $-\Delta P = \rho v\,\Delta v$, and dividing both sides by v^2 we get

$$v^2 = \frac{-\Delta P}{\rho\,\Delta v/v} = \frac{-\Delta P}{\rho\,\Delta V/V_0}$$

For compression waves in liquids

$$v = \sqrt{\frac{B}{\rho}} \qquad (12.22)$$

It takes a large increase in pressure to decrease the volume of a solid as compared to a liquid, and B is generally larger for dense media—it increases as the medium becomes increasingly rigid (Table 10.7). Thus, v usually increases as ρ increases, which is not obvious from Eq. (12.22).

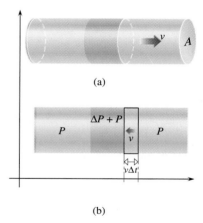

Figure 12.42 (a) A pulse of compression-wave energy traveling with a speed v through a medium with an ambient pressure P. (b) The pulse as seen in a system moving along with it at speed v. The surrounding medium enters the pulse where its pressure is changed to $P + \Delta P$.

Example 12.12 An explosion occurs not far beneath the surface of the Atlantic Ocean. Compute the speed of the resulting compression wave measured by instruments several meters below a ship.

Solution: The problem asks for the speed of a compression wave in seawater. The source is not very deep, hence the density can be taken from Table 10.1. [Given: $\rho = 1.03 \times 10^3$ kg/m³ and from Table 10.7 $B = 2.1$ GPa. Find: v.] Substituting into Eq. (12.22), we obtain

$$v = \sqrt{\frac{B}{\rho}} = \sqrt{\frac{2.1 \times 10^9 \text{ Pa}}{1.03 \times 10^3 \text{ kg/m}^3}} = \boxed{1.4 \times 10^3 \text{ m/s}}$$

▶ **Quick Check:** $B = v^2\rho \approx (1400 \text{ m/s})^2 \times (1000 \text{ kg/m}^3) \approx 2 \times 10^9 \text{ kg/m·s}^2$.

The density and therefore the speed of compression waves in the body of the sea depends on the temperature, salinity, and depth. Which is why this example was careful not to have the waves propagating very deep. Thus, while the speed is 1.4 km/s at the surface, at a depth of 5 km it's about 1.5 km/s. Moreover, a rise in temperature of 1°C increases the speed by 3.7 m/s, while an increase of 1% in salinity adds about 1.2 m/s to it—both of which would be of great interest to a dolphin or a sonar operator in a submarine. In the rock mantle of the Earth, compression waves reach speeds of 13 km/s (Fig. 12.43).

Ultrasound and Infrasound

Dolphins operate their underwater tracking system by radiating compression wave-trains, high-frequency chirps with carrier wavelengths (Fig. 12.31) of about 1.4 cm. This echo ranging, which is quite similar to sonar, determines the round-trip time it takes for a pulse once emitted to be reflected back to the dolphin. *Waves reflect effectively off objects that are at least as large as about one wavelength;* here, the carrier wavelength is small enough to still "see" rather little fish. To produce emissions of that wavelength (since $v = \lambda f = 1.4 \times 10^3$ m/s), the frequency has to be up around 10^5 Hz. These waves are well beyond the human audible range (which is less

An ordinary fairly low resolution sonogram of a left kidney (the white C-shaped form is facing down toward 5 o'clock). The speed of compression waves in the body averages $\approx 1.5 \times 10^3$ m/s, and so frequencies of 1 MHz to 10 MHz have wavelengths of 1.5 mm to 0.15 mm and can reveal fine details.

Figure 12.43 Earthquake compression waves are called primary waves, or P-waves, because they get to seismic stations before the S-waves. They predominate in the disturbances set up by underground nuclear explosions and are used to distinguish between natural quakes and weapons tests.

A Wellington airplane under 210 m of water at the bottom of Loch Ness, revealed by side-scanning sonar.

than 20 kHz) and are therefore called **ultrasonic**. Bats play much the same game in air emitting at frequencies of from 16 kHz to 150 kHz. The lower speed of the waves in air and the higher frequencies yield shorter wavelengths; bats hunt small insects and have to dodge narrow branches.

Example 12.13   While heading for a moth, a horseshoe bat emits a stream of ultrasonic wavetrains having a carrier frequency of 83 kHz, each lasting about 5.0 ms. Given that the speed of the emission is 340 m/s, determine the length in space of a single wavetrain. How many wavelengths of the carrier are there in one such train?

Solution: [Given: the pulse time $t = 5.0$ ms, the carrier frequency $f = 83$ kHz, and $v = 340$ m/s. Find: the pulselength L and the number of waves per pulse.] The pulselength is given by

$$L = vt = (340 \text{ m/s})(5.0 \times 10^{-3} \text{ s}) = \boxed{1.7 \text{ m}}$$

To find the number of carrier waves in such a train, we need first to compute the wavelength of the carrier. Since $v = f\lambda$

$$\lambda = \frac{v}{f} = \frac{340 \text{ m/s}}{83 \times 10^3 \text{ Hz}} = 4.1 \text{ mm}$$

Hence, the number of waves per pulse is

$$\frac{L}{\lambda} = \frac{1.7 \text{ m}}{4.1 \times 10^{-3} \text{ m}} = \boxed{4.2 \times 10^2}$$

▶ **Quick Check:** 4.1 mm is the size of a small bug, so λ is reasonable.

$$v = f\lambda \approx (83\,000 \text{ Hz})(0.004 \text{ m}) \approx 330 \text{ m/s}$$
$$t = 4.2 \times 10^2 \lambda/v = 5 \text{ ms}$$

Some autofocus cameras imitate bats, sending out ultrasonic pulses and using the round-trip time to determine target range. Ultrasonic techniques are routinely used in medicine in both diagnosis and treatment. Intense concentrations of energy can be directed into the body to destroy tumors and stones. In yet another application, low-energy ultrasonic waves are used to probe the body in a way that is widely believed to be far less hazardous than X-rays.

Infrasound corresponds to compression waves of subaudible frequencies (< 20 Hz). A good deal of this energy is present in the environment, especially in the vicinity of airplanes, thunderstorms, fast-moving cars, and overamplified rock bands. Really loud music booming in a relatively small bistro can impart enough compression-wave energy to set the furniture vibrating and create wave patterns on the surface of every glass of beer in the place. It can even oscillate your internal organs, which have low-frequency resonances—that's the "pleasant" buzz you may have felt in the presence of one of these bands. Human chest walls go into resonance at around 8 Hz. If the energy levels are extraordinarily high (especially around 7 Hz), dizziness, nausea, and internal bleeding can occur. It was recently discovered that elephants communicate over long distances with rumbling infrasonic calls. Similarly, the subaudible noise from the cooling systems of old Soviet nuclear submarines is transmitted so well by seawater that U.S. surveillance systems can pinpoint their locations from thousands of miles away.

A high-resolution sonogram of the head of a fetus (facing upward) in the womb. If you can't see the baby's profile, turn the book 90° clockwise.

Core Material

The time it takes for a system to complete a cycle is its **period** (T). The reciprocal of that is the **frequency** (f), represented by

$$f = \frac{1}{T} \qquad [12.1]$$

The SI unit of frequency is the *hertz* (Hz). The **angular frequency** is given by

$$\omega = 2\pi f = \frac{2\pi}{T} \qquad [12.2]$$

Simple harmonic motion (SHM) corresponds to the situation where the periodicity has a single frequency and the motion is simply sinusoidal. In that case, the displacement is

$$x = A \cos \theta = x_0 \cos \omega t = x_0 \cos 2\pi ft \qquad [12.3]$$

The speed of a harmonic oscillator is

$$v_x = -A\omega \sin \omega t = -2\pi f x_0 \sin 2\pi ft \qquad [12.7]$$

Furthermore

$$v_x = - \mp\omega\sqrt{A^2 - x^2} = \mp v_0 \sqrt{1 - \frac{x^2}{A^2}} \qquad [12.8]$$

which is the speed of the oscillator at any position x in terms of its maximum speed. Its acceleration is

$$a_x = -A\omega^2 \cos \omega t = -x_0(2\pi f)^2 \cos 2\pi ft \qquad [12.9]$$

$$a_x = -\omega^2 x \qquad [12.10]$$

The *natural frequency* of a spring carrying a load (m) is

$$f_0 = \frac{1}{2\pi} \sqrt{\frac{k}{m}} \qquad [12.12]$$

and the period is

$$T = 2\pi \sqrt{\frac{m}{k}} \qquad [12.13]$$

while the natural frequency of a *simple pendulum* is

$$f_0 \approx \frac{1}{2\pi} \sqrt{\frac{g}{L}} \qquad [12.16]$$

A progressive wave is a self-sustaining disturbance of a medium. A periodic wave propagates with a speed

$$v = f\lambda \qquad [12.18]$$

The profile of a **harmonic wave** is given by

$$y = A \sin \frac{2\pi}{\lambda} x \qquad [12.19]$$

The speed of a *transverse wave* on a stretched string is

$$v = \sqrt{\frac{F_T}{\mu}} \qquad [12.20]$$

For *compression waves* in fluids

$$v = \sqrt{\frac{B}{\rho}} \qquad [12.22]$$

Suggestions on Problem Solving

1. A system undergoing oscillation is not moving with a constant acceleration and the kinematic equations in Chapter 3 *do not apply!*

2. The basic notions to keep in mind are the definition of the spring constant ($F = ks$), the reference circle, and the fact that $\omega_0 = \sqrt{k/m}$ for a spring, and $\omega_0 = \sqrt{g/L}$ for a pendulum; most everything else can be quickly derived from these ideas, as required. If you remember an equation for ω, there's no need to also store one for $f = \omega/2\pi$ or $T = 1/f$. In general we know that $v = R\omega$, so the maximum speed, $v_0 = v_{max}$, corresponds to the maximum radius $R = A$; hence, $v_0 = A\omega$. Equations (12.3), (12.7), and (12.9) should be gotten directly from the reference circle picture without memorizing them. Physicists don't attempt to commit to memory every equation they encounter—it's much easier to *understand and retain the analysis* than the individual formulas derived from it.

3. Don't get upset by the minus signs in Eqs. (12.7), (12.9), or (12.10)—they only indicate the direction of the vectors. Remember that A is always a positive quantity.

4. Some spring problems (of middle-level difficulty) ask for f or T, thereby requiring that you have k, even though it's not explicitly given. The idea is to first determine k from the initial conditions ($F = ks$).

5. When writing the equation for the displacement, speed, or acceleration of an object in SHM, do not assume that $\varepsilon = 0$. Furthermore, *it may not be the case that A equals the initial displacement;* when the initial speed is nonzero, A is not the initial displacement.

6. The most basic equation in wave theory is $v = f\lambda$. It is likely to be encountered, in one way or another, in the solution of many of the problems. We have equations for the speeds of the different types of waves and, in conjunction with those, $v = f\lambda$ provides a means of determining either f or λ when the other is given.

7. Another fundamental relationship is $T = 1/f$. This is easy enough to remember, although the tendency at first is not to associate it with $v = f\lambda$, but, of course, $v = \lambda/T$.

8. The speed of a transverse wave on a stretched string depends on the *linear* mass-density (mass/length). Be especially careful when μ is given in units of g/m or g/cm, as it often is—convert to kg/m. All the other wave-speed equations containing density refer to the usual notion of mass/volume.

Discussion Questions

1. Draw a simple swinging pendulum and label the locations of the maxima and minima of (a) speed, (b) acceleration, (c) kinetic energy, and (d) potential energy.

2. Consider the behavior of a diving board. Compare the resulting natural frequencies if you first stood on the board at its free end and bounced up and down and then repeated the process at its middle. What would happen to its elastic constant if you shortened the board by chopping off a piece of it?

3. Imagine a hanging lightweight spring with an elastic constant k supporting a small mass m a distance L from the point of suspension. Suppose the support is free to pivot so the spring can swing in a plane. When displaced and released, the mass oscillates up and down as usual, but it soon begins to sway as the oscillation diminishes. In time, the spring oscillations vanish, and the system swings like a pendulum until that motion dies out as the oscillations again build. Explain what's happening.

4. Figure Q4 shows a spring with a mass attached to it resting on a belt. The belt is then set into constant motion counterclockwise. Assuming ordinary dry friction, describe the circumstances under which the mass will go into unstable oscillation, cycling back and forth in a way resembling a sinusoid *growing* in amplitude. Remember that friction increases as the relative speed decreases.

5. Suppose you clamp one end of a hacksaw blade to the edge of a table so it overhangs like a diving board. Now displace the free end a little and let go—twang it. What will happen to the frequency if you stick a lump of clay to the free end? Inasmuch as a tuning fork is essentially a bar bent into a U, the same thing would happen to it.

6. Why is it perfectly reasonable for the period of a pendulum to decrease as g increases? Why is it also reasonable for the period of a pendulum to increase as L increases?

Figure Q4

7. When a person in a parachute descends, vortices are shed alternately up over one side and then the other. This phenomenon periodically changes the pressure on the chute. How might that become dangerous? (Hint: What resonance effect is of concern? Incidentally, some chutes have central holes that are there precisely to break up the vortices.)

8. Figure Q8 shows a horizontal wire supporting several simple pendulums. When the bob on the far left is displaced and

Figure Q8

set oscillating, it twists the wire a little with each swing. What do you think will happen?

9. What is the general relationship between the mass of someone jumping up and down on a trampoline (or a mattress) and the frequency of the oscillation?

10. Imagine an antiroll device for a ship: a mass on horizontal springs or a tank of water tuned to resonate at the natural rolling frequency of the ship. If waves begin to rock the ship near its resonant frequency, a very dangerous situation will develop. How does the device avert catastrophe?

11. The mythical planet Mongo is a homogeneous, nonrotating sphere whose inhabitants have drilled a hole running through the center from one side to the other. Show that a ball dropped into the hole would execute SHM.

12. Imagine a chisel held up to a piece of marble and then struck on its head with a mallet. Describe what happens in light of this chapter. Consider the question on the atomic level.

13. Imagine two symmetrical pulses that are identical in every way, except that one is inverted. These pulses are moving toward each other on a rope in opposite directions. What happens to the energy of the rope at the instant of complete cancellation?

14. In the real world, a pulse sent down a very long, taut rope diminishes in amplitude and ultimately vanishes. Describe what's happening in terms of Conservation of Energy.

15. Imagine a long string of beads hung vertically from a high ceiling with a weight fixed to its lower end. The bottom is then rapidly displaced, generating a transverse pulse. Describe the motion of the disturbance, paying particular attention to how it might differ if the string were held taut horizontally.

16. Picture a long row of standing dominos, one a little behind the other so that knocking over the first will knock over the second and so on. Now imagine an idealized tube an inch or two in diameter extending from New York to California. Suppose this tube is filled with greased, essentially frictionless marbles and you push one more in at the East Coast end only to have one pop out at the West Coast end. What's happening, and how do these arrangements of dominos and marbles relate to this chapter?

17. The accompanying list is a brief accounting of the speeds of compression waves in various materials. Considering the physical characteristics of each of the materials, discuss the relative speeds and explain why they are reasonable in light of your knowledge of the process of wave propagation.

Material	Speed (km/s)
Clay	1
Sandstone	2
Limestone	4
Granite	5
Salt	6

18. Extracorporeal shock wave lithotripsy (ESWL) is a technique for pulverizing kidney stones within the body. The patient is placed in a bath of water. An intense electric spark within the tank below the patient rapidly vaporizes a small amount of water, producing a shock wave that is reflected from a curved mirror that focuses it onto the stone. After a thousand or two blasts, a typical stone is destroyed. How is the shock wave formed and what kind of wave is it? What is it used for in a physical sense? Why isn't the skin destroyed as well as the stone?

19. Figure Q19 shows two different pulses heading toward each other. Draw several pictures in sequence, illustrating the result as they superimpose and separate.

Figure Q19

Multiple Choice Questions

1. The end of a whip antenna is oscillating in SHM. Its maximum acceleration is given by the expression
   (a) $2\pi f_0^2 A$ (b) $4\pi^2 f_0^2 A$ (c) $4\pi^2 f_0^2 A^2$ (d) $4\pi f_0 A^2$
   (e) none of these.

2. Two coiled springs are made of the same wire such that they are identical in all respects, except the first is half the length of the second. Compared to the natural frequency of the first (f), the natural frequency of the second is (a) $2f$
   (b) $\frac{1}{2}f$ (c) f (d) $f/\sqrt{2}$ (e) none of these.

3. The free end of a clamped saw blade vibrates 12.8 times in 19 s. Its frequency is (a) 0.67 Hz (b) 12.8 Hz
   (c) 19 Hz (d) 1.48 Hz (e) none of these.

4. In SHM, the acceleration _____ the velocity by _____.
   (a) leads, $\frac{1}{2}\pi$ (b) lags, $\frac{1}{2}\pi$ (c) leads, $\frac{1}{4}\pi$ (d) lags, $\frac{1}{4}\pi$
   (e) none of these.

5. An adult and a child are sitting on adjacent identical swings. Once they get moving, the adult, by comparison to the child, will necessarily swing with (a) a much greater period (b) a much greater frequency (c) the same period (d) the same amplitude (e) none of these.

6. A pendulum clock is set to run accurately at sea level. It is then brought to the top of a high mountain, where it is found to (a) function unchanged (b) run slow (c) run fast (d) stop running (e) none of these.

7. A tuning fork is struck and so set vibrating. Its handle is then touched to the handle of an identical fork held near the ear and this second one immediately begins to vibrate at the same frequency. This transfer of energy is an example of (a) contact potential (b) critical damping (c) resonance (d) radiation (e) none of these.

8. Sometimes a really heavy truck will go by and set the windows of the house vibrating for a moment with a low-frequency buzz. They oscillate (a) in-phase with the truck due to critical damping (b) at one of their natural

frequencies   (c) at a random changing frequency   (d) in an undamped mode   (e) none of these.

9. An oscillator at a frequency of 1.25 Hz will make 100 vibrations in   (a) 125 s   (b) 12.5 s   (c) 8.0 s   (d) 80 s   (e) none of these.

10. A pendulum clock in a free-falling elevator   (a) runs normally   (b) runs a little fast   (c) runs a little slow   (d) runs very fast   (e) none of these.

11. Astronauts in space took a light, coiled spring of known elastic spring constant, attached a mass to it, and set it oscillating. Measuring the period, they could determine   (a) the time of day   (b) the acceleration due to gravity   (c) the mass of the bob   (d) the weight of the bob   (e) none of these.

12. The tension in the wire of a simple pendulum is at maximum when the bob is at   (a) its highest point   (b) its lowest point   (c) halfway down   (d) any location since the tension is actually constant   (e) none of these.

13. A swell of ocean waves of wavelength 1.0 m and frequency 1.25 Hz has a speed of   (a) 1.25 m/s   (b) 0.8 m/s   (c) 125 m/s   (d) 8.0 m/s   (e) none of these.

14. The speed of a wave on a stretched string _____ when the tension in it is doubled.   (a) doubles   (b) increases by a factor of 4   (c) increases by a factor or 1.414   (d) decreases by a factor of 2   (e) none of these.

15. Tie a little piece of paper to the middle of a long horizontal taut string. Now send a transverse pulse down the string. The piece of paper will suddenly rise upward as the wave passes, proving that the disturbance transports   (a) mass   (b) weight   (c) density   (d) momentum   (e) none of these.

16. Two strings of the same length are being used in a wave experiment. The first, which has twice the mass of the second, is stretched twice as taut as the second. The speed of the first's wave is   (a) twice that of the second   (b) the same as the second   (c) greater than the second by $\sqrt{2}$ (d) less than the second by $\sqrt{2}$ (e) none of these.

17. A periodic wave passes an observer, who records that there is a time of 0.5 s between crests.   (a) the frequency is 0.5 Hz   (b) the speed of 0.5 m/s   (c) the wavelength is 0.5 m   (d) the period is 0.5 s   (e) none of these.

18. A string of a certain linear mass-density is attached end-to-end to a rope whose linear mass-density is 9 times greater. The combination is stretched taut, and a pulse is sent down the string with a speed v. Its speed in the rope is   (a) $9v$ (b) v (c) $3v$ (d) $v/9$ (e) none of these.

19. For a harmonic wave of a certain type in a given medium, doubling the frequency has the effect of   (a) halving the speed   (b) halving the wavelength   (c) doubling the amplitude   (d) doubling the period   (e) none of these.

20. A stretched rope is dipped in water and thereby made wet. Without changing the tension, a pulse is then sent down its length. Compared to when the rope was dry, we can now expect the speed of the pulse to have   (a) decreased   (b) increased slightly   (c) remained the same   (d) doubled   (e) none of these.

21. Quite generally, doubling the amplitude of a wave   (a) doubles the frequency   (b) halves the period   (c) quadruples the energy   (d) doubles the speed   (e) none of these.

Problems

HARMONIC MOTION

1. [I] What is the period of a phonograph record that turns through $33\frac{1}{3}$ rotations per minute?

2. [I] The respiratory system of a medium-sized dog resonates at roughly 5 Hz so that it can pant (in order to cool off) very efficiently at that frequency. How many breaths will the dog be taking in a minute? For comparison, the dog would ordinarily breathe at around 30 breaths per minute.

3. [I] A record is turning uniformly at 78 rpm while an ant sitting at rest on its rim is being viewed by a child whose eyes are in the plane of the record. Describe the ant's motion as seen by the child. What is the frequency of the ant? What is its angular frequency?

4. [I] A hovering fair-sized insect rises a little during the downstroke of its wings and essentially free-falls slightly during the upstroke. The end result is that the creature oscillates in midair. If it typically falls about 0.20 mm per cycle, what is the wingbeat period and frequency? Compare that to the ≈ 10 Hz oscillation of a butterfly, which cannot hover.

5. [I] A body oscillates in SHM according to the equation

$$x = 5.0 \cos (0.40t + 0.10)$$

where each term is in SI units. What is (a) the amplitude,

(b) the frequency, and   (c) the initial phase?   (d) What is the displacement at $t = 2.0$ s?

6. [I] A body oscillates in SHM according to the equation

$$x = 8.0 \cos (1.2t + 0.4)$$

where each term is in SI units. What is the period?

7. [I] Show that a value of $\varepsilon = +\frac{3}{2}\pi$ is equivalent to a value of $\varepsilon = -\frac{1}{2}\pi$ in Eq. (12.5). Draw a sketch of the functions.

8. [I] A point on the very end of one-half of a tuning fork vibrates approximately as if it were in SHM with an amplitude of 0.50 mm. If the point returns to its equilibrium position with a speed of 1.57 m/s, find the frequency of vibration.

9. [I] A point at the end of a spoon whose handle is clenched between someone's teeth vibrates in SHM at 50 Hz with an amplitude of 0.50 cm. Determine its acceleration at the extremes of each swing.

10. [I] A light hacksaw blade is clamped horizontally, and a 5.0-g wad of clay is stuck to its free end, 20 cm from the clamp. The clay vibrates with an amplitude of 2.0 cm at 10 Hz.   (a) Find the speed of the clay as it passes through the equilibrium position.   (b) Determine its acceleration at maximum displacement and compare that with g.

11. [I] A small mass is vibrating in SHM along a straight line. Its acceleration is 0.40 m/s^2 when at a point 20 cm from the zero of equilibrium. Determine its period of oscillation.

12. [I] A 1.0-N bird descends onto a branch that bends and goes into SHM with a period of 0.50 s. Determine the effective elastic force constant for the branch.

13. [I] Suppose Eq. (12.5) is to be applied to the mass in Fig. 12.14, where the cycle is to begin at $t = 0$ with m released with zero speed at -3 m. What must ε equal?

14. [I] Suppose Eq. (12.5) is to be applied to the mass in Fig. 12.14, where the timing clock is started arbitrarily and it is found that at $t = \frac{1}{4}T$, the mass is at $x = +\frac{1}{2}A$. Find ε.

15. [I] A particle is oscillating with SHM along the z-axis with an amplitude of 0.50 m and a frequency of 0.20 Hz. If it is at $z = +0.50$ m at $t = 0$, where will it be at $t = 5.00$ s, $t = 2.50$ s, and $t = 1.25$ s?

16. [I] A spot of light on a computer screen oscillates horizontally in SHM along a line 20 cm long at 50 Hz. The spot reaches the center of the line at $t = T/8$. Show that $v_x = -10\pi \sin(100\pi t + \frac{1}{4}\pi)$ is the expression for the speed of the spot as a function of time.

17. [I] A bug having a mass of 0.20 g falls into a spider's web, setting it into vibration with a dominant frequency of 18 Hz. Find the corresponding elastic spring constant.

18. [I] A 5.0-kg mass resting on a frictionless air-table is attached to a spring with an elastic constant of 50 N/m. If this mass is displaced 10 cm (compressing the spring) and is then released, find its maximum speed.

19. [I] Two kilograms of potatoes are put on a scale that is displaced 2.50 cm as a result. What is the elastic spring constant? If the scale is pushed down a little and allowed to oscillate, what will be the frequency of the motion?

20. [I] A 250-g mass is attached to a light helical spring with an elastic force constant of 1000 N/m. It is then set into SHM with an amplitude of 20 cm. Determine the total energy of the system.

21. [I] Consider the frictionless cart in Fig. 12.16. If the elastic spring constants are k_1 and k_2, respectively, determine the frequency of vibration in terms of these quantities.

22. [I] What is the speed of the mass in Problem 18 as it reaches a point 5.0 cm from its undisplaced position?

23. [I] Find the frequency of a simple pendulum of length 10.0 m.

24. [I] A small lead ball is attached to a light string so that` the length from the center of the ball to the point of suspension is 1.00 m. Determine the natural period of the ball swinging through a small displacement in a vertical plane.

25. [I] How long must a simple pendulum be if it is to have a period of 10.0 s?

26. [I] One might guess that the period of a pendulum is proportional to the mass m, the length L, and the acceleration due to gravity g. Consequently, assume that

$$T = Cm^aL^bg^c$$

where C is a unitless constant. Now solve for a, b, and c using the fact that the dimensions on both sides of the equation must be the same. Of course, if this calculation produces nonsense, C wasn't unitless. Although you cannot get it from this analysis, what is the correct value of C?

27. [II] Write general expressions for both the initial displacement x_i and speed v_i of a point moving in SHM at $t = 0$. Show that

$$\tan \varepsilon = \frac{-v_i}{\omega_0 x_i}$$

What does this imply about ε if the initial speed is zero? Show that

$$A^2 = x_i^2 + \frac{v_i^2}{\omega_0^2}$$

Notice that because there is an initial speed imparted to the system at $t = 0$, $A \neq x_i$.

28. [II] A light 3.0-m-long helical spring hangs vertically from a tall stand. A mass of 1.00 kg is then suspended from the bottom of the spring, which lengthens an additional 50 cm before coming to a new equilibrium configuration well within its elastic limit. The bob is now pushed up 10 cm and released. Write an equation describing the displacement y as a function of time.

29. [II] An essentially weightless helical spring hangs vertically. When a mass m is suspended from it, it elongates an amount ΔL. When let loose, it oscillates with a measured period of T. Show how you might use this arrangement to determine g.

30. [II] A 200-g mass hung on the bottom of a light helical spring stretches it 10 cm. This mass is then replaced with a new 500-g bob that is set into SHM. Compute its period.

31. [II] An object is hung from a light vertical helical spring that subsequently stretches 2.0 cm. The body is then displaced and set into SHM. Compute the frequency at which it oscillates.

32. [II] A 5.0-kg block of wood is floating in water. It is found that a downward thrust of 10.0 N submerges the block an additional 10 cm, at which point it is let loose to oscillate up and down. Determine the frequency of the oscillation. Assume SHM.

33. [II] A uniform block with sides of length a, b, and c floats partially submerged in water. It is pushed down a little and let loose to oscillate. Given that the vertical edge has length b and the density of the block is ρ, show that the motion is SHM and determine its period. Check the units of your result.

34. [II] Derive an expression for the period of oscillation of the frictionless system shown in Fig. P34. Remember that the displacement of the mass m is the sum of the displacements of the two springs.

Figure P34

35. [II] A 5.0-g bullet is fired horizontally into a 0.50-kg block of wood resting on a frictionless table. The block, which is attached to a horizontal spring, retains the bullet and moves forward, compressing the spring. The block-spring system goes into SHM with a frequency of 9.0 Hz and an amplitude of 15 cm. Determine the initial speed of the bullet.

36. [II] Consider a simple pendulum of length L. Show that it gains an amount of potential energy

$$\Delta PE_G \approx \tfrac{1}{2} mgL\theta^2$$

when the bob is raised through an angle θ. (Hint: You may need to know that $1 - \cos\theta = 2\sin^2\tfrac{1}{2}\theta$.)

37. [II] If the period of a simple pendulum is T, what will its new period be if its length is increased by 50%?

38. [II] Extending the ideas of Problem 36, show that the pendulum's maximum angular speed is given by $\omega_0 = \theta_0\sqrt{g/L}$.

39. [II] Using energy considerations, derive an expression for the period of oscillation of a mass m. It is fixed to a horizontal spring having an elastic constant k, and is free to move on a frictionless table.

40. [III] A vibration platform oscillates up and down with an amplitude of 10 cm at a controlled variable frequency. Suppose a small rock of mass m is placed on the platform. At what frequency will the rock just begin to leave the surface so that it starts to clatter?

41. [III] Use Fig. P41 to prove that the reciprocating motion of the piston is oscillatory but that it is not SHM.

Figure P41

42. [III] A small mass is attached to the end of a long vertical string. It is displaced through an angle θ and then swung so that the mass moves in a horizontal circle. Show that if you looked at the resulting conical pendulum in the plane of motion, you would see the bob moving back and forth with a period given by

$$T = 2\pi\sqrt{\frac{L\cos\theta}{g}}$$

MECHANICAL WAVES

43. [I] Figure P43 shows two views—(a) optical and (b) acoustical—of a computer circuit. The acoustic microscope has the highly desirable ability to see several micrometers below the surface. The device focuses ultrasonic waves at a frequency of 3 GHz through a droplet of water onto the object. The speed of the waves in the water, which couples the microscope to the specimen, is 1.5 km/s. Compute the wavelength of the radiation and compare it to the wavelength of green light in air ≈ 500 nm.

44. [I] A long metal rod is struck by a vibrating hammer in

(a)

(b)

Figure P43

such a way that a compression wave with a wavelength of 4.3 m travels down its length at a speed of 3.5 km/s. What was the frequency of the vibration?

45. [I] An A note of 440 Hz is played on a violin submerged in a swimming pool at the wedding of two scuba divers. Given that the speed of compression waves in pure water is 1498 m/s, what is the wavelength of that tone?

46. [I] A wavepulse on a string travels the 10-m length in 2.0 s. A harmonic disturbance of wavelength 0.50 m is then generated on the string. What is its frequency?

47. [I] Show that, for a periodic wave, $\omega = (2\pi/\lambda)v$.

48. [I] A transverse harmonic wave on a beaded string has an amplitude of 2.5 cm and a wavelength of 160 cm. If at $t = 0$ a life-sized photo shows that the height of the wave at $x = 0$ is zero and at $x = 40$ cm it is $+2.5$ cm, then what is the string's displacement at $x = 10$ cm?

49. [I] A member of the Vespertilionidea family of bats typically emits a sequence of chirps, wavetrains lasting about 3.0 ms and having a carrier frequency that varies from 100 kHz to about 30 kHz. Generally, there is a time between individual chirps of 70 ms. Assuming the air speed of one of these wavetrains is 330 m/s, how far away can an object be and still be detected without being masked by the next outgoing chirp?

50. [I] A radar beam transmitted with a carrier frequency of 9.8 GHz is composed of a stream of nearly harmonic wavetrains each lasting 1.0 μs. Determine the number of wavelengths of the carrier that are in each pulse and show that this number is independent of v, provided the speed of the pulse equals the speed of the carrier wave.

51. [I] The A string of a violin has a linear mass-density of

0.59×10^{-3} kg/m and is stretched to a tension of 10 N. What is the speed of the transverse waves it can sustain? Incidentally, the four violin strings, when mounted and properly tuned, produce a net tension on the instrument of about 220 N (50 lb).

52. [I] A heavy rope is tied to a tree 30 m away, and it is then pulled taut with a force of 360 N. The rope is struck near its end with a stick, thus setting up a transverse wavepulse. If the rope has a linear mass-density of 0.06 kg/m, how long will it take for the pulse to make the round trip from end to tree to end?

53. [I] If the speed of a wavepulse set up on the longest string of a grand piano is 129 m/s and if the linear mass-density of the string is 64.9×10^{-3} kg/m, what is the tension? Typically, the net tension on the frame of a concert grand due to all the strings is around 2.7×10^5 N or 30 tons (60 000 lb).

54. [I] What is the speed of a compression wave in ethyl alcohol?

55. [I] Figure P55 shows a rather idealized square water wave. Show that doubling its amplitude would increase its energy by a factor of 4, thus again making the point that the energy of a wave is proportional to its amplitude squared.

Figure P55

56. [I] Figure P56 depicts two identical symmetrical transverse wavepulses, each traveling at a speed of 5.0 m/s heading toward one another. Draw the resultant disturbance 2.0 s later and again 2.0 s after that.

Figure P56

57. [I] Figure P57 depicts two identical symmetrical transverse wavepulses, each traveling at a speed of 5.0 m/s heading

Figure P57

toward one another. Draw the resultant disturbance 2.0 s later and again 2.0 s after that.

58. [II] Write an expression for the profile of a transverse harmonic wave of amplitude 10 cm and wavelength 1.2 m traveling along a rope in the positive z-direction and oscillating in the yz-plane. The height of the wave is zero at $t = 0$ and $z = 0$.

59. [II] Show mathematically that shifting the phase of a sinusoidal wave profile by $\frac{1}{2}\pi$ rad turns it into a cosine profile. Begin with

$$y = A \sin \left(\frac{2\pi}{\lambda} x \mp \tfrac{1}{2}\pi \right)$$

60. [II] The profile of a transverse harmonic wave on a long beaded string is described in SI units by the function

$$y = 0.020 \sin (6.28x)$$

Given that the wave has a speed of 5.0 m/s, determine the maximum *transverse* speed of any bead on the string.

61. [II] The profile of a transverse harmonic wave on a long taut nylon thread is described in SI units by the function

$$y = 0.040 \sin 2\pi x$$

Given that the wave travels at a speed of 2.0 m/s, determine the maximum *transverse* acceleration of any point on the thread.

62. [II] Use the method of dimensional analysis to "derive" an expression for the speed of a compression wave in a solid. We guess that the answer depends on Young's Modulus and the density, and write

$$v = KY^a \rho^b$$

where a and b are to be determined and K is a dimensionless constant (like 2π). Finish out the calculation by balancing the units.

63. [II] A string has a linear mass-density of 2.50 g/m and is put under a tension of 25.0 N as it is stretched taut along the z-axis. The free end is attached to a tuning fork that vibrates at 50.0 Hz, setting up a transverse wave on the string having an amplitude of 5.00 mm. Determine the speed, angular frequency, period, and wavelength of the disturbance.

64. [II] Using Problem 63, write an expression for the profile of the wave given that, at $t = 0$, the end of the rope ($x = 0$) is at $y = +5.00$ mm. The wave travels in the z-direction.

65. [II] A heavy nylon guitar string with a linear mass-density of 7.5 g/m is stretched to a tension of 80 N. What is the speed of the transverse wave that can be generated on the string?

66. [II] The speed of a transverse wave on a wire is 90 m/s when the tension in the wire is 120 N. What will be the speed if the tension is reduced to 80 N?

67. [II] A homemade telegraph system sends transverse pulses along a stretched string. It operates between two neighboring houses using 12 m of a string having a total weight of 0.20 N. What should be the tension in the string if the

signals are to travel at least as fast as they would were the users simply to yell? Take the speed of sound in air to be 333 m/s.

68. [II] Determine the profile of the wave function

$$y = A \sin 2\pi \left(\frac{x}{\lambda} - \frac{t}{T} \right)$$

at $t = 0$. Find the values of y at $x = 0$, $\lambda/8$, $\lambda/4$, $3\lambda/8$, $\lambda/2$, $5\lambda/8$, $3\lambda/4$, $7\lambda/8$, and λ. Make a plot of the profile. Now do the same thing for the wave function

$$y' = A \sin 2\pi \left(\frac{t}{T} - \frac{x}{\lambda} \right)$$

and compare your results.

69. [III] Given the equation

$$y = A \sin 2\pi \left(\frac{t}{0.01} - \frac{x}{40} \right)$$

describing the displacement of a string of beads in SI units, write an expression for the motion of a bead at $x = 5.0$ m as a function of time. What is the value of y at $t = 0$ and $t = 0.01$ s? When is $y = 0$? Graph your results.

70. [III] A transverse harmonic wave on a long beaded string is described in SI units by the function

$$y = 0.02 \sin (6.28x - 15t)$$

If we have a detector at $x = 0$, what will be the speed of the bead at that location at a time $t = 1.2$ s?

71. [III] A uniform rope of length L and mass m hangs freely straight down from a hook in the ceiling. Taking the bottommost point as $y = 0$, write an expression for the mass of a segment of arbitrary length Δy. What is the tension in the rope at a height Δy? Derive an expression for the speed of a transverse wave set up by wiggling the lower end. What is the maximum speed? Determine an expression for the speed of the wave if a mass M is hung on the bottom of the rope.

72. [III] The very end of a wire 14.5 m long having a mass of 0.12 kg is clamped in a vise. The other end is tugged on, putting a tension of 50 N in the wire. That end is then driven transversely by a tuning fork oscillating at 440 Hz, and the wire vibrates with a maximum displacement of 0.15 mm.   (a) Find the wavelength and period of the resulting disturbance.   (b) Determine the maximum transverse speed of any point on the wire.

13

Sound

THE WORD **SOUND** HAS two meanings of interest to us. There is the psychological and physiological concept of sound as that which humans hear—the perception produced by the ear-brain detector. And there is the physical concept of sound as **any compression wave in a material medium that has frequency content in the range from 20 Hz to 20 kHz**. Although this chapter concentrates on sound as a longitudinal elastic wave, the primary ideas are grounded in our common human experience and, consequently, we also examine how the ear works, as well as the significance of the physiological-psychological notions of pitch, loudness, intensity-level, and timbre.

The specific *frequency bandwidth* from 20 Hz to 20 kHz is the spectrum heard by the so-called "average" person, who is a statistical fiction. Anyone in our society who can hear across that entire range is actually something of a rarity. Even so, the frequency response of the human ear-brain system diminishes with age beyond about 20 years. Moreover, the exposure to loud sounds, as is common these days, desensitizes the ear at both frequency extremes.

ACOUSTICS

The insight that sound is a wave phenomenon is remarkably ancient. It seems to have been suggested initially by the Roman architect Marcus Vitruvius Pollio, who lived about 2000 years ago. While considering the design of amphitheaters, where it's crucial to control echoing, he likened sound to water ripples moving through space "upward, wave after wave." In a sense, this wonderful image marked the beginning of the study of the physics of sound—that is, *acoustics* (from the Greek *akoustikos,* relating to hearing). The recognition that a sound wave is longitudinal follows from the fact that it propagates in air, which, like all fluids, has no stiffness, cannot resist shear, and cannot sustain a transverse wave. Because a wave does not transport the medium, its speed can be extraordinarily large as is clearly the case with sound—a bolt of lightning a mile away produces a clap of thunder heard only a moment (≈ 5 s) later. For millennia, sound was recognized as the most rapidly propagating phenomenon (exclusive of light, which many thought was instantaneous).

Sound propagates in any medium that can respond elastically and thereby transmit vibrational energy. Air at ordinary density is certainly springy (squeeze an empty plastic soda bottle capped and uncapped, and the point will be made). The fact that the region connecting the source and detector must contain an adequate amount of matter to sustain the disturbance was established experimentally (1672) by von Guericke. He produced a partial vacuum within a chamber containing an alarm-clock mechanism. As the air was removed, the sound became weaker until it vanished altogether. Only after air was readmitted could the ringing bell again be heard. Sound does not propagate in vacuum. All the old science fiction movies that depicted roaring explosions in space have missed the point.

13.1 Sound Waves

Consider a loudspeaker being driven by a sinusoidal signal, a *pure tone.* The flexible cone vibrates harmonically, pumping the air in front of it into a series of condensations and rarefactions (Fig. 13.1). The resulting pressure variations that constitute the sound wave are really quite small. A very loud sound corresponds to a pressure change of less than about 10 Pa (10^{-5} atm), and quiet music exists at a gauge pressure of $< \pm 0.002$ Pa. The speaker cone usually moves back and forth a fraction of a centimeter, and any given air molecule moves an equally small amount (Fig. 12.41). That distance determines the amplitude of the wave and, therefore, its energy and the loudness of the sound. In the same way, a large drum head moves through a relatively large distance and makes a large sound. The time it takes the speaker cone to pass through a cycle fixes the period and frequency of the sound. The distance in air between successive condensations (or rarefactions) is the wavelength (Table 13.1). The wavelength, too, is determined by how fast the cone moves through a cycle. Energy streams into the air all the while the cone is in motion. A slow progression, even though it is through a small cone displacement (< 1 cm), can launch a wave with a λ of as much as ≈ 17 m at 20 Hz.

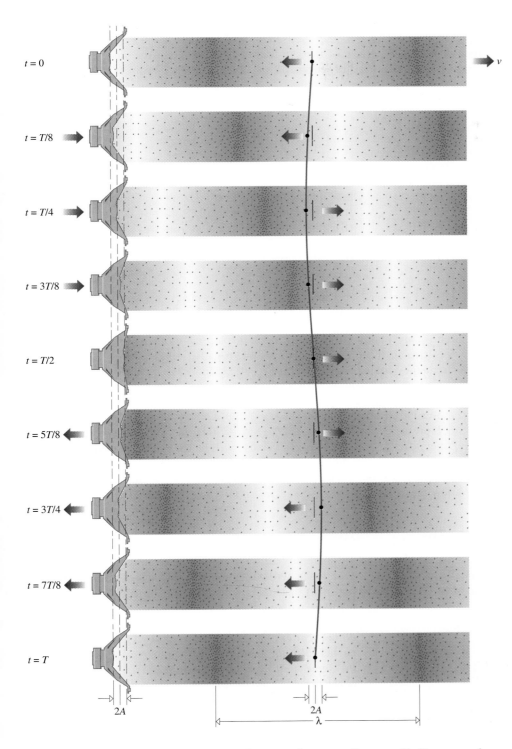

Figure 13.1 A sound wave coming from a loudspeaker. Note that the speaker cone sweeps through a distance of 2A. Each atom oscillates through a distance of 2A as well. Compare that to the wavelength λ.

This same kind of thing happens in more dense media as well. For example, punch a hole in the bottom of a topless tin can, pass a string through it, and make a large knot at the end so it will not come out the hole. Then do the same thing several meters away at the other end of the string with a second can (bottoms toward each other). With one can at your mouth and the other at your listener's ear, pull the string tight and talk into the open end—you've made a kid's "telephone." Your voice vibrates the bottom of the can, pumping energy into the string and sending compression waves down its length, ultimately setting the bottom of the receiver can vibrating just like the speaker cone.

Example 13.1 By international agreement most orchestras tune to a frequency of 440 Hz, which is called A440 (the A note above middle C). Given that the speed of sound in air at room temperature is 343.9 m/s, what is the wavelength of A440?

Solution: [Given: $f = 440$ Hz and $v = 343.9$ m/s. Find: λ.] The problem deals with frequencies and wavelengths, and that should always bring to mind

Eq. (12.18), $v = f\lambda$; hence

$$\lambda = \frac{v}{f} = \frac{343.9 \text{ m/s}}{440 \text{ Hz}} = \boxed{0.782 \text{ m}}$$

▶ **Quick Check:** For 100 Hz, $\lambda = 3.4$ m, which is 4.4 times smaller; hence, 4.4(0.78 m) = 3.4 m. It's a good thing to remember that sound waves extend from ≈20 mm to ≈20 m. Thus, we know 0.78 m is at least in the right range.

Unlike a wave on a rope, compression waves are a nuisance to picture directly. Instead, we could draw a curve with an axis along the propagation direction, whose magnitude at each point corresponds to the displacement of the particles of the medium (positive in the propagation direction, negative opposite to it). Thus, for a single-frequency compression wave, the resulting curve is a sinusoid. This representation is helpful when dealing with wind instruments, and we will use it later. Alternatively, we could plot either the density or the gauge pressure within the medium and these would also be sinusoidal (Fig. 13.2).

Since pressure can be measured directly with a microphone and readily displayed on an oscilloscope, plotting P_G versus t will be the preferred approach. As an example, Fig. 13.3 shows the sound generated by a tuning fork. By contrast, a struck wooden stick (Fig. 13.4) will produce a *transient* sound, a short-lived wavetrain only a few meters long.

13.2 The Superposition of Waves

A fascinating characteristic shared by all waves is that two or more of them moving through the same region of space will superimpose and produce a well-defined combined effect. Waves maintain their integrity upon overlapping; they cohabit the medium without themselves being permanently changed. They do not interact with each other in the sense that they do not scatter one another as would two crossing beams of particles. *In the region where two or more disturbances of the same kind overlap, the resultant is the algebraic sum of the various contributions at each point.* This is called the **Superposition Principle** and it was introduced by Thomas Young. As we will come to see, it's among the most far-reaching and fundamental insights in wave theory.

Figure 13.5 shows the effect of superimposing two harmonic waves of the same frequency and amplitude. At every value of x, we simply add the corresponding heights of the two sine curves, taking measures above the axis as

TABLE 13.1 Wavelengths in air at 20°C			
Source	**Frequency**	**Wavelength**	
	(Hz)	**(m)**	**(ft)**
Organ, lowest note	16	21	70
Piano, lowest note	28	12	40
Bass voice, low C	65	5.3	17
Middle C	262	1.3	4.3
Alto voice, high F	698	0.49	1.6
Soprano voice, high C	1047	0.33	1.1
Piano, highest note	4186	0.08	0.27

(a)

(b)

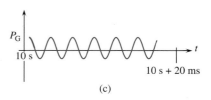

(c)

Figure 13.2 A sinusoidal sound wave. Where the pressure is maximum, above and below atmospheric, the displacement is zero. When the pressure is positive, as it is in condensations, atoms are shifted in the direction of motion of the wave.

Figure 13.3 Oscilloscope traces corresponding to gauge pressure produced by a tuning fork: (a) after the fork is initially struck; (b) after oscillating about 1 s; (c) after oscillating about 10 s. As time goes on, the high-frequency, short-wavelength components vanish, leaving a smooth, nearly sinusoidal wave. Precision tuning forks can maintain their frequency with an accuracy of 1 part in 100 000.

positive and below it as negative. Interestingly, *the sum of any number of harmonic waves of the same frequency traveling in the same direction is also a harmonic wave of that frequency,* and that's true independent of the amplitudes. When sine waves with different frequencies are added to one another, the composite disturbance is not harmonic (Figs. 13.6 and 13.7).

Fourier Analysis

A beautiful mathematical technique for synthesizing waveforms was devised in 1807 by the French physicist Jean Baptiste Joseph, Baron de Fourier. He proved that *a periodic function having a wavelength λ can be synthesized by a sum of harmonic functions whose wavelengths are integral submultiples of λ* (that is, λ, $\lambda/2$, $\lambda/4$, and so forth). In fact, any wave profile encountered in nature—pulse or periodic—can be envisioned as the result of overlapping harmonic functions. Any waveform can be

Overlapping water waves pass each other and sail away unruffled.

Figure 13.4 Gauge pressure produced by striking a wooden stick with a metal rod, as shown on an oscilloscope. The short-lived vibration dies out in about 10 ms.

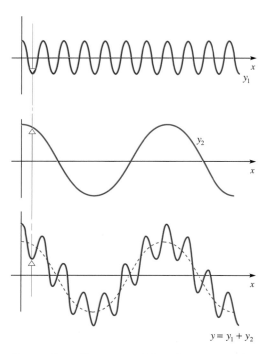

Figure 13.5 The superposition of waves y_1 and y_2, which combine to yield y. The two component waves have the same amplitude and differ in phase by ε, which corresponds to a shift in x of ε/k. In (a) the two are nearly in-phase, and the resultant is large. In (b) they are nearly out-of-phase by 180°, and the resultant is small. In (c) they are out-of-phase by exactly 180°, and the two waves cancel completely.

Figure 13.6 The superposition of waves y_1 and y_2. At each value of x, we add the magnitude of y_1 to that of y_2. Values above the axis are positive, those below are negative. The component waveforms have different frequencies, and the resultant is therefore not sinusoidal.

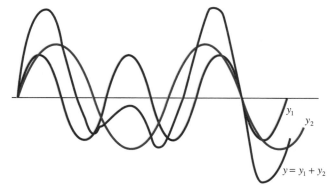

Figure 13.7 The superposition of two harmonic waves of different frequency. The resultant $y = y_1 + y_2$ is periodic, but it certainly is not harmonic. Of course, to be harmonic a disturbance must have only one frequency, and y doesn't.

decomposed into harmonic components, and this can be accomplished with sound by the human ear (or a spectrum analyzer) and with light by a prism.

According to Fourier, a periodic function of frequency f can be created out of an appropriate sum of sinusoidal terms, the first of which will also have a frequency f. This is the so-called **fundamental** or *first harmonic* ($f_1 = f$). The next term has a frequency of $f_2 = 2f$ and is called the *second harmonic,* and the next has a frequency of $f_3 = 3f$ and is called the *third harmonic,* and so on. Notice that in the time the fundamental takes to complete one cycle ($1/f_1$), every higher harmonic will complete a whole number of cycles, and the entire set will be back where it started. Thus, the frequency of the combined waveform is the frequency of the fundamental, which was made equal to f, the frequency of the periodic wave being analyzed (Fig. 13.8). If the several pure tones in Fig. 13.9a are combined, they produce the acoustic wave in Fig. 13.9b. Electronically synthesized music and talking computer chips do exactly this sort of Fourier addition (Fig. 13.10).

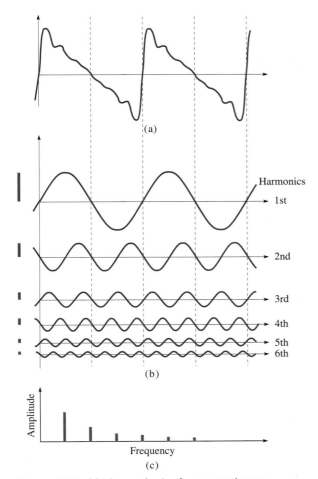

Figure 13.8 (a) The synthesis of a sawtooth wave. (b) Here, the first 6 harmonics combine to form a curve that clearly resembles a sawtooth. Another dozen terms will make it still better. (c) A plot of the amplitudes of the contributing harmonic terms plotted against frequency. This is called the *frequency spectrum.*

Figure 13.9 (a) Six sinusoidal wave-forms. (b) The result of combining these waveforms. (c) The associated spectrum. Each vertical line represents the amplitude of the corresponding harmonic in (a). Note that $f_2 = 2f_1$, $f_3 = 3f_1$, $f_4 = 4f_1$, and so forth.

13.3 Wavefronts and Intensity

In two dimensions, the harmonic disturbance spreading out on the surface of a pool of water from a rhythmically tapping toe (Fig. 13.11) forms a pattern of expanding circular ripples. The lines connecting all the points where the water is rising or falling in-step form a series of concentric circles. At every point on any one of these circles, the ripple traveling out radially is at the same stage, or phase, of its essentially sinusoidal cycle, whether that's a crest or a trough or anything between. *The phase of the disturbance is constant at every point on a given circle.* Each curve connecting all of the neighboring points of a wave that are in-step is known as a **wavefront**.

Sound waves are three-dimensional. Imagine a *point source* radiating waves uniformly in all directions, as in Fig. 13.12. Here, **the surfaces of constant phase**, the wavefronts, are perfectly spherical (yet another mathematical idealization that does not actually exist, although it is often approximated). If we examine the com-

Figure 13.10 Each button on a Touchtone telephone produces a pair of nearly pure, widely separated tones. Push the 7-button, and you can hear the superposition of 852 Hz and 1209 Hz.

Figure 13.11 A toe wiggling in the water generates circular surface waves.

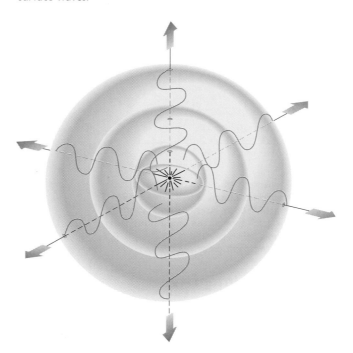

Figure 13.12 A wave from a point source spreads out uniformly in all directions. Energy flows outward in the form of a spherical wave. As the area $(4\pi R^2)$ gets larger, the concentration of energy diminishes and the amplitude of the wave decreases as $1/R$.

pression wave from an exploding firecracker ≈10 m away, the portion of the wave-front intercepted by an ear or a microphone will be essentially spherical.

A sound wave represents a moving change in the distribution of atoms in a medium, and that change is associated with the flow of energy. The rate of transfer of energy is the *acoustic power*. A person speaking at a normal conversational level emits power at about 10^{-5} J/s and shouts at about 1 mJ/s. The energy of a mechanical wave is distributed over its wavefront and, in the case of a *homogeneous* spherical wave, that distribution is uniform in all directions. This raises a practical point: when a detector intersects a portion of a wavefront, it records an amount of energy that depends on its own receiving area and on the time during which it is receptive. Both of these depend on the particular detector, and that's inappropriate. Thus, what we must know is not the total energy arriving on the detector, but the energy per unit time, per unit area—the concentration of power. The **intensity** (I) of a wave is therefore defined as *the average power divided by the perpendicular area across which it is transported:*

$$I = \frac{P_{av}}{A} \qquad (13.1)$$

A steady drip of rainwater produced these circular waves.

Intensity has the units of W/m². When a 2-m² area receives a total of 1 W of wave energy flowing perpendicularly onto it, the incident intensity is 0.5 W/m².

As a spherical wave expands outward, its area ($4\pi R^2$) increases, and since the same power flows through ever-increasing areas, its concentration—the intensity—diminishes *inversely with the square of the radius;* this is the *Inverse Square Law.* It is the reason why you cannot read a book by starlight or hear what the players on the field are saying from up in the stands. Remember that the amplitude of a *spherical wave* must diminish with distance because the energy it carries is spreading thinner and thinner while the wavefront gets larger.

Example 13.2 An underwater explosion is detected 100 m away, where the intensity is recorded to be 1 GW/m². About 1 s later, the sound wave is recorded 1.5 km away from ground-zero. What will its intensity be at that distance? Ignore absorption losses.

Solution: Assuming that the compression wave is spherical, the distances provided represent radii. [Given: $R_1 = 100$ m, $I_1 = 1$ GW/m², $\Delta t = 1$ s, and $R_2 = 1.5$ km. Find: I_2.] Whenever you have energy considerations involving spherical waves of different radii, think about the Inverse Square Law. The power flowing through the first sphere will equal the power through the

second; therefore

$$I_1(4\pi R_1^2) = I_2(4\pi R_2^2)$$

and

$$I_2 = \frac{I_1 R_1^2}{R_2^2} = \frac{(1 \times 10^9 \text{ W/m}^2)(100 \text{ m})^2}{(1.5 \times 10^3 \text{ m})^2}$$

Thus

$$I_2 = 4.4 \times 10^6 \text{ W/m}^2 \text{ or } \boxed{4 \text{ MW/m}^2}$$

▶ **Quick Check:** I goes as $1/R^2$ and 100 m is 15 times closer than 1.5 km; therefore, $I_1 = 15^2 I_2 \approx 225(4 \text{ MW/ m}^2) \approx 0.9$ GW/m².

By allowing a spherical wave to expand (Fig. 13.13) over a great distance, the wavefronts will get larger and flatter, ultimately resembling planes. The light entering a telescope from a star is usually indistinguishable from a perfectly flat wavefront.

13.4 The Speed of Sound

By the seventeenth century, Galileo, Huygens, and others were already studying the finer points of acoustics, and the wave theory was widely accepted. In Paris, Father Mersenne (ca. 1636) used a crude echo technique to first measure the speed of sound. He found a value of about 1000 ft/s, which is not far off from the present value of 331.45 m/s (1087.4 ft/s or 741 mi/h) in air at 0°C and 1 atm (Table 13.2). In

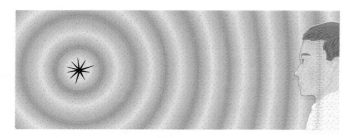

Figure 13.13 As a spherical wave expands, its radius increases and the wavefronts flatten. At very large distances, the wavefronts are quite flat and the disturbance resembles a plane wave.

a confined sample, the speed of sound increases with temperature. The springiness of a gas (for example, in an inflated ball) comes from the motion of its atoms and is measured via the pressure—the higher the temperature, the more violent the motion and the more the gas resists compression.

We have already formulated a theoretical expression for the speed of a compression wave in an elastic fluid; it is

$$v = \sqrt{\frac{B}{\rho}}$$ [12.22]

which certainly ought to tell us something about the speed of sound in air. Indeed, the equation was first derived by Newton, although he called the term represented by B the "elastic force." Sir Isaac argued that this elastic force equaled the pressure of the atmosphere. That line of thought follows from Boyle's Law, Eq. (11.5), which says that *if the temperature is constant, PV = constant*. Hence, if the pressure increases by a tiny amount ΔP, the volume will decrease by a tiny amount ΔV, where

$$PV = (P + \Delta P)(V + \Delta V) = PV + \Delta P \Delta V + V \Delta P + P \Delta V$$

Neglecting the very small term $\Delta P \Delta V$, this expression yields $P = -\Delta P/(\Delta V/V)$, which from Eq. (12.23) is indeed the Bulk Modulus. So Newton concluded that

$$v = \sqrt{\frac{P}{\rho}}$$ (13.2)

which gives sound a speed of only 979 ft/s. Although Eq. (12.24) works fine for liquids where the compressibility is very small, this application of it to gases is clearly missing something.

The discrepancy was explained in 1816 by Pierre Simon Laplace, who recognized that Newton's derivation required the temperature in the gas to be constant, and that it wasn't constant. The regions that are significantly compressed increase in temperature, while those that are rarefied decrease in temperature (p. 495). Because the waves are long and travel quickly, there is not enough time for heat to be conducted from one region to the other, and so the effect is not neutralized. Newton had overlooked the change in elasticity due to the temperature changes induced by the wave itself. What happens can be appreciated by referring to Fig. 12.40. The springs in the rarefactions are cooled, and there k decreases; the springs in the condensations are warmed, and there k increases. The greater the difference in k on either side of an atom, the more rapidly it propagates the disturbance. There is no overall gain or loss of heat and under such conditions, Boyle's Law does not hold; instead, PV^γ = constant, where $\gamma = 1.4$ for diatomic gases like air (p. 560) and

$$v = \sqrt{\frac{1.4P}{\rho}}$$ (13.3)

which produces excellent agreement with observation (Table 13.2).

TABLE 13.2 The speed of sound

Material	Temperature (°C)	Speed (m/s)
Vulcanized rubber	0	54
Carbon disulfide	0	189
Air	0	331.45
Water vapor	0	401
Cork		500
Carbon tetrachloride	23	929
Helium	0	970
Chloroform	24	1001
Lead	20	1227
Hydrogen	0	1270
Mercury	≈25	1450
Fresh water	25	1493
Seawater	20	1513
Glycerin	22	1986
Platinum	20	2690
Brass		3500
Copper	20	3560
Brick		3652
Oak		3850
Aluminum		5104
Granite		6000

Example 13.3   Use Eq. (13.3) to determine the speed of sound in air at STP.

$$v = \sqrt{\frac{1.4P}{\rho}} = \sqrt{\frac{1.4(1.013 \times 10^5 \text{ Pa})}{1.29 \text{ kg/m}^3}} = \boxed{332 \text{ m/s}}$$

Solution: [Given: STP, from Table 10.1, $\rho = 1.29$ kg/m^3, and $P = 1$ atm $= 1.013 \times 10^5$ Pa. Find: v.] From Eq. (13.3)

▶ **Quick Check:** This compares well with $v = 331.45$ m/s at STP.

One thing to observe from Eq. (13.3) is that the speed of sound does not depend on frequency. This fact is confirmed every time you listen to music in a large hall or stadium and all the sounds stay together: the high frequencies reach your ears along with the low frequencies; otherwise only the people in the front rows would hear anything recognizable.

If the pressure of some region of the atmosphere changed while the temperature remained fixed, the density would change in proportion, via Boyle's Law. That change would leave P/ρ constant and v unaltered: the speed of sound in the thin air on a mountaintop is the same as it is at sea level, provided the temperature is the same. Changing the temperature will alter the density of the air, but barometric pressure is determined by the weight of the column of air, which is independent of temperature. Since the density decreases as the temperature increases, the speed will increase with the square root of the temperature. *Over the usual range of temperatures encountered at sea level, the speed of sound changes by about ±0.60 m/s per change of ±1.0°C.*

Example 13.4   What is the speed of sound at room temperature (20°C) and normal atmospheric pressure?

$$v = 331 \text{ m/s} + 20(0.60 \text{ m/s}) = \boxed{343 \text{ m/s}}$$

Solution: [Given: a temperature of 20°C. Find: the speed of sound.] The speed increases from its 0°C value of 331 m/s by 0.60 m/s for each degree increase; hence

▶ **Quick Check:** The density of air at 20°C can be approximated from Table 10.1 as 1.2 kg/m^3; using Eq. (13.3), $v \approx 343.8$ m/s.

In deriving Newton's equation for v, it was assumed that the pressure change was small and that the $\Delta P \Delta V$ term could properly be neglected. For example, during ordinary speech, the air just in front of the mouth will experience a pressure change ($\pm \Delta P$) of no more than about one-millionth of atmospheric pressure, or about 0.1 Pa. Still, for extraordinarily loud sounds where ΔP is appreciable compared to P (which in most cases is 1 atm), $\Delta P \Delta V$ cannot be omitted. Then the above derivation leads to a Bulk Modulus of $(P + \Delta P)$, which must go into Eqs. (13.2) and (13.3). The change (ΔP) from ambient pressure in the sound wave at the mouth of a cannon can exceed atmospheric. Thus, if ΔP happens to equal atmospheric pressure, the speed of that sound wave will increase by a factor of $\sqrt{2}$ over its ordinary low-volume rate. This accounts for the peculiar fact that under the right conditions (you have to be far enough away), one can actually hear the report of a cannon *before* one hears the order to fire it. *Very loud sounds travel slightly faster than quiet ones.*

13.5   Hearing Sound

The human sound receiver, though its operation is not fully understood, is composed of three subsystems: the outer, middle, and inner ears (Fig. 13.14). Sound intercepted

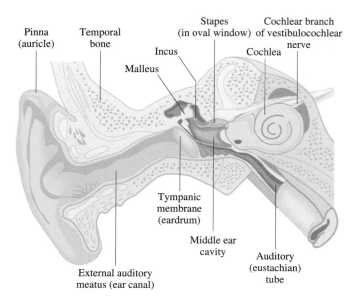

Pinna (auricle)

Temporal bone

Malleus

Incus

Stapes (in oval window)

Cochlear branch of vestibulocochlear nerve

Cochlea

Tympanic membrane (eardrum)

Middle ear cavity

Auditory (eustachian) tube

External auditory meatus (ear canal)

Figure 13.14   The human ear.

by the external ear is channeled via the auditory canal to the eardrum, which is driven into oscillation. This three-part structure is the *outer ear*. Besides aiding in determining the direction of sound, it provides a resonant cavity (p. 459) that amplifies (by about 100%) sounds in the range from ≈3 kHz to ≈4 kHz, making the entire system especially sensitive to this mid-frequency band. The ear is 1000 times more responsive at 1 kHz than at 100 Hz. Were the ear more sensitive to the low-frequency range, we would have to contend with a constant drone of internal body noises.

The *middle ear* links the eardrum, via three pivoted bones, to the oval window. This arrangement increases the force exerted on the flexible window into the inner ear. In addition, since the effective area of the eardrum is about 20 times larger than that of the oval window, there is another increase in pressure.

The *inner ear* is the transducer that converts a mechanical pressure input into an electrical nerve output. It begins at the oval window, which is the entrance to a fluid-filled coiled cavity known as the *cochlea*. The cochlea contains the *basilar membrane* on which is a structure of over 20 000 strands of hairlike receptor cells leading to nerve fibers. Hydraulic compression waves generated at the oval window induce vibrations in the basilar membrane, which flutters transversely along its length. At the front end of the cavity where the membrane is narrow and stiff, it vibrates at high frequencies; toward the rear where the membrane is slack, its resonant frequency is lower. As the membrane oscillates, the receptor cells in the different regions flutter and trigger impulses at a rate that depends on both the amplitude and frequency of the motion. The extent of the displacement of the basilar membrane determines the firing rate of the receptor cells, which, in turn, manifests itself in the perception of loudness.

Pitch

Pitch is a human perception resulting from the sensing of acoustic energy. It is an ear-brain response mainly to the frequency of sound. Certainly we know what we mean by the term: a soprano voice is high-pitched, a bass voice is low-pitched. The pitch of a *pure tone* (a sinusoidal wave with a single fixed wavelength, as approximated by a tuning fork) more or less corresponds to frequency—the higher the frequency, the higher the pitch (Fig. 13.15). Nonetheless, the ear-brain system is far from simple, and there is no one-to-one relationship between frequency and pitch. For example, a high-frequency tone will be heard to increase in pitch as its intensity increases, while a low-frequency tone will cause the opposite response.

Timbre

Another auditory sensation that allows us to distinguish between sounds is *timbre*. Listening to a flute, a trumpet, a saxophone, a violin, or a tuning fork, each producing the same note (same pitch) at the same loudness, there would be little difficulty telling one from the other. The catchall attribute that allows for this distinction is the timbre, which depends primarily on the waveform—that is, the frequencies present, their relative phases, and amplitudes. When the same tones made by these various instruments are picked up by a microphone and displayed on an oscilloscope

Figure 13.15 The frequency ranges of voices and instruments.

(Fig. 13.16), they are seen to be substantially different. Similarly, the same note produced by a piano and a singer have distinctly different frequency spectra (Fig. 13.17), and the ear immediately distinguishes the vocalist from the accompaniment, even when the sounds are made at the same moment.

An actual tone will contain frequencies higher than the fundamental (or first harmonic), and these are called *overtones*. An overtone need not be a whole-number multiple of the fundamental; that is, it need not be a harmonic. Thus, it is the number of overtones (whether they are harmonics or not) and their relative amplitudes that more than anything else determines timbre. Probably the most important difference between a fine musical instrument and an ordinary one is its timbre, or tone color.

An adult male generates basic voicing vibrations at ≈80 Hz to ≈240 Hz for speech and upwards of ≈700 Hz for song. By comparison, the adult female voice range is ≈140 Hz to ≈500 Hz in speech and up to ≈1100 Hz in song. Given enough control over a wide spectrum of pure-tones (≈80 Hz to ≈1100 Hz), we presumably could artificially duplicate all human vocal sounds—a process just beginning with "talking" alarm clocks, elevators, computers, and cameras.

Figure 13.16 Waveforms produced by several instruments all playing the same A_4 note ($f = 440$ Hz): (a) a flute; (b) a trumpet; (c) a soprano saxophone; (d) a violin. Each portion shows roughly $3\frac{1}{2}$ cycles lasting about 8 ms, corresponding to a frequency of 0.44 kHz.

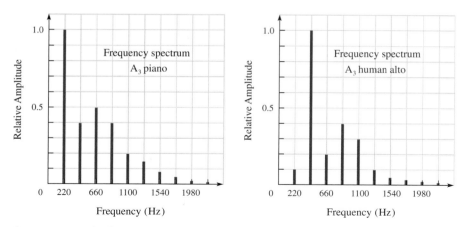

Figure 13.17 The frequency spectra for the A_3 (220 Hz) note sounded by a piano and an alto voice.

Loudness

The auditory sensation of loudness depends on the frequency spectrum, duration, and, most importantly, on the intensity of the sound. In general, the intensity of an elastic wave is proportional to the square of both the frequency and amplitude ($I \propto f^2 A^2$). The intensity of sound is proportional to the square of the amplitude (P_0) of the pressure wave ($I \propto P_0^2$).

The breadth of sound intensities that occur in nature, from the faint buzz of a mosquito to the roar of a volcano, is tremendous, and the human ear is remarkably sensitive to much of it. We can bear intensities in excess of around 1 W/m², which is so loud it can be quite uncomfortable. In fact, at this level, one begins to feel the sound as opposed to just hearing it. That *threshold of pain* corresponds to a gauge pressure change of only about 10^{-4} atm, or roughly 10 Pa. At the other extreme, the ear can just detect sounds at intensities as low as 10^{-12} W/m². That *threshold of hearing* is associated with a pressure change of about 10^{-10} atm, or 10^{-5} Pa. Lord Rayleigh showed that, at 1 kHz, this change is equivalent to air molecules vibrating with an amplitude of only about 10^{-6} mm (see Problem 35).

The ear functions over a pressure range that corresponds to a factor of a million (10^{-4} atm to 10^{-10} atm), and the intensity range—the square of that—extends a factor of 10^{12}. This is an amazing breadth of operation for any sensor—it's the equivalent of using the same device to measure both the diameter of an atom and the length of a football field. Figure 13.18 depicts the limits of sensitivity as a function of frequency.

We judge the relative loudness of a sound not by the difference in intensity between it and some reference, but by their ratio. Doubling the intensity of a faint sound will produce a sensation of some increase in loudness. Doubling the intensity of an already loud sound will nevertheless produce the sensation of the same increase in loudness. In fact, **multiplying any intensity by 10 will generally be perceived as approximately doubling it in loudness**; 10^{-8} W/m² sounds twice as loud as 10^{-9} W/m², and 0.1 W/m² is heard to be twice as loud as 0.01 W/m². As we will see presently, that's equivalent to saying that the ear responds logarithmically.

13.6 Intensity-Level

The *intensity-level* (or *sound-level*) of an acoustic wave is defined as *the number of factors of 10 that its intensity is above the threshold of hearing:* $I_0 = 1.0 \times$

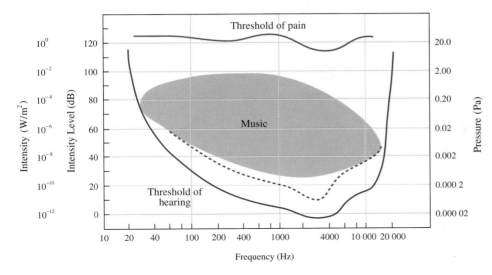

Figure 13.18 About 1% of the people in the United States can hear sound below the levels of the lower curve. Perhaps 50% can hear below the dashed curve. The shaded region corresponds to the levels of normal music.

10^{-12} W/m². To honor Alexander Graham Bell, who did important research in acoustics, the unit of sound intensity-level is called the *bel*. The performance of audio amplifiers, tuners, loudspeakers, and so forth, is given in terms of the smaller unit, 1/10 of a bel, which is a *decibel* (abbreviated dB). The decibel, like the radian, is not a physical unit expressible in terms of meters, kilograms, or seconds. It is simply a reminder of the meaning of the number preceding it: 10 dB means something very different from just 10 or from 10 rad.

By definition, the logarithm to the base 10 of any number X equals the power to which 10 must be raised to equal X. In other words, if $X = 10^Y$, then the $\log_{10} X = Y$. Hence, $1000 = 10^3$ and so $\log_{10} 1000 = 3$. If we form the ratio of the intensity of a wave I to the reference level I_0 (remembering how the ear responds to such ratios), and then take its logarithm, we will get the number of factors of 10 that I is above I_0: this number is the intensity-level in bels, and 10 times that gives it in decibels. Thus, the **intensity-level** β in dB of any sound is

$$\beta = 10 \log_{10} \frac{I}{I_0} \qquad (13.4)$$

For example, a sound wave with an intensity of 10^{-6} W/m² (about the level of normal conversation with a person 1 m away) has an intensity-level of

$$\beta = 10 \log_{10} \frac{10^{-6} \text{ W/m}^2}{10^{-12} \text{ W/m}^2} = 10 \log_{10} 10^6 = 10(6) = 60 \text{ dB}$$

The sound has an intensity of 1 million times the threshold intensity and therefore has a sound-level of 6 bel, or 60 dB—it's 6 factors of 10 or 10^6 above threshold (see Table 13.3).

Observe that since $\log_{10} 1 = 0$, the intensity-level at the threshold of hearing is $\beta = 10 \log_{10} 1 = 0$ dB. The entire intensity range from 1.0 W/m² to 10^{-12} W/m² corresponds to 12 bels (12 multiplicative factors of 10) or spreading the scale out a bit so it can be graphed more conveniently, to 120 dB. A range of 1 to 1 million mil-

TABLE 13.3 Intensity ratios and intensity-level	
Ratio of intensities	**Intensity-level (dB)**
100 000 000 000:1	110
1 000 000:1	60
10 000:1	40
100:1	20
10:1	10
4:1	6
2:1	3
1:1	0
1/2:1	−3
1/4:1	−6
1/10:1	−10
1/100:1	−20
1/1000:1	−30

TABLE 13.4 Approximate intensity-levels in dB

Source	Intensity-level (dB)	
Large rocket	≈180	
Jet engine	140	
Jet takeoff (30 m − 60 m)	≈125	intolerable
Rock concert (1.0 W/m^2)	≈120	painful
Yell into the ear (20 cm)	120	(immediate danger)
Pneumatic hammer	110	
Passing subway train (10^{-2} W/m^2)	100	(damage after
Car without muffler	100	2 hours)
Shout (1.5 m)	100	very loud
Car horn, loud	95	
Heavy truck (15 m)	90	
City street	80	(damage after
Hair dryer	80	8 hours)
Loud music	80	
Freeway traffic	75	loud
Automobile interior	≈70	
Toilet flushing	≈67	
Noisy store (10^{-6} W/m^2)	60	
Conversation, average (1 m)	60	moderate
Office	50	
Living room, city	40	
Bedroom (10^{-9} W/m^2)	30	quiet
Library	30	
Broadcast studio (10^{-10} W/m^2)	20	very quiet
Whisper	20	
Cat purring	15	
Rustling leaves	≈10	barely audible
Threshold (10^{-12} W/m^2)	0	

lion is thus converted into a range of 0 dB to 120 dB (Table 13.4). *A young healthy ear can distinguish between sound-levels that differ by as little as 1 dB; 2 dB or greater is more often the case, however.*

Several logarithm identities will be useful in performing sound-level calculations. (See Appendix A-3.) These are

$$\log_{10} \frac{A}{B} = \log_{10} A - \log_{10} B \qquad (13.5)$$

$$\log_{10} AB = \log_{10} A + \log_{10} B \qquad (13.6)$$

$$\log_{10} A^n = n \log_{10} A \qquad (13.7)$$

and reversing the process

$$10^{\log_{10} A} = A \qquad (13.8)$$

For example, suppose we increase the intensity of a sound by some multiplicative factor; by how much would the intensity-level change? In general, for a change from I_1 to I_2

$$\Delta\beta = \beta_2 - \beta_1 = 10 \log_{10}\left(\frac{I_2}{I_0}\right) - 10 \log_{10}\left(\frac{I_1}{I_0}\right)$$

and so from Eq. (13.5)

$$\Delta\beta = 10 \log_{10}\left(\frac{I_2}{I_1}\right) \qquad \text{in dB} \qquad (13.9)$$

Increasing the intensity by a factor of 10 changes the sound-level by 1 bel or 10 dB; increasing it by 100 changes β by 2 bels, or 20 dB, and so on. A 60-dB sound compared to a 20-dB sound is 40 dB (4 bels) greater, which means— from Eq. (13.9)—that $\log_{10}(I_2/I_1) = 4$ and the intensity of one is 10^4 times the other. Since every increase of 10 dB corresponds to a doubling in loudness, an 80-dB sound is louder than a 60-dB sound by (two 10-dB steps or) a factor of 4.

Example 13.5 Two public address systems are being compared, and one is perceived to be 32 times louder than the other. What will be the difference in intensity-levels between the two when measured by a dB-meter?

Solution: [Given: a factor of 32 in *loudness*. Find: $\Delta\beta$.] We are given the change in loudness, which correlates to a change in intensity-level by way of the fact that a dou-

bling in loudness is roughly a change of 10 dB. Hence, since $32 = 2^5$, and $\Delta\beta = 5(10 \text{ dB}) = \boxed{50 \text{ dB}}$.

▶ **Quick Check:** $\Delta\beta = 10 \log_{10}(I_2/I_1) = 50$; hence, $\log_{10}(I_2/I_1) = 5$ and $(I_2/I_1) = 10^5$, the intensities differ by 5 factors of 10, each one doubling the loudness, which increases 2^5, or 32-fold.

Example 13.6 Imagine that 10 identical violins are each about to play something at 70 dB. If they join in one at a time, what will be the sound-level as each contributes?

Solution: [Given: 10 sources at 70 dB each. Find: the corresponding sound-levels.] The intensity goes from I to $2I$ to $3I$ up to $10I$. We could solve the problem using Eq. (13.9), but let's go back to Eq. (13.4) to compute β

(continued)

(continued)
and use Eq. (13.6). For 2 violins

$$10 \log_{10} \frac{2I}{I_0} = 10 \log_{10} 2 + 10 \log_{10} \frac{I}{I_0}$$

and the sound-level becomes $\boxed{3 \text{ dB} + 70 \text{ dB}}$. *Doubling the intensity produces a 3-dB sound-level increase.* With 3 instruments playing, the sound-level rises to $\boxed{75 \text{ dB}}$; with 4, $\boxed{76 \text{ dB}}$; with 5, $\boxed{77 \text{ dB}}$, and so on

up to 10, when it reaches $\boxed{80 \text{ dB}}$. Ten violins produce an intensity 10 times that of 1 and therefore 1 bel (10 dB) greater, and that essentially corresponds to a doubling of the perceived loudness.

▶ **Quick Check:** $\Delta\beta = 10 \log_{10}(I_2/I_1) = 10 \log_{10} 2 = 3.0$ and $\beta + \Delta\beta = 70 \text{ dB} + 3 \text{ dB}$.

It follows that if we have a 12-W audio system and wish to replace it by one that is equally efficient but which is capable of being twice as loud, we will need to increase its power, not twofold but by a factor of 10, raising it to 120 W.

PRODUCTION AND PROPAGATION OF SOUND

Anything that will cause compression waves in the audible range in a material medium is a source of sound. Smack your hands together or tap a hammer and a sound will be made—it will not be sustained very long, but it certainly will be an acoustical wave (Fig. 13.4). The source itself does not have to vibrate to produce sound, even though that is most often the case. Physically, **noise** refers to an unrelated jumble of disturbances (Fig. 13.19), a *nonperiodic,* randomly changing wave with an essentially continuous frequency spectrum. When the spectrum has a *broad bandwidth encompassing the entire audible range with equal intensity everywhere,* it is called *white noise,* and we hear it as a mix of wind, pouring water, and continuous radio static.

Suppose that the tapping of a hammer occurs at a slow, regular pace. A series of separate impulses will be heard, increasing in rate as the tapping increases, until at roughly around 20 impacts per second, the ear-brain will perceive the clatter as a continuous sustained hum. Galileo noticed that he could produce a tone by drawing a knife blade over a serrated edge, generating a rapid succession of taps. The periodic nature of the process was apparent to him, and he inferred that the pitch was proportional to the rapidity of the tapping.

Between 1793 and 1801, J. Robison devised a technique for producing puffs of air at specific frequencies. A pipe (Fig. 13.20) carried a continuous stream of air to a rotating disc having a circle of evenly spaced holes in it. By adjusting the speed so that the device produced 720 puffs per second, a sound "was most smoothly uttered, equal in sweetness to a clear female voice." When de la Tour (1819) found that it would sing underwater, he gave it the name *siren* after the legendary sea nymphs. The importance of the apparatus was that it provided future researchers, notably Dr. Hermann Helmholtz, with a sound source having a precisely controlled frequency that could be operated across a wide spectrum.

The metal reeds in a music box mechanism are displaced by rotating pins and set vibrating. Note the different lengths of the reeds.

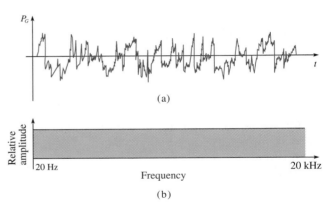

Figure 13.19 (a) Random aperiodic noise is called *white noise* when (b) its frequency spectrum has a constant amplitude all across the audible frequency range.

Figure 13.20 Professor Robison's siren. Puffs of air blowing through the holes in a rotating disc created a sound "equal in sweetness to a clear female voice."

13.7 Beats

With a standard source, we can determine the frequencies of other sound-generating devices, such as tuning forks, vibrating wires, and so forth. The most common technique for the purpose (still used by some traditional piano tuners) relies on the phenomenon of **beats**. Mersenne was the first to report the effect, and it was he who gave it that name. Describing the result of playing two organ pipes that differ slightly in pitch, he wrote that they "produce a throbbing or thrumming sound similar to that of beating a tambourine." The effect is a manifestation of a basic wave property called **interference**: two or more waves overlap and produce a resultant wave.

As was seen earlier (Fig. 13.5), two waves of the same frequency superimpose to form a stable resultant wave. When they are in-phase, or nearly so, crests fall on crests, troughs on troughs, and the resultant is large. When they are 180° out-of-phase, or nearly so, crests fall on troughs, and the resultant is small or even nonexistent. Figure 13.21 depicts the overlapping and interference of two pure tones having slightly different frequencies, traveling in the same direction. The two waves, because they are not quite able to stay in-step, gradually (and periodically) go in- and out-of-phase; at one instant, rising and falling together, they reinforce one another to form a large amplitude sound. Then, one rising while the other is falling, they subtract to cancel each other altogether. The two waves of frequency f_1 and f_2 combine to produce a carrier wave having the average frequency $(f_1 + f_2)/2$ that is *amplitude modulated,* rising and falling with the **beat frequency** of $(f_1 - f_2)$. What you hear is a tone—the carrier—building and fading in intensity at the beat frequency. (A mathematical analysis of the effect is explored in Problem 75.)

Example 13.7 A siren is used as a frequency reference during the fine adjustment of a middle-A tuning fork. The siren is slowly reduced in frequency and, together with the tuning fork, produces a quavering tone. This warbling also decreases in frequency until the siren reaches and holds at 440 Hz. At that point, the tone varies in loudness from maximum to minimum and back to maximum in $\frac{1}{4}$ s. What is the frequency of the fork?

Solution: The quavering tone tells us that this is a problem in beats. [Given: $f_1 = 440$ Hz and the period of the beats is $\frac{1}{4}$ s. Find: the frequency of the fork, f_2.] The beat frequency is 1 over the beat period; hence

$$\frac{1}{(\frac{1}{4}\,\text{s})} = (f_1 - f_2) = 440\,\text{Hz} - f_2$$

and $\boxed{f_2 = 436\,\text{Hz}}$

▶ **Quick Check:** 1/0.25 s = 4 Hz and since the siren was coming down, its frequency must exceed that of the tuning fork by 4 Hz.

A variety of vibrating objects can serve as acoustical sources that will produce sustained sounds with discernable pitches. An unburdened honeybee fluttering its wings at about 440 Hz generates a buzz at what has come to be the musician's standard A_4 (middle-A) frequency. Strings and reeds, membranes, rods, bars, lips, air columns, and tongues can all be set vibrating, all making sounds, even ones we call music, whatever that is.

13.8 Standing Waves

When a wavetrain of any kind is created in some real, finite medium (whether a string, a drum, or the Earth itself), it will propagate outward until it encounters an end or boundary. There, some fraction of the wave-energy will usually be reflected backward (p. 426), and if the original disturbance is sustained, the medium will quickly fill with waves traveling back and forth over one another. These disturbances will combine or interfere to form a steady-state distribution of energy known somewhat paradoxically as a *standing* or *stationary wave*. This situation is remarkably commonplace. It happens within every sort of musical instrument from the piano to the triangle (and, for that matter, within every kind of laser as well). It occurs in the head, throat, and mouth when we speak, in the ear canal when we listen, and in the shower stall when we sing. Standing waves exist when you ring a bell, strike a tuning fork, run water from the tap into an open sink, or blow across an empty soda bottle.

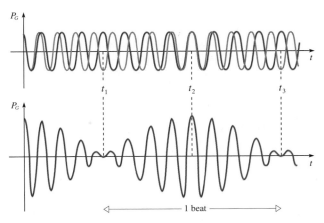

Figure 13.21 When two signals of different but nearly equal frequencies (f_1 and f_2) superimpose, they create a beat pattern, which is true of sound as well as light. The intensity rises and falls with a frequency equal to the difference ($f_1 - f_2$). Note that at t_1 and t_3 the signals are 180° out-of-phase, while at t_2 they are in-phase.

The phenomenon was first appreciated in 1821 by the brothers Ernst and Wilhelm Weber in Leipzig. They recognized that the effect (which they observed while pouring mercury) arose when "two wavetrains of the same wavelength and intensity but [traveling] in opposite sense encounter each other." Figure 13.22 shows pictorially how the process occurs for two harmonic waves passing over one another. What results is a fixed pattern of **nodes** where the resultant is zero and, midway between them, **antinodes** where the resultant is a maximum. Every point on each separate portion of the wave between successive nodes oscillates in-phase, with its own amplitude. There is no transporting of energy and, in that sense, one might not even consider it a wave at all. Still, there are good mathematical reasons to call the sum of two waves a wave even if it isn't going anywhere, and so, along with the brothers Weber, we distinguish between progressive (traveling) waves and stationary (standing) waves. Here, *the wavelength is twice the node-to-node distance*.

Standing Waves on Strings

One of the most common arrangements is the taut string with both ends held immovable, which is the basic setup for the piano and all stringed instruments (Fig. 13.23). Energy can be continuously pumped into such a system by attaching one end to a source, like a driven tuning fork or vibrator, which has such a small amplitude that it might well be considered a fixed point. Envision a harmonic wave with a tiny amplitude continuously introduced at the left, moving to the right, down the length of the string L at a speed v, as in Fig. 13.24 (p. 462). Presume that it will be reflected back to the left without any loss in amplitude but with a 180° phase shift (p. 427). This reflected wave will soon arrive back at the origin (that is, at the vibrator), there to again be phase-shifted another 180° as it is reflected (from that effectively fixed end) off to the right. What conditions have to be met if this twice-reflected wave is to move off exactly in-step with the continuously flowing original disturbance? We want both waves traveling to the right to be in-phase, meaning that the round-trip time $2L/v$ must equal the period T since the two 180° phase shifts due to reflection add up to 360° and so essentially cancel. Inasmuch as $v = f\lambda$, $2L/v = T$ yields

$$L = \tfrac{1}{2}\lambda$$

and the round trip is one wavelength long. This being the case, each trip down and back will bring the reflected wave to the fork just in time to combine in step with the newly emitted wave. Each such trip will cause the wave in the system to grow slightly larger.

A pail used to wash a floor contained a suspension of fine dirt particles in water. When placed in a curved sink, the pail gently rocked along a fixed axis, setting up standing waves and distributing the particles in ridges as they settled.

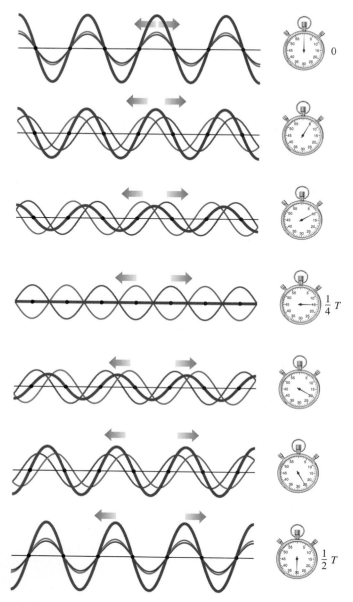

Figure 13.22 The creation of standing waves. Two waves of the same amplitude and wavelength traveling in opposite directions form a stationary disturbance that oscillates in place.

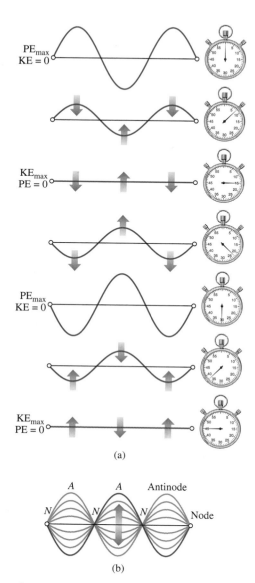

Figure 13.23 (a) A standing wave on a string that has both ends fixed. (b) represents a composite of all the configurations demonstrated in (a).

As can be seen in Fig. 13.24, if the length of the string is half a wavelength, its two ends will remain motionless as the center oscillates as an antinode. This synchronization of the waves, which depends on a matching of the vibrator's frequency with the size of the medium, is a resonance effect. Initially, the disturbance on the string will continue to build as energy is pumped in until the increasing frictional losses stabilize the system. At that point, the energy provided by the vibrator at each oscillation equals the sum of the losses, and there is no more growth of the wave, whose amplitude is nonetheless much larger than that of the source. *The ability to amplify the input is an extremely important feature of standing-wave systems.*

Similarly, if the ends are fixed and the string is displaced, it will naturally oscillate in this configuration, or *mode.* Yet, unless every point on the string is carefully displaced along a smooth sine curve, there will be higher-order standing-wave modes present as well. These occur when the round-trip time corresponds to $2T$, $3T$, $4T$, and so forth, or NT where N is a whole number (Fig. 13.25). Then

$$L = \tfrac{1}{2} N\lambda \qquad (13.10)$$

and the wavelength of the Nth mode is $\lambda_N = 2L/N$. Accordingly, the pattern shown in Fig. 13.23 comes into being when $N = 3$. Stretch a long rubber band between your two hands and then pluck it. Now repeat the process holding it first horizontally and then vertically in front of the screen of a TV set or computer—these flicker rapidly and so serve as stroboscopes.

We can find the resonant frequencies f_N of the system using the fact that speed equals frequency times wavelength, along with Eq. (13.10), which becomes $L = \tfrac{1}{2} Nv/f_N$, and $f_N = Nv/2L$. Substituting from Eq. (12.20) for the speed of a transverse wave on a string, we get

$$f_N = \frac{N}{2L}\sqrt{\frac{F_T}{\mu}} \qquad (N = 1, 2, 3, 4, \ldots) \qquad (13.11)$$

Notice that each frequency is a whole-number multiple of the *fundamental,* or lowest resonant frequency: $f_N = Nf_1$. Thus, each higher resonant frequency is a harmonic. Keep in mind that the Nth harmonic will have N antinodes, so it is easy to recognize when you see it. Usually each string of an instrument is played in its fundamental mode. In the case of a guitar, for example, the strings sound different, even though they all have the same length, because the tensions and linear mass-densities are different. By "fingering" a string; that is, by pressing it down onto a fret, thereby fixing a node at that point and effectively shortening the length of string that is able to vibrate, the player increases the fundamental frequency.

Galileo understood that an oscillating string causes "the air immediately surrounding the string to vibrate and quiver," thus producing a sound of the same frequency. A plucked or struck string actually vibrates in several modes at once and generates several tones (a fact observed by Mersenne). Strike a key on a piano, and the string will generate a strong fundamental along with a number of much quieter overtones (Fig. 13.17), which is what gives the instrument its characteristic tone color.

Example 13.8 What must be the tension in the E_5-string of a violin if it is to be tuned to 660 Hz? The length of the string from bridge to peg is 330 mm, and its mass-per-unit length is 0.38 g/m.

Solution: [Given: $f_1 = 660$ Hz, $L = 0.330$ m and $\mu = 0.38$ g/m. Find: F_T.] The violin is tuned to a fundamental of 660 Hz; hence, $N = 1$ in Eq. (13.11), and so

$$F_T = \mu(2Lf_1)^2 = (0.38 \times 10^{-3}\ \text{kg/m})4(0.330\ \text{m})^2 \times (660\ \text{Hz})^2$$

and

$$\boxed{F_T = 72\ \text{N}}$$

or about 16 lb.

▶ **Quick Check:** $f_1 = \dfrac{1}{2L}\sqrt{\dfrac{72\ N}{0.38 \times 10^{-3}\ \text{kg/m}}} = 660$ Hz.

Because vibrating strings are not good at pushing large amounts of air, they are not, by themselves, very loud. In instruments, strings are therefore coupled to a large section of the body in such a way as to transmit vibrations to it—the sounding board in pianos or the sounding box in violins and guitars. An alternative is to couple the strings to a pick-up and amplify the sound electronically.

Figure 13.24 A standing wave. A wave enters from the left (a) and is reflected (dashed) from the termination point on the right (d). Now there are two overlapping waves that combine to form a standing wave (red), oscillating (g through k) in place.

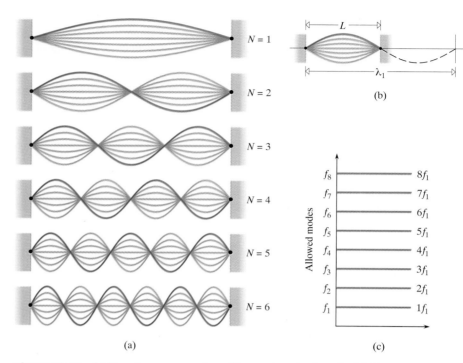

Figure 13.25 (a) Standing-wave modes with a node at both ends. (b) The wavelength of the $N = 1$ fundamental equals $2L$. (c) Allowed modes of oscillation.

Standing Waves in Air Columns

Standing waves can be sustained by the air within almost any shaped chamber from the vocal tract to the shower stall. Indeed, we speak and hear and make music via the acoustical resonance of air-filled chambers of one sort or another that amplify initial sounds. Most musical instruments driven by streams of air (organs, woodwinds, and brasses) utilize this same kind of standing-wave mechanism for increasing intensity.

There are many ways to excite a column of air into oscillation: most woodwinds use a small vibrating reed; the brasses rely on the vibration of the player's lips; and the flute and organ pipe are driven by a fluttering jet of air that pumps the column (Fig. 13.26). Similarly, blowing across the top of a soda bottle sets the air within it vibrating. Whatever the source, the air column initially oscillates with a wide range of frequencies, but only those corresponding to the standing-wave modes of the particular chamber will be sustained and amplified. There are three basic configurations for tubular enclosures: closed at both ends, closed at one end, and open at both ends.

Figure 13.27 shows a chamber closed off at the right end by a wall and sealed at the left end with a slightly movable piston or diaphragm, something like the situation that exists when you hum with your throat or mouth closed. Because the ends are essentially fixed, they will always be the locations of displacement nodes—the air molecules cannot move beyond these points (and, as we saw on p. 445, that means these are also pressure antinodes). Small oscillations of the piston will excite the air column, and a standing-wave pattern will occur. Longitudinal sound waves do not shift phase on reflection from a fixed surface as do the transverse waves on a string—*a rarefaction is reflected from a wall as a rarefaction*. Even so, because there are two 180° shifts for a string with both ends fixed, its standing-wave dis-

Ultrasonic levitation. A drop of water floating in a so-called energy well. Ultrasonic waves, one traveling up, the other down, create a standing-wave pattern. The suspended object is confined to a nodal region.

(a)

(b)

(c)

(d)

(e)

Figure 13.26 An organ pipe. The primary driving mechanism is a wavering, sheetlike jet of air from the flue-slit, which interacts with the upper lip and the air column in the pipe to maintain a steady oscillation.

Figure 13.27 (a) A sealed pipe. The dark blue regions are at density or pressure antinodes. (b) The fundamental mode. (c) The second harmonic at the instant a positive peak is at $x = \frac{1}{4}\lambda_2$. (d) The direction of motion of the molecules at that moment. (e) The associated pressure, which is 90° out-of-phase with the displacement represented in (c).

placement pattern will be the same as that of the sound wave in a closed chamber. Here again

[*both ends closed*]
$$L = \tfrac{1}{2}N\lambda \qquad\qquad [13.10]$$

and the wavelength of the Nth mode is $\lambda_N = 2L/N$. This is all fine, but if you want to really make some intense sounds, you must allow the energy to escape the instrument more efficiently: you must open your mouth and sing out, or open the end of the chamber.

The resonant chamber or cavity can be open either at one end or both, and there are organ pipes of each kind, just as there are musical instruments of each kind: the soda bottle and the trumpet versus the flute and piccolo. The organ pipe open at one end can be thought of as half the closed chamber of Fig. 13.27. At the hole where it is open to the atmosphere, the gauge pressure is zero and a pressure node (Fig. 13.28)

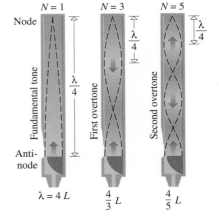

Figure 13.28 A few standing-wave modes for an organ pipe open at its lower end. The arrows show the direction in which air moves for half the cycle, whereupon the modes reverse. The dashed lines indicate the displacement nodes and antinodes. All brass instruments are closed at one end by the mouth of the player.

always exists there. Unimpeded at that point, the molecules of air can move freely and a displacement antinode exists there. This situation happens because *sound waves are phase-shifted by 180° when they reflect off an open end.* It is as if the compression wavefront travels more easily in the open air, stretching the bonds with the layers behind it and sending back a reflected rarefaction. *At an open end, a compression is reflected as a rarefaction and vice versa.* It should be mentioned that the antinodes do not actually occur precisely at the open end, but this effect is a small one for narrow pipes, and we can overlook it here. For the singly-open pipe

[*one end open*] $$L = \tfrac{1}{4}N\lambda \qquad (13.12)$$

where N is an *odd integer.* Again, since $v = f\lambda$, it follows that $L = \tfrac{1}{4}Nv/f_N$ and the resonant frequencies are given by

$$f_N = \frac{Nv}{4L} \qquad (N = 1, 3, 5, \dots) \qquad (13.13)$$

Here v is the speed of sound in air. Pressing on a valve of a trumpet opens a loop of tubing into the air stream, effectively adding to the length of the pipe and increasing the wavelength and decreasing the pitch.

Example 13.9 As we saw earlier, the auditory canal, which is essentially a narrow chamber closed at one end by the eardrum, is a standing-wave cavity. Knowing that it is about 2.7 cm long determines the fundamental frequency that will be amplified at room temperature.

Solution: [Given: $L = 2.7$ cm and $v = 343$ m/s. Find: f_1.] From Eq. (13.12) with $N = 1$, $\lambda = 4L = 4(0.027 \text{ m}) = 0.108$ m and so

$$f_1 = \frac{v}{\lambda} = \frac{343 \text{ m/s}}{0.108 \text{ m}} = \boxed{3.2 \text{ kHz}}$$

▶ **Quick Check:** 3 kHz is in the mid-frequency audible region where it should be (p. 452). As sound near that frequency flows into the canal, it rapidly builds in amplitude until it is many times greater than the ambient value.

The doubly-open pipe of Fig. 13.29 must have displacement antinodes at both ends and so must have a node at its middle. Clearly, the pipe has to be half a wavelength long; that is

[*both ends open*] $$L = \tfrac{1}{2}N\lambda \qquad [13.10]$$

and $$f_N = \frac{Nv}{2L} \qquad (N = 1, 2, 3, \dots) \qquad (13.14)$$

Except for the location of the nodes, this result is the same one we obtained for the closed pipe and the string fixed at both ends. The flute is an open pipe, open to the atmosphere both at the mouthpiece or embouchure and at the far end. Like many wind instruments, it contains a row of holes that can be covered by the fingers. Opening any one has the effect of cutting the tube off at that length and therefore shortening the wavelength and raising the pitch (Fig. 13.30). As a rule, blowing gently on a pipe (or soda bottle) primarily excites the fundamental. When air is blown much more forcefully across the opening of a flute, the column can resonate at the second harmonic, and the entire scale will be raised one octave corresponding to this doubling.

The human vocal tract—larynx, pharynx, and nasal and oral cavities—can crudely be thought of as a singly-open pipe with the vocal cords at the closed end and the mouth at the open end. It therefore sustains a series of standing-wave modes

Figure 13.29 A few standing-wave modes for an organ pipe open at both ends. The arrows show the direction in which air moves for half the cycle, whereupon the modes reverse. This time, both ends of the tube are at atmospheric pressure ($P_G = 0$) and there are pressure nodes at both.

Figure 13.30 Organ pipes and a flute. The length of the resonant chamber determines the pitch.

that amplify the sounds made primarily by the vocal cords. Unlike other musical instruments, this one changes its shape. When the fundamental generated by the vocal cords is matched by a vocal tract resonance at that frequency, the amplitude of the note will be a maximum. Such is the business of a skilled singer.

The pipes of a large organ.

13.9 Sound Waves

The means of propagation of optical and acoustical energy, though totally different in physical nature, are remarkably similar in many regards. Our present emphasis will be qualitative, leaving the more quantitative treatment for the discussion of light. Here, we briefly introduce some aspects of the reflection, refraction, interference, and diffraction of sound that are applicable to all waves.

Reflection

Waves are reflected when they encounter the boundary between two physically different media. Figure 13.31 shows a succession of plane wavefronts reflecting from a smooth, flat surface. Notice how the sound reflects off at the same angle at which it impinges—that behavior has been known from observations for centuries. As da Vinci put it, "the sound of an echo will strike and rebound to the ear at equal angles"—light does the same thing.

In addition to redirecting sonic wavefronts, reflecting surfaces (sonic mirrors) can reshape them as well. When a small source such as a ticking watch is positioned at the focus of a parabolic surface (Fig. 13.32), the diverging

(a)

θ_i | θ_r

(b)

Figure 13.31 (a) Plane waves reflecting off a surface. (b) Sound reflects from a smooth surface such that the angle of incidence equals the angle of reflection.

Figure 13.32 The sound from the watch is reflected by the parabolic mirror. It is beamed out as a plane wave and then focused down to a microphone. The modern equivalent is used for eavesdropping and picking up the sounds of the play at football games.

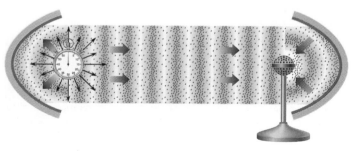

spherical sound waves that are emitted will be converted into plane waves. This beam then travels to the other mirror, where it is detected. The arrangement makes it possible to hear the ticking at up to about 10 m. In fact, by using two ordinary soup bowls as mirrors and tilting the receiver so it points toward your ear, you can hear the watch from several feet away—try it.

Refraction

A wave that begins as a sphere in a homogeneous isotropic medium will continue to propagate out in all directions at the same speed and will stay a sphere. If, on the other hand, the medium is not uniform, the speed will not be constant in all directions and the wavefronts will be altered accordingly, a situation known as **refraction**. Figure 13.33 pictures the effects of a wind (increasing in altitude) blowing (a) *with* and (b) *against* a sound wave. Refraction occurs because the speed of sound is increased when it moves along with a wind and decreased when it moves against a wind. The same thing happens (Fig. 13.34) when the density of the medium varies as a result of temperature variations. As a consequence, it is possible to hear things out of doors from farther away on cold days (especially over a body of water, where the air temperature is more likely to increase with altitude) than on warm days. Most often, the air near the ground is warmer than that above it, and distant sounds tend to be deflected upward. That's why you can often see a flash of lightning (beyond around 23 km or so away) and never hear the associated clap of thunder. A thunderhead is typically about 4 km high, and the sound from it heading toward the ground slowly climbs upward and away, never reaching a distant observer.

Interference

As with all waves, the overlapping of two sounds of the same frequency can manifest interference. A stereo audio system (Fig. 13.35) driven by a generator at 1 kHz will fill a room with waves about 34 cm long. Where the two waves overlap, an interference pattern will exist. A listener moving around in the room (with one ear covered) would hear a dramatic variation in intensity going from almost zero, where the waves are out-of-phase by 180°, to loud maxima, where they are in-phase.

When music is playing, a room is filled with many such patterns, each corresponding to a different frequency and each with all its maxima (except for the central one) and minima located at different places. This, along with the inevitable reflections, tends to obscure the interference pattern. Still, if you want the best sound in the middle of the room, the speakers have to pump in-phase, making compressions and rarefactions in step. Reversing the leads to one speaker will cause it to be 180° out-of-phase with the other. That creates a line of nodes running down the middle of the room, along which the intensity is reduced.

Diffraction

Two people walking through a forest can hold a normal conversation even though there might be a line of trees between them. There are never substantial "shadow" regions behind the trees where no sound reaches. Sound waves, like all waves, more or less bend around obstructions, which is in part why you can hear around corners, and why you can be heard even if you talk with a hand directly in front of your mouth. *Whenever a portion of a wavefront is obstructed in any way (such that any part of it changes its*

(a)

(b)

Figure 13.33 Windspeed increases with altitude. Moreover, when sound moves in moving air its speed increases in the direction of the wind and decreases against the wind. Since the frequency of sound is unaltered, λ must change, and that bends the wavefronts and alters the direction of propagation.

Max Min Max Min Max Min Max Min Max Min

Water ripples formed by two in-phase oscillators. The waves combine to form a region up the center where the disturbance is a maximum. There, we see peaks (bright) and troughs (dark) forming a long wedge shape. On either side of this wedge is a narrow gray band along which the disturbance is canceled. These are interference minima and are highlighted with magenta lines. The gold lines lie along the maxima.

(a)

Figure 13.34 The speed of sound is greater in warm air and the wavelength is longer, which causes a bending of the wavefronts. The same idea has been used to make lenses that reconfigure sound waves. Lenses will be discussed later when light is considered.

(b)

Figure 13.35 The interference of two sound waves. Where a crest (dark region) overlaps a crest, or a trough (light region) overlaps a trough, there will be reinforcement (●). Where a trough overlaps a crest, there will be cancellation (○). Compare this with the photo on p. 467.

phase or amplitude or both), there will be some alteration in the wave's direction of propagation as it progresses beyond the obstruction. This deviation from straight-line propagation is known as **diffraction** (Ch. 27).

A general rule maintains that diffraction will be pronounced when the obstacle around which the wave is passing or the aperture through which it is traveling is small compared to the wavelength. That's why a tree will cast a substantial shadow in light (which has a tiny wavelength) and not in sound (which has a comparatively large wavelength). Sound waves quickly close up as they advance past the tree.

When most of the sound energy of a wave diffracts beyond an obstacle, little of it will remain to be reflected backward. Therefore reflection, in a sense, is the complement of diffraction. If the reflection from a body is to be enhanced, λ should be smaller than the target—a conclusion already discussed as it relates to bats, moths, and sonar (p. 432).

Imagine plane waves impinging on a hole in an opaque screen (Fig. 13.36)—the sound of distant traffic pouring indoors through an open window. If the wavelength is small compared to the size of the hole, there will be little diffraction, and the sound will propagate forward essentially along a straight line. On the other hand, if λ is comparable to or larger than the hole, the transmitted wave will "bend" well beyond the straight perpendicular projection of the aperture. You can hear outside noises way off to the side of an open window. The same thing will happen to a loudspeaker—it behaves as if it were an illuminated hole (Fig. 13.37). A speaker that is, say, 0.15 m (6 in.) in diameter projects waves of 100 Hz (λ = 3.4 m) in all forward directions almost as if it were a point source. By comparison, at 10 kHz (λ = 0.03 m), the radiation from the same speaker will be predominantly in a narrow (< 30°) cone in the forward direction (Fig. 13.36). That is one reason why tweeters—high-frequency speakers—are made small (and it is something to consider when finding a place to sit and listen to recorded music).

13.10 The Doppler Effect

Every kind of wave moves through a homogeneous medium at a constant speed that depends only on the physical properties of the medium. That's true irrespective of any motion of the source—it launches the waves and away they go. Still, the perception of a wave's frequency and wavelength can be significantly altered by a relative motion between the observer and the source. Almost everyone has heard a shift in frequency when a car blowing its horn passes by. The pitch while it is approaching is somewhat higher than when at rest, and as it passes, the pitch drops. The result is a kind of eeeeoooo sound. This phenomenon is known as the **Doppler Effect** after the Austrian physicist Johann Doppler, who first worked out the analysis for sound in 1842. He also correctly suggested that the effect would apply to light. The predictions of the theory were soon confirmed (1845) by Ballot, who arranged an acoustical extravaganza that must have been a joy to behold. For two days, a locomotive, flatcar, and band of trumpeters performed—coming and going—for a group of musician observers.

Figure 13.38 indicates how a frequency shift arises for a source moving toward a stationary observer at a rate of half the speed of sound. The motion of the source essentially closes up the gap between successive wavefronts. Thus, the observer will measure a wavelength that is halved and a frequency that is doubled. In the same way, Fig. 13.39 depicts a stationary source radiating sound to the right at speed v and an observer moving left, this time also at v. The listener rushing toward the source overtakes the wavefronts, intercepting more of them per second.

(a)

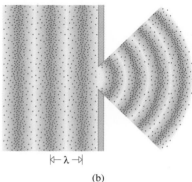

|← λ →|

(b)

Figure 13.36 (a) When the wavelength is smaller than the aperture, the wave passes through in a fairly straight path. (b) When the wavelength is comparable to or larger than the aperture, the wavefronts bend and the disturbance propagates into regions beyond the straight-ahead path.

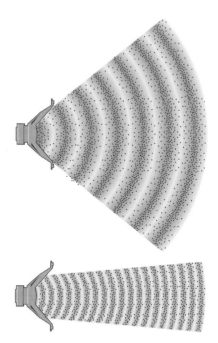

Figure 13.37 A small source behaves like an illuminated aperture. The amount of diffraction increases as the wavelength approaches the size of the aperture.

The simplest analytic situation is a medium at rest, with a point source moving at speed v_s and an observer moving at v_0, both traveling along the same line, in either direction. Figure 13.40 is a multiple exposure showing the point source's location at five consecutive moments and the associated wavefronts. Let's start time ($t = 0$) when the source was at point 0 whereupon it moved a distance $v_s t$ in a time t. Since its frequency is f_s, it has generated a number of waves equal to $f_s t$. Propagating at the speed of sound, the front emitted at $t = 0$ is now expanded out to a radius of vt, where it has encountered the observer who is moving to the right at speed v_0. Notice that the wavelength on either side of the source is different. On the left, where the observer is, λ equals the distance between the source and the observer ($vt + v_s t$) divided by the number of waves that span that space ($f_s t$), and so

$$\lambda = \frac{v + v_s}{f_s} \tag{13.15}$$

On the other side, the same number of waves is squeezed into a space of ($vt - v_s t$), and so λ in the forward direction equals ($v - v_s$)/f_s. Once we have the wavelength, we can find the frequency by dividing the relative speed by λ.

As far as the observer is concerned, she moves right at v_0 (with respect to the medium) as the sound travels toward her at v (with respect to the medium), and so she sweeps through the wavefronts at a rate of ($v + v_0$). Consequently, the observer will perceive a frequency of

$$f_0 = \frac{v + v_0}{\lambda}$$

Using Eq. (13.15), this expression becomes

$$\frac{f_0}{v + v_0} = \frac{f_s}{v + v_s} \tag{13.16}$$

In this derivation, v_0 and v_s are both positive in the directions shown. Changing those directions simply changes either sign. Accordingly, if the observer is at rest ($v_0 = 0$) and the source is moving toward it, v_s must enter Eq. (13.16) as a negative number and $f_0 = v f_s/(v + v_s)$ so that $f_0 > f_s$. If instead the source moves away, v_s is positive and $f_0 < f_s$, as observed. When the source is at rest, $v_s = 0$ and $f_0 = f_s(v + v_0)/v$. If the observer moves toward it (as in Fig. 13.39), v_0 is positive ($f_0 > f_s$) and if it moves away, v_0 is negative and $f_0 < f_s$.

Example 13.10 An automobile traveling at 20.0 m/s (45 mi/h) blows a horn at a constant 600 Hz. Determine the frequency that will be perceived by a stationary observer both as it approaches and recedes. Take the speed of sound to be 340 m/s.

Solution: [Given: $v_s = \pm 20.0$ m/s, $v = 340$ m/s, and $f_s = 600$ Hz. Find: f_0.] Referring to Fig. 13.40, we set $v_0 = 0$ and then calculate the approach from Eq. (13.16) with $v_s = -20.0$ m/s and the recession with $v_s = +20.0$ m/s. Thus, as the car approaches the observer

$$f_0 = \frac{v f_s}{v + v_s} = \frac{(340 \text{ m/s})(600 \text{ Hz})}{(340 \text{ m/s}) + (-20.0 \text{ m/s})}$$

which equals $\boxed{638 \text{ Hz}}$. As it recedes

$$f_0 = \frac{v f_s}{v + v_s} = \frac{(340 \text{ m/s})(600 \text{ Hz})}{(340 \text{ m/s}) + (+20.0 \text{ m/s})}$$

which equals $\boxed{567 \text{ Hz}}$.

▶ **Quick Check:** $v/(v + v_s) = (340 \text{ m/s})/(340 \text{ m/s} \pm 20 \text{ m/s})$—namely, 0.944 and 1.063; 0.944(600 Hz) = 567 Hz; 1.063(600 Hz) = 638 Hz.

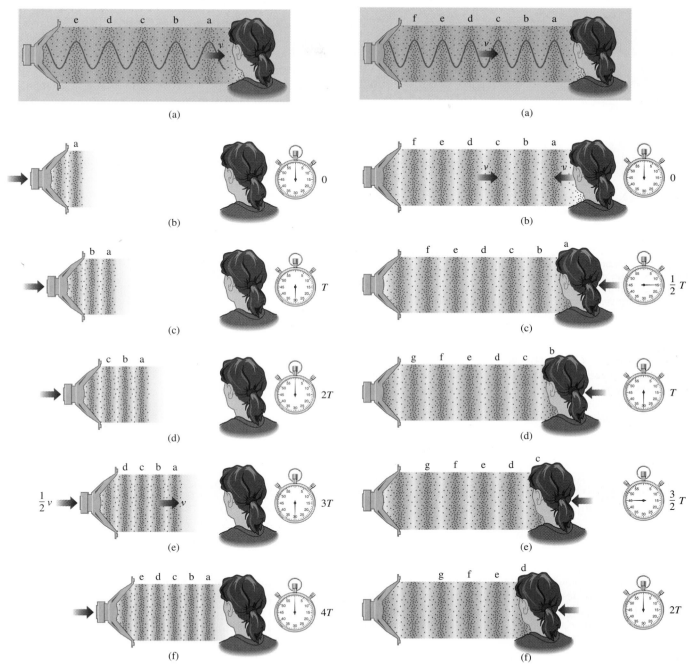

Figure 13.38 The Doppler Effect for a source moving toward a stationary observer at half the speed of sound (v). Compare the wavelength in (a), where the source is at rest, to the wavelength in (f), where it is moving at $\frac{1}{2}v$. Note how the source advances $\frac{1}{2}\lambda$ during the emission of one complete wave.

Figure 13.39 The Doppler Effect for an observer moving at speed v toward a stationary source of sound of speed v. Notice how the moving observer receives fronts a, b, c, and d in a time of two periods. Had she remained at rest as in (a), she would have received only two wavefronts a and b in that time.

All this traveling around is taking place with respect to the stationary medium, and the amount of the frequency shift actually depends on who is doing the moving—it is *not* symmetrical. The relative motion of source to observer is important

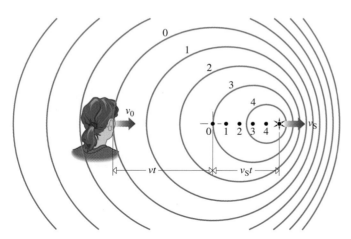

Figure 13.40 The source on the right is moving to the right at speed v_s, which is taken to be positive. The observer is also moving to the right, this time with a speed v_0, which is taken as positive.

here, but so, too, is the motion of each with respect to the medium. Whichever is moving relative to the medium will experience a wind blowing. Thus, if we recompute the first part of Example 13.10 for an observer moving toward a stationary horn at 20.0 m/s, instead of getting a frequency of 638 Hz, we get $f_0 = (v + v_0)f_s/v = (340 \text{ m/s} + 20 \text{ m/s}) \times (600 \text{ Hz})/(340 \text{ m/s}) = 635 \text{ Hz}$.

When waves are sent out from a stationary transmitter and reflect back off some moving target, that target functions as though it were a source of waves. The resulting Doppler shift (which is usually determined by beating the returning wave against a signal having the original frequency) provides the speed of the target. The police do this little trick with microwaves to clock passing cars on the highway, and the same approach is used to track both satellites and dangerous weather conditions. Likewise, Doppler-shifted ultrasonic waves (8 MHz) are being used to monitor the flow of blood, aiding in the diagnosis of conditions such as deep-vein thrombosis.

Example 13.11 Derive a general expression for the Doppler shift seen when a wave is bounced off an approaching target and returned to the stationary source, where it is viewed by an observer at rest.

Solution: There are two Doppler shifts involved in this process: the wave received by the target is shifted and then reemitted, and the returning wave is shifted when viewed by the final observer. Thus, the target (the observer in Fig. 13.40) receives the wave at a frequency $f_0 = f_t$ where Eq. (13.16), with $v_s = 0$, yields

$$f_t = \frac{(v + v_t)f_s}{v}$$

This same frequency is next reflected, becoming the source wave to be received by the final observer. But now this reflecting source is receding (Fig. 13.40); to keep track of the signs, v_s becomes $-v_t$ in Eq. (13.16).

Thus, $f_s = f_t$ in Eq. (13.16) where the observer is at rest ($v_0 = 0$), and

$$f_0 = \frac{vf_t}{v - v_t}$$

Notice that both of these shifts raise the frequency because the target is approaching. Substituting for f_t, we have

$$\boxed{f_0 = \frac{(v + v_t)f_s}{v - v_t}} \qquad (13.17)$$

If the target were receding, the signs in front of v_t would have to be changed accordingly.

▶ **Quick Check:** First, the units are right. Second, the observed frequency (f_0) is greater than the source frequency (f_s), which is correct. Third, as $v_t \to 0$, $f_0 \to f_s$, which is as it should be.

Example 13.12 A sound wave at 1000 Hz is sent out from a stationary source toward an approaching target. If the reflected wave is measured to have a frequency of 1200 Hz, what is the speed of the target? Take the speed of sound to be 332 m/s.

Solution: [Given: $f_s = 1000$ Hz, $f_0 = 1200$ Hz, and $v = 332$ m/s. Find: v_t.] Since this is a bounce-off-the-target Doppler problem, Eq. (13.17) is immediately suggested; thus

(continued)

(continued)

$$f_0 = \frac{(v + v_t)f_s}{v - v_t}$$

and $\quad v_t = \dfrac{(332 \text{ m/s})(200 \text{ Hz})}{2200 \text{ Hz}} = \boxed{30.2 \text{ m/s}}$

Hence $\qquad f_0(v - v_t) = (v + v_t)f_s$

▶ **Quick Check:** $f_0/f_s = 1.2 = (332 \text{ m/s} + 30 \text{ m/s})/$

and $\qquad v_t(f_s + f_0) = v(f_0 - f_s)$

$(332 \text{ m/s} - 30 \text{ m/s})$.

Thus $\qquad v_t = v\dfrac{f_0 - f_s}{f_0 + f_s}$

As applied to light, the Doppler Effect provides a key insight into the creation of the Universe. The light from the vast majority of galaxies is seen to be shifted toward the low-frequency end of the spectrum. This shift implies that if the Doppler Effect is the cause (and no other possible mechanism is known), then all these thousands of millions of galaxies must be moving away from us. A notion that can only be reasonable if the Universe is expanding so that every galaxy is moving away from every other one and ours is neither at the center nor flying off on its own. If that picture is correct, then the "observed" expansion must have begun back at some moment of creation. From today's recession rates and the apparent size of the Universe, we conclude that it came into being ≈ 17 thousand million years ago in a cosmic fireball known as the *Big Bang*. One can almost hear the trumpeters on the flatcar celebrating it all.

Core Material

Sound is *a compression wave in any material medium oscillating within the frequency range of 20 Hz to 20 kHz.* The **intensity** (I) of a sound wave is defined as *the average power radiating divided by the perpendicular area across which it is transported:*

$$I = \frac{P_{av}}{A} \qquad [13.1]$$

The speed of sound in a diatomic gas such as air is given by

$$v = \sqrt{\frac{1.4P}{\rho}} \qquad [13.3]$$

Sound travels in air at 331.45 m/s at STP. Over the usual range of temperatures encountered at sea level, the speed of sound changes by about 0.60 m/s per change of 1.0 degree Celsius.

The **intensity-level** (or *sound-level*) of an acoustic wave is defined as *the number of factors of 10 that its intensity is above the threshold of hearing:* $I_0 = 1.0 \times 10^{-12}$ W/m^2. The intensity-level is measured in *decibels* (abbreviated dB) and is given by

$$\beta = 10 \log_{10} \frac{I}{I_0} \qquad [13.4]$$

For a change from I_1 to I_2

$$\Delta\beta = 10 \log_{10} \frac{I_2}{I_1} \quad \text{in dB} \qquad [13.9]$$

When two wavetrains of the same wavelength and amplitude traveling in opposite directions overlap, they create a **standing wave** (p. 458). For a system with zero displacement at both ends

$$L = \tfrac{1}{2}N\lambda \qquad (N = 1, 2, 3, \ldots) \qquad [13.10]$$

and the wavelength of the Nth mode is $\lambda_N = 2L/N$. The resonant frequencies of a string fixed at both ends are

$$f_N = \frac{N}{2L}\sqrt{\frac{F_T}{\mu}} \qquad (N = 1, 2, 3, \ldots) \qquad [13.11]$$

For a pipe open at one end

$$L = \tfrac{1}{4}N\lambda \qquad [13.12]$$

and

[*antinode at one end*] $\quad f_N = \dfrac{Nv}{4L} \qquad (N = 1, 3, 5, \ldots) \qquad [13.13]$

For a pipe open at both ends

$$L = \tfrac{1}{2}N\lambda \qquad [13.10]$$

and

[*antinode at both ends*] $\quad f_N = \dfrac{Nv}{2L} \qquad (N = 1, 2, 3, \ldots)$

$$[13.14]$$

The Doppler Effect is a shift in frequency when the source and/or the observer are in motion (p. 469):

$$\frac{f_0}{v + v_0} = \frac{f_s}{v + v_s} \qquad [13.16]$$

The Doppler shift seen when a wave is bounced off an approaching target and returned to the stationary source where it is viewed by an observer at rest is given by

$$f_0 = \frac{(v + v_t)f_s}{v - v_t} \qquad [13.17]$$

Suggestions on Problem Solving

1. There are several facts associated with the notion of sound-level that are worth committing to memory. An intensity-level that is +10 dB higher or −10 dB lower than another will be heard to be twice as loud, or twice as soft, respectively. Doubling (or halving) either the acoustic power or intensity will increase (or decrease) the sound-level by 3 dB. A change of ≈1 dB is the smallest audible increment in loudness you are likely to be able to hear. Knowing this kind of thing helps to establish the physical meaningfulness of numerical answers and that should always be a primary concern.

2. When a physical entity, a string or a bar, vibrates and makes a sound, it pumps back and forth and sets the air around it vibrating at the same rate, the same *frequency*. The corresponding wavelength of the string depends on its physical conditions and the generated wavelength of the sound depends on its physical conditions, as manifested by the speed of the wave. The two wavelengths will generally *not* be the same. The natural emphasis is therefore on frequency and speed. **The energy of a wave is associated with its frequency** (p. 454).

3. Bear in mind that the standing-wave modes for all sys-

tems will have the same basic form; namely, the frequency will equal the speed over the wavelength. For a system bounded by two nodes (or two antinodes), the wavelength is some fraction of the fundamental $\lambda_1 = 2L$; that is, $2L/N$ where ($N = 1, 2, 3, \ldots$). Hence, $f_N = v/(2L/N)$. When the system is bounded at its ends by a node and an antinode, respectively, $\lambda_1 = 4L$ and only odd fractions of that (namely, $4L/N$ where $N = 1, 3, 5, \ldots$) are possible. Each different type of wave will have its own speed, but these frequency expressions hold for all of them. Try to think in terms of $f_N = v/\lambda_N$, where the form of λ_N can be determined from a mental picture of the particular mode pattern, such as Figs. 13.25, 13.28, or 13.29—that may be easier for you than memorizing the equations. Professionals store the basics and derive the rest as needed.

4. If you are having difficulties with the signs of any of the terms when doing Doppler Effect problems, remember that if the source and observer are approaching each other there must be a shift up in pitch and, hence, down in wavelength. If they are receding from one another, the shift is down in pitch and up in wavelength.

Discussion Questions

1. The behavior of sound waves can be studied using electronic frequency generators, microphones, and oscilloscopes. Suppose, then, that the speaker of Fig. Q1 sends out a single short (5 ms) compression pulse that travels into the doubly-open tube, which is 20 m long. What signal will appear on the scope as provided by the pressure microphone at the middle of the pipe?

Microphone

Signal generator   Oscilloscope

Figure Q1

2. In December 1916, a major disaster befell a contingent of the Austrian army in the Alps fighting in World War I. Almost

as soon as they began a cannon barrage, an avalanche of snow came roaring down, burying thousands of soldiers. What does that suggest about sound?

3. As an orchestra literally warms up, its sound changes. What do you think happens to the pitch of the strings and the wind instruments as a result of their being played for a while? Explain.

4. According to Galileo, "a glass of water may be made to emit a tone merely by the friction of the fingertip upon the rim." What more can you now say about the ringing goblet or any bell?

5. The ear has a protective overload mechanism called the *acoustic reflex*. A sound in excess of about 85 dB causes muscles attached to the eardrum and ossicles to engage, which provides a safety margin of about 20 dB or 30 dB (the equivalent of ear plugs). This reflex takes about 30 to 40 ms to cut in, and maximum effect only occurs after about 150 ms. What is good and bad about such a system in our modern world? Why do many hunters show loss of hearing (especially in the right ear)?

6. Suppose you inhale a little helium (too much will displace oxygen and you will get into trouble even though helium is otherwise nontoxic) and speak while exhaling it. The resulting voice will have a comic high-pitched sound. Why? What would happen to an organ pipe blown with carbon dioxide?

7. When a string on, say, a violin is plucked, particularly toward one end, it initially takes on a nearly triangular shape

Figure Q11

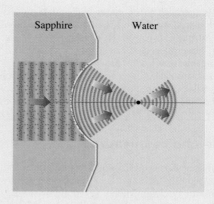

Figure Q15

rather than a sinusoidal one. Why does that result in a more "metallic" sound? Similarly, plucking the strings of a guitar with the fingers produces a more mellow sound than using a pick. Remember that the banjo is played with a pick and the harp with the fingers. Explain. Keep in mind that a signal with fine details, such as sharp corners, will require high-frequency Fourier components in order to synthesize it.

8. Suppose you sing or yell into an open piano with its dampers lifted off the strings. You are likely to hear the instrument "speak" back in response. Explain. Imagine that you have two identical tuning forks and you tap one and then place it near the other. What, if anything, will you hear if you now silence the first fork? "If one bows the bass string on a viola rather smartly," wrote Galileo, "and brings near it a goblet of fine, thin glass having the same tone as that of the string, this goblet will vibrate and audibly resound." Explain.

9. As a rule, instruments such as the piano that utilize struck strings will arrange to have the striking take place at a point 1/7th of the way down the string. Considering the fact that the 7th harmonic is generally unpleasantly dissonant, explain why the strings are struck where they are.

10. Imagine two people turning a jump rope while a third hops up and down. Describe every aspect of the process using the ideas developed in this chapter.

11. The tuning fork, which was invented in 1712 by John Shore, Handel's trumpeter, can vibrate (Fig. Q11) briefly in a mode that produces its first overtone, or clang tone, at a frequency of 6.27 times the fundamental. If you wish to have the fork produce a nearly pure tone and so suppress the high-pitched clang tone, where should it be struck? Explain. Compare the tuning fork to the rod in the previous question.

12. Discuss two ways in which temperature changes will affect the pitch of an organ pipe.

13. A tuning fork is mounted to a hollow open-ended wooden sounding box that amplifies the sound. Explain how it manages to do that. What can you anticipate would be the difference between the sounding boxes for two such forks at, say, 1000 Hz and 50 Hz?

14. In light of the Question 13, what can you say about the reasons for the difference in size between a violin (60 cm long) and a double bass (200 cm long)?

15. Fig. Q15 shows the lens assembly for an acoustic microscope. Plane waves from a piezoelectric transducer enter from the left. Explain how the device works.

16. Fig. Q16 shows an early phonograph or "Speaking Machine" devised by Thomas Alva Edison (1877, patent number 200,521). The part labeled B is the "mouthpiece" and a clockwork mechanism advances it and the output cone C relative to the turning cylinder A. The latter has "a helical indenting groove, cut from end to end, say, ten grooves to the inch. Upon this is placed the material to be indented, preferably metallic foil. . . . It is obvious that many forms of mechanism may be used. . . . a revolving plate may have a volute spiral cut on both its upper and lower surfaces." How does the thing work?

17. The long glass tube (Kundt's tube) in Fig. Q17 is sealed at the lower end with a

Figure Q16

Figure Q17

movable plunger and at the upper end with a loosely fitted piston attached to a stationary rod (itself fixed at the middle). The tube contains an even layer of light, dry, very fine powder (lycopodium, precipitated silica, or cork dust). When the rod is stroked lengthwise with a rosined cloth or chamois, it gives out a high-pitched sound, and the powder flies up in violent agitation. When the tone is stopped, the powder settles back into heaps, and by adjusting the location of the plunger, these piles become

well defined, uniform, and periodic. The device, which was invented by A. E. Kundt of Germany in 1876, can be used to measure the speed of sound (Problem 66). Discuss what is happening to the rod, to the air in the tube, and to the powder.

18. If you strike a tuning fork, hold it vertically near your ear, and rotate it slowly around its axis, the sound will rise and fall in intensity. Why does this happen?

Multiple Choice Questions

1. The higher the frequency of a sound wave, the   (a) shorter the wavelength   (b) greater the speed   (c) lower the pitch   (d) greater the amplitude   (e) none of these.

2. When the loud music in the living room causes the doors on the china cabinet to buzz, the effect is due to (a) interference   (b) diffraction   (c) superposition (d) resonance   (e) none of these.

3. The speed of sound in air at 40°C by comparison to the speed at 0°C is   (a) much lower   (b) a little lower (c) the same   (d) higher   (e) none of these.

4. Given: two sources of sound waves having the same amplitude, one with frequency 250 Hz and the other with frequency 500 Hz. A comparison of their relative loudness would reveal that   (a) the first was twice as loud   (b) the second was louder   (c) the two were of equal loudness (d) the second was one-quarter as loud   (e) none of these.

5. A dB-meter placed in front of an audio system reads 62 dB. The volume is turned up and the meter now reads 82 dB. Evidently, the sound got   (a) twice as loud (b) half as loud   (c) 20 times louder   (d) 10 times louder (e) none of these.

6. A 60-dB sound-level corresponds to an intensity of (a) 10^{-10} W/m^2 (b) 10^6 W/m^2 (c) 10^{-12} W/m^2 (d) 10^{12} W/m^2 (e) none of these.

7. A sound that has an intensity-level of 100 dB is how many times more intense than a sound of 20 dB?   (a) 5   (b) 8 (c) 1000   (d) 10^8 (e) none of these.

8. On a day when you are listening to music at a sound-level of 20 dB above the threshold of audibility, an average person would say that it was   (a) very loud   (b) of normal listening level   (c) very soft   (d) slightly loud (e) none of these.

9. A sound-level meter placed in front of the loudspeaker of a 60-W audio system reads 70 dB. All else being equal, when placed in front of a 120-W system, the meter will read (a) 120 dB   (b) 140 dB   (c) 63 dB   (d) 73 dB   (e) none of these.

10. In order to cause the loudness of a sound to appear half its original value, it must be   (a) increased by 50 dB (b) decreased by 50 dB   (c) decreased by 10 dB (d) decreased by 3 dB   (e) none of these.

11. A doubly-open organ pipe, in comparison to a singly-open pipe of the same length, has a fundamental frequency that is   (a) half as great   (b) twice as great   (c) three times larger   (d) $1/\pi$ times smaller   (e) none of these.

12. The flute and piccolo are in the same family of instruments, and each functions as a pipe open at both ends. Both have the same fingering but different lengths. The flute is about 66 cm long and is an octave lower than the piccolo—that is, the piccolo plays at twice the frequency of the flute. It follows that the piccolo, in comparison to the flute, is about   (a) twice as long (b) half as long   (c) eight times as long   (d) the same length   (e) none of these.

13. Fig. MC13 shows the frequency spectrum of a flute playing a D$_4$ note (294 Hz). It's clear from the diagram that the flute is likely to be   (a) made of metal   (b) 66 cm long (c) open at both ends   (d) cylindrical   (e) none of these.

Figure MC13

14. What will happen to the fundamental frequency of an organ pipe open at both ends if a cap is placed over one end sealing it? The frequency will   (a) go up 20%   (b) be halved   (c) stay the same   (d) double   (e) none of these.

15. If one stereo sounds 16 times louder than another, the difference in their sound-levels is about   (a) 40 dB (b) 30 dB   (c) 16 dB   (d) 160 dB   (e) none of these.

16. A clarinet has a single vibrating reed at one end coupled to a cylindrical resonance tube that is flared open at the other end. The coupling of reed and air column produces an interesting spectrum (Fig. MC16) even though there is a node at the reed. Typically, there is a strong first harmonic, little second, a strong third, some fourth, a strong fifth, and an appreciable sixth. It appears that the instrument functions as if it were   (a) purely closed at both ends   (b) purely open at one end   (c) a mix of somewhat open at both ends and yet effectively closed at one   (d) purely open at both ends   (e) none of these.

Frequency spectrum
Clarinet F₃

Figure MC16

17. If you stick a small lump of clay to the end of one or both tines of a tuning fork, you can expect the fundamental frequency to (a) remain unchanged (b) go down (c) go up (d) vary from moment to moment (e) none of these.

18. Two tuning forks of frequency 380 Hz and 384 Hz are excited simultaneously. The resulting sound will waver in intensity with a frequency of (a) 382 Hz (b) 0 (c) 4 Hz (d) 2 Hz (e) none of these.

19. One is likely to hear low-pitched sounds coming from behind the head better than high-pitched ones because of (a) diffraction (b) dispersion (c) interference (d) refraction (e) none of these.

20. A vehicle heading down a highway at 80 km/h is being pursued by a police car that is right behind it, also traveling at 80 km/h. If the cop's siren puts out a 500-Hz wail, what frequency will the perpetrator hear? (a) < 500 Hz (b) > 500 Hz (c) 500 Hz (d) 80 Hz (e) none of these.

Problems

ACOUSTICS
HEARING SOUND

1. [I] In music, the standard A_4 note has a frequency of 440 Hz. What is its period and wavelength at room temperature?

2. [I] A person standing at one side of a playing field on a cold winter night emits a brief yell. The short acoustical wavetrain returns 1.00 s later as an echo having "bounced off" a distant dormitory. Approximately how far away was the building?

3. [I] A groove on a monophonic phonograph record wiggles laterally such that its amplitude and frequency more or less correspond to the sound that is recorded. If at a given moment the needle is moving through the groove at 0.50 m/s, what would be the wiggle-wavelength for a 1.5-kHz tone?

4. [I] A low-frequency loudspeaker is called a woofer. At what frequency will a tone have a wavelength equal to the diameter of a 15-in. woofer? Take the speed of sound to be 344 m/s.

5. [I] A string vibrating at 1000 Hz produces a sound wave that travels at 344 m/s. How many wavelengths will correspond to 1 m?

6. [I] The intensity of the acoustic wave from an underground explosion measured 5.0 km away is 1.6×10^4 W/m². What intensity will be recorded at a site 50 km away? Assume no losses.

7. [I] A point source of sound waves emits a disturbance with a power of 50 W into a surrounding homogeneous medium. Determine the intensity of the radiation at a distance of 10 m from the source. How much energy arrives on a little detector with an area of 1.0 cm² held perpendicular to the flow each second? Assume no losses.

8. [I] At what temperature will the speed of sound in air at standard pressure equal 320 m/s?

9. [I] If a tuning fork puts out a tone at 440 Hz, what is its wavelength in air at 25°C?

10. [I] By what percentage does the speed of sound change when the air temperature rises from 0°C to 30°C?

11. [I] The contact time between the hammers and strings of a piano is one of the determining factors of the tone of the instrument; if it's too long, the higher overtones will be damped out. Though contact time varies across the keyboard, an average is about $\frac{1}{2}$ of a period. To two significant figures, what is the contact time for middle-C at 261.6 Hz?

12. [I] One of the most important acoustical characteristics of a room is its *reverberation time,* the time it takes for a sound to decrease 60 dB. What does that mean as far as the sound's intensity is concerned? For a concert hall, the reverberation time is typically 1 to 3 s.

13. [I] A radio playing quietly produces a sound intensity of about 10^{-8} W/m². What is the corresponding intensity-level?

14. [I] Someone playing a CD at 60 dB wants to make the music twice as loud. At what intensity-level should it be played?

15. [I] Two (otherwise identical) audio systems at a demonstration are blasting away, with one putting out 10 times the acoustic power of the other. What is the difference in their sound-levels?

16. [I] Two people having a normal conversation are 1.0 m from a sound-level meter. Approximately what will the meter read?

17. [I] Two audio systems each produce 50 W of acoustic power at the location of a microphone. What is the difference in their intensity-levels in dB at that point?

18. [I] Does a 0-dB sound-level mean there is no sound? Explain.

19. [I] The intensity of a sound is tripled. By how many decibels does it increase?

20. [I] On occasion, it turns out that we know the intensity-level of a sound and wish to find out the associated intensity. Show that

$$I = 10^{\beta/10} I_0$$

21. [I] The noise of traffic registers 77 dB on a meter. What intensity does that correspond to?

22. [I] A cannon produces a 90-dB sound-level at a certain distance from a detector. What will the device read when two such cannons at that same distance are fired?

23. [I] Twenty-five singers all singing with the same intensity produce a 65-dB sound. What is the intensity-level of the average chorus member?

24. [II] A bar of aluminum alloy 10 m long with a cross-sectional area of 1.0 cm^2 has a mass of 2.7 kg and a Young's Modulus of 7.0×10^{10} N/m^2. If the end is tapped at a rate of 100 Hz, how long will it take for the sound wave to reach the other end of the bar, given that $v = \sqrt{Y/\rho}$? What will be its wavelength?

25. [II] What is the sound-level 10 m away from a point source radiating 1.2 W of acoustic power?

26. [II] Two sounds have intensities of 10^{-2} W/m^2 and 10^{-10} W/m^2. What will be the difference in their intensity-levels as read by a sound-level meter?

27. [II] A small but noisy printer produces an acoustic intensity of 56×10^{-5} W/m^2 at a point 5.0 m away. Approximately what value will a dB-meter read at that location and, again, at 20.0 m from the printer?

28. [II] With Problem 27 in mind, what will be the *change* in intensity-level measured for any point source at 5.0 m and then at a 20.0-m distance?

29. [II] A small source emits nearly spherical waves with an acoustic power of 60 W. How far away must a sound-level meter be if it is to read 60 dB?

30. [II] The intensity-level of a large engine is measured to be 130 dB at a distance of 10.0 m. Approximately what intensity will exist at a point 100 m away?

31. [II] The sound-level 2.0 m from a pneumatic chipper is 120 dB. Assuming it radiates uniformly in all directions, how far from it must you be in order for the level to drop 40 dB down to something more comfortable?

32. [II] How much acoustic power impinges on a 10-cm^2 detector when the intensity-level is 70 dB?

33. [II] Given that the standard reference pressure for sound is $P_0 = 2 \times 10^{-5}$ Pa, write an expression for the intensity-level in terms of the pressure of the disturbance P. (Hint: p. 454.)

34. [II] Someone turns on a radio at 65.0 dB while vacuuming the floor at 80.0 dB. What will be the total intensity-level in the room?

35. [III] If the displacement of the oscillating air molecules in a sound wave is written in terms of a cosine function of amplitude A, then (with a bit of calculus and some effort) the corresponding gauge pressure can be expressed as

$$P = \frac{2\pi}{\lambda} BA \sin\left(\frac{2\pi}{\lambda}x - \omega t\right)$$

(a) Show that both sides of this expression have the same units. (b) What can you say about the displacement as compared to the pressure? (c) Show that the maximum value of the change in pressure (P_0) from the no-wave ambient value, the so-called *pressure amplitude*, is given by

$$P_0 = \frac{2\pi}{\lambda} \rho v^2 A$$

(d) Discuss, in general terms, why it is reasonable to have B and A in the equation for P.

36. [III] The pressure amplitude associated with the loudest sounds tolerable by human beings is roughly 30 Pa. Such pressure waves vary ± 30 Pa with respect to atmospheric (10^5 Pa). Determine (using the results of Problem 35) the maximum displacement of the air molecules in such a disturbance having a frequency of 1000 Hz at human body temperature. Take the density of the air to be 1.22 kg/m^3.

37. [III] If one audio system sounds 6 times louder than another, what is the difference in their intensity-levels?

PRODUCTION AND PROPAGATION OF SOUND
SOUND WAVES

38. [I] Two sound waves of angular frequencies 900.0 rad/s and 896.0 rad/s overlap. What is the resulting beat frequency?

39. [I] The sound from a tuning fork of 1000 Hz is beat against the unknown emission from a vibrating wire. If beats are heard at a frequency of 4 Hz, what can be said about the frequency of the wire?

40. [I] With Problem 39 in mind, what can you say if a small piece of tape is fixed to the tuning fork and the beat frequency now increases?

41. [I] On tapping two tuning forks, an observer hears a succession of intensity maxima arriving at a rate of one every 0.99 s. What is the difference in frequency between the two forks?

42. [I] A thin wire is stretched between two posts 50 cm apart. It is then bowed and thereby set into oscillation. What are the wavelengths of the fundamental and the first overtone of the system?

43. [I] A taut string is fixed at both ends, which are 0.50 m apart. It is then set into resonance at its 6th harmonic. Determine the wavelength of the oscillation and draw the standing-wave pattern.

44. [I] A string stretched between fixed posts is 250 cm long and oscillates in its fundamental mode at 100 Hz. Determine the speed of a transverse wave on the string.

45. [I] A piece of steel piano wire is held fixed at both ends under a tension of 100 N. The free length of wire is 1.00 m, and it has a mass of 2.5 g. What is its fundamental frequency?

46. [I] A narrow tube 1.00 m long is closed rigidly at one end and with a piston at the other. Given that the speed of sound is 335 m/s, what is the frequency of the tube's fundamental oscillatory mode?

47. [I] Show theoretically that for a string fixed at both ends, its standing-wave frequencies are given by

$$f_N = \frac{1}{2} N \sqrt{\frac{F_T}{Lm}}$$

where L is its length and m its mass.

48. [I] A trumpet is a bent tube roughly 140 cm long and closed at one end by the player's mouth. Determine the fundamental (which will be quite difficult to blow) and the first three overtones. Take the temperature to be 20°C.

49. [I] Two loudspeakers separated by 2.0 m are fed a 688-Hz signal that drives them both in-phase (Fig. 13.35). A microphone is placed at a point 6.75 m from one and 7.00 m from the other. The magnitudes of the pressure

variations, above and below atmospheric, at that point are essentially the same when either speaker acts alone. What can you say about the intensity there when both are radiating? Take the speed of sound to be 344 m/s.

50. [I] A police car, its siren blaring at 1000 Hz, is traveling at 20.00 m/s while chasing a garbage truck moving at 15.00 m/s a block in front of it. What apparent frequency will the garbage collectors hear? The speed of sound is 330.0 m/s.

51. [I] The setup in Problem 49 is changed slightly. By reversing the wires going to one of the speakers, it is made to move forward, creating a compression, just as the other speaker is moving backward, creating a rarefaction. What, if anything, will happen to the intensity at the microphone now?

52. [I] A train whistle is blown by an engineer who hears it sound at 650 Hz. If the train is heading toward a station at 20 m/s, what will the whistle sound like to a waiting commuter? Take the speed of sound to be 340 m/s.

53. [I] A car with its horn blaring at 500 Hz passed a woman standing on the street at 25 m/s. What frequency did she hear as the car receded into the distance? The speed of sound that day was 344 m/s.

54. [I] An ultrasonic wave at 80 000 Hz is emitted into a vein where the speed of sound is about 1.5 km/s. The wave reflects off the red blood cells moving toward the stationary receiver. If the frequency of the returning signal is 80 020 Hz, what is the speed of the blood flow?

55. [I] A man running toward the stage in a theater hears an A_4 note from a stationary tuning fork to have a frequency of 441 Hz instead of its more normal 440 Hz. About how fast is he going?

56. [II] Galileo determined experimentally that the fundamental frequency of a string fixed at both ends was inversely proportional to both its diameter and to the square root of its density. Prove that this description is true theoretically.

57. [II] A B-string from a guitar is held fixed at both ends under tension with a vibrating length of 33 cm. It oscillates at its fundamental frequency of 246 Hz. What are the wavelengths on the string, and in the air at room temperature?

58. [II] Imagine a hypothetical piano with all strings made of the same material and all under the same tension. The piano extends from 27.5 Hz to 4186 Hz, which is over seven octaves—that is, seven doublings of frequency. If the highest note corresponds to a string 15 cm long, how long will the lowest string have to be? What can you conclude about this approach to piano design?

59. [II] Two identical piano strings are both tuned to 440 Hz. The tension in one is then increased by 1.00%, and both strings are activated so that they sound their fundamental frequencies. What will be the resulting beat frequency?

60. [II] A narrow glass tube 0.50 m long and sealed at its bottom end is held vertically just below a loudspeaker that is connected to an audio generator and amplifier. A tone with a gradually increasing frequency is fed into the tube, and a loud resonance is first observed at 170 Hz. What is the speed of sound in the room?

61. [II] An organ pipe that ordinarily sounds at 600 Hz at 0°C is connected to a source of helium at that temperature. At what frequency will it now operate?

62. [II] A speaker (Fig. P62) attached to a generator emits a constant 340-Hz tone, which is soon reflected off a board. The microphone is moved along a perpendicular line from source to board, and it is found that there is an intensity maximum at point P_1 and a minimum at P_2, which is 25 cm closer to the speaker. What is the speed of sound?

Figure P62

63. [II] A quartz tube open at both ends has a fundamental resonant frequency of 200 Hz at 0°C. Neglecting any changes in length, by how much will the fundamental change when the tube is sounded in a chamber at 40°C?

64. [II] A C-flute with all its holes covered plays a middle C (262 Hz) as its fundamental. Assuming room temperature (20°C) and overlooking end corrections, how long should the flute be from embouchure hole to end?

65. [II] A wire stretched between two posts and under a tension of 200 N oscillates at a fundamental frequency of 420 Hz. At what tension would it oscillate instead at 430 Hz?

66. [II] Referring to Kundt's tube (Fig. Q17), if v_a and v_r are the speeds of sound in the air and rod, respectively, and if L is the length of the rod and l the node-to-node distance, show that

$$v_r = \frac{v_a L}{l}$$

In other words, knowing the speed of sound in air, we can determine it in the material of the rod.

67. [II] A copper bar 1.00 m long is clamped at its middle and set vibrating. Given that $v = \sqrt{Y/\rho}$ and the bar has a density of 8.9×10^3 kg/m^3 and a Young's Modulus of 11×10^{10} N/m^2, what will its frequency be?

68. [II] Two small loudspeakers separated by a distance of 3.00 m emit a constant tone of 344 Hz (Fig. P68). A microphone is moved along a line parallel to and 4.00 m from the line connecting the speakers. It is found that intensity maxima exist on the center line and directly opposite each speaker. From this information, calculate the speed of sound. (Hint: The waves must arrive in-phase if they are to interfere to form a maximum.)

69. [II] Two small loudspeakers are set up as in Fig. P68. They emit a tone of 2000 Hz in-phase with one another. Point O is equidistant from the two and a microphone

Figure P68

Figure P71

placed there reads a maximum intensity. Now speaker S_1 is slowly moved away to the left along line S_1O, and the output of the mike gradually drops to near zero and then goes back to a maximum when the displacement reaches 0.172 m. What is the speed of sound?

70. [II] A point source of sound on top of a police car emits a signal at 1000 Hz. If the car is traveling in a straight line at 30 m/s, what will be the wavelength perceived by people standing on the road both directly in front of and behind the car? Take the speed of sound to be 335 m/s. What is the wavelength as measured in the car?

71. [II] Figure P71 shows a loudspeaker emitting a 340-Hz sine wave toward a pressure microphone 5.00 m away. The mike is 1.146 m above the table, and its output at this location is found to be a minimum. The temperature is not known, but it's not far from room temperature. Assume there are no phase shifts on reflection and determine the speed of sound.

72. [II] A sound wave at 1000 Hz is sent out from a stationary source toward a target approaching at 20.0 m/s. What will be the frequency of the returning signal? Take the speed of sound to be 340 m/s.

73. [II] An observer with a stationary source of sound sends out a signal directly toward an approaching target. Write an expression for the frequency of the beats heard by the observer.

74. [III] Consider a long, narrow rod of length L clamped at both ends. If the rod is rubbed with a rosined cloth, a compression wave will travel its length. Show that the expression for the resulting standing-wave modes is

$$f_N = \frac{N}{2L}\sqrt{\frac{Y}{\rho}}$$

where $N = 1, 2, 3, \ldots$. A novice playing with a violin is likely to move the bow lengthwise along a string. Explain why this movement will generally produce a shrill squeaking sound.

75. [III] Consider the expressions for two waves of the same amplitude with slightly different frequencies traveling in the same direction:

$$y_1 = A \sin(\omega_1 t) \quad \text{and} \quad y_2 = A \sin(\omega_2 t)$$

where we simplify things a little by just looking at them at the point $x = 0$. Now, supposing the two waves were to overlap, derive the following equation for the combined waveform:

$$y = 2A \cos[\tfrac{1}{2}(\omega_1 - \omega_2)t] \sin[\tfrac{1}{2}(\omega_1 + \omega_2)t]$$

and interpret each term in reference to Fig. 13.21.

76. [III] The two loudspeakers in Fig. P68 are emitting sounds of the same frequency in-phase. Show that the separation between them (which equals a) must exceed $\tfrac{1}{2}\lambda$ if cancellation via interference is to be observed anywhere in the region in front of the speakers.

14

Thermal Properties of Matter

CHAPTERS 14, 15, AND 16 are about internal energy, the energy stored via the motion and interaction of the atoms of a system. The discipline is called *thermal physics* and it deals, broadly speaking, with temperature (Chapter 14), and the transfer (Chapter 15) and transformation (Chapter 16) of energy. The study begins with our rather crude bodily appreciation of hot and cold; with the what-can-you-put-in-your-mouth-without-burning-it understanding of temperature. Our sensory perception of temperature, however unreliable, must correspond to some unseen aspect of the physical world. We will find that, *even though temperature may well be a basic quantity like*

Man has no Body distinct from his
    Soul;
  for that called Body is a portion of
    Soul
  discern'd by the five Senses, the
    chief
  inlets of Soul in this age.

Energy is the only life and is from
    the Body;
  and Reason is the bound or
    outward
  circumference of Energy.

Energy is Eternal Delight.
                WILLIAM BLAKE (1793)
          The Marriage of Heaven and Hell

Figure 14.1 Galileo's thermoscope is not called a thermometer because the scale was arbitrary. The egg-sized globe at the top is the sensor. The gas within it expands or contracts, and the liquid level rises and falls.

mass and time, it can usually be associated with the concentration of thermal energy in a material system. This chapter focuses on a variety of temperature-dependent properties of bulk matter.

The firebox is at the center of our civilization. Out of sight, in power plants everywhere on the planet, there are furnaces and reactors generating thermal energy that ultimately ends up driving our TVs and electric toothbrushes. Most cars, trains, ships, and planes burn fuels in heat engines of one sort or another. The First and Second Laws of Thermodynamics (pp. 550 and 570) are about energy and change. These ideas transcend the discipline, providing us with profound insights into life itself and the very workings of the Universe.

TEMPERATURE

Like so many other physical quantities, temperature was measured long before it was understood. Galileo appears to have invented (ca. 1592) the first device for indicating "degrees of hotness." He simply placed the end (Fig. 14.1) of an inverted narrow-necked flask, warmed in his hands, into a bowl of water (or wine). As it cooled, the liquid was drawn up, partially filling the neck. Air captured in the bulb at the top either expanded or contracted when subsequently heated or cooled, and the column fell or rose proportionately. Much the same effect can be seen by putting an inflated balloon in the freezer. The gas molecules inside the balloon keep it puffed up by constantly bombarding the inner walls. Cooling the gas, removing kinetic energy from its molecules, lessens their bouncing and the balloon collapses.

The medical applications of the thermometer were recognized almost immediately, and normal body temperature became a focus of interest. In 1631, J. Rey, a French physician, inverted Galileo's device, filling the bulb with water and leaving much of the stem with air in it. In that configuration, which is more like today's thermometer, the liquid's expansion operates the device. Within 70 years of Galileo's invention, sealed pocket-sized thermometers containing either alcohol or mercury were in use (both freeze at much lower temperatures than water, and sealing the devices isolated them from atmospheric pressure changes).

By exploring the behavior of matter with the aid of the thermometer, it soon became evident that there were a number of physical occurrences that happened at fixed temperatures. Boyle, Hooke, and Huygens independently recognized (1665) that that fact could provide a reliable reference point for any thermometer. Hooke suggested using the freezing point of water, and Huygens offered its boiling point. C. Renaldini (1694) used both the freezing and boiling points of water to standardize two widely spaced points on his thermometer (something Newton would do independently seven years later).

The scale we still use in the United States is a gift of the instrument maker G. D. Fahrenheit (1717). Not wanting to deal with negative values, he set the 0 mark at the coldest temperature he could produce, that of a mixture of water, ice, and sea salt. Following Newton's lead, Fahrenheit set the upper reference at normal human body temperature, which he took as 96 (probably because that number was divisible by 12 and easily halved and quartered). Upon dividing the distance between these two fixed points into 96 equal-sized divisions, or degrees, he could then extend the scale as high or low as he wished. That done, water froze at what turned out to be 32°F (which is read *degrees Fahrenheit*) and boiled at 212°F.

The scale that is still used in modern scientific work (although it is gradually being replaced) is associated with the name of a man who neither first proposed it nor ever actually produced it. Anders Celsius (1742) used the freezing and boiling of water for the reference points and then divided the distance between them into 100 equal parts—a convenient scheme for the user. Strangely enough, Celsius set the temperature of freezing water at 100° and boiling at 0°. Some years later, with these two numbers more reasonably interchanged (Fig. 14.2) the arrangement came to be known as the *centigrade scale* (from the Latin *centum,* meaning 100, and *gradus,* degree). The Tenth International Conference on Weights and Measures (1954) changed the name to the *Celsius scale.*

Since the Fahrenheit scale goes from 32° to 212° (a total of 180 divisions), and the Celsius scale goes from 0° to 100°, each Fahrenheit degree is smaller than a Celsius degree. A change of 1 C° is equivalent to a larger change of (180/100)F°, or (9/5)F°. If we move a certain number of Celsius degrees, T_C, from freezing (0°C), that corresponds to a change from freezing (32°F) of (9/5)T_C Fahrenheit degrees. In short

$$T_F = 32° + \tfrac{9}{5}T_C \quad \text{or} \quad T_C = \tfrac{5}{9}(T_F - 32°) \tag{14.1}$$

Rather than memorizing these equations, it might be easier to make the transformations logically. For example, human body temperature is 98.6°F, which is 66.6 Fahrenheit degrees above freezing—equivalent to (5/9)66.6 = 37.0 of the larger Celsius degrees above freezing (0°C). Hence, body temperature is 37.0°C.

Figure 14.2 A comparison of the Fahrenheit and Celsius temperature scales.

Example 14.1   Determine the temperature at which the Fahrenheit and Celsius thermometers have the same value.

Solution: [Find the value at which $T_F = T_C$.] Using Eq. (14.1) and setting $T_F = T_C$, we have

$$T_F = 32° + \tfrac{9}{5}T_F$$

hence

$$-\tfrac{4}{5}T_F = 32°$$

and

$$\boxed{T_F = T_C = -40°}$$

The two scales are the same at −40°.

▶ **Quick Check:** $T_C = (5/9)(T_F - 32°) = (5/9) \times (-72°) = -40°.$

As we will see, the manner in which any material—mercury, alcohol, water, glass, whatever—expands on heating is characteristic of that material. What this means practically is that thermometers that differ in their physical construction and yet have the same two fixed reference points will necessarily agree exactly only at those two points.

The range of temperatures we deal with is extensive, and there is now a whole arsenal of different kinds of thermometers, each with its own virtues. The familiar mercury-in-glass instrument is only useful between the points where mercury freezes and glass melts. Moreover, it can be used reliably only when its presence doesn't affect the temperature being determined. If you want to measure the temperature of a flea or a thermonuclear fireball, the old mercury-in-glass standby will not be of much help. Accordingly, there are electrical resistance thermometers, optical thermometers, thermocouples, and constant-volume gas thermometers, to name a few. Of these, the standard for accuracy and reproducibility is still the constant-volume gas instrument (Fig. 14.3), though it is large, slow, delicate, and inconvenient.

At present we know of no purely mechanical quantity—that is, one expressible in terms of mass, length and time only—which can be used, however inconveniently, in place of temperature. We are inclined to conclude that temperature probably is itself a basic concept.

A. G. WORTHING (1940)

Figure 14.3 A constant-volume gas thermometer. The chamber at the left is the sensing bulb. The gas in the bulb changes volume when it experiences a temperature change. That causes the levels of mercury in the U-tube manometer to shift, and h gives us the pressure in the probe. Lifting the movable mercury column on the right shifts the other two mercury levels and allows the volume of gas to be kept constant.

14.1 Thermodynamic Temperature and Absolute Zero

The Fahrenheit and Celsius scales have little or nothing to do with the fundamental nature of the concept of temperature. After all, the freezing point of water at Earth's atmospheric pressure has no obvious relationship to any basic aspect of the Universe. Presumably, an alien scientist on a planet that has no water will devise a thermometer that measures temperature equally well. More to the point, is there a universal zero of temperature linked to the very essence of matter and energy, a zero that all scientists (human or otherwise) might discover? The answer is yes, and it is called the *absolute zero of temperature.*

Sometime around 1702, G. Amontons devised an improvement on Galileo's thermoscope that has evolved into the modern constant-volume gas thermometer (Fig. 14.3).

Figure 14.4 shows the behavior of a typical gas as its temperature is lowered. The important feature is that the graph is a straight line—the pressure falls linearly until the gas liquefies and the process abruptly ends. Remarkably, all gases behave nearly the same way. Moreover, when the straight line for any gas is extended downward, it crosses the zero-pressure axis at a temperature of $-273.15°C$. Similarly, if the pressure is kept constant, the volume of a gas will collapse linearly, again approaching the zero-volume axis at the same $-273.15°C$ ($-459.7°F$), as in Fig. 14.5. This point is now believed to be the very limit of low temperature, **Absolute Zero**.

Amontons, studying the behavior of a number of gases, showed that each volume changed by the same percentage for a given change in temperature. Astutely, he recognized that a decrease in pressure down to zero would accompany a decrease in temperature down to some limiting value of coldness. Later, J. H. Lambert repeated these experiments with greater accuracy, concluding (ca. 1779) that there was indeed a temperature limit. "Now a degree of heat equal to zero," wrote Lambert, "is really what may be called absolute cold."

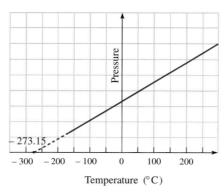

Figure 14.4 Pressure versus temperature for a gas at constant volume. As the gas is cooled, its atoms lose some of their thermal energy; they travel more slowly and collide with the chamber walls less frequently and less forcefully. As a result, the pressure in the chamber drops. All gases behave essentially the same way. Their P versus T graphs have different slopes, but they are all straight lines, and they all head toward $-273.15°C$.

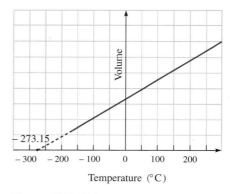

Figure 14.5 Volume versus temperature for a gas at constant pressure. As the temperature of a gas is lowered, its atoms are less energetic. To keep the pressure constant, the volume must decrease, which will make it easier for atoms to strike the chamber walls frequently, thereby sustaining the pressure. All gases shrink toward zero volume at $T = -273.15°C$.

Absolute Zero certainly seems the place to begin a temperature scale. Still, the idea received little attention for well over half a century. Then, in 1848, William Thomson (later to become Lord Kelvin), formalized the idea theoretically. On the basis of very different thermodynamic considerations (Sect. 16.5, p. 568), he proposed an **absolute temperature scale** that turned out to be in perfect agreement with the results from the constant-volume gas thermometer.

Today, the practical laboratory temperature reference is the *triple point* of water: the unique circumstance of temperature (0.01°C) and pressure (610 Pa) where solid, liquid, and vapor coexist. Absolute Zero (−273.15°C) and the triple point (0.01°C) are separated by 273.16 Celsius degrees.

Temperature is now recognized as a fundamental physical quantity like mass, length, and time, so the old degree symbol is discarded. As with all other quantities, the **thermodynamic** or **absolute temperature** is given a unit, the *kelvin* (abbreviated K). The SI value of 1 K is defined as 1/273.16 of the temperature of the triple point of water. Note that 1 kelvin is effectively the same size as 1 degree Celsius. For practical reasons (not the least of which are all the volumes of research data already in existence), the scheme is constructed to mesh with the Celsius system (Fig. 14.6). Using an arrow to mean "equivalent to," Absolute Zero is 0 K → −273.15°C, therefore 0°C → 273.15 K, and room temperature is just about 293 K (which is read 293 kelvins). The numerical equivalent is summarized symbolically by

$$T \rightarrow T_C + 273.15 \qquad (14.2)$$

This expression, awkward as far as units are concerned, is more a statement of the logic of making numerical conversions than it is an equation—the 273.15 can either be degrees or kelvins.

Figure 14.6 A comparison of the Celsius and Kelvin scales. All scientific work is done in these two systems, with the Kelvin scale being preferred.

Example 14.2 Gold freezes at 1337.58 K. What is the equivalent Celsius temperature?

Solution: [Given: $T = 1337.58$ K. Find: T_C.] To go from the thermodynamic temperature to Celsius, Eq. (14.2) suggests we subtract, and therefore

$$T_C = (1337.58 - 273.15)°C = \boxed{1064.43°C}$$

▶ **Quick Check:** This result is 1064.43°C above 0°C, which is ≈273°C above 0 K. Hence, it is ≈1337 K above 0 K.

14.2 The Range of Temperatures

The Universe sustains an incredible temperature range that is completely alien to our lukewarm intuition: we live out our delicate lives within a tiny band of hot and cold (see Table 14.1).

The highest temperatures likely to exist at this moment are to be found deep within certain stars: ≈4 × 10⁹ K seems to be the theoretical extreme. At a temperature only about 10 times higher, matter fragments into subatomic particles. A hydrogen bomb ignites at roughly 40 × 10⁶ K, and blazes up to nearly 10 times that, whereas the innards of the Sun churn away at a mere 15 × 10⁶ K. Atoms can form, if only momentarily, in the seething body of a star even at 10⁵ K. Such a raging mix of atoms and atomic fragments (of ion cores and electrons) is *plasma,* typical star-stuff.

At temperatures lower by a factor of 10, matter exists in clouds of atoms, ions, electrons, and an occasional molecule. There is evidence for the presence of a vari-

Lord Kelvin, born William Thomson (1824–1907), was one of the most accomplished scientists of the nineteenth century.

TABLE 14.1   Temperatures of several physical phenomena

°C		°F
−273.2	Absolute Zero	−459.7
−269	Helium boils	−436
−196	Nitrogen boils	−320
−183	Oxygen boils	−297
−79	Dry ice (CO_2) freezes	−109
−39	Mercury freezes	−38
0	Water freezes	32
3.8	Heavy water freezes	38.9
≈20	Room temperature	≈68
31	Butter melts	88
≈37	Body temperature	≈98.6
≈54	Paraffin melts	≈129
78	Alcohol boils	172
100	Water boils	212
101.4	Heavy water boils	214.6
108	Saturated salt solution boils	226
232	Tin melts	449
327	Lead melts	621
445	Sulfur boils	833
657	Aluminum melts	1215
801	Salt (NaCl) melts	1473
961	Silver melts	1762
1063	Gold melts	1945
1083	Copper melts	1981
1000–1400	Glass melts	1830–2550
1300–1400	Steel melts	2370–2550
1530	Iron melts	2786
1620	Lead boils	2948
1774	Platinum melts	3225
1870	Bunsen burner	3398
2450	Iron boils	4442
≈3410	Tungsten melts	≈6170
3500	Oxyacetylene flame	6332
5500	Carbon arc	9932
6000	Surface of the Sun	10 832
6020	Iron welding arc	10 868

ety of molecules in cool spots (<5000 K) on the surface of the Sun. If we take a cloud of steam up to ≈3000 K, about one-quarter of the water molecules rupture into separate atoms of oxygen and hydrogen. The hottest thing you are likely to find around the house is a tungsten light bulb filament, which operates at ≈2800 K (2500°C). By comparison, lava is molten at ≈2000 K. Lead melts at 327°C (621°F), paper burns at ≈230°C (450°F), and now we are down to the temperatures of baking apple pie and a hot cup of tea.

Ice melts at 273.15 K (0°C), but if salt (NaCl) is added to icy slush, the temperature of the mix drops as low as 252 K (−21°C). That's the way people traditionally cool homemade ice cream during churning. Mercury freezes at 234.29 K (−38.87°C). Adding 4 parts by weight of crystalline calcium chloride to 3 parts of water will produce a mix at 218 K ($-55°C$). Dry ice, solid CO_2, exists at 194 K ($-79°C$), and by adding ether (as was first done by M. Thilorier in 1835) or alcohol, the resulting rapid evaporation lowers the temperature further. By pumping on the mixture, carrying off the vapor, temperatures as low as ≈163 K ($-110°C$) can be achieved.

The mainstay of the modern low-temperature laboratory is liquid nitrogen, a clear, colorless fluid that looks like water and boils away at about 77.4 K ($-196°C$). It is easy to handle and relatively inexpensive (about the price of beer). Hydrogen gas can also be liquefied, but not until it is brought down to 20.4 K ($-253°C$). The atoms within the molecules of any substance at that penetrating cold are still capable of motion, but they do not have enough energy to rearrange themselves into new compounds; all chemical and biological activity has long ceased. Still, the atoms vibrate and a piece of frozen oxygen carries compression waves like any other solid.

The most exotic of all cryogenic materials is liquid helium. Clear, colorless, and incredibly cold, it boils in an open container at 4.25 K ($-268.9°C$). Difficult to handle, it must be stored in a special thermos bottle surrounded by liquid nitrogen. It vaporizes easily and, since liquid helium sells at the price of a decent champagne, that's to be avoided. By pumping away the evaporating helium, the remaining liquid is further cooled, and the temperature can be lowered appreciably. At 2.18 K, helium undergoes a dramatic change, becoming not a solid, but a so-called *superfluid*. In that state, it is the best-known conductor of heat, though it remains an electrical insulator.

Dropping the temperature of a system becomes increasingly difficult and more costly in terms of energy expended, the lower we go. In order to cool a sample much below 1 K, a technique called *adiabatic demagnetization* is generally used; with it, solid specimens have been taken down to about 0.003 K. To plunge still lower, an even more subtle method of nuclear adiabatic demagnetization is employed. In the quest for Absolute Zero, no one has ever gotten bulk matter any closer than ≈0.000 000 02 K.

At these low temperatures, only a faint rustling of atoms persists. Yet, enough energy remains in the atoms to keep superfluid helium from solidifying, making it the only liquid that does not become a solid under ordinary pressure. To form crystalline helium, the atoms must be forced close together. Even below 1 K, where there is hardly any disruptive jiggling left, it still takes about 30 times atmospheric pressure to solidify helium. *Absolute zero corresponds to a state of minimum but finite atomic motion:* at 0 K, a tiny **zero-point energy** associated with the vibration of atoms remains—presumably enough to keep helium fluid, however close it approaches 0 K.

The fountain effect. A tiny amount of thermal energy introduced by the spiral heater causes superfluid helium at ≈1.2 K to spurt into the air.

THERMAL EXPANSION

The realization that most materials expand when heated is an ancient one: centuries before the thermometer was invented, blacksmiths were putting red-hot iron rims on wooden wagon wheels (so they would cool and shrink tight). Nevertheless, it was the ability to measure temperature that made possible a quantitative study of **thermal expansion**. Early thermometers themselves operated via thermally induced changes in the volumes of gases and liquids (by 1747 there were even metal-rod thermometers), which was all the more reason to study the effect.

14.3 Linear Expansion

For the moment, let's limit ourselves to considering elongated solids so that one dimension predominates: bars, pipes, wires, girders, and so forth. Experimentally, it

(a)   (b)   (c)

A sprig of pachysandra frozen in a bath of liquid nitrogen. Holding the plant in the liquid for a minute or so drops its temperature down to −196°C (−320°F). The leaves are then as brittle as glass and shatter into sharp fragments when squeezed. The sprig is so cold in (b) it is able to condense the water vapor in the air into streamers. The hand isn't harmed because the mass of the sprig is small, as is its specific heat capacity (p. 521).

Years ago trains rattled along because fairly large spaces had to be left between the lengths of steel track. This was done so that the rails could expand in warm weather. The photo shows what would happen when that precaution wasn't taken. New low-expansion steel alloys have smoothed the ride considerably.

is found that the change in length (ΔL) of a solid bar of original length L_0 depends on the change in temperature (ΔT) it experiences when heated uniformly; that is, $\Delta L \propto \Delta T$. The process is similar to the one encountered in Chapter 10 where we considered what would happen when a solid was stretched. Heating the bar causes its atoms to vibrate more actively with greater amplitudes, and *they also put more distance between each other*. At any one temperature, each atom vibrates in three dimensions about a more or less fixed equilibrium point in the solid. As thermal energy is added to the material, the equilibrium positions of the rapidly oscillating atoms separate, and the solid gradually expands.

The weaker the interatomic cohesive force is, the greater is the excursion of the atoms from their previous equilibrium separations, and the more the material expands as a whole. Since even in the liquid state the atoms are not much farther apart, it can be expected that this increase in separation will be small. A typical metal expands about 7% when its temperature rises from near 0 K to its melting point. If for some small ΔT the atoms or molecules on average separate from one another by an additional 0.001%, we can expect the rod to increase in length by that same 0.001%. Thus, a long rod will increase more than a short one, but both will change by the same percentage. As a consequence, $\Delta L \propto L_0$ and we can anticipate that $\Delta L \propto L_0 \Delta T$. The change in length depends on both the original length and the change in temperature.

Once again, we have a proportionality that can be made into an equation by introducing a constant of proportionality, in this case α; thus

$$\Delta L = \alpha L_0 \Delta T \qquad (14.3)$$

Here, ΔT can be in either °C or kelvins, since *the change in temperature is numerically the same for both*. At this point, we go into the laboratory and measure α (the **temperature coefficient of linear expansion**) and find that it has a specific value for every material tested (Table 14.2). The vast majority of materials have positive values of α; rubber under tension is a notable exception, and there are a number of ceramics (having strong atomic bonding) whose expansion coefficients at room temperature are nearly zero or even negative.

It is important to realize the logic of this process because it's used over and over in the development of physics. We do not yet have a theory complete enough to write out exactly what ΔL equals on an atomic level. Instead, we determine which macroscopic parameters it depends on (L_0 and ΔT) and then lump the effects of everything else into an empirical constant specific to each material, which we assume takes into account all the atomic details. If we knew more physics, we could presumably write an equation that would show exactly why lead expands 73 times more than quartz.

We can say that, because of its relatively weak interatomic bonds, lead is soft (has a low Young's Modulus), melts at a relatively low temperature (327°C), and has a high coefficient of expansion. The idea that the melting point and α are interrelated can be seen in Fig. 14.7, where a number of metals fall on the same smooth curve. The reason tin (Sn) and silicon (Si) are not closer to that curve is that they each experience a certain amount of strong *covalent* bonding. Covalent bonds, where electrons are shared, are stronger than metallic bonds (Table 10.6), making tin and silicon less able to expand for a given increase in vibrational (thermal) energy; hence they have a lower value of α than would otherwise be the case.

TABLE 14.2 Approximate values* of coefficients of linear expansion

Material	Coefficient (α) (K^{-1})
Aluminum	25×10^{-6}
Brass (yellow)	18.9×10^{-6}
Brick	10×10^{-6}
Diamond	1×10^{-6}
Cement and concrete	$10–14 \times 10^{-6}$
Copper	16.6×10^{-6}
Glass (ordinary)	$9–12 \times 10^{-6}$
Glass (Pyrex)	3×10^{-6}
Glass (Vycor)	0.08×10^{-6}
Gold	13×10^{-6}
Granite	8×10^{-6}
Hard rubber	80×10^{-6}
Invar (64% Fe, 36% Ni)	1.54×10^{-6}
Iron (soft)	$9–12 \times 10^{-6}$
Lead	29×10^{-6}
Nylon (molded)	81×10^{-6}
Paraffin	130×10^{-6}
Platinum	8.9×10^{-6}
Porcelain	4×10^{-6}
Quartz (fused)	0.55×10^{-6}
Steel (structural)	12×10^{-6}
Steel (stainless)	17.3×10^{-6}

*At temperatures around 20°C.

Example 14.3 A structural steel beam holds up the scoreboard in an open-air stadium. If the beam is 12 m long when it's put into place on a winter day at 0°C, how long will it be in the summer at 32°C? Given that it has a cross-sectional area of 100 cm², how much force would it take to stretch the beam that much?

Solution: [Given: $L_0 = 12$ m, $T_i = 0$°C, $T_f = 32$°C, and $A = 100$ cm². Find: ΔL and F.] From Table 14.2, $\alpha = 12 \times 10^{-6}$ K^{-1}, so

$$\Delta L = \alpha L_0 \Delta T \qquad [14.3]$$

Hence

$$\Delta L = (12 \times 10^{-6} \text{ K}^{-1})(12 \text{ m})(32 \text{ K}) = 4.6 \text{ mm}$$

or almost $\frac{1}{2}$ cm. To find the force, remember $Y =$ stress/strain. From Eq. (10.10) and Table 10.4

$$F = YA \frac{\Delta L}{L_0}$$

$$F = \frac{(200 \times 10^9 \text{ Pa})(100 \times 10^{-4} \text{ m}^2)(4.6 \times 10^{-3} \text{ m})}{12 \text{ m}}$$

and $\boxed{F = 77 \times 10^4 \text{ N}}$ or 17×10^4 lb. This is also the force exerted by the beam on whatever might be resisting its change in length. Evidently, provision will have to be made in the design of the structure to allow for thermal expansion and contraction.

▶ **Quick Check:** Steel changes length by 0.001 2% per kelvin. Here, ΔT is 32 K, so the length change is 0.038 4% $L_0 = 4.6$ mm.

Figure 14.7 The higher the melting point of a material, the stronger the interatomic binding, and the smaller the amount of expansion.

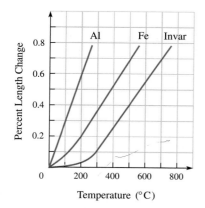

Figure 14.8 A plot of $\Delta L/L_0$ versus T. The slopes of the curves, $(\Delta L/L_0)/T$, equal α. For aluminum, over the temperature range shown, α is constant. That's not quite true for iron, and it's certainly not true for the alloy invar. The very low value of α for invar from 0°C to 200°C makes it useful for bimetallic devices and precision applications where expansion can be troublesome.

Equation 14.3 is a useful, practical relationship, even if it is not a fundamental law of nature. Provided we are careful about the circumstances under which it is applied, it works nicely. Something to watch out for is that α really depends on T, increasing slightly as T increases (Fig. 14.8). The technical literature contains extensive tabulations of its values for a variety of substances over various ranges of temperature. Moreover, materials that have their atoms arranged differently in different directions (anisotropic materials, such as crystalline quartz and bismuth) expand and contract differently in those different directions. Indeed, when heated, graphite expands along one axis and contracts along another. But we need not worry about that, beyond realizing that none of this is as simple as it might appear.

14.4   Volumetric Expansion: Solids and Liquids

Using the same procedure as above, we can create an expression for the change in volume of a substance that has undergone a change in temperature ΔT; thus

$$\Delta V = \beta V_0 \Delta T \qquad (14.4)$$

Here, V_0 is the original volume, and β is known formally as the **temperature coefficient of volume expansion** (see Table 14.3). Notice that the coefficients for liquids are about 50 times greater than for solids, which is to be expected. The weak intermolecular bonds inherent in that state account for the fact that liquids are both more compressible and more readily expanded thermally than are solids. As a rule, β *decreases as the thermodynamic temperature decreases, approaching zero as* 0 K *is approached.* For example, alcohol expands roughly 20% more per kelvin change near 373 K (100°C) than at 273 K (0°C). Fortunately, β for mercury, over the same range, varies by less than 0.01%, making it a superior thermometric substance.

Much of what was said before regarding α pertains here. The atoms again separate, but now that happens in three dimensions, not one. Indeed, a comparison of Tables 14.2 and 14.3 suggests that $\beta \approx 3\alpha$ is generally the case for isotropic solids. That relationship between α and β can be understood by imagining a rectangular solid with sides a, b, and c. If this block is changed in temperature by ΔT, we can expect each side to change its length proportionately. Accordingly, it will go from an initial volume of (abc) to a final one of $(a + \alpha a \Delta T)(b + \alpha b \Delta T)(c + \alpha c \Delta T)$. Multiplying out this expression provides the new volume, but α is a very small number and we can drop any terms containing α^2 or α^3 because they will be still smaller. The result is a final volume approximately equal to $(abc + 3\alpha abc \Delta T)$, and a change in volume $\Delta V \approx 3\alpha V_0 \Delta T$, from which it follows that $\beta \approx 3\alpha$.

Suppose that we had a solid, homogeneous object whose shape is of no particular interest. Picture anywhere inside of it a portion of the material enclosed within a smaller imaginary sphere. When the big object is heated, it expands uniformly and, since no holes form inside of it, we can assume that the little imaginary sphere of material expands at the same rate. If the little sphere of matter is removed, the resulting cavity must expand at the same rate as the sphere itself. When an object expands or contracts, all its protuberances and cavities expand or contract proportionately.

Thus, when a bowling ball is heated the finger holes get bigger, as does the cavity within a mercury-in-glass thermometer. Similarly, a red-hot steel ring cools and contracts to fix tightly on a wagon wheel. That's why running hot water on a jar lid loosens it—the coefficient of expansion for steel is greater than for glass. A jar is larger when hot than cold and it holds more. In fact, because ordinary glass is a poor conductor of heat, putting a thick-walled glass vessel into an already hot oven is usually disastrous. The exposed faces of the vessel get hot while the region within the glass stays cool. Uneven expansion can crack the glass unless it is heated or cooled slowly. The advantage of Pyrex glass is that its expansion coefficient is 3 times smaller than that of ordinary glass.

Whenever materials are exposed to large variations in temperature, we must be concerned about expansion. If spaces are not left between slabs of ordinary concrete on streets and roads, they can buckle. Most exposed outdoor steel structures (overpasses, trestles, and so forth) have expansion joints to provide for changes in length. During a particularly cold week in 1937, the Bay Bridge in San Francisco contracted a total of 4 ft, 5 in. Often, small bridges are fastened only at one end, while the other rests on rollers to permit expansion. It has even been necessary, when finishing a large bridge in summer, to pack the final spans in ice for several hours to shrink them so they fit in place. Dental fillings and caps can experience substantial temperature variations in the mouth and should be made of materials that closely match the β for teeth.

Thermal expansion is utilized to advantage in the *bimetallic* strip, where flat lengths of two metals with different values of α are bonded together. When heated, the strip bends toward the side with the lower coefficient. These strips are used to make thermometers and circuit breakers or to open and close electrical switches in thermostats.

TABLE 14.3 Approximate values* of coefficients of volumetric expansion

Material	Coefficient (β) (K^{-1})
Solids	
Aluminum	72×10^{-6}
Asphalt	$\approx 600 \times 10^{-6}$
Brass (yellow)	56×10^{-6}
Cement and concrete	$\approx 36 \times 10^{-6}$
Glass (ordinary)	$\approx 26 \times 10^{-6}$
Glass (Pyrex)	9×10^{-6}
Invar	2.7×10^{-6}
Iron	36×10^{-6}
Lead	87×10^{-6}
Paraffin	590×10^{-6}
Porcelain	11×10^{-6}
Quartz (fused)	1.2×10^{-6}
Steel (structural)	36×10^{-6}
Liquids	
Acetone	1487×10^{-6}
Ethyl alcohol	1120×10^{-6}
Gasoline	950×10^{-6}
Glycerin	505×10^{-6}
Mercury	182×10^{-6}
Turpentine	973×10^{-6}
Water	207×10^{-6}

*At temperatures around 20°C.

Example 14.4 A driver pulls into a gas station and casually says "fill it up." The attendant does just that, filling the steel tank to the very brim with 56 liters of gasoline at 10°C. The trip home is a short one, and the heated garage is at 20°C. How much gasoline will overflow onto the floor?

Solution: [Given: $V_0 = 56$ liters, $\Delta T = +10$ K and that the liquid is gasoline and the tank is steel. Find: the amount of overflow.] First, convert the volume from liters to m^3: 56 liters $= 56 \times 10^3 \text{ cm}^3 = 0.056 \text{ m}^3$. The change in volume of the gasoline is

$$\Delta V_g = \beta_g V_0 \Delta T = (950 \times 10^{-6} \text{ K}^{-1})(0.056 \text{ m}^3)(+10 \text{ K})$$

and

$$\Delta V_g = +5.32 \times 10^{-4} \text{ m}^3$$

But the tank's capacity also increases; the tank's 0.056-m^3 cavity expands at the same rate as a solid steel mass in that shape. Hence, the tank's volume increases by

$$\Delta V_t = \beta_t V_0 \Delta T = (36 \times 10^{-6} \text{ K}^{-1})(0.056 \text{ m}^3)(+10 \text{ K})$$

and

$$\Delta V_t = +0.202 \times 10^{-4} \text{ m}^3$$

Therefore, about $\boxed{0.51 \times 10^{-3} \text{ m}^3}$, or 0.51 liter overflows.

▶ **Quick Check:** $\Delta V_g / \Delta V_t = \beta_g / \beta_t = 26.4$; $(+5.32 \times 10^{-4} \text{ m}^3)/(+0.202 \times 10^{-4} \text{ m}^3) = 26.3$.

The Expansion of Water

Water is especially important to us, and that makes its physical behavior of great interest. Like the majority of liquids, water contracts slightly as it is cooled (Fig. 14.9),

A common device that allows an elevated roadway to expand and contract without affecting traffic. This one is part of an old wooden bridge over the Seine River in Paris.

This thick-walled glass mug cracked when it was filled with hot water.

A bimetallic strip bends when it's heated.

becoming a trifle more dense. In a unique way, however, this contraction gradually diminishes until at 277.13 K (3.98°C), it stops altogether. Between 373.15 K (100°C) and 277.13 K (3.98°C), water contracts about 4%, which is not very much, but still about 20 times more than would occur over that range for most solids. Water

Temperature (°C)	Density (g/cm³)
0	0.999 84
3.98	0.999 973
10	0.999 70
20	0.998 20
30	0.995 65
40	0.992 22
50	0.988 05
60	0.983 22
70	0.977 79
80	0.971 81
90	0.965 32
100	0.958 36

Figure 14.9 (a) The density of mercury plotted against temperature in °C. Mercury is a more representative liquid than water. (b) The density of water is a maximum at 3.98°C.

line. In other words, *when the pressure is kept constant, the volume of a given amount of any gas varies directly with the thermodynamic temperature.* This is known as **Charles's Law**, after Citizen Jacques Alexandre César Charles of Paris (ca. 1787)—the man who inadvertently made the first solo balloon ascent when his assistant accidentally let go while climbing out of the gondola. Putting it another way, $V \propto T$, or

[*constant P and m*]
$$\frac{V}{T} = \text{constant} \qquad (14.5)$$

All gases increase in volume when the temperature increases. Doubling the temperature of a gas at constant pressure doubles its volume (Fig. 14.11).

The third basic gas relationship, called **Gay-Lussac's Law** after Joseph Louis Gay-Lussac (1802), maintains that *when the volume is kept constant, the absolute pressure of a given amount of any gas varies directly with the thermodynamic temperature;* that is, $P \propto T$, or

[*constant V and m*]
$$\frac{P}{T} = \text{constant} \qquad (14.6)$$

The pressure in an automobile tire (V = constant) rises in the summer as T rises, and it's dangerous to throw aerosol cans in a fire because P increases so much they explode.

Multiplying these three relationships yields

[*constant m*]
$$\frac{PV}{T} = \text{constant} \qquad (14.7)$$

for a fixed quantity of gas. Thus, if V is decreased rapidly as in Fig. 14.12, P will rise and so too must T. Note that each of the constants in the last four equations is different.

Figure 14.11   The weight on the piston keeps the pressure constant at P. Heating the gas, and thereby doubling T, must double V.

Figure 14.12   The fire syringe is a small cylinder with a tight-fitting piston. When the piston is rapidly driven in, the gas is compressed, V decreases while both P and T increase. A small piece of tissue paper will burst into flames within the syringe.

Example 14.5   A tank having a volume of 1.00 m^3 is filled with air at 0°C to 20.0 times atmospheric pressure. How much volume will that gas occupy at 1.00 atm and room temperature?

Solution: [Given: $V_i = 1.00 \text{ m}^3$, $T_i = 273.15$ K, $P_i = 20.0$ atm, $P_f = 1.00$ atm, and $T_f = 293.15$ K. Find: V_f.] Since the ratio in Eq. (14.7) is constant

$$\frac{P_i V_i}{T_i} = \frac{P_f V_f}{T_f}$$

hence
$$V_f = \frac{P_i V_i T_f}{P_f T_i}$$

(Divide pressures first so we need not convert units)

$$V_f = \frac{20(1.00 \text{ m}^3)(293.15 \text{ K})}{273.15 \text{ K}} = \boxed{21.5 \text{ m}^3}$$

There is a slight increase in volume due to the increase in temperature, but the predominant increase is due to the 20-fold drop in pressure.

▶ **Quick Check:** The rise in temperature increases the volume by a factor of $\approx 293/273 \approx 1.07$; the pressure change increases it by a factor of 20; hence, $(1.07)20 \approx 21.5$ and $V_f \approx 21.5(1 \text{ m}^3)$.

We still have one quantity that can vary to account for, and that is the mass of the gas. Experience tells us that with all else unchanged, the volume of a gas must depend directly on the *amount* of gas present. Consider two identical bottles of gas, each with the same pressure, temperature, volume, and number of molecules. It seems reasonable that we could attach them together, remove the separating wall, and nothing significant would change on a macroscopic level—twice the number of gas molecules, twice the volume. Similarly, if V and T are fixed, increasing the number of molecules will increase the number of their collisions with the walls (p. 365) and therefore will increase the pressure. Hence $PV \propto m$, or

[*constant T*]
$$\frac{PV}{m} = \text{constant} \qquad (14.8)$$

Example 14.6   A 100-liter storage tank is slowly being filled with gas. At 5.00 times atmospheric pressure, the tank holds 0.60 kg of gas. If the temperature is kept constant, how much gas will be in the tank when the pressure is raised to 10.00 atm?

Solution: [Given: $V = 100$ liters = constant, P_i is 5.00 atm, $m_i = 0.60$ kg, and $P_f = 10.00$ atm. Find: m_f.] Since 1 liter = 10^{-3} m^3, $V = 100 \times 10^{-3} \text{ m}^3$. It follows from Eq. (14.8) that the initial and final values of the ratio PV/m must be equal; consequently

$$\frac{P_i V_i}{m_i} = \frac{P_f V_f}{m_f}$$

hence
$$m_f = \frac{P_f m_i}{P_i}$$

and
$$m_f = \frac{(1.013 \text{ MPa})(0.60 \text{ kg})}{0.506 \, 6 \text{ MPa}} = \boxed{1.2 \text{ kg}}$$

▶ **Quick Check:** To double the pressure, double the mass: $\Delta m = 0.60$ kg. (Notice that it would have been easier to divide the two pressures before converting their units.)

14.5 The Ideal Gas Law

The most commonly used equations of state express the amount of gas in terms of either the number of moles (n) present (p. 320) or the number of molecules present (N); that is, $PV \propto nT$ or $PV \propto NT$. The masses of the individual particles play a less central role than the total number of particles. We saw this earlier when we found that the same number of molecules of any gas at the same temperature and pressure occupies the same volume. Thus, it seems reasonable that if a single equation is to describe both H_2 and CO_2, we should deal with equal numbers of molecules, not equal masses of gas.

A mole is the mass of any substance in grams, equal numerically to the molecular mass of that substance: CO_2 has a molecular mass of 12 u + 2 × 16 u = 44 u and one mole of it corresponds to 44 g, or 0.044 kg. More to the point, a mole of any substance contains the same number of molecules. Introducing a constant of proportionality R, the equation of state becomes

$$PV = nRT \qquad (14.9)$$

Remarkably, experiments reveal that R is nearly the same for all gases at low pressures; $R = 8.314\,41$ J/mol·K, and so it is called the **universal gas constant**. Still, we must limit the circumstances under which this formula is applied—it does not work well at high values of density for real gases. Accordingly, we say that a model theoretical gas for which this expression works perfectly is an *ideal gas,* and therefore Eq. (14.9) is known as the **Ideal Gas Law**. Despite its limitations, the concept has very practical applications. For example, helium at low pressure acts like an ideal gas almost to the point where it liquefies.

Example 14.7 Determine the volume of 1.00 mole of any gas at STP.

Solution: [Given: $T = 273.15$ K, $P = 1.013 \times 10^5$ Pa, and $n = 1.00$ mol. Find: V.] From the Ideal Gas Law

$$V = \frac{nRT}{P} = \frac{(1.00 \text{ mol})(8.314 \text{ J/mol·K})(273.15 \text{ K})}{1.013 \times 10^5 \text{ Pa}}$$

and

$$\boxed{V = 0.022\,4 \text{ m}^3}$$

▶ **Quick Check:** This result is confirmed by the discussion on p. 320.

Avogadro's number $N_A = 6.022 \times 10^{23}$ is the number of molecules per mole, and so the number of molecules N present in a sample of gas is just nN_A. Hence

$$PV = nRT = \frac{N}{N_A}RT$$

Accordingly, we define a new constant, called **Boltzmann's Constant**, $k_B = R/N_A = 1.380\,6 \times 10^{-23}$ J/K and thereby obtain the final form of the Ideal Gas Law

$$PV = Nk_BT \qquad (14.10)$$

This expression may seem to be just another way of saying the same thing, but Boltzmann's Constant is actually one of the fundamental constants of nature.

Example 14.8 Estimate the number of air molecules there are in a room 4.00 m by 6.00 m by 3.00 m at standard pressure and room temperature (20°C). How many moles is that? How much volume would the gas

(continued)

(continued)

occupy if the temperature were 273 K (0°C), all else constant? Roughly how much does the air weigh? Compare that to your own weight. (See Table 14.4.)

Solution: [Given: $T = 293$ K (20°C), $P = 0.101$ MPa, $V = 4.00 \times 6.00 \times 3.00$ m³. Find: N.] From Eq. (14.10)

$$N = \frac{PV}{k_B T} = \frac{(0.101\,3 \times 10^6 \text{ Pa})(72.0 \text{ m}^3)}{(1.381 \times 10^{-23} \text{ J/K})(293 \text{ K})}$$

and $N = 1.803 \times 10^{27}$ molecules.

To find the number of moles, we divide by Avogadro's number, thus

$$n = \frac{N}{N_A} = \frac{1.803 \times 10^{27}}{6.022 \times 10^{23}} = \boxed{2.99 \times 10^3 \text{ moles}}$$

At STP, each mole occupies 22.414 liters; hence

$$V_{STP} = (22.414 \text{ liters/mol})(2994 \text{ mol})$$

and $\quad V_{STP} = 67.1 \times 10^3$ liters $= \boxed{67.1 \text{ m}^3}$

as compared to 72 m³ at room temperature. From Table 14.4, each mole of air has a mass of 0.029 kg; hence, the total mass of air is

TABLE 14.4	The masses of one gram mole of several gases	
Material		**Mass (kg)**
Hydrogen		0.002
Helium		0.004
Nitrogen		0.028
Air (average)		0.029
Oxygen		0.032
Argon		0.040
Carbon dioxide		0.044
Ozone		0.048

$$M = (0.029 \text{ kg/mol})(2994 \text{ mol}) = 86.8 \text{ kg}$$

which weighs $\boxed{0.85 \text{ kN}}$ or 1.9×10^2 lb.

▶ **Quick Check:** Using Eq. (14.10) at STP, $V_{STP} = N k_B (273 \text{ K})/P_A = (1.8 \times 10^{27})(1.381 \times 10^{-23} \text{ J/K}) \times (273 \text{ K})/(0.101 \times 10^6 \text{ Pa}) = 67$ m³.

14.6 Real Gases: Liquefaction

A given quantity of an ideal gas at a fixed temperature obeys the relationship $PV =$ constant, which is the equation of a hyperbola. Plotting P against V at any constant T produces a smooth curve (Fig. 14.13), symmetrical with the 45° line, known as an **isothermal**, or *isotherm* (from the Greek *isos*, meaning equal). When we construct the same sort of *PV*-diagram from data for real gases, a number of important differences appear. These were first observed by T. Andrews (ca. 1863) who used carbon dioxide under great pressure, but his results are broadly applicable (Fig. 14.14).

Each curve in the diagram is an isotherm and at high temperatures they are nearly hyperbolic. At lower temperatures, however, the isotherms drop below the corresponding hyperbolas (dashed curves) in the region of small volume and high pressure. It's here that the gas is compact enough for the intermolecular cohesive force to begin to be effective. The isotherms continue to distort as T is reduced until a special *critical temperature* is reached: the 304.3-K (31.1°C) isotherm, whose left-hand portion (beginning at the point of inflection C) marks the boundary between liquid and gas. At or below that line, there can be a change in the *phase,* or state, of the substance—it may go from liquid to gas or vice versa.

Andrews called point C the *critical point,* and the pressure there is known as the *critical pressure.* Here, the molecules are moving around, but not rapidly enough to overcome the now-potent cohesive force (Table 14.5). At C, *liquid and gas coexist at the same density* and are indistinguishable. Above the critical temperature, no amount of pressure will liquefy the gas; it can become denser and denser, but there will never be a clear gas-liquid transition. For example, above 647 K (374°C), water can exist only as a gas no matter what the pressure is. Evidently, the critical temperature is a measure of the strength of the cohesive force of the substance. The relatively high value (647 K) for water is a reflection of the sizable attraction between its polar molecules.

Let's follow along one of the isotherms below the critical temperature. That can be done by filling a cylinder with a large volume of the gas (Fig. 14.15). It is customary to call the substance *vapor,* rather than gas, when it is below the critical temperature—thus *vapors can be condensed via compression, gases cannot.* Compressing the vapor without allowing its temperature to change carries it from point 1 to point 2. Compressing it further increases the pressure and brings the vapor to point 3, where it is just about to condense. Further compression results in liquefaction at a constant pressure, bringing the system first to point 4 and then, when all is liquid, to point 5. Beyond that it takes considerably more pressure (note the slope of the curve) to compress the liquid, bringing it to point 6.

At point 4, the chamber contains both liquid and vapor, and the amount of each is stable. Some molecules that happen to be moving fast enough up near the surface will leave the liquid altogether, in the usual process of evaporation. Others in the vapor above it will crash into the liquid, becoming part of it. When the rates of evaporation and condensation are equal, the system remains in a state of *dynamic equilibrium.* If the concentration of vapor above the liquid is stable, it exerts, on average, a stable pressure on the liquid known as the **equilibrium vapor pressure**, or the *saturated vapor pressure,* or, when there is no ambiguity, just the vapor pressure. That's what the pressure gauge in the chamber (Fig. 14.15) will be reading because the piston is being moved very slowly. This pressure depends both on the nature of the liquid and on the temperature. The larger the cohesive force, the lower the vapor pressure. The higher the temperature, the higher the vapor pressure. Thus, at 293 K (20°C), strongly interacting water has a vapor pressure of 2.3 kPa, while volatile, weakly bound chloroform has a vapor pressure of 21.3 kPa—it can keep evaporating much longer before equilibrium sets in.

Systematic research on the liquefaction of gases began when Michael Faraday took up the work. In 1823, using a mixture of ice and salt as the cooling agent, Faraday was able to liquefy a number of gases, including carbon dioxide (Fig. 14.16). Still, despite all efforts, the most common gases of the air defied liquefaction, even to pressures in excess of 3000 atm. In 1869, Andrews pointed out that the failure to

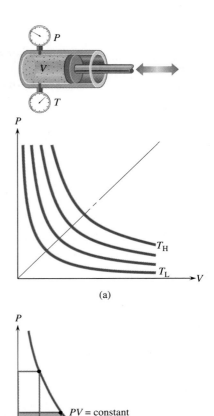

Figure 14.13 (a) A family of isothermal curves of P versus V for an ideal gas: $T_H > T_L$. (b) The area of each rectangle is $P \times V$, and that's constant.

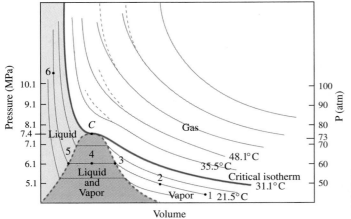

Figure 14.14 Isothermals for carbon dioxide. Below the critical isotherm and to the right, CO_2 exists as an unsaturated vapor. Liquid and saturated vapor coexist in the middle region, and only liquid exists in the region on the left.

P

T

Point 1

Point 2

Point 3

Point 4

Point 5

Point 6

Figure 14.15 The isothermal compression of a real gas below its critical temperature. These are the six points on the isothermal shown in Fig. 14.14.

TABLE 14.5   Critical constants

Material	Critical Temperature (K)	Critical Pressure (MPa)	Critical Density (kg/m³)
Ammonia	405.5	11.28	235
Argon	150.7	4.86	531
Bromine	584.2	10.34	1180
Carbon dioxide	304.3	7.38	468
Carbon monoxide	133.0	3.50	301
Chlorine	417.2	7.71	573
Deuterium	38.4	1.66	67
Helium	5.21	0.23	69
Hydrogen	43.24	1.30	31
Hydrogen sulfide	373.2	9.01	349
Neon	44.4	2.73	484
Nitrogen	126.3	3.39	311
Nitrous oxide	309.7	7.27	459
Oxygen	154.8	5.08	410
Water	647.4	22.12	326
Xenon	289.8	5.88	1110

liquefy these gases occurred because the early attempts used compression without cooling below the critical temperature—in short order, oxygen (1877), hydrogen, and nitrogen were liquefied.

14.7   Phase Diagrams

Figure 14.17 is a typical *PT*-diagram, also known as a *phase diagram* because it summarizes the relationships between the solid, liquid, and gaseous states of a substance, in this case, water. Two different phases coexist at temperatures and pressures lying along each of the three boundary lines. The curve from O to C corresponds to the equilibrium vapor pressure and is, as was recognized by Dalton (1804), logarithmic. It is called the *boiling-point curve*. Again, C corresponds to the critical point. If the line from O to C is crossed, there will be a clear liquid-vapor phase transition. By contrast, if the system is carried (by altering P and T) from the region on the lower right where it is a gas (up around and over C), it gradually blends into a liquid, without there being any phase transition. That's what would be seen if

Figure 14.16 Gas is generated by heating the substance on the left, creating a high pressure in the sealed bent tube. Cooling the far end below the critical temperature causes the gas to condense out as a liquid.

Low temperature bath

we could descend into the hydrogen-helium atmosphere of the giant planet Jupiter. Under its own crushing pressure, the atmosphere gets thicker and thicker, imperceptibly merging into the liquid body of the planet.

For every pressure above that of point O, there is a temperature at which a pure solid will be in equilibrium with its liquid state. The locus of all such melting (or freezing) points is the line rising up from O on the left, which is the *melting-point curve*. Notice that this solid-liquid transition line leans to the left. Raising the pressure by 1 atm lowers the melting temperature by about $0.0072°C$ (which is one reason for packing snowballs as hard as possible—it fuses the snow together). Along with friction, a rise in pressure causes the ice under the blades of a skate to momentarily melt and lubricate the surface. A person on skates might exert a pressure of 50 atm on the ice, lowering its melting point by about $0.4°C$. Still, friction seems to produce most of the melting (Fig. 14.18).

At point O in Fig. 14.17, which is the **triple point**, all three phases coexist in equilibrium (Table 14.6). Like liquids, solids also have a vapor pressure, though it may be extremely small. Thus, raising the temperature of a solid at a fixed pressure below that of the triple point will cause it to pass directly from solid to vapor, a process known as *sublimation*. The line descending from O with decreasing temperature corresponds to the vapor pressure and is often called the *sublimation curve*. The intermolecular bonds in dry ice are so weak that even at room temperature it will transform directly to the vapor state, as will camphor and naphthalene (mothballs). In this spirit, take a block of frozen coffee, ice cream, or even blood at a temperature below the triple point; lower the pressure and pump off the vapor as the solid is gently heated. The ice will sublime, leaving a light porous residue, which is said to be *freeze-dried*.

When the process is run the other way via a reduction in temperature, the vapor condenses out directly to the solid state. This condensation is what produces snow in the upper atmosphere and hoarfrost down at the surface of the Earth. The latter is the feathery crystalline deposit of ice that can be seen on the inside of a window pane in winter.

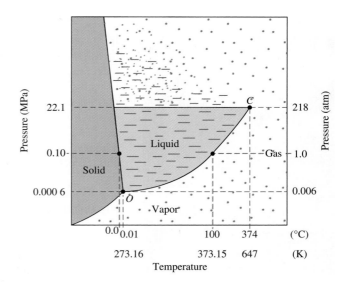

Figure 14.17 Phase diagram for water. If we take a block of ice at atmospheric pressure and slowly raise its temperature, it melts completely at 273 K (0°C) and remains liquid until 373 K (100°C), whereupon it completely vaporizes.

Figure 14.18 A wire that supports a load passes through a block of ice because the melting point drops due to the high pressure just under the wire. Above the wire, the water refreezes.

TABLE 14.6 Triple-point data		
Material	**Temperature (K)**	**Pressure (kPa)**
Water	273.16	0.610
Carbon dioxide	216.55	517.00
Ammonia	195.4	6.07
Nitrogen	63.18	12.5
Oxygen	54.36	0.152
Neon	24.57	43.2
Deuterium	18.63	17.1
Hydrogen	13.84	7.04

Hoarfrost.

14.8 Kinetic Theory

In ancient times, the Greek philosopher Plato observed that "heat and fire . . . are themselves begotten by impact and friction: but this is motion." That insight, that thermal phenomena arise from motion, was echoed almost 2000 years later by Francis Bacon, who maintained that "the very essence of heat . . . is motion and nothing else." The picture of a gas as a tumult of "very minute corpuscles which are driven hither and thither" was used as early as 1738 by Bernoulli (p. 365) to explain Boyle's Law. By the middle of the 1800s, this motion theory, now known as **Kinetic Theory**, was one of several competing formalisms concerning the nature of heat and temperature. Keep in mind that the reality of the atom was not established until the early twentieth century.

Kinetic Theory, as further developed mainly by James Clerk Maxwell and Ludwig Boltzmann, is a mathematical description of how a vast number of minute, rapidly moving particles can manifest itself macroscopically in the observed properties of bulk matter. We will examine only the basics of the theory and use it to gain some insights into the nature of gases. Following Maxwell, we model a gas as "an indefinite number of small, hard, and perfectly elastic spheres acting on one another only during impact." Clearly, we are talking about an ideal gas. To make the treatment still easier, we assume that the gas particles are essentially points having no extension, that do not collide with one another. This is certainly not the case, nor is it a necessary condition. (The inclusion of interparticle collisions would bring us into the study of Statistical Mechanics.)

Figure 14.19 depicts a piston in a cylinder of length L and cross-sectional area A. The pressure exerted on the piston originates from the countless millions of impacts made on it by the gas molecules each and every second. Inasmuch as the system is in dynamic equilibrium, this is the pressure P everywhere within the cylinder. The motion of a molecule can be resolved into its x-, y-, and z-components (Fig. 14.19a). The collisions are perfectly elastic (conserving KE) and so each time a particle bounces off the piston, its velocity's y- and z-components, parallel to the face of the piston, remain unaltered; only the x-component changes, reversing itself from $+v_x$ to $-v_x$. Thus, each impact produces a change in the momentum of a single particle of

$$\Delta p_x = 2mv_x$$

Each molecule moves along the length of the cylinder at a speed of v_x, which is constant (even though it may well be bouncing around from side to side in the process). Hence, it travels from the piston to the far face and back a distance of $2L$ in a time

$$\Delta t = \frac{2L}{v_x}$$

This is the time between collisions on the piston: the number of seconds per collision. The reciprocal of that value is the number of collisions per second, or the rate of collision: $1/\Delta t = v_x/2L$. Multiplying the number of collisions per second ($1/\Delta t$) by the change of momentum per collision (Δp) gives us the *rate of change of momentum:*

$$\frac{\Delta p_x}{\Delta t} = \frac{2mv_x}{2L/v_x} = \frac{mv_x^2}{L}$$

From Newton's Second Law, this equals the average force exerted by that one particle on the piston. Assuming, just for a moment, all N of the molecules in the cylinder behave the same way, the total average force on the piston due to the hail of

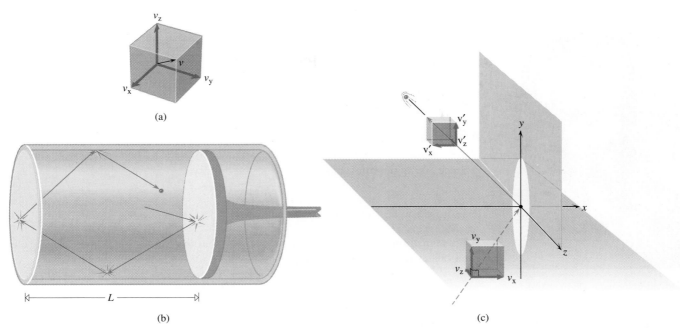

Figure 14.19 Atoms colliding with the face of a piston. Only the v_x component changes its direction on collision with the surface.

particles is

$$F_{av} = \frac{Nmv_x^2}{L}$$

By definition, the force per unit area is the pressure:

$$P = \frac{Nmv_x^2}{LA}$$

This equation can be expressed a little more conveniently using the three-dimensional equivalent of the Pythagorian Theorem, which tells us that the total speed squared equals the sum of the squares of its components; that is

$$v^2 = v_x^2 + v_y^2 + v_z^2$$

Since there are no external forces acting on the system, on average, we can expect that the gas particles will be moving in all directions with equal likelihood: $v_x^2 = v_y^2 = v_z^2$, and so

$$v_x^2 = \tfrac{1}{3}v^2$$

Accordingly

$$P = \frac{Nmv^2}{3LA} \tag{14.11}$$

This would be the pressure if each molecule had the same speed, but in fact they don't. Not surprisingly, there is a temperature-dependent distribution of speeds, something Maxwell and Boltzmann each worked out theoretically. For us, it will suffice to recognize that fact, and we will use an average speed in the pressure equation in place of v^2. The sum of the squares of the speeds of each of the N particles divided by N is the average, or *mean,* of the speed squared; consequently

Ludwig Boltzmann (1844–1906). Despondent and disturbed by the idea that his life's work was a waste, Boltzmann killed himself in 1906. Within a few decades, the atom was established as a reality, and Boltzmann's brilliant achievement was recognized as seminal.

$$(v^2)_{av} = \frac{v_1^2 + v_2^2 + \cdots + v_N^2}{N}$$

Thus

$$P = \frac{Nm(v^2)_{av}}{3LA}$$

and noticing that the volume of the cylinder is $V = LA$, we get

$$P = \frac{Nm(v^2)_{av}}{3V}$$

Since Nm is the total mass of the gas, the density is $\rho = Nm/V$ and

$$P = \tfrac{1}{3}\rho(v^2)_{av} \qquad (14.12)$$

Moreover

$$PV = \tfrac{1}{3}Nm(v^2)_{av}$$

This equation is Boyle's Law. Remembering that the translational kinetic energy of an object is $\tfrac{1}{2}mv^2$, it follows that

$$PV = \tfrac{2}{3}N(KE)_{av} \qquad (14.13)$$

Pressure depends on the number of molecules per unit volume as well as on their average translational KE.

Comparing this situation with the Ideal Gas Law, Eq. (14.10), we see that

$$(KE)_{av} = \tfrac{3}{2}k_B T \qquad (14.14)$$

The temperature of an ideal gas is proportional to the average translational kinetic energy of its molecules. The microscopic average translational KE of the molecules manifests itself macroscopically in the form of temperature.

Still, real gases do interact, and that must be treated via Quantum Mechanics and not by Newtonian physics. Thus, although the relationship between translational kinetic energy and temperature is undeniably an important insight, one that can fruitfully guide our thinking, it has its shortcomings. Remember that a substance must have a zero-point energy even at absolute zero, something not in accord with Eq. (14.14), which therefore cannot be the whole truth at low temperatures. All of this underscores the conclusion that temperature is a fundamental quantity and, like length, time, and mass, is not expressible in terms of other quantities.

The Speeds of Molecules in a Gas

It follows from Eq. (14.14) and the fact that $(KE)_{av} = \tfrac{1}{2}m(v^2)_{av}$ that

$$(v^2)_{av} = \frac{3k_B T}{m}$$

which is the mean of the speed squared. If we take the square root of it, we get something called the *root of the mean of the square* (*rms* for short):

$$v_{rms} = \sqrt{(v^2)_{av}} = \sqrt{\frac{3k_B T}{m}} \qquad (14.15)$$

This is not quite the average speed, but a statistical analysis shows that $v_{av} = 0.92v_{rms}$, and so Eq. (14.15) gives us a measure of the typical speed of a gas particle.

Example 14.9 What is the *rms*-speed of a hydrogen molecule (H_2) at STP? Take the molar mass of hydrogen molecules to be 2.0 g/mol.

Solution: [Given: $T = 273$ K. Find: v_{rms}.] The mass of a molecule m is the molar mass divided by the number of molecules per mole, and therefore

$$m = \frac{0.002\,0 \text{ kg/mol}}{6.022 \times 10^{23} \text{ molecules/mol}}$$

and $m = 3.3 \times 10^{-27}$ kg

Hence, from Eq. (14.15)

$$v_{rms} = \left[\frac{3(1.38 \times 10^{-23} \text{ J/K})(273 \text{ K})}{3.3 \times 10^{-27} \text{ kg}} \right]^{\frac{1}{2}}$$

and so $\boxed{v_{rms} = 1.9 \text{ km/s}}$ or about 6.3×10^3 ft/s, which is faster than a bullet.

▶ **Quick Check:** m is correctly just about the mass of two protons. From Eq. (14.12) and Table 10.1, $v_{rms}^2 = 3P/\rho = 3(1.01 \times 10^5 \text{ Pa})/(0.090 \text{ kg/m}^3)$; $v_{rms} = 1.9$ km/s.

We can expect sound waves to travel at speeds less than that of the individual molecules. The speed of sound in hydrogen (Table 13.2) is 1.27 km/s $= 0.7v_{av}$, and that's reasonable.

Our present understanding is that the molecules of a gas undergo a vast number of collisions with each other—about 5×10^9 per second for each molecule in the air around us, which is about 10^5 collisions for each centimeter of path traveled. The uninterrupted distance between collisions—the *mean free path*—is about 10^{-5} cm (Table 14.7), which is ≈ 100 atomic-diameters. That's why it takes time for the scent of a perfume to fill a room after the bottle is opened. The molecules, even though they move at great speeds, travel zigzag paths, only gradually diffusing outward.

The effect of the tremendous number of intermolecular collisions does *not* even out the speeds of the particles, but distributes them in a stable pattern over a broad range from zero up toward infinity. The corresponding curves represent the number of molecules in each interval of speed (Fig. 14.20). They peak at a speed that is likely to be possessed by more particles than any other, and that's roughly 13% below the average speed v_{av} and 23% below v_{rms}. Most of the particles are grouped around this peak speed; in fact, there will probably be no more than 1 out of about 10 000 molecules with speeds in excess of $3v_{av}$. Nonetheless, the immense number of particles present ensures that small but finite quantities will be traveling very slowly

TABLE 14.7 Molecular parameters at STP

Property	Hydrogen (H$_2$)	Oxygen (O$_2$)
Number of molecules per cm^3	2.7×10^{19}	2.7×10^{19}
Diameter of molecule	0.24 nm	0.32 nm
Molecules per mole	6.06×10^{23}	6.06×10^{23}
Mass per cm^3	8.99×10^{-8} kg	1.43×10^{-6} kg
Average speed	18.4×10^2 m/s	4.6×10^2 m/s
Collisions per second	10×10^9	4.6×10^9
Mass of molecule	3.3×10^{-27} kg	53×10^{-27} kg
Mean free path	1.8×10^{-7} m	1.0×10^{-7} m
Volume per gram	11.2×10^3 cm^3	700 cm^3

Figure 14.20 The Maxwell-Boltzmann speed distributions for molecular oxygen at $T = 73$ K and $T = 273$ K. The area under either curve between any two speeds, say, 400 m/s to 600 m/s, is the percentage of the total number of particles N in that range. The total area under either curve is 100%. Multiply the percentage in any speed range by the total number of molecules present, and you obtain the number with that range of speeds.

Figure 14.21 At the same temperatures, oxygen and hydrogen molecules have the same average kinetic energy; therefore, the less massive hydrogen molecules, on average, move faster.

and very rapidly. Realize, too, that the curve represents a stable pattern even though, from moment to moment, collision to collision, the speed of any one molecule will change drastically. These are the *Maxwell-Boltzmann distributions,* each displaying the most probable spectrum of molecular speeds available to the system at a particular temperature. Though purely theoretical when proposed, the distribution received direct confirmation in the late 1920s and early 1930s via molecular beam experiments.

In the derivation that led to Eq. (14.11), we assumed that the squares of the three speed components were equal. That suggests that the three possible means of sharing the total KE of the system are also equal. In other words, each mode of action, or *degree of freedom,* of the system possesses the same amount of energy. This is really a guess, but it's a guess that has been successfully generalized (and not without controversy) into the *Equipartition Theorem.* The idea was proposed by J. J. Waterston when considering a mix of two different gases held at the same temperature in a chamber. He suggested that each gas would have the same average translational KE. This same conclusion can also be seen to follow from Eq. (14.14). The implication is that massive particles will travel slower than light ones, as affirmed by Eq. (14.15) and illustrated in Fig. 14.21.

Diffusion

Presumably, if two gases of different mass are retained in a chamber that has a small hole in it, the lighter, more rapidly moving ones would escape sooner. That conclusion was confirmed experimentally (1896) by Lord Rayleigh, who partially separated two gases by allowing them to diffuse through a porous barrier into a region of vacuum.

This is the idea behind one of the processes devised during World War II to produce the atomic bomb. Bomb-stuff, fissionable uranium-235, is found in the ore mixed with heavier, but chemically identical, uranium-238. What was needed was a mechanical scheme that could separate the two kinds of atoms. Several different methods were pursued to ensure ultimate success—one of these was gaseous diffusion. Uranium was combined with fluorine, producing gaseous uranium hexafluoride. This gas was then cycled, over 4000 times, through barriers each containing thousands of millions of tiny holes smaller than one-tenth of a mean free path ($<10^{-5}$ mm). Because it was less massive, uranium-235 passed through the barriers more effectively, separating from the uranium-238. The scheme worked surprisingly well, and by the summer of 1945 we entered the age of nuclear weapons.

Core Material

Temperature is a fundamental quantity that describes an aspect of the physical state of a system. The two common temperature scales, Fahrenheit and Celsius, are related by

$$T_F = 32° + \tfrac{9}{5}T_C \qquad \text{or} \qquad T_C = \tfrac{5}{9}(T_F - 32°) \qquad [14.1]$$

The **thermodynamic temperature** is measured in units of *kelvins* (K):

$$T \to T_C + 273.15 \qquad [14.2]$$

and the Absolute Zero of temperature corresponds to 0 K (−273.15°C).

Most solids expand on having their temperatures raised and

$$\Delta L = \alpha L_0 \Delta T \qquad [14.3]$$

Here, ΔT can be in either °C or kelvins and α is the *temperature coefficient of linear expansion*. The change in volume of a substance that undergoes a change in temperature ΔT is

$$\Delta V = \beta V_0 \Delta T \qquad [14.4]$$

Here, V_0 is the original volume and β is the *temperature coefficient of volume expansion* (p. 490).

Boyle's Law is

[*constant T and m*] $PV = \text{constant}$ [11.5]

Charles's Law is

[*constant P and m*] $\dfrac{V}{T} = \text{constant}$ [14.5]

Gay-Lussac's Law is

[*constant V and m*] $\dfrac{P}{T} = \text{constant}$ [14.6]

These three relationships can be combined; thus, *for a fixed quantity of gas*

[*constant m*] $\dfrac{PV}{T} = \text{constant}$ [14.7]

When the mass of gas can change but the temperature is held constant, $PV \propto m$, or

[*constant T*] $\dfrac{PV}{m} = \text{constant}$ [14.8]

The equation of state for an ideal gas is

$$PV = nRT \qquad [14.9]$$

where $R = 8.314\,41$ J/mol·K is called the **universal gas constant**. We define a new constant, called **Boltzmann's Constant**, $k_B = R/N_A = 1.380\,6 \times 10^{-23}$ J/K and thereby obtain the final form of the Ideal Gas Law:

$$PV = Nk_B T \qquad [14.10]$$

The Kinetic Theory of gases leads to Boyle's Law, in the form

$$PV = \tfrac{2}{3}N(\text{KE})_{av} \qquad [14.13]$$

where $(\text{KE})_{av} = \tfrac{3}{2}k_B T$ [14.14]

The average translational KE of the molecules of an ideal gas is related to the temperature of that gas. Furthermore

$$v_{rms} = \sqrt{(v^2)_{av}} = \sqrt{\dfrac{3k_B T}{m}} \qquad [14.15]$$

Suggestions on Problem Solving

1. When confronted by a problem on gases, note any physical parameters that are held constant. Use the Ideal Gas Law in one of its two forms to summarize all the gas laws. Thus, if T and m (that is, n) are constant, the problem deals with variations in pressure and volume and $PV = \text{constant}$; if V and m (that is, n) are constant, the problem deals with variations in pressure and temperature and $P/T = \text{constant}$, and so on. When given or asked for N, use $PV = Nk_B T$.

2. To determine the mass of any molecule, either consult the Periodic Table or Table 14.4. Atomic masses are usually given in atomic mass units (u), where $1.000\,00$ u $= 1.660\,54 \times 10^{-27}$ kg. The numbers in the Periodic Table are averages for naturally occurring mixes of isotopes (something we will come back to later), so there is no need to carry around all those figures.

3. In the problems concerning variations in the size of an object, the change in temperature is central. Whether you use °C or K, ΔT will be the same; however, you cannot use Fahrenheit when α or β is given in SI. **All the gas equations are formulated in terms of thermodynamic temperature, and using anything other than kelvins will produce wrong answers.**

4. The gas equations often lead to a symmetrical form like $P_i V_i / T_i = P_f V_f / T_f$, in which case the units cancel from both sides. It is then allowable to enter mixed units into the equation, but be very careful. Like it or not, there will always be a body of historical data in the literature that uses all the old, now shunned, units, and we should be able to deal with that, too. When one unit can be converted to another by a multiplicative constant, then the ratios of quantities in either set of units will be the same: P_i / P_f is the same whether it is in Pa or mm Hg. *But* when the units are converted one to the other by the addition of a numerical factor, this will not be true: T_i / T_f is *not* the same in degrees Celsius as in kelvins. (Just think about the chaos that would arise in the equations at 0°C.) So, yes you can use liters on both sides, or atmospheres, but you must use kelvins for temperature.

5. Remember that $\Delta L/L_0 = \alpha \Delta T$ and so α contains no length unit. It is the fractional change per kelvin (or per °C, which is the equivalent) and, hence, ΔL and L_0 can be entered in any unit you like.

Discussion Questions

1. List all the possible sources of errors you can think of that might limit the accuracy of a temperature reading made with a typical mercury-in-glass thermometer. Be as imaginative as you can and, as Descartes put it, "doubt of everything"; that is, take nothing for granted.

2. Plunge a sensitive mercury-in-glass thermometer in a bath of hot water. The temperature reading will first drop for a moment and then gradually rise. Explain.

3. Type metal, which is poured into molds to make printer's type, is an alloy of lead, antimony, and tin. Similarly, cast iron, which is an alloy of iron and carbon, is also used in making castings. What uncommon physical property do you think these materials might share with water? Explain.

4. A demonstration device known as a cast-iron bomb is a thick-walled sphere that is filled to the top with water and then sealed with a tightly fitting iron plug. The thing is then submerged in a pail of ice, water, and salt. What happens next? Explain. How does this same process, occurring naturally, affect rocks and mountains?

5. The two steel plates in Fig. Q5 have been attached to one another with a rivet, which was slipped into the hole running through the plates while red hot and "headed over" with a hammer. Why were the rivets put in hot? (Incidentally, the usual approach was to throw the red-hot glowing rivets, almost the size of your fist, through the air from the fire, up to the installing team, which caught them in buckets.)

Figure Q5

6. Suppose that you are visited by an inventor who has found a liquid that has exactly the same β as glass. It is proposed that the wonder fluid be dyed red so it can be seen easily and used to make inexpensive, but incredibly accurate thermometers. Would you invest in such a scheme? Explain the physical reasons behind your decision.

7. The old-fashioned way to break boulders is to heat them up for a long time in a big fire and then quickly douse them with cold water. It is a lot easier than hitting at them with sledge hammers. Explain.

8. Anyone who has ever worked in a chemistry lab knows that such places invariably house a multitude of beakers and test tubes. Why is it that such equipment is generally made of Pyrex

(or some other special glass), and why are they so thin-walled and delicate? (You never see a good, thick, sturdy test tube.)

9. In light of Question 8, why are glass incandescent bulbs so thin? What might happen to a hot light bulb if a drop of cold water fell on it? *Don't try it!* (This can be a problem to sloppy house painters.) What does that suggest about outdoor lighting? How does this relate to traditional photographic flashbulbs?

10. Though it was no doubt observed countless times before, Michael Faraday was the first person to report that two pieces of ice can be frozen together as one by simply pressing them against each other. Explain. What does this phenomenon have to do with snowballs?

11. I remember days as a kid when there was snow everywhere, but the stuff just would not pack into snowballs no matter how we tried. Although we concluded that it was just "terrible snow," what might have been the actual cause of the problem?

12. Professor James Thomson, Lord Kelvin's older brother, was the one who suggested that the three transition curves in the phase diagram (Fig. 14.17) actually meet at a single point—the triple point. Show logically that this must be the case (Fig. Q12).

13. The thick-walled sealed tube in Fig. Q13 shows liquid carbon dioxide in equilibrium with its vapor at room temperature and a pressure of about 6.2 MPa (61 atm). Describe what will happen in the tube as the temperature is slowly raised to 31.1°C.

Figure Q12   **Figure Q13**

14. Under what circumstances, if any, can water freeze and boil at almost the same time? What would happen to a dish of water in an evacuated region where the vapor is efficiently drawn away?

15. Can we expect the ordinary gases of the air (oxygen, hydrogen, nitrogen, and so forth) to behave like an ideal gas? Explain.

16. Sir James H. Jeans, a physicist who made a number of important contributions to modern physics during the early part of the twentieth century, wrote the following in *The Growth of Physical Science:*

In warmer air, the molecules must, of course, move more rapidly, in cooler air more slowly. In general the energy of motion per molecule is proportional to the temperature, measured from the absolute zero at which the energy of motion is nil.

Which equations derived in the chapter confirm his comment? What, if anything, might you take exception to?

17. Given: two 1.0-m³ sealed evacuated chambers surrounded by the same constant temperature bath. Suppose we put two moles of hydrogen in one of them and two moles of nitrogen in the other. What can be said about the pressures in the two chambers? Undeniably, the nitrogen molecules are much more massive than the hydrogen molecules. How can that difference be understood in light of your conclusion about the pressures? What, if anything, can be said about the speeds of the different molecules?

18. If we should ever be visited by some advanced extraterrestrial society, is it likely that they, too, would have arrived at the same notion of an absolute zero of temperature? How reasonable is it to expect that any thermometer they might pull out would have a scale we would be familiar with? If you answered affirmatively, which scale?

19. The air in an ordinary room heated by a stove of some sort will rise in temperature but may not increase its net amount of molecular translational energy. Explain. Be careful about the assumptions you make.

Multiple Choice Questions

1. The Celsius equivalent of 90°F is   (a) 90°   (b) 32°   (c) 194°   (d) 45°   (e) none of these.
2. At the temperature known as absolute zero   (a) all motion ceases   (b) time ceases   (c) all matter is in the solid state   (d) atomic motion is at a minimum   (e) none of these.
3. The surface of a very hot star has a temperature of 2×10^5 °C. Very roughly, how many times hotter is that than room temperature (293 K)?   (a) 10^3 (b) 10^5 (c) 10^{-3} (d) 10^{-5} (e) none of these.
4. What is the order of magnitude of the fractional change in the length of a metal bar when it is altered in temperature by 1 K?   (a) 10^{-6} (b) 10^{-4} (c) 10^{-5} (d) 10^6 (e) none of these.
5. Water has its maximum density at a temperature of about   (a) 0°C   (b) 32°F   (c) 4°C   (d) 273 K   (e) none of these.
6. If the temperature of a sealed chamber full of gas is increased, the pressure will   (a) remain constant   (b) decrease proportionately   (c) increase proportionately   (d) first decrease and then increase   (e) none of these.
7. Considering an ideal gas in equilibrium at a fixed temperature, volume, and pressure, we can expect that the velocity of any one of its molecules on average is   (a) large   (b) small   (c) variable   (d) zero   (e) none of these.
8. Fahrenheit observed that water could be cooled below its freezing point without solidifying, and Gay-Lussac took it down to −20°C in the liquid state. This is an example of   (a) superconductivity   (b) supercooling   (c) superfluidity   (d) supersaturation   (e) none of these.
9. What is the approximate mass of 11.11 liters of molecular hydrogen gas at STP? Each hydrogen atom has an average mass of 1.007 97 u.   (a) 1.0 g   (b) 1.0 kg   (c) 11.1 g   (d) 22.4 liters   (e) none of these.
10. The density of carbon dioxide at STP is   (a) 1.96 kg/m³   (b) 44 g/liter   (c) 44 kg/m³   (d) 6.02×10^{23} g/liter   (e) none of these.
11. Given that we have 17.0 g of ammonia gas (NH_3), the volume it occupies at STP is   (a) 17 liters   (b) 22.4 liters   (c) 22.4×10^6 m³   (d) 17 m³   (e) none of these.

12. According to Kinetic Theory, the molecules of a gas at a given temperature   (a) all move with a speed v_{rms} (b) all move at speeds in excess of v_{rms} (c) all move at speeds less than v_{rms} (d) move at speeds above, at, and below v_{rms} (e) none of these.
13. In a mixture of oxygen and hydrogen gas, we can expect, on average, that   (a) the hydrogen molecules will be moving faster than the oxygen molecules   (b) both kinds of molecules will be moving at the same speed   (c) the oxygen molecules will be moving more rapidly than the hydrogen molecules   (d) the KE of the hydrogen will exceed that of the oxygen   (e) none of these.
14. A sample of gas is held at a constant pressure in a cylinder closed by a movable piston. If the volume is halved, how will the new *rms*-speed of the molecules compare with the original *rms*-speed? It will be   (a) $\sqrt{2}$ times greater   (b) the same   (c) 2 times greater   (d) 4 times greater   (e) none of these.
15. A sample of gas is held at a constant pressure in a cylinder closed by a movable piston. If the volume is doubled, how will the new *rms*-speed of the molecules compare with the original *rms*-speed? It will be   (a) $\sqrt{2}$ times greater   (b) the same   (c) 2 times greater   (d) 4 times greater   (e) none of these.
16. Molecular hydrogen has a mass per mole of 2 g/mol, while the equivalent value for molecular oxygen is 32 g/mol. That means that the ratio of the *rms*-speed of a hydrogen molecule to that of an oxygen molecule will be   (a) 2:1   (b) 4:1   (c) 8:1   (d) $\sqrt{2}$:1   (e) none of these.
17. An atom of argon ($^{40}_{18}A$) in the air at 20°C has an average translational kinetic energy of   (a) 6.0×10^{21} J   (b) 6.0×10^{-21} J   (c) 6.0×10^{-21} m/s   (d) 21×10^{-6} J/s   (e) none of these.
18. The molecules of a gas have a certain average translational kinetic energy at 10°C. They will have twice that average value at   (a) 20°C   (b) 283 K   (c) 566 K   (d) 14.1°C   (e) none of these.
19. According to Kinetic Theory, the molecules of a gas at a given temperature   (a) all have the same speed   (b) all

have the same direction of motion   (c) all have the same kinetic energy   (d) all have the same momentum   (e) none of these.

20. If the temperature of an ideal gas is increased from 100 K to 400 K, the *rms*-speed of the particles will change by a multiplicative factor of   (a) 2   (b) $\frac{1}{2}$ (c) $\frac{1}{4}$ (d) $\sqrt{2}$ (e) none of these.

21. A flat metal disc has a small off-center hole drilled in it. When heated,   (a) the disc will expand and so will the hole   (b) the disc will expand, but the hole will remain unchanged   (c) the disc will expand, but the hole will shrink   (d) the disc will remain unchanged, but the hole will enlarge   (e) none of these.

Problems

TEMPERATURE
THERMAL EXPANSION

1. [I] What is normal body temperature (98.6°F) in degrees Celsius?

2. [I] Helium boils at −268.9°C and melts at −272°C (at 26 atm). What are the corresponding thermodynamic temperatures?

3. [I] A block of steel is heated from 0°F to 100°F. What is the change in its thermodynamic temperature?

4. [I] Hydrogen melts at 14.01 K. What is that in °C?

5. [I] If the temperature at the center of a star is 10 MK, what is the equivalent in °C and °F?

6. [I] A uranium fuel rod in a reactor operates at about 4000°F. What is that temperature in kelvins?

7. [I] The temperature of steam in a typical fossil-fuel plant is about 770 K. What would the temperature be in Fahrenheit degrees?

8. [I] For lead, what is the unitless ratio of its boiling point to its melting point computed in kelvins and in °C?

9. [I] By how much would a 1.00-m-long aluminum rod at 20°C increase if its temperature were raised 1.00°C?

10. [I] By how much would a 1.00-m-long brass rod shrink if its temperature were lowered 1.00°C?

11. [I] By how much would a 10-m-long aluminum bar change its length in the process of going from 30°C to 50°C?

12. [I] A steel girder 10 m long is installed in a structure on a day when the temperature is 5°C. By how much will it increase in length when it warms up to 35°C?

13. [I] The roadway of the Golden Gate Bridge is 1280 m long, and it is supported on a steel structure. If the temperature varies from 0°C to 35°C, how much does the length change?

14. [I] A thin sheet of copper 50.00 cm by 20.00 cm at 30.00°C is heated to 60.00°C. What will be the new area of one face? (See Problem 15.)

15. [I] Redo Problem 14 using the equation derived in Problem 18. Compare the two results.

16. [II] A steel television tower is 150.00 m tall at 10 degrees below zero Fahrenheit. What is its height at 95°F?

17. [II] A carpenter's steel measuring tape is 10.000 m long and 1.00 cm wide. Assuming that it was calibrated in the factory at room temperature (20°C), what is its length on a summer day when the temperature is 96.8°F? (For this steel, $\rho = 7860$ kg/m^3, $\alpha = 12.15 \times 10^{-6}$ K^{-1}, $\beta = 36.46 \times 10^{-6}$ K^{-1}.) Comment on the practical aspects of your results.

18. [II] You have a sheet of material of area A_0 whose temperature is changed by an amount ΔT. Prove that the area will change by

$$\Delta A = 2\alpha A_0 \Delta T$$

19. [II] The mercury in a thermometer at 32°F has a volume of 0.50 cm^3. What will be its volume at 212°F?

20. [II] Imagine that we have a mercury-in-Pyrex thermometer. What is the *apparent* volume coefficient of expansion of mercury in Pyrex? That is, what is the effective coefficient taking into consideration the expansion of the body of the device?

21. [II] A soft glass beaker with an inside diameter of 10.00 cm is filled to the very top with 800.0 cm^3 of pure mercury at 95°C. How much will the level change if the temperature is lowered to 0°C?

22. [II] A thermometer has a quartz body within which is sealed a total volume of 0.400 cm^3 of mercury. The stem contains a cylindrical hole with a bore diameter of 0.10 mm. How far does the mercury column extend in the process of rising from 10°C to 90°C? Neglect any change in volume of the quartz.

23. [II] A beaker made of Pyrex is filled to the very brim with 100 cm^3 of water at 10°C. How much will overflow when it's raised to 50°C?

24. [II] A sheet of brass has a 2.000-cm-diameter hole drilled through it while at a temperature of 20°C. What will be the diameter of that hole if the sheet is surrounded by boiling water?

25. [II] A block of iron has an old label stuck to its side that gives its density as 7.85 g/cm^3 (at 20°C) and its coefficient of volume expansion as 36×10^{-6} °C^{-1}. What will its density be at 0°C?

26. [II] When one cubic meter of water freezes at 0°C, it becomes how many cubic meters of ice?

27. [III] A pendulum clock is made (at 20.00°C) by supporting the *c.g.* of a bob at the very end of a 1.000 0-m-long aluminum rod. Determine its period. Now find its period on a winter's day when the heater goes out and the temperature drops to 0.000°C. Did the clock run fast or slow that day?

28. [III] A 1-second pendulum clock is made at 20.000°C by suspending a mass at its *c.g.* from a steel wire with a linear coefficient of expansion of 12.1135×10^{-6} K^{-1}. What will its period be at 40.000°C? Will it run fast or slow?

29. [III] Given that β is small, show that the final density of a solid specimen that has experienced a small temperature change of ΔT is

$$\rho \approx \rho_0(1 - \beta \Delta T)$$

30. [III] Show that the *thermal stress* (the stress due to an expansion or contraction of the object as a result of a temperature change) is given by

$$\text{Thermal stress} = Y\alpha\,\Delta T$$

31. [III] With Problem 30 in mind, suppose two concrete slabs are butted flat up against one another with their far ends fixed. Each is 10 m long and the contact area is 1000 cm². If the concrete was poured on a day when the temperature was 5.0°C, what will be the stress in them when the temperature is 42°C? Will they fracture? Take $\alpha = 10 \times 10^{-6}$ K⁻¹ and $Y = 25$ GPa.

THE GAS LAWS

32. [I] A container of helium gas at STP is sealed and then raised to a temperature of 730 K. What will be its new pressure?

33. [I] The needle is removed from a hypodermic syringe, and that end is sealed. The air inside of it is slowly compressed (so that it remains at room temperature) to one-tenth its original volume. What is the final pressure (to two significant figures) in the syringe?

34. [I] A container of gas is kept at a constant pressure (which is close to atmospheric) by supporting a glob of mercury as in Fig. P34. If its volume is initially 500×10^{-6} m³ at a temperature of 273 K, what will it be at 300 K?

Figure P34

35. [I] Given that we have 1200 cm³ of helium at 15°C and 99 kPa, what will be its volume at STP?

36. [I] Three moles of a gas are to be stored at a temperature of 60°C and a pressure of 93.0 kPa. How much volume will the gas occupy?

37. [I] A quantity of helium gas at atmospheric pressure occupies the 300-cm³ volume of a cylinder fitted tightly with a piston. If the temperature is initially 25.0°C, what will be its final value when the gas occupies 200 cm³ at atmospheric pressure?

38. [I] A chamber sealed with a movable piston contains 5.50 liters of hydrogen at 28°C and 81.3 kPa. What would its new volume be at STP?

39. [I] What phase is water in at (a) 1.0 atm and 102°C (b) 2.0 atm and 50°C (c) 0.006 atm and 5°C (d) 0.001 atm and 1.0°C (e) 0.0001 atm and 0.01°C?

40. [I] Determine the density of oxygen (O_2) at STP.

41. [I] What is the mass of a water molecule in kilograms? Assume the mass is simply the sum of the constituent masses. The masses of hydrogen and oxygen atoms are 1.008 u and 15.999 u, respectively.

42. [I] One mole of ammonia gas has a mass of 17.03 g. What is its density at STP to two significant figures?

43. [I] An expandable chamber contains 16.0 kg of molecular oxygen at STP. How many molecules are in it? How many moles?

44. [I] An expandable chamber contains 16.0 kg of molecular oxygen at STP. What's its volume?

45. [I] If 1.40 moles of ammonia gas are to be put into a 10.0-liter container at a pressure of 202.6 kPa, at what temperature should the gas be maintained?

46. [I] A very good vacuum system can pump a chamber down to about 10^{-15} atm. Assuming a temperature of 273 K in the evacuated region, how many molecules of air remain in each cubic centimeter?

47. [I] With Problem 46 in mind, how many molecules are in a cubic centimeter of air at STP?

48. [I] A spacesuit with a volume of 0.10 m³ contains air at a pressure of 0.01 MPa at a temperature of 20°C. How many gas molecules are in the suit?

49. [I] Given five molecules with speeds of 1.0 m/s, 2.0 m/s, 3.0 m/s, 4.0 m/s, and 5.0 m/s, find their average speed.

50. [I] Given five molecules with speeds of 1.0 m/s, 2.0 m/s, 3.0 m/s, 4.0 m/s, and 5.0 m/s, find their *rms*-speed and compare it to the results of Problem 49.

51. [I] Given that the mass of an average oxygen molecule is 5.3136×10^{-26} kg, what is its *rms*-speed at 293.15 K?

52. [I] What is the average translational kinetic energy of an oxygen molecule in the air at 0°C?

53. [II] Taking the normal lung capacity to be 500 cm³ and the pressure therein to be the equivalent of 761 mm Hg (which is still the way the medical texts are listing it), estimate the number of molecules per breath.

54. [II] An automobile tire is pumped up to an absolute pressure of 33 lb/in.², with air at a temperature of 40.0°F. After driving for several hours, the temperature in the tire reaches 120°F. Find the pressure in the tire at that point in SI units.

55. [II] A large, aluminumized-Mylar™ balloon contains 1.000 m³ of helium when filled out of doors at 0°C on a winter's day. What will its volume be at home at 20°C?

56. [II] Hydrogen gas at 273 K fills an expandable chamber to a volume of 30.0 liters at a pressure of 2.00 atm. The gas is then compressed to 15.0 liters at a pressure of 3.00 atm. What will its new temperature be?

57. [II] At an altitude of about 12.5 km, the temperature of the Earth's atmosphere is roughly −55°C, and the pressure is around 19.4 kPa. How many kilograms of hydrogen gas (H_2) should be put in a balloon to fill it to 2000 m³ at that altitude?

58. [II] If the density of nitrogen in a chamber at a pressure of 1.00 atm is found to be 1.245 kg/m³, what is the temperature of the gas?

59. [II] Show that for a given amount of gas

$$\frac{P_1}{T_1\rho_1} = \frac{P_2}{T_2\rho_2}$$

60. [II] Determine the density of hydrogen gas at STP. What is its density at 1.00 atm and at a temperature of 273°C?

61. [II] What is the average amount of translational kinetic energy possessed by all the molecules of oxygen contained in 0.50 m^3 of gas at STP?

62. [II] Prove that the pressure produced by an ideal gas in a sealed chamber is proportional to both the number of molecules per unit volume and their average translational kinetic energy.

63. [II] Derive an expression for the *rms*-speed of an ideal gas in terms of its density and pressure.

64. [II] What is the ratio of the *rms*-speeds of nitrogen in the air at 0°C and 100°C?

65. [III] A gas is being collected in a beaker over mercury. This is done by inverting a graduated beaker filled with mercury and then partially submerging it in a mercury-filled bowl. Gas is introduced into the beaker from below, bubbles up through the mercury and displaces it. Suppose, at room temperature, 214 cm^3 of gas is collected, leaving the surface of the mercury 3.90 cm above the level in the bowl. A barometer on the wall reads 101.05 kPa. If the gas is removed and slowly cooled to 0.00°C at a pressure of 101.32 kPa, what will be its new volume?

66. [III] Hydrogen is being collected in an inverted beaker over water. The gas occupies a volume of 370 cm^3, and the surface of the water inside the beaker is raised 5.00 cm while a barometer reads 759 mm. The temperature of the system is 17°C. Given that water has a vapor pressure at that temperature of 1.92 kPa, find the pressure exerted by the hydrogen alone.

67. [III] Suppose 3.0 liters of oxygen at a pressure of 5.0 atm are added to 1.0 liter of nitrogen at 2.0 atm in a chamber with a volume of 4.0 liters. What is the pressure in the chamber if the temperature is kept constant throughout?

68. [III] Prove that the volume coefficient of expansion (keeping P constant) of an ideal gas equals $1/T$.

69. [III] Considering the discussion on p. 505, what is the percentage difference between the *rms*-speeds at room temperature of the two different forms of uranium hexafluoride (UF_6)—namely, those containing either $^{238}_{92}U$ or $^{235}_{92}U$?

15

Heat and Thermal Energy

THIS CHAPTER IS ABOUT internal energy, heat, and the associated behavior of bulk matter. It provides the basis for the study of Thermodynamics, which will be treated in Chapter 16. That discipline developed in the 1800s completely independently of atomic theory, and that's both a virtue and a failing. Here we will pursue an integrated approach that seeks to explain the macroscopic in terms of the microscopic.

We will attempt to answer such questions as: What is heat? What is the relationship between heat and temperature? What happens to bulk matter as thermal energy is added or removed? How is energy transferred from one

A temperature map of the surface of the east coast of the United States produced by a space vehicle orbiting the planet. Florida is outlined by warm water (deep red) at the lower left. Long Island juts into the cool blue-green coastal waters of the Atlantic Ocean. The great lakes are at the upper left. Notice the warm vortices spun off by the Gulf Stream (at the center of the picture).

Energy will remain in some sense the lord and giver of life, a reality transcending our mathematical descriptions. Its nature lies at the heart of the mystery of our existence as animate beings in an inanimate universe.

FREEMAN DYSON
(contemporary physicist)

place to another? The discourse is not just about boiling water and barbecues. As with all the creatures on the planet, thermal concerns are of vital personal interest at every moment of our lives. On a far grander level, these ideas lay the groundwork for a more complete understanding of the concept of energy and with that, a more complete understanding of the way change occurs throughout the Universe.

THERMAL ENERGY

The atoms of a solid, constantly colliding with one another, forever vibrate every-which-way about their fixed equilibrium positions (Fig. 15.1). In a liquid those positions shift, and much of the long-range order of the solid state vanishes, but the oscillations persist. By comparison, the gas is a tumult of impacts and recoils. Gas atoms are essentially free. There are no equilibrium positions, no oscillations; the motion is a perpetual high-speed zigzag dance punctuated by collisions. Much of this chapter treats the macroscopic manifestations of this hidden atomic motion.

Since a good deal of today's nomenclature and imagery came by way of the erroneous caloric theory of heat, we do well to examine it, if only for the roots of our present understanding.

15.1 Caloric versus Kinetic

Rarely is a hypothesis that is once decisively set aside ever resurrected, but that is exactly what happened with the Kinetic Theory. It was widely accepted, in one form or another, throughout the seventeenth century. Moreover, it was correctly believed by many that *heat* was a manifestation of this atomic motion. Galileo had said as much and Plato, long before, appreciated the equivalence of motion and heat.

The eighteenth century brought with it the gradual acceptance of a new and incorrect theory of combustion. This conception maintained that there was an invisible, weightless "matter of fire." By the 1760s, that unfortunate invisible-fluid metaphor was being extended to describe heat as well. In France and Germany especially, the consensus held that heat was an imponderable, self-repellent, indestructible fluid, which the chemist Lavoisier (ca. 1787) christened **caloric**.

The caloric theory proved to be highly useful for understanding and predicting a variety of thermal effects, much more so than Kinetic Theory, which no one at the time was able to apply practically. After all, wrong hypotheses can explain the world too, even if erroneously. We still say things like "pour on the heat" and "soak up the heat." Armed with one of the new thermometers, clever philosophers could measure the amount of caloric poured into a system—at least they thought they could.

The minority opinion, of heat-as-motion, would ultimately win out, but only later on, when the new concept of energy had matured to the point where heat could be recognized as a different kind of subtle "stuff"—random KE. This saga, the rise of the energy formalism and the understanding of heat as a process within that picture, was among the greatest intellectual achievements of the nineteenth century.

The Repudiation of Caloric

The first important attack on the caloric theory was delivered in 1798 by one of the most fascinating characters ever to be involved in science. The American, Benjamin

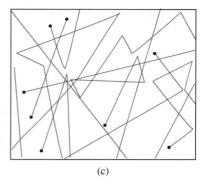

(a)   (b)   (c)

Figure 15.1 (a) In the solid, each atom oscillates about a fixed lattice site and collides (interacting electromagnetically) with its immediate neighbors. (b) In the liquid, the equilibrium positions move, and each atom vibrates over a larger region. (c) In the gas, the oscillations vanish, and the motion is essentially unconstrained until two atoms collide.

Thompson, a shrewd and elegant opportunist, was a professional soldier who saw little action yet rapidly rose to the rank of major general. Relying on an audacious blend of charm and cunning, he managed to be knighted by the King of England, became minister of war in Bavaria, and ultimately was made Count Rumford of the Holy Roman Empire. A tall, handsome, philandering, blue-eyed, likeable rogue, he fathered three children, of whom only one was legitimate. Thompson was a spy and a scoundrel, a quasi-beloved benefactor of the poor, a social reformer, a superb administrator (who even seems to have provided for a Bavarian brothel), and the founder of the British Royal Institution.

After numerous escapades our semihero, the Imperial Count Rumford, was all but running the tiny state of Bavaria. In 1787, he performed a series of very fine experiments (using his Most Serene Highness's best balance) from which he concluded that caloric, if it existed at all, had to be weightless—a fact that would become a primary objection to the theory and one that convinced him, at least, of the validity of the heat-as-motion hypothesis.

Years later, having newly outfitted the arsenal at Munich, he was watching the cannons being bored one day and realized that there was a tremendous amount of heat liberated in the process. He immediately arranged to have a brass gun-barrel blank, surrounded by several gallons of cold water, rotated against a dull steel borer (Fig. 15.2). Gradually, the temperature rose, until after $2\frac{1}{2}$ hours "the water *actually boiled!*" Heating by friction was certainly well known—aborigines and Boy Scouts still start fires that way, and people at the time even believed that forest fires began when the leaves on the trees rubbed too vigorously. The crucial point was that most calorists would have said that caloric was squeezed out of the brass by the boring tool. Moreover, the central postulate of the heat-as-matter theory was that caloric was indestructible and uncreatable. And yet Rumford showed that heat could apparently continue to be generated indefinitely as long as work was done turning the cannon barrel. It "appeared evidently to be *inexhaustible*" and could not therefore "possibly be a *material substance.*"

Though no single experiment can prove a theory, and alternative explanations were offered on behalf of caloric, Rumford's work weighed heavily on the side of the Kinetic Theory of Heat.

15.2 Heat and Temperature

In previous chapters, we studied mechanical energy, the energy associated with an object that moved or interacted with other objects as a whole. In that sense, we have thus far been dealing with organized, or *ordered energy*. On the other hand, it is possible to impart uncoordinated motions to the individual constituent atoms—

Count Rumford. As a young man, he had seen the Boston Massacre, but that didn't keep him from becoming a major in the King's American Dragoons. When British General Gage marched on Lexington and Concord, starting the Revolutionary War, it was on the strength of secret intelligence provided in a letter written (in invisible ink) by Benjamin Thompson.

Figure 15.2 Rumford's diagrams of his experiment. A solid brass cannon blank (Fig. 1) was turned down to form a short, wide cylinder at its front end (Fig. 2), then encased (Fig. 3) in a wooden box (Fig. 4). A blunt borer (Figs. 5 and 6) was forced into the cylinder and the cannon rotated by a team of horses (not shown). The box was filled with water, which subsequently boiled.

disorganized random motion, not of the body as a whole, but within the body. **Thermal energy is the net disordered KE (rotational, translational, and vibrational) associated with any group of particles, usually the atoms of bulk matter.**

If we throw an apple, it will not experience an increase in thermal energy, although it will have an increased amount of KE (with respect to us). A thermometer stuck in the apple will show no change due to the motion. All the apple's atoms will move together as a whole in an ordered fashion. By contrast, if the apple crashes into a wall, it deforms and some of that organized KE will invariably be converted into an increase in thermal energy. Surprisingly, this energy perspective was essentially understood as early as 1695. Leibniz pointed out that in an inelastic collision the seeming loss of *vis viva* (which is equivalent to KE) only appeared to happen. In fact, it was "not destroyed, but dissipated among the parts. That is not to lose it, but to do like those who change large money into small." The ordered energy of the whole is shifted into the disordered energy of the parts.

In a sense, **temperature** *is a measure of the ability of randomly moving particles, usually atoms, to directly impart thermal energy to a thermometer or any other object. It reflects not the net amount of random* KE, *but its average value; the concentration of thermal energy.* The temperature of a substance is independent of the total number of atoms present. The net amount of thermal energy in the Atlantic Ocean is huge, but the average random KE of its molecules is low, and its temperature is low. By contrast, thermal energy depends on both the total number of atoms present and on their average random KE (which depends, in large measure, on temperature). The more atoms present and the more KE each possesses on average, the more thermal energy is present in the system. Because the individual atomic motions are random, altering the thermal energy of a system has no effect on its net momentum.

We now consider three basic ways in which the thermal energy of a body can be increased (or, if reversed, decreased), though much of this chapter will focus only on the last of these mechanisms.

First, we can do work on the body: rub it, compress it, or in any way *distort it; that is, physically displace some of its atoms against an internal restoring force.* Blacksmiths sometimes kindled a fire using a soft iron rod made red hot by pounding on it with rapid hammer blows. **Work** *is the organized mechanical energy transferred into or out of a system by way of a force acting through a distance.* If a force is applied to overcome interatomic forces (for example, friction) thereby displacing atoms, the energy introduced is rapidly distributed within the body through atomic collisions, and much of it will appear in the disordered form of thermal energy. That's what happens when you warm your hands by rubbing them together. Increasing the total amount of thermal energy of an object (like your hand) increases the average random KE, which means an increase in temperature.

The second means of increasing thermal energy involves shining electromagnetic radiation (light, infrared, ultraviolet, X-rays, etc.) onto the body. In other words, we can transfer radiant energy directly to atoms, which thereupon become agitated, collide with one another, and redistribute that energy into a disordered thermal form. That's what happens when you warm your face in the sunshine: radiant energy is absorbed and transformed into thermal energy, the average KE of the molecules rises, and your skin temperature goes up.

The third mechanism is perhaps the one that comes to mind first: we know from experience that the "hotness" of an object can be increased by placing it in contact with something at a higher temperature. A spoon at room temperature positioned in a flame will soon heat up. The high-temperature flame consists of atoms randomly moving about with a high average KE. This violent storm of glowing gas bombards the atoms in the surface of the spoon, which have a much lower average KE. The lower temperature spoon-atoms take up energy on impact from the more energetic flame-atoms. The former are thus increased in their random motion, and thermal energy is thereby imparted to the spoon. Energetic spoon-atoms that are touching the flame strike quiescent spoon-atoms just behind them, and the jostling continues back until the entire object becomes hot—a process we ordinarily describe by saying that "heat flows" down the spoon. More formally, **heat** (*Q*) *is the thermal energy transferred, via atomic collisions, from a region of high temperature* *to a region of lower temperature.* It is the energy of random motion transported by way of contact from one object to another, from one group of atoms to another—exclusively as a result of a temperature difference.

A body contains or stores thermal energy, not heat. Heat is thermal energy in transit, and once transferred it is no longer called heat. (We blow wind into a balloon, but once inflated, what is stored is air.) As before, we are again more concerned with the change in energy of a system than with the net amount of energy present at any moment. An agency of change of thermal energy is heat, just as an agency of change of mechanical energy is work. Once transmitted, the abiding mechanical energy of a system is no longer called work—what is stored is KE or PE, not work. In general, *a system contains energy when it can undergo spontaneous change.*

15.3 Quantity of Heat

Four friends dominated the intellectual life of Scotland in the latter half of the 1700s. They were Adam Smith, the political economist; David Hume, the renowned philosopher; his personal physician, Dr. Joseph Black, professor of chemistry at Glasgow; and that university's resident scientific instrument maker, James (Jamy) Watt. Both Black and Watt were reserved advocates of the caloric theory, and that was reasonable enough; caloric was a practical contrivance, and these were practical men, more concerned with using heat than with hypothetical abstractions.

Black was the great pioneer in thermal physics and probably the first to distinguish between "quantity of heat" and temperature. He saw temperature as reflecting the "intensity," or concentration, of what we would call thermal energy and he called "the matter of heat."* A stew tastes salty because of the concentration of salt, and it is hot because of the concentration of thermal energy. The bigger the potful, the more salt must be sprinkled in and the more heat must be added. The ocean, cold as it is, contains far more thermal energy than the scalding stew, and no matter how salty the dish, the ocean will contain far more salt as well. The fact that its thermal energy is spread out thin, not concentrated, accounts for the ocean's low temperature, but the large amount of random KE is there nonetheless.

Sometime around 1760, Black introduced a simple scheme for quantifying heat that was operational and hence totally independent of what heat actually is—caloric or KE, it didn't really matter. One only needed to have a thermometer and a scale to weigh out some water. Following Black's lead, the **calorie** (cal) ultimately became the pre-SI metric unit of heat. It is defined as *the amount of heat that must be added to raise the temperature of 1 gram of water 1 degree Celsius*—from 14.5°C to 15.5°C. This particular temperature range is now specified because the heat required is very slightly different at other temperatures. As H. A. Rowland discovered (1878), it differs by less than 1% over the range from 0°C to 100°C which, for our purposes, is negligible. There we are then: we can say exactly what a calorie is, even without knowing what heat is, just as we can say exactly what a kilogram is, without knowing what mass is.

By definition, 1 cal will raise 1 g of water 1°C. We can assume (and experiment confirms) that 2 cal will raise the temperature of 1 g of water 2°C, and so on. In other words, the resulting temperature change is proportional to the heat-in: $\Delta T \propto Q$. Similarly, 1 cal will raise the temperature of 2 g of water only $\frac{1}{2}$°C. That is,

James Prescott Joule (1818–1889) was a well-to-do brewer's son from Manchester, England, and a pupil of John Dalton. He was greatly influenced by Rumford's research at the cannon factory.

*Physicians were very involved in research in the early days of thermal science. Indeed, the word *temperature* appears to derive from medicine, where the goal was to temper the extremes of chill and fever.

$\Delta T \propto 1/m$; the more water, the more heat is needed to raise the temperature, $\Delta T \propto Q/m$. Introducing a constant of proportionality c (to take care of the units), we have

$$Q = cm \, \Delta T = cm(T_\text{f} - T_\text{i}) \tag{15.1}$$

In this particular case of water, $c = 1$ cal/g·C°, exactly.

Example 15.1 The equivalent of a glass of water (that is, 270 g of liquid) at 20.0°C receives 1000 cal from a heater. Assuming that all the energy goes into the water without any losses, what is the final temperature of the liquid?

Solution: [Given: $m = 270$ g, $T_\text{i} = 20.0°C$, and $Q = 1000$ cal. Find: T_f.]

$$Q = cm \, \Delta T = cm(T_\text{f} - T_\text{i})$$

$$1000 \text{ cal} = (1.00 \text{ cal/g·C°})(270 \text{ g})(T_\text{f} - 20.0°C)$$

and

$$T_\text{f} = \frac{1000 \text{ cal}}{(1.00 \text{ cal/g·C°})(270 \text{ g})} + 20.0°C = \boxed{23.7°C}$$

▶ **Quick Check:** 270 cal will raise 270 g by 1°C; 1000 cal/270 cal = 3.7, hence the new temperature is 20.0°C + 3.7°C.

The calorie turns out to be inconveniently small, especially for weight watchers. So the world of dietetics (Table 15.1) uses the **kilocalorie** (kcal), sometimes cleverly called the Calorie (1 kcal = 1 Cal = 1000 cal). More often, the fact that "calorie" is supposed to mean kilocalorie is just kept as a little secret (Table 15.2). When you

TABLE 15.1   Energy provided by various foods

Food	Kilocalories
Martini	145
Peanut butter sandwich	330
Buttered popcorn (1 cup)	55
Vanilla ice cream ($\frac{1}{3}$ cup)	145
Whiskey (1 shot)	105
Brownies (1)	140
Jelly doughnut	225
Dry red wine (1 glass)	75

TABLE 15.2   Approximate calories consumed per hour

Body weight lb	100	150	200	250
Body mass kg	45	68	90	113
Sleeping it off	40	60	80	105
Sitting on the grass	65	95	130	165
Standing in line	70	100	140	170
Strolling through a park	130	195	260	320
Running from a mugger	290	440	580	730

Just 1 kilocalorie per glass. Note that the Europeans use commas in their numbers where we use periods (1 kcal = 4.18 kJ).

read on the box that your favorite breakfast cereal "contains" 110 calories per serving, that's really 110 kilocalories, and what's meant is that that amount of energy will be liberated upon the complete combustion of one serving of cereal. Some engineers and aspiring air-conditioner salesmen (mostly in the United States) hold fast to another fading anachronism, the Btu, or *British thermal unit*. One Btu is the amount of heat needed to raise 1 lb of water 1°F.

The realization that heat is energy only came gradually in the mid-nineteenth century. To the advocates of this unifying perspective, the pivotal question then became: how much heat, as defined by its effect on the temperature of water, is equal to a unit of energy, as defined by its ability to do work? If, indeed, heat is energy, then *calories* must be directly convertible into *joules*.

15.4 The Mechanical Equivalent of Heat

The German physician who took the next step was twenty-eight years old when his first essay appeared (1842). Julius Robert Mayer reinterpreted some long-forgotten results on expanding gases to determine a rough value for what came to be called the "mechanical equivalent of heat"—namely, the numerical relationship between work and heat. Mayer's bold speculative essay lacked direct experimental confirmation and was ridiculed as unprofessional.

As if the time was ripe for change, within one year of Mayer's paper there appeared in print the first report of the remarkable researches of James Prescott Joule. Unlike Mayer, Joule was a gifted experimenter who devised a whole range of apparatus using various means (for example, electric currents, friction, and compression of gases) to generate "heat." His most famous was a paddle wheel device driven by falling weights (Fig. 15.3). By equating the work done by the weights to the thermal energy imparted to the water, Joule concluded that 773 ft·lb was equivalent to 1 Btu (which is not far from the modern value of 778 ft·lb per Btu). Nowadays, we use SI units and say the **mechanical equivalent of heat** is

$$4.186 \text{ J} = 1 \text{ cal}$$

or

$$1 \text{ kcal} = 4186 \text{ J}$$

By international agreement, the calorie is no longer the recognized unit of heat. As with all forms of energy, the preferred unit is the *joule*. Thus, we can summarize much of what has been discussed so far with the statement that *the temperature of 1 kilogram of water can be increased (or decreased) by 1 kelvin through the addition (or removal) of 4186 joules of heat.*

Example 15.2 The following statement is taken from Mayer's paper of 1842: "The warming of a given weight of water from 0°C to 1°C corresponds to the fall of an equal weight from the height of about 365 metres." Compare his value of the mechanical equivalent of heat with the modern value.

Solution: [Given: $\Delta T = 1°C$ and $h = 365$ m. Find: Mayer's mechanical equivalent of heat.] By "a given weight," he means any value of mg, so let's take $m = 1.00$ kg for convenience. We must find the corresponding energy associated with a fall of 365 m in SI units. The *gravitational*-PE is given by

$$\text{PE}_G = mgh \qquad [9.13]$$

hence

$$\text{PE}_G = (1.00 \text{ kg})(9.81 \text{ m/s}^2)(365 \text{ m}) = 3.58 \text{ kJ}$$

That's the mechanical energy associated with the fall of a 1-kg mass (of water, or anything else), and it is supposed to equal the heat needed to raise the temperature of that mass (1 kg) of water 1°C, which, from Eq. (15.1), is just 1000 cal = 1 kcal. According to Mayer, then, $\boxed{1 \text{ kcal} = 3.58 \text{ kJ}}$, which is about $\boxed{14\% \text{ too small}}$.

15.5 The Zeroth Law

Joseph Black began one of his lectures with an observation so central to any complete understanding of the behavior of heat that it has since come to be called the **Zeroth Law of Thermodynamics**:

> We can perceive a tendency of heat to diffuse itself from any hotter body to the cooler ones around it, until the heat is distributed among them in such a manner that none of them is disposed to take any more from the rest. The heat is thus brought into a state of equilibrium . . . . The temperature of all of them [i.e., all the bodies] is the same.

In other words, when two bodies at different temperatures come into contact, thermal energy flows from the higher to the lower until the temperatures of the two equalize and they are said to be in **thermal equilibrium**. The idea that matter, all by itself, spontaneously tends to a uniform-temperature equilibrium state is based squarely on observation—the rest of the statement about the flow of energy is hypothesis.

Today, this equilibrium insight is often used as the starting point in a formal theoretical development that introduces the idea of thermodynamic temperature as a unique variable describing the state of the system. After all, that's really the wonderful part of the process, the fact that a cold beer left on the kitchen table will ultimately reach the same temperature as the table (and anything else sitting on it) reveals something fundamental about nature. Thus, when system 1 is placed in thermal contact with system 2 and no net energy flows from one to the other, the two exist in a special physical condition, or state, that can be identified via the concept that they have the same thermodynamic temperature.

The Zeroth Law of Thermodynamics is often stated as follows: *when two systems are at the same temperature as a third, they are all three at the same temperature as each other.* It might seem at first reading that this is a trivial tautology that must be true, but that's not quite the case. Certainly, two people can be in a state of equal and mutual respect for a third person without necessarily respecting each other. The conclusion that temperature does nonetheless behave that way is an experimentally verified law. The reason for emphasizing the notion is that we tacitly assume it's true whenever we compare the temperatures of two samples using a thermometer as an intermediary third body. Thus, it is unnecessary to place our test sample in contact with a standard-temperature bath to see if they are in equilibrium. We calibrate the thermometer with the bath, and if the sample is subsequently found to be at that temperature along with the thermometer, it is necessarily at the temperature of the bath.

Figure 15.3 Joule's apparatus for determining the mechanical equivalent of heat. As the weights fall, the paddle turns, raising the temperature of the water. *Gravitational*-PE is converted into the KE of the paddle and the water. That KE is ultimately converted into thermal energy within the isolated chamber.

15.6 Specific Heat Capacity

We know that 4.2 kJ (1 kcal) will raise the temperature of 1 kg of water 1°C, but there is no reason to assume it will do the same thing to 1 kg of iron or peanut butter or pizza; still, that was the common wisdom before Dr. Black put things right. He found, quite to the contrary, that *every substance changed temperature by a characteristic amount when infused with a fixed quantity of heat.* It is informative to see how he actually managed to transport the "fixed quantity" of heat to the different substances under study.

In one scheme, he placed equal weights of different substances in identical vessels that were subsequently put on the same burner, or in front of the same fire, for the same length of time. The amount of heat transferred was then presumably proportional to the time the sample was exposed to the source (Fig. 15.4). In this way,

Figure 15.4 Imagine some water and an equal mass of iron at, say, 25°C. If an equal amount of heat is added to both, the temperature of the iron increases 9.1 times as much as the water. If the water rises to 35°C, the iron rises to 116°C.

it was confirmed that an equal mass of iron will get hotter than water by a factor of almost 10 in the same amount of time.

Another more controllable approach, which came to be known as the *method of mixing*, evolved into the modern technique of calorimetry. It was fairly common knowledge in the community of scientists that if a given mass of cold water at, say, 5.0°C is mixed with an equal amount of hot water at, say, 95°C, the blend will come to an equilibrium temperature of 50°C. The hot system loses a certain amount of heat to the cold system; the former drops 45°C, the latter gains 45°C—*equal changes, because equal amounts of the same material are being mixed*. We can see as much from Eq. (15.1): the *heat lost* by hot system 1 is

$$Q_1 = cm_1(T_{f1} - T_{i1})$$

where $T_{i1} > T_{f1}$. Similarly, the *heat gained* by cold system 2 is

$$Q_2 = cm_2(T_{f2} - T_{i2})$$

where $T_{f2} > T_{i2}$. But these are plainly equal; that is, $-Q_1 = Q_2$; **heat-out is negative, heat-in is positive.** Moreover $m_1 = m_2$; hence

$$-c(T_{f1} - 95°C) = c(T_{f2} - 5.0°C)$$

At equilibrium, $T_{f1} = T_{f2} = T_f$ and $T_f = \frac{1}{2}(95°C + 5.0°C) = 50°C$ as expected. The crucial points here are that the final temperatures of the two samples are equal, and that the heat lost by one is gained by the other (that is, Conservation of Energy).

This approach of mixing different systems became rather important, especially after Fahrenheit performed much the same experiment using water and mercury. The hot mercury decreased in temperature more than the water rose in the process of their coming together at equilibrium. As Black interpreted the results, an equal mass of mercury required less heat to change its temperature 1 degree than did water. He concluded that "quicksilver [mercury] has less *capacity for heat*" than does water. In time, the *heat capacity* (c) of a substance was defined to be *the number of joules of heat that must be added to raise the temperature of 1 kg of the material 1 K* (that is, *1°C*). It follows from Eq. (15.1), $Q = cm\,\Delta T$, that each different substance has its own characteristic value of c (Table 15.3).

Because there were several different measures of heat in the mid-1800s, it was common to tabulate a unitless quantity called the *specific heat*. Taking a lead (as

TABLE 15.3   Specific heat capacity for some materials*

Material	Specific heat capacity	
	kJ/kg·K	kcal/kg·K
Solids		
Aluminum	0.90	0.21
Clay (dry)	0.92	0.22
Copper	0.39	0.093
Glass	0.84	0.20
Gold	0.13	0.031
Human body (average)	3.47	0.83
Ice (water, −5°C)	2.1	0.50
Iron	0.47	0.11
Polyamides (e.g., Nylon)	1.7	0.4
Polyethylenes	2.3	0.55
Polytetrafluorethylene (e.g., Teflon)	1.0	0.25
Lead	0.13	0.031
Marble	0.86	0.21
Platinum	0.14	0.032
Protein	1.7	0.4
Silver	0.23	0.056
Stainless steel (type 304)	0.50	0.12
Wood	1.8	0.42
Liquids		
Acetone	2.2	0.53
Alcohol (ethyl)	2.4	0.57
Ammonia	4.71	1.13
Mercury	0.14	0.033
Nitrogen (−200°C)	1.98	0.474
Oxygen (−200°C)	1.65	0.394
Sulfuric acid	1.4	0.34
Water	4.186	1.000
Gases		
Air (100°C)	1.0	0.24
Argon	0.52	0.13
Carbon monoxide	1.0	0.25
Hydrogen	14.2	3.39
Methane	2.2	0.53
Steam (110°C)	2.01	0.481

*At ≈20°C.

well as its name) from the idea of specific gravity, which is a unitless ratio, specific heat was defined as the ratio of the heat capacity of the substance to the heat capacity of water. That was convenient because the heat capacity of water was defined to be unity. Today, with our standardization of units and because the SI heat capacity of water is not unity, this quantity is of less practical significance, though it persists. Regrettably, the terminology has blurred with time, and one has to be careful when consulting different sources. It's fairly common now to speak of c as the **specific heat capacity** and to give it the appropriate units. In the literature, those units are likely to be (cal/g·C°) or (kcal/kg·C°) or (J/kg·C°), though (J/kg·K) is to be preferred. Since specific heat capacity is the amount of heat (Q) that must be added or removed per kilogram to change the temperature one kelvin, it follows that

$$c = \frac{Q}{m \, \Delta T} \qquad (15.2)$$

Example 15.3   A copper pot has a mass of 0.50 kg and is at 100°C. How much heat must be removed from it if its temperature is to be lowered to precisely 0°C?

Solution: [Given: Cu, m = 0.50 kg, T_i = 100°C, and T_f = 0°C. Find: Q.] Here Q is the heat-out of the system, which is negative since $\Delta T = -100$ K. From Eq. (15.1)

$$Q = c_{Cu} m \, \Delta T = (390 \text{ J/kg·K})(0.50 \text{ kg})(-100 \text{ K})$$

and so $Q = -19.5$ kJ

or to two significant figures $\boxed{-20 \text{ kJ}}$.

▶ **Quick Check:** $\Delta T = -100$ K. From Table 15.3, the specific heat capacity of copper is 9.3% that of water; if the pot were made of water, we would have to remove $\frac{1}{2}$ 100 kcal; for copper we need only remove 9.3% of that, or 4.65 kcal = 19.5 kJ.

Water has a high specific heat capacity, a fact that is crucial to all life on the planet. A quantity of water will change temperature only when a great deal of heat is added or removed; it heats up slowly and cools off slowly. The legendary hot-water bag is the ultimate testament. A potato or a lima bean, fished out of a hot stew, can be eaten with a little caution, but a juicy onion (mostly water) will have to be treated with respect. To lower its temperature from 100°C in the pot to something you can deal with will require removing a lot more heat first. And if you don't bother, that considerable amount of heat will be deposited in your mouth as it descends down to equilibrium with your tongue. We are mostly water ourselves, and that high value of c is one factor that helps us keep from rapid temperature changes. Cities on the coast of large bodies of water get that same benefit. It is also the reason water is used to carry thermal energy in radiators that heat houses and in cooling systems that maintain automobile engines.

Example 15.4   A quantity of what looks like lead shot, having a mass of 100 g, is poured into a test tube. The tube is then partially submerged in a beaker of boiling water and kept there until the shot reaches 100.0°C. The hot shot, as it were, is then transferred to a vessel containing 100 g of water at 20.00°C. The contents are gently stirred until they come to an equilibrium temperature of 22.41°C. What is the specific heat capacity of the shot?

Solution: [Given: m_S = 100 g, m_w = 100 g, T_{is} = 100.0°C, T_{iw} = 20.00°C, and $T_{fs} = T_{fw} = T_f$ = 22.41°C. Find: c_S.] We can do this problem in several different ways. All are equivalent, and all rely on the fact

that the heat lost by the shot is gained by the water and that both are at the same final temperature. Using Eq. (15.2),

$$c_S = \frac{Q_S}{m_S \, \Delta T_S}$$

and we need Q_S and ΔT_S. The heat lost by the shot is the heat gained by the water, which we can calculate. The water has changed temperature by (22.41°C − 20.00°C) = +2.41°C. There being 100 g of water present, each 1 degree change corresponds to 100 × (4.186 J); hence, the water has gained +1008.8 J. The shot changed temperature by (22.41°C − 100°C) =

(continued)

(continued)
−77.59°C; thus

$$c_S = \frac{-1008.8 \text{ J}}{(0.100 \text{ kg})(-77.59 \text{ K})} = \boxed{130 \text{ J/kg·K}}$$

▶ **Quick Check:** From Eq. (15.1) and the fact that the

heat-out equals the heat-in, $-Q_S = Q_w$; $-c_S m_S \Delta T_S = c_w m_w \Delta T_w$; hence, $c_S = c_w \Delta T_w / \Delta T_S = -(4186 \text{ J/kg·K}) \times (22.41°C - 20.00°C)/(22.41°C - 100°C)$, and $c_S = 130 \text{ J/kg·K}$.

In all these thermal experiments, it is tacitly assumed that there is no significant heat loss. To make that a reality, we use a *calorimeter* as pictured in Fig. 15.5. It's simply a well-insulated chamber whose vacuum or styrofoam walls keep the contents from exchanging heat with the outside. A thin metal cup with a low specific heat and a small mass is used to hold a measured amount of water. The cup changes temperature easily and can store very little thermal energy in the process. Thus, it tends to remove very little from the system. That's a lot like the behavior of aluminum foil used in cooking—it can be handled, albeit carefully, right out of the oven in spite of its high temperature because the quantity of heat given off in coming to equilibrium with your fingers is necessarily small (*mc* is small). Typically, a sample whose specific heat is to be measured is raised to a known temperature and then immersed in the water. The change in temperature of the water allows the required specific heat to be computed.

Figure 15.5 A water calorimeter.

Example 15.5 A calorimeter consisting of a thin copper cup of mass 150 g containing 500 g of water is at a temperature of 20.0°C. A 225-g sample of an unidentified material at 508°C is lowered into the bath, and the device is sealed. After a few minutes, the system reaches a constant temperature of 40.0°C. Determine the specific heat capacity of the sample. Overlook any losses due to the thermometer, stirrer, or insulation.

Solution: [Given: $m_c = 0.150$ kg, $m_w = 0.500$ kg, $T_{ic} = T_{iw} = 20.0°C$, $m_S = 0.225$ kg, $T_{iS} = 508°C$, and $T_f = 40.0°C$. Find: c_S.] The temperature change in kelvins equals that in Celsius, so there is no need to change units. The heat transferred out of the sample (which is a negative number) equals the heat transferred into the water and cup

$$-Q_S = Q_w + Q_c$$
$$-m_S c_S \Delta T_S = m_w c_w \Delta T_w + m_c c_c \Delta T_c$$

Using Table 15.3, we obtain

$$-(0.225 \text{ kg})(c_S)(40°C - 508°C) =$$
$$(0.500 \text{ kg})(4186 \text{ J/kg·K})(20 \text{ K}) +$$
$$(0.150 \text{ kg})(390 \text{ J/kg·K})(20 \text{ K})$$

and $\quad 105.3 \, c_S = (41860 + 1170) \text{J/kg·K}$

hence $\qquad \boxed{c_S = 409 \text{ J/kg·K}}$

▶ **Quick Check:** $[(0.15 \text{ kg})(0.093 \text{ kcal/kg·K}) + (0.5 \text{ kg})(1.0 \text{ kcal/kg·K})] (20 \text{ K}) = 10.3 \text{ kcal} = c_S(0.225 \text{ kg})(468 \text{ K})$; $c_S = 0.098 \text{ kcal/kg·K}$.

The Calorie content, or better still the *energy content,* of your breakfast crunchies can be found with a modified device known as a *bomb calorimeter.* A weighed sample of dried material is placed into a sealed chamber containing a heating coil and an over-pressure of oxygen. That chamber is then positioned in the water bath of an otherwise traditional calorimeter. An electric current heats the coil

TABLE 15.4 Approximate heats of combustion

Material	Heat of combustion (MJ/kg)
Alcohol, methyl fuel	22
(denatured)	27
Anthracite (hard coal)	33
Bituminous coal	30
Bread	10
Butter	33
Carbohydrates	17
Charcoal	28
Diesel oil	45
Dung	17
Fats	38
Gasoline	48
Methane	56
Oil, furnace	44
Propane	50
Proteins	17
TNT	5
Wood	15

TABLE 15.5 Approximate energy content of various materials

Source	Energy (J)
Ton of coal	3×10^{10}
Barrel of oil	6.3×10^{9}
Ton of TNT	4.2×10^{9}
Gallon (U.S.) of oil	1.5×10^{8}
Gallon (U.S.) of gasoline	1.3×10^{8}
Kilogram of fat	3.8×10^{7}
Pound of coal	1.5×10^{7}
Pound of wood	8.6×10^{6}
Cubic foot (1 atm, 16°C) of natural gas	1.1×10^{6}

(with a known amount of energy), and the crunchies burn to dust. The same scheme can be used to determine the energy content of fuels, the so-called *heats of combustion* (see Tables 15.4 and 15.5).

An Atomic View of Specific Heat

Joseph Black could simply explain the differences in specific heat by supposing that water had more space between its atoms than iron and so could contain more caloric. The actual explanation is a good deal more subtle and depends on the structure of the molecules and on their interactions. Roughly put, the heat once entered into a sample is energy that is distributed among the molecules of the system. The more molecules there are, the less the increase in thermal energy per molecule. That means less increase in the average random KE and a smaller resulting temperature increase. A kilogram of many low-mass atoms will increase less in temperature, for a given input of heat, than will a kilogram of fewer high-mass atoms and so have a higher specific heat capacity.

That's the case for almost all the elements, provided they remain in the solid state. In fact, *the product of the specific heat capacity and the atomic mass for each of the solid elements is roughly constant*—something known as **Dulong and Petit's Law** (1819). The unified atomic mass of an atom (u) corresponds numerically to the number of grams there are of it per mole. Hence, if we multiply specific heat capacity in cal/g·K by g/mol, we get cal/mol·K, or J/mol·K, which is called the *molar specific heat capacity* (*C*). What Dulong and Petit found experimentally was that the molar specific heat capacity—the heat needed to raise the temperature of 1 mol, 1 K—is about 25 J/mol·K for most solid elements.

Classical theory explains this result by maintaining that the atoms of a monatomic solid behave like those of a gas, this time oscillating about a mean position with a distribution of energies similar to the Maxwell-Boltzmann distribution. Unlike the noninteracting gas, here there is an ongoing exchange between KE and PE, as there is with an oscillating spring-mass system. Remember that specific heat capacity reflects the medium's ability to manifest a change in the energy it possesses as a change in temperature. But not all forms of energy stored internally do that—PE **does not affect temperature**.

For a gas, the average KE of each atom, as given by Eq. (14.14), is $\frac{3}{2}k_B T$. Since any atom can be imagined as moving along three perpendicular axes, it is assumed that the KE is, on average, distributed equally among the three motions—that's the Equipartition Theorem. Each mode of behavior, or *degree of freedom,* is assumed to correspond to an average energy of $\frac{1}{2}k_B T$. In the case of a solid or liquid, where there is appreciable interaction among the particles, a portion of the entering energy also goes into overcoming these forces and is stored as PE. Again, there are three basic directions and three more degrees of freedom. Extending the Equipartition Theorem, it is assumed that each PE mode has the same energy as each KE mode; namely, $\frac{1}{2}k_B T$. Thus, an atom in a solid should have 6 modes of freedom, each able to take up $\frac{1}{2}k_B T$ in energy, for a total of $3k_B T$. Each mole of material, containing N_A atoms, will then have a net **internal energy** (U = KE + PE), which is the sum of the random atomic KE (the thermal energy) and the interatomic PE. At any absolute temperature T, the internal energy is given by

$$U = N_A 3 k_B T$$

where $N_A k_B$ equals R, the universal gas constant (p. 000). Hence

$$U = 3RT$$

and if an amount of heat Q is introduced, it represents a change in the internal energy ΔU such that

$$Q = \Delta U = 3R\,\Delta T$$

The specific heat capacity per mole is then

$$C = \frac{Q}{\Delta T} = 3R \tag{15.3}$$

Because the atoms interact, half the energy entering the system goes into PE, and there is that much less of a temperature change. Since $R = 8.31$ J/mol·K, it follows that $C = 24.9$ J/mol·K, which is just what Dulong and Petit found. Well, this result is wonderful, but unhappily not the whole story. What is found experimentally is that C is actually temperature dependent, generally approaching zero at 0 K, and $3R$ at temperatures around and above room temperature (Fig. 15.6).

Evidently, the interactions between atoms will be different for the various elements, and we can expect some differences in C. Substances such as diamond that are bound very strongly (are hard and have high melting points) are found to have appreciably lower room-temperature values of C than predicted—a small input of Q causes a large ΔT. It would seem that at low-to-moderate temperatures not much energy goes into PE. Why that should be cannot be understood from classical theory. It is also unclear classically what role the free electrons play in determining the specific heat of metals since they, too, should be moving around and picking up energy. A completely satisfactory theory of specific heat can only be produced using Quantum Theory, but we won't attempt that here.

Liquids allow a greater freedom of their atoms (they can both rotate and move around), which are nonetheless bound to one another. In general, liquids have higher specific heats than either solids or gases.

Figure 15.6 The specific heat capacity per mole for several solids. The curves approach a value of 25 J/mol·K as the temperature increases.

The Specific Heat of Water

Water stands out for its high specific heat capacity (4.2 kJ/kg·K, or 1.0 kcal/kg·K). Even in the solid state (that is, ice) water has a comparatively high specific heat capacity (2.1 kJ/kg·K, or 0.50 kcal/kg·K). The same is true for steam, which at a constant 1 atm has a specific heat capacity of roughly (2.0 kJ/kg·K, or 0.48 kcal/kg·K) that increases with pressure, and above 200°C, with temperature as well.

Water is molecularly rather lightweight; 1 kg of it contains 3 times as many molecules as there are atoms in 1 kg of iron. Additionally, those molecules effectively interact with one another, storing energy in the form of *electrical*-PE.

The hydrogen atoms in a water molecule behave like partially naked protons because their electrons tend to spend more time near the oxygen atom. As a result, there is a powerful intermolecular force binding the liquid together that forms a short-ranged, ordered structure of clusters of molecules. An individual molecule that can nonetheless slowly meander around (with relatively little energy) also oscillates about its moving equilibrium position (with a good deal more energy). Besides increasing this oscillation, which directly increases temperature, adding heat to a liquid partially breaks up the clusters, reducing their size and liberating molecules. Thus, it takes a certain amount of energy to free up the molecules and that energy does not enhance the temperature. The result of all these mechanisms is a substantial specific heat.

The Specific Heat of Gases

A similar relationship to that found by Dulong and Petit also exists among the elemental gases, though here the molar specific heat capacity is different for the different types of gas. In part, that is why hydrogen (at constant pressure and room temperature) has so large a value of c: 14.2 kJ/kg·K, or 3.4 kcal/kg·K (it's such a light molecule). In addition to this mass effect, specific heat capacity is again influenced by the various ways energy is imparted to or stored by the molecules. Monatomic gases such as helium, argon, and neon can manifest their thermal energy only as translational KE; they are not set rotating or vibrating through collisions. They are not molecular, so there can't be any internal vibration of constituent atoms. As for rotation, the moment of inertia of an atom about an internal axis is very small, and that precludes it (for quantum mechanical reasons) from transferring rotational energy via a collision. A monatomic gas easily changes temperature and has a comparatively low molar specific heat—about half that of a typical metal.

With its volume kept constant, such a simple noninteracting gas has 3 degrees of freedom due to its translational KE. Consequently, the internal energy per mole is $U = 3(\frac{1}{2}k_B T)N_A = \frac{3}{2}RT$ and $\Delta U = \frac{3}{2}R\,\Delta T$. The molar specific heat capacity at constant volume is then $C_V = \frac{3}{2}R = 12.5$ J/mol·K, or 2.98 cal/mol·K. This value is almost precisely the molar specific heat capacity of monatomic helium and argon.

Gas molecules with more than one atom can support rotations of the entire molecule about its center-of-mass, as well as internal vibrations of the atoms with respect to each other. These modes of motion are transferable via collisions and presumably can take up appreciable amounts of energy while having little effect on changing the temperature of the system. The more complicated the gas, the more the number of degrees of freedom, the greater the specific heat.

CHANGE OF STATE

The transport of heat into or out of a substance can alter it in a number of ways. For instance, it can change the temperature of the sample, and it can change the physical structure, or *phase*, of the sample.

15.7   Melting and Freezing

When a solid is transformed into a liquid by the addition of thermal energy, the process is known as *melting*, though at one time it was widely called *fusion*. The common wisdom in the mid-1700s (and a notion still afoot today) was that once at the melting point, it took only a slight addition of heat to cause a further small rise in temperature and thereafter the rapid melting of the body. Joseph Black understood that this was wrong; he knew how long it often took for the piles of winter snow to melt, even on warm days. Black was convinced that when a body melts, "a large quantity of heat enters into it . . . without making it apparently warmer. . . . This heat must be added in order to give it the form of a liquid . . . ." Imagine that we take two thin glass bottles, put a kilogram of pure water in each, and cool them both to as close to 0°C as possible, leaving one liquid and freezing the other. Now bring them both into a warm room and observe the time it takes for the water to rise 1°C (on gaining 4.2 kJ, or 1 kcal). Next, observe how long it takes the ice to completely melt. What you will find is that the melting takes about 80 times longer. Black established experimentally that *when a solid is warmed to its melting point, the slow continuing addition of heat results in the material's gradual and complete liquefaction at that fixed temperature.* Only then does the temperature again begin to rise.

The heat added to a system while it melts apparently has no effect on the temperature, and so Black called it *latent* (meaning unrevealed) *heat.* Experiments have established that 1 kg of ice at 0°C will be transformed into 1 kg of liquid water at 0°C when 334 kJ (79.7 kcal) of heat are added. The amount of heat that must be added to a kilogram of any solid at its melting temperature in order to liquefy it (or, alternatively, the same amount of heat that must be removed from a kilogram of liquid at its freezing temperature in order to solidify it) has come to be called the *latent heat of fusion,* or just the **heat of fusion**. It's an unfortunate term simply because it is energy per unit mass (J/kg) and not energy. Be that as it may, the heat of fusion (L_f) for water is

$$L_f = 334 \text{ kJ/kg}$$

which is just about 80 kcal/kg. Table 15.6 lists the heats of fusion for a representative selection of substances at a pressure of 1 atm. As we will see presently, L_f depends on pressure. It follows that to change the phase of a mass m already at its melting point requires the addition (or removal) of an amount of heat Q such that

$$Q = \pm mL_f \qquad (15.4)$$

where the positive sign corresponds to heat-in (melting) and the negative sign to heat-out (solidification).

The element gallium has a heat of fusion of only 80 kJ/kg and a melting temperature of 29.8°C (85.6°F)—it melts in the hand.

TABLE 15.6 Approximate heats of fusion and vaporization

Material	Melting point (°C)	Heat of fusion (kJ/kg)	Boiling point (°C)	Heat of vaporization (kJ/kg)
Antimony	630.5	165	1380	561
Alcohol, ethyl	−114	104	78	854
Copper	1083	205	2336	5069
Gold	1063	66.6	2600	1578
Helium	−269.65	5.23	−268.93	21
Hydrogen	−259.31	58.6	−252.89	452
Lead	327.4	22.9	1620	871
Mercury	−38.87	11.8	356.58	296
Nitrogen	−209.86	25.5	−195.81	199
Oxygen	−218.4	13.8	−182.86	213
Silver	960.8	109	1950	2336
Water	0.0	333.7	100.0	2259

Example 15.6 A container holding 0.250 kg of water at 20°C is placed in the freezer compartment of a refrigerator. How much energy must be removed from the water to turn it into ice at 0°C?

Solution: [Given: $m_w = 0.250$ kg, $T_{iw} = 20$°C, and $T_{fw} = 0$°C. Find: Q_{out}.] The heat-out (that is, the heat that must be removed) must be enough to drop the water at 20°C down to 0°C and then transform it to ice (also at 0°C). Both processes correspond to negative amounts of heat; accordingly

$$Q_{out} = c_w m_w (T_{fw} - T_{iw}) + (-mL_f)$$

Substituting in the numerical values, we have

$$Q_{out} = (4.2 \text{ kJ/kg·K})(0.250 \text{ kg})(0°C - 20°C)$$
$$- (0.250 \text{ kg})(334 \text{ kJ/kg})$$
$$Q_{out} = -21 \text{ kJ} - 83.5 \text{ kJ} = -104.5 \text{ kJ}$$

and

$$\boxed{Q_{out} = -1.05 \times 10^5 \text{ J}}$$

▶ **Quick Check:** To drop the temperature, we must remove $(\frac{1}{4}$ kg$)(20$ K$)(1$ kcal/kg·K$) = 5$ kcal. To freeze it, another $(80$ kcal/kg$)(\frac{1}{4}$ kg$) = 20$ kcal must be removed, for a total of 25 kcal = 105 kJ.

From an atomic perspective, it is apparent that the liquid state has more internal energy than the solid state at the same temperature; after all, heat is transferred into the sample as it melts. Still, no corresponding rise occurs in temperature, and that is true for monatomic systems, like the metals, as well as molecular systems, like water. It would seem that the energy input must be going into an increase in PE. The energy associated with the heat of fusion liberates the atoms or molecules from the rigid bonding of the solid state, allowing them to move apart slightly into a less ordered, more fluid configuration.

Since heat must flow into a body when it melts, fusion is a cooling process: it removes thermal energy from the immediate environment in contact with it. A block of ice in a picnic cooler slowly melts, drawing thermal energy from the surrounding warm cans of beer. After a while, the beer is cold and the ice is melted. The same is true for ice cubes in a glass of warm soda—that's why the ice is added, and that's why it always melts. In reverse, freezing is a warming process: it exhausts thermal energy into the immediate environment. Years ago, people used to protect their fruit cellars and greenhouses from freezing by storing large vats of water in them. As the water began to freeze, it liberated thermal energy, making it that much harder to further cool the room. In part, the thermal energy exhausted into the room by a refrigerator comes from the water freezing in the ice cube trays.

15.8 Vaporization

The process of transformation of a liquid or solid into the gaseous state is known as **vaporization**. The resulting product is called a *vapor* because it deviates radically from the behavior predicted by Boyle's Law, the behavior of an ideal gas. The change from a liquid to a gas that occurs continuously at a free surface (that is, one not in contact with a solid or other liquid) is known as **evaporation**.

A liquid is composed of a large number of swarming particles, atoms, or molecules that move around with a distribution of kinetic energies much like that of a gas. Even at low temperatures, a few of those traveling in the vicinity of the surface move up toward it with speeds great enough to overcome the cohesive forces and escape the liquid altogether (Fig. 15.7). As we have already seen (p. 499), some molecules will return to the liquid as new ones leave it. As long as the vapor pressure above the liquid is less than the saturated value for that temperature, evaporation continues. And this is essentially independent of the presence of other gases (such as, air). If the vapor pressure above the liquid is made to exceed the saturated value (for example, by pressing down on the vapor with a piston), there will be condensation. Similarly, a drinking glass filled with ice cubes can cool the surrounding moist air, thereby lowering its saturated vapor pressure enough to allow water to condense onto the cold surface.

The rate of evaporation can be increased by altering the physical condition of the liquid in two ways: First, by increasing the surface area (a wet shirt dries slowly if it's balled up, and brandy evaporates from a wide snifter a lot faster than from an uncorked bottle). Second, increasing the temperature of the liquid accelerates evaporation because at greater temperatures the molecules have higher average kinetic energies and more likelihood of escaping the surface. Warming the clothes in a dryer, heating the back window of your car to drive off the condensation, or holding the bowl of the brandy snifter in a warm hand are each examples of this effect.

Just as with melting, energy must be supplied to the molecules being liberated from the cohesion of the liquid state. Vaporization is an even more drastic physical change than melting: the molecules are both torn free and appreciably separated, a process requiring the input of a substantial amount of energy. The vapor contains far more energy than an equal mass of liquid at the same temperature. Accordingly, we define the **heat of vaporization** (L_v) as *the amount of thermal energy required to evaporate 1 kilogram of a liquid at a constant temperature,* or if removed from a kilogram of vapor, to condense it under the same conditions. As we will presently, that temperature is usually the boiling point, but it certainly need not be. Moreover, the heat of vaporization generally decreases as T increases. For example, L_v for water at 0°C is 2.49×10^3 kJ/kg (595.4 kcal/kg), whereas at skin temperature, ≈ 33°C, it's down somewhat to 2.42×10^3 kJ/kg (578 kcal/kg). Since the density of the liquid is greater than the density of the vapor, it is evident that energy is being supplied to each escaping molecule to overcome the attraction of the other molecules. Yet this density difference decreases as the sample approaches its critical temperature (p. 498). There, the two states are indistinguishable, and the heat of vaporization is zero.

As Table 15.6 shows, each substance has a characteristic heat of vaporization. Thus, the amount of heat Q that must be supplied (or removed) in order to vaporize (or condense) a mass m of liquid (or vapor) at a specific constant temperature is given by

$$Q = \pm mL_v \qquad (15.5)$$

With this equation, we could compute how much energy had to be supplied to your wet bathing suit the last time you impatiently waited around for it to dry. Clearly, one of the synthetic fabrics that does not absorb much water in the first place (so that m is small) requires the least amount of thermal energy and dries the most quickly, especially if it's a warm, dry day.

Since this rather substantial amount of thermal energy must be supplied to the system (often drawn from the immediate environment), vaporization can be considered a cooling process: the molecules that escape are the energetic ones, leaving behind their lower-average-KE brethren in the liquid, which therefore must have a reduced temperature (unless the system is being supplied heat from outside). That's in part why we perspire and why dogs pant. It's why you might lick a finger to test the wind or find your unattended bath water colder than the room even though it started out warm a few hours before.

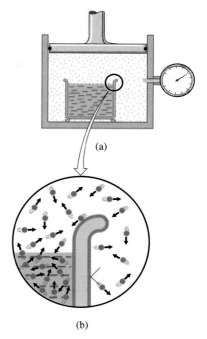

Figure 15.7 (a) A liquid in equilibrium with its vapor at a constant pressure. (b) On the average, as many molecules leave the surface from below as enter it from above.

Example 15.7 At what rate would you have to perspire to rid your body of the 75 W of thermal power you generate metabolically just sitting around resting? Assume a skin temperature of 33°C. As we will see, this is not our only cooling technique, and we generally don't have to sweat that much. Still, an athlete may lose several liters of water to perspiration during a good workout. What happens when you dry off the perspiration with a towel?

Solution: [Given: a power output of 75 W. Find: the evaporation rate.] First, 75 W = 75 J/s. From the above remarks, the heat of vaporization of water at skin temperature is 2.42×10^3 kJ/kg. Thus, you must lose energy at a rate of 75 J/s = $Q/t = mL_v/(1 \text{ s})$, and $m/(1 \text{ s}) = (75 \text{ J/s})/L_v = (75 \text{ J/s})/(2.42 \times 10^6 \text{ J/kg}) = 3.1 \times 10^{-5}$ kg/s. If this were your only available cooling mechanism, you would sweat away 3.1×10^{-2} g/s, or about $\boxed{0.11 \text{ kilogram per hour}}$.

Normally we sweat about $\frac{1}{2}$ liter per day, or about the equivalent of 1.2 MJ. Toweling perspiration off the skin or letting it drip off is of no use—it has to evaporate to do any cooling.

▶ **Quick Check:** The power equals L_v times the mass evaporated per second; $(2.4 \times 10^3 \text{ kJ/kg})(3.1 \times 10^{-5} \text{ kg/s}) = 74$ W.

The primary mechanism that causes hot, moist food to cool is evaporation. That's why you blow on a hot cup of soup—without air currents to carry away the vapor, it builds up above the soup and becomes saturated. That build-up decreases the evaporation rate and slows down the cooling process. To increase evaporation, you blow away the vapors every now and then. The hot food at a barbecue always seems to cool too fast because of the active air currents outdoors.

15.9 Boiling

Evaporation takes place at a liquid's free surface at any temperature, but under special circumstances it can also occur throughout the body of the liquid. Since vapor generated at any point within the liquid has a much lower density than the surrounding medium, it pushes the liquid in the immediate vicinity out of the way, creating (under hydrostatic pressure) a little gaseous sphere. The formation of such a bubble of vapor signals the beginning of the familiar process of **boiling**. To be precise, any liquid (water, olive oil, or molten gold) will boil at a specific temperature where its saturated vapor pressure equals the surrounding pressure. After all, to exist for an appreciable time, a bubble must be essentially in equilibrium with the surrounding liquid. That means the pressure inside of it—the vapor pressure—must equal the ambient hydrostatic pressure, and that's usually ≈1 atm. The vapor pressure of water, which is 0.006 atm at 0°C and rises to 0.2 atm at 60°C, equals 0.1 MPa (1 atm) at 100°C. That should be no surprise; the Celsius scale was fixed via the temperature of boiling water at atmospheric pressure, so we have just gone full circle.

When a pot of water has been heated for a while, bubbles begin to form around the walls and bottom of the vessel. The nucleation sites are centered on minute pockets of air or particles of dust. Some of the bubbles are air, driven out of solution. Once created at the hot bottom, air bubbles remain intact until they break loose. By contrast, bubbles of water vapor rapidly form, vanish, and reform. When they break loose and rise into the colder water above, they are cooled and suffer a decrease in vapor pressure, collapsing, condensing, and vanishing before reaching the surface. This compression can occur violently and is often noticeably noisy. Only when the upper regions of the water reach 100°C will true boiling take place throughout the liquid, and then bubbles reach and burst at the surface. That bursting process liberates a great deal of vapor and with it a great deal of energy—the heat of vaporization of water at 100°C is 2.259 × 10³ kJ/kg (539.6 kcal/kg). If heat is continuously fed into the water, it will continue to boil, and raising the flame under the pot will only increase the rate of boiling without changing the temperature of the water. More heat-in generates more vapor, which at temperatures ≥100°C is known as **steam**.

Each kilogram of steam at 100°C can conveniently be used to carry that 2.3 MJ from one place to another, from a basement boiler to a third floor radiator. On condensing in the radiator, the steam liberates its 2.3 × 10³ kJ/kg to the environment. Of course, steam can be raised to temperatures well in excess of the 1-atm boiling point, the ultimate limitation being the breaking up of the water molecules. In most modern applications, steam above 500°C is used. This *superheated steam* (as it's called above 100°C) whirls the great turbines that drive the majority of large ships and electric generators.

Example 15.8 How much heat must be added to a 1.0-kg mass of water ice at −10°C and atmospheric pressure in order to transform it into superheated steam at 110°C? Compare the energy associated with each stage of the process and confirm Fig. 15.8.

Solution: [Given: $m = 1.0$ kg, $T_i = -10°C$, and $T_f = 110°C$. Find: Q_{in}.] We have to raise the temperature of the ice to 0°C, melt it, raise the temperature of the water to 100°C, vaporize it, and raise the temperature of the steam to 110°C. Remember that each state has its own value of specific heat capacity. Using

$L_f = 334$ kJ/kg, $L_v = 2.26 \times 10^3$ kJ/kg, $c_w = 4.2$ kJ/kg·K, $c_i = 2.1$ kJ/kg·K, $c_s = 2.0$ kJ/kg·K, and the relationship

$$Q_{in} = mc_i \Delta T_i + mL_f + mc_w \Delta T_w + mL_v + mc_s \Delta T_s$$

we get

$$Q_{in} = (1.0 \text{ kg})(2.1 \text{ kJ/kg·K})[0°C - (-10°C)]$$
$$+ (1.0 \text{ kg})(334 \text{ kJ/kg}) + (1.0 \text{ kg})$$
$$\times (4.2 \text{ kJ/kg·K})[100°C - 0°C]$$
$$+ (1.0 \text{ kg})(2.26 \times 10^3 \text{ kJ/kg}) + (1.0 \text{ kg})$$
$$\times (2.0 \text{ kJ/kg·K})[110°C - 100°C]$$

Figure 15.8 A graph of temperature versus thermal energy added to, or removed from, 1 kg of H₂O. The sample is raised from −10°C to above 100°C and passes from ice, to ice plus water, to water, to water plus steam, to steam.

(continued)

(continued)

Accordingly

$$Q_{in} = 21\ kJ + 334\ kJ + 420\ kJ + 2260\ kJ + 20\ kJ$$

thus $Q_{in} = 3055\ kJ$

and to two significant figures $\boxed{Q_{in} = 3.1 \times 10^3\ kJ}$. Notice that most of this energy was required to make the steam.

▶ **Quick Check:** Because the calorie was defined via the behavior of water, problems that deal exclusively with water are easier to treat in the old units:

$$Q_{in} = (1\ kg)(0.50\ kcal/kg{\cdot}K)(+10\ K) + (1\ kg)$$
$$\times (80\ kcal/kg) + (1\ kg)(1.00\ kcal/kg{\cdot}K)$$
$$\times (+100\ K) + (1\ kg)(540\ kcal/kg)$$
$$+ (1\ kg)(0.48\ kcal/kg{\cdot}K)(+10\ K)$$
$$Q_{in} = 5.0\ kcal + 80\ kcal + 100\ kcal + 540\ kcal$$
$$+ 4.8\ kcal$$

and $Q_{in} = 729.8\ kcal = 3.1 \times 10^3\ kJ$

The boiling point of a substance depends on the external pressure. Lower the surface pressure, and bubbles can form at a lower vapor pressure, which corresponds to a lower temperature (they are squeezed less and form easier). A pot of water on a campfire atop Mont Blanc (5 km) will boil at only 83°C, while on Mt. Everest (9 km) making a decent cup of tea on an open fire would be impossible (the pressure is ≈37 kPa; that is, 0.4 atm, and the boiling point is ≈74°C). This effect has been used commercially to boil off water from delicate foodstuffs like milk and syrups (in the manufacture of sugar) without cooking them. They're boiled in a low-pressure vessel where the vapor is continuously pumped off. If we put a small sample of water in a vacuum system and vigorously pump off the vapor, dropping the saturated pressure, the liquid will boil at room temperature. This being a cooling process, the sample will rapidly decrease in temperature and freeze into ice.

The opposite process is employed in the pressure cooker (invented by Denis Papin in the late 1600s) in which the liberated vapor builds up pressure inside a heated sealed vessel. The domestic pressure cooker usually operates at a gauge pressure of 1 atm and a boiling-water temperature of 121°C (250°F). The chemical reactions of cooking roughly double their rate with every 10°C increase beyond 100°C. Thus, the 21°C rise is significant, and the pot will cook things much more quickly than the traditional open-to-the-atmosphere culinary approach. The hospital autoclave is a large-scale pressure cooker for sterilizing instruments.

THE TRANSFER OF THERMAL ENERGY

Put a hot cup of coffee on a table in your kitchen and come back in an hour and it will certainly be cooler, just as a cold bottle of soda left out will get warmer. Objects transfer thermal energy to and from the environment in three basic ways: radiation, convection, and conduction.

15.10 Radiation

Figure 15.9 Thermal radiation: the electromagnetic energy radiated by a body at various temperatures as a function of wavelength. The narrow band from ≈350 nm to ≈700 nm is the visible region of the spectrum.

All bodies radiate electromagnetic energy (Ch. 24) as a natural result of their constituent atoms randomly oscillating. This **thermal radiation** is characterized by a broad, continuous range of frequencies (Fig. 15.9) that arises out of the electromagnetic interactions among the atoms of solids, liquids, and dense gases. At this very instant, because of the warmth of your body, you are emitting copious amounts of invisible infrared (IR) electromagnetic radiation (which can easily be observed by IR cameras and pit vipers). The reverse process of absorption becomes evident whenever you put your face in the sunshine (or near an open furnace) and feel a rush of warmth as the radiant energy, especially the IR, is converted to thermal energy in your skin.

As long ago as 1791, P. Prevost maintained that matter at a constant temperature was in dynamic equilibrium, absorbing as much radiant energy as it emitted. *The amount of thermal radiation emitted by a body depends on its surface condition (color, texture, exposed area, etc.) and on its temperature.* The higher the temperature, the more energy it radiates per second (Sect. 30.1). On the other hand, *the amount of thermal radiation absorbed by a body depends on its surface condition and on the nature of the incident radiant energy (wavelength, intensity, etc.), which, in turn, depends on the temperature of the source.*

All else equal, black rough materials emit radiant energy effusively, whereas polished metals such as silver and copper are (20 or 30 times) less given to radiate, and white surfaces are usually between the two. As a rule, *a good emitter is a good absorber,* and vice versa (it would follow that a poor absorber is a good reflector or a good

Much of the Hubble space telescope is wrapped in a highly reflecting metalized plastic covering. It reflects radiant energy from the Sun and helps maintain an appropriate range of operating temperatures.

transmitter—if it cannot absorb the energy that lands on it, it has to get rid of it some other way). Were it not so, we could put our great little absorber out in the sunshine, and it would get hotter and hotter indefinitely. As is, an object bathed in radiant energy from a higher temperature source (but otherwise isolated from the environment) will absorb energy and rise in temperature. It will correspondingly increase its initially (low T) low rate of emission until, at an appropriately elevated temperature, emission precisely balances absorption. At that fixed equilibrium temperature, the body will thereafter continue to radiate as much as it absorbs. That's what happens when you park your car in the sunshine: if it is black, it heats up a lot and quickly; if white, it heats up a little and slowly. In neither case does the temperature keep rising until the car melts. The more surface exposed to the radiation, the more energy will be absorbed and the more rapidly the temperature will rise— that's why we sit in the shade when it's hot and why, if there is no shade, a camel will sit facing the Sun rather than expose the length of its body to the Sun's rays.

If it were possible to put our hot cup of coffee off alone in an empty universe, it would continue to radiate away energy, slowly approaching Absolute Zero. But that process is not what happens in our far-from-empty Universe. Here, any object is actually immersed in a local sea of electromagnetic radiation, and it continuously exchanges radiant energy with that environment. The rate at which it radiates is determined by its thermodynamic temperature; the rate at which it absorbs is determined by the character of the incident radiation and therefore by the effective temperature of the surroundings. That holds true even in the distant reaches of space where the ambient radiation has a thermal spectrum consistent with a temperature of about 3 K (indicating to physicists that the radiation was once in equilibrium with matter in a denser, hotter, earlier time, presumably near the beginning of the Universe). Space is awash with randomly fluctuating electromagnetic fields that correspond to thermal radiation.

This photo of a dog was made using only the infrared radiant energy emitted by the animal. Such images are essentially temperature mappings.

Putting it rather simplistically, when an object (like our coffee) is hotter than the local radiation environment, it radiates more than it absorbs and tends to cool down to the surrounding temperature. Your kitchen is filled with a spectrum of radiant energy that is typical of a room-temperature thermal source, and the hot coffee loses energy to it. On the contrary, our cold bottle of soda still radiates, but much less than it absorbs from the radiation environment of the kitchen, and so it slowly warms up. This mechanism has nothing to do with any contact the bottle might have with the table or the air. It would not be affected if the bottle were suspended inside of a vacuum. Radiant energy passes through vacuum unimpeded, which is obvious in the case of solar radiation.

Because we humans constantly generate thermal energy internally, we must continuously exhaust it to the environment if a constant body temperature is to be maintained. Actually, we radiate rather well—light-skinned folks emit at about 60% of the maximum rate for an ideal (black) body at that temperature, while dark-skinned people approach about 80%. The difference is not unreasonable given that lighter skins probably evolved in colder climates. Sitting quietly reading a book you lose upwards of 50% of your exhausted thermal energy to the room by way of radiation.

15.11 Convection

When a region of a fluid is heated (for example, the air around a candle flame), its density decreases via thermal expansion and it rises. Displacing cooler portions in its path, the resulting fluid current constitutes a flow of thermal energy, up and away from the source. Driven by gravity, this process is known as *free* or *natural convection* (Fig. 15.10). The household radiator warms a space via convection currents that swirl up and around the room. A hot cup of coffee will produce the same effect in the air on a smaller scale. During the day on the coast, the land (heated more quickly by the Sun because of its low specific heat) heats the air above it. The warm air circulates up and around to create a cooling breeze off the sea (Fig. 15.11). At night, when the sea tends to be warmer than the land, the air current reverses. Large birds such as eagles and vultures (as well as sailplane pilots) seek out warm air updrafts. They conserve their energy by riding convection currents, hovering effortlessly or gaining altitude in the uprush of these so-called thermals. On a larger scale, the trade winds are caused by atmospheric convection currents rising up from the tropics. Ocean currents, such as the Gulf Stream, are also the result of free convection.

A reason to wear clothing (if you need one) is to prevent convection currents from carrying off too much thermal energy from your body on cold days. After all, we are tropical animals who could not survive naked very much below 20°C. Stopping convection currents is also a good reason to insulate water heaters, put up storm windows, and sleep under blankets when it's cold. The blanket, unless it's electric, does not supply heat; it just cuts down on the loss of the thermal energy you generate—covering a shut-off cold automobile engine with a blanket will not warm it up.

Figure 15.10 (a) Convection currents. Heated liquid rises, cooler liquid descends. The effect is due to a decrease in density as *T* increases and is driven by gravity—no gravity, no free convection. (b) With a tilting match, convection currents carry thermal energy up to the unburned wood, which fuels the process.

(b)

(a)

Figure 15.11   Convection currents coming off the sea.

An eagle soaring on updrafts of warm air corresponding to convection currents.

15.12   Conduction

We began this chapter with a discussion of what heat is, and now we end with a few remarks about how it is conducted—namely, by atomic collisions. At the center of the concept is the idea of a temperature difference: ΔT. Heat is only conducted from a high to a low temperature. The measure of the ability of a substance to transmit heat is known as its **thermal conductivity** (k_T), and it depends on the atomic structure of the material. Metals are roughly 400 times better conductors of heat than other solids, which are a trifle better than most liquids, which, in turn, are about 10 times better than the gases (Table 15.7).

It is usually true that a good conductor of heat is a good conductor of electricity, a parallel that is no coincidence. A metal can be envisioned as a lattice of positive ions immersed in a tenuous sea of essentially free electrons. Both heat and electricity are transported largely by the motion of these unbound electrons, which behave like a highly mobile fluid within the metal. In poor conductors of electricity, the transport of heat is primarily by the slower mode of molecular collisions. Thus, amorphous nonmetallic solids are only modest thermal conductors. But the ability to conduct energy is enhanced a bit when the substance is crystalline and the lattice of atoms can also vibrate as a continuous structure (as it does when it carries a sound wave). Whether it is via the random motion of atoms or molecules or electrons, a temperature difference propels heat much as a gravity difference propels a river.

The transfer of energy always occurs from the action of one of the fundamental Four Forces. In the case of thermal conduction, the force is electromagnetic. A group of atoms with a high average thermal energy transfers random KE to an adjacent group of atoms having a lower average thermal energy, and we observe the process as heat conduction across a temperature difference.

Body tissue without blood moving through it is a relatively poor thermal conductor (it's very slightly better than cork or felt), and the simple conduction of heat

A candle burning in the Microgravity Laboratory of the *Space Shuttle*. While in Earth-orbit, convection currents are far less pronounced than they are under normal gravitational conditions. Ordinarily, hot air, being less dense than cool air, rises. Convection currents feed oxygen to the flame, which also rises.

TABLE 15.7 Approximate values* of *thermal conductivities*

Material	Thermal conductivity, k_T (W/m·K)	Material	Thermal conductivity, k_T (W/m·K)
Metals		Linen	0.088
Aluminum	210	Paper	0.13
Brass (yellow)	85	Paraffin	0.25
Copper	386	Plaster of Paris	0.29
Gold	293	Polyamides (e.g., Nylon)	0.22–0.24
Iron	73	Polyethylenes	0.3
Lead	35	Polytetrafluroethylene	
Platinum	70	(e.g., Teflon)	0.25
Silver	406	Porcelain	1.1
Steel	≈46	Rubber, soft	0.14
Other solids		Sand, dry	0.39
Asbestos	0.16	Silk	0.04
Brick, common red	0.63	Snow, compact	0.21
Cardboard	0.21	Soil, dry	0.14
Cement	0.30	Wood, fir, parallel to grain	0.13
Chalk	0.84	*Liquids*	
Concrete and cement mortar	1.8	Acetone	0.20
cinder block	0.7	Benzene	0.16
Down	0.02	Alcohol, ethyl	0.17
Earth's crust	1.7	Mercury	8.7
Felt	0.036	Oil, engine	0.15
Flannel	0.096	Vaseline	0.18
Glass	0.7–0.97	Water	0.58
fiberglass	0.04	*Gases*	
Granite	2.1	Air	0.026
Human tissue (no blood)	0.21	Carbon dioxide	0.017
fat	0.17	Nitrogen	0.026
Ice	2.2	Oxygen	0.027
Leather	0.18		

*Near room temperature.

from inside the body to the skin's surface, where it can be expelled, would be impractical. Instead, thermal energy conducted into the blood is brought near to the body's surface by the circulatory system. When it is cold and thermal energy must be retained, the capillaries near the suface contract, cutting off the flow of blood, thereby creating an insulating layer of tissue.

The molecules of a gas are widely separated, which makes them much less able to communicate thermal energy via collisions. Accordingly, air is a poor thermal conductor, and you can hold your hand beside a candle flame (as opposed to above it in the convection current) and barely feel it. Anything that will trap a layer of air will function as a thermal insulator. That's what a storm window does and why wearing many layers of clothing works nicely in the winter. Your body generates thermal energy, and the clothing, along with the air it captures, keeps it from escaping. In part, the containment of air is also what makes fibrous materials such as wool, fur, and fiberglass such effective thermal insulators. Trapped layers of motionless or "dead" air make excellent insulation. Because the thermal conductivity of air is low, our bodies don't lose much heat to it via conduction. We do exhaust heat by convection, radiation, and evaporation. A swim in a cold ocean will quickly make the point that liquids are better conductors than gases.

The steady-state amount of heat Q that flows per unit time Q/t through a sample has been studied extensively, both theoretically and in the laboratory. Because it's actually a complicated problem, we will consider only the simplest geometry and use the classic approach of compressing all the subtle atomic details into a single material constant. The amount of heat that is conducted per second through a uniform slab or rod of material (Fig. 15.12) is found to be proportional to the *fixed* temperature difference ΔT across its two faces, $Q/t \propto \Delta T$. We expect that our sample slapped against a blazing furnace wall would transmit more heat to your hand than if it were pressed against a lukewarm radiator. Moreover, it is reasonable to assume that the amount of molecular agitation that can be propagated down the sample should depend directly on its cross-sectional area A: $Q/t \propto A$.

If the sample has a thickness or length d, the amount of heat that will traverse it per second in steady-state flow is found to be inversely proportional to d. In other words, Q/t is proportional to the **temperature gradient** $\Delta T/d$ (which, if we were looking at the rate at which balls were rolling down a hill, would be the equivalent of how steep the hill was, and not just how high it was). Remember the end-face temperatures (Fig. 15.12) are held constant, and the closer those faces are, the greater the gradient and the more rapidly the collision energy is moved along. In summary, then, $Q/t \propto A(\Delta T/d)$, and if we introduce a material constant of proportionality, namely, the thermal conductivity (k_T), then

Figure 15.12 The rate (Q/t) of transport of heat is proportional to the temperature gradient ($\Delta T/d$) and the area (A).

$$\frac{Q}{t} = k_T A \frac{\Delta T}{d} \qquad (15.6)$$

where k_T has the units of J/s·m·K or W/m·K. This expression is usually called *Fourier's Conduction Law*, after J. B. J. Fourier (1807). It provides the rate of flow of heat across an area A in a perpendicular direction in terms of the temperature gradient in that direction. For a given ΔT, the greater d is, the lower Q/t is (the thicker your clothing is on a winter's day, the lower the rate of heat loss). When the human body is exposed to cold, the surface blood vessels contract, and that contraction goes on deeper and deeper, providing a thicker (larger d) layer of low k_T insulation as needed.

In the steady state, the amount of heat flowing into a conducting system traverses it and emerges; there is no loss or gain of energy, no storage. What that suggests (for Q/t = constant) is that regions of low conductivity correspond to large gradients and often to large temperature changes. Thus, a thin layer of insulating material generally experiences a large change in T across it—that's what a blanket does.

This material was developed for the heat-shield tiles of the *Space Shuttle*. It has both an extremely low thermal conductivity and small specific heat capacity; so much so, that a cube of it can be held at its edges even when red hot.

Example 15.9 A pane of window glass is 0.90 m wide by 1.5 m high and 4.0 mm thick. It's a cold blustery winter's day in the Midwest. The temperature of the inside face of the window is 10°C, and the outside face is at −9.0°C. How much thermal power is being transported through the windows (k_T = 0.84 W/m·K)?

Solution: [Given: A = 0.90 m × 1.5 m, d = 4.0 mm, and ΔT = 10°C − (−9.0°C). Find: Q/t.] Using

(continued)

(continued)

we obtain

$$\frac{Q}{t} = \frac{k_T A \, \Delta T}{d} \qquad [15.6]$$

$$\frac{Q}{t} = \frac{(0.84 \text{ W/m·K})(1.35 \text{ m}^2)(19 \text{ K})}{4 \times 10^{-3} \text{ m}} = 5386.5 \text{ W}$$

and to two significant figures $\boxed{Q/t = 5.4 \text{ kW}}$. Notice that the inside face of the window is a lot colder than room temperature. That's the result of the blanketlike quality of a thin layer of stationary air that usually

forms against the window. Even more power would be wasted if the inside air were moving across the inner face of the window, keeping its temperature closer to that of the room. This underscores the virtues of having drawn curtains or window shades (in addition to storm windows), especially at night when little radiation enters via the windows.

▶ **Quick Check:** $\Delta T = Qd/tk_T A = (5386.5 \text{ W}) \times (0.004 \text{ m})/(0.84 \text{ W/m·K})(1.35 \text{ m}^2) = 19 \text{ K}.$

Streaks of snow remain on the roof where the shingles stay cold, above the thick roof beams. This roof is obviously uninsulated; heat flows up from the warm room below, passing out between the beams.

When you awake on a chilly morning, miss the rug and step onto the "cold" bathroom floor, it's a lesson in conductivity. The floor and the rug must have the same temperature—they have been in equilibrium all night. But they certainly don't feel equally warm. The floor tiles have about 10 times the conductivity the rug has and will draw a great deal of heat per second from your warmer feet. That loss of heat is what produces the perception of coldness, more so than the temperature difference—we are not very good thermometers.

Core Material

Thermal energy is the disordered KE of any group of particles, usually the atoms constituting bulk matter. **Temperature** is essentially a measure of the concentration of thermal energy (p. 515). **Heat** (Q) is the thermal energy transferred, via atomic collisions, from a region of high temperature to a region of lower temperature. The amount of heat that must be added to raise the temperature of 1 gram of water 1 degree Celsius—from 14.5°C to 15.5°C—is a calorie, where

$$4.186 \text{ J} = 1 \text{ cal}$$

or

$$1 \text{ kcal} = 4186 \text{ J}$$

The **Zeroth Law of Thermodynamics** maintains that *when two systems are at the same temperature as a third, they are all three at the same temperature as each other* (p. 521). Two systems are in thermal equilibrium only when their temperatures are equal.

The **specific heat capacity** (c) of a substance is the number of joules of heat that must be added to raise the temperature of 1 kg of the material 1 K; thus

$$Q = cm \, \Delta T = cm(T_f - T_i) \qquad [15.1]$$

where the SI units of c are J/kg·K. Water stands out for its high specific heat capacity: as liquid (4.2 kJ/kg·K, or 1.0 kcal/kg·K), as ice (2.1 kJ/kg·K, or 0.50 kcal/kg·K), and as steam (2.0 kJ/kg·K, or 0.48 kcal/kg·K).

The amount of heat that must be added to a kilogram of any solid at its melting temperature in order to liquefy it is the **latent heat of fusion**. For water $L_f = 334$ kJ/kg ≈ 80 kcal/kg. Changing the phase of a mass m requires the addition (or removal) of an amount of heat Q:

$$Q = \pm mL_f \qquad [15.4]$$

The **heat of vaporization** (L_v) is the amount of thermal energy required to evaporate 1 kilogram of a liquid at a constant temperature, thus

$$Q = \pm mL_v \qquad [15.5]$$

The heat of vaporization of water at 100°C is 2.259×10^3 kJ/kg = 539.6 kcal/kg.

Objects transfer thermal energy to and from the environment in three basic ways: radiation, convection, and conduction. The amount of heat conducted per second through a uniform slab or rod of material is

$$\frac{Q}{t} = k_T A \frac{\Delta T}{d} \qquad [15.6]$$

where k_T is the **thermal conductivity**.

Suggestions on Problem Solving

1. Common practice is to represent zero degrees Celsius as 0°C with no reference to significant figures. *Take 0°C to have the same number of significant figures as the other temperatures given.*

2. Because we are generally dealing with temperature changes, we need not transform the data from C° to K. Thus, when using the equation $Q = cm\,\Delta T$, you can express the temperatures in Celsius degrees or kelvins, whichever is more convenient. Still, if you use Celsius and calculate T_f, the answer will be in Celsius.

3. Be careful with calculations when mixing various substances in *different states* (for example, water and ice, or steam and water) in a calorimeter. Whenever there is a change of state, it is entirely possible that not all of the substance will actually make the transition. The governing equation $-Q_{out} = Q_{in}$ is correct, but saying that $Q_{in} = mL_f + cm\,\Delta T$ assumes that the entire mass m has melted, and that need not be the case. We must compute the thermal energy that would be made available on descending to the transition temperature and then check to see if that is enough to accomplish the change of state for the whole mass m. If it isn't, only part of the ice melts (or part of the steam condenses, or whatever).

4. Another common error is to use L_v when you should be using L_f, and vice versa. Double check that you have made the correct substitution before carrying out the calculation.

5. The analyses of many of the mixing problems begin with $-Q_{out} = Q_{in}$. The minus sign is there because the object providing the thermal energy decreases in temperature: $\Delta T = (T_f - T_i)$ is negative.

6. In the old cgs system of units, the specific heat capacity of water is 1 cal/g·C°, which in mks is equivalent to 1 kcal/kg·C°. Watch out for the units; those are kilocalories and kilograms.

Discussion Questions

1. Discuss the physics involved in making a soft piece of steel red hot by hammering on it. Will that work as well for hard steel?

2. Suppose you leisurely stirred two identical cups of very hot coffee, one with a spoon made of steel and the other of glass, each having the same mass. If you absentmindedly put one of the spoons in your mouth, which would you rather it was? Explain.

3. The rate of energy consumed by the human body in the process of doing something, divided by the area of the body, is found to be fairly constant for different weights and body types; it's called the *metabolic rate* (usually expressed in kcal/m²·h). Measurements show that the metabolic rates associated with bicycling and shivering are about equal at 250 kcal/m²·h. Considering that muscles are only about 20% efficient, the rest being lost as thermal energy, compare these activities from an energy perspective. What function does shivering serve?

4. Imagine that we place a wooden rod and a brass rod, which are otherwise identical, in a freezer. Which will cool faster? Compare what they would feel like after being in the freezer for a few hours. Explain.

5. Why are steel saucepans and tea kettles sometimes covered with copper on their outside bottoms? Why are high-quality saucepans thick and tea kettles thin-walled?

6. Suppose we wrap an ice cube in aluminum foil (bright side out) and another in black cloth and place them both in the sunshine. What will probably happen and why?

7. Having evolved an expression for the rate of heat transfer via conduction, Eq. (15.6), see if you can do the same thing for convection. Explain each assumption.

8. Ordinary tungsten incandescent light bulbs are evacuated. Explain why, and discuss the effect it has on the amount of power consumed and the amount of heat wasted. (Hint: Why do light bulbs often get blacker inside the longer they are used?)

9. Is it likely that you will feel more or less comfortable in a low-humidity environment at 80°F than in a high-humidity one at 70°F? Explain. Why do many people become "flushed" when they get overheated?

10. Suppose you had to remove a large hot pot from the stove, and to protect your hands you could use either of two cloth towels that are identical, except one is dry and the other is thoroughly wet with cool water. Which would you use and why?

11. Place some water in a metal ice cube tray in a freezer (or

Figure Q13

outside if it's below 0°C) and periodically observe what's happening. Where does the ice first begin to grow? Why is there often a region of water at the center of the cube? Now try the same thing with two trays, one with warm water and one with cool. Explain how it might be possible for the warm water to freeze first.

12. Most modern automobile cooling systems operate at a gauge pressure of 1 atm. The cooling fluid is usually a 50% mixture of water and ethylene glycol. What purpose does the ethylene glycol serve as regards the cooling process? Why is the system pressurized?

13. Figure Q13 depicts an experiment on conductivity performed by Forbes in 1864. Two metal containers at fixed temperatures of 100°C and 0°C are connected to each other by a metal rod. In the first case, the rod is exposed; in the second, it's insulated. Steady-state temperature readings are made all along the rod and the values plotted as the two *T*-versus-*d* curves in the diagram. Interpret the significance of the arrows and explain the curves. What is the meaning of the slope at each point on each curve? Does what is happening agree with the picture of heat as an incompressible liquid?

14. Given a wet shirt hanging on a clothesline, what will determine the rate at which the water evaporates from it? Why will it dry faster in the sunshine?

15. Figure Q15 shows a demonstration devised by the English physicist John Tyndall. Four 1-kg cylinders made respectively of aluminum, iron, copper, and lead, each with the same

Figure Q15

cross-sectional area, are heated to 100°C in boiling water. They are then all placed upright upon a block of paraffin. Describe what happens and why.

16. How do storm windows work? Should they be sealed tight or left open a little? It's a hot, sunny day and you have the central air conditioning running full blast. Should you lower the blinds and draw the curtains?

17. If your goal is to keep warm in a very cold climate, should you wear a fur coat with the hair inside or out? Explain. If your goal is to keep cool in a hot environment where drinking water is plentiful, what should you wear outdoors? What if water is scarce?

18. We saw earlier that when a stream of air flows over a body, there is usually a thin layer in contact with the surface that is at rest. This same effect should exist when a flame carries hot gas over the bottom of a tea kettle. Explain how it is possible to boil water held in a paper milk container that is placed on an open flame. Why should you avoid turbulence in the hot air?

19. Imagine that you are floating around in space one or two hundred thousand miles from Earth with a thermometer. What will determine the temperature of your suit (assuming no active air conditioning), and what will the thermometer read floating nearby?

20. Leslie's cube is a hollow copper box whose outer surfaces are finished differently. One face is smooth, shiny, and gold-plated; one is roughened with sandpaper; and another is coated with black soot. The box is filled with boiling water, and the emanations from it are examined with a detector. What will be found? Should household heating system radiators be painted shiny silver, as is most often done?

Multiple Choice Questions

1. Two bodies that are not initially in thermal equilibrium are placed in intimate contact. After a while the (a) temperature of the cooler one will rise the same number of kelvins as the temperature of the hotter one drops   (b) amount of thermal energy contained by both bodies will be equal   (c) specific heats of both bodies will be equal   (d) thermal conductivity of each body will be the same   (e) none of these.

2. An open beaker of pure water is gently boiling at atmospheric pressure. A thermometer held deep in the water will likely read a temperature   (a) equal to 100°C   (b) a little less than 100°C   (c) a little greater than 100°C   (d) equal to 212°C   (e) none of these.

3. Two blocks of aluminum, one with a 1.0-kg mass, the other with a 2.0-kg mass, are in thermal equilibrium with a third block of brass at 100°C. The two aluminum blocks are at temperatures, respectively, of   (a) 100°C and 50°C   (b) 50°C and 100°C   (c) 100°C and 100°C   (d) 200°C and 100°C   (e) none of these.

4. Referring to Question 3, the ratio of the net thermal energy of the 1.0-kg aluminum block to that of the 2.0-kg block is   (a) $\frac{1}{2}$ (b) 4   (c) 2   (d) $\frac{1}{4}$ (e) none of these.

5. An open pot of water is boiling on a gas stove when someone raises the flame. The result will be   (a) a substantial increase in the temperature of the water   (b) a tiny decrease in the rate of evaporation   (c) an increase in the rate of boiling   (d) an appreciable increase in both the rate of boiling and in the temperature of the water   (e) none of these.

6. Figure MC6 shows a glass flask in which water was boiled until the steam drove out much of the air. Thereafter it was tightly corked and inverted. Cold water was then sprinkled over the spherical bottom, and the water inside   (a) froze because of the evaporation   (b) boiled because of the condensation of steam   (c) froze because of conduction   (d) boiled because of convection of thermal energy   (e) none of these.

7. Everyone who has ever walked barefoot on a beach in summer has noticed how fast the dry sand gets hot in the

Figure MC6

morning. That's because sand has a   (a) light color   (b) fairly low specific heat   (c) high thermal conductivity   (d) great deal of convection   (e) none of these.

8. Figure MC8 shows a test tube containing water and some ice (held down by a weight). As shown, the water at the top is boiling. Is that possible and, if so, why?   (a) yes, because water has a high heat of vaporization   (b) no, because water has a low specific heat capacity   (c) yes, because water has a low thermal conductivity   (d) no, because water has a low heat of fusion   (e) none of these.

9. Figure MC9 depicts a Dewar flask, or thermos bottle. The space between the two glass walls is evacuated (via a tube that is then sealed). This is done to   (a) decrease conduction   (b) decrease the specific heat capacity

Figure MC8

(c) decrease radiation   (d) decrease weight   (e) none of these.

Glass

Shiny metal

Vacuum

Figure MC9

10. Referring to the Dewar flask of Fig. MC9, why is the outer surface of the inner wall covered with a shiny layer of metal?   (a) to decrease conductivity   (b) to decrease radiation to and from the inner chamber   (c) to increase convection currents in the glass   (d) to increase the loss of coldness   (e) none of these.

11. Given that a thermos bottle (Fig. MC9) contains a cold drink, what function does the metal film on the inner surface of the outer wall serve?   (a) it keeps heat from being conducted inward beyond that point   (b) it looks nice shiny   (c) it reflects radiant energy incident from outside   (d) it cools the inner chamber by emitting radiation   (e) none of these.

12. Smoke is drawn out of a fireplace by an updraft that is initiated by wind blowing across the top of the chimney. To help matters along, it is recommended that a burning piece of newspaper be placed up the exhaust hole in the fireplace for a minute or two before lighting the fire. This is done in order to   (a) increase the conductivity of the air   (b) increase the updraft via convection currents   (c) decrease the specific heat of the air   (d) warm up the air so it burns better   (e) none of these.

13. Suppose you pour a hot cup of coffee and the phone rings so you can't drink it. To keep it hot as long as possible   (a) add the cool milk and sugar immediately   (b) don't add the cool milk until you're ready to drink it   (c) stir it once gently without the milk, but don't add the sugar   (d) don't add the milk and use a black mug if possible   (e) none of these.

14. Which of the following materials has the highest thermal conductivity?   (a) wood   (b) water   (c) air   (d) gold   (e) none of these.

15. The purpose of a cover on a soup tureen is to   (a) increase conductivity   (b) decrease convection   (c) increase radiation   (d) decrease conduction   (e) none of these.

16. The thermal effect of dissolving sugar in a cup of hot tea   (a) is to lower the temperature   (b) is to raise the temperature   (c) is to leave the temperature unchanged   (d) cannot be predicted   (e) none of these.

17. Metal pots are often made shiny on the outside especially on the top and side, and that makes sense thermally because   (a) this conducts heat better   (b) this radiates less energy out from the pot   (c) this lowers the loss to conduction   (d) this appreciably decreases convection losses   (e) none of these.

18. Fanning yourself on a hot day (say, 100°F, or 38°C) cools you by   (a) increasing conductivity   (b) decreasing the mean free path of the air   (c) increasing the radiation rate of the skin   (d) increasing the evaporation of perspiration   (e) none of these.

19. Franklin's experiment (Fig. MC19) consists of a glass bulb connected to a wide, initially open tube. The whole thing is heated until steam drives the air from the cylinder, whereupon it is sealed. Then, by holding the bulb in the hand, the water is warmed and boils in the tube because   (a) the temperature can quickly reach 100°C   (b) the water is a good conductor and the heat travels to the cylinder where it is concentrated   (c) the vapor pressure in the cylinder is quite low   (d) radiation is trapped in the connecting tube and brought to bear on the cylinder   (e) none of these.

Figure MC19

20. Contemporary radiators of all sorts consist of the heated body—a pipe or transistor, for example—to which are affixed a series of spaced flat vanes, often painted black. The function of the vanes is to   (a) increase the convection currents and so decrease conduction   (b) increase the heated surface area to increase radiation and convection   (c) increase the ability to block air currents thereby increasing conduction   (d) conduct more heat to the stagnant air layer and so increase radiation   (e) none of these.

Problems

THERMAL ENERGY

1. [I] If 15 kcal of heat are added to a system, how many joules of energy are added?

2. [I] A wrapped sugar cube is labeled to produce 92 kJ. How many kilocalories is that?

3. [I] A person just standing around consumes energy at a rate of about 100 kcal/h. How many watts is that?

4. [I] The human brain operates with an average power consumption of about 20 W. What is that in kcal/h?

5. [I] Alcohol has a rather high caloric value of about 7 kcal/g, which is completely converted into energy in the body. Because it contains no vitamins or minerals, people who drink to excess often have nutritional problems. How much energy is liberated by the body (as thermal energy) via one shot or drink, about 30 g?

6. [I] How much thermal energy must be added to 5.0 kg of pure water at 0°C to raise the temperature to 37°C?

7. [I] The cooling system of an engine contains 30 liters of water. If, during a brief run, the water carries off 500 kcal, by how much will its temperature rise?

8. [I] How much energy does it take to raise the temperature of 500 g of mercury from −19°C to 61°C? Give the answer in both joules and kilocalories.

9. [I] How much thermal energy must be removed from 30 g of tin with a specific heat of 0.060 to drop its temperature from 373 K to 283 K?

10. [I] How much heat does it take to raise the temperature of 0.40 kg of aluminum from 50°C to 60°C given that it has a specific heat of 0.217?

11. [I] If a bar of pure copper is found to absorb 16 kJ in the process of having its temperature raised from 293 K to 353 K, what is the mass of the metal?

12. [I] Show that the molar specific heat capacity of water is 75.3 J/mol·K.

13. [I] How much energy will be given out upon the complete combustion of 1 gallon of gasoline? One gallon equals 3785 cm³ and the density is 0.68×10^3 kg/m³.

14. [I] Two beakers of water, one at 15°C and the other at 95°C, each contain 1.52 kg of the liquid. If the water is combined, what will its final temperature be, assuming no heat losses from the liquid to the environment?

15. [I] It is found experimentally that by adding 100 g of iron, with a specific heat capacity of 0.113 kcal/kg·K at 80°C, to a quantity of water at 25°C, the new equilibrium temperature is 30°C. What is that amount of water?

16. [I] Suppose we combine 100 g of aluminum with a specific heat capacity of 920 J/kg·K at a temperature of 495°C with 99.9 g of water at 0.010°C. What will be the final temperature?

17. [I] A quantity of aluminum shot ($c = 910$ J/kg·K) at 473 K is mixed with 4.95 kg of water at room temperature, and in a little while, the whole thing comes to equilibrium at 300 K. How much aluminum was used?

18. [I] A 340-g glass mug at 20.0°C is filled with 250 milliliters of water at 96.0°C. Assuming no losses to the external environment, what is the final temperature of the mug?

19. [I] The Frenchman Gustave Hirn, in the 1800s, was the first person to make a quantitative study of the thermal energy involved in collisions. He crushed a lead ball with a steel cylinder of known mass dropping through a known height and quickly transferred it to a calorimeter. Hirn essentially found that dropping a 423-g mass a distance of 1.0 m produced 1.0 cal. Explain this result and comment on its accuracy.

20. [II] Suppose we could convert 1.00 kcal completely into work, thereby raising a 1.00-kg mass in a uniform gravitational field ($g = 9.807$ m/s²). How high would it get?

21. [II] Water at 20°C flows through a 4.0-kW heater at a rate of 1.5 liter/min. Assuming no losses, at what temperature will it emerge?

22. [II] A person who has a surface area of 1.65 m² consumes energy at about 70 kcal/h while resting. Determine the corresponding metabolic rate (see Discussion Question 3) and compute the approximate amount of heat that will be generated in 2 h.

23. [II] According to the *Shooter's Bible,* a 357-magnum bullet has a kinetic energy of 540 ft·lb at 50 yards from the pistol. If it strikes and comes to rest inside a 1.0-kg block of wood (with a specific heat capacity of 1700 J/kg·K), by how much would the block's temperature change? The bullet is lead and weighs 158 grains (1.0 grain is equivalent to 0.064 8 g).

24. [II] A 1000-kg car traveling at 60 km/h is brought to a stop by its braking system. How much thermal energy is evolved in the brakes? Friction with the ground does no work unless there is slipping.

25. [II] A 50-W immersion heater is placed in a beaker containing 1.0 kg of water at 20°C. How long will it take to raise the temperature to 100°C? Neglect any heat losses to the air or beaker.

26. [II] Suppose we have 0.30 kg of unknown material. It is heated to 371 K and placed into 0.60 kg of room temperature water. The water is in a copper cup of mass 0.14 kg surrounded by insulating material, and the final equilibrium temperature is 295 K. Compute the specific heat capacity.

27. [II] A 20-g dried sample of food is placed in an aluminum bomb calorimeter and burned completely. The device consists of an aluminum chamber (0.50 kg) containing the sample, a surrounding water bath (2.00 kg), and an aluminum cup (0.60 kg) holding the water. If the calorimeter changes temperature from 20°C to 32°C, how many kilocalories did the food make available? Neglect the mass of the ash.

28. [II] Let's model the human body as being composed of 75% water and 25% protein. Given that the specific heat capacity of protein is 0.4 cal/g·C°, approximate the specific heat capacity of the body in units of kJ/kg·K.

29. [II] Due to the presence in the body of both minerals (in the bone) and fat, its specific heat capacity is closer to 0.83 cal/g·C° than the value computed in Problem 28. Given that a 150-lb human at a temperature of 98.6°F must be dropped to a hypothermic value of 93°F, how much heat must be removed?

30. [II] A 1-kg sample of a material is surrounded by insulation material as in Fig. P30. An electric heater is

Figure P30

inserted into a hole in the sample, as is a thermometer. Given that 15 W are supplied by the heater for 16.667 min and the thermometer changes temperature by 16.67 K, what is the specific heat capacity of the sample? Ignore any losses.

31. [II] The specific heat of a 0.210-kg sample is to be determined. It's first heated to 373 K and then placed into a bath of 0.082 kg of water at 288 K, which is held in a 0.121-kg copper cup at that same temperature. If the final temperature of the system is 310 K, what is the specific heat capacity of the sample?

32. [II] A handful (405 g) of lead shot is removed from boiling water and dropped into a 99.0-g glass beaker containing 198 g of water at 20°C. If the shot and glass have specific heat capacities of 0.031 kcal/kg·K and 0.20 kcal/kg·K, respectively, find the equilibrium temperature of the system.

33. [II] A 70-kg person engaged in moderate physical activity produces thermal energy internally at a rate of 200 kcal/h. If all the cooling mechanisms become inoperable so there is no dissipation of this energy, how long would it take before the person collapses with a body temperature of 43°C (109°F)? (A body temperature of about 44°C begins to cause irreversible protein damage.)

34. [II] How much excess heat is produced when a 65-kg person with the flu experiences a rise in temperature from 98.6°F to 102°F? Take the average specific heat capacity of the human body to be 0.83 cal/g·C°.

35. [II] How much hard coal must be burned to raise the temperature of 1.00 kg of water from 0°C to 100°C?

36. [III] A house is heated electrically with a system that is 100% efficient; all the power consumed is converted into thermal energy via baseboard heaters. Suppose that it takes 2500 kW·h of energy per month to heat the place. How much wood would be needed to do the same job? Assume the wood stove, which must be exhausted to the outside, is 35% efficient. Given that a cord of wood is a stack of roughly about 1000 kg, how many cords would have to be burned each month?

37. [III] Figure P37 shows a device for measuring the specific heat of a solid rod of material by mechanical means. The rod is rotated, and the rope wrapped around it slides on its surface, producing a frictional drag. The crank is turned at a rate that takes all the tension off the rubber band holding the rope's far end. A thermometer is set in a hole in the specimen that is separated from the rest of the system by

Figure P37

thermal insulators. Explain how the apparatus works and how it is used to find c. If the load is 6.00 kg, the mass of the sample 250 g, its diameter 3.00 cm, the number of rotations 240, and the corresponding temperature increase of the rod is 12.0 K, determine its specific heat capacity.

38. [III] Marsh gas [that is, methane (CH_4)] has a heat of combustion of 212 kcal/mol. Show that this value agrees with Table 15.4. If 10 liters of it at STP are burned completely, and

$$CH_4 + 2O_2 \rightarrow CO_2 + 2H_2O$$

how much energy is released in the process? What is the net energy released per molecule burned? Note that the formation of methane gives off an amount of energy of (−)17.9 kcal/mol, whereas the formation of carbon dioxide and water releases (−)94.0 kcal/mol and (−)68.3 kcal/mol, respectively. The heat of formation of the elements in their natural states is zero. From this fact, determine the heat of combustion of methane.

CHANGE OF STATE
THE TRANSFER OF THERMAL ENERGY

39. [I] Determine the thermal energy that must be added to a 200-g block of copper at 1356 K to melt it without raising its temperature.

40. [I] Determine the thermal energy that must be added to 200 g of molten copper at 2336°C to completely vaporize it.

41. [I] Silver vapor at atmospheric pressure condenses at a temperature of 2466 K. How much thermal energy must be removed from 1.5 kg of silver vapor if it's to form a liquid at that temperature?

42. [I] A layer of ice 2.0 mm thick covers the 0.75-m² windshield of a car. How much power must be supplied by a heater if the ice is to be melted in 2 min and the air temperature is 0°C? Assume no losses.

43. [I] What is the final state of the system consisting of 500.0 g of liquid mercury at its boiling temperature given that 16.244 kcal is added and there are no losses?

44. [I] Large air conditioners are often rated in tons. A 1-ton unit has a cooling capacity equivalent to changing 2000 lb

of water per day at 0°C into ice at 0°C. Determine (to two significant figures) the thermal power removed from a building by a 1-ton air conditioner.

45. [I] A 0.50-kg quantity of mercury, with an average specific heat capacity of 138 J/kg·K, at 240 K is to be brought to its boiling point and completely vaporized while being maintained at that temperature. How much thermal energy must be provided?

46. [I] A cup of water, 0.25 kg, at room temperature is set on the stove in an open pot to make some tea. The phone rings and you get distracted only to return to the pot as the last drops of water boil off completely. How much heat was transferred to the water?

47. [I] A 500-g lead mass is heated to 150°C and placed on a block of ice at 0°C. How much, if any, ice will melt?

48. [I] Compare the amount of heat it takes to raise an arbitrary amount of water from 0°C up to 100°C with the amount of heat needed to vaporize that water.

49. [I] How much ice at 0°C must be added to 0.50 kg of water at 20°C if it is all to melt, leaving only water at 0°C? (Do the problem using SI units and see Problem 50.)

50. [I] Redo Problem 49 using kilocalories rather than joules and compare your answers.

51. [I] Suppose we combine 100 g of aluminum with a specific heat capacity of 920 J/kg·K at a temperature of 495°C with 99.9 g of ice at 0.00°C. What will be the final temperature?

52. [I] Body tissue without blood flowing through it has a thermal conductivity of about 18 Cal·cm/m²·h·C°. It's often given in those units in the life sciences because the surface area of the body is generally determined in m² and the thickness of the tissue in cm. Convert the conductivity to SI units.

53. [I] The following provides some idea of the inadequacy of conduction as a mode of heat transfer for the human body (see Problem 52). Suppose a person has a surface area of 1.4 m² and an average tissue thickness of 2.0 cm between what we might consider the inner and outer body. If the skin temperature is 33°C and the inner body temperature is 37°C, how much heat will be conducted to the surface per hour? Compare that to the fact that about 100 kcal/h are developed while just standing around staying alive.

54. [I] Determine the amount of heat that is conducted through a 4.0-m² portion of a brick wall 15 cm thick in the course of 1.0 h if the inside temperature is 20°C and the outside temperature is 0°C.

55. [I] A sheet of ice 1.00 cm thick with an area of 2.00 m² covers a piece of sidewalk. Since it's a nice clear day, we simply sprinkle a thin layer of black soot on the ice exposed to the Sun's rays. The idea is to melt the stuff rather than going out to break it up with a chopper. If the air temperature is 0°C and the solar energy per unit area per unit time impinging perpendicularly on the ice is 350 W/m², how long will it take before it melts? Assume 100% absorption and no losses.

56. [II] One gram of carbohydrate burned in a calorimeter liberates 4.10 Cal, and it is estimated that in the body about 98% of that energy finds its way to the cells. If a woman consumes 150 Cal/h jogging, how long must she exercise to burn off 100 g of carbohydrate?

57. [II] Knowing that the heat of vaporization of water at STP is 2492 kJ/kg, approximate the amount of energy required by one molecule to escape from the liquid.

58. [II] Compute the heats of fusion and vaporization of water in electron volts per molecule and compare those with the energy needed to tear apart a water molecule into oxygen and hydrogen (2.96 eV/molecule).

59. [II] How much steam at 100°C must be fed into 1.00 kg of water at 10°C to bring it up to 50°C?

60. [II] A lead bullet is heated up to 600 K and then fired at a stone wall, where it comes to rest in a completely molten state. Assuming no energy losses, compute its impact speed.

61. [II] A 10.0-g lead bullet at 23.0°C slams into a stone wall and, squashing, comes to rest. Assuming no loss of energy to the environment, how fast must the bullet be traveling if it is to totally melt?

62. [II] A 10-kg block of ice at 0°C falls, impacting in a way that brings it to rest with no loss of thermal energy. From what height should it drop if 0.010% of the ice is to melt?

63. [II] How many grams of steam at 110°C must be added to 1 kg of ice at 0°C to melt it without raising its temperature? Do this one in both kilocalories and joules and check your answer.

64. [II] An unheated garage in New Jersey in the winter is maintained inside at 0°C by the freezing of a tank of 1000 kg of water, despite the fact that it's colder than that outside. The garage loses heat at a rate of 10⁴ cal/min. How long will it be able to stay at 0°C?

65. [II] A calorimeter contains 398 g of water in a 102-g copper cup at 5.1°C. To this is added 40.5 g of crushed ice at −7.6°C. What is the final temperature of the system?

66. [II] As little as 60 years ago, most people kept food in iceboxes that were simply well-insulated cabinets. A large block of ice, delivered roughly once a week, was placed inside above a drip pan. The remaining space was packed with food, the stuff to be colder kept nearer the ice. How does the system work? What's the equilibrium temperature inside the icebox? How much heat would a typical 100-lb block remove from the food before it melted?

67. [II] A block of ice at 0°C is placed on a balance, and an equal mass of water at 80°C is measured off. The water and ice are then combined in an isolated chamber. Determine the equilibrium temperature of the system and describe in detail what it consists of.

68. [II] How much wood would have to burn completely to provide enough energy (assuming no losses) to vaporize 1.00 kg of water boiling away at 100°C?

69. [II] What thickness of brick, with a thermal conductivity of 0.60 W/m·K, will conduct heat at the same rate as a 10-cm layer of dead air under the same conditions?

70. [II] A wall is made up of a 10-cm layer of common red brick in contact with a 20-cm layer of cinder block concrete. If the inside concrete face has a temperature of 20°C and the outside brick face is at 0°C, how much heat flows through a 1.0-m² area each hour? What is the power loss?

71. [III] How much thermal energy must be added to 1.00 kg of water to change its temperature from 5.0°C to 100°C? If that same quantity of energy is added to 1.20 kg of silver at 5.0°C, what will its final temperature be?

72. [III] In 1899, Callendar and Barnes devised a new method for measuring the specific heat of water as a function of temperature. It utilized a scheme (Fig. P72) in which only

Thermometer

Thermometer

Water in

Water out

Figure P72

constant temperatures occurred that could be measured slowly and with great accuracy using platinum resistance thermometers. Water was made to flow at a constant rate through a tube that was thermally insulated by a vacuum jacket. The tube contained an electrical heating coil that generated a constant known amount of thermal power (P). In steady-state operation, the heat generated electrically is carried off by the water that enters the tube at a tempera-

ture T_i and leaves at T_o. Write an expression for the specific heat of water in terms of the experimental variables. Explain your reasoning.

73. [III] How does the system in Problem 72 lose energy and thus introduce error? Suppose the power loss is P_l. Repeat the previous experiment at a new flow rate such that a mass of water m' passes through in time t, but also change the power input to P' so that T_o is the same. Show that the thermal losses can be eliminated from the equations, whereupon

$$c_w = \frac{P - P'}{(m - m')\Delta T}$$

Explain your reasoning.

74. [III] A double-glazed window consists of two sheets of glass (each 3.0 mm thick) separated by a 5.0-mm layer of air. Determine the ratio of the thermal conductivity per unit area to that of a single 3.0-mm pane under the same conditions.

16

Thermodynamics

NOW THAT WE KNOW a little about the basics of thermal energy, temperature, and heat, we can turn our attention to **Thermodynamics**, *the study of thermal energy, its transfer, transformation, degradation, and dispersal*. Although it was conceived (1824) out of narrow pragmatic concerns about steam engines, Thermodynamics has evolved into a grand discourse on energy and change. Energy is a measure of change (the change in position, speed, mass, temperature, etc.), and change informs time; at base, Thermodynamics is the study of change, the study of the unfolding of events, the progression of the Universe.

Thermodynamics treats the thermal behavior of matter, and when we consider some specific entity or group of entities, it's called a *system:* it might be a bottle of gas, a jet engine, or a yard full of chickens. Whatever else there is beyond the system (that is, the rest of the Universe) is called the *surroundings.* A system might be connected to its surroundings in any number of ways: for example, it might allow heat, or electromagnetic radiation, to cross its boundaries, or it might even emit acoustical energy. On the other hand, we can imagine a system to be completely insulated from the rest of the environment—in which case it is *isolated.* The ultimate isolated system is the entire Universe, there being no surroundings for it to interact with. Each of us can be considered a system in rather elaborate contact with our surroundings, transferring heat to and from it, exhausting and intaking gases, radiating and absorbing electromagnetic energy, ingesting and excreting material, and so forth. Living creatures are just as much the subject of Thermodynamics as steam engines.

THE FIRST LAW OF THERMODYNAMICS

Perhaps the first person to grasp the nature of heat and its relationship to, and unity with, all the various forms of energy was Dr. J. Mayer (p. 520). His work contained the broad notion of Conservation of Energy, but it generated little interest beyond ridicule. Before the decade passed, Mayer was overwhelmed by a deep depression fed by derision and made more intense with the death of two of his children. Lost in despair, he leapt from a second-story window (1849) in a painfully unsuccessful attempt at suicide. Within two years, he was a straitjacketed inmate in an insane asylum.

Though Mayer was first (1842) to recognize that all the various forms of energy were one, Hermann von Helmholtz, a renowned German physiologist (1847), independently formulated the idea of Conservation of Energy, as did Joule. In 1847, Joule gave a short uninvited talk at a scientific meeting. He was cautioned to be brief, and discussion was neither expected nor encouraged, but a young man in the audience began to raise some interesting questions and soon there was a lively discussion that ultimately caused a sensation in the scientific community. That young Scotsman who was so moved by what he had heard was Professor William Thomson, the future Lord Kelvin.

The laws of Thermodynamics "control, in the last resort, the rise and fall of political systems, the freedom or bondage of nations, the movements of commerce and industry, the origins of wealth and poverty, and the general physical welfare of the race."

FREDRICK SODDY (1877–1956)
Nobel Prize-winning chemist

16.1 Conservation of Energy

The most complete conception of the Law of Conservation of Energy (p. 297) includes *all* kinds of energy and is known as the **First Law of Thermodynamics**:

Energy can neither be created nor destroyed, but only transferred from one system to another and transformed from one form to another.

Remember that, whatever energy is, it is associated with the change imparted to a system as the result of interactions. If a positive amount of work is done *on* a sample of matter (on the worked), simultaneously an equal amount of negative work will be done *by* that sample of matter (on the worker). In that sense, change is transferable and insofar as it is transferred, energy (ΔE) is transferred.

The transformation of energy from one form to another is mediated by matter and occurs in the region of that matter. Energy is a property of matter; it does not have a separate existence. Thus, it cannot spontaneously vanish at one place in the Universe and simultaneously appear at another, even though that would not explicitly violate the First Law. *The transformation of energy occurs through the action of the Four Forces.*

The First Law, fleshed out with all of the associated quantitative ideas, such as heat, work, *gravitational*-PE, *electrical*-PE, KE, and so on, allows us to build an understanding of the operation of the natural world. (Fig. 16.1). It was the contribution of the nineteenth century to add heat into the energy balance, and it was the work of Einstein in the twentieth century that showed that matter itself was part of the balance, too—but we will come back to this later, in Chapter 28, since its effect here is negligible.

When a 1-N apple falls 10 m from a tree into your hand, the original 10 J of *gravitational*-PE is transformed into 10 J of KE at the instant before it lands (minus a tiny bit transferred to the air via friction). When the apple comes to rest (neglecting a minute amount of energy lost via sound as it thumps down), the 10 J are shared as thermal energy between hand and apple, so that both increase in temperature slightly. A meteorite with much more KE may even melt on impact, but all the energy present at any moment is constant, whatever its form.

If we do 10 J of work compressing a spring, then tie it up and lower it into an insulated acid bath, what happens to the *elastic*-PE as the spring dissolves? The energy stored via the interatomic forces must be liberated when the spring vanishes. We can expect the temperature of the bath to rise slightly more when dissolving the compressed spring than when dissolving a relaxed one.

To slow down a spaceship reentering the atmosphere, some of its KE is deliberately converted into thermal energy in the air and heat shield. For that matter, every time you activate the brakes on a moving car, you are doing the same thing: converting KE—yours and the car's—to random thermal energy via friction and the First Law. As Mayer suggested, friction that occurs during the tidal action of the oceans, the back and forth rubbing against the ground, heats the waters a bit. And that ther-

Figure 16.1 The First Law of Thermodynamics established Conservation of Energy as a central principle in physics. Now thermal energy joins other forms of energy, broadening the concept of conservation. (See Figures 1.4, 4.20, 8.26, and 9.21.)

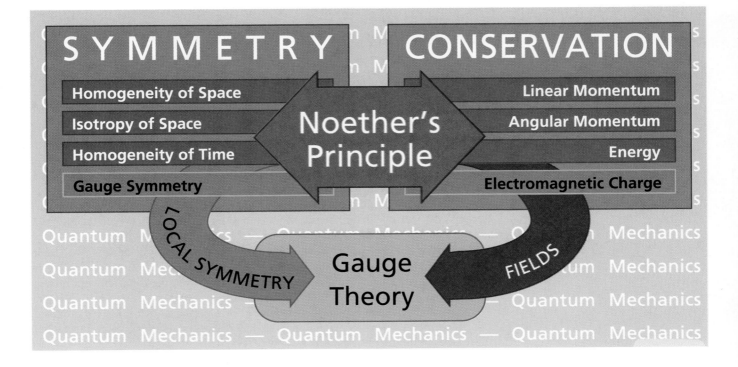

mal energy comes from the rotational KE of the planet—the Earth experiences drag and slows down.

Work, Heat, and Internal Energy

Consider the particles of an ideal gas, which are effectively noninteracting (PE = 0) point-masses. These are endowed with neither rotational nor vibrational KE, and the internal energy U is purely random translational KE. That being the case, it follows from Kinetic Theory that the internal energy of the ideal gas depends only on its temperature [Eq. (14.14)]. Because of its mathematical simplicity, we shall be using the ideal gas as a guide to the behavior of real gases, which generally respond in more or less the same way.

We already know that both heat Q and work W correspond to the transfer of energy by very specific means. The early application of these ideas was to the steam engine, where it was natural to "pour" heat in and get work out. Consequently, it made sense to define both heat-in and work-out as positive quantities, and that usage persists. Hence, work done *by* the system on the environment (work-out) is positive, while work done by the environment *on* the system (work-in) is negative. If an amount of heat Q enters the system, it could manifest itself as either an increase in internal energy, or as a resulting quantity of work performed by the system on the surroundings, or a bit of both. Consequently, the First Law of Thermodynamics becomes

$$Q = W + \Delta U \tag{16.1}$$

The definition of the internal energy of a system might then be taken as

$$\Delta U = Q - W \tag{16.2}$$

Example 16.1   A system in contact with a furnace receives 5000 J of heat while doing 2700 J of work on its surroundings. What, if any, is the change in the internal energy of the system?

Solution: [Given: $Q = +5000$ J = heat-in, $W = +2700$ J = work-out. Find: ΔU.] The relationship be-

tween Q, W, and ΔU is given by the First Law

$$\Delta U = Q - W = (+5000 \text{ J}) - (+2700 \text{ J}) = \boxed{2300 \text{ J}}$$

▶ **Quick Check:** $Q = W + \Delta U = 2700$ J + 2300 J = 5000 J.

Here, for the most part, we limit the discussion to the transfer of energy via work and heat. A more general treatment would simply add on terms to Eq. (16.2) as needed; for example, if the system were heated by putting it in a microwave oven, we would add in a radiation term, and so on.

Whenever there are forces acting, it is necessary to decide where the boundaries of the system will be before beginning the analysis (just as it was when applying the idea of Conservation of Momentum). The choice affects the details of the description but does not change the end results. For example, Fig. 16.2 depicts two chambers that are in contact and can exchange heat, but otherwise are isolated from the rest of the environment. An amount of heat Q flows from the high-temperature liquid into the low-temperature gas. There are three possible systems: 1 and 2, which are open, and 3, which is closed; Q flows out of 1 into 2, and nothing flows into or out of 3. Since no work is being done, ΔU_1 decreases by an amount equal to Q, ΔU_2 increases by an amount equal to Q, and $\Delta U_3 = 0$.

By definition, an isolated system is thermally insulated ($Q = 0$) and can neither do work on the surroundings nor have work done upon it ($W = 0$). It follows, then, that $\Delta U = 0$. Another equivalent statement of the First Law is then

The internal energy of an isolated system is constant (even though it may be transformed from one type of energy to another).

This applies to all isolated systems, including the Universe itself. That's a presumptuous statement to make considering our meager role in the Cosmos, but to date, the First Law has never failed, so we keep saying it. For an isolated system, the internal energy is constant regardless of what processes go on within it—be they chemical, nuclear, or biological.

The same sort of thing happens in the special case of a system that is not isolated, provided that the net amount of heat and work into and out of it turns out to be zero: $Q - W = 0$, $\Delta U = 0$, and $U =$ constant. This notion actually applies to the Earth, provided we add on a radiation term to Eq. (16.2). Take the system to be everything within a great sphere enclosing the atmosphere. Practically speaking, the amount of work done on or by the Earth is negligible ($W = 0$), and since it is thermally insulated, being surrounded by vacuum, $Q = 0$. Still, energy pours in in the form of electromagnetic radiation (mostly from the Sun and mostly visible) at a rate of about 1.7×10^{17} J/s. The Earth reflects some of this and furthermore radiates energy on its own (mostly infrared). The result is that it emits just about at a rate of -1.7×10^{17} J/s. Thus, the total internal energy of the Earth (that is, everything on it and everything in it) is more or less constant. Indeed, as long as you are not changing your weight, much the same thing can be said about you—energy goes in, changes form, and goes out.

The internal energy of a finite system is finite. Clearly, then, if work is to be extracted from it, an equivalent amount of energy must be supplied, or the internal energy will diminish and ultimately no longer be available to drive the system. A *perpetual-motion machine* is one that puts out more work than its input of energy and can presumably continue to do so forever. We call such a mythical device a *perpetual-motion machine of the first kind* because it violates the First Law. Despite the fact that people are still attempting to patent such machines, they cannot possibly work. Recently, a Los Angeles jury acquitted two men of 50 counts of fraud, theft, and conspiracy. They had raised several hundred thousand dollars from backers of a compressed air turbine that was supposedly a perpetual-motion machine. According to the deputy district attorney, "The jurors told me they believed this machine produces more energy than goes into it."

All of this is not to say that the First Law precludes the existence of isolated systems that could, for example, spin forever or bounce forever—provided they do not output work or any other form of energy, they are not forbidden by the First Law.

Processes

A system is in equilibrium when there is no tendency for it to undergo spontaneous change. Despite the fact that its atoms are moving about randomly, it manifests constant measurable macroscopic *physical properties,* such as pressure, volume, temperature, and so on. A configuration of the system, corresponding to a specific set of values of these physical properties (P, V, T, etc.), is called a *state*. As long as they are constant, the system is said to be in a particular state, even though its constituent atoms are continuously moving around. Accordingly, an ideal gas in equilibrium can be described by Eq. (14.9), $PV = nRT$, which is its *equation of state*. A

Figure 16.2 Because of the action-reaction nature of interactions, whenever forces act, it is necessary to establish the boundaries of the system being analyzed.

process occurs when the system changes from one state (one set of values of its physical properties) to another state. The system returns to its original state when all of its macroscopic physical properties resume their original values.

When a process is taking place, there is some observable alteration of the system: something measurable (for example, P, V, or T) changes. A system can be varied in a variety of ways, but there are four basic modes that are especially simple to deal with:

isothermal A process is isothermal when the temperature of the system is constant. The word comes from the Greek *isos*, equal, and *therma*, heat.

isobaric A process is isobaric when the pressure of the system is constant. The word comes from the Greek *baros*, weight.

adiabatic A process is adiabatic when no heat is transferred to or from the system. The word comes from the Greek *adiabatos*, not passable.

isovolumic A process is isovolumic (or isochoric) when the volume of the system is constant.

The thermal processes involving vapors and gases are particularly important: they play a central role, for example, in the operation of the electric-power–producing steam engine, the internal combustion automobile engine, and the jet propulsion airplane engine.

Work and the First Law

Let's now express the work W in Eq. (16.1) in terms of directly measured variables. Imagine a system that expands against an external pressure, changing its volume and doing work. The primary features are the existence of an applied force and the volume change. What that system is really doesn't matter much—it could be a solid, liquid, or gas. To keep things simple, let the system consist of a cylinder of gas sealed with a weightless, frictionless piston of area A. Some influence in the surroundings pushes down on the piston with a constant force F, as shown in Fig. 16.3. The gas and piston are in equilibrium, and the downward force F is exactly opposed by an upward force produced by the gas—namely, PA. The gas is now somehow made to expand very slowly (quasi-statically) so that it always remains essentially in equilibrium with the piston (here, that means P is constant)—the process is isobaric. This could be accomplished by slowly providing heat to the gas, which will increase the average speed of the molecules thereby sustaining P, even as V increases (p. 504). If the displacement of the piston is Δs, the work done *on the surroundings by the expanding gas* is

$$W = F\Delta s = PA\,\Delta s$$

Inasmuch as the change in volume is $V_f - V_i = \Delta V = A\,\Delta s$, *the work done in an isobaric process* becomes

$$W = P\Delta V \qquad (16.3)$$

When the system expands, ΔV is positive, work is done *by* the system, and W is positive. When the system contracts, ΔV is negative, work is done *on* the system, and W is negative.

Figure 16.3 A gas sealed in a cylinder by a weightless, frictionless piston. The constant downward-applied force F equals PA, and when the piston is displaced, downward work is done on the gas.

Example 16.2 A cylinder closed off with a movable piston contains 10.0 g of steam at 100°C. The system is heated and its temperature rises 10.0°C as the steam expands 30.0×10^{-6} m^3 at a constant pressure of 0.400 MPa. Determine (a) the work done by the steam and (b) the change in its internal energy. Take the specific heat capacity of steam to be 2.02 kJ/kg·K.

Solution: [Given: $P = $ constant $= 0.400$ MPa, $m = 10.0$ g (of water), $\Delta T = 10.0$ K, $\Delta V = 30.0 \times 10^{-6}$ m^3, and $c = 2.02$ kJ/kg·K. Find: W and ΔU.] (a) The process is isobaric, and so W is given by Eq. (16.3)

$$W = P\Delta V = (0.400 \times 10^6 \text{ Pa})(30.0 \times 10^{-6} \text{ m}^3)$$

and
$$\boxed{W = 12.0 \text{ J}}$$

(b) Since $\Delta U = Q - W$, we will need Q, the heat-in. $Q = cm\,\Delta T$, and so $\Delta U = cm\,\Delta T - W$, hence

$$\Delta U = (2.02 \times 10^3 \text{ J/kg·K})(0.010\,0 \text{ kg})(10.0 \text{ K}) - 12.0 \text{ J}$$

and
$$\Delta U = 202 \text{ J} - 12.0 \text{ J} = \boxed{190 \text{ J}}$$

▶ **Quick Check:** $Q = cm\,\Delta T = \Delta U + W = 190$ J $+ 12.0$ J; $\Delta T = (202 \text{ J})/cm = 10$ K.

Example 16.3 An aluminum cube 20 cm on a side is heated from 50°C to 150°C in a chamber at atmospheric pressure. Determine the work done by the cube and the change in its internal energy. If the same process was carried out in vacuum, what would the change in internal energy be then?

Solution: [Given: aluminum cube, $L = 20$ cm, $T_i = 50$°C, $T_f = 150$°C. Find: ΔU when $P = 1$ atm and when $P = 0$.] This isobaric process involves ΔU and W, and that immediately suggests Eq. (16.2): the block expands against the hydrostatic pressure of the atmosphere and does work on it. We have to determine Q and W to find ΔU. To get $W = P\Delta V$, we need ΔV, and that follows from Eq. (14.4) and Table 14.3:

$$\Delta V = \beta V_0 \Delta T = (72 \times 10^{-6} \text{ K}^{-1})(8.0 \times 10^{-3} \text{ m}^3)$$
$$\times (100 \text{ K}) = 5.76 \times 10^{-5} \text{ m}^3$$

Hence

$$W = P\Delta V = (0.101\,3 \text{ MPa})(5.76 \times 10^{-5} \text{ m}^3) = \boxed{5.8 \text{ J}}$$

Now, to find Q, we use Eq. (15.1), $Q = cm\,\Delta T$, but we will need m first. Since $m = \rho V$ and from Table 10.1 (p. 321), it follows that

$$m = (2.7 \times 10^3 \text{ kg/m}^3)(8.0 \times 10^{-3} \text{ m}^3) = 21.6 \text{ kg}$$

Using Table 15.3:

$$Q = cm\,\Delta T = (0.90 \text{ kJ/kg·K})(21.6 \text{ kg})(100 \text{ K})$$
thus
$$Q = 1.94 \text{ MJ}$$

Hence
$$\Delta U = Q - P\Delta V \qquad (16.4)$$

and
$$\Delta U = 1.94 \text{ MJ} - 5.8 \text{ J} = \boxed{1.9 \text{ MJ}}$$

The amount of work done is negligible by comparison to Q, so it doesn't matter if this process is carried out in vacuum or not; the internal energy change will still be 1.9 MJ.

▶ **Quick Check:** Using rounded off numbers, $\Delta V \approx (70 \times 10^{-6} \times 10 \times 10^{-3} \times 100)$ m$^3 \approx 7 \times 10^{-5}$ m^3. $W \approx 0.1$ MPa $\times 6 \times 10^{-5}$ m$^3 \approx 6$ J. $Q \approx (1 \times 10^3 \times 20 \times 100)$ J ≈ 2 MJ.

The PV-diagram for an isobaric expansion is shown in Fig. 16.4. From an initial state I, the system was transformed at a constant pressure to a different final state F. Notice that **the work done, $P\Delta V$, is the area under the curve**. If the pressure is now decreased (Fig. 16.5) while maintaining a constant volume $\Delta V = 0$, the system arrives at a new state F with no additional work done (the area under the curve is unchanged). In the case of an ideal gas, we can imagine the various processes of constant P, V, or T as occurring on the surface of a three-dimensional PVT-diagram (Fig. 16.6). The points A, B, and C in Fig. 16.5 correspond to those in Fig. 16.6, where the system goes from A to B to C.

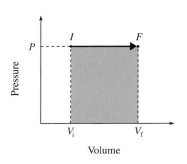

Figure 16.4 The volume of a gas is changed while the pressure is kept constant. The work done by the gas in expanding is $P\Delta V$, and that equals the area under the curve.

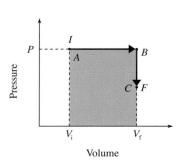

Figure 16.5 No work is done in decreasing the pressure carrying the system from B to C and a final state F. The work done by the gas in going from I to F is equal to the area under the curve.

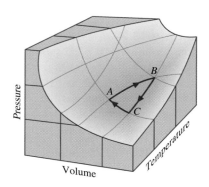

Figure 16.6 *PVT*-diagram for an ideal gas.

Example 16.4 A cylinder containing a gas is sealed with a nearly frictionless piston at a pressure of 0.20 MPa. The cylinder is placed in contact with a source of heat and the gas very slowly expands, moving the piston, whose area is 1000 cm^2, a distance of 5.0 cm. Assuming the process is isobaric, and given that 300 J of heat enters the system, determine its change in internal energy.

Solution: [Given: $P = 0.20$ MPa = constant, $A = 1000$ cm^2, $\Delta s = 5.0$ cm, and $Q = 300$ J. Find: ΔU.] The only formula we have for internal energy is the First Law, Eq. (16.2), $\Delta U = Q - W$, but because this process is isobaric

$$\Delta U = Q - P\Delta V \qquad [16.4]$$

Here, $\Delta V = (1000$ cm$^2)(5.0$ cm$) = 5000$ cm$^3 = 5.0 \times 10^{-3}$ m^3. Hence

$$\Delta U = (300 \text{ J}) - (0.20 \times 10^6 \text{ Pa})(5.0 \times 10^{-3} \text{ m}^3)$$

and

$$\boxed{\Delta U = -0.70 \text{ kJ}}$$

The gas does much of the work of expansion at the expense of internal energy to the extent of a 700-J reduction in U.

▶ **Quick Check:** $\Delta V = 5 \times 10^3$ cm$^3 \approx 10^{-2}$ m^3; $P\Delta V \approx 10^5$ Pa $\times 10^{-2}$ m$^3 \approx 10^3$ J; $Q = \Delta U + P\Delta V \approx -700$ J $+ 10^3$ J ≈ 300 J.

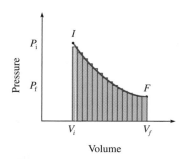

Figure 16.7 The area under a *PV*-curve equals the work done on or by the gas. Here, that area is divided into a large number of small rectangles so that it can be computed.

If the pressure varies continuously as the volume changes (Fig. 16.7), the area under the *PV*-curve would still correspond to the work done. That area can be found by dividing it up into small rectangular segments (as in Fig. 9.6, p. 284), each terminated by the average value of the pressure over that interval. The work done during the i-th constant-P segment ΔW_i is given by Eq. (16.3) as $P_i \Delta V_i$, and the sum of all such contributions, as they get narrower and their number approaches infinity, approaches the total work exactly. The only proviso is that the system be in equilibrium everywhere. If it isn't, PA will not equal F over some of those vanishingly narrow area segments, a point we will return to presently.

Depending on what we do to it, the system can be transformed from its initial condition (state I) to its final condition (state F) by various routes, each corresponding to a different amount of work done (Fig. 16.8). Clearly, first dropping P at constant V and then changing the volume encompasses less area (and so does less work) than initially keeping P constant. *The work done on or by a system depends on the manner in which it is transformed from its initial to its final state—work is not a state variable, it does not reflect the state of the system*. Remember that heat may also be entering or leaving the system, and the final state is determined by a balance of both these contributions. Thus, the path taken in the *PV*-diagram must be fully specified if the work is to be determined. Still, whatever path is taken to get there, the final state is the same—its atoms presumably should not remember the sequence of events that brought them there. Consequently, we can expect that U_f will depend on P_f and V_f but will be the same for all paths leading to that final state. Experiments bear out this conclusion: ΔU is always found to be path-independent. **The internal energy of a system is a unique measure of the state of the system**. Although U is path-independent, W is not, and it follows from the First Law that Q is not. Both W and Q depend on the history of the system and combine to sustain U so that it is independent of that history.

Isothermal Change

Suppose that the device in Fig. 16.9a contains an ideal gas that is to expand isothermally. Accordingly, imagine that the cylinder is made of a highly conducting material, is thin-walled, and surrounded by a large constant temperature bath (a so-called *thermal reservoir*). Now, if the gas is allowed to expand very slowly, raising the piston, heat will be able to enter the system at a rate such that the gas remains at a constant temperature. Quite generally, an expanding gas does work on the movable wall of its container, the piston. When a molecule collides with a wall that is moving away and rebounds off, it has a final speed that is less than its initial precollision speed (Fig 16.9b). The decrease in molecular KE equals the work done on the piston. That work is therefore done at the expense of internal energy. Thus, *unless heat is supplied to the gas, its temperature will drop*. Here, the cylinder is in contact with a thermal reservoir, and any tendency for the temperature to change is negated by a flow of heat across the boundary.

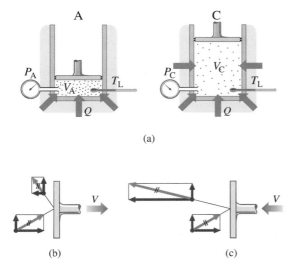

Figure 16.9 (a) A gas expanded isothermally (see Fig. 16.10) from state *A* to state *C*. As heat enters, the piston rises and pressure drops. (b) When molecules bounce off a receding piston, they lose momentum. (c) When they bounce off an advancing piston, they gain momentum.

(a)

(b)

(c)

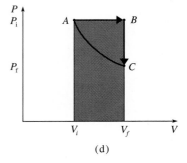

(d)

Figure 16.8 A system (a) may go from an initial state *I* to a final state *F* along several different routes, each corresponding to the doing of a different amount of work. Parts (b), (c), and (d) depict different procedures, areas, and amounts of work.

Figure 16.10 (a) An isothermal expansion of a gas. (b) The area under the *PV*-diagram equals work done.

In an isothermal expansion of an ideal gas, since *U* depends only on the translational KE, which depends on *T*, which is constant, it follows that $\Delta U = 0$ and $Q = W$: The work done by the gas is provided by the heat-in from the reservoir. In effect, the gas serves as a conduit that carries energy from reservoir to piston. On the other hand, when a gas is compressed, the molecules bounce off an approaching wall (Fig. 16.9c), picking up speed like a tennis ball colliding with an advancing racket. The KE of the gas increases; the piston does work on the gas, and unless heat can escape the system, its temperature will rise. Figure 16.10a depicts an ideal gas being moved along an isotherm from point *A* to point *C*, and the same path is pictured in Fig. 16.6, as well. Notice how going from *A* to *B* entails raising the gas from T_L to a higher temperature T_H, crossing several isotherms along the way. It's helpful when interpreting *PV*-diagrams to envision a grid of isothermals ranging from a low to a high temperature. Here, since $P = nRT/V = \text{(constant)}/V$, the isothermals are hyperbolas.

Consider the work done by the isothermal expansion of the ideal gas. The area under the curve in Fig. 16.10b is gotten by summing up all of the tiny ΔW area elements (or, via calculus, by integration). Here, $\Delta W = P\Delta V$, but if the process is done slowly so that the system never deviates appreciably from equilibrium, $P = \text{(constant)}/V$ and the pressure can be determined in terms of the volume at each element. Thus, $\Delta W = \text{(constant)}\Delta V/V$; more work is done (that is, the area element is taller) initially when the volume is small and the pressure high than later, when the volume is large and the pressure low. This particular functional dependence is extremely common in the real world: ΔW is proportional to both ΔV and to $1/V$. The situation where the change in something depends both on the change in another quantity and, inversely, on how much of that other quantity exists at each step is encountered quite frequently in physics. That kind of dependence is called *logarithmic* (Appendix A-3), and we define the natural logarithm (abbreviated ln) by way of it. We encountered the $\log_{10}$ when we dealt with sound intensity. This *natural* logarithm (so called because it occurs in so many natural processes) is the log to a special base: rather than 10, it is the number known as *e* equal to 2.718 . . . . In other words, $\ln x = \log_e x = 2.30 \log_{10} x$, as shown in Fig. 16.11, and the corresponding values of $\ln x$ are to be found in standard tables or computed via electronic calculators.

The area under the isotherm varies logarithmically with volume—that is, *W* varies logarithmically with *V*. *The work done on or by an ideal gas during an isothermal volume change from V_i to V_f is*

$$W = nRT \ln \frac{V_f}{V_i} \tag{16.5}$$

where $W = Q$. At least for the purpose of analysis, many processes are essentially isothermal and, as a rule, they occur slowly—a very small leak in a tank of compressed gas at room temperature, an open pot of water boiling away on a low flame.

Example 16.5 Determine the amount of work needed to compress 4.0 g of oxygen at STP down to one-third its original volume, keeping the temperature constant. Assume it behaves as an ideal gas.

Solution: [Given: oxygen at STP, $m = 4.0$ g, and $V_f = V_i/3$. Find: *W*.] The work for an isothermal volume change follows from Eq. (16.5) and so we need *n*, the number of moles. The atomic mass of O_2 is $2(16) = 32$;

(continued)

(continued)

hence, there are 32 g/mol and $n = (4.0 \text{ g})/(32 \text{ g/mol}) = 0.125$ mol. Hence

$$W = nRT \ln \frac{V_f}{V_i}$$

$$W = (0.125 \text{ mol})(8.314 \text{ J/mol·K})(273 \text{ K})(\ln \tfrac{1}{3})$$

and $\boxed{W = -0.31 \text{ kJ}}$. The minus sign shows that work was done on the gas, and since the temperature did not change, it had to lose the same amount of energy as heat-out.

▶ **Quick Check:** Redoing the calculation using simplified numbers, we get 4 g $= \tfrac{1}{8}$ mol; $nRT \approx \tfrac{1}{8}(8 \text{ J/mol·K})(273 \text{ K}) \approx 273 \text{ J/mol}$; $\ln \tfrac{1}{3} \approx -1$; $W \approx -0.3 \text{ kJ}$.

Reversibility

If the isothermal process that carried the gas from point A to point C in Fig. 16.10 can be retraced so that the system returns from P_f, V_f, and T to P_i, V_i, and T along the same curve, then we say that that process was **reversible**. Another way to put it is that the system must always be infinitesimally close to thermodynamic equilibrium even as it passes from one state to the next. Imagine a vertical, thermally insulated cylinder containing some gas sealed in with an upright frictionless piston on which is piled a mass of fine sand. The sand provides the downward force that is balanced by the trapped gas, leaving the piston motionless in equilibrium. If one miniscule sand grain is added to the pile, the piston will descend through a tiny displacement, in an essentially quasi-static fashion. One grain at a time, the compression can continue until we choose to stop at any moment. If sand is now slowly removed, the process will be reversed into an expansion that carries it through the same states it has already experienced. The fact that a process can be stopped anywhere along the way and turned around by a minute change in the operating conditions is the hallmark of reversibility.

In truth, reversible processes are idealizations: they can be approximated but never actually attained. For an isothermal process such as the one above to be reversible, the piston must be weightless and frictionless; moreover, the volume change would have to be done very slowly, keeping the gas in equilibrium and not generating any organized motion (such as eddies or turbulence) that would take up energy. Any process that entails friction losses is irreversible. Consequently, real processes can, at best, only resemble being reversible, but even so, such behavior is of great practical interest.

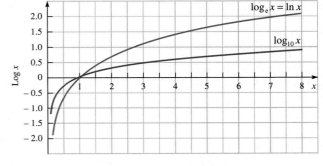

Figure 16.11 The function ln x equals $-\infty$ at $x = 0$, 0 at $x = 1$, and continues to rise slowly as x increases.

Adiabatic Change

Imagine an *adiabatic* process occurring in a system; that is, one where no heat crosses the boundary. This can be accomplished, with varying degrees of success, by thermally insulating the system. Alternatively, processes that occur suddenly tend to be adiabatic because heat takes a fair amount of time to flow. Thus, sound waves in air are adiabatic since there isn't enough time for heat to flow from the compressions to the rarefactions (see p. 450). Adiabatic processes are important in many situations, not the least of which is the operation of the internal combustion engine.

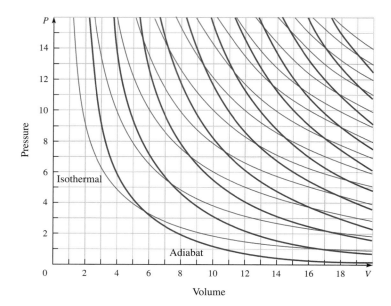

Figure 16.12 *PV*-diagram for an ideal gas; isothermals and adiabats.

Consider an ideal gas, sealed in a cylinder closed off with a piston, all of which is thermally insulated so no heat can enter or leave. The process about to be performed will be adiabatic, but it may or may not be reversible—that depends on how it is carried out. Let's stay with a reversible adiabatic process and so do everything nice and slowly to allow equilibrium to always obtain. Now, compressing the gas ($W < 0$), it follows that U must increase as the work done on the gas goes directly into internal energy ($\Delta U = -W$), and so the temperature increases. Accordingly, PV, which equals nRT, is no longer constant. The curve corresponding to this process in a *PV*-diagram is called an *adiabat* (Fig. 16.12) and is evidently everywhere steeper than an isotherm. That's the case because as the volume is decreased, the temperature rises, sustaining a higher pressure than would be if T were constant. On the other hand, as the gas expands adiabatically, it cools, sustaining a lower pressure at any value of V than would the isothermal process.

For an isothermal process, $PV = constant$ by way of Boyle's Law. For an adiabatic process, P drops more quickly with increasing V, and it is found experimentally that

[*reversible adiabatic*] $$PV^\gamma = \text{constant} \qquad (16.6)$$

where γ is a constant that equals 1.67 for monatomic gases, ≈ 1.4 for diatomic gases, and ≈ 1.3 for polyatomic gases. It can be shown theoretically, though we shall not, that γ is actually the ratio of the specific heat of the gas at constant pressure to that at constant volume. Another way to write Eq. (16.6) is

$$P_i V_i^\gamma = P_f V_f^\gamma \qquad (16.7)$$

and, since the gas is still ideal, the resulting temperature can be determined from

$$\frac{P_i V_i}{T_i} = \frac{P_f V_f}{T_f} \qquad (16.8)$$

Example 16.6 Argon gas (which, like the other Noble Gases, is monatomic) is very slowly compressed adiabatically in a well-insulated cylinder to half its original volume of 0.100 m^3. If it was at atmospheric pressure and 27.0°C to start, what will be the final pressure and temperature of the gas?

Solution: [Given: an adiabatic process, $V_i = 0.100$ m^3, $P_i = 1$ atm, $V_f = \frac{1}{2}V_i$, and $T_i = 27.0$°C. Find: P_f and T_f.] The process is adiabatic and involves P, V, and T, so we begin with Eq. (16.7). The gas is monatomic; hence, $\gamma = 1.67$ and

$$P_f = P_i \left(\frac{V_i}{V_f}\right)^\gamma = (0.101 \text{ MPa})(2)^{1.67}$$

and $P_f = (0.101 \text{ MPa})(3.182) = \boxed{0.322 \text{ MPa}}$

To find the temperature, use Eq. (16.8), and so

$$T_f = T_i \left(\frac{P_f}{P_i}\right)\left(\frac{V_f}{V_i}\right) = (300 \text{ K})(3.19)\left(\frac{1}{2}\right) = \boxed{479 \text{ K}}$$

▶ **Quick Check:** Using Eq. (16.7), $P_i V_i^\gamma = (0.1 \text{ MPa})(0.1 \text{ m}^3)^{1.67} = 2 \times 10^3$, whereas $P_f V_f^\gamma = (0.3 \text{ MPa})(0.05 \text{ m}^3)^{1.67} = 2 \times 10^3$.

A common instance of an adiabatic process is the rapid compression of air with a hand pump on the downstroke. You do (negative) work on the gas and $Q = 0$; it rises in temperature ($\Delta U = -W$), heating the pump quite noticeably. Stretch a fat rubber band, doing work on it in one quick adiabatic jerk, and its temperature will rise appreciably—you can easily sense the result with your lips.

An example of the reverse process, where the system does work, is the discharge of a carbon dioxide fire extinguisher. This happens so fast, it's again adiabatic, but now the temperature drops. Initially at room temperature, the CO_2 blasts out into a region of atmospheric pressure and does (positive) work on the surroundings. Since $\Delta U = -W$, the temperature of the gas drops, often to the point of forming dry ice crystals. Meanwhile, the canister is cooler, too, because the CO_2 remaining in it consists of the slower molecules (lower KE) left behind. The same sort of thing happens when you fill a spare tire from a can of compressed gas—the can gets cold. Hold a fat rubber band stretched for a while and then let it quickly contract, and it will cool appreciably. The elastic force does work as the band contracts that decreases the internal energy, and the temperature drops.

CYCLIC PROCESSES: ENGINES AND REFRIGERATORS

Let's suppose that we are working with reversible processes and, furthermore, once having carried out several of them, we wish to return the system to its original state. The PV-diagram would then represent a *cycle* and would appear as a closed figure that could be swept through without change, over and over again. The simplest program is to put an ideal gas in a cylinder closed by a piston (Fig. 16.3), surround it with a constant temperature bath, and expand it along an isothermal, as in Fig. 16.13a. An amount of heat Q_{AC} flows in as the system expands from A to C: $W_{AC} > 0$, and because $\Delta T = 0$, $\Delta U = 0$ for an ideal gas (p. 558); hence, $Q_{AC} > 0$. The work done by the gas is the area under the curve. If we return to point A via the same isothermal in reverse, as in Fig. 16.13b, work is done on the gas and heat flows out of it, where $W_{AC} = -W_{CA}$ and $Q_{AC} = -Q_{CA}$, and the system is energetically back to its starting point. Notice that if we add the negative work done on the gas during the compression leg to the positive work done during the expansion leg, the result is the net positive work done over the cycle, or in this case, zero. That result is represented by the *area enclosed by the closed curve*—zero.

When the cap is removed from a beer bottle, gas in the neck expands rapidly and adiabatically. It does work on the air and also loses thermal energy, overcoming its own internal interactions. As a result, the temperature in the neck of the bottle can drop from ≈41°F to ≈−31°F. The sudden change causes water vapor in the neck to condense into a cloud of tiny droplets.

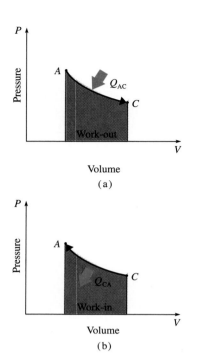

Figure 16.13 (a) The *PV*-diagram for an isothermal expansion during which a quantity of heat Q_{AC} enters the system as it goes from *A* to *C*. (b) During the isothermal compression back to *A*, the same amount of heat Q_{CA} is ejected.

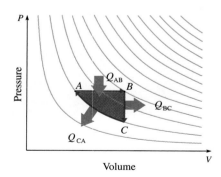

Figure 16.14 A cycle corresponding to a nonzero amount of work done by the system. The work done on the system during compression from *C* to *A* is negative.

Figure 16.14 shows a cycle where the net work done by the gas (the area enclosed) is nonzero and positive. In going from *A* to *B*, work is done by the gas. Its temperature rises, so *U* increases and heat enters the system. In going from *B* to *C*, no work is done and the temperature drops, so *U* decreases and heat leaves the system. The work-out (that is, the net work done by the gas) equals the net heat-in, and $\Delta U = 0$.

This cycle suggests the operation of an engine. *An engine is a device that converts some form of energy (electrical, gravitational, chemical, thermal, etc.) into work.* In the broadest sense, both a cannon and a solid-fuel rocket motor are engines. But these are essentially one-shot affairs, and we will do better to limit the term *engine* to cyclic devices that can continue to operate for as long as we provide energy (fuel) to them. *A heat engine is a cyclic device that converts thermal energy into work-out.*

16.2 The Carnot Cycle

When Nicholas Léonard Sadi Carnot (pronounced car-no) published his masterwork *Reflections on the Motive Power of Heat* in 1824, he was twenty-eight years old. Carnot's father, Lazare, had been minister of war to Napoleon Bonaparte, and Sadi had fought in the outskirts of Paris in 1814. Recognizing that military power came with industrial mastery, Carnot turned his special ingenuity to bear on the great symbol of the age, the fire-chariot of the Industrial Revolution: the steam engine.

Carnot began his analysis with something more manageable; a simple reversible cycle that corresponds to a purely hypothetical device. Remarkably, the conclusions he drew were applicable, in one way or another, to all heat engines from the gas turbine to the power system that propels the family car. The Carnot engine (Fig. 16.15) is just a gas-filled cylinder and piston that can alternately be brought into contact with, or isolated from, either a high-temperature reservoir (steam) that serves as a source of heat or a low-temperature reservoir (cooling water) that serves as a sink into which heat is exhausted. The Carnot cycle is a four-stage reversible sequence (Fig. 16.16) consisting of: (1) an isothermal expansion (from *A* to *B*) with the reception of heat-in (Q_H) at a high temperature; followed by (2) an adiabatic expansion (from *B* to *C*); followed by (3) an isothermal compression (from *C* to *D*) with the rejection of heat-out ($-Q_L$) at a low temperature; followed by (4) an adiabatic compression (from *D* to *A*). Observe that Q_H and Q_L are traditionally positive so that the net flow of heat to the system is ($Q_H - Q_L$).

Figure 16.17 shows the corresponding stages of the engine. The *A-B-C* portion, the expansion, is the *power stroke,* during which positive work is done. For the first part of it (*A-B*), the work is provided for by an input of heat. During the second part (*B-C*), the work is provided for by draining some of the internal energy of the gas—that's why its temperature drops. The piston may transmit work to the environment by way of a crankshaft. And it would be the inertia of the rotating crankshaft that drives the system back from *C* to *D* to *A*, doing work on the gas and returning it to its original condition.

There are several crucial points here: by using both a high-temperature source (steam) and a low-temperature sink (cooling water), the cycle rides out and back on isothermals but now encloses a nonzero area, unlike Fig. 16.13; less heat is exhausted into the sink than entered from the source (since $\Delta U = 0$, $Q_H - Q_L = W_o$). The lower the sink temperature, the more work would be done, and the less heat would be exhausted.

Figure 16.15 A Carnot engine consists of a cylinder and piston filled with gas. The device is able to be brought into contact with either a high-temperature or a low-temperature reservoir.

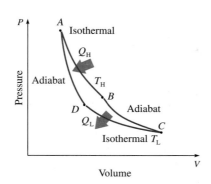

Figure 16.16 The four-stage reversible sequence representing a Carnot cycle.

Figure 16.17 The various stages of the Carnot cycle.

Efficiency

The primary reason the Carnot engine is so important is that it represents a class of idealized devices (reversible engines) that can be shown to be the most efficient possible. Thus, it sets an upper limit on the attainable efficiency of any real heat engine, no matter how cleverly contrived, and that's a marvelous practical tool to have.

Most generally, the energy **efficiency** (e) of a process is defined as

$$\text{Efficiency} = \frac{\text{useful energy-out}}{\text{energy-in}} \quad (16.9)$$

which is the ratio of what you get out (that is of value) to what you put in (to accomplish the process). With heat engines, the useful energy output is work, and the energy input is heat; thus, over one cycle

$$\text{Efficiency} = \frac{\text{work-out}}{\text{heat-in}}$$

This expression is applicable to all heat engines, real or imagined. For an idealized cyclic heat engine (which is both frictionless and lossless), the First Law demands that $W_o = Q_H(\text{in}) - Q_L(\text{out})$, and so

$$e = \frac{W_o}{Q_i} = \frac{Q_H - Q_L}{Q_H} = 1 - \frac{Q_L}{Q_H} \quad (16.10)$$

where both Q_H and Q_L are positive numbers.

Figure 16.18 schematically depicts a heat engine operating between a source and sink at high and low temperatures, respectively. Note that the efficiency increases as

Nicolas Léonard Sadi Carnot
(1796–1832).

Q_L decreases, becoming 1 (that is, 100%) only when there is no exhaust heat, a condition that has never been achieved (and one that is expressly forbidden by the Second Law of Thermodynamics, p. 570). Actual engines dissipate energy via friction and lose appreciable amounts of energy to the environment by convection, conduction, and radiation. That's why the hood of a car gets hot after a few minutes of operation, and the hotter it gets, the more wasted energy there is (Table 16.1).

The form of Eq. (16.10) in terms of heat is really not very convenient; it would be far simpler to measure if it were expressed in temperatures instead, and Carnot did just that. "According to established principles at the present time," he wrote, "we can compare with sufficient accuracy the motive power of heat to that of a waterfall." The idea was simple enough: consider a waterfall and an appropriate ideal harness-the-rushing-water-machine located somewhere in the cataract, as shown in Fig. 16.19. The maximum amount of energy available that can be converted to work by the machine is ΔPE, which is proportional to the height of the descent ($h_H - h_L$). Thus, the fractional amount of energy available from the mill is ($h_H - h_L$)/h_H. Carnot wrongly believed heat was a liquid, and so with caloric flowing like water, he argued by analogy that the theoretical efficiency of an ideal heat engine should be ($T_H - T_L$)/T_H. Notice how the discussion has built into it a zero-temperature reference level—the denominator is the maximum change in temperature ($T_H - 0$). Heat is naturally propelled down a temperature gradient just as water is naturally propelled down a gravity gradient. Moreover, $Q = cm\Delta T$ and $W = gm\Delta h$. Delightfully, even though there is no such heat fluid, Carnot's answer is correct. Before he died at age thirty-six in the cholera epidemic that swept through Paris in 1832, Carnot (his unpublished papers reveal) had already rejected caloric and even anticipated the First Law.

Figure 16.18 A schematic representation of a heat engine operating between a high- and a low-temperature reservoir. Q_H flows in, Q_L flows out, and the engine does work. Energy-in equals energy-out.

TABLE 16.1	Approximate efficiencies of various devices
Device	**Efficiency (%)**
Electric generator	70–99
Electric motor	50–93
Dry cell battery	90
Domestic gas furnace	70–85
Storage battery	72
Hydrogen-oxygen fuel cell	60
Liquid fuel rocket	47
Steam turbine	35–46
Fossil-fuel power plant	30–40
Nuclear power plant	30–35
Nuclear reactor	39
Aircraft gas turbine engine	36
Solid-state laser	30
Internal combustion gasoline engine	20–30
Gallium arsenide solar cells	>20
Fluorescent lamp	20
Silicon solar cell	12–16
Steam locomotive	8
Incandescent lamp	5
Watt's steam engine	1

The preceding temperature statement of the efficiency of a Carnot engine (e_c) can be established more rigorously, but we would gain little from the effort. Let's instead focus on understanding the concept, represented by

$$e_c = 1 - \frac{T_L}{T_H} = \frac{T_H - T_L}{T_H} \qquad (16.11)$$

Keep in mind that this expression is independent of the engine's working fluid—it could be a hot gas in an internal combustion engine or steam in a steam turbine. Remember, too, that **these are absolute temperatures**.

As we shall see presently, the **Carnot engine has the highest allowable efficiency for a heat engine operating between any two fixed temperatures**. Nonetheless, that efficiency can only equal 1 (that is, 100%) when the cold-sink temperature is 0 K. In other words, *even the Carnot engine operates at less than 100% efficiency*. One way to ensure that the limiting value of Eq. (16.11) is as large as possible is to make T_L as low as possible, but there are practical restraints on that, too. In most real situations, the cold-sink is the open air or cool water from a river, so there isn't very much latitude on the low-temperature end. Of course, you could cool the low-sink with some sort of refrigerator, but that will dissipate additional amounts of energy and negatively affect the overall efficiency of the system. On the other hand, increasing T_H to increase e_c is far more practical. That's why power plants operate with superheated steam and why modern jet engines and auto engines alike are being designed to run at increasingly higher temperatures.

Figure 16.19 A waterfall analogy for the heat engine.

Example 16.7 Determine the maximum possible efficiency for a steam engine operating between 200°C and 27.0°C.

Solution: [Given: $T_H \to 200°C$ and $T_L \to 27.0°C$. Find: e_c.] We need only substitute into Eq. (16.11), being sure to use kelvins. Hence

$$e_c = 1 - \frac{300 \text{ K}}{473 \text{ K}} = 1 - 0.634 = 0.365 = \boxed{36.5\%}$$

In practice, losses will cut that down by about one-third, but this efficiency is the very best we can hope for operating between those two temperatures.

▶ **Quick Check:** $(473 \text{ K} - 300 \text{ K})/(473 \text{ K}) = 37\%$.

A modern fossil-fuel electric-generating power plant uses superheated steam to provide a high-temperature source at perhaps 500°C (≈ 800 K). High-pressure steam violently expands into the turbine, impacting upon and driving the spinning blades (Fig. 16.20). The steam is expelled into a cooled condenser at ≈ 373 K. The maximum theoretical efficiency follows from Eq. (16.11) and is 53%, although thermal losses generally reduce that to about 40% in modern facilities (much of the untransformed energy goes up the smokestack).

The automobile engine functions at approximately 3300 K, exhausting its spent fuel at about 1400 K. That gives it a theoretical efficiency of roughly 58%. Even though the exhaust gas escapes into the surrounding cool air, it leaves the engine, where it can do work, at a much higher temperature T_L. In reality, most of the fuel energy fed into the system is lost by way of thermal energy escaping to the air around the engine and out with the still hot exhaust gas.

Figure 16.20 A modern fossil-fuel electric-generating power plant. Steam drives the turbine, transferring energy to it. The turbine powers a high-voltage electric generator. A large pressure difference is maintained across the turbine by condensing the steam. The steam turns back to water in the condenser (the cold-sink). Water of extreme high purity is recirculated in the sealed system.

16.3 Internal Combustion

Carnot discussed the possibilities of powering an engine by igniting an inflammable gas within a cylinder, but it was J. Lenoir (1859) who produced the first workable internal combustion engine. A traveling salesman and mechanic named Nikolaus Otto happened upon a newspaper account of Lenoir's work (1876) and within a year built and patented the first efficient four-stroke internal combustion engine. The automobile, airplane, and countless other noisy paraphernalia from the power lawnmower to the motorcycle and chainsaw all owe their existence to Otto.

Figure 16.21a shows the sequence of events occurring in each cylinder of the four-cylinder engine depicted in Fig. 16.21b. Each piston is connected by a pivoted arm to the rotating crankshaft. Starting at *A*, a mixture of air and gasoline (blended in a carburetor) at atmospheric pressure is drawn into a cylinder through the intake valve as the piston moves down to *B*. This is the *intake stroke*. The piston now rapidly (and more or less adiabatically) compresses the mix to point *C* with both valves closed in the *compression stroke*. The spark fires at point *C*, igniting the fuel, and raising the temperature and pressure up to *D*, while the volume stays almost constant. The piston then rushes down on what is supposed to be the adiabatic *power stroke,* carrying the system to *E*. The exhaust valve now opens, and the temperature and pressure drop as the volume stays constant and the system moves to *F*, having vented much of the waste gas. Mechanical energy stored in a rotating flywheel pushes the piston back to *A*, forcing the remaining waste products out of the exhaust valve and into the once fresh air. This stage is the *exhaust stroke*. That valve closes, the intake valve opens, and the cycle begins again.

16.4 Refrigeration Machines and Heat Pumps

A heat engine uses thermal energy flowing from a high-temperature reservoir to a low-temperature reservoir to produce work. A *refrigeration machine* does the reverse—it uses work to transfer thermal energy from a low-temperature reservoir to a high-temperature reservoir. When the region to be cooled is a food chest, the device, an ordinary household refrigerator, simply expels the extracted heat into the room. That's fine in the winter, but it's admittedly a little dumb in the summer. On the other hand, to cool the room itself, the refrigeration machine, known as an air conditioner, simply dumps the hot exhaust outside. If we turn the air conditioner around and cool the outdoors, regardless of its temperature, blowing the exhaust energy into the house, we have a heating system: such a reversible device is called a *heat pump*.

Figure 16.22 schematically depicts a continuously operating refrigeration machine, which is simply an engine run in reverse. If it were a Carnot refrigerator, the *PV*-diagram would be identical to Fig. 16.16, but run through counterclockwise. From the First Law it follows that (heat-out) = (heat-in) + (work-in)

or

$$Q_H(\text{out}) = Q_L(\text{in}) + W_i \tag{16.12}$$

A — Intake of fuel and expansion
B — Compression starts
C — Ignition and pressure rise
D — Isovolumic pressure rise completed
E — Power stroke completed
F — Exhaust stroke starts

(a)

Water

Flywheel

Crankshaft

Connecting rod

Intake — Compression — Explosion — Exhaustion

(b)

Figure 16.21 The internal compression engine. (a) The complete sequence of operations. (b) A four-cylinder engine with each cylinder undergoing a different part of the cycle.

The effectiveness of a refrigeration machine cannot be measured by way of the efficiency, Eq. (16.10), because there is no work-out. Hence, we introduce the *coefficient of performance* (η), which is defined as the ratio of the amount of heat removed from the cold reservoir (the heat-in to the system) to the amount of work done on the system, represented symbolically as

$$\eta = \frac{Q_L}{W_i} = \frac{Q_L}{Q_H - Q_L} \qquad (16.13)$$

The larger is η, the more effective is the refrigeration machine, and values of about 5 are typical.

The best possible coefficient will correspond to a Carnot engine running in reverse. Accordingly, set Eq. (16.10) equal to Eq. (16.11), whereupon

$$e_c = 1 - \frac{Q_L}{Q_H} = 1 - \frac{T_L}{T_H}$$

and so

$$\frac{Q_L}{Q_H} = \frac{T_L}{T_H} \qquad (16.14)$$

The *coefficient of performance for an ideal system* in terms of the operating temperatures can now be gotten from Eq. (16.13) using Eq. (16.14); accordingly

$$\eta_c = \frac{T_L}{T_H - T_L} \qquad (16.15)$$

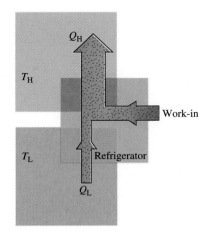

Q_H

T_H

Work-in

T_L — Refrigerator

Q_L

Figure 16.22 A schematic representation of a refrigeration machine. It takes heat Q_L from a low-temperature region and dumps it in a high-temperature region, at the cost of a certain amount of work-in.

Example 16.8 Determine the best possible coefficient of performance that can be gotten for an air conditioner maintaining a room at 21°C (70°F) when the outside temperature is 36°C (97°F). Suppose 5.0 MJ of heat leaks into the room in an hour. The machine exhausts the waste heat it generates in operating, outside the room via a cooling fan. How much work would that air conditioner have to do per hour to maintain the temperature of the room? How much heat in total is exhausted outside per hour?

Solution: [Given: $T_L \to 21°C$, $T_H \to 36°C$, and $Q(\text{in}) = 5.0$ MJ. Find: η_c and W_i.] Remember that $\Delta U = 0$, and so ΔQ equals the work. Using Eq. (16.15), we obtain

$$\eta_c = \frac{T_L}{T_H - T_L} = \frac{294 \text{ K}}{(309 \text{ K}) - (294 \text{ K})} = 19.6$$

The hourly heat leak, $Q(\text{in})$, enters the system at the low temperature and equals Q_L. The work that must be done each hour to remove Q_L joules of thermal energy follows from Eq. (16.13); thus

$$W_i = \frac{Q_L}{\eta_c} = \frac{5.0 \text{ MJ}}{19.6} = \boxed{0.26 \text{ MJ}}$$

To remove 5.0 MJ, this machine does only 0.26 MJ of work. In the process, that 0.26 MJ is converted to thermal energy, which is also exhausted outside. From Eq. (16.12), we have

$$Q_H = Q_L + W_i = 5.0 \text{ MJ} + 0.26 \text{ MJ} = \boxed{5.3 \text{ MJ}}$$

This is the net heat exhausted to the outside.

▶ **Quick Check:** For every 19.6 J it removes, the machine does 1 J of work and produces 1 J of waste thermal energy. When it does 0.26 MJ of work, it removes 5 MJ of thermal energy from the room and produces an additional 0.26 MJ itself.

Many refrigeration machines used in the home operate on what's called the vapor compression cycle (Fig. 16.23). A liquid under high pressure (≈ 7.4 atm) passes through a very small aperture, a needle valve, into a region of lower pressure (≈ 1.4 atm). The restriction drastically slows the speeds of the molecules and what emerges from the valve is a wet low-temperature ($\approx -8°C$) vapor (see Fig. 14.14). A typical *refrigerant* is an easily vaporized liquid, one with a low boiling point at atmospheric pressure, such as ammonia ($-33°C$) or freon ($-30°C$). The partially vaporized refrigerant next enters the evaporator (usually a set of coils in the freezer), where it absorbs heat from the surroundings (the food) as it isothermally and isobarically vaporizes. The low-pressure vapor (at $\approx 16°C$) is then compressed roughly adiabatically via a cylinder-and-piston and emerges as a hot high-pressure vapor (Fig. 16.24). A fan blows away enough thermal energy so that the vapor condenses back to a high-pressure liquid. This is the source of the warm breeze that you may have noticed coming from the bottom or back of a refrigerator. The warm high-pressure liquid freon ($\approx 38°C$) again encounters the needle valve and the process begins again. Amusingly, early electric refrigerators cleverly had their compressors mounted on top of the machine, where the hot air could rise up and away. For aesthetic reasons, the compressor was moved to the bottom, which is precisely where you would not want a constant supply of heat.

16.5 Absolute Temperature

A Carnot engine takes in a certain amount of heat (Q_H) at a high temperature (T_H) and exhausts a lesser amount (Q_L) at a lower temperature (T_L), the energy difference going into work-out. The amount of heat exhausted, and therefore essentially wasted, is directly proportional to T_L: the lower the sink temperature, the less the value of Q_L, the more work is done, and the higher the efficiency. That much has

Figure 16.23 (a) A refrigeration system for either an air conditioner or a home refrigerator. Evaporation of the wet, cold vapor in the coils in the evaporator draws heat from the material to be cooled. (b) The *PV*-diagram shows the irreversible expansion through the needle valve.

been considered already:

$$\frac{Q_L}{Q_H} = \frac{T_L}{T_H} \qquad [16.14]$$

Kelvin proposed that this ratio be used to define an absolute thermodynamic temperature scale (p. 484) that utilized the Carnot engine as a thermometer. This would

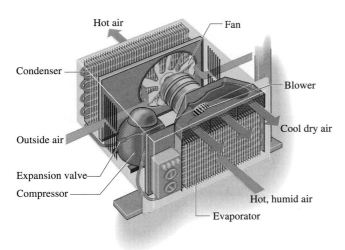

Figure 16.24 A window-mounted air conditioner. Hot, humid room air is drawn over the cold evaporation coils, where it is cooled. Water vapor condenses, runs off, and collects in the base pan, dripping away outside.

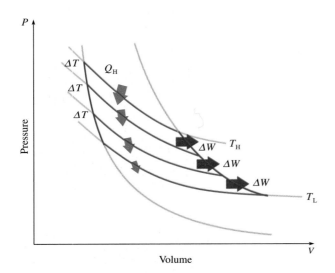

Figure 16.25 A Carnot cycle divided into a number of little Carnot cycles. Each one expels less heat than the one before.

have the virtue of eliminating the dependence on the working substance that plagues all ordinary thermometers. Kelvin assumed the validity of Eq. (16.14) and used it to define the ratio of two thermodynamic temperatures. Inasmuch as Eq. (16.14) is known to be satisfied when an ideal gas is used in a gas thermometer (since we used the ideal gas to derive the equation in the first place), the gas thermometer must agree with the Kelvin thermometer. We can arbitrarily set the difference between the temperatures of the steam and ice points of water as 100 kelvins. In that way, the thermodynamic scale will match up with the long-established Celsius scale.

Suppose, then, that we have a Carnot cycle operating between the temperatures of boiling and freezing water. Now divide the cycle (Fig. 16.25) into 100 equal-area little Carnot subcycles, each doing the same work ΔW and each residing between isotherms 1 K apart. As we progress down from one cycle to the next, more and more net work is being done, and less and less heat exhausted to the next little engine. It follows that if we continue to add more of these little 1-K cycles below the ice-point temperature, sooner or later we will reach a point where the heat-in to one of them all goes into the work-out (ΔW), and no heat is exhausted to the last low-temperature sink, which must be the Absolute Zero of temperature: 0 K. At that point the total amount of heat-in, Q_H, is transformed entirely into work-out, $\Sigma \Delta W$. Presumably there could be no temperature lower than 0 K, since that would imply an efficiency greater than 1, via Eq. (16.11), and more work-out than heat-in, which would violate the First Law.

As we will see later, the notion that all the heat flowing into the last little engine is converted to work violates the Second Law of Thermodynamics, and that suggests that it may be impossible to actually bring bulk matter to a temperature of Absolute Zero.

THE SECOND LAW OF THERMODYNAMICS

The Second Law of Thermodynamics was introduced into physics by way of a rather mundane statement about heat flow. But it evolved into a principle of great scope and significance, as far-reaching as the First Law, and perhaps even more revealing. Things happen via change, and energy is the measure of that change. Everything is tied together by a web of interactions, and change tends to spread out—one entity influencing another—and so energy tends to disperse. Strike a match and the smoke, sound, thermal energy, and light will all disperse. The Second Law is about the dispersal of energy, the unfolding of events. Time itself is given direction by the Second Law.

The germ of the idea was already there when scientists realized from countless observations that:

Heat flows naturally from a region at high temperature to a region at low temperature. By itself, heat will not flow from a cold to a hot body.

And this is one of several equivalent but different statements of the Second Law.

Once the concept of internal energy was formulated, this observation was enriched with the insight that the spontaneous flow of heat from one object to another is independent of the amount of internal energy in either entity. Heat is propelled by a temperature gradient—no gradient, no heat. From Carnot's work, we see that a cyclic heat engine must operate between high- and low-temperature reservoirs. If both temperatures are the same, its efficiency is zero and no heat enters the engine, across which there must be a gradient.

Furthermore, with any low-sink temperature other than zero, even for an ideal engine, there will be some unavoidable loss of heat Q_L to the reservoir: the efficiency must be less than 100%. That much follows from Eq. (16.14), if $T_L \neq 0$, $Q_L \neq 0$. There is an energy equivalence between heat and work, but there is also a *dissymmetry*, a difference on the microscopic level, that has its roots in the difference between order and disorder. We can convert all the ordered energy we wish *entirely* into thermal energy—the brakes on your car do it and Rumford's cannon borer (p. 516) did it. Nature allows the continued spontaneous transformation of work entirely into thermal energy, but it does not allow the reverse. The transformation of disordered heat into ordered work must be paid for, as it were, by an additional loss of heat Q_L (Fig. 16.17). The formal wording of the Second Law in terms of the above ideas is known as the *Kelvin statement:*

No process is possible for which the sole result is the removal of heat from a source and its complete transformation into work.

In other words, *it is quite impossible to produce a cyclic engine that will generate work by extracting heat from a reservoir without expelling some waste heat to a lower-temperature sink* (Fig. 16.26). An anti-Kelvin device, which could continuously do work extracting thermal energy from a single reservoir with no concern about temperature gradients, is a so-called **perpetual motion machine of the second kind**. Although such a device would not violate the First Law, it would violate the Second. It could not create energy or, for that matter, run forever, but it could do the next best thing—namely, draw off energy at no cost from the vast low-temperature sources of atmosphere and ocean. No such machine capable of solely producing ordered work out of disordered heat is possible.

It is not impossible to convert thermal energy directly into work. What is impossible is to do it with no other change occurring: A pistol can transfer the thermal energy of the expanding gases into the KE of the bullet, but it will also blast out those gases into the air. The work done on the bullet will not be the sole result. Understand that that "sole result" stipulation means that the system is otherwise unchanged; the system is back in the state it was in originally, and the process can be repeated and run continuously.

16.6 The Clausius Formulation

The next major figure to appear on the thermodynamic scene was a young man in Germany, a member of the Gottlieb family. Somewhere along the way that surname received a classical rebirth, and his father was called Gottlieb Clausius. Rudolf Julius Emmanuel Clausius was a contemporary of Joule and Kelvin and, like them, was caught up in the ferment that surrounded the creation of Thermodynamics in the nineteenth century. Inspired by Carnot, Clausius put aside caloric, speculated about the relationship of heat and motion in the microworld, and recognized that

Figure 16.26 A schematic representation of an impossible anti-Kelvin device. It converts heat-in to work with no wasted heat-out.

Rudolf Clausius (1822–1888).

Figure 16.27   If an anti-Kelvin engine could exist, it could be attached to an ordinary engine, and the combination would violate the Clausius statement of the Second Law. It could take in Q_L and dump it as Q_H and do nothing else.

Figure 16.28   If an anti-Clausius engine could exist, it could be attached to an ordinary engine, and the combination could violate the Kelvin statement of the Second Law. It could remove heat Q_H from a source, converting it entirely into work.

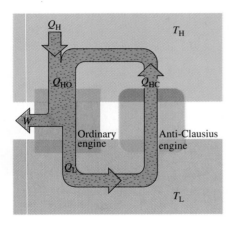

there were actually *two* energy principles that spoke, in turn, to the quantity and quality of energy.

It's commonplace for hot objects to cool spontaneously, all by themselves, losing heat to a colder environment, but we presumably never observe cold objects spontaneously getting colder by losing heat to a warmer environment. And that's true even when heat flows from a warm body with a tiny amount of internal energy (a cup of tea) into a cold body with an immense amount of internal energy (the atmosphere). In the same way, an object can start at the top of a hill and spontaneously roll down it, but never does one begin at the bottom and roll uphill unaided by any external agent. There is a specific way in which all processes (not just those involving heat) naturally unfold, and the Second Law, at its heart, is about just that: the natural direction in which energy flows as it irreversibly redistributes itself.

Clausius proposed an alternative statement of the Second Law that at first glance seems quite different from Kelvin's:

> **No process is possible in which the sole result is the transfer of a given amount of thermal energy from a body at low temperature to a body at a higher temperature.**

Refrigerators are not forbidden, provided they produce some other result in addition to the transfer of thermal energy—like the consumption of electrical power. The above statement may imply that an input of work is necessary to transport heat "uphill," but it doesn't say so explicitly. The Kelvin statement talks about the conversion of heat to work. The Clausius statement talks about the natural direction of the flow of heat. Both suggest a hidden asymmetry in the manner in which processes spontaneously unfold.

To prove that these two perspectives are in fact equivalent, refer to Fig. 16.27, which depicts an ordinary reversible engine being driven in reverse by a hypothetical anti-Kelvin engine (one that violates Kelvin's statement). The latter converts an amount of heat (Q_{HK}) directly into work, which drives the ordinary engine as a refrigerator. It, in turn, exhausts heat (Q_{HO}), some of which can be imagined circulating back to the anti-Kelvin device. The net effect is that the combined system has the sole result of taking heat from the low-temperature reservoir and dumping it into the high-temperature reservoir in direct conflict with the Clausius statement. Thus, if an anti-Kelvin device were possible, it could be used to make an anti-Clausius device. Similarly, the ordinary engine on the left in Fig. 16.28 is effectively being supplied heat by a hypothetical anti-Clausius engine that carries heat from the low to the high reservoir. The ordinary engine draws off heat (Q_{HO}), expels heat (Q_L), and performs work. The net effect of the combined system is solely to convert an amount of heat Q_H into work in direct conflict with the Kelvin statement. Thus, if an anti-Clausius engine were possible, it could be used to make an anti-Kelvin engine. We must conclude that if either statement is false, the other is false—they might be wrong, but they must be equivalent.

16.7 Entropy

In 1865 Clausius introduced a new concept, *entropy*, in order to distinguish between the ideas of *conservation* and *reversibility*. When a match burns, energy is conserved, but that's not the whole story of the change. The match will never spontaneously reform itself from the ash and smoke—the process is irreversible and the energy naturally disperses. This is very unlike the idealized occurrences of (lossless) Newtonian dynamics that are reversible, and conservation tells us all we need to know. Dynamics is independent of the direction of time: a lossless bouncing ball bounces forever and is independent of whether time runs forward or backward—the ball falls, climbs, and falls again or climbs, falls, and climbs again, either way. By contrast, heat conduction and friction are irreversible changes; indeed, without irreversible processes, life would be impossible.

When an engine operates, converting heat-in into work-out, some fraction of the input energy is irreversibly lost as heat-out. In effect, the engine takes in heat and puts out two distinct channels of energy: highly useful ordered energy (work) and not so useful disordered energy (heat). The expelled waste thermal energy is at a low temperature and will thereafter be quite difficult to extract and use again. To be sure, the more efficient the engine, the lower the sink temperature is likely to be, and the harder it will be to get at that exhausted thermal energy in the future. It's like water pouring over the falls in Fig. 16.19; once it has descended down to sea level, it is a lot less useful. In a real sense, the *quality of thermal energy (that is, its ability to do work on a macroscopic scale*) is dependent on its temperature.* When an engine operates, it does not use up energy—energy is conserved—it transforms energy, and it does that at the cost of degrading a certain amount of high-quality, high-temperature energy into low-quality, low-temperature energy. Energy does not vanish, it is dissipated. The family car is an example of how highly ordered available chemical energy (in the fuel) is converted into highly disordered, low-temperature unavailable thermal energy. Drive for 20 miles and you will convert a gallon of gasoline into a ribbon of heated air. Then try to drive back by reclaiming those 1.3×10^8 (now essentially useless) joules that are still there warming the sky.

Suppose there is an ideal gas in a chamber and it's released through a valve into an insulated evacuated vessel. The gas spontaneously expands, filling the vessel, but there is no change in internal energy because no work is done. The pressure and volume both change; the temperature remains constant. The system in its initial state transforms, all by itself, to a new final state where the energy is the same. And yet, if we want to return the gas to the original chamber, we will have to do work on it—the gas will not spontaneously squeeze back through the valve. Once the valve is opened, the gas is no longer in equilibrium, and the spontaneous rush toward equilibrium is irreversible: there is no infinitesimal change in any state variable (P, V, T) that will reverse the process. The system is directed by a hidden imperative (other than maintaining the total energy) to seek a new dispersed state from which it will not return on its own.

In the process of spontaneously expanding, the gas loses some of its ability to do work, though its energy is unchanged. The amount of work it takes to recompress the gas would seem to be equivalent to the ability to do work it lost. In any event, the final state of the gas is profoundly different from the initial state. What we want now is a new property of the system that will reflect that difference; a property that will allow us to know if spontaneous change will take place between any two states, a property that governs the direction of natural change in all systems from galaxies to embryos. Clausius proposed that there was such a measure, and he gave it the symbol S and the name **entropy**.

---

*The familiar definition of energy as "the ability to do work" (and that presumably means work on a macroscopic scale, an organized phenomenon) is inconsistent with the idea that energy is both conserved (First Law) and degraded (Second Law).

Experience suggests that the entropy of a system be defined such that it increases as the system degrades. In other words, natural changes (that is, spontaneous ones) carry an isolated system toward disorder and are accompanied by an increase in entropy. Unnatural changes, which must be forced on a nonisolated system by an outside agency, can bring it more order and decrease *its* entropy. **Reversible changes are entropy neutral**, they leave S for an isolated system unaltered. Since entropy has something to do with degradation and disorder, it's not surprising that a reversible process carried out by an isolated system (which must be able to return to exactly the state from which it started) should have $\Delta S = 0$. The entropy equivalent of the Second Law, which is often called the *Entropy Principle,* is as follows:

> **When an isolated system undergoes a change, passing from one state to another, it will do so in such a way that its entropy will increase, or at best remain the same.**

This hidden imperative gives direction to the natural unfolding of events. Remember that entropy describes the state of a system. It is a property, like internal energy and temperature, of engines and ice cubes and cats.

Whenever a process occurs in an isolated system, its entropy increases, or at best ideally remains the same. If the system is not isolated, then its entropy can be made to decrease by impressing order upon it from outside via external agencies. In that case (according to the Entropy Principle), the entropy of the surroundings will have to increase even more, so that the entropy of the ultimate isolated system, the entire Universe, nonetheless increases.

> **Every process increases (or at best ideally leaves unchanged) the entropy of the Universe.**

The essential measure of the difference between reversible and irreversible processes is the change in entropy of the Universe. The essential measure of the difference between what occurs naturally and what does not occur naturally is the change in entropy of the Universe.

The Second Law and Entropy

We can get some insight into the behavior of entropy by using the Second Law as a guide. The Kelvin statement of the Second Law is subsumed into the entropy version provided that *the entropy of a system increases when it receives heat, decreases when it loses heat, and remains unchanged by work done in the absence of friction.* In that context, work is an organized mechanical change in energy and, as such, does not affect the microscopic disorder of a system. (By contrast, any mechanism—such as friction—that disperses energy into the agitation of the system's particles will increase entropy.) Consider an isolated system composed of a reservoir of thermal energy, an anti-Kelvin engine, and its immediate surroundings. Running the engine off the reservoir will have only one effect; the entropy of the reservoir will decrease, and therefore the engine is forbidden.

Similarly, the Clausius statement will be subsumed if we assume that *the higher the temperature of the heat transferred, the less entropy change results.* Imagine an isolated system consisting of two objects, one hot, the other cold, each with the same internal energy. Bring them into contact—heat spontaneously flows out of the hot one into the cold one and they both reach a new equilibrium temperature. The same amount of heat is lost by one as gained by the other. But the high-temperature heat is "high quality" and only slightly decreases the entropy of the high-temperature object. Whereas the heat delivered at a low temperature is of "low quality" and increases the entropy of that body appreciably—the net effect of this spontaneous transfer of heat is properly an increase in the entropy of the system and, hence, the Universe.

Thus, we know that entropy is directly proportional to the heat supplied and inversely proportional to the temperature at which it's supplied. Moreover, for a Carnot cycle, it follows from Eq. (16.14) that

$$\frac{Q_L}{T_L} - \frac{Q_H}{T_H} = 0 \qquad (16.16)$$

We interpret this expression to represent the zero net change in entropy experienced by a Carnot engine over its reversible cycle. Entropy "flows" in from the high-temperature source, "flows" out to the low-temperature sink, and the engine returns to its original state. Accordingly, Clausius defined the change in the entropy of a system (ΔS) experienced during a *reversible* process as the ratio of the heat-in to the absolute temperature at which it enters:

[*reversible process*] $$\Delta S = \frac{Q}{T} \qquad (16.17)$$

When heat enters a system, its entropy increases; when heat leaves, its entropy decreases. Observe that the entropy of a system will increase by the same 1 unit when 1000 J enters at 1000 K or, alternatively, when 0.001 J enters at 0.001 K. To use Eq. (16.17) for finite changes (ΔS), the temperature has to be constant.

Example 16.9 By how much does its entropy change when 10 kg of ice at 0°C is completely melted into water at 0°C?

Solution: [Given: $m = 10$ kg, ice melting at 0°C. Find: ΔS.] Here T is absolute; namely, $T = 273$ K. To find the entropy for this isothermal process, we use Eq. (16.17), first determining Q for the melting. From Eq. (15.4),

$Q = mL_f = (10 \text{ kg})(334 \text{ kJ/kg}) = 3.34 \text{ MJ}$. Thus

$$\Delta S = \frac{3.34 \text{ MJ}}{273 \text{ K}} = \boxed{12 \text{ kJ/K}}$$

▶ **Quick Check:** $Q = \Delta S T = (12 \text{ kJ/K})(273 \text{ K}) = 3.3$ MJ. We can expect an increase in entropy corresponding to the increase in molecular disorder as the crystalline solid gives way to the liquid.

Suppose the isolated system is a Carnot engine with the attendant source and sink (Fig. 16.16). In isothermally expanding from A to B, the source loses entropy, $\Delta S_H = (Q_H/T_H)$, and the engine gains the same amount—the net change is zero. The expansion to C is adiabatic ($Q = 0$) and therefore **isentropic** ($\Delta S = 0$). Over the isothermal compression to D, the engine loses entropy, $\Delta S_L = (Q_L/T_L)$, and the sink gains the same amount. The final adiabatic compression back to A is isentropic ($\Delta S = 0$), and it closes the cycle. Since $\Delta S_H = \Delta S_L$, from Eq. (16.16), the net change in entropy of the entire system during this reversible cycle is indeed zero. The Carnot engine is just efficient enough to exhaust the smallest allowable amount of waste heat. Had it been still more efficient, expelling less low-temperature heat, the entropy change would have been negative and that's forbidden by the Second Law. All real engines will exhaust more heat and produce a net increase in entropy.

Of course, entropy increases in irreversible processes even though we don't have a formula that spells out exactly how. Still, *we can get around that and use*

Eq. (16.17) to calculate the entropy change using a hypothetical reversible process between the same initial and final states. Since the entropy change only depends on these two states, ΔS for the reversible process will be the same as for the irreversible process. For example, imagine that an ideal gas undergoes a free expansion to a new state. Now suppose that the gas is returned to its initial state via a quasi-static iso-thermal compression, which decreases its entropy by an amount ΔS, given by Eq. (16.17). Since the system is back where it started, ΔS must be the increase in entropy it experienced during the irreversible expansion.

Example 16.10   A hot object at 573 K is put in contact with a cool object at 273 K, and 20.0 kJ of heat flows irreversibly from one to the other. By how much is the entropy of the Universe changed?

Solution: [Given: $T_H = 573$ K, $T_L = 273$ K, and $Q = 20.0$ kJ. Find: ΔS.] The entropy change of the Universe equals the net entropy change of the high- and low-temperature objects; thus

$$\Delta S = \Delta S_L + \Delta S_H = \left(\frac{Q}{T_L}\right) + \left(\frac{Q}{T_H}\right)$$

Remembering that heat-in is positive

$$\Delta S = \frac{+20.0 \times 10^3 \text{ J}}{273 \text{ K}} + \frac{-20.0 \times 10^3 \text{ J}}{573 \text{ K}}$$

$$\Delta S = 73.26 \text{ J/K} - 34.90 \text{ J/K}$$

and to three figures $\boxed{\Delta S = +38.4 \text{ J/K}}$.

▶ **Quick Check:** $\Delta S_H < \Delta S_L$ is proper since high-temperature heat has less entropy than low-temperature heat. Moreover, $T_H/T_L = \Delta S_L/\Delta S_H = (573 \text{ K})/(273 \text{ K}) \approx 2.1 \approx (73.3 \text{ J/K})/(34.9 \text{ J/K})$.

16.8 Order and Disorder

It was not until 1878 that Boltzmann reformulated entropy, giving it a new fundamental clarity. He recognized that entropy was a measure of the disorder of the Universe on an atomic level. Once lit, the neatly arranged molecular pattern of a match spontaneously transforms into the mayhem of gas, smoke, and flame. The disorder of the Universe increases, entropy increases; energy once concentrated is soon spread out with an evenness that is the essence of thermodynamic disorder. *Maximum disorder corresponds to a random tumult, an undifferentiated uniformity.*

We tend, at first, to think of disorder as a scene with everything tossed about, a shoe here, a sock there. But that's only crude disorder. Splendid disorder requires a fine, dispersed dust—no shoes, no socks, but a uniform cloud of agitated atoms. A bull in a china shop will spontaneously powder the pretty cups, crush the delicate vases, and increase the uncompromising entropy of the Universe. The disorder of a distribution of particles is reflected by both their separations and their speeds. If all the particles in a perfect crystal were at rest, that would correspond to its state of maximum order, only approached at 0 K.

We never measure entropy, but only entropy change, ΔS, which accompanies physical change. And a system naturally changes (energy naturally flows) so as to increase universal entropy ($\Delta S \geq 0$). **The quantity ΔS measures the degree of disorder associated with any change (ΔE) in a system**. *An isolated system in a maximum entropy configuration cannot of itself change macroscopically; it is*

OZYMANDIAS

I met a traveller from an antique land
Who said: "Two vast and trunkless legs of stone
Stand in the desert. Near them, on the sand,
Half sunk, a shattered visage lies, whose frown
And wrinkled lip, and sneer of cold command,
Tell that its sculptor well those passions read
Which yet survive, stampt on these lifeless things,
The hand that mockt them and the heart that fed;
And on the pedestal these words appear:
'My name is Ozymandias, king of kings:
Look on my works, ye Mighty, and despair!'
Nothing besides remains. Round the decay
Of that colossal wreck, boundless and bare
The lone and level sands stretch far away."
                                        P. B. SHELLEY

in equilibrium. Every physical and chemical process in nature occurs in such a way that the net entropy of all the matter involved increases—the disorder of the Universe inexorably moves toward a maximum.

Mixtures of any kind are more disordered than were their separate constituents. Two gases introduced into a chamber will "always" mix, even though there is no energy incentive to do so. Colored grains of sand shaken in a bag will "always" mix. Yet neither the gases nor the sands are likely to spontaneously unmix, to neatly separate, to decrease in entropy. For that matter, you can shake the bag of sand for centuries, and though it is possible, it is most improbable that the different colored specks will ever separate perfectly again. True, you could reach into the system (then no longer isolated) and sort out the sand—you could reestablish order and decrease entropy. But you would spend more order in the effort; you would disorder your breakfast more than you would order the sand—the Universe would lose again. It must lose again.

Wherever entropy is decreasing—in a factory where someone is stringing beads, among the ice cubes in a freezer, or with the cells forming a living embryo—order is rising out of disorder. But whenever that occurs locally, there will always be a hidden harvest of mayhem that overbalances the scales on the side of universal disorder. The freezer will make its ice cubes, but the system isn't isolated. The cost will be in electrical energy converted to heat and in the end, $\Delta S > 0$.

The Arrow of Time

Of all the laws we have studied, only the Second gives a direction to the unfolding of events, to the progression of processes. It points the arrow of time in the "forward" direction—toward maximum disorder. Time stops when all events stop, when nothing happens, and the Second Law provides the direction to processes and to time. Have you ever seen the gases come back, the smoke descend, the light return, the heat pour in, the flame collapse, and a match reform, unburnt and whole? Will the gray hairs spontaneously darken, the skin grow tight and supple, the sags unsag? Not likely! And yet energy would be conserved. If everything reversed and processes ran toward order, would time be going backward? All by itself the game runs downhill, and it is the Second Law that shows the way.

16.9 Entropy as Probability

Boltzmann found the mathematical fingerprint of disorder in a statistical relationship describing the likelihood that a particular molecular arrangement would occur. Were we to take all the separate molecules that make up a match and put them in a box and shake them, the chance of finding them all bound together, neatly arranged as a match, is incredibly small. It is faintly possible but, given the astounding number of possible combinations that do not lead to a match, exceedingly unlikely. Unwrap a deck of playing cards and observe the arrangement. That ordered state is just one out of 80 million, million, million, million, million, million, million, million, million, million, million possible arrangements of those 52 cards. Order is far less likely than disorder. Once the molecules in an egg are scrambled, the chance that all of them will simultaneously find their way back to so unlikely an arrangement as the original ordered ovum is slim indeed. Every process irreversibly scrambles a bit of the cosmic egg.

A system of continually agitated molecules moves from one state to the next because the new configuration is more disordered and therefore statistically more

TABLE 16.2   Possible arrangements of four coins

Number of heads	Equivalent microstates	Number of microstates
0	TTTT	1
1	HTTT, THTT, TTHT, TTTH	4
2	HHTT, THHT, TTHH, HTTH, THTH, HTHT	6
3	HHHT, THHH, HTHH, HHTH	4
4	HHHH	1
	Total number of microstates	16

probable. Throw four coins on the table and it is most likely that two will be heads and two tails. There are a total of 2^4 or 16 different ways the four coins can land (16 so-called *microstates*), as shown in Table 16.2. Of these, 6 equivalent microstates correspond to 2-heads-2-tails (that is, 1/2-heads): That's 6/16, or 38%, which is the most likely state to occur. By comparison, there is only one way (1 microstate) where all the coins will be heads, and the probability of that highly ordered state happening is only 1/16, or 6%. Clearly, the probability of finding the system in some particular state, say 2-heads-2-tails, is proportional to the number of possible equivalent microstates ($\mathcal{W}$) composing that state—in this case, 6.

A crystal is a regular structure in which, especially at low temperatures, there is very little variation in the arrangement of its atoms. Each such configuration is a microstate of the system, and there are relatively few of them possible. The crystal is highly ordered and possesses little entropy. Still, as its internal energy increases, the number of ways that energy can be distributed among the constituent particles increases, which increases the number of available microstates and, hence, the entropy. In contrast, a mass of gas has a tremendous number of possible equivalent microstates that will produce the same macrostate with the same values of temperature, volume, and pressure. The gas is highly disordered and possesses a large amount of entropy.

If there were 100 coins in our system, there would still be only 1 microstate corresponding to all-heads, but now there would be a total of 2^{100} or $\approx 10^{30}$ microstates. That makes the chance of throwing all-heads about 1 out of 10^{30}. Since there are roughly 32×10^6 seconds per year, you can throw down the coins once every second for 3×10^{22} years, and you are not likely to hit all-heads more than once,

Entropy.

though it certainly could happen on the first shot. By comparison, there are about 10^{29} ways to get 1/2-heads and that's by far the most likely state, occurring 10% of the time. About 90% of all throws will result in between 45 and 55 heads.

Sequential probabilities build by multiplication. The chance of getting a head with 1 coin is 1/2 per throw (50%), while the chance of following that with another head is $1/2 \times 1/2$ and so on. Thus, the probability of throwing 4 heads in a row is $1/2 \times 1/2 \times 1/2 \times 1/2 = 1/16$, just as it was when we threw them all at once, which is reasonable since they don't talk to each other as they fall. By contrast, sequential processes would progressively *add* entropy changes. A system usually increases in entropy every time something happens to it.

The more microstates a state has, the more likely it is that that state will be observed, and the more disordered it must be. Hence, we can guess that $S \propto \mathcal{W}$, but not directly, because one is additive and the other multiplicative. This suggests using a function like the logarithm to overcome the difference because $\ln ab = \ln a + \ln b$; taking the logarithm turns products into sums.

Boltzmann (1877) proposed that the entropy could be explicitly connected to the notion of disorder by defining it in a new way, which was nonetheless consistent with Clausius's definition:

$$S = k_B \ln \mathcal{W} \qquad (16.18)$$

where the k_B is Boltzmann's Constant (p. 497). Keep in mind that the logarithm (Fig. 16.11) rises rapidly at low values and levels off at high values. That would reflect the greater disordering effect of heat-in at lower temperatures. Even though fewer microstates are available at low temperatures, S is more sensitive to changes in internal energy at small T.

The probabilistic version of the Second Law does not forbid a state from spontaneously moving to a lower entropy; rather, it maintains that such a case would be most unlikely. If this statistical formulation is true to life, and it certainly seems to be, anything is possible—any process that can occur via the First Law may occur, if only remotely, via the Second. It is possible for the egg to unscramble, but it's unthinkably improbable.

To live at all is to churn through spontaneous processes, to be out of equilibrium, to pour forth entropy (while usually exhausting just about as much energy as we take in). Life feeds on order, and even as it builds, it disperses.

Core Material

Thermodynamics (or thermo, as it's affectionately called) is the study of thermal energy—its relationship to heat, mechanical energy, and work, and the conversion of one to the other.

The **First Law of Thermodynamics** is

$$Q = W + \Delta U \qquad [16.1]$$

Hence

$$\Delta U = Q - W \qquad [16.2]$$

An equivalent statement of the First Law is

The internal energy of an isolated system is constant (even though it may be transformed from one type of energy to another).

A system is in **equilibrium** when there is no tendency for it to undergo spontaneous change (on a macroscopic level). If a system is always infinitesimally close to thermodynamic equilibrium, even as it passes from one state to the next, the process that is occurring is said to be **reversible** (p. 559).

The work done in an isobaric process is

$$W = P\Delta V \qquad [16.3]$$

The work done on or by an ideal gas during an isothermal volume change from V_i to V_f is

$$W = nRT \ln \frac{V_f}{V_i} \qquad [16.5]$$

For a reversible adiabatic process

$$PV^\gamma = \text{constant} \qquad [16.6]$$

where γ is a constant that equals 1.67 for monatomic gases, ≈ 1.4 for diatomic gases, and ≈ 1.3 for polyatomic gases.

A *heat engine* is a cyclic device that converts thermal energy into work-out. A *Carnot engine* is a reversible device operating in a cycle composed of two adiabats and two isothermals. The efficiency of an engine is

$$e = \frac{W_o}{Q_i} = \frac{Q_H - Q_L}{Q_H} = 1 - \frac{Q_L}{Q_H} \qquad [16.10]$$

where both Q_H and Q_L are positive numbers. The efficiency of a Carnot engine is

$$e_c = 1 - \frac{T_L}{T_H} = \frac{T_H - T_L}{T_H} \qquad [16.11]$$

A *refrigeration machine* uses work to transfer thermal energy from a low-temperature reservoir to a high-temperature reservoir.

The *coefficient of performance* (η) is defined as the ratio of the amount of heat removed from the cold reservoir (the heat-in to the system) to the amount of work done on the system; consequently

$$\eta = \frac{Q_L}{W_i} = \frac{Q_L}{Q_H - Q_L} \qquad [16.13]$$

For a **Carnot engine**

$$\frac{Q_L}{Q_H} = \frac{T_L}{T_H} \qquad [16.14]$$

The coefficient of performance for an ideal system is

$$\eta_c = \frac{T_L}{T_H - T_L} \qquad [16.15]$$

The **Kelvin statement** of the Second Law is

No process is possible for which the *sole result* is the removal of heat from a source and its complete transformation into work.

The **Clausius statement** of the Second Law is

No process is possible in which the *sole result* is the transfer of thermal energy from a body at low temperature to a body at a higher temperature.

A Carnot engine has the same efficiency as any other reversible engine operating between the same two temperatures. No engine can exceed the efficiency of a reversible engine (p. 565).

Clausius proposed that the measure of the disorder of a system is the **entropy** S. The Second Law in terms of entropy is

When an isolated system undergoes a change, passing from one state to another, it will do so in such a way that its entropy will increase, or at best remain the same.

Clausius defined the change in entropy of a system (ΔS) experienced during a *reversible* process as

$$\Delta S = \frac{Q}{T} \qquad [16.17]$$

A system with maximum entropy is in equilibrium. Boltzmann proposed that entropy could be explicitly connected to the notion of disorder by defining it as

$$S = k_B \ln \mathcal{W} \qquad [16.18]$$

where $\mathcal{W}$ is the number of microstates.

Suggestions on Problem Solving

1. Keep in mind that all the calculations involving temperature in this chapter use kelvins. Look out for data given in °F; convert to degrees Celsius and then to kelvins early in the analysis.

2. Whether Q_H and Q_L are flowing into or out of the device in a particular problem can be confused, so it's a good idea to draw an energy flow diagram, like Fig. 16.18. This is especially true when dealing with refrigerators.

3. When an engine is reversible, the values of Q_L, Q_H, and W must remain the same in either mode of operation. Of course, when it's run as an engine, Q_H is the heat-in whereas, as a refrigerator, Q_H is the heat-out, but numerically the values stay the same. In other words, the energy channels in schematic diagrams like Fig. 16.28 don't change width on reversing a reversible engine; they do change if the engine is irreversible.

4. Adiabatic processes utilizing gases require values for γ, and most often air is involved, for which $\gamma = 1.4$. If not explicitly stated, you can assume that a gas behaves adiabatically if the process occurs rapidly (p. 559).

5. The heat leaving an object is negative, and so the change in entropy it experiences is also negative. Keep careful track of the signs throughout entropy calculations. If the temperature drops $\Delta T < 0$, $Q = cm\,\Delta T < 0$ and $\Delta S = Q/T$.

6. It often happens that a quantity you might think would be needed explicitly to solve a problem cancels out of the equations. That sort of thing happens especially with before-and-after volumes and pressures of gases. Moreover, there are situations involving the gas laws where it's useful to work on a per-unit-volume basis. Again, because the equations for gases often have before-and-after pressures and volumes (that is, P and V on both sides), you can use non-SI units such as atm or cm^3. That will save the time to convert, but it is risky if you are not careful. It's better to play it safe and convert everything to SI right away. To illustrate the point, a few of the solutions given in the back of the book use mixed units.

Discussion Questions

1. Figure Q1 is the *PV*-diagram for a diesel engine. It uses the internal combustion of a gaseous fuel in a cylinder-and-piston arrangement. Here, the fuel is sprayed into the cylinder at C. Explain each part of the cycle. Why do you think only air is compressed from B to C with the fuel injected only at C? Is work done? When does heat enter and leave the system? What kind of processes are occurring? At what point does the cycle end?

Where in the cycle does the exhaust valve open to eject spent fuel? Use the gasoline engine as a guide since it's fairly similar to the diesel.

2. The gas in the combustion chamber of a rocket is in violent random motion at high temperature. It leaves the engine through a nozzle at high speeds. Describe what's happening in terms of energy. Considering that the exhaust emerges in a well-

Figure Q1

Figure Q3

defined direction, what is the temperature of the exhaust relative to that of the combustion chamber?

3. Figure Q3 depicts a reciprocating steam engine. Explain how it works.

4. The so-called *Third Law of Thermodynamics,* which to some physicists does not seem as secure as the other two, states that:

> It is impossible to bring a body to a temperature of Absolute Zero in a finite number of steps.

In other words, it is widely believed that Absolute Zero is unattainable. Discuss what this means and determine whether or not it fits in with the Second Law.

5. Discuss the thermodynamics of a possible mechanism for the formation of clouds as a result of rising columns of moist, warm air coming from near the ground.

6. You are offered the opportunity to "buy in on the ground floor" of a company selling a new machine. It is a container designed to keep drinks at a fixed temperature indefinitely. One loads it with hot liquid; the heat that inadvertently leaks off is cleverly converted into work that is fed to a stirrer that continually churns the liquid putting the thermal energy back and maintaining the temperature. Do you buy? Explain. Discuss the process from an entropy perspective.

7. What is the ultimate source of the energy stored in coal, wood, and oil? Explain. Does this arrangement violate the entropy version of the Second Law?

8. Suppose someone is placed in a totally isolated room with enough food for one day. What will inevitably happen to him? Explain your answer in terms of both the First and Second Laws.

9. What is the ultimate source of the energy extracted via a hydroelectric power plant? Explain.

10. A red-hot coal is hung on an insulating fiber inside a room-temperature metal box that is sealed, evacuated, and completely isolated from the rest of the environment with massive amounts of insulation. Discuss the situation from an entropy perspective. What happens to the coal and the box?

11. Imagine a thermally insulated cylinder-and-piston containing an ideal gas at a certain initial volume *V*. Suppose the

gas, pushing on the piston, is allowed to expand to a volume 2*V*. Discuss, in thermodynamic terms, every aspect of the process. What kind of process is it? What happens to the internal energy of the gas? What happens to *P*, *V*, and *T*? Is work done? By what, on what? Does the entropy change?

12. Suppose you fly in a jet plane from New York to London. What is the net effect on the energy and entropy of the planet?

13. Suppose an automobile engine burns its hydrocarbon fuel at about 2400 K while the outside air temperature is 300 K. Explain why the actual efficiency is far less than the corresponding Carnot efficiency between those two temperatures.

14. What would you say was the overall effect of the rise of technology on the entropy of the Universe?

15. A large room may have some 10^{28} air molecules filling it. Is it possible for these molecules to suddenly all end up in one corner of the room, leaving us gasping for breath? Do we have to worry about that happening and if not, why not? How would matters change if there were only 10 molecules in the room?

Multiple Choice Questions

1. A cylinder-and-piston contains an ideal gas whose volume is to be halved. The work done on the gas will be

(a) greater if the compression is isothermal rather than adiabatic   (b) greater if the compression is adiabatic rather than isothermal   (c) the same if it's either adiabatic or isothermal   (d) equal to the change in internal energy in all cases   (e) none of these.

2. A cylinder-and-piston contains a gas that is made to expand quasi-statically from A to B at a constant temperature as shown in Fig. MC2. The process is (a) adiabatic and heat leaves the gas   (b) isothermal and heat enters the gas   (c) adiabatic and heat enters the gas (d) isothermal and heat leaves the gas   (e) none of these.

Figure MC2

3. Referring to the previous question and Fig. MC2, when the gas is carried from B to C, the process is   (a) isobaric and no heat is transferred   (b) isovolumic and no heat is transferred   (c) adiabatic and no heat is transferred (d) isovolumic and heat leaves the gas   (e) none of these.

4. Referring to the previous question and Fig. MC2, when the gas is carried from C to D, the process is (a) isothermal and no heat is transferred   (b) isovolumic and no heat is transferred   (c) adiabatic and heat enters the gas   (d) isovolumic and heat leaves the gas   (e) none of these.

5. Referring to the previous question and Fig. MC2, when the gas is carried from D to A, the process is   (a) isothermal and no heat is transferred   (b) isovolumic and no heat is transferred   (c) adiabatic and heat enters the gas (d) isovolumic and heat enters the gas   (e) none of these.

6. When Joule wrote "the grand agents of nature are, by the Creator's fiat, indestructible," (1843) he was talking about (a) the F.B.I.   (b) the Second Law   (c) an automobile (d) the First Law   (e) none of these.

7. An ideal gas is contained in a cylinder-and-piston that is 0.5 m long, and 50 J of heat is added while the piston is held in place. The work done by the gas is   (a) 50 J (b) 0   (c) 25 J   (d) 250 J   (e) none of these.

8. If a system does work adiabatically,   (a) the temperature stays the same   (b) the internal energy stays the same (c) the internal energy decreases   (d) the internal energy increases   (e) none of these.

9. Fig. MC2 shows the PV-diagram for a gas. In reference to the work done by the gas in going from state A to state C   (a) the route A-C corresponds to a minimum (b) the route A-D-C is equivalent to route A-B-C   (c) all

the routes are equivalent   (d) route A-D-C corresponds to a maximum   (e) none of these.

10. When this book is given a push across a desktop, it comes to rest losing its initial KE in a(n)   (a) reversible process (b) irreversible process   (c) process that is isothermal (d) process that is isentropic   (e) none of these.

11. A gas is compressed adiabatically by a force of 500 N acting through a distance of 5.0 cm. The net change in its internal energy is   (a) +500 J   (b) +25 J   (c) −500 J (d) −25 J   (e) none of these.

12. In an adiabatic operation, the value of γ for mercury vapor is likely to be about   (a) 1.67   (b) 1.4   (c) 1.3   (d) 1 (e) none of these.

13. The statement that heat flows spontaneously only from a high-temperature object to a low-temperature object is   (a) not always true   (b) the First Law of Thermodynamics   (c) the Second Law of Thermodynamics   (d) only true for isothermal processes (e) none of these.

14. An ice-filled glass of soda is so cold, water from the air condenses out on it in beads. As a result, the entropy of that water must have   (a) increased   (b) remained unchanged   (c) first increased then decreased (d) decreased   (e) none of these.

15. As it regards the ordinary processes of everyday life, it can be said that the Second Law of Thermodynamics (a) distinguishes the possible from the impossible   (b) is quite inapplicable   (c) demands that energy be conserved, no matter what   (d) requires that the sum of the heat-in and the heat-out be zero   (e) none of these.

16. The addition of heat to a system   (a) decreases its order and so decreases its entropy   (b) decreases its order and its internal energy   (c) decreases its order and increases its entropy   (d) increases its disorder and decreases its entropy   (e) none of these.

17. The disordering effect of heat   (a) increases with increasing temperature   (b) decreases with increasing entropy   (c) is constant   (d) increases with decreasing temperature   (e) none of these.

18. The area under a process curve on a TS-diagram is the (a) heat flowing into or out of the system   (b) the work done on or by the system   (c) the change in pressure of the system   (d) the change in entropy of the system (e) none of these.

19. The entropy associated with a particular state of a system is   (a) independent of probability of any sort (b) dependent on the logarithm of the probability of the state occurring   (c) inversely proportional to the number of microstates associated with that state   (d) independent of the number of microstates associated with that state (e) none of these.

20. The statement that processes take place in such a way that the entropy of the Universe tends toward a maximum should be interpreted as saying   (a) that the Universe is moving inextricably toward equilibrium   (b) that nothing much is happening in the Universe   (c) that energy is approaching a minimum   (d) that order is rising out of chaos   (e) none of these.

Problems

THE FIRST LAW OF THERMODYNAMICS
CYCLIC PROCESSES: ENGINES AND REFRIGERATORS
THE SECOND LAW OF THERMODYNAMICS

1. [I] An amount of heat equal to 500 J flows into a system that does 250 J of work on its surroundings. What is the change in the internal energy of the system?

2. [I] A quantity of gas in a cylinder sealed with a piston absorbs 1.00 kcal of heat as it expands by 15 liters at constant pressure. Determine that pressure if the internal energy is found to increase 286 J.

3. [I] A system has 150 J of work performed upon it while its internal energy is made to decrease by 300 J. Determine the heat transferred to or from the system.

4. [I] A brass sphere having a 15-cm diameter is heated from 0°C to 280°C. How much work does it do on the atmosphere in the process?

5. [I] A 100-g sample of gas is contained in a sealed insulated tank. Heat is slowly transferred to the gas (via an electric heater), and its internal energy increases by 500 J. Determine the amount of heat added to the gas.

6. [I] When a chunk of ice melts, it changes from a density of 920 kg/m³ to 1000 kg/m³, and work is done on it by the atmosphere. Compare the work done per kilogram on the ice to the heat-in per kilogram.

7. [I] An amount of heat equal to 100 cal is transferred to a system while 100.4 N·m of work is done on it. What is the net change in its internal energy?

8. [I] An ideal gas is compressed to one-half its original volume by a piston moving in a cylinder. During the process, 500 J of heat leave the gas, but its temperature remains fixed. How much work is done on or by the gas?

9. [I] An ideal gas in a cylinder-and-piston is made to follow the cycle shown in Fig. P9 from A to B to C to A. What is the work done in terms of the initial volume and pressure?

Figure P9

10. [I] A cylinder-and-piston containing an ideal gas is placed in contact with a thermal reservoir. The volume of the gas is very slowly changed from 50 liters to 10 liters as 40 J of work is done on it by an external agency. Determine the change in the internal energy of the gas and the amount of heat flowing into or out of the system.

11. [I] An ideal gas in a cylinder-and-piston is made to follow the cycle shown in Fig. P11 from A to B to C to D. What is the work done in terms of the initial volume and pressure?

Figure P11

12. [I] An ideal gas contained by a cylinder-and-piston is made to double its volume isothermally, at 27°C. If the chamber contains 5.0 moles of gas, how much work is done in the process?

13. [I] A gas experiences a change from state I to state F as indicated in Fig. P13. Determine the work done.

Figure P13

14. [I] An engine doing work receives 100 kJ and exhausts 75 kJ. Determine its efficiency.

15. [I] A person consumes food at 98.6°F and exhausts waste heat at 20°C. What is the maximum theoretical efficiency for a heat engine operating at the same temperatures?

16. [I] What is the maximum theoretical efficiency of a steam engine operating between 100°C and 400°C?

17. [I] To improve the theoretical efficiency of a heat engine, would it be better to lower T_L by 10 K or raise T_H by 10 K? Explain.

18. [I] In a given amount of time, a refrigerator extracts 75 kJ from a cool chamber while exhausting 100 kJ. Determine its coefficient of performance.

19. [I] A 1.0-ton air conditioner has the capacity to freeze

2000 lb of ice at 0°C in a day. What is the equivalent rate of transport of heat in watts?

20. [I] What is the coefficient of performance of a Carnot refrigerator operating between 0°C and 85°C?

21. [I] What is the change in entropy of a quantity of boiling water in a covered pot when 1.00 kJ of heat is added?

22. [I] By how much does its entropy change when 10 g of water at 0°C is completely frozen into ice at 0°C? Make it clear whether your answer is an increase or decrease in entropy.

23. [I] Draw a *T*-versus-*S* diagram for the Carnot cycle and label it to match Fig. 16.16.

24. [I] A thick metal bar, insulated everywhere but on its ends, is placed between two thermal reservoirs at 600 K and 300 K, respectively, such that heat flows and a stable temperature distribution exists along the bar. If 10.0 kcal of heat traverses the bar, what is the attendant change in entropy of (a) both reservoirs, (b) the bar, (c) the Universe?

25. [I] A 500-g piece of copper melts at 1083°C. Determine its change in entropy in the process.

26. [I] By how much does the entropy of 1.00 kg of water change when it is completely converted to steam?

27. [I] With Problem 26 in mind, compare the changes in entropy that result when 1.00 kg of water ice is melted at 0°C with that when the liquid is vaporized at 100°C. Account for the great difference in these two values.

28. [II] An ideal gas is compressed from 20 liters to 2.0 liters at a constant pressure of 3.0 atm by drawing heat from it so that its temperature drops. At that point, its volume is held constant, and heat is added to it so the pressure rises and it returns to its original temperature. Draw a *PV*-diagram. Compute the net work done on or by the gas. Determine the net heat flow.

29. [II] A piston with a cross-sectional area of 0.50 m² seals a gas within a cylinder at a pressure of 0.30 MPa. Heat is very gradually added to the cylinder, and the piston rises 6.5 cm as the gas expands isobarically. What work is done by the gas?

30. [II] A quantity of 5.0 kg of water in a vacuum chamber is supplied with heat, and its temperature increases by 10°C. What is the change in its internal energy?

31. [II] A large thermally insulated cylinder is sealed with a piston. Steam at 100°C enters the cylinder and pushes the piston outward against a pressure of 99 kPa, increasing the occupied volume by 4950 cm³. How much water condenses inside the cylinder?

32. [II] A 2.0-mole sample of a gas at 0°C, which behaves like an ideal gas, is compressed to half its original volume isobarically. How much work must be done on the gas?

33. [II] A bicycle tire filled with air to a pressure of 4.5 atm is rapidly emptied into a large bag at atmospheric pressure. If the air in the tire was initially at 27°C, what will be its final temperature in the bag? Assume ideal gas behavior.

34. [II] An ideal diatomic gas at STP is adiabatically compressed to half its original volume. Determine its final temperature and pressure.

35. [II] A cylinder-and-piston with a volume of 22.4 liters contains 1.00 mole of diatomic hydrogen at a pressure of 1.00 atm. The gas is allowed to expand adiabatically to twice its original volume, whereupon it is compressed back

to its initial volume isothermally. (a) What is the lowest pressure the system attains? (b) What is the lowest temperature the system attains? Draw a *PV*-diagram showing everything that's happening.

36. [II] Considering Problem 35, how much work is done on isothermally returning the system to its original volume? What is the final temperature of the system? What is the final pressure?

37. [II] An ideal gas contained in a cylinder-and-piston assembly undergoes a reversible isothermal expansion. Derive an expression for the work done by the gas in terms of the number of moles present, the temperature, the gas constant, and the initial and final pressures.

38. [II] With the results of Problem 37 in mind, compute the amount of work done by 2.00 moles of an ideal gas in an isothermal expansion at 400 K from an initial pressure of 12 atm down to 1.5 atm.

39. [II] Figure P39 shows two reversible cycles *A*-*B*-*C*-*A* and *C*-*D*-*E*-*C* carried out on an ideal gas (the curves are isothermals). Prove that the work done is the same for each cycle.

Figure P39

40. [II] A reversible heat engine operates between 325°C and 25°C taking in 610 kcal of heat per cycle. Determine the efficiency of the system, the work done per cycle, and the heat exhausted per cycle. If the engine runs at 5.0 cycles per second, what is the power delivered by it?

41. [II] It is claimed that a heat engine has been built that takes in 102 kJ and produces 26 kJ of useful work. It supposedly operates between 144°F and 5°F. Is such a device physically possible?

42. [II] The air conditioner in my bedroom has a little plaque on it that says 18 000 Btu/h, 230 V, 9.0 amps. That means that it consumes electrical power at a rate of (230 V) × (9.0 amps) = 2070 W while removing 18 000 Btu each hour from the room. Given that 1 Btu = 1.055 × 10³ J, determine the coefficient of performance of the machine.

43. [II] An old-fashioned commercial 1-ton refrigerator removes heat from water in a chamber at 32°F at such a rate as to produce 1.0 ton (2000 lb) of ice (at 32°F) per day. The heat is exhausted to the environment at 88°F. At its very best, how much power would be needed to drive this system?

44. [II] Suppose that in the course of a day you consume one peanut butter sandwich (330 kcal), a jelly doughnut

(225 kcal), 1.0 kg of milk (650 cal/g) and two martinis (145 kcal each). You spend a quiet day averaging an output of about 60 watts of useful work for 6 hours, above and beyond the energy consumed just staying alive. What's your efficiency?

45. [II] A heat pump, which removes thermal energy from the outdoors and exhausts it inside to heat the house, has a coefficient of performance of 5.5. What is the ratio of the heat supplied to the house, to the work done? How much energy do you pay for as compared to the amount you get pumped into the house? How does this compare with heating with electricity directly?

46. [II] A refrigerator maintains food at 40°F when the room temperature is 80°F. Ideally, how much heat will be removed from the food per joule of work done by the compressor on the refrigerant?

47. [II] You are designing an electric refrigerator to remove 360 J of heat per minute from a cold chamber at −4.0°C and exhaust that energy into a room at 27°C. At the very best, how much electrical power will it require?

48. [II] We have designed a 150-W electrical refrigeration system that removes the 360 J/s of heat leaking into a test chamber and expels that thermal energy (the so-called *heat load*) into the room (which is at 23°C). Determine the lowest temperature we can expect to reach in the chamber.

49. [II] A Carnot engine operates at an efficiency of 42.2% with a high-temperature reservoir at 473 K. If the efficiency is to be raised to 50% using a new high-temperature reservoir, what will its temperature have to be?

50. [II] A Carnot refrigerator is driven by 2.00 J of mechanical work each second, from an electric motor. It, in turn, removes heat from a cold region (0°C), expelling it into the room (27°C). What is the change in entropy of the cold region?

51. [II] During each cycle of operation of a heat engine, it takes in 1200 J of heat from a hot reservoir and exhausts 450 J of heat to a cool reservoir. What is its efficiency? The engine is now run backward between the same reservoirs, and while delivering 1200 J per cycle to the high-temperature reservoir, it is driven by 1800 J of applied work. Determine whether this engine is reversible or irreversible in a thermodynamic sense.

52. [II] A 100-g chunk of ice at 0°C is placed on a large marble table, which is at 27°C. What, if any, is the change in entropy of the ice-table system when the ice melts completely?

53. [II] A 20-kg sample of pure water at 40.0°C is mixed with a 20-kg sample of pure water at 32.0°C. Using the average temperatures of each sample, determine the approximate net change in entropy of the total quantity of water.

54. [II] An ideal gas expands isothermally from V_i to V_f. Prove that the associated entropy change is given by

$$\Delta S = nR \ln \frac{V_f}{V_i}$$

55. [II] Using the data of Fig. P55, which is a TS-diagram for

Figure P55

a thermal engine, determine Q_H, Q_L, the net work done by the system, and its efficiency.

56. [III] A 1.0-m^2 panel of solar cells absorbs 35% of the total radiation (of intensity 0.83 kW/m^2) impinging on it from the Sun. It generates electrical energy, which is fed to a motor that raises a load, doing work at a rate of 150 J each second. Assuming no other losses, determine the rate at which the internal energy of the system (panel and motor) is changing.

57. [III] Please show that for an adiabatic volume change of an ideal gas

$$\frac{T_i}{T_f} = \left(\frac{V_f}{V_i}\right)^{\gamma-1} \qquad (16.19)$$

58. [III] A diesel engine adiabatically compresses an air-fuel mixture from its initial volume at atmospheric pressure and a temperature of 25.0°C to 1/20 that volume. What is its final temperature? (Take $\gamma = 1.35$.)

59. [III] The compression ratio (V_i/V_f) of a gasoline engine is 7.5. Determine the ratio of the temperature of the gas before and after the compression. (Take $\gamma = 1.35$.) Explain your results.

60. [III] Derive an expression for the entropy change experienced by a gas during a free expansion from a volume V_i to a volume V_f.

61. [III] Two identical thermally insulated chambers are connected by a small valve. One chamber contains 1.0 mole of an ideal diatomic gas; the other chamber is evacuated. The valve is then opened, and the gas rushes through the valve so that it occupies both chambers. What is the resulting change in entropy?

62. [III] A 200-kg boulder rolls, slides, and slips down the side of a mountain, descending 200 m vertically before coming back to rest. It's a nice cool winter's day, and the temperature of the forest is 7.0°C. What is the change in entropy of the Universe as a result of the tumble?

Appendixes:
A Mathematical Review

APPENDIX A   ALGEBRA

Algebra is essentially a body of rules and procedures for logically exploring the relationships between concepts using symbols to provide a concise and easily read format. Our concern is usually with equations describing interdependencies among physical quantities. Thus, the length traveled (l) in a time (t) by a bird moving at a speed (v) is given by $l = vt$. *When we have one equation we can solve for only one unknown quantity.* Thus, suppose the speed is known to be 2 m/s and the length traveled is 500 m (ignoring units for the moment), the equation becomes $500 = 2t$; one equation, one unknown (namely, t). How long did it take the bird to make the trip? That is, solve for t—*get the unknown all by itself on one side of the equation.* To do that, remember the following three logical rules:

1. The same quantity (constant or variable) can be added to or subtracted from both sides of an equation without changing the equality: $100 = 100$; $100 - 2 = 100 - 2$.
2. Both sides of an equation can be multiplied or divided by the same quantity (constant or variable) without changing the equality: $100 = 100$; $100/2 = 100/2$.
3. Both sides of an equation can be raised to the same power (squared or square rooted, cubed or cube rooted) without changing the equality: $100 = 100$; $\sqrt{100} = \sqrt{100}$.

Wherever possible, apply rule 1 first followed by rule 2. Isolate the variable to whatever power it is raised to and if that power is other than one, use rule 3 to make it one.

Back to the bird: $2t = 500$. To get the t alone, remove the 2 (using rule 2) by dividing both sides by 2; hence, $t = 250$ seconds.

Let's solve $8x + 2 = 42$ for x. To get the x alone, first remove the 2 via rule 1, $8x = 42 - 2 = 40$. Next, use rule 2 to remove the 8; $x = 40/8 = 5$.

Now solve $5x^2 + 12 = 57$ for x. Apply rule 1: $5x^2 = 57 - 12 = 45$. Next isolate x^2 using rule 2 to divide both sides by 5: $x^2 = 9$. Now use rule 3, taking the square root of both sides: $x = \pm 3$. Note that there are two solutions: $+3$ and -3.

Solve $\frac{3}{4}t^2 - 6 = 0$ for t. Use rule 1 to move the 6, yielding $\frac{3}{4}t^2 = 6$. Now, using rule 2, multiple both sides by 4 to get $3t^2 = 24$. Use it again, dividing both sides by 3 to get $t^2 = 8$. Now, using rule 3, take the square root of both sides: $t = \sqrt{8}$.

Some Trouble Spots

1. Dividing by fractions can be troublesome. $8/\frac{1}{4}$ is *not* 2. *A fraction is not changed when the top and bottom are multiplied by the same quantity.* Here, multiply top and bottom by 4 to get 32/1. Similarly, given $\frac{1}{4}/\frac{1}{2}$ multiply top and bottom by 2 to get $\frac{1}{2}/1 = \frac{1}{2}$.
2. $(a + b)^2$ is *not* equal to $a^2 + b^2$! $(a + b)^2 = (a + b)(a + b) = aa + ab + ba + bb = a^2 + 2ab + b^2$.

3. $\dfrac{1}{a} + \dfrac{1}{b}$ is *not* equal to $\dfrac{1}{a+b}$. We can only add terms with the same denominators. To get a common denominator (ab), multiply top and bottom by whatever is needed: $\dfrac{b}{ba} + \dfrac{a}{ab}$ and add. Thus

$$\frac{1}{a} + \frac{1}{b} = \frac{a+b}{ab}$$

Similarly
$$\frac{1}{2} + \frac{2}{5} = \frac{5}{10} + \frac{4}{10} = \frac{9}{10}$$

4. Remember that $\sqrt{a}\,\sqrt{b} = \sqrt{ab}$; thus, $\sqrt{4}\,\sqrt{9} = \sqrt{36}$, or $2 \times 3 = 6$. Similarly, $\sqrt{8} = \sqrt{4}\,\sqrt{2} = 2\sqrt{2}$; hence, $\sqrt{25gt^2} = 5t\sqrt{g}$.

5. Given the relationship $F = GmM/R^2 = 2GmM \neq R^2$, how does F change if m is doubled? The new F is $G(2m)M/R^2$, which is twice the original F. How does F change when R is doubled? The new F is $GmM/(2R)^2 = GmM/4R^2$, which is one-quarter the original F.

A-1   Exponents

A square with sides of length a has an area of $a \times a$, which is more concisely written as a^2 (a raised to the second power). Because of that, a^2 is read "a squared." In the same way, a cube with sides of a has a volume of $a \times a \times a = a^3$ or "a cubed." In general, a raised to the nth *power* is a^n, where the *exponent* n can be any number, fractional or whole, positive or negative. Quantities raised to various powers are often multiplied and divided. Thus, $a^2 \times a^3 = (a \times a) \times (a \times a \times a) = a^2 a^3 = a^5$ and, in general

$$(a^n)(a^m) = a^{n+m} \tag{A.1}$$

The base numbers (a) being raised to the powers n and m are the same. Alternatively, if the base numbers (a and b) are different and the powers are the same, then

$$(a^n)(b^n) = (ab)^n \tag{A.2}$$

Now, for division: $2^3/2^2 = 8/4 = 2 = 2^1$; the two 2s on the bottom cancel two 2s on the top or, equivalently, the exponent on the bottom subtracts from that on the top. More generally

$$\frac{a^3}{a^2} = \frac{a \times a \times a}{a \times a} = a^{3-2} = a$$

As a rule
$$\frac{a^n}{a^m} = a^{n-m} \tag{A.3}$$

which suggests that we define

$$\frac{1}{a^m} = a^{-m} \tag{A.4}$$

Hence, $1/5 = 5^{-1}$, $1/3^2 = 3^{-2}$, and $1/4^{-2} = 4^2$. Inasmuch as $a/a = 1$, it follows from Eq. (A.3) that $a^{1-1} = a^0 = 1$. *Any quantity raised to the zero power is 1.*

Fractional exponents correspond to roots, thus

$$a^{1/n} = \sqrt[n]{a} \tag{A.5}$$

Thus, using Eq. (A.1), we have

$$\sqrt{a}\,\sqrt{a} = a^{\frac{1}{2}} a^{\frac{1}{2}} = a^1 = a$$

Since powers undo roots and vice versa

$$(a^{1/n})^n = 1$$

For example, $(25^{\frac{1}{2}})^2 = 5^2 = 25$. In general

$$(a^n)^m = a^{nm} \tag{A.6}$$

where n and m can be either whole or fractional numbers.

A-2   Powers of Ten: Scientific Notation

We deal with numbers that are extremely small (like the mass of an electron, 0.000 000 000 000 000 000 000 000 000 000 911 kg) and extremely large (like the number of stars in the Universe, $\approx 10\,000\,000\,000\,000\,000\,000\,000$). The *scientific notation* is a shorthand way of writing numbers in terms of powers of ten:

$10^0 = 1$	$10^0 = 1$
$10^1 = 10$	$10^{-1} = 1/10 = 0.1$
$10^2 = (10 \times 10) = 100$	$10^{-2} = 1/(10 \times 10) = 0.01$
$10^3 = (10 \times 10 \times 10) = 1000$	$10^{-3} = 1/(10 \times 10 \times 10) = 0.001$
$10^4 = (10 \times 10 \times 10 \times 10) = 10\,000$	$10^{-4} = 1/(10 \times 10 \times 10 \times 10) = 0.0001$

and so forth. For positive exponents, *the number of zeros corresponds to the power of ten.* Thus, the number of stars in the Universe is a 1 with 22 zeros, or 1×10^{22}, or just 10^{22}.

Suppose we want to express a number greater than 1.0 given in ordinary form (such as the number of seconds in a year—31 560 000) in scientific notation. Start by indicating the decimal at its reference position (31 560 000.0). Then decide where you would like to move it to (for example, 31*560 000.0). Here the decimal is to be relocated six places to the left yielding 31.56, and multiplying by 10^6 will move it back six places to the right, where it started. Thus, $31\,560\,000 = 31.560 \times 1\,000\,000 = 31.56 \times 10^6$, or 3.156×10^7, or $0.315\,6 \times 10^8$. To write a number less than 1.0 in scientific notation (for example, 0.000 000 000 000 000 000 000 000 000 000 911 kg), again locate where you would like the decimal to appear (0.*000 000 000 000 000 000 000 000 000 9*11 kg). That's 31 places to the right, so multiplying 9.11 by 10^{-31} will shift the decimal back 31 places to the left where it started. The mass of an electron is 9.11×10^{-31} kg.

When multiplying or dividing numbers in scientific notation, process the numerical terms separately from the exponents; as

$$(1.1 \times 10^{12})(5.0 \times 10^{17}) = (1.1 \times 5.0)(10^{12} \times 10^{17}) = 5.5 \times 10^{29}$$

$$\frac{(1.1 \times 10^{12})}{(5.0 \times 10^{17})} = \frac{(1.1)}{(5.0)} \times \frac{(10^{12})}{(10^{17})} = 0.22 \times 10^{-5}$$

Scientific calculators have an "exp" key that allows you to enter exponents, and they will keep track of the decimal automatically.

A-3 Logarithms

Suppose we have a positive number y expressed as a power of b where $b > 0$ and $b \neq 1$; accordingly

$$y = b^x$$

Let's define x to be the logarithm of y, to the base b; thus

$$x = \log_b y$$

The logarithm x is the exponent, and the two equations are just different ways of saying the same thing. For example, if $2^3 = 8$, then $3 = \log_2 8$; if $8^{-4/3} = 1/16$, then $-4/3 = \log_8 1/16$. Because logarithms are exponents, the laws governing the two are very similar. Thus, independent of the base

$$\log(ab) = \log a + \log b$$

$$\log \frac{a}{b} = \log a - \log b$$

$$\log a^n = n \log a$$

where

$$\log_b b = 1$$

$$\log 1 = 0$$

and

$$b^{\log_b a} = a$$

There are two widely used bases, 10 and $e = 2.718\,281\,828 \cdots$. The *common logarithms* arise when $y = 10^x$, whereupon $x = \log_{10} y$. Before the advent of the electronic calculator, common logs were used to carry out complicated calculations. Our interest in them arises out of the idea of intensity-level (p. 455) and the dB. The *natural logarithms* arise when $y = e^x$, whereupon $x = \ln y$ (it's customary to write $\ln y$ rather than $\log_e y$). The exponential function e^x (Fig. A1) and the natural log (Fig. A2) occur in a great many physical situations in which the rate-of-change of some quantity depends on that quantity. We can see this dependence in Fig. A1: the rate-of-change of the curve $y = e^x$ equals e^x. When the curve has small values, the slope is small; when the curve has large values, the slope is large. This property is very special and it means that if some quantity N changes in time exponentially, then $\Delta N / \Delta t \propto N$.

It follows from the above that

$$\ln e^x = x \qquad \text{and} \qquad e^{\ln x} = x$$

Notice that with $y = e^x$ when $y = 1$, $x = 0$; hence, $y = 2$ when $2 = e^x$, whereupon $x = \ln 2 = 0.693$. Doubling y yields $y = 4$ when $x = \ln 4 = 1.386 = (0.693 +$

Figure A1

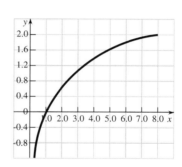

Figure A2

0.693), and again $y = 8$ when $x = \ln 8 = 2.079 = (0.693 + 1.386)$. Every time x increases by 0.693, y doubles. Indeed, any time x increases by a fixed amount, y will increase by a fixed multiplicative factor—y increases by the same factor in equal intervals of x. When x increases by $\ln 10 = 2.302\,6$, y increases by a factor of 10.

A-4   Proportionalities and Equations

We frequently find that one physical quantity depends on another: the force exerted by a spring depends on the amount it's stretched; the current through a resistor depends on the voltage across it. Suppose the quantities A and B depend on one another such that doubling either doubles the other. They are said to be **directly proportional** to each other or, symbolically, $A \propto B$. On the other hand, it's possible that doubling one could halve the other, in which case $A \propto 1/B$, and they are **inversely proportional**. Table A1 gives a list of values of A and B. At first, there may seem no relationship between them, but form the ratio B/A and we see that there is a consistent pattern: $B/A = 3$. Thus, $B = 3A$ and 3 is known as a **constant-of-proportionality**. When two quantities are proportional, we can always turn the proportionality into an equality using an appropriate constant-of-proportionality. The circumference of a circle (C) is directly proportional to the diameter (D), where the constant of proportionality is π: $C = \pi D$. The area of a sphere (A) is directly proportional to the square of the radius (r), where the constant-of-proportionality is 4π: $A = 4\pi r^2$.

When the variables in an equation occur only to the first power, it's called a **linear equation**; for example, $y = mx$, where m is a constant. Here, when $x = 0$, $y = 0$, and the curve (Fig. A3) is a straight line passing through the origin. For any two points on the line, $\Delta y / \Delta x$ is its **slope**, or tilt, and it equals m. If a constant (b) is added to y such that $y = mx + b$, the line shifts parallel to itself; b is the place where the line crosses the y-axis, the y-intercept (Fig. A4).

The Quadratic Equation

When an object is uniformly accelerating, it travels a distance s given by Eq. (3.9); namely, $s = v_i t + \frac{1}{2} a t^2$. This expression is typical of a class of equations that contains a variable (t) raised to the second power, and it's called a **quadratic equation**. The standard form has everything on the left set equal to zero on the right:

$$ax^2 + bx + c = 0$$

TABLE A1	
A	B
3.1	9.3
18.0	54
5.2	15.6
71.0	213

Figure A3

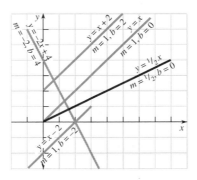

Figure A4

where a, b, and c are constants. This equation can be solved directly, but it's common practice to memorize the solution in the form

$$x_{\pm} = \frac{-b \pm \sqrt{b^2 - 4ac}}{2a}$$

where, provided $b^2 > 4ac$, there are two real solutions. For example, $2x^2 + 6x - 20 = 0$ corresponds to $a = 2$, $b = 6$, and $c = -20$; therefore

$$x_+ = \frac{-6 + \sqrt{36 - 4(2)(-20)}}{2(2)} = \frac{-6 + \sqrt{196}}{4} = +2$$

$$x_- = \frac{-6 + \sqrt{36 - 4(2)(-20)}}{2(2)} = \frac{-6 - \sqrt{196}}{4} = -5$$

It may happen in a physical analysis that only one of the two solutions corresponds to a possible situation; we then simply ignore the other. For example, we might solve Eq. (3.9) for the time and find one of the solutions to be negative—it works mathematically, but not physically.

Simultaneous Equations

For every unknown, we must have an independent equation (one that isn't equivalent to any other equation in the set). Situations with one and two unknowns are common throughout the discipline. Indeed, we will encounter problems in circuit analysis where there are as many unknowns as there are loops in the circuit, but three will be our limit. Suppose there are two unknowns, x and y, and two equations

$$4x - 2y = 16$$
$$3x + 4y = 23$$

Lots of numbers satisfy each of these equations (for example, in the first one, $x = 4$, $y = 0$ or $x = 0$, $y = -8$). We want to solve these *simultaneously* so that the solutions satisfy both equations at the same time. There are two ways to accomplish this goal. (1) *Solve either equation for either unknown in terms of the other unknown, and substitute that into the remaining equation.* From the first equation, $4x = 16 + 2y$, $x = 4 + \frac{1}{2}y$. Now put this result into the second equation so it has only one unknown $3(4 + \frac{1}{2}y) + 4y = 23$ and so $12 + 3y/2 + 4y = 23$, $3y/2 + 8y/2 = 23 - 12$, $11y/2 = 11$, $y/2 = 1$, $y = 2$. Put that back into the first equation: $4x - 2(2) = 16$, $4x = 20$, and $x = 5$. (2) Alternatively, *multiply either or both of the equations by whatever numbers it takes to make the same coefficient appear for either one of the unknowns. Then add or subtract the two, thereby removing one unknown.* Here, to get rid of y, multiply the first equation $4x - 2y = 16$ by 2

getting $\qquad\qquad\qquad\qquad\qquad 8x - 4y = 32$

and add $\qquad\qquad\qquad\qquad\qquad \underline{3x + 4y = 23}$

$$\qquad\qquad\qquad\qquad\qquad 11x \qquad\quad = 55$$

or $x = 5$. Substitute this result into either equation and solve for y.

A-5   Approximations

It's often desirable to approximate the solution to a problem, either as a quick check or because the exact solution is too elaborate to deal with. Newton's *Binomial Theo-*

rem, expressed as

$$(a + b)^n = a^n + na^{n-1}b + \frac{n(n-1)}{2 \times 1}a^{n-3}b^2 + \frac{n(n-1)(n-2)}{3 \times 2 \times 1}a^{n-3}b^3 + \cdots$$

(A.7)

produces some very helpful approximations. In general, if n is a positive integer, the series ends in a finite number of terms, otherwise it has an infinite number. Expressions of the form $(1 + x)^n$ arise frequently and can be approximated using the Binomial Theorem ($a = 1$, $b = x$); thus

$$(1 + x)^n = 1 + nx + \frac{n(n-1)}{2}x^2 + \cdots$$

(A.8)

When x is very small ($x \ll 1$), the x^2 and higher terms will be negligibly small and

$[x \ll 1]$	$(1 + x)^n \approx 1 + nx$
with $n = 2$	$(1 + x)^2 \approx 1 + 2x$
with $n = 3$	$(1 + x)^3 \approx 1 + 3x$
with $n = \frac{1}{2}$	$(1 + x)^{\frac{1}{2}} = \sqrt{1 + x} \approx 1 + \frac{1}{2}x$
with $n = -\frac{1}{2}$	$(1 + x)^{-\frac{1}{2}} = \dfrac{1}{\sqrt{1 + x}} \approx 1 - \frac{1}{2}x$
with $n = -1$	$(1 + x)^{-1} = \dfrac{1}{(1 + n)} \approx 1 - x$
with $n = \frac{1}{2}$	$(1 - x)^{\frac{1}{2}} = \sqrt{(1 - x)} \approx 1 - \frac{1}{2}x$

Try taking the square root of $1.000\,000\,001\,0$ on your calculator. Now, let $x = 0.000\,000\,001\,0$ and the square root of $(1 + x)$ is $\approx 1.000\,000\,000\,5$. For $a \gg b$

$$(a + b)^n = a^n \left[1 + \frac{b}{a} \right] \approx a^n \left[1 + n\frac{b}{a} \right]$$

(A.9)

What is the value of $a/(a + b)^2$ when $a \gg b$? Since b is negligible compared to a, $(a + b)^2 \approx (a)^2$ and $a/(a + b)^2 \approx 1/a$. Check this expression with Eq. (A.9), $n = 2$ and $(a + b)^2 \approx a^2(1 + 2b/a) \approx a^2$. Suppose $b \gg a$, $(a + b)^2 \approx (b)^2$ and $a/(a + b)^2 \approx a/b^2$. We frequently know how a system behaves at its extremes and can confirm the analysis at those extremes when a key quantity is very large or very small. For example, any equation for the force between two magnets must go to zero as the separation between them gets very large.

A useful means of improving the likelihood that an analysis is correct is to simplify the numbers by rounding them off and then run quickly through the calculation to get a crude answer. When a number is rounded off to the nearest power of 10, we say it's an **order-of-magnitude** figure. Thus, the order-of-magnitude of the acceleration due to gravity on Earth (9.81 m/s^2) is 10 m/s^2; the great mountains have heights of the order-of-magnitude of 10 km; the electron's mass ($9.109\,389\,7 \times 10^{-31}$ kg) is of the order-of-magnitude of 10^{-30} kg; the number of atoms per cubic centimeter of a solid is of the order-of-magnitude of 10^{23}.

A-6   The Change in a Quantity

We will often encounter quantities that are formed of products such as $z = xy$. The question arises, what happens to z when either x or y or both change? The initial value of z is $z_i = xy$. Its final value, after a change in x of Δx, and a change in y of Δy, is

$$z_f = (x + \Delta x)(y + \Delta y)$$

$$z_f = xy + x\,\Delta y + y\,\Delta x + \Delta x\,\Delta y$$

Hence $$\Delta z = z_f - z_i = x\,\Delta y + y\,\Delta x + \Delta x\,\Delta y$$

When Δx and Δy are small, $\Delta x\,\Delta y$ is negligibly small and

$$\Delta z \approx x\,\Delta y + y\,\Delta x$$

This is an important relationship that we will make use of often.

APPENDIX B   GEOMETRY

Figure B1 provides a number of useful relationships for the angles formed by intersecting lines and triangles. Of special interest is the Pythagorean Theorem, which relates the sides of any right triangle; consequently

$$C = \sqrt{A^2 + B^2}$$

as in Fig. B2. There are a number of right triangles with whole number sides—the 3–4–5 and 5–12–13 are the most commonly encountered (Fig. B3).

Figure B1

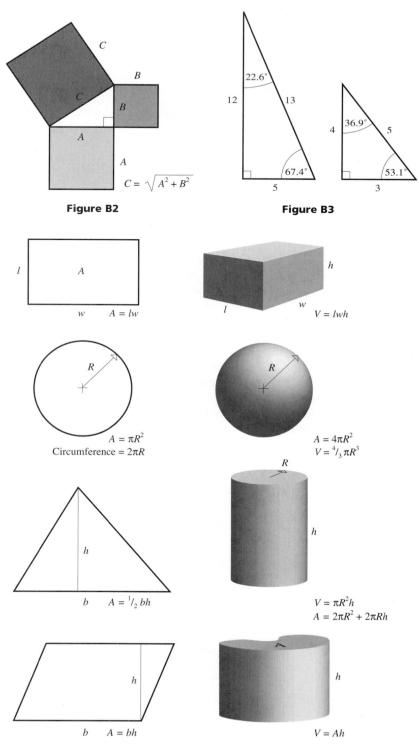

Figure B2

Figure B3

$C = \sqrt{A^2 + B^2}$

$A = lw$

$V = lwh$

$A = \pi R^2$
Circumference $= 2\pi R$

$A = 4\pi R^2$
$V = {}^4/_3 \pi R^3$

$A = {}^1/_2 bh$

$V = \pi R^2 h$
$A = 2\pi R^2 + 2\pi Rh$

$A = bh$

$V = Ah$

Figure B4

Figure B4 displays the areas and volumes of shapes that are frequently used in this text. Notice that the area of the curved surface of a circular cylinder ($2\pi Rh$) is that swept out by a line of height h moved around the circumference of the base

$(2\pi R)$. In a similar way, the volume of a cylinder is the area of the base times the height (that is, the volume swept out by a vertical line of height h moved to every point on the base area).

APPENDIX C TRIGONOMETRY

Figure C1a shows a right triangle with an acute angle θ and the sides (A) adjacent and (B) opposite to it. A particular value of θ fixes a set of triangles, a few of which are shown in Fig. C1b. Although the triangles are different, the ratios of the sides in any one of them is the same for all of them. In other words, the ratio of the opposite over the adjacent is the same for each triangle and is indicative of θ. The various possible ratios are given names, and electronic calculators are programmed to compute them. Knowing θ we can determine any ratio and vice versa. Thus

$$\sin\theta = \frac{\text{opposite}}{\text{hypotenuse}} \qquad \cos\theta = \frac{\text{adjacent}}{\text{hypotenuse}} \qquad \tan\theta = \frac{\text{opposite}}{\text{adjacent}}$$

For example, for an angle of $30°$, the ratio of the opposite to the adjacent side will always be 0.5774. To determine as much on a calculator, first make sure it's in the *degree* mode, then punch in [3][0][tan]. Alternatively, knowing the ratio, you can find the angle by entering [0][·][5][7][7][4][tan^{-1}].

Referring to Fig. C1a, observe that $\sin\theta = B/C = \cos\phi$ and that $\cos\theta = A/C = \sin\phi$; don't confuse these relationships. In Fig. C1c, increasing θ, with the adjacent side constant, increases both the opposite side and the hypotenuse. When $\theta \approx 0$, $B \approx 0$, $A \approx C$ and so $\sin\theta = 0$, $\cos\theta = 1$ and $\tan\theta = 0$. Similarly, when $\theta \approx 90°$, $C \approx B$ and both are tremendously large. Thus, $\sin 90° = 1$, $\cos 90° = 0$, and $\tan 90° = 1$. These results are summarized in Fig. C2, which provides $\sin\theta$, $\cos\theta$, and $\tan\theta$ for all values of θ between 0 and $360°$.

Some useful relationships are

$$\tan\theta = \frac{\sin\theta}{\cos\theta}$$
$$\sin^2\theta + \cos^2\theta = 1$$
$$\sin 2\theta = 2\sin\theta\cos\theta$$

For all triangles (Fig. C3)

Law of Sines $\dfrac{\sin\alpha}{A} = \dfrac{\sin\beta}{B} = \dfrac{\sin\gamma}{C}$

Law of Cosines $C^2 = A^2 + B^2 - 2AB\cos\gamma.$

Figure C1

(a)

(b)

(c)

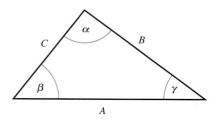

Figure C2

APPENDIX D   VECTORS

Vectors are discussed at length in Chapter 2. Here, we'll go over some of the trouble spots often encountered when dealing with them. Figure D1 shows a vector **C** in each of the four quadrants and the corresponding x and y components. In each case

$$C_x = C \cos \theta \quad \text{and} \quad C_y = C \sin \theta$$

where θ is measured either up or down from the x-axis and is always acute. Now, suppose we add several vectors along the x-axis and get C_x and add several along the y-axis and get C_y. The next step in determining **C** is to use $\tan \theta = C_y/C_x$ to find θ. It's best to write the tangent as $\tan \theta = |C_y|/|C_x|$, using the absolute values of the components so that $\tan \theta$ is always positive and θ is always between 0° and 90°. Figure C2c is a plot of $\tan \theta$ against θ from 0° to 360°, where θ is always measured from the positive x-axis. Thus, the inverse tangent of -1.732 is both 120° and 300°, lying in either the second or fourth quadrants, as expected from the figure. A calculator provides another form of the answer—namely, $-60°$ (equivalent to 300°). It neglects 120° altogether, which means that if we take the ratio of the components when finding the tangent and one or both of them are negative, there will be some ambiguity in the resulting angle. Use the ratio of absolute values; determine which quadrant the resultant is in by drawing C_x and C_y. Then measure $\theta < 90°$ up or down from the horizontal axis, as in Fig. D1.

We now wish to analytically add the two vectors **A** and **B** in the four situations depicted in Fig. D2. In each case, find $C_x = A_x + B_x$ and $C_y = A_y + B_y$. Since the angles are between 0 and 90°, the sines and cosines will be positive. Thus, oppositely directed components must be assigned plus and minus signs. Enter any component as plus ($+$) if it is in the positive x- or y-direction, and minus ($-$) if it is in

Figure C3

Figure D1

$C = A + B$

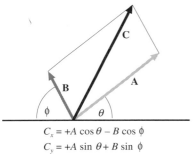

$C_x = +A \cos \theta + B \cos \phi$
$C_y = +A \sin \theta + B \sin \phi$

$C_x = +A \cos \theta - B \cos \phi$
$C_y = +A \sin \theta + B \sin \phi$

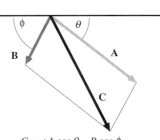

Figure D2

$C_x = +A - B \cos \phi$
$C_y = -B \sin \phi$

$C_x = +A \cos \theta - B \cos \phi$
$C_y = -A \sin \theta - B \sin \phi$

the negative x- or y-direction. The appropriate expressions are shown in each case in Fig. D2.

APPENDIX E   DIMENSIONS

To determine a distance, we must specify the units in which the measurement will be made:—feet, yards, light-years, meters, and so forth. But all such measurements have something in common—they are all lengths, regardless of the specific unit. The type of unit is the *dimension* of the quantity; in this case, the dimension of length indicated as $[L]$. Area is a distance times another distance and so has the dimensions of $[L][L] = [L^2]$. We can assign dimensions to all the quantities in mechanics using the basic dimensions of *length* $[L]$, *time* $[T]$, and *mass* $[M]$. Thus, speed, which is measured in units of m/s, ft/s, mi/h, km/h, and so on, has the dimensions of length/time or $[L]/[T]$.

The volume of a sphere of radius R is $\frac{4}{3}\pi R^3$, whereas that of a cube of side a is a^3. Each has units of m^3 or ft^3 and each has the dimensions of volume $[L^3]$, suggesting that a technique of dimensional analysis, with appropriate logical rules, can be used to check our equations. To that end, consider the equation representing three physical quantities A, B, and C, as

$$C = A + B$$

There are three rules. The first is: *dimensions can be treated as algebraic quantities*—added, subtracted, multiplied, and divided accordingly. Furthermore, A and B must have the same units (we can't add seconds to meters). Our second rule is: *only quantities having the same dimensions can be added or subtracted*. Whatever it might be, it's clear that C must have the same units as $A + B$. More generally, our third rule is: *the dimensions on either side of an equal sign must be the same*. All physical equations must obey these rules.

If you come up with an expression that obeys the rules, it *may* be correct; if it doesn't obey them, it's *definitely* incorrect. For example, in Chapter 3 we find that the distance traveled s by an object that was initially moving at speed v_i after it accelerates at a constant rate a for a time t is given by $s = v_i t + \frac{1}{2}at^2$. Knowing that the dimensions of speed and acceleration are $[L]/[T]$ and $[L]/[T^2]$ respectively, apply dimensional analysis. The expression in dimensional form is

$$[L] = \frac{[L]}{[T]}[T] + \frac{[L]}{[T^2]}[T^2]$$

Cancelling time on the right, the equation obeys the second rule since each term being added has the dimension $[L]$. And it obeys the first rule as well. Conclusion: the equation may be correct.

Answer Section

CHAPTER 1
Answers to Odd-Numbered Multiple Choice Questions
1. a 3. a 5. c 7. d 9. d 11. d 13. d 15. b 17. a 19. c

Answers to Odd-Numbered Problems
1. $10\,000\,000\,000 = 10^{10}$ 3. (a) 0.010 s (b) 0.001 s (c) 10 000 s (d) 100 000 000 s (e) 0.000 001 s 5. 1.00×10^7 mm 7. 76.2 mm 9. 5×10^2 nm 11. 2×10^2 13. 86 400 s 15. 9.46×10^{15} m 17. 8.4×10^{10} bacteria 19. 13 times greater 21. (a) 0.000 001 00 g (b) 0.000 000 000 001 g (c) 0.100 0 g (d) 0.010 000 g (e) 10 000 g 23. 3×10^8 m/s, 3.00×10^8 m/s, 2.998×10^8 m/s, $2.997\,924\,6 \times 10^8$ m/s 25. $(1.602\,177\,33 \pm 0.000\,000\,48) \times 10^{-19}$ C 27. $0.092\,90$ m^2 29. $V = 2.2 \times 10^{19}$ m^3 31. 0.393 7 in./cm 33. 1.6×10^3 m 35. $0.092\,90$ m^2 37. 237×10^{-6} m^3 39. 1×10^{-7} g; 0.3×10^2 kg 41. 1806.8 m 43. 1.00 kg 45. 6.1×10^{-3} g^3

Solutions to Selected Problems
3. (a) 0.010 s; (b) 0.001 s; (c) 10 000 s; (d) 100 000 000 s; (e) 0.000 001 s. 7. 1.00 in. = 2.54 cm = 25.4 mm; 3.00 in. = 76.2 mm. 11. 1 kg = 1000 g; (1000 g)/(5 g) = 200 = 2×10^2. 15. (5.88×10^{12} mi)(1.609 km/mi) = 9.46×10^{15} m. 21. (a) 0.000 001 00 g; (b) 0.000 000 000 001 g; (c) 0.100 0 g; (d) 0.010 000 g; (e) 10 000 g. 25. $(1.602\,177\,33 \pm 0.000\,000\,48) \times 10^{-19}$ C. 29. $V = (4/3)\pi R_\mathbb{C}^3 = 2.1991 \times 10^{19}$ m$^3 = 2.2 \times 10^{19}$ m^3. 33. (45 hairs/d) $\times$ (365 1/4 d/y) = 16 436 hairs/y; (16 436) $\times$ (0.10 m) = 1.6×10^3 m. 37. 1 liter = 1000 cm$^3 = 10^{-3}$ m^3; hence, 1.00 cup = 237×10^{-6} m^3. 41. These add up to 1806.818 m, and the least number of decimal places is one, so the answer is 1806.8 m. 45. $(0.002\,1$ g)(655.1 $\times 10^3$ g)(4.41 $\times 10^{-6}$ g) = 6.1×10^{-3} g^3.

CHAPTER 2
Answers to Odd-Numbered Multiple Choice Questions
1. d 3. b 5. c 7. c 9. c 11. d 13. b 15. a 17. c

Answers to Odd-Numbered Problems
1. 29.5 km/s 3. (a) 1.0×10^{-9} s (b) 3.336×10^{-6} s 5. 1.000 m/s = 3.600 km/h; 1.000 m/s = 2.237 mi/h; 1.000 m/s = 3.281 ft/s 7. 2.0 m; $v_{max} = 4.0$ m/s, $v_{min} = 0$; from 1.5 to 3.3 s and from

4.0 s to 5.0 s 9. (a) 60 km/h, 20 km/h (b) 30 km/h 11. 15 m 13. 0.33 m/s 15. 1.7 km 17. 120 km/h 19. 6.7 km/h 23. 1.4 m 25. 15 km 27. $D = 10$ m 29. zero, 36 km 31. $\mathbf{s} = 24$ m–straight up 33. $\mathbf{C} = 5$ units–due N 35. $v_{av} = 4.3$ m/s; $\mathbf{v}_{av} = 0.30$ m/s–south 37. $\mathbf{s} = 63$ m–72° up from the horizontal 39. (a) zero (b) $\mathbf{v} = 0.4$ m/s–south 41. $v_{av} = 150$ m/s; $\mathbf{v}_{av} = 150$ m/s–upward: round trip $v_{av} = 150$ m/s; $\mathbf{v}_{av} = 0$ 43. $\mathbf{v}_{av} = 2.1$ m/s–southeast 45. (a) 2×10^2 m/s–north (b) 2×10^2 m/s–north 47. 100 s 49. 10×10^2 m 51. $\theta_1 = 67.38°$; $\theta_2 = 22.62°$ 53. 10 m 55. 12 m 57. 74 km/h 59. (a) $v_H = 254$ m/s (b) $v_V = 159$ m/s (c) $v = 300$ m/s 61. $v_{av} = 1.3 \times 10^3$ km/h 63. $\theta = 11°$; $v = 10$ m/s 65. (a) no (b) $\theta = 42°$ (c) 35 km/h 67. 22 km/h–3.0° W of N 69. (a) $v_{RS} = 1470$ km/h; $v_{RC} = 0$ (b) $v_{RC} = 2940$ km/h (c) eastward (d) Florida is closer to the equator.

Solutions to Selected Problems
1. $v_{av} = l/t = (885$ km)/(30.0 s) = 29.5 km/s. 5. 1.000 m/s =

$$\frac{(1.000 \text{ m})/(1000 \text{ m/km})}{(1.000 \text{ s})/(60.00 \text{ s/min})(60.00 \text{ min/h})}$$
$= 3.600$ km/h: 1.000 m/s =
$$\frac{(1.000 \text{ m})(3.281 \text{ ft/m})/(5280 \text{ ft/mi})}{(1.000 \text{ s})/(60.00 \text{ s/min})(60.00 \text{ min/h})}$$
$= 2.237$ mi/h: 1.000 m/s =
$$\frac{(1.000 \text{ m})(3.281 \text{ ft/m})}{(1.000 \text{ s})} = 3.281 \text{ ft/s}.$$
9. (a) $v_{av} = l/t = (15$ km)/(0.25 h) = 60 km/h; $v_{av} = l/t = (15$ km)/ (0.75 h) = 20 km/h; (b) $v_{av} = l/t = (30$ km)/(1.0 h) = 30 km/h. 15. Assuming the time for the light to reach you is negligible, $l = (3.3 \times 10^2$ m/s)(5.0 s) = 1.7 km. The error incurred by not including the time for light to travel is a few millimeters. 19. L is the one-way distance; $v_{av} = l/t = 2L/[L/(10$ km/h) + $L/(5.0$ km/h)] = 2/[1/(10 km/h) + 1/(5.0 km/h)] = 20/(3.0 km/h) = 6.7 km/h. 21. $l = Ct^2$; $(l + \Delta l) = C(t + \Delta t)^2 = C[t^2 + 2t\Delta t + (\Delta t)^2]$; $\Delta l = Ct^2 + 2Ct\Delta t + C(\Delta t)^2 - l$; $v_{av} = \Delta l/\Delta t = [2Ct\Delta t + C(\Delta t)^2]/\Delta t = 2Ct + C\Delta t$. Letting $\Delta t \to 0$, in $v_{av} = 2Ct + C\Delta t$, the second term vanishes and $v = 2Ct$. 25. This is a 3-4-5 right triangle, so the displacement is 15 km. 29. Zero, 36 km. 35. $v_{av} = l/t = (43$ m)/ (10 s) = 4.3 m/s; $\mathbf{v}_{av} = \mathbf{s}/t = (3.0$ m–south)/ (10 s) = 0.30 m/s–south. 39. (a) The slopes

are zero, as are both velocities. (b) The slope is -2.0 m in 5.0 s or $v = (-2.0$ m)/(5.0 s) = -0.4 m/s and so $\mathbf{v} = 0.4$ m/s–south. 43. From Eq. (2.6), $\mathbf{v}_{av} = (\mathbf{s}_f - \mathbf{s}_i)/\Delta t$; $\Delta \mathbf{s} = \sqrt{(3.0 \text{ m})^2 + (3.0 \text{ m})^2}$–southeast = 4.24 m–southeast; $\mathbf{v}_{av} = 2.1$ m/s–southeast. 45. (a) $\mathbf{v}_{av} = (600$ m)/(3 s)–north = 2×10^2 m/s–north; (b) also 2×10^2 m/s–north. 47. Each runs 500 m in time t at speed 5.00 m/s; (500 m)/(5.00 m/s) = 100 s. Alternatively, together they cover the track at a speed of 5.00 m/s + 5.00 m/s = 10.0 m/s; hence, (1000 m)/(10.0 m/s) = 100 s. 51. $\tan \theta_1 = 12/5 = 2.400$, $\theta_1 = 67.38°$; $\tan \theta_2 = 5/12 = 0.416\,7$, $\theta_2 = 22.62°$. 55. $\tan 25° = h/(25$ m) = 0.466; $h = 11.66$ m, or 12 m. 61. $\cos \theta = R_\theta/R$; $R_\theta = R \cos \theta = (6400$ km) cos 41° = 4830 km; the circumference is $2\pi R_\theta = 3.0 \times 10^4$ km; $v_{av} = (3.0 \times 10^4$ km)/(24 h) = 1.3×10^3 km/h, or 7.9 $\times 10^2$ mi/h. 65. (a) No. Those velocity vectors cannot form a right triangle because the hypotenuse ($\mathbf{v}_{BW}$) must be larger than either of the other sides, and here $v_{BW} = v_{WE}$. (b) $\mathbf{v}_{BE}$ cuts directly across the river. Head at $\mathbf{v}_{BW}$ somewhat upstream at an angle θ such that $\mathbf{v}_{BE} = \mathbf{v}_{BW} + \mathbf{v}_{WE}$, $\sin \theta = (20$ mi/h)/ (30 mi/h) = 0.667, $\theta = 42°$. (c) $v_{BE} = [(30$ mi/h)$^2 - (20$ mi/h)$^2]^{1/2} = 22$ mi/h = 35 km/h. 69. (a) $v_{RS} = 1470$ km/h; $\mathbf{v}_{RC} = \mathbf{v}_{RS} + \mathbf{v}_{SC} = 1470$ km/h–west + 1470 km/h–east = 0. (b) $\mathbf{v}_{RC} = 1470$ km/h–east + 1470 km/h–east = 2940 km/h–east. (c) Eastward. (d) Florida is closer to the equator and is moving faster than Maine.

CHAPTER 3
Answers to Odd-Numbered Multiple Choice Questions
1. a 3. c 5. b 7. b 9. e 11. a 13. d 15. c 17. d

Answers to Odd-Numbered Problems
1. 10 m/s^2 3. 0.17 m/s^2–north 5. 0.23 m/s^2 7. 18 m/s^2 9. 5.0 m/s^2 11. 20 m/s^2 13. 3.5 m/s^2 15. (a) $a_{av} = 0$ (b) $a = -1.1$ m/s^2 17. -17 km/s^2 19. $a = A + 2Bt$ 21. 1×10^2 m/s^2 23. $v_f = 4.7$ m/s 25. $v_i = 48$ m/s 27. $a = -0.9 \times 10^6$ m/s^2 29. 0.2 s 31. $s_3 - s_2 = 1.6$ m 33. 244.5 m and they don't collide 35. (a) 70 m (b) 90 m 37. 0.1×10^2 m 39. 10 m/s^2 41. 74 m 43. 0.12 km 45. 3.1 m/s; 10 ft/s; 7.0 mi/h 47. -9.8 m/s, -20 m/s, -49 m/s, -98 m/s; -4.9 m,

−20 m, −0.12 km, −0.49 km   **49.** $v_f =$ 423 m/s = 947 mi/h   **51.** s versus t graph is a parabola. s versus t^2 graph is a straight line   **53.** proof   **55.** bullet falls 49.0 cm below target center   **57.** 20 m   **59.** the equation must be wrong   **61.** $v_i = 4.9$ m/s   **63.** 127.5 m   **65.** $a = -gs_{max}/s_a$ **69.** 11 m/s   **71.** 24.6 m, $v_f = 50.8$ m/s   **73.** proof   **75.** $s_B = 6.1$ m   **77.** $24t^4 - 255t^3 + 901t^2 - 400 = 0$

Solutions to Selected Problems

3. Taking north as positive, $a_{av} = \Delta v/\Delta t = [(+10$ m/s$) - (-10$ m/s$)]/$ $(120$ s$) = 0.17$ m/s^2–north.   **7.** $a_{av} = \Delta v/\Delta t =$ $(244 \times 0.4470$ m/s$)/(6.2$ s$) = 18$ m/s^2.   **11.** Draw a line on the curve from $t = 0$ to $t = 3$ s: the rise is 6(10 m/s); the run is 3.0 s; the slope is $(60$ m/s$)/(3.0$ s$) = 20$ m/s^2.   **17.** $a_{av} = \Delta v/\Delta t = (-60$ km/h $\times$ 1000 km/m $\div$ 3600 s/h$)(1$ s/1000$) =$ -17 km/s^2.   **19.** $v = At + Bt^2$; $(v + \Delta v) = A(t + \Delta t) + B(t + \Delta t)^2 =$ $At + A\Delta t + Bt^2 + B(\Delta t)^2 + B2t\Delta t$; $\Delta v = A\Delta t + B(\Delta t)^2 + 2Bt\Delta t$; $a_{av} = \Delta v/\Delta t = A + 2Bt + B\Delta t$; as $\Delta t \rightarrow 0$, $a = A + 2Bt$.   **23.** $v_f^2 = v_i^2 + 2as =$ $(1.5$ m/s$)^2 + 2(1.0$ m/s$^2)(10$ m$) = 22$ m^2/s^2; $v_f = 4.7$ m/s.   **29.** From Eq. (3.12), $s =$ $v_i t_R - v_i^2/2a$; $t_R = (23.3$ m$)/(16.7$ m/s$) +$ $(16.7$ m/s$)/2(-7$ m/s$^2) = 0.2$ s.   **33.** To stop, the Express needs $s_E = -v_i^2/2a =$ $-(26.67$ m/s$)^2/2(-4$ m/s$^2) = 88.9$ m; the Flyer needs $s_F = -v_i^2/2a = -(30.56$ m/s$)^2/$ $2(-3$ m/s$^2) = 155.6$ m; for a total of 244.5 m and they don't collide.   **39.** It covered 20 m in 2.0 s; $s = v_i t + \frac{1}{2}at^2$; the initial relative speed was zero; $a = 2s/t^2 = 10$ m/s^2.   **43.** His initial speed with respect to the train is 50 km/h, and he passes it as if it were standing still. Hence $s = v_i t + \frac{1}{2}at^2$, 500 m = (13.89 m/s)$t + \frac{1}{2}(10$ m/s$^2)t^2$, and using the quadratic formula, $t = [-(13.89$ m/s$) \pm$ $\sqrt{(13.89 \text{ m/s})^2 - 4(5.0 \text{ m/s}^2)(-500 \text{ m})]/}$ $2(5.0$ m/s$^2) = (-13.89$ m/s ± 100.96 m/s$)/$ $(10$ m/s$^2) = 8.7$ s; the train travels $s =$ $(13.89$ m/s$)(8.7$ s$) = 0.12$ km.   **49.** Taking down as positive, $v_f^2 = v_i^2 + 2as =$ $0 + 2(9.81$ m/s$^2)(9144$ m$)$; $v_f = 423$ m/s = 947 mi/h.   **53.** $v_{iH} = v_i \cos\theta$, $v_{iV} = v_i \sin\theta$, hence Eq. (3.16) yields the desired expression.   **57.** (Taking down as positive) $s_V = \frac{1}{2}gt^2 =$ $\frac{1}{2}(9.81$ m/s$^2)(2.0$ s$)^2 = 19.6$ m, or to two figures, 20 m.   **61.** $s = v_i t + \frac{1}{2}at^2$; 0 = $v_i(1.0$ s$) + \frac{1}{2}(-9.81$ m/s$^2)(1.0$ s$)^2$; $v_i =$ 4.9 m/s.   **65.** The lift-off speed is v_I, where $v_I^2 = 2as_a$. During the deceleration, $0 = v_I^2 + 2gs_{max}$ or $v_I^2 = -2gs_{max} = 2as_a$; $a = -gs_{max}/s_a$.   **71.** To avoid the quadratic, use Eq's. (3.10) and (3.8) applied to the vertical motion, where $v_{iV} = (40.0$ m/s$) \times$ $\sin 60.0° = 34.64$ m/s. Taking down as positive, $v_{fV}^2 = (34.64$ m/s$)^2 +$ $2(9.81$ m/s$^2)(50.0$ m$)$; and $v_{fV} = 46.70$ m/s. $s_V = 50.0$ m $= \frac{1}{2}(v_{iV} + v_{fV})t = (40.67$ m/s$)t$ and $t = 1.23$ s. It strikes the water a distance downrange of $v_H t = (40.0$ m/s$)(\cos 60.0°) \times$

(1.23 s) = 24.6 m. It hits at a speed of $v_f = (v_{fH}^2 + v_{fV}^2)^{1/2} = 50.8$ m/s.   **75.** Down is positive; over his height $s = v_i t + \frac{1}{2}at^2$; 2 m = $v_i(0.20$ s$) + \frac{1}{2}(9.81$ m/s$^2)(0.20$ s$)^2$; $v_i = 9.02$ m/s at top of his head; at ground $v_f = v_i + gt = 9.02$ m/s + (9.81 m/s^2) $\times$ $(0.20$ s$) = 10.98$ m/s; for total fall, $v_f^2 =$ $v_i^2 + 2as_B$; $(10.98$ m/s$)^2 = 0 + 2(9.81$ m/s$^2)s_B$; $s_B = 6.1$ m.   **77.** Using the Pythagorean Theorem, we obtain $(20$ m$)^2 = (v_H t)^2 +$ $(v_{iv}t + \frac{1}{2}gt^2)^2$, $v_{iH} = v_i \cos\theta = (30$ m/s$) \times$ $\cos 60°$ and $v_{iV} = v_i \sin\theta = (30$ m/s$) \sin 60°$. $24t^4 - 255t^3 + 901t^2 - 400 = 0$.

CHAPTER 4

Answers to Odd-Numbered Multiple Choice Questions

1. a   **3.** a   **5.** c   **7.** c   **9.** e   **11.** c   **13.** b   **15.** c   **17.** a   **19.** c

Answers to Odd-Numbered Problems

1. $v_M = 10$ m/s–horizontal   **3.** 1.000 0 m   **5.** 8.36 km compared to 110 m   **7.** 16.7 km   **9.** 100 N   **11.** 100 N on each person   **13.** $F = 298$ N at $\theta = +35°$ **15.** 0.12 kN at $\theta = 54°$ **17.** (a) 24.0 m   (b) 96.0 m   **21.** 500 N at 76.9° up from x-axis   **23.** one due east; the other two, $\pm 60°$ **25.** 400 N–straight up   **27.** 37 m   **29.** -10 N·s   **31.** unchanged, 10 kg·m/s   **33.** -2 MN   **35.** 0.012 s   **37.** 3.3 m/s–south   **39.** -0.12 m/s   **41.** 1.53 kN   **43.** 0.3 N   **45.** 203 N   **49.** $s_h = 1.7$ m   **51.** 42 MN   **53.** 1.0 kN   **55.** -1.00 kN   **57.** 18 m/s   **59.** 0.020 2 m/s opposite to snowball   **61.** 83.9 kN   **63.** -0.12 MN   **65.** $v_{Nf} =$ -0.100 m/s, $v_{Sf}' = 0.198$ m/s, $v_{Sf}'' =$ 0.400 m/s

Solutions to Selected Problems

1. At maximum altitude, $v_V = 0$, hence $\mathbf{v}_M = 10$ m/s–horizontal.   **5.** Using Eq. (3.15), $s_p = -v^2/2g = 8.36$ km, which is tremendously different from 110 m! The difference is due to drag, which is large because of the high speed and effective because of the small mass.   **11.** The force is 100 N on each person.   **13.** $F_{1x} = (100$ N$) \times$ $\cos 45° = 70.7$ N; $F_{2x} = (200$ N$) \cos 30° =$ 173.2 N; $F_{1y} = (100$ N$) \sin 45° = 70.7$ N; $F_{2y} = (200$ N$) \sin 30° = 100$ N; $F^2 =$ $(243.9$ N$)^2 + (170.7$ N$)^2$; $F = 298$ N; at $\theta =$ $\tan^{-1} 170.7/243.9 = +35°$.   **17.** Just to use something other than Eq. (3.15), $v_{iV} =$ $(30.7$ m/s$) \sin 45° = 21.71$ m/s, take up as positive; $v_{fV} = v_{iV} + gt$, 0 = 21.71 m/s + $(-9.81$ m/s$^2)t_p$; $t_p = 2.21$ s. $s_{Hp} = v_H t_p =$ $(21.7$ m/s$)(2.21$ s$) = 48.0$ m. (a) $s_p =$ $\frac{1}{2}(v_{iV} + v_{fV})t_p = \frac{1}{2}(21.7$ m/s$)(2.21$ s$) = 24.0$ m above the launch height. (b) $s_R = 2s_{Hp} =$ 96.0 m.   **21.** There is 90° between the two vectors, which form a 3-4-5 right triangle, so the resultant is 500 N at an angle of $36.9° + 40°$ up from the x-axis.   **25.** By symmetry, the horizontal components of the forces cancel, leaving only $4(200$ N$) \times$ $\cos 60° = 400$ N acting straight up.   **29.** Taking the direction of motion as

positive, $\Delta p = m\Delta v = (1.0$ kg$) \times$ $(-10$ m/s$) = -10$ N·s.   **33.** To find F_{av}, the force exerted on the gas, take the direction of the exhaust as positive and use $F_{av}\Delta t = m\Delta v$; $F_{av} = (1000$ kg$)(2$ km/s$)/(1.000$ s$) = 2$ MN. Thrust is -2 MN.   **37.** The initial momentum of the boat-person system is zero, $\mathbf{p}_i = \mathbf{p}_f = 0$, and $m_p v_{pf} + m_b v_{bf} = 0$; taking motion north as positive, $(50$ kg$)(+10$ m/s$) = -(150$ kg$)v_{bf}$; $v_{bf} = -3.3$ m/s; or 3.3 m/s–south.   **43.** 2 oz = 2/16 lb; $m = (2/16$ lb$) \times$ $(0.453\,6$ kg/lb$) = 0.056\,7$ kg; $F_{av} =$ $m\Delta v/\Delta t = (0.056\,7$ kg$)(0.50$ m/s$)/$ $(0.1$ s$) = 0.284$ N, or 0.3 N.   **47.** $F_{av} =$ $m\Delta v/\Delta t$; impact speed from Eq. (3.10), $v_I = \sqrt{2gs_h}$. The duration of the collision follows from Eq. (3.8), $s_c = \frac{1}{2}v_I\Delta t$, and so $\Delta t = 2s_c/v_I$; hence, $F_{av} = m\Delta v/\Delta t =$ $mv_I/(2s_c/v_I) = mv_I^2/2s_c = mgs_h/s_c$.   **53.** Taking the initial direction as positive, $F_{av} = m\Delta v/\Delta t$; from Eq. (3.8), $s = \frac{1}{2}v_I\Delta t$, $\Delta t = 2(0.20$ m$)/(200$ m/s$) = 2.0$ ms; $F_{av} =$ $(0.010$ kg$)(200$ m/s$)/(0.002\,0$ s$) = 1.0$ kN.   **57.** $p_i = (10\,000$ kg$)(20$ m/s$) = 2.0 \times$ 10^5 kg·m/s $= p_f = (10\,000$ kg + 900 kg$)v_f$; $v_f = 18.3$ m/s, or 18 m/s.   Let's see what a is first, then t, and then F. $v_f^2 = v_i^2 + 2as$; $a = (69.5$ m/s$)^2/(914.4$ m$) = 2.64$ m/s^2; $v_f = v_i + at$; $t = (69.5$ m/s$)/(2.64$ m/s$^2) =$ 26.3 s; m = 31 751 kg; $F = m\Delta v/\Delta t =$ $(31\,751$ kg$)(69.5$ m/s$)/(26.3$ s$) = 83.9$ kN. We could have started with Eq. (3.8).   **63.** 120 mi/h = 53.64 m/s; if v_T is his terminal speed and s_D the depth of the crater, Eq. (3.8) yields $\Delta t = 2s_D/v_T = 0.039\,8$ s. $F_{av} = m\Delta v/\Delta t = (90$ kg$)(-53.64$ m/s$)/$ $(0.039\,8$ s$) = -0.12$ MN.   **65.** $p_{Ni} + p_{ai} =$ $0 = p_{Nf} + p_{af} = (100$ kg$)v_{Nf} + (0.500$ kg$) \times$ $(20.0$ m/s$)$; $v_{Nf} = -0.100$ m/s—he moves in the opposite direction to the asteroid, whose direction is taken as positive; she catches it and so $p_{ai}' = p_{af}' + p_{Sf}'$; $(0.500$ kg$) \times$ $(20.0$ m/s$) = v_{Sf}'(m_a + m_S) = v_{Sf}'(50.5$ kg$)$; $v_{Sf}' = 0.198$ m/s. Now she throws it back, $p_{af}'' + p_{Sf}'' = p_{Si}'' + p_{ai}''$; $(50.5$ kg$) \times$ $(0.198$ m/s$) = (0.500$ kg$)(-20.0$ m/s$) +$ $(50.0$ kg$)v_{Sf}''$; $v_{Sf}'' = 0.400$ m/s.

CHAPTER 5

Answers to Odd-Numbered Multiple Choice Questions

1. c   **3.** d   **5.** e   **7.** c   **9.** b   **11.** c   **13.** d   **15.** d   **17.** a   **19.** c

Answers to Odd-Numbered Problems

1. -71 N   **3.** >4.9 N   **5.** $1.5 F_w$ **7.** 592 N, 392 N   **9.** $0.586\,g$ **11.** 508 N, upward   **13.** 4.81 m/s^2 **15.** $\theta = 22°$ **17.** $F_{T1} = 1.00$ kN, $F_{T2} = 475$ N, $F_f =$ 125 N   **19.** $v_f = 12$ m/s   **21.** $a = 1.7$ km/s^2, 7.8×10^{-4} N   **23.** $F_{T1} = 0.03$ kN, $F_{T3} =$ 0.02 kN, $a = \frac{1}{4}g$, the tensions are internal to the 3-body system, the weights are external   **25.** $g_M = 3.4$ m/s^2 **27.** 17 m/s   **29.** $F_C = 6.6 \times 10^2$ N, friction   **31.** $F_C =$ 2.1 kN   **33.** $R = 9.5 \times 10^2$ m   **35.** $a_C =$ 1.8×10^6 m/s^2 **37.** 0.12 kN   **39.** $F_N =$ $m(v^2/r + g)$ **41.** 19.2 m/s   **43.** (a) time to go once around, $T = 55$ s, or 0.018 rotations per second   (b) $a_C = 4\pi^2 r/T^2$ **45.** 99.7 N   **47.** $\mu_s = 0.53$ **49.** $a = \mu_s g$ **51.** 0.31   **53.** crate slides toward back of truck   **55.** 0.10   **57.** 20 m   **59.** 3 s   **61.** $\mu_k = 0.070$ **63.** $\mu_k = 0.30$ **65.** (a) $a = 0$ (b) $a = 1$ m/s^2 **67.** $\theta = 0.2 \times 10^2$ degrees   **69.** $\mu_k = 0.11$

Solutions to Selected Problems

1. $F = ma = 0$; $F_f = -F \cos 45° = -(100 \text{ N})(0.707) = -71$ N.   **7.** $+\!\uparrow\Sigma F_V = ma = F_N - F_W$; $F_N = ma + F_W = m(a + g) = (40.0 \text{ kg})(14.81 \text{ m/s}^2) = 592$ N. $F_W = mg = 392$ N.
11. $+\!\downarrow\Sigma F_V = ma = F_W + F_f$; $m(2.00 \text{ m/s}^2) = mg + F_f$; $F_f = -(7.81 \text{ m/s}^2)(65.0 \text{ kg}) = -508$ N, or 508 N, upward.   **17.** The tension in the front rope is 1.00 kN. For the first barge (taking the pull of the tug to be positive and to the right), $\xrightarrow{+}\Sigma F_{H1} = m_1 a = 1.00 \text{ kN} - F_{T2} - F_f = (4.00 \times 10^3 \text{ kg})(0.100 \text{ m/s}^2)$; for the second barge, $\xrightarrow{+}\Sigma F_{H2} = m_2 a = F_{T2} - F_f = (3.50 \times 10^3 \text{ kg})(0.100 \text{ m/s}^2)$. Adding the two equations, $-2F_f = (7.50 \times 10^3 \text{ kg}) \times (0.100 \text{ m/s}^2) - 1.00 \text{ kN}$; $F_f = +125$ N, opposing the motion; $F_{T2} = 475$ N.
21. $s_R = 12$ in. $= 0.304\,8$ m $= [2v_i^2/(9.8 \text{ m/s}^2)] \cos \theta \sin \theta$; $v_i = 1.729$ m/s; (a) $a = \Delta v/\Delta t = (1.729 \text{ m/s})/(1.0 \times 10^{-3} \text{ s}) = 1.7$ km/s^2; (b) $F = ma = (4.5 \times 10^{-7} \text{ kg}) \times (1.73 \text{ km/s}^2) = 7.8 \times 10^{-4}$ N.   **23.** For mass-1, $+\!\downarrow\Sigma F_{V1} = m_1 a = F_{W1} - F_{T1}$; for mass-2, $\xrightarrow{+}\Sigma F_{H2} = m_2 a = F_{T1} - F_{T2}$; for mass-3, $+\!\uparrow\Sigma F_{V3} = m_3 a = F_{T2} - F_{W3}$; adding all three of these, $a(m_1 + m_2 + m_3) = F_{W1} - F_{W3} = g(m_1 - m_3)$; (b) $a = g(2 \text{ kg})/(8 \text{ kg}) = \frac{1}{4}g$; (c) the tensions are internal to the 3-body system; the weights are external; (a) $m_1 a = m_1 g - F_{T1}$; $F_{T1} = m_1(g - a) = 29.4$ N $= 0.03$ kN. $m_3 a = F_{T3} - F_{W3}$, $F_{T3} = m_3(a + g) = 24.5$ N $= 0.02$ kN.
27. $F_C = mv^2/r$, $v = \sqrt{(20 \text{ N})(0.355\,6 \text{ m})/(0.025 \text{ kg})} = 17$ m/s.   **31.** The distance traveled once around is $2\pi r$, hence $v = 2(2\pi r)/(1.0 \text{ s})$; $r = 1.828$ m; $v = 22.98$ m/s; $F_C = mv^2/r = (7.257 \text{ kg})v^2/r = 2096.8$ N, or 2.1 kN.   **35.** $a_C = v^2/r$; $r = (0.100 \text{ m}) \sin 30° = 0.050$ m $= 0.10$ m; $v = 2\pi r(40\,000)/60 = 418.88$ m/s; $a_C = 1.8 \times 10^6$ m/s^2.   **39.** Using Fig. 5.24 with F_T replaced by F_N, we get $F_N = m(v^2/r + g \cos \theta)$; at the bottom, $\theta = 0$, $\cos \theta = 1$; $F_N = m(v^2/r + g)$.   **43.** (a) $a_C = v^2/r = 1.0g$; $v^2 = \frac{1}{2}(1500 \text{ m})g$; $v = 85.76$ m/s; the cylinder must go once around a distance $(2\pi r)$ in a time $(2\pi r)/v = T = 54.9$ s, or 55 s; (b) $a_C = v^2/r = (2\pi r/T)^2/r = 4\pi^2 r/T^2$, where the time to go once around is the same throughout the station. Hence, the simulated gravity varies linearly with r, decreasing with distance "up" from the floor.   **45.** $+\!\downarrow\Sigma F_V = ma_C = F_G - F_N$; F_N is equal and opposite to the effective weight; F_G is the actual gravitation force downward, namely, 100 N. $F_N = F_G - ma_C = F_G - (F_G/g)v^2/R_\oplus = 100 \text{ N} - 0.345 \text{ N} = 99.7$ N.

49. $F = ma$; $a = F_f/m = mg\mu_s/m = \mu_s g$.   **53.** For no slipping, $+\!\swarrow\Sigma F_\parallel = 0 = F_W \sin \theta - F_f(\text{max})$; and $\mu_s = \tan \theta$, $0.3 = \tan \theta$; $\theta = 16.7°$; hence, the crate slides toward the back of the truck.   **57.** $v_i = 50.0$ km/h $= 13.89$ m/s; to find the stopping distance, $v_f^2 = v_i^2 + 2as$; $s = -v_i^2/2a$; where $a < 0$ is determined from $F_f(\text{max}) = \mu_s mg = ma$; hence, $a = \mu_s g$ and so $s = -v_i^2/2(-\mu_s g) = 20$ m.
61. $+\!\swarrow\Sigma F_\parallel = 0 = F_W \sin \theta - F_f$; $\tan \theta = \mu_k = 0.070$. At $-10°$C, the water film that usually forms via melting under the skis would not.   **65.** (a) $F_H = F \cos \theta = (300 \text{ N})(0.866) = 259.8$ N, which must exceed $F_f(\text{max}) = 0.5[(50 \text{ kg})g + (300 \text{ N}) \times \sin 30°] = 320$ N, but it does not, so there is no acceleration;   (b) $F_f(\text{max}) = 256$ N—it moves! $\xrightarrow{+}\Sigma F_H = ma = 259.8$ N $- 0.3(640.3 \text{ N}) = 67.7$ N and $a = 1.4$ m/s^2, or because the friction coefficients are known to only one figure, $a = 1$ m/s^2.   **69.** $v_f^2 = 2as$, $a = (8.33 \text{ m/s})^2/2(30.0 \text{ m}) = 1.158$ m/s^2; $v_f = at$, $t = (8.33 \text{ m/s})/(1.158 \text{ m/s}^2) = 7.196$ s; the box accelerates with respect to the truck at a_b, where $s_b = \frac{1}{2}a_b t^2 = 1.00$ m, and $a_b = 0.038\,6$ m/s^2. If there were no friction, it would have accelerated back at 1.158 m/s^2; hence, $(1.158 \text{ m/s}^2 - 0.038\,6 \text{ m/s}^2)m = F_f = \mu_k mg$, $\mu_k = 0.11$.

CHAPTER 6

Answers to Odd-Numbered Multiple Choice Questions

1. d   **3.** d   **5.** c   **7.** e   **9.** b   **11.** a   **13.** e   **15.** d   **17.** d   **19.** c

Answers to Odd-Numbered Problems

1. (a) 250 N (b) 200 N   **3.** (a) 30 N (b) 38 N, 2.0 N   **5.** they slide equally   **7.** $F_{Tb} = 0$, $F_{Tl} = 98$ N   **9.** $\theta = 38°$, $F_T = 16$ N   **11.** 39.7 N   **13.** $F_T = 50.0$ N, 100 N   **15.** $F_{BC} = 0$, $F_{DE} = 0$, $F_{FG} = 0$ **17.** 6.7 N   **19.** $F_T = 3.1$ kN   **21.** 75°, 193 N   **23.** $\mu_s = 0.93$ **25.** $F_{RB} = 2.00$ kN, $F_{RA} = 1.73$ kN $= F_{RC}$, $F_{RD} = 3.00$ kN   **27.** 100 N·m   **29.** 140 N   **31.** $F_R = 355.9$ N, $x = 0.731\,7$ m   **33.** $F_{RB} = 6000$ N, $F_{RA} = 2000$ N   **35.** $F_{N1} = F_{N2} = 69.3$ N   **37.** 1.0 kg   **39.** $F_{RD} = 360$ N, $F_{RC} = 80.0$ N, $F_{RA} = 40$ N   **41.** $F_n = 0.83$ kN   **43.** $F_f = 11$ N   **45.** $F_{AH} = -1.5$ kN (to the left); $F_{AV} = -0.50$ kN (down)   **47.** midface in the glue 8 cm from the end   **49.** $x_{cg} = 0.063$ m, $y_{cg} = 0.063$ m   **51.** $F_{T2} = 100$ N, $F_W = 173$ N   **53.** yes   **55.** yes   **59.** $F_{RG} = 0.31$ kN at 74°   **61.** $F_{Nh} = 0.21$ kN, $F_{Nt} = 0.11$ kN   **63.** $x_{cg} = F_N h/(F_{N1} + F_{N2})$ **65.** $F_T = 90.4$ N

Solutions to Selected Problems

3. (a) $(2.0 \text{ N/m})(20 \text{ m})(3/4) = 30$ N. (b) 1.0 m from the top, the tension is 38 N; 1.0 m from the bottom, it's 2.0 N.   **5.** For each one, $+\!\swarrow\Sigma F_\parallel = F_W \sin \theta - \mu_s F_W \cos \theta = 0$; and $\mu_s = \tan \theta$ independent of weight, so they

tend to slide equally.   **9.** $+\!\uparrow\Sigma F_V = (20 \text{ N}) \cos 30° + F_T \cos \theta - 30 \text{ N} = 0$; $\xrightarrow{+}\Sigma F_H = F_T \sin \theta - (20 \text{ N}) \sin 30° = 0$; hence, $F_T \cos \theta = 12.679$ N and $F_T \sin \theta = 10$ N. We could divide these and get $\sin \theta/\cos \theta = 0.788\,7 = \tan \theta$ and $\theta = 38°$, or square and add them, since $\sin^2 \theta + \cos^2 \theta = 1$, and get $F_T^2 = (12.697 \text{ N})^2 + (10 \text{ N})^2$; $F_T = 16$ N.   **15.** Since BC is perpendicular to AC and CE at joint C, there can be no vertical force in BC; hence, force in BC is zero. The same is true at joint D (so $F_{DE} = 0$) and at joint G (so $F_{FG} = 0$).   **19.** $+\!\uparrow\Sigma F_V = 2F_T \sin 5.0° - 533.8 \text{ N} = 0$; $F_T = 3.1$ kN.   **25.** On the small ball, $\xrightarrow{+}\Sigma F_H = F_{RA} - F_{RB} \cos 30° = 0$ and $+\!\uparrow\Sigma F_V = F_{RB} \sin 30.0° - 1.00 \text{ kN} = 0$; $F_{RB} = 2.00$ kN, $F_{RA} = 1.73$ kN $= F_{RC}$; $F_{RD} =$ the net weight $= 3.00$ kN.
31. $+\!\uparrow\Sigma F_V = 533.8 \text{ N} - 889.66 \text{ N} + F_R = 0$; $F_R = 355.9$ N; $(+\Sigma \tau_S = (355.9 \text{ N}) \times (1.829 \text{ m}) - (889.66 \text{ N})x = 0$; $x = 0.731\,7$ m from Selma.   **35.** The reaction forces are each normal to the surfaces because there is no friction. $F_{N1} = F_W \sin 45.0° = F_W \cos 45.0° = F_{N2} = 69.3$ N.   **41.** $(+\Sigma \tau = (300 \text{ N})(0.25 \text{ m}) - F_n(0.09 \text{ m}) = 0$; $F_n = 0.83$ kN.   **45.** Take the horizontal bar out and draw a free-body diagram; $(+\Sigma \tau_A = (1000 \text{ N})(3.0 \text{ m}) - (2.0 \text{ m})F_C \sin 45° = 0$; $F_C = 2121.6$ N. At A, the horizontal force is $F_{AH} = -F_{CH} = -1.5$ kN (force is to the left). $+\!\uparrow\Sigma F_V = F_C \sin 45° - 1000 \text{ N} - F_{AV} = 0$; $F_{AV} = -0.50$ kN (force is down).
49. Taking the upright side as $y = 0$, $\Sigma mx = (0.50 \text{ kg/m})(0.25 \text{ m})(0.125 \text{ m}) = 0.015\,6$ kg·m and $x_{cg} = (0.015\,6 \text{ kg·m})/(0.25 \text{ kg}) = 0.063$ m, and, by symmetry, $y_{cg} = 0.063$ m.   **55.** If each is of length L, the $c.g.$ of the topmost block is at the edge of the one beneath it. Measured from the left end, for the top two blocks, $x_{cg} = (mL + mL/2)/2m = 3L/4$, and that is located just above the edge of the third block. For the top three blocks, $x_{cg} = (2mL + mL/2)/3m = 5L/6$. For the top four blocks, $x_{cg} = (3mL + mL/2)/4m = 7L/8$. The overhangs are $L/8$, $L/6$, and $L/4$, which is more than $L/2$.   **59.** $+\!\uparrow\Sigma F_V = F_N - g(30 \text{ kg}) = 0$; $F_N = 0.29$ kN. $(+\Sigma \tau = 0 = F_N(8.0 \text{ m}) \cos 60° - F_f(8.0 \text{ m}) \sin 60° - g(30 \text{ kg})(4.0 \text{ m}) \cos 60°$; $F_{RG} = 0.31$ kN at 74°.   **63.** $(+\Sigma \tau_1 = F_W x_{cg} - F_{N2}h = 0$; $(+\Sigma \tau_2 = F_{N1}h - F_W(h - x_{cg}) = 0$; $F_W = F_{N2}h/x_{cg}$; $F_{N1}h - F_{N2}h(h - x_{cg})/x_{cg} = 0$; $x_{cg} = F_{N2}h/(F_{N1} + F_{N2})$.   **65.** Taking O at the base of the wall, $(+\Sigma \tau_o = (4.00 \text{ m})F_{N2} - (3.00 \text{ m})F_{N1} - F_W(2.00 \text{ m}) = 0$; and $\xrightarrow{+}\Sigma F_H = F_T \cos \theta - F_{N1} = 0$, where θ is the angle the rope makes with the ground. From the geometry (we are dealing here with a 3-4-5 right triangle), $\theta = 10.62°$. $+\!\uparrow\Sigma F_V = F_{N2} - F_W - F_T \sin \theta = 0$. Eliminating F_{N1} from the first and second equations, $-(3.00 \text{ m}) \times F_T \cos \theta + F_W(2.00 \text{ m}) + F_{N2} = 0$; and substituting F_{N2} into the third equation, $(0.75 F_T \cos \theta + 0.50 F_W) - F_W - F_T \sin \theta = 0$. Hence, $F_T(0.75 \cos \theta - \sin \theta) = 0.50 F_W$; $F_T = 90.4$ N.

CHAPTER 7

Answers to Odd-Numbered Multiple Choice Questions

1. c   3. d   5. c   7. c   9. b   11. d   13. c
15. b

Answers to Odd-Numbered Problems

1. would be halved   3. $m = 1.22 \times 10^5$ kg
5. 4 times larger   7. 2.5×10^{16} N
9. $s_\female = 1.1$ m   11. $F_E = 2 \times 10^{39} F_G$
13. $0.11\ M_\oplus$ 15. 4.4 m   19. $0.379\ g_0$
21. $g_n = 1 \times 10^{12}$ m/s^2 23. $F_G = GmMr/R^3$
25. $T = 1 \times 10^3$ s   29. 1.62 m/s^2
31. 1.65×10^3 m/s   35. $T_2 = 2.8\ T_1$
37. 3.32×10^{-5} m/s^2 39. 5.9×10^{-3} m/s^2
41. 8.000 Earth-years   45. 1.078 AU,
1.12 Earth-years   47. $GM/2R^2$ 49. 1.05×10^3 kg/m^3 51. 9.7 d, 1.440 km/s

Solutions to Selected Problems

3. $F_G = Gm^2/r^2$; 1.00 N = (6.67 $\times$
10^{-11} N·m^2/kg^2)m^2/(1.00 m)^2; $m = 1.22 \times$
10^5 kg.   9. $v_f^2 = 0 = v_i^2 - 2gs$; $s = v_i^2/2g$,
$s_\female/s_\oplus = g_\female/g_\oplus$, (1.00 m)/$s_\female$ = 0.88, $s_\female =$
1.1 m.   13. $g_\delta = GM_\delta/R_\delta^2$, $M_\delta =$
(3.7 m/s^2)($\frac{1}{2}6.8 \times 10^6$ m)^2/(6.67 $\times$ 10^{-11} N·m^2/
kg^2) = 6.4 $\times$ 10^{23} kg = 0.11 $M_\oplus$.   15. To
find $g_\female$, we need the mass; $M_\female = \frac{4}{3}\pi r^3 \rho =$
4.82 $\times$ 10^{24} kg; $g_\female = GM_\female/R_\female^2$ = 8.79 m/s^2;
$s = \frac{1}{2}g_\female t^2$ = 4.4 m.   19. $g_\delta = GM_\delta R_\delta^2 =$
$G(0.108\ M_\oplus)/R_\oplus^2(0.534)^2 = 0.379\ g_0$.
23. $F_G = GmM_r/r^2$, where $M_r = \frac{4}{3}\pi r^3 \rho$, and
$\rho = M/(4\pi R^3/3)$; hence, $M_r = Mr^3/R^3$ and
$F_G = GmMr/R^3$.   27. $g_p = GM/(R + h)^2 =$
$GM/R^2(1 + h/R)^2 = (GM/R^2)(1 + h/R)^{-2}$.
From the binomial expansion, the bracketed
term becomes $1 + (-2)(1)h/R + \frac{1}{2}(-2)(-3) \times$
$(1)(h/R)^2 + \dots$, which is $\approx(1 - 2h/R)$ since
the $(h/R)^2$ term is very small.   29. $g_\mathbb{C} \approx$
$GM_\mathbb{C}(1 - 2h/R)/R^2 \approx GM_\mathbb{C}[1 - (200\ m)/$
$(1.74 \times 10^6\ m)]/R_\mathbb{C}^2 \approx 0.999885\ GM_\mathbb{C}/$
$R_\mathbb{C}^2 \approx 1.62$ m/s^2. The difference is
negligible.   33. From Eq. (7.7),
$T^2 = 4\pi^2 r_\oplus^3/M_\oplus G = 9.90 \times 10^{-14} r_\oplus^3$;
$T = 3.15 \times 10^{-7} r_\oplus^{3/2}$.   37. $g_\mathbb{C} = GM_\mathbb{C}/r_{\mathbb{C}\oplus}^2 =$
3.32 $\times$ 10^{-5} m/s^2.   41. From Eq. (7.6),
$T^2 = r_\odot^3/C_\odot = (4.000$ AU)^3/(1 AU^3/
Earth-year^2) = 64.00 Earth-years^2;
T = 8.000 Earth-years.   45. $r_{\odot I} =$
$\frac{1}{2}$(0.186 AU + 1.97 AU) = 1.078 AU.
$r_{\odot I}^3/T_I^2 = 1 = (1.078$ AU)$^3/T_I^2$; $T_I = 1.12$
Earth-years.   51. From Eq's. (7.6) and (7.7),
you find that $C_\oplus = GM_\oplus/4\pi^2 =$
1.010×10^{13} m^3/s$^2 = r_\oplus^3/T_s^2 = (\frac{1}{3}3.844 \times$
10^8 m)$^3/T_s^2$ and $T_s = 0.8385 \times 10^6$ s =
9.7 days; $v = 2\pi r_\oplus/T_s$ = 1.440 km/s. Check
this answer using Eq. (7.8).

CHAPTER 8

Answers to Odd-Numbered Multiple Choice Questions

1. b   3. e   5. b   7. c   9. b   11. b   13. d
15. a   17. d   19. b

Answers to Odd-Numbered Problems

1. 0.017453 rad   3. 1.6 rad, 0.02 rad,
0.02 rad, 2.0 rad   5. $D = 3 \times 10^5$ m

7. 2.0 m   9. 69 rev   11. 12.2 rad/s
13. 2.3 m   15. 105 rad/s^2 17. 4.2 s
19. 2.97×10^4 m/s, 1.99×10^{-7} rad/s
21. 0.2 m   23. 0.029 rad/s, 0.26 m/s^2
25. 2.7 m/s   27. $v_E = 0.1$ m/s, upward
29. $\alpha = 9.82$ rad/s^2, 16.0 s
31. 4.84×10^3 km, $\theta = 40.6°$
33. $\omega_4 = 1200$ rpm, $\omega_5 = 400$ rpm,
$\omega_n = \omega_1 N_1/N_n$; if n is odd, the direction is
unchanged   35. 1.4 m   37. 0.75 m/s—upward
39. 25 rev   41. -2.4×10^{-22} rad/s^2,
6.51×10^{-5} rad/s, 27 h   43. MR^2
45. 450% decrease   47. $\frac{1}{3}ml^2$
49. 3×10^{-3} kg·m^2/s   51. 0.20 rad/s^2
53. 426 rad/s^2 55. $a = mg/(m + I/R^2)$
57. $a = 1.5$ m/s^2 59. (a) 1.6×10^{-4} kg·m^2
(b) 8.0×10^{-2} kg·m^2 (c) ≈ 0.080 kg·m^2
(d) 0.010 kg·m/s (e) 4.0×10^{-3} kg·m^2/s
(f) 5.0×10^{-2} rad/s   61. $a = \frac{1}{2}g \sin \theta = g/4$
65. 1.2 m/s, 1.8×10^4 kg·m^2/s, 24 m/s,
17×10^3 N   69. $I \approx \frac{1}{3}Ml^2$

Solutions to Selected Problems

3. (ii) 90° = 1.6 rad; tan α = 1/100,
$\alpha = 5.7 \times 10^{-1}$, $2\alpha = 1.15° = 0.0201$ rad;
$\theta = l/r = (2.00$ m)/(100 m) = 0.0200 rad.
(i) $\theta = l/r = 2.00$ rad.   5. $l = r\theta =$
$(3.8 \times 10^8$ m)(8 $\times$ 10^{-4}) = 3 $\times$ 10^5 m, which
is the diameter.   9. $2\pi r = 2\pi(0.15$ m) =
0.942 m, (65 m)/$2\pi r$ = 69 rev.
13. $\omega = 2\pi(1800$ rpm)/(60 s/m) =
188.5 rad/s, $v = R\omega = (0.12$ m) $\times$
(188.5 rad/s) = 22.62 m/s; $l = vt =$
(22.62 m/s)(0.10 s) = 2.3 m.   17. $\omega_i =$
$200(2\pi)/(60$ rad/s) = 20.94 rad/s; $0 = \omega_i +$
$(-5$ rad/s$^2)t$; $t = 4.2$ s.   21. $\omega^2 = 0 + 2\alpha\theta$,
$(20 \times 2\pi/60)^2 = 2(5$ rad/s$^2)\theta$, θ = 0.4 rad, or
0.07 turns. (0.4 rad)(0.5 m) = 0.2 m.
25. $v = 2\pi R(100$ rpm)/(60 s/min) =
8.7 ft/s = 2.7 m/s.   29. $\omega_f^2 = \omega_i^2 + 2\alpha\theta$,
$(1500 \times 2\pi/60)^2 = 2\alpha(200 \times 2\pi)$,
α = 9.82 rad/s^2. $\omega_f = \omega_i + \alpha t$,
$t = (1500 \times 2\pi/60)/(9.82$ rad/s^2) = 16.0 s.
33. 1500 rpm $\times$ 20 = $\omega_4 25 = \omega_5 75$;
ω_4 = 1200 rpm; ω_5 = 400 rpm.
$\omega_n = \omega_1 N_1/N_n$, where N is the number of
teeth and n is the number of gears. If n is odd,
the direction is unchanged.   35. $a_C = R\omega^2$,
$a_T = R(4.0$ rad/s^2), $a = \sqrt{a_C^2 + a_T^2}$,
$(8.0$ m/s$^2)^2 = R^2\omega^4 + R^2(4.0$ rad/s$^2)^2$; $R^2 =$
2.0 m^2, R = 1.4 m.   37. $l_3 = (0.24$ m)θ_1,
$l_5 = (0.12$ m) $\theta_1 = (0.20$ m)θ_2,
$l_4 = (0.15$ m)θ_2, so $\theta_2 = 0.60\theta_1$ and
$l_4 = (0.15$ m)(0.60 θ_1) = (0.90 $\times$ 10^{-1}) $\times$
$\theta_1 = 0.90 \times 10^{-1}$ $(l_3/0.24) = 0.375 l_3$,
$v_4 = 0.375 v_3 = 0.375(2.0$ m/s) = 0.75 m/s,
upward.   45. 450% decrease; $L_i = L_f$.
49. $L = I\omega$; $I \approx mr^2 = (2 \times 10^{-3}$ kg) $\times$
$(0.5$ m)$^2 = 5 \times 10^{-4}$ kg·m^2; $L \approx (5 \times$
10^{-4} kg·m^2)(2π rad/s) = 3 $\times$ 10^{-3} kg·m^2/s.
53. $\Sigma \tau = 50.0$ ft·lb = $I\alpha$, $I = mR^2 =$
$(F_W/g)R^2$, 50.0 ft·lb = [(3.22 lb)/(32.2 ft/s^2)] $\times$
[(13 in.)/(12 in./ft)]$^2\alpha$; α = 426 rad/s^2.
57. $+\downarrow \Sigma F_v = ma = F_W - F_T$; $\zeta + \Sigma\tau = I\alpha =$
$F_T R - 0.060$ N·m; $a = R\alpha$; $F_T = (I\alpha +$
0.060 N·m)/$R = Ia/R^2 + (0.060$ N·m)/R;
$F_W - Ia/R^2 - (0.060$ N·m)/$R =$

ma, $a(m + I/R^2) = Mg - (0.060$ N·m)/R,
a = 9.2/6.0 = 1.5 m/s^2, as compared to
1.6 m/s^2 previously.   61. $+\nwarrow\Sigma F_\| =$
$F_W \sin \theta - F_f = ma$; $\Sigma \tau_{cm} = F_f R =$
$I_{cm}\alpha = mR^2\alpha$; $a = R\alpha$; $F_f R = mRa$;
$F_W = mg$; $a = \frac{1}{2}g \sin \theta = g/4$.   65. $v =$
$a_T t = 1.0 \times 10^{-3} g(120$ s) = 1.2 m/s;
$L = rmv = 1.8 \times 10^4$ kg·m^2/s.
$L = (5.0$ m)(150 kg)$v = 1.8 \times 10^4$ kg·m^2/s,
v = 24 m/s, or 54 mi/h. $F_C = mv^2/r =$
17×10^3 N, or 4×10^3 lb!   67. $+\nwarrow\Sigma F_\| =$
$mg \sin \theta - F_f = ma$; $\zeta + \Sigma\tau_{cm} = F_f R = I\alpha$,
$\alpha = a/R$, $I_{cm} = \frac{1}{2}mR^2$, $F_f = \frac{1}{2}ma$;
$mg \sin \theta = (3/2)ma$, $a = (2/3)g \sin \theta$;
$s = h/\sin \theta$, $v_f^2 = 2as = \frac{2(2/3)(g \sin \theta)h}{\sin \theta} = (4/3)gh$. $v_f = 2\sqrt{gh/3}$.

CHAPTER 9

Answers to Odd-Numbered Multiple Choice Questions

1. a   3. a   5. a   7. c   9. c   11. a   13. a
15. c   17. e   19. c   21. d

Answers to Odd-Numbered Problems

1. 23 J   3. 2.0 J   5. 20 MJ   7. 4.00 N
9. 5.0 W   11. 10 m/s   13. 5.0 m
15. 0.50 kN·m   17. 1.2×10^2 J
19. (a) 0.20 kN·m (b) 2.5 N·m/cm^2
(c) 0.20 kN·m   21. 5.3 s   23. 60 W/m^2
25. 74 W/m^2 27. 90 J   29. 10^{15} W
31. 0.98 m/s, $KE_G = 0.95$ J, $KE_B = 0.29$ kJ
33. 6×10^2 m   35. 1.2×10^6 J
37. (a) 2.9×10^5 J (b) 7.2×10^4 J/s
39. 8 min   41. (a) 100 J (b) no rope
(c) +100 J   43. (a) 0.16 $g_\oplus$ (b) 1.6×10^5 J
45. 1.7×10^3 m   47. 5.2 m   49. both
51. 4.85×10^3 J, h = 9.00 m or 10.0 m above
ground   53. 0.46 km/s   55. -6.2×10^7 J,
-3.1×10^7 J, -2.1×10^7 J, -1.6×10^7 J,
-1.2×10^7 J, -0.62×10^7 J,
E $= -1.0 \times 10^7$ J   57. 2.4 km/s
59. v_{esc}^2(Moon)/v_{esc}^2(Earth) = 0.046
61. 2121 m/s   63. 0.070 m/s
65. v_{2f} = 15 m/s, $v_{1f} = -10$ m/s
67. $v = \sqrt{2(m_2 - m_1)gy/(m_1 + m_2)}$
69. $v_B = [(m_B + m_C)/m_B]\sqrt{2gh}$

Solutions to Selected Problems

3. $W = Fs = (20$ N)(0.10 m) = 2.0 J.
5. The thrust is constant at 10×10^3 N;
hence, $W = (2.0 \times 10^3$ m)(10^4 N) = 20 MJ.
11. P $= Fv = 1.0 \times 10^4$ J/s $= (1.0 \times 10^3$ N)v;
v = 10 m/s, or 22 mi/h.   15. $\sin \theta = 5/13$,
$W = (F_W \sin \theta)s = (100$ N)(5/13)(13 m) =
500 N·m. Straight up, $W = (100$ N)(5.0 m) =
0.50 kN·m.   19. (a) $W = (100$ N)(2.0 m) =
0.20 kN·m; (b) (1.0 cm)(1.0 cm) =
(10 N)(0.25 m) = 2.5 N·m/cm^2; (c) 10 cm $\times$
8.0 cm = 80 cm^2; (80 cm^2)(2.5 N·m/cm^2) =
0.20 kN·m.   23. $\Delta W/\Delta t = (10^7$ J)/
(24 h $\times$ 3600 s/h) = 1.2×10^2 W. BMR =
$(90$ W)/(1.5 m^2) = 60 W/m^2.   29. KE =
$\frac{1}{2}mv^2 = \frac{1}{2}(0.5 \times 10^{-3}$ kg)(200 $\times$ 10^3 m)^2 =
1×10^7 J = $\Delta KE/\Delta t = (1 \times 10^7$ J)/
$(10^{-8}$ s) $= 10^{15}$ W.   33. $PE_G = mgh$;
$h = (6 \times 10^9$ J)/(10^6 kg) $\times$
(9.8 m/s^2) $= 6 \times 10^2$ m.   39. P$t = W =$
$\Delta KE = \frac{1}{2}(2 \times 10^3$ kg)(268 m/s)$^2 =$

7.18×10^7 J; $t = 8$ min. **43.** (a) $g_{\mathbb{C}} = GM_{\mathbb{C}}/R_{\mathbb{C}}^2 = (1/100)/(1/4)^2 g_{\oplus} = 0.16\, g_{\oplus}$. (b) $\Delta PE = mg\Delta h = (1000\text{ kg})(0.16) \times (9.8\text{ m/s}^2)(100\text{ m}) = 1.6 \times 10^5$ J. **45.** Let θ be the incline angle; $\tan\theta \approx \sin\theta = \frac{1}{2}\%$; the net work done by all forces acting on the train equals its ΔKE, $W = [22.7 \times 10^4\text{ N} - (35\text{ N})(1.8 \times 10^3) - (1.8 \times 10^7\text{ N})\sin\theta]s = (7.4 \times 10^4\text{ N})s$; $\Delta KE = \frac{1}{2}(45)[(4.0 \times 10^5\text{ N})/(9.8\text{ m/s}^2)][(13.4\text{ m/s})^2 - (6.7\text{ m/s})^2] = 1.237 \times 10^8$ J; hence, $W = \Delta KE$ and $s = (1.237 \times 10^8\text{ J})/(7.4 \times 10^4\text{ N}) = 1.7 \times 10^3$ m. **49.** For one car, $KE = \frac{1}{2}mv^2 = \frac{1}{2}[(7.12\text{ kN})/(9.81\text{ m/s}^2)] \times (26.67\text{ m/s})^2 = 258$ kN·m; and for the other, $KE =$ twice that; viz., 516 kN·m. For one car on top of the mountain, $PE = mgh = (7.12\text{ kN})(33.5\text{ m}) = 238.5$ kN·m; and for the other, $PE =$ twice that; viz., 477 kN·m. Thus, both cars have enough energy to coast over the top. **53.** $p_i = (1.00\text{ kg})v_i + 0 = p_f = (91\text{ kg})(5.0\text{ m/s})$; $v_i = 0.46$ km/s. **57.** $v_{esc} = (2GM/R)^{1/2} = [2(6.67 \times 10^{-11}\text{ N·m}^2/\text{kg}^2)(7.4 \times 10^{22}\text{ kg})/(1.74 \times 10^6\text{ m})]^{1/2} = 2.4$ km/s. It can be launched in any direction that does not intersect the Moon. **63.** Taking east as positive, $p_i = (+8.0\text{ kg}) \times (0.15\text{ m/s}) - (2.0\text{ kg})(0.25\text{ m/s}) = 0.70$ kg·m/s $= p_f = (10.0\text{ kg})v_f$; $v_f = 0.07$ m/s. **65.** $p_i = m(15.0\text{ m/s}) + m(-10\text{ m/s}) = mv_{1f} + mv_{2f}$, $+5.0 = v_{1f} + v_{2f}$. But $v_{2f} - v_{1f} = v_{1i} - v_{2i} = 25$ m/s; hence, adding the last two equations, $2v_{2f} = 30$ m/s, $v_{2f} = 15$ m/s, and $v_{1f} = -10$ m/s. **67.** $PE_i = 0$; $E_i = 0$; hence, $\frac{1}{2}m_1v^2 + m_1gy + \frac{1}{2}m_2v^2 - m_2gy = 0$ a moment after release. $v = \sqrt{2(m_2 - m_1)gy/(m_1 + m_2)}$. **69.** $m_Bv_B = (m_B + m_C)v_C$. After impact, $\frac{1}{2}(m_B + m_C)v_C^2 = (m_B + m_C)gh$, $v_C^2 = 2gh$, $v_B = [(m_B + m_C)/m_B]\sqrt{2gh}$.

CHAPTER 10
Answers to Odd-Numbered Multiple Choice Questions
1. d **3.** d **5.** b **7.** d **9.** c **11.** b **13.** b **15.** a **17.** c **19.** b

Answers to Odd-Numbered Problems
1. 6.02×10^{23} **3.** 6×10^7 kg/m³ **5.** 49 kg **7.** 19 GJ **9.** 12.0×10^{23} **11.** 303 u **13.** 6×10^{77} **15.** 3.3×10^{28} molecules/m³ **17.** 3.14 kg **19.** 0.46 nm **21.** 3.3 nm **23.** 1.0 kN/m **25.** 2 MPa **27.** 0.15% **29.** 0.399 8 m **31.** same **33.** 1×10^{-2} m **35.** 1.7×10^2 N/m² **37.** 5.0×10^2 N/m **39.** 16 J/m² **41.** 0.77 mm **43.** 29.5 kN **45.** 0.22% **47.** $R = 1.01$ cm, 0.173% **49.** 2 cm **51.** 0.911 MN

Solutions to Selected Problems
1. 6.02×10^{23} **5.** (13 gallons)(3.785 liters/gallon) = 49.2 liters; 1 liter = 1000 cm³; hence, the human body contains $\approx 49 \times 10^3$ cm³, or $\approx 49 \times 10^3$ g, or 49 kg. **9.** $C_{12}H_{22}O_{11}$: 12(12) + 22(1) + 11(16) = 342; hence, 684 g is 2 gram-moles and, therefore,

12.0×10^{23} molecules. **13.** (10^{33} g per star) $\times$ (10^{11} stars per galaxy)(10^{11} galaxies)/(10 g per mole) = number of moles = 1×10^{54} moles; (10^{54} moles)(6.02×10^{23} atoms/mole) = 6×10^{77} atoms. **17.** The mass of a molecule is 18 u, of which 2 u is hydrogen; that is, 11.1%; hence, 11.1% of 62.4 lb is 6.93 lb $\to$ 3.14 kg. **21.** 22.4 liters contain 6.0×10^{23} molecules; hence, the volume "occupied" by each one is $(22.4 \times 10^3\text{ cm}^3) \times (10^{-6}\text{ m}^3/\text{cm}^3)/6.0 \times 10^{23} = 3.73 \times 10^{-26}\text{ m}^3$; and the cube root of this is 3.3×10^{-9} m, or 3.3 nm, which is roughly 10 times the nucleus-to-nucleus separation in a solid. **23.** $F = ks$; $k = F/s = (50\text{ N})/(0.05\text{ m}) = 1.0$ kN/m. **25.** $\sigma = F/A = (200\text{ N})/(10^{-4}\text{ m}^2) = 2 \times 10^6\text{ N/m}^2 = 2\text{MPa}$. **29.** $\Delta L/L_0 = \varepsilon$; $\Delta L = (0.400\ 0\text{ m})(5.000 \times 10^{-4}) = 2.000 \times 10^{-4}$ m and $L = L_0 - \Delta L = 0.399\ 8$ m. **33.** $Y = (F/A)/(\Delta L/L_0)$; $\Delta L = (F/A)(Y/L_0) = FL_0/YA$; the mike is already on the rod and so only the weight of the *sumatori* enters; $\Delta L = (9.8\text{ m/s}^2)(181.4\text{ kg})(4\text{ m})/(200 \times 10^9\text{ Pa})\pi(\frac{1}{2} 2 \times 10^{-3}\text{ m})^2 = 1 \times 10^{-2}$ m. **37.** $k = F/s = (10\text{ N})/(2.0 \times 10^{-2}\text{ m}) = 5.0 \times 10^2$ N/m. **41.** $F/A < 435$ MPa; (200 N)/(435 MPa) $< A$; $A > 4.598 \times 10^{-7}$ m²; $\pi R^2 > 4.598 \times 10^{-7}$ m²; $R > 3.8 \times 10^{-4}$ m; minimum diameter is 0.77 mm. **45.** $Y = 50$ GPa = (110 MPa)/ε_R; $\varepsilon_R = (110\text{ MPa})/(50\text{ GPa}) = 2.2 \times 10^{-3} = 0.22\%$. **49.** $F/A = S\gamma$; $F/S\gamma = A = (200\text{ N})/(0.6\text{ GPa}) \times (0.001) = 3.33 \times 10^{-4}\text{ m}^2 = \pi R^2$; $D = 2R = 2$ cm. **53.** $PE_e = F^2\frac{1}{2}L_0/2AY + F^2\frac{1}{2}L_0/2(2A)Y = 3F^2L_0/8AY$. **55.** From p. 341, $k = YA/L_0$, $Y = (F/A)/(\Delta L/L_0) = \sigma_R/(\Delta L/L_0)$; $PE_e = \frac{1}{2}k(\Delta L)^2 = \frac{1}{2}(YA/L_0)(\sigma_R L_0/Y)^2 = \frac{1}{2}AL_0\sigma_R^2/Y$.

CHAPTER 11
Answers to Odd-Numbered Multiple Choice Questions
1. b **3.** d **5.** a **7.** a **9.** a **11.** d **13.** b **15.** c **17.** c **19.** d

Answers to Odd-Numbered Problems
1. 3×10^4 Pa **3.** 50.7 N/cm² **7.** 1.0×10^4 Pa **9.** 10.3 m **11.** 1.1×10^8 Pa **13.** 4.1×10^5 Pa **15.** 10.5 **17.** 13 mN **19.** 3.9×10^3 Pa **21.** -1.1×10^4 Pa **23.** 1.1×10^5 Pa **25.** 21.5, platinum, $V = 3.00 \times 10^{-8}$ m³ **27.** 10% is visible, 8% is above **29.** 300 cm³ **31.** 1.1×10^5 N **33.** 72.7 kg **35.** 4.6 kN, 0.15 m **37.** 2.7×10^{-2} N **41.** 2×10^5 N **43.** $\rho_1 = (F_{wa} - F_{wl})\rho_w/(F_{wa} - F_{ww})$ **45.** 16×10^{-3} m³/s **47.** -27 kPa **49.** 8.8 kPa **51.** $v = \sqrt{2P_G/\rho}$ **53.** $x = 2\sqrt{yh}$ **55.** 12 m/s **57.** $v = \sqrt{2\Delta P/\rho}$ **59.** $v_p = \sqrt{2g\Delta y/[(A_p^2/A_t^2) - 1]}$ **61.** 82 m/s **63.** $v_2 = \sqrt{2gh}$, $v_3 = \sqrt{2g(h + Y)}$ **65.** $v_2 = \sqrt{2gH}$ **67.** $y = h\sin^2\theta$

Solutions to Selected Problems
3. $P_1 = 5(1.013 \times 10^5\text{ N/m}^2) = 5.07 \times 10^5$ N/m² = $(5.07 \times 10^5\text{ N})/(10^4\text{ cm}^2) = 50.7$

N/cm². **7.** (3.0 in.)(2.54 cm/in.) = 7.62 cm; $\Delta P = \rho g\Delta h = (13.6 \times 10^3\text{ kg/m}^3) \times (9.8\text{ m/s}^2)(7.62 \times 10^{-2}\text{ m}) = 1.0 \times 10^4$ Pa. **11.** $P = \rho gh = (1.025 \times 10^3\text{ kg/m}^3)(9.8\text{ m/s}^2) \times (11 \times 10^3\text{ m}) = 1.1 \times 10^8$ Pa, or 1.6×10^4 psi. **15.** From Eq. (11.9), (10.0 g)/(0.952 g) = 10.5, and it's likely silver. **19.** $P = \rho gh = (1.0 \times 10^3\text{ kg/m}^3)(9.8\text{ m/s}^2)(0.40\text{ m}) = 3.9 \times 10^3$ Pa. **23.** At the level of the lower surface, $P = P_A + \rho g\Delta h = (1.013 \times 10^5\text{ Pa}) + (1.00 \times 10^3\text{ kg/m}^3)(9.8\text{ m/s}^2)(0.61\text{ m}) = 1.1 \times 10^5$ Pa, or 1.1 atm. **25.** From Eq. (11.9), $gm/(gm - gm_r) = (6.327 \times 10^{-3}\text{ N})/(6.327 \times 10^{-3}\text{ N} - 6.033 \times 10^{-3}\text{ N}) = 21.5$; platinum; weight of water displaced is $(6.327 \times 10^{-3}\text{ N} - 6.033 \times 10^{-3}\text{ N}) = 2.94 \times 10^{-4}$ N, equivalent to 3.00×10^{-5} kg, or 3.00×10^{-8} m³. **29.** $P_iV_i = P_fV_f$; (228 cm Hg)(100 cm³) = (76 cm Hg)V_f; $V_f = 300$ cm³. **33.** mass of air displaced = mass of helium + mass of balloon; $V(1.29\text{ kg/m}^3) = V(0.178\text{ kg/m}^3) + 454$ kg; $V = 408.3$ m³, for a mass of $\rho V = (0.178\text{ kg/m}^3)(408.3\text{ m}^3) = 72.7$ kg. **37.** There are two surfaces here, so $F = 2\gamma L = 2\gamma\pi D = 2\pi(72.8 \times 10^{-3}\text{ N/m})(6.0 \times 10^{-2}\text{ m}) = 2.7 \times 10^{-2}$ N, which is the weight of a 2.8-g mass. **39.** The area of the left face is $A_a = L(L\sin\phi)$, the area of the top face is $A_c = L^2$, and that of the bottom face is $A_b = L^2\cos\phi$. So the corresponding forces are $F_a = P_aL^2\sin\phi$, $F_c = P_cL^2$, and $F_b = P_bL^2\cos\phi$. Taking the sum of the horizontal forces equal to zero, we have $F_a = F_c\sin\phi$; $P_aL^2\sin\phi = P_cL^2\sin\phi$. Thus, $P_a = P_c$. The weight of the fluid wedge is the volume, $\frac{1}{2}L(L\sin\phi)(L\cos\phi)$, times ρg; hence, the sum of the vertical forces yields $F_b = F_c\cos\phi + F_W$ or $P_bL^2\cos\phi = P_cL^2 \times \cos\phi + \frac{1}{2}\rho g L^3\sin\phi\cos\phi$ and $P_b = P_c + \frac{1}{2}\rho gL\sin\phi$. As $L{\to}0$, we get $P_b = P_c$, which equals P_a. **41.** Because the pressure varies uniformly with depth, the average pressure on the wall is $\frac{1}{2}$ the pressure at the bottom. From Problem 11.1, $P_1 = 3 \times 10^4$ Pa; hence, $F = (1.5 \times 10^4\text{ Pa}) \times (5\text{ m} \times 3\text{ m}) = 2 \times 10^5$ N. **43.** $F_{wa} - F_{ww} = \rho_wgV$; $V = (F_{wa} - F_{ww})/\rho_wg$; $V = (F_{wa} - F_{wl})/\rho_lg$; setting these equal, $(F_{wa} - F_{ww})/\rho_wg = (F_{wa} - F_{wl})/\rho_lg$; and $\rho_l = \dfrac{(F_{wa} - F_{wl})}{(F_{wa} - F_{ww})}\rho_w$. **47.** From Eq. (11.18), $P_1 - P_2 = \rho g(y_2 - y_1) = (0.68 \times 10^3\text{ kg/m}^3)(9.8\text{ m/s}^2)(-4.0\text{ m}) = -27$ kPa. $P_2 > P_1$. **51.** $v = \sqrt{2gh}$, $P_G = \rho gh$; hence, $v = \sqrt{2P_G/\rho}$. **55.** Consider a slug of gas of volume $V = AL$, where the length $L = v\Delta t$. The mass per unit time (Δt) flowing is 1.0 kg/s $= V\rho/\Delta t = Av\Delta t\rho/\Delta t = \frac{1}{4}\pi D^2\rho v$; $v = 4(1.0\text{ kg/s})/\pi D^2\rho = 12$ m/s. **63.** At the surface and the spigot using gauge pressure, $0 + \rho g(h + Y) + 0 = 0 + \rho gY + \frac{1}{2}\rho v_2^2$, hence, $v_2 = \sqrt{2gh}$. Then, between points 2 and 3, $0 + \rho gY + \frac{1}{2}\rho(2gh) = 0 + 0 + \frac{1}{2}\rho v_3^2$; $v_3 = \sqrt{2g(h + Y)}$. These are just what we expect if the water free-falls.

67. $P_2 + \frac{1}{2}\rho v_2^2 + \rho g y_2 = P_3 + \frac{1}{2}\rho v_3^2 + \rho g y_3$; $0 + \frac{1}{2}\rho v_2^2 + 0 = 0 + \frac{1}{2}\rho v_3^2 + \rho g y$; since there is no horizontal acceleration, $v_3 = v_2 \cos \theta$, $y = \frac{1}{2}(v_2^2 - v_2^2 \cos^2 \theta)/g$, but $\cos^2 \theta + \sin^2 \theta = 1$; hence, $y = (\frac{1}{2}v_2^2 \sin^2 \theta)/g$ and, since $v_2^2 = 2gh$, $y = h \sin^2 \theta$ and it cannot exceed h.   **69.** Imagine a flow tube from some distant point 1 in front of the plane at the level of the top of the wing; then $P_1 + \frac{1}{2}\rho v_1^2 + \rho g y_1 = P_\alpha + \frac{1}{2}\rho v_\alpha^2 + \rho g y_\alpha$. Now do the same thing for a tube starting at point 2 just below the wing and running to the bottom of the wing as $P_2 + \frac{1}{2}\rho v_2^2 + \rho g y_2 = P_\beta + \frac{1}{2}\rho v_\beta^2 + \rho g y_\beta$. Since $P_1 \approx P_2$, $v_1 \approx v_2$, $y_1 \approx y_2$, $P_\alpha + \frac{1}{2}\rho v_\alpha^2 + \rho g y_\alpha = P_\beta + \frac{1}{2}\rho v_\beta^2 + \rho g y_\beta$, and with $y_\alpha \approx y_\beta$, $F_L = A(P_\beta - P_\alpha) = \frac{1}{2}\rho(v_\alpha^2 - v_\beta^2)A$.

CHAPTER 12

Answers to Odd-Numbered Multiple Choice Questions
1. b   **3.** a   **5.** c   **7.** c   **9.** d   **11.** c   **13.** a   **15.** d   **17.** d   **19.** b   **21.** c

Answers to Odd-Numbered Problems
1. 1.8 s   **3.** 1.3 Hz, 2.6π rad/s   **5.** (a) 5.0 m (b) 0.064 Hz (c) 0.10 rad (d) $x = 3.1$ m   **9.** 4.9×10^2 m/s^2 **11.** T = 4.4 s   **13.** $\varepsilon = \pi$ **15.** $z = +0.50$ m, $z = -0.50$ m, $z = 0$ **17.** $k = 2.6$ N/m   **19.** $k = 784$ N/m, $f = 3.15$ Hz   **21.** $f_0 = (1/2\pi)\sqrt{(k_1 + k_2)/m}$ **23.** 0.16 Hz   **25.** $L = 24.8$ m   **29.** $g = 4\pi^2 \Delta L/T^2$ **31.** 3.5 Hz   **33.** $T = 2\pi\sqrt{b\rho/\rho_w g}$ **35.** $v_b = 0.86$ km/s   **37.** 1.2 T **39.** $T = 2\pi\sqrt{m/k}$ **43.** 500 nm   **45.** $\lambda = 3.40$ m   **49.** 12 m   **51.** 1.3×10^2 m/s   **53.** 1.08×10^3 N

57.

Fig. AN 57.

61. 6.3 m/s^2 **63.** 100 m/s, 314 rad/s, 0.020 0 s, 2.00 m   **65.** $v = 1.0 \times 10^2$ m/s   **67.** $F_T = 1.9 \times 10^2$ N   **69.** $y = -0.7 A$, $y = -0.7 A$ **71.** $v_{max} = \sqrt{gL}$, $v = \sqrt{g\Delta y + MgL/m}$

Solutions to Selected Problems
1. $T = 1/f = 1/[(33\frac{1}{3} \text{ rpm})/(60 \text{ s/min})] = 1.8$ s.   **5.** By comparison with Eq. (13.3), (a) $A = 5.0$ m (b) $\omega = 2\pi f = 0.40$ rad/s; $f = 0.064$ Hz (c) $\varepsilon = 0.10$ rad (d) $x = 3.1$ m.   **9.** $a_0 = -A(2\pi f)^2$; dropping the sign, $a_0 = (0.005 \text{ m})4\pi^2(50 \text{ Hz})^2 = 4.9 \times 10^2$ m/s^2.   **13.** $x = A \cos \theta = A \cos(\omega t + \varepsilon) =$

$A \cos(2\pi f t + \varepsilon)$; hence, at $t = 0$, $x = -3 = A \cos \varepsilon$, and since $A = 3$, $\cos \varepsilon = -1$ and $\varepsilon = \pi$.   **17.** $f = (1/2\pi)\sqrt{k/m}$; $k = (2\pi f)^2 m = 2.6$ N/m.   **21.** Since $x = x_1 = x_2$ and $F = F_1 + F_2$; hence, $kx = k_1 x_1 + k_2 x_2$ and $k = k_1 + k_2$. Now, simply substitute k into Eq. (12.12).   **25.** From Eq. (12.17), $T = 2\pi\sqrt{L/g}$; hence, $L = gT^2/4\pi^2 = 24.8$ m.   **29.** $k\Delta L = mg$; hence, $m/k = \Delta L/g$ and, from Eq. (12.13), $T = 2\pi\sqrt{\Delta L/g}$; hence, $g = 4\pi^2\Delta L/T^2$.   **33.** The additional buoyant force if the block is pushed down a distance y is $-acy\rho_w g$, which is the restoring force $-ky$, so the motion is SHM, since $F \propto -y$ and $k = ac\rho_w g$. Here, $m = abc\rho$, so $T = 2\pi\sqrt{(abc\rho)/(ac\rho_w g)} = 2\pi\sqrt{(b\rho)/(\rho_w g)}$.   **37.** From Eq. (12.17), $T = 2\pi\sqrt{L/g}$, and if $L' = L + 50\% L = 1.50 L$, $T' = 2\pi\sqrt{L'/g} = \sqrt{1.50} T = 1.2 T$.   **41.** $x = L \cos \phi + R \cos \theta$; $\cos \phi = 1 - \sin^2 \phi$, but $L \sin \phi = R \sin \theta$; $x = R \cos \theta + L\sqrt{1 - (R/L)^2 \sin^2 \theta}$. Although the first term corresponds to SHM, the presence of the second term negates that.   **45.** $v = f\lambda$; $\lambda = (1498 \text{ m/s})/(440 \text{ Hz}) = 3.40$ m.   **49.** The round-trip distance is $2x = vt = (330 \text{ m/s})(70 \text{ ms}) = 23.1$ m, $x = 12$ m.   **51.** $v = \sqrt{F_T/\mu} = \sqrt{(10 \text{ N})/(0.59 \times 10^{-3} \text{ kg/m})} = 1.3 \times 10^2$ m/s.   **53.** $F_T = v^2\mu = 1.08 \times 10^3$ N.   **59.** Using the trigonometric identity $\sin(\alpha \mp \beta) = \sin \alpha \cos \beta \mp \cos \alpha \sin \beta$, where $\alpha = k(x - vt)$ and $\beta = \mp\frac{1}{2}\pi$, we get $y = \mp A \cos\frac{2\pi}{\lambda}(x - vt)$.   **63.** $v = \sqrt{F_T/\mu} = 100$ m/s; $\omega = 2\pi f = 314$ rad/s; $v = f\lambda$, $\lambda = 2.00$ m; $T = 1/f = 0.020$ 0 s.   **67.** $m = F_w/g = (0.20 \text{ N})/(9.81 \text{ m/s}^2) = 0.020$ 4 kg; $\mu = 1.7 \times 10^{-3}$ kg/m; $v = \sqrt{F_T/\mu}$; $F_T = v^2\mu = 1.9 \times 10^2$ N, or 42 lb.   **69.** The phase is $2\pi[(t/0.01) - (5.0/40)] = 2\pi[(t/0.01) - (1/8)] = (200\pi t - \frac{1}{4}\pi)$; hence, $y = A \sin(200\pi t - \frac{1}{4}\pi)$ is the displacement of the bead. At $t = 0$, $y = A \sin(-\frac{1}{4}\pi) = -0.7 A$, and at $t = 0.01$ s, which is one period later, $y = A \sin(2\pi - \frac{1}{4}\pi) = -0.7 A$.

CHAPTER 13

Answers to Odd-Numbered Multiple Choice Questions
1. a   **3.** d   **5.** e   **7.** d   **9.** d   **11.** b   **13.** c   **15.** a   **17.** b   **19.** a

Answers to Odd-Numbered Problems
1. $T = 2.27$ ms, $\lambda = 0.780$ m   **3.** 3.3×10^{-4} m   **5.** 2.91   **7.** 4.0×10^{-2} W/m^2, $E = 4.0$ MJ   **9.** 0.79 m   **11.** 1.9 ms   **13.** 40 dB   **15.** 10 dB   **17.** 0.79 m   **19.** 4.8 dB   **21.** 5.0×10^{-5} W/m^2 **23.** 51 dB   **25.** 89.8 dB   **27.** 87 dB, 75 dB   **29.** $R = 2.2$ km   **31.** 200 m   **33.** $\beta = 20 \log_{10} (P/P_0)$ **35.** (b) displacement and pressure are $\frac{1}{4}\lambda$ out-of-phase   **37.** 25.8 dB   **39.** 996 Hz or 1004 Hz   **41.** 1.0 Hz   **43.** $\lambda_6 = \frac{1}{6}$ m   **45.** $f_1 = 100$ Hz   **49.** zero   **51.** intensity increases by a factor of 4 times   **53.** 466 Hz   **55.** 0.782 m/s   **57.** air, $\lambda = 1.4$ m; string, 66 cm   **59.** 2 Hz   **61.** 1.76 kHz

63. $\Delta f = 14$ Hz   **65.** 210 N   **67.** 1.8 kHz   **69.** 344 m/s   **71.** $v = 340$ m/s   **73.** $f_0 - f_s = 2v_t f_s/(v - v_t)$ **75.** $2A \cos [\frac{1}{2}(\omega_1 - \omega_2)t]$ is a cosinusoidally modulated amplitude

Solutions to Selected Problems
1. $T = 1/f = 1/440$ Hz $= 2.27$ ms. $\lambda = v/f = (343 \text{ m/s})/(440 \text{ Hz}) = 0.780$ m.   **5.** $v = f\lambda = (1000 \text{ Hz})\lambda = 344$ m/s; $\lambda = 0.344$ m; 2.91 waves per meter.   **11.** $\tau = \frac{1}{2}T = 1/2f = 1/2(261.6 \text{ Hz}) = 1.9$ ms.   **15.** Ten times the power means 10 times the intensity; therefore $\Delta\beta = 10 \log_{10} 10 = 10$ dB.   **19.** $\Delta\beta = 10 \log_{10} 3 = 4.8$ dB.   **23.** The increase over one voice is $\Delta\beta = 10 \log_{10} 25 = 14$ dB; hence, each singer puts out 65 dB $-$ 14 dB $= 51$ dB.   **25.** $I = (1.2 \text{ W})/4\pi R^2 = 0.955$ mW/m^2; $\beta = 10 \log_{10} 0.955 \times 10^9 = 89.8$ dB.   **27.** $\beta = 10 \log_{10} (56 \times 10^{-5} \text{ W/m}^2)/(10^{-12} \text{ W/m}^2) = 87$ dB. The intensity drops with the inverse square of the distance; $(5.0 \text{ m})^2/(20 \text{ m})^2 = I/(56 \times 10^{-5} \text{ W/m}^2)$; at 20.0 m, $I = 3.5 \times 10^{-5}$ W/m^2; and there $\beta = 75$ dB.   **31.** -40 dB means a change of -4 bel or 10^{-4}; hence, from the Inverse Square Law, $R^2/(2.0 \text{ m})^2 = 1/10^{-4}$ of whatever the original intensity was. $R = 200$ m.   **35.** (a) N/m$^2 = (\text{N/m}^2)(1/\text{m})(\text{m})$; (b) the displacement and pressure are $\frac{1}{4}\lambda$, or $90°$, out-of-phase; (c) $P_0 = \frac{2\pi}{\lambda}BA$; $v = \sqrt{B/\rho}$, $P_0 = \frac{2\pi}{\lambda}v^2\rho A$. The Bulk Modulus relates pressure change to volume change in the gas, and that volume change, in turn, is related to the displacement of the molecules. P is a pressure change from ambient, and so we can expect it to be associated with B and displacement.   **39.** The beat frequency is 4 Hz; hence, the wire is vibrating at *either* 996 Hz or 1004 Hz.   **43.** $\frac{1}{2}\lambda = L = \frac{1}{2}$ m, $\lambda = 1.0$ m; $\lambda_6 = 1/6$ m.   **47.** $f_N = (N/2L) \times \sqrt{F_T/\mu}$; but $\mu = m/L$; hence, $f_N = (\frac{1}{2}N) \times \sqrt{F_T/L^2(m/L)}$, which equals the desired expression.   **51.** The addition of a 180°, or $\frac{1}{2}\lambda$, phase shift will bring the waves into phase and they will now reinforce one another, creating a peak pressure 2 times as great and an intensity 4 times that of either speaker alone.   **55.** $f_0 = f_s(v + v_0)/(v + v_s)$; here $v_s = 0$ and $v_0 = v(f_0 - f_s)/f_s = (344 \text{ m/s})(1/440) = 0.782$ m/s.   **59.** $f \propto \sqrt{F_T}$; hence, the increased frequency is given by $f/(440 \text{ Hz}) = \sqrt{1.010 F_T/F_T}$; $f = 442$ Hz and the beat frequency is 2 Hz.   **63.** $f_1/f_1' = 200 \text{ Hz}/f_1' = (331.5 \text{ m/s})/(331.5 + 0.60 \times 40) = 0.932$; $f_1' = 214$ Hz; $\Delta f = 14$ Hz.   **67.** $v = \sqrt{Y/\rho} = 3516$ m/s; $v = f\lambda$, $\lambda = 2L = 2.00$ m; $f = v/2L = (3516 \text{ m/s})/(2.00 \text{ m}) = 1.8$ kHz.   **71.** There is a tiny residual signal because the paths are different; hence, the amplitudes are slightly different and the direct and reflected waves cancel, but not completely. The angle between these two beams is $\tan \theta = 1.146/2.5$, $\theta = 24.62°$; the speaker-table distance is $1.146/\sin 24.62° = 2.75$ m. The direct beam travels 5.00 m, the reflected beam travels 5.50 m, and there is cancellation; the path difference (0.50 m)

must be an odd multiple of $\frac{1}{2}\lambda$. Speed is ≈ 344 m/s, and since $f = 340$ Hz, $\lambda \approx 1$ m; hence, 0.50 m cannot be $3\lambda/2$ or $5\lambda/2$ and must be $\frac{1}{2}\lambda$, ergo $\lambda = 1.00$ m and $v = 340$ m/s. **75.** $y = y_1 + y_2 = A \sin(\omega_1 t) + A \sin(\omega_2 t)$. Using the identity $\sin \alpha + \sin \beta = 2 \sin \frac{1}{2}(\alpha + \beta) \cos \frac{1}{2}(\alpha - \beta)$, we get the desired result. Remember $\omega_1 \approx \omega_2$; there is a relatively high-frequency sinusoidal carrier oscillating at the average frequency $\frac{1}{2}(\omega_1 + \omega_2)$, and this is multiplied by an amplitude term $2A \cos[\frac{1}{2}(\omega_1 - \omega_2)t]$. This amplitude varies slowly in time, and so the modulated carrier has maxima whenever $\cos[\frac{1}{2}(\omega_1 - \omega_2)t] = \pm 1$; hence, beats occur at twice that frequency, namely, at $(\omega_1 - \omega_2)$ or $(f_1 - f_2)$.

CHAPTER 14

Answers to Odd-Numbered Multiple Choice Questions

1. b **3.** a **5.** c **7.** d **9.** a **11.** b **13.** a **15.** a **17.** b **19.** e **21.** a

Answers to Odd-Numbered Problems

1. 37°C **3.** 55.6 K **5.** $10 \times 10^{6\circ}$C $\approx 18 \times 10^{6\circ}$F **7.** 927°F **9.** $\Delta L = 25 \times 10^{-6}$ m **11.** 5.0 mm **13.** 0.54 m **15.** 1.008 cm^2 **17.** 10.002 m **19.** 0.51 cm^3 **21.** $\Delta H = -1.5$ mm **23.** 0.79×10^{-6} m^3 **25.** 7.9×10^3 kg/m^3 **27.** 2.006 s, ran fast **31.** 9.3 MPa **33.** 1.0 MPa **35.** 1.1×10^3 cm^3 **37.** $T_f = 199$ K **39.** (a) vapor (b) liquid (c) vapor (d) vapor (e) vapor **41.** 2.992×10^{-26} kg **43.** 3.01×10^{26} molecules **45.** 174 K **47.** N $= 27 \times 10^{18}$ **49.** 3.0 m/s **51.** 478.03 m/s **53.** 1.19×10^{22} **55.** 1.1 m^3 **57.** 43 kg **61.** 76 kJ **63.** $v_{rms} = (3P/\rho)^{1/2}$ **65.** 189 cm^3 **67.** 4.3 atm **69.** 0.427%

Solutions to Selected Problems

1. 37°C **5.** 10 000 000 K $\rightarrow$ 10 000 273°C $\approx$ $10 \times 10^{6\circ}$C $\approx 18 \times 10^{6\circ}$F. **9.** $\alpha = 25 \times 10^{-6}$ K^{-1}; hence, $\Delta L = 25 \times 10^{-6}$ m. **13.** $\Delta L = \alpha L_0 \Delta T = (12 \times 10^{-6}$ K$^{-1}) \times (1280$ m$)(35$ K$) = 0.54$ m. **17.** $\Delta L = \alpha L_0 \Delta T = (12.15 \times 10^{-6}$ K$^{-1})(10.000$ m$) \times (16.0$ K$) = 1.9 \times 10^{-3}$ m. $L = 10.002$ m. 2 mm is an appreciable error, but it's not likely to trouble a carpenter. **21.** $\Delta V = \beta V_0 \Delta T = (182 \times 10^{-6}$ K$^{-1})(800.0 \times 10^{-6}$ m$^3) \times (-95$ K$) = -1.383 \times 10^{-5}$ m^3 for the mercury. $\Delta V = \beta V_0 \Delta T = (26 \times 10^{-6}$ K$^{-1}) \times (800.0 \times 10^{-6}$ m$^3)(-95$ K$) = -1.98 \times 10^{-6}$ m^3 for the glass. Hence, the total $\Delta V = -1.185 \times 10^{-5}$ m^3. $\Delta V = A \, \Delta H = \pi r^2 \Delta H$; $\Delta H = -1.5$ mm. **25.** $\rho_i = m/V_i$; $\rho_f = m/(V_i + \Delta V)$; $\rho_i V_i = \rho_f (V_i + \Delta V)$; $\rho_f V_i (1 + \beta \Delta T)$; $\rho_i (1 + \beta \Delta T) = \rho_f = \rho_f V_i (1 + \beta \Delta T)$; $\rho_f = (7.85 \times 10^3$ kg/m$^3)/[(1 + (36 \times 10^{-6}$ K$^{-1}) \times (-20$ K$)] = 7.9 \times 10^3$ kg/m^3. **29.** $\rho_0 = m/V_0$; $\rho = m/(V_0 + \Delta V)$; $\rho_0 V_0 = \rho(V_0 + \Delta V) = \rho V_0 (1 + \beta \Delta T)$; $\rho_0/(1 + \beta \Delta T) = \rho$: from the Binomial Theorem, $(1 + x)^n \approx (1 + nx)$ for $x \ll 1$; here, $n = -1$, and so $\rho \approx \rho_0 (1 - \beta \Delta T)$. **31.** Stress $= Y\alpha\Delta T = (25$ GPa$)(10 \times 10^{-6}$ K$^{-1})(37$ K$) =$

9.3 MPa. Concrete is strong in compression (Table 10.3), so it's more likely to buckle than break. **33.** $P_i V_i = P_f V_f$; $P_f = 10 P_i =$ 1.0 MPa. **37.** No change in P; $V_i/T_i = V_f/T_f$; $(300$ cm$^3)/(298$ K$) = (200$ cm$^3)/T_f$; $T_f = 199$ K. **41.** $2H \rightarrow 2(1.008$ u$)$; $0 \rightarrow 15.999$ u; $(18.015$ u$)(1.660\,6 \times 10^{-27}$ kg/u$) = 2.992 \times 10^{-26}$ kg. **45.** $PV = nRT$; $(202.6 \times 10^3$ Pa$)(10.0 \times 10^{-3}$ m$^3) = (1.40$ mol$) \times (8.314$ J/mol·K$)T$; $T = 174$ K. **49.** $v_{av} = (1.0$ m/s $+ 2.0$ m/s $+ 3.0$ m/s $+ 4.0$ m/s $+ 5.0$ m/s$)/5 = 15/5$ m/s $= 3.0$ m/s. **53.** $PV = Nk_B T$; $T = 98.6$°F $= 37$°C $\rightarrow 310$ K; $PV/k_B T = (101.46$ kPa$)(500 \times 10^{-6}$ m$^3)/(1.380\,7 \times 10^{-23}$ J/K$)(310$ K$) = N = 1.19 \times 10^{22}$ molecules, which is the same as the number of stars there are in the entire Universe. **57.** $n = PV/RT = (19.4 \times 10^3$ m$^3)(2000$ m$^3)/(8.314$ J/mol·K$)(218$ K$) = 21.4 \times 10^3$ moles. Each mole has a mass of 0.002 kg; hence, the total mass is $(0.002$ kg/mol$)(21.4 \times 10^3$ mol$) = 43$ kg. **61.** $(KE)_{av} = (3/2)k_B T = (3/2)(1.380\,662 \times 10^{-23}$ J/K$)(273.15$ K$) = 5.66 \times 10^{-21}$ J/molecule. 0.50 m$^3 = 500$ liter; $(500$ liter$)/(22.4$ liter/mol$) = 22.3$ mol; $(22.3$ mol$)(6.022 \times 10^{23}$ molecules/mol$) \times (5.66 \times 10^{-21}$ J/molecule$) = 76$ kJ. **65.** To find the pressure on the gas, 3.90 cm Hg $= 5.199\,6$ kPa, $P_i = (101.05 - 5.199\,6)$ kPa $= 95.85$ kPa; $P_i V_i/T_i = P_f V_f/T_f$; $(95.85$ kPa$) \times (214$ cm$^3)/(293.15$ K$) = (101.32$ kPa$)V_f/(273$ K$); V_f = 189$ cm^3. **69.** $v_{rms} = \sqrt{3k_B T/m}$; hence, the ratio of the rms-speeds varies as the inverse of the square roots of the masses. The masses are 238 u $+ 6(18.998$ u$) = 351.99$ u and 235 u $+ 6(18.998$ u$) = 348.99$ u. $(v_{rms}^{235} - v_{rms}^{238})/v_{rms}^{235} = [(348.99$ u$)^{-1/2} - (351.99$ u$)^{-1/2}]/(348.99$ u$)^{-1/2} = 0.427\%$.

CHAPTER 15

Answers to Odd-Numbered Multiple Choice Questions

1. e **3.** c **5.** c **7.** b **9.** a **11.** c **13.** a **15.** b **17.** b **19.** c

Answers to Odd-Numbered Problems

1. 63 kJ **3.** 116 W **5.** 9×10^5 J **7.** 17 K **9.** 0.68 kJ **11.** 0.68 kg **13.** 1.2×10^8 J **15.** 113 g **17.** 0.921 kg **19.** W $=$ 4.148 J ≈ 1 cal **21.** 58°C **23.** $+0.43$ K **25.** 1.9 h **27.** 27 kcal **29.** 0.74 MJ **31.** $c = 0.65$ kJ/kg·K **33.** 1.8 h **35.** 12.7 g **37.** $c = 443$ J/kg·K **39.** 41.0 kJ **41.** 3.5 MJ **43.** 0.230 kg vaporizes **45.** 0.17 MJ **47.** 0.029 kg **49.** 0.125 kg **51.** $T_f = 24$°C **53.** 2.1×10^5 J **55.** 2.65 h **57.** 7.4×10^{-20} J **59.** 0.068 kg **61.** 354 m/s **63.** 1×10^2 g **65.** $T_f = 0$°C **67.** 0°C **69.** 2.3 m **71.** 0.36 MJ, $T_f = 1234$ K

Solutions to Selected Problems

1. 15 kcal $= 63$ kJ. **5.** $(7$ kcal/g$)(30$ g$) = 210$ kcal $= 879$ kJ, or to 1 significant figure, 9×10^5 J. **9.** $Q = cm\Delta T = (0.060)(1.0$ kcal/kg$)(0.030$ kg$)(-90$ K$) = 0.162$ kcal $= 162$ cal $= 0.68$ kJ. **13.** $m = (3785 \times 10^{-6}$ m$^3)(0.68 \times 10^3$ kg/m$^3) = 2.574$ kg;

$(2.574$ kg$)(48$ MJ/kg$) = 124$ MJ $= 1.2 \times 10^8$ J. **17.** $-(910$ J/kg·K$)m(-173$ K$) = (4186$ J/kg·K$)(4.95$ kg$)(7.00$ K$)$; $m(157\,430$ J/kg$) = 145\,044.9$ J; $m = 0.921$ kg. **21.** $Q/t = cm\Delta T/t$; 1.5 liters/min $= 1.5$ kg/min $= 0.025$ kg/s; 4000 W $= (4186$ J/kg·K$)(0.025$ kg/s$)\Delta T$; $\Delta T = 38$ K; $T_f = 58$°C. **25.** $Q = cm\Delta T = (4186$ J/kg·K$)(1.00$ kg$)(80$ K$) = 334\,880$ J; $(334\,880$ J$)/(50$ J/s$) = 6697.6$ s $= 1.9$ h. **29.** 150 lb $\rightarrow 68.04$ kg; $Q = cm\Delta T$; $\Delta T = 33.89$°C $- 37$°C $= -3.11$ K; $Q = (3474.4$ J/kg·K$)(68.04$ kg$)(3.11$ K$) = 0.74$ MJ. **33.** The average specific heat capacity of the body is 3.5 kJ/kg·K $(0.83$ kcal/kg·C°$)$, and normal temperature is 37°C. From Eq. (19.1), $Q = cm\Delta T = (3.5$ kJ/kg·K$)(70$ kg$)(6$ K$) = 1470$ kJ will raise the temperature 6 K. Since 200 kcal/h $= 837.2$ kJ/h, it will take $(1470$ kJ$)/(837.2$ kJ/h$) = 1.8$ h. **37.** The tension in the rope is $g(6$ kg$) = 58.8$ N, and this is the force exerted (via friction) on the rope by the rod, and it does it over a distance of $2\pi R(240) = 22.62$ m. The work thus done is $(58.8$ N$)(22.62$ m$) = 1330$ J, and this result equals the heat-in: $Q = cm\Delta T$; 1330 J $= c(0.250$ kg$)(12.0$ K$)$; $c = 443$ J/kg·K, which is what we might expect for some sort of steel. **41.** $L_v = 2336 \times 10^3$ J/kg; $Q = mL_v = (1.5$ kg$)(2336 \times 10^3$ J/kg$) = 3.5$ MJ. **45.** $Q = cm\Delta T + mL_v = (138$ J/kg·K$) \times (0.50$ kg$)(630$ K $- 240$ K$) + (0.50$ kg$) \times (296 \times 10^3$ J/kg$) = 26\,910$ J $+ 148\,000 = 0.17$ MJ. **49.** $-m(334\,000$ J/kg$) = (0.50$ kg$)(4186$ J/kg·K$)(0° - 20$°C$)$; $m = 0.125$ kg. **53.** $Q/t = \lambda_T A \Delta T/d = (18$ Cal·cm/ m^2·h·C°$)(1.4$ m$^2)(4$°C$)/(2.0$ cm$) = 50.4$ Cal/h $= 0.014$ kcal/s $= 0.059$ kJ/s $= 0.059$ kW. Therefore, 2.1×10^5 J per hour. **57.** 1 mol $= 18$ g; at STP there are 6.02×10^{23} molecules/mol; to vaporize 1 mol requires $Q = mL_v = (0.018$ kg$)(2492$ kJ/kg$) = 44.86$ kJ. Hence, per molecule the energy is $(44.86 \times 10^3$ J$)/(6.02 \times 10^{23}$ molecules/ mol$) = 7.4 \times 10^{-20}$ J. **61.** $Q = mL_f$; $mL_f = (23 \times 10^3$ J/kg$)m$; to which must be added $Q = cm\Delta T = (130$ J/kg·K$)m \times (327$ K $- 23$ K$) = (39\,520$ J/kg$)m$; the total is then $(62\,520$ J/kg$)m = \frac{1}{2}mv^2$; $v = 354$ m/s. **65.** $-Q_{out} = Q_{in}$; if it dropped to zero, the calorimeter would provide an amount of heat equal to $(1.00$ kcal/kg·K$)(0.398$ kg$)(5.1$ K$) + (0.093$ kcal/kg·K$)(0.102$ kg$)(5.1$ K$) = 2.078$ kcal; the amount of heat needed to melt the ice is $(0.040\,5$ kg$)(80$ kcal/kg$) = 3.24$ kcal; hence, not all the ice melts (!), and $T_f = 0$°C. **69.** $Q/t = k_{TB}A\Delta T/d_B = k_{TA}A\Delta T/d_A$; $k_{TB}/d_B = k_{TA}/d_A$; $k_{TB}/k_{TA} = 0.60/0.026 = 23 = d_B/d_A$; $d_B = 2.3$ m. **73.** In the first set-up, $Pt = c_w m\Delta T + P_l t$ and, in the second, $P't = c_w m'\Delta T + P_l t$; note that the mean temperature is the same in both cases, and so P_l is the same. Thus, solve for P_l in both equations and set them equal. $P - c_w m\Delta T/t = P' - c_w m'\Delta T/t$; and $(P - P')t = c_w \Delta T(m - m')$.

CHAPTER 16

Answers to Odd-Numbered Multiple Choice Questions

1. b 3. d 5. d 7. b 9. e 11. b 13. c 15. a 17. d 19. b

Answers to Odd-Numbered Problems

1. $+250$ J 3. -450 J 5. $+500$ J 7. 519 J 9. $W = \frac{1}{4}P_i V_i$ 11. $W = \frac{1}{2}P_i V_i$ 13. $+12$ J 15. 5.5% 17. decrease T_L 19. 3.5 kW 21. 2.68 J/K 25. $+75.6$ J/K 27. $\Delta S_f = 1.22$ kJ/K, $\Delta S = 12$ J/K 29. $+9.8$ kJ 31. 0.22 g 33. $T_f = -78°C$ 35. (a) $0.038\ 4$ MPa (b) 207 K 37. $W = nRT \ln(P_i/P_f)$ 41. impossible 43. 399 W 45. $Q_H/W_i = 6.5$, pay for 1 unit of work and get 6.5 units of heat 47. 0.69 W 49. $T_H = 547$ K 51. $e = 62.5\%$, irreversible 53. $+0.014$ kJ/K 55. $W = 6.0$ MJ, $e = 67\%$ 59. $T_i/T_f = 0.49$ 61. 5.8 J/K

Solutions to Selected Problems

1. From Eq. (16.2), $\Delta U = Q - W = (+500\ \text{J}) - (+250\ \text{J}) = +250$ J. 5. Since $\Delta V = 0$, $W = 0$, and from Eq. (16.1), $Q = \Delta U = +500$ J. 9. $W = \frac{1}{2}(\frac{1}{2}P_i) \times (2V_i - V_i) = \frac{1}{4}P_i V_i$. 13. The work done is the area under the curve, which is $\frac{1}{2}(3.0\ \text{MPa}) \times (2.0 \times 10^{-6}\ \text{m}^3) + (4.0 \times 10^{-6}\ \text{m}^3) \times (2.0\ \text{MPa}) + (1.0\ \text{MPa})(1.0 \times 10^{-6}\ \text{m}^3) = 3.0\ \text{J} + 8.0\ \text{J} + 1.0\ \text{J} = 12$ J, and this is positive, since the gas expanded and it did work on the surroundings. 17. $e_c = 1 - (T_L/T_H) = (T_H - T_L)/T_H$; increasing T_H increases both the numerator and the denominator; decreasing T_L increases the numerator only and therefore has the more profound effect. 21. $\Delta S = (1.00\ \text{kJ})/(373\ \text{K}) = 2.68$ J/K. 25. $\Delta S = Q/T = mL_f/T = (0.500\ \text{kg})(205\ \text{kJ/kg})/(1356\ \text{K}) = +75.6$ J/K. 29. $W = P\Delta V = (0.30\ \text{MPa}) \times (0.50\ \text{m}^3)(+0.065\ \text{m}) = +9.8$ kJ. 33. $P_i V_i^\gamma = P_f V_f^\gamma$, where we assume the rapid expansion means an adiabatic process; final pressure is 1 atm; we don't have initial volume, so let's work per-unit-volume, $(4.5\ \text{atm})(1\ \text{m}^3) = (1\ \text{atm})\ V_f^\gamma$; $V_f = 4.5^{1/\gamma} = 2.93\ \text{m}^3$ per m³, where $\gamma = 1.4$. To find T_f, use Ideal Gas Law; $P_i V_i/T_i = P_f V_f/T_f$; $(4.5\ \text{atm})(1\ \text{m}^3)/(300\ \text{K}) = (1.0\ \text{atm})(2.93\ \text{m}^3)/T_f$; $T_f = 195$ K, or $-78°C$. Notice that we didn't need to know the actual initial volume. 37. From Eq. (16.5), $W = nRT \ln(V_f/V_i)$, but $P_i V_i = P_f V_f$; hence, $W = nRT \ln(P_i/P_f)$. 41. The maximum efficiency from Eq. (16.9) requires the temperatures in kelvins; $T_H \to 144°F = 62.2°C$ or 355.2 K; $T_L \to 5°F = -15°C$ or 258 K; $e_c = 1 - (258/335.2) = 23\%$. As compared to the general expression for efficiency, Eq. (16.10), $e = (26\ \text{kJ})/(102\ \text{kJ}) = 25.4\%$. The engine exceeds the efficiency of a Carnot engine and is therefore impossible. 45. $\eta = $ heat-in/work-in $= Q_L/W_i = 5.5$. Moreover, $Q_H = Q_L + W_i = 5.5\ W_i + W_i = 6.5\ W_i$; hence, the ratio we want $Q_H/W_i = 6.5$. You pay for 1 unit of work and get 6.5 units of heat into the house, which compares very well with electrical heating (6.5 times better), where 100% of the electrical energy is converted 'into heat and you have to pay for each unit

of energy that comes into the house. 49. $e_c = 1 - T_L/T_H = 42.2\%$ and $T_L = 273.4$ K. $T_H = 547$ K. 53. Since the masses are equal, $\Delta T = \pm 4.0$ K; thus, $Q = mc\Delta T = (20\ \text{kg})(4.186\ \text{kJ/kg·K})(\pm 4.0\ \text{K}) = \pm 334.88$ kJ. Average temperatures are $38°C$ and $34°C$; $\Delta S = (-334.88\ \text{kJ})/(311\ \text{K}) + (+334.88\ \text{kJ})/(307\ \text{K}) = (-1.077\ \text{kJ/K}) + (+1.091\ \text{kJ/K}) = +0.014$ kJ/K. 57. Using Eq. (16.7), $P_i V_i^\gamma = P_f V_f^\gamma$, we have $P_i/P_f = (V_f/V_i)^\gamma$. $P_i V_i = nRT_i$ and $P_f V_f = nRT_f$ and, dividing these, $P_i/P_f = T_i V_f/T_f V_i$; getting rid of the pressures, $(V_f/V_i)^\gamma = T_i V_f/T_f V_i$, from which Eq. (16.19) follows immediately. 61. From Problem 60, $\Delta S = nR \ln(V_f/V_i) = (1.0\ \text{mol})(8.314\ 4\ \text{J/mol·K}) \ln 2 = 5.8$ J/K.

CHAPTER 17

Answers to Odd-Numbered Multiple Choice Questions

1. c 3. b 5. c 7. a 9. a 11. d 13. b 15. c 17. d 19. e

Answers to Odd-Numbered Problems

1. 5.7×10^{-18} kg 3. $\pm 1.1 \times 10^{-5}$ C 5. 2.3 N 7. $5.3\ \mu$C 9. zero 11. $8.0\ \mu$N 13. zero 15. 0.58 N 17. 8.3 N, $\theta = 49°$ 19. 6.6 N, $\theta = 74°$ 21. $0.11\ \mu$C 23. $q = -0.34Q$, $a = 0.41$, where Q and q are separated by ad 25. 8.0×10^2 N/C 27. $-1/10$ N/C due east 29. 5.6×10^{-11} N/C straight down 31. 5.1×10^{11} N/C 33. 7.8×10^4 N/C 35. 6.4×10^2 N/C straight upward 37. $E = \sigma/\varepsilon_0$ 39. 4.4 nC 41. 9.6×10^{10} m/s² 43. $E = R\sigma/r\varepsilon$ 45. $E = \sigma/\varepsilon_0$ 47. 2.6×10^{15} m/s² 49. $E = Q/4\pi R^2 \varepsilon_0$ within the space, zero beyond the shells 51. $E = rQ/4\pi\varepsilon_0 R^3$ 53. 1.4×10^{21} N/C

Solutions to Selected Problems

1. $N = (10^{-6}\ \text{C})/(-1.6 \times 10^{-19}\ \text{C}) = 6.2 \times 10^{12}$ electrons. Each has a mass of 9.109×10^{-31} kg; hence, the mass increase is 5.7×10^{-18} kg. 5. $F = kqq/r^2 = (9.0 \times 10^9\ \text{N·m}^2/\text{C}^2)(1.6 \times 10^{-19}\ \text{C})^2/(1.0 \times 10^{-14}\ \text{m})^2 = 2.3$ N. 9. By symmetry, the net force is zero. 15. $F_3 = F_{31} + F_{32} = (9 \times 10^9\ \text{N·m}^2/\text{C}^2)(-12.5 \times 10^{-6}\ \text{C}) \times (-10.0 \times 10^{-6}\ \text{C})/(3.0\ \text{m})^2 + (9 \times 10^9\ \text{N·m}^2/\text{C}^2)(-5.0 \times 10^{-6}\ \text{C})(-10.0 \times 10^{-6}\ \text{C})/(1.0\ \text{m})^2 = 0.125\ \text{N} + 0.450\ \text{N} = 0.58$ N, in the increasing x-direction. 17. See Fig. AN 17. $F_{12} = kq_1q_2/r_{12}^2 = 3.6$ N; $F_{13} = kq_1q_3/r_{13}^2 = 4.5$ N; $F_{14} = kq_1q_4/r_{14}^2 = 1.8$ N; $F_{x1} = (1.8\ \text{N}) + (4.5\ \text{N})4/5 = 5.4$ N; $F_{y1} = -(3.6\ \text{N}) - (4.5\ \text{N})3/5 = -6.3$ N; $F_1 = \sqrt{5.4^2 + 6.3^2} = 8.3$ N; $\theta = \tan^{-1}(6.3/5.4) = 49°$. 23. The repulsive force between the two known charges is $F = kQ2Q/d^2$; put a negative charge $-q$ on the line connecting the two positive charges and between them at a distance ad from $+Q$ and $(1 - a)d$ from $+2Q$. The attraction between $+Q$ and $-q$ will cancel the repulsive force on $+Q$ when $k2Q^2/d^2 = kqQ/(ad)^2$ or when $a^2 = q/2Q$; similarly $+2Q$ will be in equilibrium when $k2Q^2/d^2 = kq2Q/[(1 - a)d]^2$, that is, $(1 - a)^2 = q/Q$;

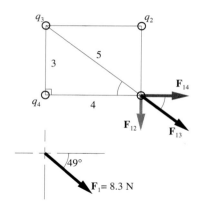

Fig. AN 17.

hence, $(1 - \sqrt{q/2Q}) = \sqrt{q/Q}$; $1 - \sqrt{q}/\sqrt{2Q} = \sqrt{q}/\sqrt{Q}$; $1 = \sqrt{q}(1 + \sqrt{2})/(\sqrt{2}\sqrt{Q})$; $q = 2Q/(1 + \sqrt{2})^2 = -0.34\ Q$ and $a = 0.41$. 27. $E = F/q = (2.0\ \text{nN})/(-20\ \text{nC}) = -1/10$ N/C; due east. 31. $E = kq_e/r^2 = (9.0 \times 10^9\ \text{N·m}^2/\text{C}^2)(1.6 \times 10^{-19}\ \text{C})/(5.3 \times 10^{-11}\ \text{m})^2 = 5.1 \times 10^{11}$ N/C. 35. The horizontal field cancels, leaving only twice the vertical contribution from each charge: $E = 2(kq/r^2) \cos \theta = 2[(9.0 \times 10^9\ \text{N·m}^2/\text{C}^2)(+50 \times 10^{-9}\ \text{C})/(1.0\ \text{m})^2](\cos 45°) = 6.4 \times 10^2$ N/C, straight upward.

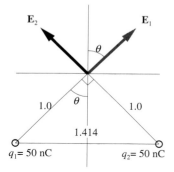

Fig. AN 35.

37. See Fig. AN 37. The field is parallel to the curved side and perpendicular to the endface; since $E = 0$ inside the metal, $EA = \sigma A/\varepsilon_0$ and $E = \sigma/\varepsilon_0$. 41. $F = qE = ma$; $a = qE/m = (1.6 \times 10^{-19}\ \text{C})(20 \times 10^3\ \text{N/C})/(3.35 \times 10^{-26}\ \text{kg}) = 9.6 \times 10^{10}$ m/s². 45. See Fig. AN 45. Use a very short perpendicular Gaussian cylindrical surface. Assume that E and σ are constant over the tiny area of the cylinder (both may vary from region to region, depending on the shape of the conductor). Embed one endface within the conductor, where $E = 0$. Hence, $EA = \sigma A/\varepsilon_0$ and $E = \sigma/\varepsilon_0$. 49. See Fig. AN 49. Surround the inner sphere with a spherical Gaussian surface of radius R ranging in the gap between the shells. The field is radial and $EA = E4\pi R^2 = Q/\varepsilon_0$; hence, $E = Q/4\pi R^2\varepsilon_0$, within the space. Now draw a Gaussian surface encompassing the large sphere anywhere beyond it. Again, if there is

Fig. AN 37.

Fig. AN 45.

a field, by symmetry it would have to be radial, but now the net charge enclosed is zero and the field is therefore zero.

Fig. AN 49.

53. $E = kq_e/r^2 = (9 \times 10^9 \text{ N·m}^2/\text{C}^2) \times (1.6 \times 10^{-19} \text{ C})/(1.0 \times 10^{-15} \text{ m})^2 = 1.4 \times 10^{21}$ N/C.

CHAPTER 18

Answers to Odd-Numbered Multiple Choice Questions

1. b   **3.** c   **5.** b   **7.** d   **9.** d   **11.** d   **13.** c
15. d   **17.** b   **19.** b

Answers to Odd-Numbered Problems

1. -25.0×10^{-7} J   **3.** ± 0.10 V   **5.** V = 0
7. -3.6×10^5 V   **9.** ΔKE = +500 eV,
ΔPE$_E$ = −500 eV   **11.** ΔKE = 4.8×10^{-15} J,
1.0×10^8 m/s   **13.** 1.4×10^{-17} J   **15.** 1.3×10^2 V   **17.** 5.58×10^{-12} V   **19.** +0.23 J
21. 11 MV/m   **23.** V = −8.3 kV, E = 0
25. $Q = 1.5 \times 10^{-10}$ C   **27.** $Q = 1 \times 10^{-5}$ C
29. R = 90 mm   **31.** 5.0 μF
33. 8.9×10^{-10} F   **35.** 2.5 pF   **37.** 110 pF
39. 0.99 μF   **41.** 10 pF   **43.** 1.4 nJ
45. 0.01 F/m^2 **47.** 7×10^{-4} C/m^2
49. 1.5 nJ/m   **51.** 9 pF   **53.** 13 μF
55. 3.6×10^{-10} C   **57.** 1.0 V   **59.** 2.2 nJ
61. 24 nF   **63.** 1.4 nJ

Solutions to Selected Problems

1. $\Delta W = \Delta$PE$_E = q\Delta V = (-25.0 \times 10^{-9}$ C)(100 V) $= -25.0 \times 10^{-7}$ J. This is the work done against the field. The negative sign means the charge has work done on it by the field. It moves from 0 to 100 V and so in the opposite direction to the field. That's the direction a negative charge moves in spontaneously, propelled by the field.   **5.** A charge of $-Q$ is induced on the inner surface of the outer sphere. At a point outside, the potential is equivalent to that due to a point charge of $+Q$ and a point charge of $-Q$, both at the center. Hence, at P, V = 0.   **13.** At the surface, $V = k_0Q/R = (8.99 \times 10^9$ N·m^2/C) $\times (1.00 \times 10^{-9}$ C)/(0.10 m) = 90 V. $W = q_e\Delta V = (1.6 \times 10^{-19}$ C)(90 V) $= 1.4 \times 10^{-17}$ J.   **19.** The potential at the point is $V = k_0Q_1/r_1 + k_0Q_2/r_2 = (8.99 \times 10^9$ N·m^2/C^2)(+10 μC)/(4.0 m) + (8.99 \times 10^9 N·m^2/C^2)(−25 μC)/(5.0 m) = −22.5 kV. $W = q\Delta V = (-10 \mu$C)(−22.5 kV − 0) = +0.23 J.
23. Each face has a diagonal of $\sqrt{2}$ m, and so there is a tilted rectangle $\sqrt{2}$ m by 1.0 m passing through the center and having four spheres at its corners. The diagonal of that square is $\sqrt{3}$ m long and half of that, $\frac{1}{2}\sqrt{3}$, is the corner-to-center distance. Hence, $V = 8kQ/r = -8.3$ kV. And E = 0.
25. $Q = CV = (100 \times 10^{-12}$ F)(1.5 V) $= 1.5 \times 10^{-10}$ C.   **31.** $C = Q/V = (20 \mu$C)/(4.0 V) = 5.0 μF.   **35.** $C = \varepsilon A/d$; $A = \pi R^2 = \pi(0.25 \times 10^{-2}$ m)$^2 = 1.96 \times 10^{-5}$ m^2; $C = (4.8)(8.85 \times 10^{-12}$ F/m) $\times (1.96 \times 10^{-5}$ m^2)/(0.33 $\times 10^{-3}$ m) = 2.5 pF.
41. The 12-pF and 4.0-pF capacitors are in series and so equivalent to 3.0 pF, which, in turn, is in parallel with 7.0 pF. Hence, C = 10 pF.   **45.** The membrane is like a rolled-up parallel-plate capacitor, so $C = \varepsilon A/d$; $C/A = \varepsilon/d = 7\varepsilon_0/d = 7(8.85 \times 10^{-12}$ C^2/N·m^2)/(6 nm) = 0.01 F/m^2.
49. ΔPE$_E/L = \frac{1}{2}CV^2/L = \frac{1}{2}(3 \times 10^{-7}$ F/m) $\times (0.1$ V)^2 = 1.5 nJ/m.   **53.** The two 6.0-μF capacitors are in series (yielding 3.0 μF), as are the two 5.0-μF capacitors (yielding 2.5 μF). Now everything is in parallel, so 3.0 μF + 7.5 μF + 2.5 μF = 13 μF.
55. The 9.0-pF capacitor is in parallel with the battery, and so the voltage is 12 V. All the capacitors are in parallel; hence, C = 30 pF. $Q = CV = (30$ pF)(12 V) $= 3.6 \times 10^{-10}$ C.
63. The diagram shows successive simplifications of the circuit. The equivalent capacitance is 20 pF; hence, PE$_E = \frac{1}{2}CV^2 = \frac{1}{2}(20$ pF)(12 V)^2 = 1.4 nJ.

CHAPTER 19

Answers to Odd-Numbered Multiple Choice Questions

1. c   **3.** d   **5.** c   **7.** e   **9.** b   **11.** b   **13.** a
15. c   **17.** d   **19.** c   **21.** a

Answers to Odd-Numbered Problems

1. 10 μA   **3.** 96.5 A   **5.** 3.6×10^2 C
7. 6.2×10^{14} **9.** 120 C   **11.** 1.5×10^3 cells
13. 36 kC   **15.** 27 h   **17.** (a) A and B

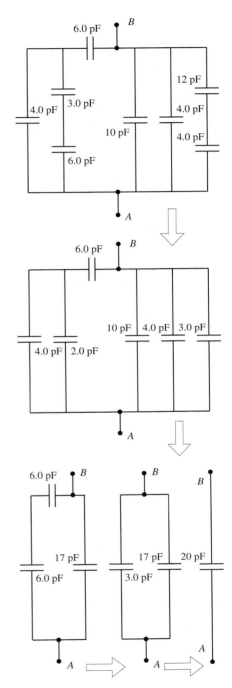

Fig. AN 63.

(b) A and E (c) A and D and C   (d) A and D   (e) A and F when the headlights are on
19. 20 cells in series form a row, and there are 20 rows in parallel   **21.** 27 h
23. 0.11 MC   **25.** 10 Ω **27.** 50×10^9 Ω
29. if it's 1 m on a side   **33.** 1.6×10^{-8} Ω·m   **35.** 1.9 mm   **37.** (a) 0.12 Ω
(b) 0.10 Ω **39.** 0.1 A   **41.** 0.1 W

43. 0.16 A **45.** 2.2 kW **47.** 0.017 V/m
49. $1 \times 10^3\ °C$ **51.** 26 Ω **53.** 220° C
55. 87 kW **57.** (a) $2.0 \times 10^{-4}\ Ω$
(b) 2.4×10^{-5} V (c) 2.9×10^{-7} W
59. 5.2 kW **61.** 30 W **63.** 1.3 W, 12 Ω
65. 12 W **67.** $\Delta R/R_0 = 2.7$, poor: 12

Solutions to Selected Problems

1. $I = \Delta Q/\Delta t = (10\ \mu C)/(1.0\ s) = 10\ \mu A$.
7. $I = \Delta Q/\Delta t$; $\Delta Q = (1.0\ mA)(0.10\ s) =$
0.10 mC. The number of protons is $(0.10 \times$
$10^{-3}\ C)/(1.602 \times 10^{-19}\ C) = 6.2 \times 10^{14}$.
11. (220 V)/(0.15 V per cell) = 1466.7 cells =
1.5×10^3 cells. Salt water, because of the
presence of ions, is a much better conductor.
15. Number of coulombs passing per second
is $(I/A)A = (1.0\ MA/m^2)(1.00 \times 10^{-6}\ m^2) =$
1.0 C/s; or 6.24×10^{18} electrons/s; hence,
$(6.02 \times 10^{23}$ electrons$)/(6.24 \times 10^{18}$
electrons/s$) = 96\ 507\ s = 27\ h$. **19.** First,
do the voltage, (9.0 V)/(0.45 V) = 20; we
need 20 cells in series forming one row. To
get the current, we must have (440 mA)/
(22 mA) = 20 such rows in parallel.
25. $R = V/I = (100\ V)/(10\ A) = 10\ Ω$.
29. If the cube is 1 m on a side, $R = \rho$.
33. $R = \rho L/A$; $\rho = AR/L = 1.6 \times$
$10^{-8}\ Ω\cdot m$. **37.** (a) From Eq. (19.8),
$R = 0.10\ Ω + (0.10\ Ω)(0.005\ 0\ K^{-1}) \times$
(30 K) = 0.12 Ω (b) since the coefficient is
essentially zero, $R = 0.10\ Ω$. **41.** P =
$IV = (16 \times 10^{-3}\ A)(9\ V) = 0.144\ W =$
0.1 W. **47.** From Eq. (19.6) for a uniform
E-field, $V = Ed$, so we must find V; since
$V = IR$, we need R; $R = \rho L/A =$
$(1.7 \times 10^{-8}\ Ω\cdot m)(1.0\ m)/(1.0 \times 10^{-6}\ m^2) =$
0.017 Ω; $V = (1.0\ A)(0.017\ Ω) = 0.017\ V$;
$E = V/d = (0.017\ V)/(1.0\ m) = 0.017\ V/m$.
51. $R_f = R_0(1 + \alpha_0 \Delta T) = (10\ Ω) \times$
$(1 + 0.003\ 9\ K^{-1} \times 400\ K) = 25.6\ Ω =$
26 Ω. **55.** The current entering the factory
must be $I = P/V = (45 \times 10^3\ W)/(110\ V) =$
409.09 A; the total power loss in the cables is
$I^2R = (409.09\ Ω)^2(2 \times 0.25\ Ω/mile \times$
0.50 mile) = 41.84 kW; hence, the net power
supplied must be 45 kW + 41.8 kW = 87 kW.
59. P = (12 A)(500 V) = 6.0 kW.
86%(6.0 kW) = 5.2 kW = 6.9 hp. **63.** P =
$IV = (1/3\ A)(3.9\ V) = 1.3\ W$. $P = I^2R =$
1.3 W = $(1/3\ A)^2R$; $R = 12\ Ω$. **69.** See
Fig. AN 69. From the diagram $R/(T' + T) =$
R_0/T'; and so $R = R_0(1 + T/T')$; comparing
this expression with Eq. (19.8), we find that T
is equivalent to ΔT and $\alpha_0 = 1/T'$.

CHAPTER 20

Answers to Odd-Numbered Multiple Choice Questions

1. d **3.** d **5.** c **7.** c **9.** a **11.** b **13.** d
15. c **17.** a **19.** b **21.** b

Answers to Odd-Numbered Problems

1. 1.5 A, goes out **3.** 0.15 A, P doubles
1.8 W to 3.6 W **5.** 1 Ω **7.** 5.0 A **11.** 2 Ω
15. 5.0 Ω **17.** 5 Ω **19.** 9 A through 17 Ω,
6 A through 3 Ω, 3 A through 4 Ω and 2 Ω
21. 1 A through 4 Ω, 6/9 A through 3 Ω,
3/9 A through 6 Ω **23.** 10 Ω **25.** $R_s =$

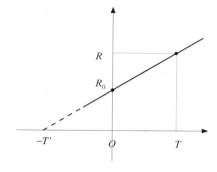

Fig. AN 69.

0.05 Ω **27.** 4.0 s **29.** 6.0 μA, 2.0 s
31. $R_e = 1.8\ Ω$ **33.** 2 A through 3 Ω and
1 Ω, 5 A through 2 Ω, 2 A through 4 Ω and
1 Ω **35.** 4 Ω, 95 W **37.** 24 W, C is 1.5 V
above A **39.** 0.89 A, 8.6 V **41.** 27 W
43. $R_s = 0.100\ Ω$ **45.** 0.12 A, 0.60 ms
47. 1.6 μA **49.** 3 A, 3 A, 6 A **51.** $I_1 =$
2 A, $I_2 = 2/3$ A, $I_3 = 4/3$ A **53.** 4 V
55. 32 V **57.** $I_1 = 2.0$ A, $I_2 = 1.0$ A,
$I_3 = 1.0$ A **59.** 8 μC **61.** $R_1 = 46\ Ω$,
$\mathcal{E}_1 = 14$ V **63.** 28 Ω **65.** (a) zero
(b) 36 μC **67.** $R = 4.0\ Ω$, $V = 30$ V,
$I_1 = 8.0$ A, $I_2 = I_6 = 2.0$ A, $I_3 = 1.0$ A,
$I_4 = I_5 = 5.0$ A **69.** A at +14 V, E at
+29 V, F at +12 V, D, C, B, and G at zero.
71. 42/5 Ω.

Solutions to Selected Problems

3. $R_e = 80\ Ω$; $I = V/R_e = 0.15$ A. After S is
closed $R_e = 40\ Ω$, and $I = 0.30$ A. P = IV,
so the power doubles from 1.8 W to 3.6 W.
7. Each lamp draws $I = P/V = (100\ W)/$
(100 V) = 1.0 A; hence, the source must
provide 5.0 A. **11.** 6 Ω in parallel with 3 Ω
is 2 Ω. **17.** The only resistance between A
and B is 5 Ω. **21.** The 3.0-Ω and 6.0-Ω
resistors in parallel equal 2.0 Ω, and that's in
series with the 4.0-Ω resistor; hence, $R_e =$
6.0 Ω. $I = V/R_e = (6.0\ V)/(6.0\ Ω) = 1.0$ A.
This current passes through the 4.0-Ω resistor
and splits 6 parts out of 9, that is, 6/9 A,
through the 3.0-Ω resistor and 3/9 A through
the 6.0-Ω resistor. **25.** $(50\ Ω)(0.001\ I) =$
$(0.999\ I)R_s$; $R_s = 0.05\ Ω$. **29.** $I_i = V/R =$
(12.0 V)/(2.0 MΩ) = 6.0 μA; $RC =$
(2.0 MΩ)(1.0 μF) = 2.0 s. **31.** 3 Ω and 9 Ω
are in parallel, and that's in parallel with the
18-Ω resistor yielding 2 Ω, which is in
parallel with 3 Ω and 6 Ω for a total of 1 Ω;
hence, $R_e = 1.8\ Ω$. **37.** See Fig. AN 37. The
equivalent resistance is 6.0 Ω; the battery
provides 2.0 A and 24 W. Working backwards,
the currents split as shown. Going from A to
B, there is a rise of $\frac{1}{2}$ V, and from B to C,
another rise of 1 V; hence, C is $1\frac{1}{2}$ V above A.
39. See Fig. AN 39. The equivalent resistance
is 10.14 Ω. The current from the battery is
0.89 A, and terminal voltage is 9.0 V −
(0.89 A)(0.50 Ω) = 8.6 V. **43.** The coil
carries 1.00 mA when the meter carries
1.00 A, and so the shunt carries a maximum
of 0.999 A; $(1.00 \times 10^{-3}\ A)r = (0.999\ A)R_s$,
$R_s = 0.100\ Ω$.

Fig. AN 37.

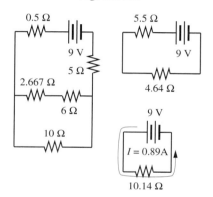

Fig. AN 39.

45. See Fig. AN 45. $I_i = V/R = (12\ V)/$
(100 Ω) = 0.12 A; $RC = (100\ Ω)(6.0\ \mu F) =$
0.60 ms.

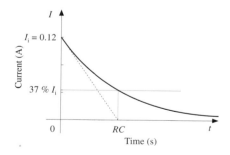

Fig. AN 45.

51. See Fig. AN 51. At A, $I_1 = I_2 + I_3$ around
loop A-B-C-D-A going clockwise: $-3I_2 -$
$3I_2 + 20\ V - 8I_1 = 0$. Around loop A-B-C-A,
clockwise: $-3I_2 - 3I_2 + 3I_3 = 0$. Around
loop A-C-D-A, clockwise: $-3I_3 + 20\ V -$
$8I_1 = 0$. Hence, $I_1 = 2$ A, $I_2 = 2/3$ A,
$I_3 = 4/3$ A. **55.** The drop across the top
branch is 16 V; with that across the middle
branch the current through it must be 1.0 A;
hence, the current in the bottom branch, from

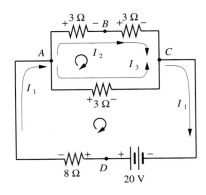

Fig. AN 51.

the Node Rule, is 1.0 A. The drop across the two bottom resistors is 16 V; hence, the voltmeter must read 32 V. **59.** The equivalent resistance in the steady state is 12 Ω; hence, the current is 1 A. The voltage across C is (1 A)(4 Ω) = 4 V, and since $CV = Q$, $Q = 8\ \mu C$. **63.** See Fig. AN 63. $I_1 = 200$ mA, $V_{AB} = +2.4$ V; hence, around the first loop clockwise: $-0.200R + 12 - 2.4 - 4 = 0$, $R = 28\ \Omega$.

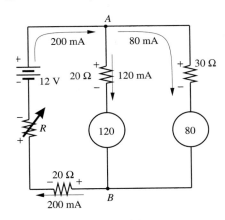

Fig. AN 63.

71. Place a source of, say, 18 V, across the group. Then solve for the six branch currents: here, 6/7 A goes through both 12-Ω resistors, 3/7 A goes through the central 6-Ω resistor, and 9/7 A passes through each outer 6-Ω resistor. Thus, the source provides 15/7 A, and so the equivalent resistance is 42/5 Ω, independent of the source.

CHAPTER 21

Answers to Odd-Numbered Multiple Choice Questions
1. d **3.** c **5.** c **7.** b **9.** b **11.** a **13.** a **15.** b **17.** c **19.** c **21.** a

Answers to Odd-Numbered Problems
3. 4.0×10^{-6} T **5.** 4.0 mT **7.** 0.10 mT **9.** 63 μT **11.** 0.63 nT **13.** 0.20 A

15. 2.5 mT **17.** 1.3×10^{-11} T clockwise looking toward the source **19.** 500 A **21.** 24×10^{-7} T **23.** 31° W of N **25.** $B_z = \mu_0 IR^2/2z^3$ **27.** $\approx 6 \times 10^9$ A **29.** $B = \mu_0 NI/2\pi r$, same for single wire **35.** $B = \mu_0 i$ inside, $B = 0$ outside **37.** $+x$-direction **39.** 1.3 T **43.** 1.8×10^7 m/s **47.** $-y$-direction **49.** 1.00 A **51.** 0.31 mN **53.** 0.019 N·m **55.** 3.8×10^{18} m/s^2 **57.** circle, $R = 14$ mm **59.** $p = qBR$ **63.** 0.5 m/s **65.** $\tau_f = 0.75\ \tau_i$ **67.** 2.0×10^{-5} N, repulsive **69.** $f = qB/2\pi m$

Solutions to Selected Problems
3. From Eq. (21.2), $B = (1.26 \times 10^{-6}$ T· m/A)(10 A)$/2\pi(0.50$ m) $= 4.0 \times 10^{-6}$ T. **7.** $V = IR$; $I = (12$ V)/(1.2 Ω) $= 10$ A; $B = \mu_0 I/2\pi r = 0.10$ mT. **11.** $B_z = \mu_0 I/2R = 0.63$ nT. **15.** $V = IR$, $I = 0.20$ A; $B_z \approx \mu_0 nI = 2.5$ mT. **19.** $B = \mu_0 I/2\pi r$; $I = 500$ A. **23.** $B = \mu_0 I/2\pi r = 0.30 \times 10^{-4}$ T, west—due to the current; the net field is at an angle of $\tan^{-1} 0.30/0.50$, or 31° west of north. **27.** We take a single circular equatorial current loop circulating with a radius of 2.3×10^3 km—it would be more realistic to assume several such loops distributed over the spherical core. We know that the field drops off as in Fig. 21.19b and the formula given. At a distance of 2.78R, it's down by a factor of about 26. $B = \mu_0 I/2R$; 0.6×10^{-4} T $= (1.26 \times 10^{-6}$ T·m/A)$I/2(2.3 \times 10^6$ m); $I = 0.22 \times 10^9$ A, but this distance is to be 6.3×10^6 m away along the z-axis, so the current should be roughly 26 times larger, or $I \approx 6 \times 10^9$ A. Since there is a south magnetic pole in the north, the current circulates from east to west. **29.** Using Ampère's Law around a circular path of radius R within the torus, $\Sigma B_\parallel \Delta l = \mu_0 \Sigma I$. If we take the field to be axial (a solenoid bent around on itself), $B2\pi r = \mu_0 NI$; $B = \mu_0 NI/2\pi r$. The field varies inversely with r—the density of the windings is not the same on the inner and outer surfaces. For a single wire, $B = \mu_0(NI)/2\pi r$, which is identical to the coil. **33.** See Fig. AN 33. For a finite sheet, the field lines would loop around the sheet counterclockwise looking into the current (along the minus z-axis). The sheet is infinite so there cannot be a preferred direction other than parallel to it—no $\pm y$-component to the field. Moreover, there's no reason for the field to be any different at different distances from the sheet because, being infinite, the sheet "looks" the same as seen from any point above it. In effect, the notion of "distance from" loses its traditional meaning in this idealized situation. Going around the Ampèrean loop in the figure, $\Sigma B_\parallel \Delta l = \mu_0 \Sigma I$; $Bl + Bl = \mu_0 il$; and $B = \frac{1}{2}\mu_0 i$. **37.** In the $+x$-direction. **41.** Using $F = qvB$; $B = F/qv$; 1 T $= 1$ N/C·m/s $= 1$ (N·m)s/C·m^2 $= 1$ (J/C)s/m^2 $= 1$ V·s/m^2. **45.** $qE = qvB$; $v = E/B$. **49.** $F = IlB$; $I = F/lB = 1.00$ A. **53.** $\tau = NIAB \sin\phi = (20)(1.5$ A)$(1.3 \times 10^{-3}$ m^2) $\times (0.90$ T)(0.529 9) $= 0.019$ N·m. **55.** $F =$

Fig. AN 33.

$qvB \sin\theta = ma$; $a = (1.60 \times 10^{-19}$ C)$(5.0 \times 10^6$ m/s)(5.0 T)0.866/$(9.11 \times 10^{-31}$ kg) $= 3.8 \times 10^{18}$ m/s^2. **59.** $qvB = mv^2/R$; $qB = mv/R$; $p = qBR$. **63.** $V = vBl$; $v = V/Bl = (1.0\ \mu V)/(0.50$ mT)(4 mm) $= 0.5$ m/s. **65.** $\tau = NIAB$; hence, $\tau_i/\tau_f = I_i/I_f = I_i/0.75I_i$; $\tau_f = 0.75\tau_i$. **69.** $qvB = mv^2/R$, $qB = mv/R$, $v = qRB/m$; if f is the frequency, $1/f$ is the period, which equals $2\pi R/v$; hence, $f = v/2\pi R$ and $f = qB/2\pi m$.

CHAPTER 22

Answers to Odd-Numbered Multiple Choice Questions
1. e **3.** c **5.** a **7.** a **9.** c **11.** c **13.** b **15.** d **17.** d **19.** b **21.** b

Answers to Odd-Numbered Problems
1. 3.0 μWb **3.** 1.2 Wb/m^2 **5.** 0.1 T, south **7.** 0.50 V **9.** 75 ms **11.** 8.0 V **13.** 0.60 V **15.** bulb would not light **17.** 90 μV **19.** $I = Blv/R$, clockwise, constant for a time l/v **21.** 4.1 μC **23.** 3.5 mV, counterclockwise **25.** P $= (Blv)^2/R$ **27.** $B = RK\theta/NA$ **29.** 0.83 mT **31.** 1.6 V **33.** 39 V **35.** 100 V, $f = 60.000$ Hz **37.** $L = 0.26$ H **39.** 0.80 mWb **41.** $N = 2.5 \times 10^3$ **43.** 8.4×10^2 **45.** 5.0 H **47.** 2.5 H **49.** 0.632 **51.** 1.2 H **53.** 21 mT **55.** 12 mV **57.** $\mathscr{E} = (1.3$ kV) $\sin 200\pi t$ **59.** 0.15 s **61.** 79 mJ/m **63.** (a) 6.0 A (b) 12 ms (c) 3.8 A **65.** (a) 2.0 A (b) 4.0 s (c) ≈ 20 s **69.** 7×10^{16} J, 7×10^8 gallons of gasoline **71.** $\Delta I/\Delta t = (V - IR)/L$, 0.89 kA/s never reaches 1.0 A **73.** 21 V

Solutions to Selected Problems
1. $\Phi_M = B_\perp A = (1.2$ mT)(0.002 5 m^2) $= 3.0\ \mu$Wb. **5.** The final field minus the initial field is the change in the field and that's 0.1 T, south. **9.** From Eq.(22.3), $\Delta t = 150 \times (0.030$ Wb)/(60 V) $= 75$ ms. **13.** $\mathscr{E} = vBl = (200$ m/s)(0.050 mT)(60 m) $= 0.60$ V. **19.** As soon as the leading edge of the loop starts cutting field lines, there is an emf $= Blv$, which produces a counterclockwise current $I = Blv/R$. The emf is constant (for a time l/v) until the trailing edge of the loop enters the field. It produces an oppositely directed emf (the flux no longer changes), and the net emf goes to zero. It remains zero until the leading edge emerges from the field at a time $3l/v$ after it entered, whereupon the emf jumps to Blv in the opposite direction. A

current $I = Blv/R$ appears clockwise and remains constant for a time l/v. **23.** From Eq. (22.1), $N\Delta\Phi_M = 1(0.500\text{ T}) \cos 30° \times (1.6 \times 10^{-3}\text{ m}^2) = 6.928 \times 10^{-4}\text{ T·m}^2$; from Eq.(22.3), emf $= 3.5$ mV. Counterclockwise. **27.** $I = V/R$; $\Delta Q = N\Delta\Phi_M/R = NBA/R = K\theta$, $B = RK\theta/NA$. **33.** For each length $\mathcal{E} = vBl = (22\text{ m/s})(0.35\text{ T})(0.10\text{ m}) = 0.77$ V. Since there are 2 lengths per turn and 25 turns, $50(0.77\text{ V}) = 39$ V. **37.** $N\Phi_M = LI$; $L = (500)(0.002\ 0\text{ Wb})/(3.8\text{ A}) = 0.26$ H. **41.** $N\Phi_M = LI$; $N = (2.5\text{ H})(1.80\text{ A})$ $(1.80\text{ mWb}) = 2.5 \times 10^3$. **45.** $\mathcal{E} = -L\dfrac{\Delta I}{\Delta t}$, $L = -(10\text{ V})/(2.0\text{ A/s}) = 5.0$ H (forget the sign, since we don't know if the current is increasing or decreasing). **49.** $1 - 0.368 = 0.632$. **55.** $\mathcal{E} = \frac{1}{2}Br^2 2\pi f = \frac{1}{2}(0.050\text{ mT} \times \cos 40°)(5.0\text{ m})^2 2\pi(240/60) = 12$ mV. **59.** $\mathcal{E} = -L\dfrac{\Delta I}{\Delta t}$; $\Delta t = (1.5\text{ H})(10\text{ A})/$ $(100\text{ V}) = 0.15$ s. **63.** (a) $I = V/R = (150\text{ V})/(25\ \Omega) = 6.0$ A; (b) $L/R = (300\text{ mH})/(25\ \Omega) = 12$ ms; (c) $0.632(6.0\text{ A}) = 3.8$ A. **67.** From Eq. (22.9), $L \approx \mu N^2 A/l = An^2 l\mu$ where An^2 is a constant, so ΔL depends on the change in μl. Once the shaft is inserted, part of the region of the coil (d long) changes from μ_0 to μ; hence, $\mu d + \mu_0(l - d)$ is the new value, $\mu_0 l$ the original value. The difference $d(\mu - \mu_0)$ is what we want; hence

$$\Delta L \approx d(\mu - \mu_0)\frac{N^2 A}{l^2}$$

71. From Eq. (22.11), $\Delta I/\Delta t = (V - IR)/L$, which (in the limit as $\Delta t \to 0$) is also the instantaneous rate of change of current, or the slope of the curve in Fig. 22.21 at any instant. At $t = 0$, $I = 0$ and $\Delta I/\Delta t = V/L = 2.4$ kA/s. At $t = L/R$, $I = 0.63V/R$ and $\Delta I/\Delta t = (0.37\ V)/L = 0.89$ kA/s. The maximum current is $V/R = 0.60$ A, so it never reaches 1.0 A.

CHAPTER 23

Answers to Odd-Numbered Multiple Choice Questions
1. b **3.** d **5.** e **7.** c **9.** b **11.** e **13.** a **15.** b

Answers to Odd-Numbered Problems
1. (a) zero (b) 84.9 V **3.** $i = (14\text{ A}) \times \sin 2\pi(50\text{ Hz})t$ **5.** 141 V **7.** 0.833 A **9.** 113 W **11.** 0.13 W **13.** 2.00 Ω **15.** 22 μF **17.** $f = 11$ Hz **19.** 94 V **21.** $f = 100$ Hz **23.** one resistor is probably shorted internally **25.** 20 A, 120 V **27.** 15 W, 240 W **31.** 2.12 A **33.** (a) something is wrong with it (b) new capacitor is shorted internally **35.** $P_m = 2P_{av}$ **37.** 10.1 Ω **39.** 219 Ω **41.** 1.30 kΩ **43.** $\theta = 26.6°$, voltage leads current **45.** 0.25 A **47.** $\theta = -63°$ **49.** $R = 1.7$ kΩ **51.** 500 Ω **53.** 5.1×10^{-9} F **55.** 52.8 mH **57.** 100:1 **59.** $V_s = 30$ V, $I_s = 4.0$ A **61.** 5.0 A, $\theta = 51°$ **63.** $V_L = 302$ V, $V = 477$ V **65.** 27 μF **69.** (a) 20:1 (b) $I_p = 0.11$ A **71.** 98.3% **73.** 20:1

Solutions to Selected Problems
3. $I_m = 1.414(10\text{ A}) = 14$ A; $i = (14\text{ A}) \sin 2\pi(50\text{ Hz})t$. **7.** $P_{av} = 100$ W $= IV$; $I = (100\text{ W})/(120\text{ V}) = 0.833$ A. **9.** $P_{av} = IV = (1.50\text{ A})(75.0\text{ V}) = 113$ W. **13.** $X_C = 1/2\pi fC$; $X_{Ci}/X_{Cf} = f_f/f_i = 100$; hence, the final reactance is 2.00 Ω. **17.** $X_L = 2\pi fL$; $f = (10\ \Omega)/2\pi(0.15\text{ H}) = 10.6$ Hz $= 11$ Hz. **21.** $v = V_m 0.951$; $0.951 = \sin 2\pi f(0.002\ 00\text{ s})$; $f = 100$ Hz. **25.** $1/R_e = 1/(10\ \Omega) + 1/(15\ \Omega)$, $R_e = 6.0\ \Omega$; $I = V/R_e = 20$ A. The voltmeter reads 120 V. **29.** $X_L = 2\pi fL$; $V = -L\Delta i/\Delta t$, so $L = -V\Delta t/\Delta i$ and 1 H $= 1$ V·s/A; hence, reactance has units of $(1/s)(H) = V/A = \Omega$. **33.** (a) $X_C = 1/2\pi fC = 132.6\ \Omega$; hence, $I = V/X_C = (120\text{ V})/(132.6\ \Omega) = 0.90$ A. The current should be about 0.9 A and, instead, the ammeter reads 0.25 A. Something is wrong with that capacitor. (b) The fuse would blow if the new capacitor was shorted internally—it would short out the first capacitor. **37.** $Z = \sqrt{R^2 + X^2} = 10.1\ \Omega$. **39.** $Z = \sqrt{R^2 + X^2} = 219\ \Omega$. **43.** $\tan\theta = X/R$; $\theta = 26.6°$, and voltage leads current. **47.** $\tan\theta = (200\ \Omega)/(100\ \Omega) = 2.0$; $\theta = -63°$, this is a capacitive reactance. **51.** $Z = 2500\ \Omega - 2000\ \Omega = 500\ \Omega$. **53.** $C = 1/4\pi^2 f_0^2 L = 5.1 \times 10^{-9}$ F—such values are not unusual. **57.** 100:1. **61.** $2\pi f = 1257$ s^{-1}; $X_L = 2\pi fL = 37.7\ \Omega$; $Z = \sqrt{R^2 + X^2} = 48.18\ \Omega$; $I = V/Z = 5.0$ A; $\tan\theta = X_L/R = 1.257$, $\theta = 51°$, voltage leads current. **63.** $2\pi f = 377$ s^{-1}; $X_L = 2\pi fL = 150.8\ \Omega$; $Z = \sqrt{R^2 + X^2} = 238.7\ \Omega$; $I = V_R/R = (370\text{ V})/(185\ \Omega) = 2.00$ A. $V_L = IX_L = 302$ V. $\tan\theta = X_L/R$; $\theta = 39.18°$; $V = V_R/\cos\theta = 477$ V. **67.** $Z = \sqrt{R^2 + X_C^2}$; $I = V_i/Z$; $V_0 = IX_C$; $V_0 = V_i X_C/Z$; $V_0/V_i = X_C/Z = (1/2\pi fC)/\sqrt{R^2 + X_C^2} = 1/\sqrt{1 + (2\pi fRC)^2}$. **69.** (a) $V_p N_s = V_p/V_s = 20:1$; (b) $2\pi f = 377$ s^{-1}; $X_L = 2\pi fL = 1.131$ kΩ; $I_p = V/X_L = 0.11$ A. **73.** $Z_p = V_p/I_p$; $Z_s = V_s/I_s$; $Z_p/Z_s = (V_p/V_s) \times (I_s/I_p)$ but, as we saw earlier, $V_p N_s = V_s N_p$ and $N_p I_p = N_s I_s$; hence, $Z_p = Z_s(N_p/N_s)^2$. $\sqrt{Z_p/Z_s} = N_p/N_s = 20:1$.

CHAPTER 24

Answers to Odd-Numbered Multiple Choice Questions
1. c **3.** b **5.** a **7.** b **9.** a **11.** d **13.** c **15.** d **17.** b **19.** e **21.** c

Answers to Odd-Numbered Problems
1. 12.6×10^6 m^{-1} **3.** 10.00×10^3 m **7.** 20 V/m **11.** radiowaves, 100 s **13.** 1.50×10^7 m **15.** 5 μH **17.** 1.50×10^6 Hz **19.** 1.2×10^{-9} J/m^2·s **21.** 6.30×10^4 J **23.** 620 nm, 484 THz, 3.20×10^{-19} J, orange-red light **27.** -20 V/m **29.** $B_0 = 6.7 \times 10^{-7}$ T **31.** 14.5 **33.** 53 W/m^2 **35.** $E_{max} = 1.00 \times 10^4$ eV, 1.24×10^{-10} m **37.** (a) 628 nm (b) 4.78×10^{14} Hz (c) positive x-direction **39.** (a) 0.600 m (b) 1.0×10^6 J/m^3

Solutions to Selected Problems
1. $k = 2\pi/\lambda = 2\pi/(500\text{ nm}) = 12.6 \times 10^6$ m^{-1}. **5.** See Fig. AN 5. At $x = \lambda/4$, $kx = (2\pi/\lambda) \times (\lambda/4) = \pi/2$.

Fig. AN 5.

Fig. AN 9.

9. See Fig. AN 9. $E = 10 \sin(kx + \varepsilon)$ mark off the axis in intervals of $\lambda/4$. Notice that shifting the phase by 90° results in a cosine function. **13.** $c = f\lambda_0$, $\lambda_0 = c/f = (3.00 \times 10^8\text{ m/s})/(20.0\text{ Hz}) = 1.50 \times 10^7$ m. **17.** $f = c/\lambda = (3.00 \times 10^8\text{ m/s})/(200\text{ m}) = 1.50 \times 10^6$ Hz. **21.** Irradiance is energy divided by area divided by time: $E/At = I$, $E = (1.05 \times 10^3\text{ W/m}^2)(1.00\text{ m}^2)(60.0\text{ s}) = 6.30 \times 10^4$ J. **23.** 1 eV $= 1.602 \times 10^{-19}$ J; $E = hf = 3.20 \times 10^{-19}$ J; $f = 484$ THz; $\lambda_0 = c/(484\text{ THz}) = 620$ nm; orange-red light. **27.** $E(0,0) = 20 \cos(0 - 0 + \pi) = -20$; $E(0, T/4) = 20 \cos(0 - 2\pi fT/4 + \pi) = 20 \cos(+\pi/2) = 0$; $E(0, T) = 20 \cos(0 - 2\pi fT + \pi) = 20 \cos(-\pi) = -20$ V/m. **31.** $\Delta l = c\ \Delta t = (3.00 \times 10^8\text{ m/s}) \times (30.0 \times 10^{-15}\text{ s}) = 9.00 \times 10^{-6}$ m. Now dividing this result by the wavelength yields the number of waves: $(9.00 \times 10^{-6}\text{ m})/(620 \times 10^{-9}\text{ m}) = 14.5$. **37.** (a) $k = 10^7 = 2\pi/\lambda$, $\lambda = 628$ nm. (b) $f = c/\lambda = (3.00 \times 10^8\text{ m/s})/(628 \times 10^{-9}\text{ m}) = 4.78 \times 10^{14}$ Hz. (c) It moves in the positive x-direction. **39.** (a) $l = c\Delta t = (3.00 \times 10^8\text{ m/s})(2.00 \times 10^{-9}\text{ s}) = 0.600$ m. (b) The volume of one pulse is $(0.600\text{ m})(\pi R^2) = 2.945 \times 10^{-6}$ m^3; hence, $(3.0\text{ J})/(2.945 \times 10^{-6}\text{ m}^3) = 1.0 \times 10^6$ J/m^3.

CHAPTER 25

Answers to Odd-Numbered Multiple Choice Questions
1. c **3.** d **5.** e **7.** a **9.** d. **11.** d **13.** d **15.** d **17.** c **19.** e **21.** d

Answers to Odd-Numbered Problems
1. 6.25% **3.** 30° **5.** $\approx$90° **7.** 10° **9.** 12 m **11.** 12.5 cm **13.** it misses the mirror **15.** 6.0 m **17.** unchanged **19.** $H = hd/(d + s_0)$ **23.** $L = d/\tan 2\phi$ **25.** 1.24×10^8 m/s **27.** 1.1 **29.** 2.25×10^8 m, $2.21 \times$

10^8 m/s **31.** 1.3 **35.** 566 THz, 353 nm
37. $\theta_c = 49.8°$ **39.** $\theta_c = 33.7°$ **41.** 2.2 mm, 4.2×10^3 waves **43.** $n_i = 1.57$ **45.** 2.2 cm
47. 0.355 mm **49.** 98° **51.** $(\overline{AB} - L)/\lambda_0$, $(\overline{AB}/\lambda_0) + L(1/\lambda - 1/\lambda_0)$, 2000π
53. 0.885 m

Solutions to Selected Problems
1. $(1/\lambda_r)^4/(1/\lambda_v)^4 = (\lambda_v/\lambda_r)^4 = 6.25\%$.
9. The statue is 16 m from the point of incidence, and since the ray-triangles are similar, 4 m:16 m as 3 m:Y and $Y = 12$ m.
15. See Fig. AN 15. $h = \frac{1}{2}$ m, $d = 1.0$ m, $D = 11.0$ m, find H. $h/d = H/(d + D)$, $h = H/(1.0 + D)$, $\frac{1}{2}(1.0 + D) = H = \frac{1}{2}(1.0 + 11.0) = 6.0$ m.

Fig. AN 15.

17. You still see the same fraction of your total body. **21.** See Fig. AN 21. There will be two images arising from single reflections from each of the single mirrors, and there will be a third image due to the double reflections, one from each mirror. The central image is *not* reversed in its handedness—you hold up a right hand, and diagonally across from it, the image will hold up its right hand.

Fig. AN 21.

23. See Fig. AN 23. Construct a normal at M_2; the angle it makes with $\overline{SM_2}$ call α. $\angle M_1 A M_2 = 45° + 2\phi$, hence $(45° + 2\alpha) + (45° + 2\phi) = 180°$, and so $\phi = 45° - \alpha$ and $\angle M_1 S M_2 = 90° - 2\alpha = 2\phi$, thus $\tan 2\phi = d/L$; hence, $L = d/\tan 2\phi$.
29. $l = vt = (c/n)t = (3.00 \times 10^8$ m/s$) \times (1.00$ s$)/1.333 = 2.25 \times 10^8$ m. $n = 1.36 = c/v$; $v = c/n = 2.21 \times 10^8$ m/s.
31. $1.00 \sin 55° = n \sin 40°$; $n = 1.27$ or 1.3.
35. $\lambda = (1.06 \ \mu m)/2 = 0.530 \ \mu m = 530$ nm, hence $f = 566$ THz. $(1.06 \ \mu m)/3 = 0.353 \ \mu m = 353$ nm in the near-UV.
41. The photo shows the trail of the pulse as it moves through the water during the 10-ps exposure time. $v = c/n = (3.0 \times 10^8$ m/s$)/1.36 = 2.2 \times 10^8$ m/s; the pulse moves $v \ \Delta t = (2.2 \times 10^8$ m/s$)(10 \times 10^{-12}$ s$) = $

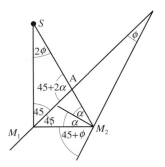

Fig. AN 23.

2.2 mm during the exposure and is about 2.2 mm long to begin with. $(2.2 \times 10^{-3}$ m$)/(530 \times 10^{-9}$ m$) = 4.2 \times 10^3$ waves.
43. $\tan \theta_i = 15/20$, $\theta_i = 36.87°$; from Snell's Law $n_i \sin 36.87° = 1.00 \sin 70°$, $n_i = 1.57$.
47. The glass will change the depth of the object from d_R to d_A, where $d_A/d_R = 1.00/1.55$; but $d_R = 1.00$ mm; hence, $d_A = 0.645$ mm, and the camera must be raised 1.00 mm $- 0.645$ mm $= 0.355$ mm.
51. The number of waves in vacuum is $\overline{AB}/\lambda_0$. With the glass in place, there are $(\overline{AB} - L)/\lambda_0$ waves in vacuum and an additional L/λ waves in glass for a total of $(\overline{AB}/\lambda_0) + L(1/\lambda - 1/\lambda_0)$. The difference in number is $L(1/\lambda - 1/\lambda_0)$, giving a phase shift $\Delta\phi$ of 2π for each wave; hence, $2\pi L(1/\lambda - 1/\lambda_0) = 2\pi L(n/\lambda_0 - 1/\lambda_0) = 2\pi L/2\lambda_0 = 2000\pi$. **53.** See Fig. AN 53. $d_{A1}/d_{R1} = 1.50/1.33$; $d_{R1} = 1.00$ m; $d_{A1} = 1.1278$ m: $d_{R2} = d_{A1} + 0.20$ m; $d_{A2}/d_{R2} = 1.00/1.50$; $d_{A2} = (1.3278)(1.00/1.50) = 0.885$ m.

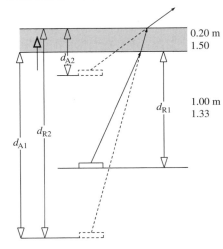

Fig. AN 53.

CHAPTER 26
Answers to Odd-Numbered Multiple Choice Questions
1. b **3.** b **5.** d **7.** b **9.** a **11.** e **13.** b **15.** d **17.** a **19.** c **21.** e

Answers to Odd-Numbered Problems
1. 100 cm **3.** 0.86 m **5.** 2.25 **7.** $s_i = 150$ cm **9.** $s_i = 0.38$ m; real, inverted, and minified **11.** $s_i = -240$ mm, virtual **13.** $f/2$ **15.** 37 mm **17.** 0.13 m **19.** 3 cm **21.** $f = -20$ cm, 40 cm in dia. **23.** $f = 10$ cm, will help **25.** $f = +4.0$ cm, yes **27.** $\mathcal{D} = +2.94$ D **29.** $s_i = -5.0$ m **31.** $L = 6.0$ cm **33.** 1.03 m, 33× **39.** (a) -3.8 cm (b) 1st—real, inverted, magnified; after jump—virtual, right-side-up, magnified **41.** $f = 6.5 \times 10^2$ mm **43.** (a) $\mathcal{D} = -1.02$ D (b) $s_0 = 30.9$ cm **45.** The greater is s_i, the greater is M_T **47.** (a) 200× (b) $f_E = 25.4$ mm, $f_0 = 8.0$ mm (c) $s_0 = 8.4$ mm **49.** R = 0.083 m **51.** $f_w = 4f_a$ **53.** 3.2 m, set should be upside down **55.** $\mathcal{D} = 5.0$ D **57.** -13 cm **59.** 1.33 m **61.** (a) -0.58 m (b) 1.5 m **63.** virtual, right-side-up, minified (0.7 times), 0.14 m behind mirror **65.** 2.0×10^{-6} m **67.** R = -0.060 m **69.** $M_T = -0.14$, real, inverted, and minified **73.** $f = -2.5$ m **75.** -1.0 m

Solutions to Selected Problems
1. From Eq. (26.4), $\sigma \approx h^2/2R$, $\frac{1}{2}$ cm $\approx (10$ cm$)^2/2R$; $R = 100$ cm. **3.** $1/f = 0.58 \times (2$ m$^{-1})$; $f = 0.86$ m. **5.** $1/f = (n - 1) \times [1/(2.00$ m$) - 1/\infty]$; $1/(1.60$ m$) = (0.500$ m$^{-1})(n - 1)$; $n = 2.25$. **9.** Given the focal length and the object-distance, we want the image-distance that follows from the Gaussian lens formula: $1/s_0 + 1/s_i = 1/f$; $1/s_i = 1/f - 1/s_0 = 2.66$ m^{-1}; $s_i = 0.38$ m. The object is three focal lengths away from the lens and the image is real, inverted, and minified. **13.** The diameter is 25 mm; hence, f-number $= (50$ mm$)/(25$ mm$) = f/2$. **17.** $M_A = d_n \mathcal{D}$, $2.0 = (0.254$ m$) \mathcal{D}$, $\mathcal{D} = 7.87$ D and $f = 0.13$ m. **21.** For the second lens $s_0 = -10$ cm and $s_i = +20$ cm; hence, $1/f = 1/(-10$ cm$) + 1/(20$ cm$)$; $f = -20$ cm. $M_T = -(20$ cm$)/(-10$ cm$) = +2$, and the image is 40 cm in diameter. **25.** Applying Eq. (26.19) to two lenses at a time, $1/f = 1/10 + 1/(-20) + 1/5.0$; $f = +4.0$ cm. Since it's positive, it can form real images. **29.** The lens makes an object at a distance s_0, corresponding to infinity, appear to be at the far-point (s_i), where it can be seen clearly. $1/(-5.0$ m$) = 1/\infty + 1/s_i$; $s_i = -5.0$ m. **33.** Putting the long focal length ($+1.00$ m) first, we'll need a tube of length $f_0 + f_E = 1.00$ m $+ 0.030$ m $= 1.03$ m. $M_A = -(1.00$ m$)/(0.030$ m$) = 33×$. **37.** $n_1 \overline{HI} + n_1 \overline{IJ} = n_m \overline{AB} + n_m \overline{BC} + n_1 \overline{CD} + n_1 \overline{DE} + n_m \overline{EF} + n_m \overline{FG}$, now divide both sides by n_m and you are back to the form of the first derivation provided $n_1/n_m \rightarrow n_1$. **41.** $M_T = y_i/y_0 = -s_i/s_0$; $s_i = (-2.0$ m$) \times (-0.024$ m$)/(0.05$ m$) = 0.96$ m. $1/(2.0$ m$) + 1/(0.96$ m$) = 1/f$; $f = 6.5 \times 10^2$ mm. **47.** (a) $20 \times 10 = 200×$. (b) $f_E = (254$ mm$)/M_{AE} = 25.4$ mm, $f_0 = -(160$ mm$)/(-20) = 8.0$ mm. (c) $1/s_0 = 1/f_0 - 1/s_i = 1/(8.0$ mm$) - 1/(L + f_0) = 1/(8.0$ mm$) - 1/(168$ mm$)$; $s_0 = 8.4$ mm. **51.** From Eq. (26.7)

$1/f_w = (n_1/n_w - 1)/(n_1 - 1)f_a = (1.5/1.33 - 1)/(1.5 - 1)f_a = 0.125/0.5 f_a; f_w = 4f_a$.
55. With her glasses on, 2.0 D $= 1/(0.80$ m $- 0.020$ m$) + 1/s_i$, which gives the distance from the eyeglass plane at which she can see clearly; that is, $(s_i + 2.0$ cm$)$ is her new near-point. The prescribed focal length of her new spectacles must satisfy the expression $1/f = 1/(0.254$ m $- 0.020$ m$) + 1/s_i$. Combining these two equations; $1/f - 4.274$ D $= 2.0$ D $- 1.282$ D; and $\mathcal{D} = 5.0$ D. **59.** $1/(2.00$ m$) + 1/(4.00$ m$) = 1/f$; $f = 1.33$ m. **63.** $1/(0.20$ m$) + 1/s_i = -2/(1.00$ m$); s_i = -0.14$ m. $M_T = -(-0.14$ m$)/(0.20$ m$) = +0.7$. The image is virtual, minified (0.7 times life-size), and 0.14 m behind the mirror. **67.** For the image to be magnified and erect, the mirror must be concave and the image virtual; $s_o = 0.015$ m; $M_T = 2 = -s_i/(+0.015$ m$); s_i = -0.030$ m and, hence, $1/f = 1/(0.015$ m$) - 1/(0.030$ m$); f = 0.030$ m and $f = -R/2; R = -0.060$ m. **71.** Take another look at your solution of Problem 68; $M_T(s_o - f) = -f$; hence, $s_o = (-f + fM_T)/M_T$, which equals what we want. Furthermore, $s_i = fs_o/(s_o - f)$, and on substituting in for s_o, we do indeed get the desired expression. **75.** The mirror projects a real, inverted, life-size image onto the empty socket. $s_o = 2f = s_i = 1.0$ m $= 2(-R/2); R = -1.0$ m.

CHAPTER 27

Answers to Odd-Numbered Multiple Choice Questions

1. c   **3.** a   **5.** b   **7.** a   **9.** c   **11.** b   **13.** c   **15.** b   **17.** a

Answers to Odd-Numbered Problems

1. 1327 W/m^2 **3.** 100 W/m^2 **5.** 188 W/m^2 **7.** 40 W/m^2 **9.** 53°   **11.** 0.19   **13.** 1.501   **15.** 39°   **17.** ±6.33 mm   **19.** 419 nm   **21.** 9.76 mm   **23.** 1.84×10^{-7} m   **25.** 107 nm, 322 nm, 537 nm, etc.   **27.** $y_{m'} = sm'\frac{1}{2}\lambda/a$ **29.** 10 km   **31.** 0.346 m   **33.** $\Delta x = \lambda_f/2\alpha$ **35.** 550 nm   **37.** 96.6   **39.** $\theta_m = \sin^{-1}(\sin\theta_i \pm m\lambda/a)$ **41.** 1474 nm, 737.0 nm, 491.3 nm, 368.5 nm   **43.** 0.53°   **45.** 12 cm   **47.** Δy doubles   **49.** 2.6 µm   **51.** 1.32×10^{-7} rad, 7.57×10^{-6} degrees, 2.72×10^{-2} s   **53.** 64.5 km   **55.** 0.12 mm   **57.** 11×10^4 Hz   **59.** 0.597 mm   **61.** 4 fringes, 2N   **63.** 1.11

Solutions to Selected Problems

1. $I = \frac{1}{2}c\varepsilon_0 E_0^2 = \frac{1}{2}(2.998 \times 10^8$ m/s$) \times (8.854 \times 10^{-12}$ C^2/N·m$^2)(1000$ V/m$)^2$; hence, $I = 1327$ W/m^2.   **5.** $I = I_1 \cos^2\theta = (250$ W/m$^2)\cos^2 30 = 188$ W/m^2.   **9.** $\tan\theta_p = n_t/n_i = 1.3/1.00, \theta_p = 53°$.   **13.** $\tan\theta_p = n_t/n_i; n_t = \tan 56.33° = 1.501$.   **17.** According to Eq. (27.4), the first minimum occurs when $m' = \pm 1$. Hence, $a\sin\theta_1 = \pm\frac{1}{2}\lambda; \theta_1 \approx \pm\frac{1}{2}\lambda/a = \pm\frac{1}{2} \times (632.8 \times 10^{-9}$ m$)/(0.100 \times 10^{-3}$ m$) = \pm 3.16 \times 10^{-3}$ rad, since $y = s\theta_1 = (2.00$ m$)(\pm 3.16 \times 10^{-3}$ rad$) = \pm 6.33$ mm.

21. $y_4 = s4\lambda/a = (1.00$ m$)4(487.99 \times 10^{-9}$ m$)/(0.200 \times 10^{-3}$ m$) = 9.76 \times 10^{-3}$ m.   **25.** There will be a 180° phase shift due to the reflections, and so the film must be a quarter-wavelength thick or odd multiples thereof. $d = (m + \frac{1}{2})\lambda_0/2n_f$; for $m = 0$, $d = \lambda_0/4n_f = (580$ nm$)/4(1.35) = 107$ nm; for $m = 1, d = 322$ nm; for $m = 2, d = 537$ nm, etc.   **29.** From Eq. (27.7), $y_1 = s\lambda/a = sc/fa = (20 \times 10^3$ m$)(3.0 \times 10^8$ m/s$)/(1.0 \times 10^6$ Hz$)(600$ m$) = 1.0 \times 10^4$ m.   **33.** The film increases half a wavelength, $\lambda_f/2$. Using the results of the previous problem, we obtain $x_{m+1} - x_m = \Delta x = \lambda_f/2\alpha$.   **35.** Each arm is traversed twice, so a motion of $\lambda/2$ causes a single fringe-pair to shift past; hence, $92\lambda/2 = 2.53 \times 10^{-5}$ m and $\lambda = 550$ nm.   **37.** If the cell has a length L and is traversed twice, then the change in optical path length with and without air is $2Ln_a - 2L$, and the number of waves changed is $2L(n_a - 1)/\lambda_0$, each corresponding to a fringe-pair, and so there are 96.6 of them.   **41.** There will be a relative phase shift on reflection of $\frac{1}{2}\lambda$; hence, the longest wave not reflected is 550.0 nm $= \frac{1}{2}\lambda_f$, or $\lambda_f = 1100$ nm, and so $\lambda_0 = (1100$ nm$)1.340 = 1474$ nm in the IR. When 550.0 nm $= \lambda_f, \lambda_0 = (550.0$ nm$) \times 1.340 = 737.0$ nm. Similarly, when 550.0 nm $= 3\lambda_f/2 = 366.6$ nm or $\lambda_0 = 491.3$ nm, there is no reflection. And, finally, when $2\lambda_f = 550.0$ nm, $\lambda_f = 275.0$ nm; and $\lambda_0 = 368.5$ nm in the UV, which is not reflected.   **45.** From Eq. (27.13), $\lambda = (20$ cm$)\sin 36.87° = 12$ cm.   **49.** From Eq. (27.8), $a = 2(550 \times 10^{-9}$ m$)/0.422\,6 = 2.6 \times 10^{-6}$ m.   **53.** $\theta_a = 1.22\lambda/D = 1.22(550 \times 10^{-9}$ m$)/(4.00 \times 10^{-3}$ m$) = 1.68 \times 10^{-4}$ rad; hence, $L/(3.844 \times 10^8$ m$) = 1.68 \times 10^{-4}$ rad; $L = 64.5$ km.   **57.** The wavelength should be less than the size of the target or else diffraction will be troublesome. $v = f\lambda, f > (330$ m/s$)/(0.003\,0$ m$) = 11 \times 10^4$ Hz.   **61.** The angular width of the diffraction maximum is obtained from Eq. (27.13), $\sin\theta_1 = 1\lambda/D = (500 \times 10^{-9}$ m$)/(0.10 \times 10^{-3}$ m$) = 0.005; \theta_1 = 0.286\,5°$, and twice this value is 0.573\,0°. The angular width of one of Young's fringes is $\theta = \lambda/a = (500 \times 10^{-9}$ m$)/(0.20 \times 10^{-3}$ m$) = 2.5 \times 10^{-3}$ rad $= 0.143\,2°$; hence, 4.0 fringes will exist. And that might have been obvious from the fact that $a = 2D$ and the number of fringes is 2N.

CHAPTER 28

Answers to Odd-Numbered Multiple Choice Questions

1. d   **3.** c   **5.** b   **7.** a   **9.** e   **11.** e   **13.** a   **15.** d   **17.** c

Answers to Odd-Numbered Problems

5. 90 s, Earth-based observer; 45 s, astronaut   **7.** 601 s   **9.** 0.499c   **11.** 0.436 m   **13.** 64.34 m   **15.** 500 m   **17.** c, classically 2c   **19.** 0.999\,9c $\approx$ c   **21.** $v = 0.404\,8$c

23. 0.436c   **25.** $v = 0.9$c   **27.** 4.45 ns   **31.** 73.8×10^3 y   **33.** 36.3°   **35.** 2.9×10^{-19} kg·m/s   **37.** KE $= 1000$ MeV, $v = 0.866\,0$c   **39.** 1.0 MeV   **41.** $E_0 = 8.988 \times 10^{10}$ J, E $= 9.173 \times 10^{10}$ J, KE $= 1.853 \times 10^9$ J   **43.** 2.2 MeV   **45.** 1.000×10^{-15} kg   **47.** 3.26 MeV   **49.** $f = 3.653 \times 10^{20}$ Hz   **51.** $(\gamma - 1)/\gamma$, 20%   **53.** $v = 0.101$ m/s for both, when $\gamma \approx 1$ **57.** $v = 0.967$c, $v = 2.4$c (classically)   **59.** 0.106\,0c   **61.** 0.866   **63.** 3.438×10^{20} Hz

Solutions to Selected Problems

1. $t_\parallel = \dfrac{2L}{c}\dfrac{1}{(1 - \beta^2)} \to \dfrac{2L\sqrt{1 - \beta^2}}{c(1 - \beta^2)} = \dfrac{2L}{c}\dfrac{1}{\sqrt{1 - \beta^2}}$.   **5.** $\Delta t_M = \Delta t_S/\sqrt{1 - v^2/c^2} = 2(45$ s$) = 90$ s for the Earth-based observer; 45 s for the astronaut.   **9.** $\Delta t_M = \Delta t_S/\sqrt{1 - v^2/c^2}$; 60.0 s $= (52.0$ s$)/\sqrt{1 - v^2/c^2}$; $\sqrt{1 - v^2/c^2} = 0.866\,7, v = 0.499$c.   **13.** $L_M = L_S\sqrt{1 - v^2/c^2} = (2$ mi$) \times (5280$ ft/mi$)(0.020\,00) = 211.1$ ft $= 64.34$ m.   **17.** Let S' be the frame of our laboratory, such that one photon moves right at $v_{PO'} =$ c. The other photon in S moves to the left; that is, S' moves right ($+x$-direction) with respect to S at $v_{O'O} =$ c. The speed of one photon with respect to the other is $v_{PO} = (c + c)/(1 + cc/c^2) =$ c. Classically: 2c.   **19.** Put the electron at rest in S, and have the proton move in S', which is approaching S at -0.992c (that is, moving left). The proton travels at -0.981c (also moving left). $v_{PO} = [(-0.981$c$) + (-0.992$c$)]/[1 + (0.981$c$)(0.992$c$)/c^2] = -1.973$c$/1.973\,2 \approx -$c.   **23.** $L_M = 0.900L_S = L_S/\gamma, \gamma = 1/0.900 = 1.111, 1.2345 = 1/(1 - \beta^2), \beta^2 = (\gamma^2 - 1)/\gamma^2 = 0.190, v = 0.436$c.   **31.** $\gamma \approx 1 + \frac{1}{2}\beta^2 = 1 + \frac{1}{2}8.585 \times 10^{-13}$; $\Delta t_M = \gamma\Delta t_S = (1 + \frac{1}{2}8.585 \times 10^{-13})\Delta t_S = \Delta t_S + (\frac{1}{2}8.585 \times 10^{-13})\Delta t_S$, and we want $(\frac{1}{2}8.585 \times 10^{-13})\Delta t_S = 1.00$ s; hence, $\Delta t_S = 2.330 \times 10^{12}$ s $\approx 73.8 \times 10^3$ y.   **35.** $p = \gamma mv = 1.155(1.675 \times 10^{-27}$ kg$) \times (0.50$c$) = 2.9 \times 10^{-19}$ kg·m/s.   **37.** E $=$ KE $+ E_0 = 2E_0$, KE $= E_0 = 1000$ MeV; E $= 2000$ MeV $= \gamma mc^2 = \gamma(1000$ MeV$)$, $\gamma = 2 = 1/\sqrt{1 - v^2/c^2}, v = 0.866\,0$c.   **41.** $E_0 = mc^2 = 8.988 \times 10^{10}$ J; E $= \gamma mc^2 = 1.021(8.988 \times 10^{10}$ J$) = 9.173 \times 10^{10}$ J; E $=$ KE $+ E_0$, KE $= E - E_0 = 1.853 \times 10^9$ J.   **43.** 938.272 MeV $+$ 939.566 MeV $= 1877.83$ MeV; this result minus 1875.6 MeV is 2.2 MeV.   **47.** 2(1876.12 MeV) $- 2809.41$ MeV $- 939.566$ MeV $= 3.26$ MeV.   **51.** $(\gamma mv - mv)/\gamma mv = (\gamma - 1)/\gamma, \gamma = 1.25, (\gamma - 1)/\gamma = 20\%$.   **55.** $E^2 = E_0^2 + (pc)^2, 4E_0^2 = E_0^2 + (pc)^2, 3E_0^2 = p^2c^2, p = \sqrt{3}$ mc.   **57.** KE $= 1.50$ Mev, E $= (1.50$ MeV $+ 0.511$ MeV$) = \gamma mc^2 = \gamma E_0 = \gamma(0.511$ MeV$)$, $\gamma = 3.935, \beta^2 = (\gamma^2 - 1)/\gamma^2 = 0.935, v = 0.967$c. KE$_c = \frac{1}{2}mv^2 = 2.40 \times 10^{-13}$ J, $v = 2.4$c.   **61.** KE $= E_0$, E $=$ KE $+ E_0 = 2E_0, \gamma mc^2 = 2mc^2, \gamma = 2 = 1/\sqrt{1 - v^2/c^2}, 4 - 1 = 4\beta^2, \beta = \sqrt{3}/2 = 0.866$.   **63.** $E^2 = E_0^2 + (pc)^2, (pc)^2 = (1.000$ MeV $+$

$0.511 \text{ MeV})^2 - (0.511 \text{ MeV})^2$, $p =$ 1.422 MeV/c, which is the momentum of the electron. For the photon $E = pc =$ 1.422 MeV $= hf$, $f = 3.438 \times 10^{20}$ Hz.

Chapter 29

Answers to Odd-Numbered Multiple Choice Questions

1. d **3.** b **5.** b **7.** c **9.** b **11.** a **13.** b **15.** c **17.** a

Answers to Odd-Numbered Problems

1. 1.0978×10^{27} **3.** 35.5 g of Cl **5.** 3.55 g **7.** Ba, 25.6 g; Cl, 13.2 g **9.** 4.90×10^{-14} kg **11.** 0.11 nm **13.** $\theta = 8.54°$ **15.** 0.21 nm **17.** $\lambda = 486.2$ nm **19.** $\lambda_\infty = 364.6$ nm **21.** $\lambda_\alpha = 656.3$ nm **23.** 2.28×10^{-10} m **27.** $B = 8.71 \times 10^{-4}$ T **29.** 2.82×10^8 m/s **31.** 0.237 nm at 25.0° **33.** 12 **35.** 1.6×10^{-14} m **37.** 397 nm **41.** $\Delta\lambda = \pm 0.001$ nm

Solutions to Selected Problems

3. (965 A)(100.0 s) = 96.5 kC = 1 faraday; hence, 1 mole of each is liberated; 35.5 g of Cl.   **7.** The valence of barium is 2 and that of chlorine is 1. Hence, for Ba, $m = (2.00 \text{ A}) \times (5.00 \text{ h})(3600 \text{ s/h})(137 \text{ g})/(96.5 \text{ kC})2 =$ 25.6 g; for Cl, $m = (2.00 \text{ A})(5.00 \text{ h}) \times (3600 \text{ s/h})(35.5 \text{ g})/(96.5 \text{ kC})1 = 13.2$ g.   **11.** $d = 1(0.090 \text{ nm})/(2 \sin 25°) = 0.11$ nm.   **15.** $\lambda = 2(0.28 \text{ nm}) \sin 22° = 0.21$ nm.   **19.** $1/\lambda = (1.097 \times 10^7 \text{ m}^{-1})(1/4 - 1/\infty)$; and $\lambda_\infty = 364.6$ nm.   **21.** $1/\lambda = (1.097 \times 10^7 \text{ m}^{-1})(1/4 - 1/9)$; and $\lambda_\alpha = 656.3$ nm.   **25.** $\tan \theta \approx \theta = v_y/v_x = a_y t/v_x = (eE/m_e)t/(E/B)$ where $t = L/v_x$; hence, $\theta \approx eLB^2/Em_e$.   **31.** $d = 1\lambda/(2 \sin 25.0°) = 0.237$ nm at 25.0°.   **33.** The largest angle is 90°; $2(1.22 \times 10^{-10} \text{ m}) \sin 90° = m \times (0.020 \ 0 \times 10^{-9} \text{ m})$; $m = 12.2$, so to nearest integer $m = 12$.   **37.** $1/(390 \text{ nm}) = (1.097 \times 10^7 \text{ m}^{-1})(1/2^2 - 1/n^2)$; and $n = 7.8$, so we must use $n = 7$; $1/\lambda = (1.097 \times 10^7 \text{ m}^{-1})(1/2^2 - 1/7^2)$, and $\lambda = 397$ nm.   **39.** $Y = y + R \tan \theta$; $\tan \theta = v_y/v_x = eLB^2/Em_e$. Using Eq. (29.2), we have $y = eL^2B^2/2m_e E$ and $Y = (e/m_e)\dfrac{B^2L}{2E}(L + 2R)$.   **41.** $m\lambda = 2d \sin \theta$; $m\Delta\lambda = 2 \Delta d \sin \theta$; $\Delta\lambda = \pm 0.001$ nm.

CHAPTER 30

Answers to Odd-Numbered Multiple Choice Questions

1. c **3.** a **5.** b **7.** b **9.** c **11.** d **13.** b **15.** c **17.** b

Answers to Odd-Numbered Problems

1. $I = 76.9$ W/m², P = 108 W **3.** 9.4 μm, IR **5.** 4.1×10^{15} Hz **9.** 496 nm **11.** 0.07 W/m² **13.** 4.61×10^{14} Hz **15.** 1.89 eV **17.** 1.46×10^{-4} nm **19.** 180° **21.** 35 keV **23.** -1.51 eV **25.** $f = 4.57 \times 10^{14}$ Hz, $\lambda = 656$ nm **27.** 476 pm **29.** $n = 137\ 467$ **31.** $T = 4.46 \times 10^3$ K **33.** 0.663×10^6 m/s **35.** 4.4×10^{14} Hz, 6.8×10^{-7} m

37. 5.82×10^{-11} m **39.** 59.6° **41.** 2.19×10^6 m/s **43.** -54.4 eV **45.** $f_n = m_e k^2 Z^2 e^4/2\pi n^3 \hbar^3$ **49.** 5.54×10^{-8} eV

Solutions to Selected Problems

1. $P/A = \varepsilon\sigma(T^4 - T_e^4) = (0.97)(5.670\ 3 \times 10^{-8} \text{ W/m}^2 \cdot \text{K}^4)(306^4 - 293^4) = I = 76.9$ W/m². P = 108 W.   **5.** From Eq. (30.4), $\lambda_{max}T = 0.002\ 898$ m·K; $\lambda_{max} = 72$ nm; $f_{max} = 4.1 \times 10^{15}$ Hz.   **9.** $E = hf = hc/\lambda$, $\lambda = hc/E = 496$ nm.   **13.** KE ≈ 0 is the threshold condition; $f_0 = c/\lambda = 4.61 \times 10^{14}$ Hz.

17. $\Delta\lambda = \lambda_s - \lambda_i = \dfrac{h}{m_e c}(1 - \cos \theta) = (0.002\ 426 \text{ nm})(1 - \cos 20°) = 1.46 \times 10^{-4}$ nm. $n = 1$ is the ground state, $n = 2$ the first excited state, $n = 3$ the second excited state; $E_3 = -(13.6 \text{ eV})/3^2 = -1.51$ eV.   **27.** $n = 3$; from Eq. (30.17), $r = 3^2(0.052\ 9 \text{ nm}) = 476$ pm.   **31.** Taking $\lambda_{max} \approx 650$ nm, $\lambda_{max} T = 0.002\ 898$ m·K; $T = 4.46 \times 10^3$ K.   **35.** In order to use $hf_0 = \phi$, we first need to find ϕ; $hf = \text{KE}_{max} + \phi$; $\phi = 2.929 \times 10^{-19}$ J; $f_0 = \phi/h = 4.4 \times 10^{14}$ Hz; $\lambda_0 = 6.8 \times 10^{-7}$ m.   **39.** $0.060\ 0(20.0 \text{ pm}) = (2.426 \text{ pm})(1 - \cos \theta)$; $\cos \theta = 0.505$; $\theta = 59.6°$.   **43.** From Eq's. (30.18) and (30.19), $E_1 = (-13.6 \text{ eV})Z^2 = -54.4$ eV.   **47.** For x-component, $p_i - p_s \cos \theta = p_e \times \cos \phi$; for y-component, $p_s \sin \theta = p_e \sin \phi$; square and add these two, noting that $\cos^2\phi + \sin^2\phi = 1$; (A) $p_i^2 + p_s^2 - 2p_i p_s \cos \theta = p_e^2$. For the electron (B), $E_e^2 = c^2 p_e^2 + m_e^2 c^4$. Conservation of Energy is (C) $E_i + m_e c^2 = E_s + E_e$. Now rewrite this equation as $E_e = E_i + m_e c^2 - E_s$ and square it and plug it into (B); also multiply (A) by c^2 and plug it into (B) to get $(E_i + m_e c^2 - E_s)^2 = c^2(p_i^2 + p_s^2 - 2p_i p_s \cos \theta) + m_e^2 c^4$. Using $E = pc$ for photons and simplifying, this equation becomes $1/E_s - 1/E_i = (1/m_e c^2)(1 - \cos \theta)$, and since $E = hf = hc/\lambda$ for photons

$$\lambda_s - \lambda_i = \frac{h}{m_e c}(1 - \cos \theta) \qquad [30.13]$$

49. $\Delta E = (-3.40 \text{ eV}) - (-13.6 \text{ eV}) = 10.2$ eV; the photon of momentum $p = E/c$, Eq. (28.16), imparts that momentum to the atom; the atom's recoil energy is then $p^2/2M = E^2/2Mc^2 = (10.2 \text{ eV})^2/2(939 \times 10^6 \text{ eV/c}^2)c^2 = 5.54 \times 10^{-8}$ eV.

CHAPTER 31

Answers to Odd-Numbered Multiple Choice Questions

1. b **3.** d **5.** a **7.** d **9.** d **11.** c **13.** d **15.** a

Answers to Odd-Numbered Problems

1. 5.5×10^{-36} m **3.** 2.4×10^{-11} m **5.** 0.147 nm **7.** 150 V **9.** $l = 0$, $m_l = 0$; $l = 1$, $m_l = -1, 0, +1$; $l = 2$, $m_l = -2, -1, 0, +1, +2$ **11.** K has 2, L has 8, M has 18, N

has 32 **13.** $\hbar/2L$ **15.** 5×10^{-27} m/s **17.** 5×10^{-27} J, 3×10^{-8} eV **19.** 75 MeV, 9.8% **21.** 0.27 nm **23.** $p \approx E/c$ **25.** $v_p = \frac{1}{2}v$ **27.** 12.1 eV **29.** 2.86 eV **31.** cannot exist **33.** $1s^2 \cdot 2s^2 2p^6 \cdot 3s^2 3p^1$ **35.** 2.64×10^{-29} m **37.** 1.16×10^6 m **39.** $R = r + (1.79 \times 10^{-13} \text{ m})/r$

Solutions to Selected Problems

1. $p = h/\lambda$, $\lambda = h/p = h/mv = 5.5 \times 10^{-36}$ m.   **5.** $\lambda = h/p = h/\sqrt{2m \text{ KE}} = 0.147$ nm.   **9.** $n = 3$; $n - 1 = 2$; hence, $l = 2, 1$, or 0. m_l goes from -1 to $+1$; that is, when

$l = 0$, $m_l = 0$
$l = 1$, $m_l = -1, 0, +1$
$l = 2$, $m_l = -2, -1, 0, +1, +2$

13. $\Delta x = L$, $\Delta p_x \approx \frac{1}{2}\hbar/\Delta x \approx \frac{1}{2}\hbar/L$.   **17.** $\Delta t = 10$ ns; $\Delta E = \frac{1}{2}\hbar/\Delta t = (5.272\ 9 \times 10^{-35} \text{ J·s})/(10 \text{ ns}) = 5 \times 10^{-27}$ J $= 3 \times 10^{-8}$ eV.   **21.** $E_0 \gg$ KE; hence, classical form is okay; $p = \sqrt{2m_e \text{KE}}$, KE = (20 eV)$(1.60 \times 10^{-19}$ J/eV); $p = 2.414\ 6 \times 10^{-24}$ kg·m/s; $\lambda = h/p = 2.7 \times 10^{-10}$ m = 0.27 nm.   **25.** $v_p = E/p = (p^2/2m)/p = p/2m$; $v_p = \frac{1}{2}v$.   **29.** From Eq. (30.19), $\Delta E = (-13.6 \text{ eV})/5^2 - (-13.6 \text{ eV})/2^2 = 2.86$ eV.   **33.** $1s^2 \cdot 2s^2 2p^6 \cdot 3s^2 3p^5$ **37.** $\Delta z \Delta p_z \geq \frac{1}{2}\hbar$; $m\Delta v_z \approx \frac{1}{2}\hbar/\Delta z$; $\Delta v_z \approx 0.578\ 8 \times 10^6$ m/s. After 2.00 s, $\Delta z = \Delta v_z t = 1.16 \times 10^6$ m.   **39.** $\Delta z = 2r$; $\Delta p_z \approx \frac{1}{2}\hbar/\Delta z \approx \frac{1}{2}\hbar/2r$; $m\Delta v_z \approx \frac{1}{2}\hbar/2r$; $v_z \approx \frac{1}{2}\hbar/2mr$; in the time τ it takes to reach the observation screen, the beam spreads a distance $Z' = v_z\tau$. To find τ, we find v_y; KE = $(3/2)k_B T$; $\frac{1}{2}mv_y^2 = (3/2)k_B(1600 \text{ K})$; $v_y^2 = 3k_B(1600 \text{ K})/m$; $v_y = 450.1$ m/s. $\tau = (1.000 \text{ m})/(450.1 \text{ m/s}) = 2.221$ ms. $Z' = (\frac{1}{2}\hbar/2mr) \times (2.221 \text{ ms}) = (1.79 \times 10^{-13} \text{ m})/r$; and $R = r + Z' = r + (1.79 \times 10^{-13} \text{ m})/r$.

CHAPTER 32

Answers to Odd-Numbered Multiple Choice Questions

1. d **3.** b **5.** b **7.** d **9.** b **11.** a **13.** d **15.** b **17.** a **19.** e

Answers to Odd-Numbered Problems

1. 111 **3.** 7 protons, 8 neutrons **5.** Fr, Pb, Ag, Pr **7.** 13 043.8 MeV/c², 13.043 8 GeV/c² **9.** $9.988\ 352 \times 10^{-27}$ kg **11.** 92.906 38 u **13.** 15 fm **15.** 27 **17.** 64.749 MeV **19.** 6.474 9 MeV/nucleon **21.** 15.663 MeV **23.** no **25.** 3.9 times **29.** 8.736 MeV **31.** Ca-41, 8.36 MeV; Ca-42, 11.48 MeV **33.** 5.40 MeV **35.** $^{27}_{13}\text{Al} + ^4_2\alpha \rightarrow ^{30}_{15}\text{P} + ^1_0\text{n}$ **37.** $^{147}_{62}\text{Sm} \rightarrow ^{143}_{60}\text{Nd} + ^4_2\alpha$ **39.** 17.346 MeV **41.** 4.1×10^{-9} s^{-1} **43.** $t_{1/2} = 12$ h **45.** 4.28×10^{-4} decays/y **47.** 5.5 d **49.** 90.1 s **51.** $N/N_0 = 4.97 \times 10^{-16}$ **53.** 5.700 74 MeV **55.** 1.191 MeV **57.** 5.7×10^{12} Bq **59.** 3×10^{10} Bq **61.** 2.8×10^{10} atoms **63.** 0.18 MBq/m³ **65.** 24.685 4 MeV

Solutions to Selected Problems

5. Fr, Pb, Ag, and Pr.   **9.** (5.603 051 GeV/c²) $\times$

$(1.782\ 663 \times 10^{-36}\ \text{kg/eV})(1.000\ 000 \times 10^9\ \text{eV/}$
$\text{GeV}) = 9.988\ 352 \times 10^{-27}\ \text{kg}.$ **13.** $R =$
$R_0 A^{1/3} = (1.2\ \text{fm})(6.17) = 7.4\ \text{fm}.$
$2R = 15\ \text{fm}.$ **19.** If $E_B = 64.749\ \text{MeV},$
$E_B/A = 6.474\ 9\ \text{MeV/nucleon}.$
21. $\Delta m = 1.008\ 665\ \text{u} + 15.003\ 065\ \text{u} -$
$15.994\ 915\ \text{u} = 0.016\ 815\ \text{u} \rightarrow 15.663\ \text{MeV}.$
27. Mass of C-12 atom is $(12\ \text{kg})/$
$(6.022 \times 10^{26}\ \text{atoms/kmole}) = 1.99 \times$
$10^{-26}\ \text{kg};$ dividing by 12 yields $1\ \text{u} =$
$1.66 \times 10^{-27}\ \text{kg}.$ **31.** For Ca-41:
$(40.962\ 278\ \text{u}) - (1.008\ 665\ \text{u}) -$
$(39.962\ 591\ \text{u}) = -0.008\ 978\ \text{u}.$ For Ca-42:
$(41.958\ 618\ \text{u}) - (1.008\ 665\ \text{u}) -$
$(40.962\ 278\ \text{u}) = -0.012\ 325\ \text{u}.$ One must add
8.36 MeV to remove a neutron from Ca-41 and
11.48 MeV to remove one from Ca-42.
37. $^{147}_{62}\text{Sm} \rightarrow ^4_2\alpha + X;$ $X = ^{143}_{60}X,$ which is
neodymium-143. **41.** $\lambda = 0.693/t_{1/2} =$
$0.693/(167 \times 10^6\ \text{s}) = 4.1 \times 10^{-9}\ \text{s}^{-1}.$
45. $\lambda = 0.693/t_{1/2} = 4.28 \times 10^{-4}\ \text{y}^{-1}.$
49. $\tau = 1/\lambda = 130\ \text{s};$ $t_{1/2} = 0.693/\lambda = 90.1\ \text{s}.$
51. $\lambda = 0.693/(1.18\ \text{min}) = 0.587\ \text{min}^{-1}.$
$N/N_0 = \exp[(-0.587\ \text{min}^{-1})(60.0\ \text{min})] =$
$4.97 \times 10^{-16}.$ **55.** Using atomic masses;
initial $-$ final mass; $(4.002\ 603\ \text{u}) +$
$(14.003\ 074\ \text{u}) - (16.999\ 131\ \text{u}) -$
$(1.007\ 825\ \text{u}) = -0.001\ 279\ \text{u};$ α must supply
1.191 MeV. **59.** $t_{1/2} = 1.236 \times 10^6\ \text{s};$ from
Eq. (32.12), $\lambda = 5.609 \times 10^{-7}\ \text{s}^{-1};$ $R_0 =$
$\lambda N_0 = 3 \times 10^{10}\ \text{Bq}.$ **63.** $(444\ \text{MBq})/$
$(50\ \text{weeks})(5\ \text{d/week})(10\ \text{m}^3/\text{d}) =$
$0.18\ \text{MBq/m}^3.$ **65.** $4(1.007\ 276\ \text{u}) -$
$(4.001\ 506\ \text{u}) - 2(0.000\ 548\ 580\ \text{u}) =$
$0.026\ 501\ \text{u};$ or 24.685 4 MeV. **67.** $N =$
$N_0 \exp(-\lambda t);$ $N_0/2 = N_0 \exp(-\lambda t_{1/2});$
$\exp(\lambda t_{1/2}) = 2;$ $\ln 2 = \lambda t_{1/2}.$

CHAPTER 33
Answers to Odd-Numbered Multiple Choice Questions
1. d **3.** e **5.** c **7.** c **9.** a **11.** a **13.** d
15. a **17.** c **19.** b

Answers to Odd-Numbered Problems
1. (a) electromagnetic (b) strong (c) weak
(d) strong **3.** (a) baryon number and
strangeness (b) lepton number (c) charge and
strangeness (d) strangeness and baryon
number **5.** no **7.** no, electron-lepton
number **9.** yes **11.** no, weakly **13.** not
forbidden, 33.9 MeV **15.** 37.7 MeV,
strangeness **17.** 1877 MeV **19.** 0.660 GeV
21. 10^{30} **23.** $E_0 = hc/4\pi R$ **25.** 70.45 MeV
27. -535.4 MeV, no—must have appreciable
KE. **29.** 2.42×10^{22} Hz, 1.24×10^{-14} m,
5.34×10^{-20} kg·m/s **31.** c$\bar{\text{u}}$ **33.** Λ° = uds
35. d$\bar{\text{s}} \rightarrow$ u$\bar{\text{d}}$ + d$\bar{\text{u}}$ **37.** 2.47×10^{20} Hz
39. $\approx 10^{15}$ GeV/c^2 **41.** 8×10^{27} K

Solutions to Selected Problems
1. (a) Only the electromagnetic force affects
photons. (b) Since there's no strangeness, this
must be a strong-force decay. (c) Λ° has
strangeness, so this must be a weak-force
decay. (d) With no strangeness, this must be a
strong-force decay. **5.** No. It does not
conserve baryon number—the p$\bar{\text{p}}$ pair can
cancel, leaving only one nucleon on the right.
Also, it doesn't conserve angular momentum.
9. Yes. It conserves electromagnetic charge,
spin, and B. There is no meson number to
worry about. Because of the strangeness of K$^\circ$,
its decay will be controlled by the weak force.
13. Not forbidden. (139.6 MeV) $-$
(105.7 MeV) $= 33.9$ MeV. **17.** $2(m_p c^2) =$
$2(938.3\ \text{MeV}) = 1877\ \text{MeV}.$ **21.** If 1 proton
takes an average of 10^{30} y to decay, in a group
of $N = 10^{30}$ protons one should decay within
1 y. $\tau = 10^{30}$ y $= 1/\lambda;$ $\lambda = 10^{-30}$ fractional
decays/y; $\lambda N = 1.$ **25.** $2(938.3\ \text{MeV}) -$
$(939.6\ \text{MeV}) - (938.3\ \text{MeV}) - (139.6\ \text{MeV}) =$
-140.9 MeV; each proton must bring in
70.45 MeV. **29.** $E = hf;$ $f = E/h =$
$2.417\ 99 \times 10^{22}$ Hz $= 2.42 \times 10^{22}$ Hz; $\lambda =$
$c/f = 1.239\ 8 \times 10^{-14}$ m $= 1.24 \times 10^{-14}$ m;
$p = h/\lambda = 5.34 \times 10^{-20}$ kg·m/s. **33.** It's
made up of three quarks (no antiquarks). To
get the strangeness, we need $S = -1$ or one
s-quark that has a $Q = -1/3,$ so we must add
a net charge of $+1/3.$ One u and one d will do
it. Λ° = uds. **37.** $E = hf = 2m_e c^2 =$
1.637×10^{-13} J; $f = 2.47 \times 10^{20}$ Hz.
41. $E = (3/2)k_B T;$ $T = 2E/3k_B = 8 \times 10^{27}$ K.

Credits

Chapter 2: p. 28, Figure 2.2 adapted from Eugene Hecht, *Physics in Perspective.* © 1980 by Addison-Wesley Publishing Company, Inc. Reprinted by permission. **Chapter 3: p. 61**, Figure 3.3 from *Road & Track,* December 1977, p. 49. Reprinted by permission. **p. 70**, Excerpt: From Associated Press News Release, July 1971. Reprinted by permission. **p. 77**, Figure 3.12 adapted from Eugene Hecht, *Physics in Perspective,* © 1980 by Addison-Wesley Publishing Company, Inc. Reprinted by permission. **Chapter 4: p. 95**, Figure 4.4 adapted from Eugene Hecht, *Physics in Perspective.* © 1980 by Addison-Wesley Publishing Company, Inc. Adapted by permission. **Chapter 5: p. 127**, Figure 5.2 adapted from Eugene Hecht, *Physics in Perspective.* © 1980 by Addison-Wesley Publishing Company, Inc. Adapted by permission. **p. 137**, Figure 5.15 from Eugene Hecht, *Physics in Perspective.* © 1980 by Addison-Wesley Publishing Company, Inc. Reprinted by permission. **Chapter 7: p. 214**, Table 7.1: Adapted from *Scientific American,* August 1992. **Chapter 9: p. 305**, Figure 9.19: Adapted from Eugene Hecht, *Physics in Perspective.* (Adapted from pg. 100). © 1980 by Addison-Wesley Publishing Company, Inc. Reprinted by permission. **Chapter 10: p. 325**, Figure 10.3: From Eugene Hecht, *Physics in Perspective.* © 1980 by Addison-Wesley Publishing Company, Inc. Reprinted by permission. **Chapter 11: p. 370**, Figure 11.22: From Eugene Hecht, *Physics in Perspective.* © 1980 by Addison-Wesley Publishing Company, Inc. Reprinted by permission. **p. 371**, Figure 11.23: From Eugene Hecht, *Physics in Perspective,* © 1980 by Addison-Wesley Publishing Company, Inc. Reprinted by permission. **p. 391**, Figure Q10: Adapted from Eugene Hecht, *Physics in Perspective.* © 1980 by Addison-Wesley Publishing Company, Inc. Reprinted by permission. **Chapter 16: p. 565**, Figure 16.19 from Eugene Hecht, *Physics in Perspective.* © 1980 by Addison-Wesley Publishing Company, Inc. Reprinted by permission. **Chapter 18: p. 627**, Figure 18.2 from Eugene Hecht, *Physics in Perspective.* © 1980 by Addison-Wesley Publishing Company, Inc. Reprinted by permission. **Chapter 19: p. 683**, Figure Q15 from *Newsday,* Wednesday, July 29, 1987. Reprinted by permission of Los Angeles Times Syndicate International. **Chapter 21: p. 729**, Figure 21.11 adapted from Eugene Hecht, *Physics in Perspective.* © 1980 by Addison-Wesley Publishing Company, Inc. Reprinted by permission. **p. 750**, Figure 21.37 adapted from Eugene Hecht, *Physics in Perspective.* © 1980 by Addison-Wesley Publishing Company, Inc. Reprinted by permission. **Chapter 23: p. 828**, Figure Q15 reprinted, with permission, from *49 Easy-to-Build Electronic Projects,* by Robert M. Brown and Tom Kneitel. Copyright 1981 by TAB Books, a Division of McGraw-Hill Inc., Blue Ridge Summit, PA 17294-0850, (1-800-233-1128). **Chapter 28: p. 1003**, Figure 28.18 adapted from W. Bertozzi, 1964, *American Journal of Physics, 32,* p. 551. Adapted by permission. **p. 1007**, Figure 28.19 adapted from R. Kollarits, 1972, *American Journal of Physics, 40,* p. 1125. Adapted by permission. **Chapter 32: p. 1104**, Figure 32.1 from A. P. Arya, *Elementary Modern Physics* (Figure 11.5), © 1974 by Addison-Wesley Publishing Company, Inc. Reprinted by permission. **p. 1105**, Figure 32.3 data from R. Hofstadter, *Annual Reviews of Nuclear Science, 7,* 231, 1957. **p. 1108**, Figure 32.6 data from G. W. Farwell and H. E. Wegner, 1954, *Physical Review, 95,* 1212.

Photo Credits

Chapter 1: **1**, J. Wray; **5**, National Optical Astronomy Observatories; **7**, Brooks/Cole; **8**, The Bryn Mawr College Archives; **9**, (top) Fotomas Index, London; (bottom) E. H.; **10**, (two at top) Stock, Boston, (two at bottom) Photo Researchers, Inc.; **12**, (top) Focus on Sports, (bottom) National Bureau of Standards; **13**, (two at top), Stock, Boston, (two at bottom) Photo Researchers, Inc.; **14**, (top) Stock, Boston, (bottom) Photo Researchers, Inc.; **16**, E. H.; **18**, (left) Stephen Frisch/Stock, Boston, (right) J. Anthony Tyson/AT&T Bell Labs

Chapter 2: **23**, Brooks/Cole; **24**, Gift of Collection Société Anonyme/Yale University Art Gallery; **28**, Ana S. Arias; **30**, © The Harold E. Edgerton 1992 Trust; **39**, Canadian Forces Photo

Chapter 3: **57**, Brooks/Cole; **59**, Duomo Photography; **71**, Vandystadt/Photo Researchers, Inc.; **73**, Biblioteca Ambosiana, Milan, Italy; **76**, Peticolas/Megna/Fundamental Photographs; **77**, Dawson Jones/Stock, Boston

Chapter 4: **89**, Brooks/Cole; **90**, Burndy Library **91**, NASA; **92**, Roger-Viollet, Paris; **93**, Biblioteca Ambrosiana, Milan, Italy; **96**, E. H.; **103**, © The Harold E. Edgerton 1992 Trust; **104**, Rick Stewart/Allsport USA; **109**, (top) E. H., (bottom) Richards/PhotoEdit; **111**, (top) Hewlett-Packard; (bottom) Yoram Lehmann/Peter Arnold, Inc.; **112**, Mickey Gibson/Animals, Animals; **118**, U.S. Air Force; **124**, Tracy Lee Didas/U.S. Navy

Chapter 5: **125**, Brooks/Cole; **130**, Gerard Vandystadt/Photo Researchers, Inc.; **132**, McDonough/Focus on Sports; **134**, Joseph Sohm/Stock, Boston; **137**, Bob Martin/Allsport; **141**, James Sugar/Black Star; **142**, Gerard Vandystadt/Photo Researchers, Inc.; **144**, David Yarrow/Allsport; **145**, Six Flags Great Adventure; **147**, Brooks/Cole; **148**, Leonard Harris/Stock, Boston; **151**, (top) David Madison/Duomo Photography, (bottom) Gerard Vandystadt/Allsport USA; **152**, Gerard Vandystadt/Allsport USA; **155**, (left) E. H., (right) E. H.; **156**, E. H.

Chapter 6: **163**, Brooks/Cole; **165**, Canadian Forces Photo; **169**, Larry Molmud/Mucking Otis Press; **170**, Armed Forces Institute of Pathology; **171**, SKA; **173**, E. H.; **179**, Vanguard Racing Sailboats; **184**, Bob Daemmrich/Stock, Boston; **187**, (left) Bob Daemmrich/Stock, Boston, (right) E. H.; **190**, Hsinhai News Agency; **192**, E. H.; **193**, (top) E. H., (bottom) Central Scientific Company

Chapter 7: **207**, Brooks/Cole; **208**, (top) SKA, (bottom) Jet Propulsion Laboratory; **209**, Burndy Library; **213**, Central Scientific Company; **214**, James Sugar/Black Star; **217**, James Sugar; **221**, Uwe Fink/Dept. of Planetary Sciences, University of Arizona; **222**, Jet Propulsion Laboratory; **226**, NASA; **230**, Raymond E. Arvidson/Washington University

Chapter 8: **239**, Brooks/Cole; **255**, (top) Central Scientific Company, (bottom) Richard Megna/Fundamental Photographs; **261**, (right) NASA, (left) National Optical Astronomy Observatories, (middle) National Optical Astronomy Observatories, (bottom) Ricoh Corporation; **262**, Thomas Eakins; **264**, Jet Propulsion Laboratory; **265**, Gordon Garradd/Photo Researchers, Inc.; **266**, Sargent Welch

Chapter 9: **277**, Brooks/Cole; **287**, James Sugar/Black Star; **288**, Smithsonian Institution; **292**, (top) Rick Rockman/Duomo Photography, (bottom) Marty Stouffer/Animals, Animals; **294**, Robert Bloomberg; **295**, Dorothy Littell/Stock, Boston; **296**, NASA; **300**, The Harold E. Edgerton 1992 Trust; **304**, (left) Erik Anderson/Stock, Boston, (right) The Harold E. Edgerton 1992 Trust; **306**, The Harold E. Edgerton 1992 Trust; **308**, Central Scientific Company

Chapter 30: **1043**, Brooks/Cole; **1044**, PhotoDisc; **1048**, Agency International de Physique Solvag/AIP Neils Bohr Library; **1049**, Courtesy, German Information Center; **1053**, Linda J. LaRosa; **1057**, AIP Emilio Segre Visual Archives; **1062**, Warren Nagourney; **1064**, Laserscope

Chapter 31: **1073**, Brooks/Cole; **1074**, (top) French Embassy, Press & Information Division, (bottom) Photo Researchers, Inc.; **1075**, (left, right) Educational Development Center, Inc.; **1077**, AIP Neils Bohr Library; **1079**, (left) Courtesy, David Sarnoff Research Center, (right) A. Tonomura/Hitachi Advanced Research Center; **1081**, A. J. G. Hey/University of Southampton; **1083**, National Optical Astronomy Observatories; **1090**, AIP Neils Bohr Library, Archives for History of Quantum Physics; **1093**, AIP Neils Bohr Library, Fankuchen Collection; **1094**, AIP Emilio Segre Visual Archives; **1095**, (left) McConnell Brain Imaging Centre, Montreal Neurological Institute, (right) Dan McCoy/Rainbow

Chapter 32: **1101**, Brooks/Cole; **1102**, Photo Researchers, Inc.; **1104**, Argonne National Laboratory; **1107**, (top, bottom) Dan McCoy/ Rainbow; **1110**, AIP Meggers Gallery of Nobel Laureates; **1115**, EG&G Mound Applied Technology; **1118**, S. M. Larson and S. J. Goldsmith/Nuclear Medicine Service, Memorial Sloan-Kettering Cancer Center; **1122**, A. Agelarakis/Adelphi University; **1123**, (top) Argonne National Laboratory; **1125**, from Armin Hermann, *The New Physics,* Munich: Heinz Moos Verlag, 1979; **1126**, *N. Y. Herald Tribune;* **1127**, *Pittsburgh Post-Gazette,* AIP Emilio Serge Visual Archives; **1129**, (top) painting by Gary Sheahan, photo courtesy Argonne National Laboratory, (bottom) U. S. Council for Energy Awareness; **1130**, (top) U. S. Navy, (bottom) UPI/Bettmann Archive; **1133**, Lawrence Livermore National Laboratory

Chapter 33: **1137**, Brooks/Cole; **1139**, Fermilab Visual Media Services; **1142**, Fermilab Visual Media Services; **1157**, Brooks/Cole; **1158**, Karl Lattrell/University of Michigan; **1161**, E. H.

Index

For entries on pp. 587–1164, please see Volume 2.

Aberrations, 898
Absolute gravitational acceleration (g_0), 214, 217, 231
of the Earth, 216
Absolute pressure (P), 361
Absolute rest, 163
Absolute simultaneity, 988, 1008
Absolute temperature (T), **485**, 565, 568–570
Absolute temperature scale, 485
Absolute zero, 484–485, 507
Absorption, 847–848
process of, 427–428
Absorption spectrum, 1031
AC capacitive circuit, 803, 804
Acceleration (**a**), 57–83
angular (α), 246–247, 249–250
average (**a**$_{av}$), **58**–60, 82
centripetal (i_C), 140–141, 143, 156, 210, 246, 247, 407
concept of, 58–62
constant, 60–61, 66–69, 71–75, 82, 247–250
down-plane, 133
due to gravity (g), 71–73, 212, 217, 254
equations of constant, **66**
in free-fall, 73
instantaneous (**a**), **60**–62
physiological effects of, 132
radial, 141
relationship to force and mass, 126–139
in simple harmonic motion, 407–408
SI unit of, 60
uniform, 58, **60**–69, 83
vertical, 134–136
Accommodation, of the human eye, 911
AC generator, 774–776
Acoustic power, 448–449
Acoustics, 3, 442–457
hearing sound, 451–454
intensity-level of (β), 454–457
sound waves, 442–444
speed of sound, 449–451
superposition of waves, 444–447
wavefronts and intensity, 447–449
Acoustic waves, 429
intensity-level of (β), 454–457
AC series circuits, behavior of, 809
AC shaded-pole induction motor, 798
Action-at-a-distance, 227, 229, 602–603, 617, 618, 724
Activity (R), defined, 1119
Acute angles, sine and cosine functions of, 44–45
AC voltages, 776
Addition
significant digits in, 15
of vectors, 32–33

Additive coloration, 884
Adhesion, role in producing friction, 154–156
Adiabat, 560
Adiabatic change, 559–561
Adiabatic demagnetization, 487
Adiabatic process, 554
Aerodynamic drag, 93
Aerodynamic effects, analysis of, 93
Aether, 834
Air
as an insulator, 593
wavelengths in, 444
Air capacitor, 644
Air columns, standing waves in, 462–465
Air-extraction pump, 362
Airfoils, 147, 384–385
Air friction, 70, 93
effect on projectiles, 82
Airy, George B., 603, 968
Airy disk, 968, 973
Albert of Saxony
on weight, 209
al-Biruni, 370
Algebra, 17
of vectors, 32–35
Alhazen, 242
Alkali metals, 1088, 1089
Alkaline earths, 1088–1089
Alpha decay, **1114**–1115
Alpha particles, 1033
in the helium atom, 1026
scattering experiment with, 1028–1031
Alpha radiation, 1025–1028
Alternating current (ac), 795–818
capacitance and, 803–805
inductance and, 800–803
L-C-R ac networks, 805–818
power (P_{av}), 810
resistance and, 798–800
Alternating-current ampere, 799
Alternator, 776
al-Zarqali, 218
Amber, 588
Ammeter, 672, 681, **699**–700
placement of, 693
Amontons, G., 484, 494
Amorphous solids, 324
Ampere (A), 8, 662, 681
as the unit of alternating current, 799
Ampère, A. M., 733, 750–752
on electromagnetic induction, 764
on magnetism, 731
Ampère's Circuital Law, 836–837, 838
Ampère's hypothesis, 737–739
Ampère's Law, 739–740, 752, 836, 838
Amp-hour rating, 667
Amplifiers, electronic, 824

Amplitude of SHM, 404
Amplitude modulation (AM), 812
Amplitude of the oscillation, 409
AM radio, 812
transmissions, 849
Analog meters, 699
Anderson, Carl, 1093
Andrews, T., 498, 499–500
Andromeda galaxy, 242
Angle of flight, determining, 44
Angles
in degrees and radians, 242
measurement of, 241
Angström, A. J., 1031
Angular acceleration (α), 246–247
determining, 249–250
Angular displacement (θ), 243, 268
Angular frequency (ω), 403, 433
Angular limit of resolution, 968
Angular magnification, 915
Angular momentum (**L**), 240, 257–259, 269
on atomic and subatomic levels, 259
conservation of, 7, 240, **259**–268, 269, 308
orbital, 262–263
precessional, 266–267
quantization of, 259, 611, 1059
similarity to linear momentum, 258
symmetry and, 267–268
Angular momentum vector (**L**), 257
Angular speed (ω), **243**–245
determining, 248–249
instantaneous, 244
Anisotropic space, 268
Anode, 669
Anomalous Zeeman Effect, 1083
Antibaryons, 1148
Anticolor charges, 1149
Antileptons, 1139
Antimatter, 1081, 1092–1095
Antineutron, 1094
Antinodes, 473, 459
Antiparallel vectors, 35, 41, 42
Antiproton ($\overline{p}$), 1094
Antiprotonic hydrogen, 1094
Antiquarks ($\overline{q}$), 1148, 1159
Antireflection coating, 958–959
Aorta, 378
Aphelion, 219
Apollo, journey to the Moon by, 91, 136, 138, 227, 311
Apparent power, 810, 825
Approximations, **17**–19
Arago, Dominique, F. J., 737, 764
on magnetism, 731
Arc (inverse) tangent function, 44
Arch, 191
Archimedes of Syracuse, 209, 368

Buoyancy Principle of, 368
Law of the Lever, 173
Architecture, stability of, 188–191
Arc-length (*l*), 241, 268
Area, computing, 11, 14–15
Area-under-the-curve analysis, 63–65
Argument of a function, 15–16
Aristotle, 24, 173
    aether conception of, 834
    definition of speed, 24
    gravitational theory of, 208
    ideas on sustained motion, 90
    Law of the Lever, 173
    theory of the Universe, 69
    on time, 990
    on vacuum, 362
Arteriosclerosis, 384
Ascending objects, effect of *g* on, 74
Aspherical mirrors, 925–926
Aspherical optical elements, 897
Aspherical surfaces, 895–897
Aston, F. W., 1102
Astronomical telescope, 924, 925
Astronomical units (AU), 220
Atmosphere, density fluctuations in, 865
Atmosphere (atm), 359, 389
    Earth's, 355
Atmospheric pressure, 358–361
Atomic bomb, 1128–1129
Atomic clocks, 13, 850, 996
Atomic emission, schematic of, 847
Atomic energy, 1025, 1061–1062
Atomic masses (u), 319, 1034
    of nuclides, 1103
Atomic mass units (u), 1103, 1131
    table of, 1103
Atomic nucleus, 5, 6, 1029, 1101–1131
    neutron-proton model of, 1035
Atomic number (Z), 1034, **1065**–1066, 1102
Atomic particles, angular momentum of, 259
Atomic quantum numbers, 1084
Atomic theory, 1056–1066
    atomic number, 1065–1066
    Bohr atom, 1057–1062
    stimulated emission, 1062–1065
Atomism, 319
Atoms, 6, 318
    as elastic systems, 318
    electromagnetic interaction of, 132, 146, 154
    ground state of, 846
    light and, **846**–847
    magnetism and, 727–730
    masses and rest energies for, 1005
    matter and, 318–327
    neutrality of, 590–591
    polarized, 602
    rotational motion within, 239–240
    size of, 322–323
    in solids and liquids, 323–324
Attack angle, 384–385
Attwood's machine, 160
Auditory canal, 464
Augustine, Saint, 1130
    on time, 993
Auroras, **745**–746
Average accelerating force, computing, 126
Average acceleration, 58–60, 82

Average angular acceleration, 246
Average angular speed, 243
Average current, 662
Average deceleration, calculating, 127–128
Average density, 214
Average dissipated power, 810
Average drift speed, 670, 681
Average induced emf, 767, 781
Average power, 286
Average speed, 24–27, 50
Average velocity, 35–36, 50
Avogadro, Amadeo, 319–320
Avogadro's number (N_A), 319–320, 1017, 1021, 1049
Axial loads, 333, 335
Axis of rotation, 258

Back-emf, 779, **780**–781, 785
Bacon, Francis, 502
Balance, stability and, 188–191
Balanced steelyard, 174, 175
Ballistic motion, 92
Ballistic projectiles, shape of the path of, 76–77
Ballot, 469
Balmer, Johann, 1031–1032
Balmer series, 1032, 1036, 1061
Banking angle, 144
Bannister, Roger, 26
Bardeen, John, 678
Barkla, C., 1022, 1065
Bar magnet, force field around, 725
Barnett, S. J., 738
Barometer, 358–361, 359, 360
Bars, 171
Barycenter, 223
Baryon number, 1141
Baryons, **1141**–1142, 1148, 1159
Basal metabolic rate (BMR), 314
Basilar membrane, 452
Batteries, 664–669, 681
    capacitors across, 644
    potential difference in, 632
    terminal voltage of, 691–692
BCS Theory, 678
Beat frequency, 458
Beats, 458
Becquerel
    discovery of radioactivity, 1024
Becquerel (Bq), 1120, 1131
Beeckman, Isaac, 66
Bel, 455
Bell, Alexander Graham, 455
Benedetti, G. B., 92, 140
Bennet, A., 594
Berkeley, George, 43, 92
Bernoulli, Daniel, 365, 379, 502
    on electric force, 595
Bernoulli's Equation, 379–384, 387, 389
Bernoulli's Theorem, 380
Beryllium, electron structure of, 1086
Beta decay, 1116–1118
Beta radiation, 1025–1028
Beta rays, as electrons, 1026
B-fields, calculating, 739–740
Big Bang, 3, 230, 473, 727
Bimetallic strip, 491, 492
Binding energy, 342, 1111–1113, 1131

Binding-energy-per-nucleon, **1112**
Biot, Jean, 732, 734
Birefringence, 948–950
Black, Joseph, 518, 521, 522, 526, 529
Blackbody radiation, **1044**–1049
Black holes, 3, 303
Blood, speed of, 378
Blood pressure, 386–389
Boccioni, Umberto, 24
Bodies, rigid, 178–183
Body-centered cubic structure, 327
Bohr, Niels, 20, 1032, 1056–1057, 1058
    atomic theory, 1057
    beta decay and, 1116
    on electron shells, 1084
    nuclear fission and, 1125
    principle of complementarity, 1076
Bohr atom, **1057**–1062
Bohr magneton (μ_B), 749–750
Bohr radius (r_1), 1060, 1081
Boiling, 532–534
Boiling-point curve, 500
Boltzmann, Ludwig, 502, 503, 504, 576, 577, 579, 1045
Boltzmann's Constant (k_B), 497, 507, 579, 1049
Bomb calorimeter, 525
Bone arch, 192–193
Bones
    broken, 341, 345
    as compression members, 173
    flexibility of, 332
Born, Max, 1078–1080
Boron, electron structure of, 1086
Bosons, 1084, 1159
Bostock, John, 668
Bottom quark, 1150
Boundary layer, 376
Bound charge, 846
Boyle, Robert, 70, 319
    and vacuum, 364
Boyle's Law, 364–365, 389, 450, 494, 504, 507
Bragg, W. H., 1023
Bragg, W. L., 1023
Bragg equation, 1023, 1036
Brahe, Tycho, 218
Brain, 21
    equipotentials in, 635
Bramah, 367
Branch, defined, 693
Branching chain reaction, 1127
Breaking point, 331
Breaking strength, 169, 337
Bremsstrahlung, **1054**–1055
Brewster, David, 947
Brewster's angle (θ_p), 946, 972
Brewster's Law, 947
Bridge circuit, 702
British Royal Society, 304
British thermal unit (Btu), 520
Brittle materials, 332, 337
Brushes, 750
Bulk Modulus (B), 344–346, 430
    and speed of sound, 450
Buller, A. H. Reginald, 990
Bullialdus, Ismaël
    on gravity, 209

Buoyancy Principle, 368, 389
Buoyant force (F_b), 368
Buridan, Jean, 101
Byron, Lord, 223

Calculus, 29, 215
Caloric, 834
Caloric theory, 514–515
Calorie (cal), 518, 519–520
Calorimeter, 525
Calorimetry, 522
Camera, structure of, **912**–915
Camera obscura, 912
Capacitance (C), **639**–651
    ac and, 803–805
    capacitors in combination, 643–648
    energy in capacitors, 648–651
    parallel plate capacitor, 641–643
Capacitive reactance (X_C), 804, 825
Capacitor banks, 651
Capacitors, **639**, 681
    air, 644
    charged, 649–650
    as charge storers, 641
    in combination, 643–648
    computing charge and energy stored in, 650
    dc and, 804
    energy in, 648–651
    parallel plate, 641–643
    radio, 642
Capillaries, 378
Carbon, electron structure of, 1087–1088
Carbon-14
    dating with, 1122
    half-life of, 1121–1122
Carbon-filament lamps, 796
Carbon-zinc cell, 664
Cardiovascular system, 389
Carnot, Nicholas Léonard Sadi, 562, 564, 571
Carnot Cycle, 562–565, 570
Carnot engine, 565, 579, 580
Carrier, 423
Carrier wave, 812
Carus, Titus Lucretius, 2, 71
Cassegrainian reflecting telescope, 931
Cathode, 669
Cathode rays, 1033
Cathode-ray tube, 610, 630, 1017–1018
    electron beam in, 744
CAT scan, 857
Cauchy, Augustin Louis, 332
    stress and strain, 333
Cavendish, Lord Henry, 212
    on the inverse-square law, 595
Celestial bodies, escape speeds from, 302
"Celestial pivot," 267
Cells, electrically responsive, 605
Celsius, Anders, 483
Celsius scale, 483
Center-of-gravity, 184–188, 194
    degrees of equilibrium and, 188–189
    locating, 185–187
Center-of-mass, **254**–255, 269
Center-seeking forces, 141–143
Center-to-center gravitational interaction, 209
Centigrade scale, 483
Centimeter, 9

Central force, 221
    planetary motion and, 222
Central generating stations, 796
Centrifuge, 144–145
Centripetal acceleration (a_C), **140**–141, 156, 246, 407
    computing, 143, 247
    computing the magnitude of, 141
    of the Moon, 210
Centripetal force (F_C), 140, 141
Cesium clock, 13, 850, 996
Cgs system, 9
    units of force, mass, and acceleration in, 127
Chadwick, James, 1034, 1035, 1116, 1128
Chain reaction, 1125–1126
Change
    adiabatic, 559–561
    energy as a measure of, 278
    as a manifestation of energy, 768
Change in quantity, notation for, 28–29
Change of state, 528–534
    boiling, 532–534
    melting and freezing, 528–530
    vaporization, 530–532
Characteristic X-rays, 1065
Charge (q), **588**, 618
    conservation of, 590, 638–639
    electromagnetic, 588–595
    free, 843–844
    interaction between, 598
    mobile carriers, 663
    particles of, 1017–1022
    potential and distribution of, 636–637
    potential of several, 635–636
    quantization of, 589
    quantum of, 1016–1017
    storing, 639–641
    units of, 597
Charged conductors, 593–595
Charge densities, 616
Charge-field-charge concept, 603–604
Chargeless objects, electrifying, 590
Charge per unit length (λ), 616
Charge-to-mass ratio, 1017, 1036
Charles, Jacques Alexandre César, 495
Charles's Law, 495, 507
Charm quark, **1150**
Chemical atomic mass, 1103
Chemical energy, 278
Chemical-PE, 296
Chisov, I. M., 72
Choke coil, 802
Circuit elements, 681
Circuits, **689**–707
    ammeters and voltmeters, 699–700
    defined, 644
    domestic, 815–818
    impedance of, 808–809, 825
    maximum power and impedance, 698–699
    network analysis, 702–707
    parallel, 643–645, 651
    power factor of, 810
    R-C, 700–702
    reactance of, 808, 825
    resistors in series and parallel, 693–698
    series, 646–648, 651, 805–812
    sources and internal resistance, 691–693

Circular motion, **143**–146
Circular polarization, 940
Circulatory system, human, 378, 387, 389
Cladding, 881
Classical field theory, 784
Classical physics, 3–4
Classical Principle of Relativity, 983–984
Clausius, Rudolf, 571
Clausius Formulation, 571–572
Clausius statement of the Second Law, 580
Clifford, W. K., 20
Cochlea, 452
Coefficient of friction, 147, 156
Coefficient of kinetic friction (μ_k), 152, 153, 156
Coefficient of performance, 567, 580
Coefficient of rolling friction (μ_r), 154, 156
Coefficient of static friction (μ_s), 147, 156
Coefficients of volumetric expansion (β), 491
Coherence, 951–952
Coherence length (l_c), 952, 972
Coherence time (t_c), 952, 972
Coherent bundle, 881
Cold welding, 155
Coleridge, Samuel Taylor, 20
Colinear forces, 131
Colinear force systems, 164–167
Colinear vectors, 59
Collagen, 335, 337
Collector, 822
Collisions, 303–308
    elastic, 306
    force-time curves for, 108
    inelastic, 303
Collision time, 103
Color, **882**–885
    frequency and wavelength ranges for, 853
    human perception of, 853
Color charge, **1149**, 1151–1152
Colored light, 883
    source of, 855
Color force, 1150–1152, 1153
Combustion
    heats of, 526
    internal, 566
    theories of, 514–515
Common emitter amplifier, 824
Commutator, 750
Complementarity, Principle of, **1076**
Complementary colors, 884
Composite solids, 324
Compounds, 318
    superconducting, 678
Compressibility, 348
Compression, 165
Compression waves, 429–433
Compressive stress, normal, 333
Compton, Arthur H., 1055
Compton effect, **1055**–1056, 1067
Compton Wavelength, 1056
Computed tomography (CT), 857
Concave lenses, 897, 909
Concurrent force systems, 164, 167–173, 193
Condensation, 429
Condensed matter physics, 3
Condenser, 639
Conduction, 537–540
Conduction band, 818

Conductors, **592**–595, 674
  fields and, 610–613
  immersed in an E-field, 611
Conjugate variables, 1076
Conservation laws, 7
Conservation of Angular Momentum, 7, 240, 260, 269, 308
  isotropy of space and, 268
Conservation of Charge, 590, 638–639
  Law of, 590, 704
Conservation of Energy, Law of, 7, 277, 297, 307, 308–309, 310, 379, 550–561
  and gravitons, 230
  the homogeneity of space and, 116
  origin of the concept, 112
Conservation of Linear Momentum, Law of, 7, 90, 112–117, 267, 308
Conservation of Mass-Energy, 1006
Conservation of Mechanical Energy, 297–309
Conservation of Momentum, Law of, 230, 306, 307
Conservative force, 296, 309
Constancy of the speed of light, 984
Constant acceleration, 60–61, 73–75
  calculating, 64–65
  equations of, 66–69, 82, 247–250
  Galileo's experiments with, 71–72
Constant-*a* equations, 66–69
Constant angular speed, determining, 245
Constant speed, 27–28
Constant-volume gas thermometer, 483–484
Constructive interference, 865, 951
Contact forces, 5, 131–133
Continuity Equation, 377–379, 382, 383, 389, 613
Continuous symmetry, 7
Convection, 536
Convection currents, 536, 537
Converging lenses, 896
  focal points for, 901
Conversion factors, 26
Convex lenses, 896
Cooper, Leon, 678
Copernicus, Nicolaus, 208
Coplanar concurrent forces, 193
Coplanar forces, 180
Coplanar force systems, 164
Coplanar vectors, multiplying, 177
Coriolis, Gaspard, 279
Cosine function, 15, **45**
Cosmic force, 218–230
  General Theory of Relativity, 229–230
  gravitational field, 227–229
  gravity and Kepler's laws, 220–224
  laws of planetary motion, 218–220
  satellite orbits, 224–227
Cosmic rays, 745–746
Coulomb (C), 597, 598
  defined, 752
Coulomb, Charles, 595
  on magnetic force law, 727
Coulomb force, 626, 1107, 1109
Coulomb interactions, summation of, 599
Coulomb's Law, **595**–600, 602, 603, 619, 732
Coulomb repulsion, 601
Coulomb scattering, 1108
Coupled motions, 138–139
Covalent bonding, 489, 819

Cowan, C. L., 1117
Crab Nebula, pulsar in, 260, 261
Crick, F. H. C., 1024
Critical angle (θ_c), 879, 886
Critical constants, 500
Critical field, 730
Critically damped system, 415
Critical mass, 1129
Critical pressure, 498
Critical size, 1128
Critical temperature (T_C), 498, 677, 681
Crookes, William, 1017–1018
Crookes tube, 1017
Crossed axes, 944
Cross product, 177
Crystalline solids, 324, 325–327
Crystallization, of water, 493
Cubic crystal form, 326
Cubits, 8
Curie, Marie, 1024–1025, 1116
Curie, Pierre, 730, 1025
Curie temperature, 730
Current (I), **662**
  direct, 667
  eddy, 778–779
  effective, 799, 825
  fields and, 731–737
  instantaneous, 662, 798, 803
  maximum, 798
  steady-state, 667
  as the transporter of energy, 679
Current balance, 752
Current-carrying coil, 735, 773
  field of, 738
  magnetic field lines produced by, 737
Current-carrying wire, 732–734
  force on, 747
Current loop, 734–735
  torque on, 748
Curvilinear motion, 139–146
  center-seeking forces, 141–143
  centripetal acceleration, 140–141
  circular motion, 143–146
  equations of, 247–248
Curvilinear translation, 239
Cyclic processes, 561–570
  absolute temperature and, 568–570
  Carnot Cycle, 562–565
  internal combustion, 566
  refrigeration machines and heat pumps, 566–568
Cyclotrons, 1104
Cylinder, determining acceleration of, 253–254

Dalton, John, 319, 500
Damped anharmonic motion, 415
Damped harmonic motion, 409
Damping, 415–416
Dashpot, 427, 428
Data, 2, 19
  collection of, 2
Daughter nucleus, 1114, 1131
da Vinci, Leonardo, 71, 93, 109, 147, 164, 174, 421, 465
  on wave theory, 839
Davisson, C. J., 1074–1075
Davy, Humphrey, 603, 1016

Daylight, phenomenon of, 852–853
DC (direct current)
  capacitors and, 804
  cell, 681
  generator, 776–778
  motor, 750, 778
De Broglie, Louis Victor Pierre Ramond, Prince, 4
  antineutrinos and, 1117
  waves, **1074**
Decay constant (λ), 1120
Decay rate (R), 1119
Deceleration, **59**
  calculating, 127–128
Decibel, 455
Decimal measurement, 8
Deflection yoke, 744
Deformable objects, friction and, 282–283
Degrees of freedom, 506, 526, 1082
de Haas, W. J., 738–739
Delta notation, 28–29
De Magnete (Gilbert), 208
Dense media, light transmission through, 866–867
Density (ρ), 320–322, 346
  Earth's, 214
  pressure-energy, 380
  relative, 369–370
Depletion layer, 819
Depolarization, 668
Derivative, 30–31
Desaguliers, John, 154
Descartes, René, 91, 101, 362, 724
  Conservation of Linear Momentum, 112
Descending objects, effect of *g* on, 74
"Despinning," 778
Destructive interference, 865, 951
Determinism, 3–4, 1092
Deuterium, 1104
Deuteron, 1105
Dewitt, Bryce, 1095
Diamagnetism, 728
Diamond, 354
Diastolic pressure, 386
Diathermy machine, 851
Dichroism, 944
Diderot
  on dimensions, 997
Dielectrics, **592**–595, 602
  in a parallel plate capacitor, 643
Dielectric breakdown, 637
Dielectric constant (K_e), **608**
Differential gravity, 232
Diffraction, 466–469, 960–972, 973
  circular holes and obstacles and, 967–969
  diffraction grating and, 965–967
  Fraunhofer, 962, 968, 974
  Fresnel, 962
  holography and, 969–972
  momentum and, 1091
  single-slit, 963–965
  of sound, 466
  X-ray, 1022–1024
Diffraction limited system, 894
Diffuse reflection, 871
Diffusion of gases, 506
Digital meters, 699

Diode laser, 820
Diodes, 819–821, 825
Diopters (D), 919, 932
Dioptric power ($\mathscr{D}$), 918, 932
Dipole, 609, 723
    field, 610
    radiation, 844
Dip poles, 726–727
Dirac, P. A. M., 1093
    on the magnetic monopole, 727
    on Quantum Electrodynamics, 1093
Dirac equation, 1093
Direct current (dc), **661–681**, 667
    currents, 662–671
    resistance, 671–681
Discharge rate (J), 378, 389
Discontinuous symmetry, 7
Disintegration energy, 1114, 1131
Dispersion, 874
Displacement (**s**), 32, 36, 37
    angular (θ), 243
    of projectiles, 93
    in simple harmonic motion, 404–406
Displacement current, 837
Displacement Law, 1046–1048
Displacement-time curve, 39
Displacement triangle, 46
Displacement vector (**s**), **32**
Dissipative absorption, 848
Distance
    varying of force with, 294–296
    versus displacement, 32
Distance-time graph, 28, 29, 30
Distributed force, 333
Diverging lenses, 897
    focal points for, 902
Division, significant figures in, 14–15
DNA structure, 1024
Dolphins, and ultrasound, 431
Domains, magnetic, 728–730
Domestic circuits, 815–818
Doping, 675, 819
Doppler, Johann, 469
Doppler Effect, 469–473, 474, 994
Doppler shift, 472, 1055
Double-slit experiment, 952–956, 1079
Down-plane acceleration, 133
Down quarks, 1138, 1148
Drag force, 70, 93, 94
Drift velocity, electric fields and, 669–671
Dry cell, 666
    electric field of, 669
    flow of charge through, 667
Dry friction, 146
Duane, W., 1055
Ductile materials, 331, 337
Dufay, C., 588
Dull, Charles E., 984, 985
Dulong and Petit's Law, 526
Dynamic equilibrium, 499
Dynamics, 24, **89**, 90, 125
    fluid, 354, 374–389
    Newton's Laws and, 250
    of rotation, 250–269
Dynamos, 774
Dyne, 597

Ear, structure of, 451–452

Earnshaw, 1028
Earth
    absolute gravitational acceleration of, 216
    angular momentum of, 258
    asphericity of, 216–218
    charge stored by, 600
    composition of atmosphere of, 355
    as a conductor, 634–635
    density of, 213–215
    gravitational interaction with objects, 130, 164–165
    gyroscopic stability of, 265
    magnetism of, 726–727
    perihelion and aphelion, 263
    rotation of, 141–142
    spin rate of, 264
Earth-centered world view, 208
Earth-Moon system, 264
Earthquake compression waves, 432
Earth satellite, 224
Eddy currents, **778**–779, 814
Edison, Thomas, 1024
    incandescent lamp and, 796
Effective current, 799, 825
Effective voltage, 799, 825
Effective weight, 130, 135, 136, 137
Effective weightlessness, 136–138, 156, 226
E-field, amplitude of, 839
E-field meter, 838
Ehrenfest, Paul, 1053, 1083
Einstein, Albert, 3, 4, 20, 297
    on Ampère's hypothesis, 738
    on atomic energy, 1127
    on blackbody theory, 1049–1050
    on Conservation of Energy, 297, 1006
    on corpuscular theory, 845
    on Electromagnetic Theory, 617
    on electromotive force, 774
    General Theory of Relativity of, 3, 6, 13, 207–208, 229–230
    on light-quanta, 1052
    on magnetism, 738–739
    Photoelectric Effect, 1049
    on the photon, 845, 1051–1054
    on Quantum Mechanics, 1094
    on the Special Theory of Relativity, 982–983
    on speed-dependent mass, 1001–1002
    on stimulated emission, 1062
Elastic collisions, 306–308
Elastic energy (PE$_e$), 278, 296
Elastic force, 5, 450
Elastic force (spring) constant (k), 328, 411
Elastic hysteresis, 348
Elasticity, 327–341
    fatigue and, 340–341
    Hooke's Law and, 327–332
    strain and, 335–337
    strength and, 337–340
    stress and, 332–335
Elastic limit, 329, 331
Elastic materials, 331–332
Elastic-PE (PE$_e$), 296, 299, 300, 329–331, 551
Elastic restoring force, 408–412
Elastic shear strain, 336
Elastic stress per unit strain, 341, 346
Elastin, 337
Elastomers, 337

"Electrical capacity," 639
Electrical energy, 278
Electrical-PE (PE$_E$), 296, 527–528, 626
Electric battery, first, 664
Electric car, 698
Electric current (I), 661–671, 681
    batteries, 664–669
    electric fields and drift velocity, 669–671
Electric field (E), **602**–619
    calculating, 606–607
    of a charged conductor, 611
    conductors and, 610–613
    defined, 604
    of a distribution of charge, 615
    drift velocity and, 669–671
    energy density of (u), 783
    lines, 609–610
    flux of, 613–614
    Gauss's Law and, 613–617
    induced, 773–774
    magnitude of, 632
    of a parallel-plate capacitor, 611
    of a point-charge, 604
    quantum field theory and, 617–619
    rationalized units and, 607–609
    shielding against, 611
    table of values, 606
    between two charged plates, 617
    of two equal positive point-charges, 609
Electric field lines, 604, 615
Electric flux (Φ_e), determining, 614
Electric force, 588, 595–602
    Coulomb's Law, 595–600
    electrostatic induction, 600–602
Electric motor, 750
Electric potential (V), 625, **626**–639, 651
    conservation of charge and, 638–639
    distribution of charge and, 636–637
    equipotentials and, 634–635
    of a point-charge, 632–634
    of several charges, 635–636
    in a uniform field, 631–632
Electric power, consumption of, 817
Electrocardiogram, 402, 679
Electrodes, 665
Electrodynamics, 730–740
    Ampère's hypothesis, 737–739
    calculating B-fields, 739–740
    currents and fields, 731–737
Electrolysis, 1016
Electrolytes, 665
Electromagnet, creation of, 737
Electromagnetic charge, 587, 588–595
    insulators and conductors of, 592–595
    positive and negative, 588–589
    as quantized and conserved, 589–590
    by rubbing, 590–591
    transfer of, 591–592
Electromagnetic force, 5, 95, 132, 323
Electromagnetic induction, **763**–785
    electromotive force (emf) and, 764–774
    generators and, 774–779
    self-induction, 779–784
Electromagnetic interaction, of atoms, 132, 146, 154
Electromagnetic-PE, 331
Electromagnetic-photon spectrum, **848**–857
    gamma rays, 857

infrared radiation, 851–852
light, 852–854
microwaves, 850–851
radio waves, 848–850
ultraviolet radiation, 854–855
X-rays, 856–857
Electromagnetic radiation (EM)
origins of, 843–845
thermal energy and, 517
Electromagnetic spectrum, 849, 858
Electromagnetic Theory, 3
Electromagnetic waves, **835–837**
electric field of, 838
Electromagnetism, 722
Electromotive force (emf), **666**, 681
induced, 764–774
motional, 770–773, 785
polarity of, 769–770
Electron-antineutrino, 1116
Electron capture, 1116, 1131
Electronics, **818–824**
pn-junction and diodes, 819–821
semiconductors, 818–819
transistors, 822–824
Electron-oscillators, 947
Electron-positron pairs, 1094, 1144
Electrons (e⁻)
affinities for, 591
angular momentum of, 259
beam of, 662
behavior of, 727
beta rays as, 1026
charge on (q_e), 589–590, 619
as dipole magnets, 724
discovery of, 1018–1020
fields of, 728
as the fundamental unit of charge, 1017
J. J. Thomson on, 1020
magnetic moment of, 1095
mass of (m_e), 14
physical characteristics of, 1021
spin of, 1083–1084
transfer of, 591–592
valence, 818
value of the charge of, 597
Electron shells, 1084–1089
Electron spin, 1083–1084
Electron volt (eV), 343, 630
Electroscope, **594–595**
inducing a charge on, 600
Electrostatic equilibrium, 648
Electrostatic generator, 598, 637
Electrostatic induction, **600**
Electrostatics, 587–619, 625–651
capacitance, 639–651
defined, 588
electric field, **602–618**
electric force, 595–602
electric potential, 626–639
electromagnetic charge, 588–595
energy, 625
Electroweak epoch, 1156
Electroweak field, 1156
Electroweak force, 1152–1156
Electroweak Theory, 1155, 1159
Elemental semiconductor, 819
Elementary particles, 4, 1138–1142
angular momentum of, 259

hadrons, 1140–1142
leptons, 1139
Elements, 318, 346
chemical atomic mass of, 1103
superconducting, 677
Ellis, C. D., 1116
Emission spectrum, 1031
Emission theory, 834
Emitter, 822
Empire State Building, 403, 404
Empty space, isotropic nature of, 268
Endoscope, 881
End-pivoted lever, 188
Endurance (fatigue) limit, 340
Energy, 277–310
atomic, 1025
binding, 342
in capacitors, 648–651
chemical, 278
current as the transporter of, 679
elastic potential, 329–331
electrostatics and, 625–651
gravitational, 278
internal (*U*), **296**–297, 527, 552–553, 557
irradiance and, 842–843
kinetic (KE), 288–291
in the magnetic field, 783–784
mechanical, 288–297, **298**
mechanical change in, 279, 288–309
moduli and, 341–346
perspective on, 299–303
as possessing inertia, 1005
potential (PE), 291–296
power and, 679–681
quantization of, 1049–1054
relativistic, 629, 1003–1008
symmetry and, 308
thermal, 278, 296, 513–540
total, 1004
of a wave, 424
zero-point, 487
Energy conservation, 7, 277, 307–310, 379,
550–561
mechanical, 297
Newton's Third Law and, 309
symmetry and, 308–309
Energy content, 525
Energy density, 783
Energy efficiency, 563
Energy elements, Planck on, 1048–1049
Energy levels, 1058
atomic, **1061**–1062
Energy quanta, **845–846**
Energy transfer, 278–288, 299
power (P), 286–288
work (W), 278–286
Engines, 561–570
Engine thrust, calculating, 112
Entropic neutrality, 574
Entropy (*S*), 572–576, 580
as probability, 577–579
Entropy Principle, 574
Epitaxial layers, 822
Equal-arm balance, 173
Equation of state, 494, 496–497, 553
Equations, **16**
uses of, 17
Equilibrium, 163–205

degrees of, 188–189
First Condition of, 164, 193
force triangle corresponding to, 170
rotational, 173–183
Second Condition of, 179, 184, 193
stable, 188, 189
static, 163
of systems, 125, 553
thermal, 521, 579
translational, **164**–173, 179
Equilibrium of extended bodies, 184–194
center-of-gravity, 184–188
stability and balance, 188–191
structures, walls, and arches, 191–193
Equilibrium vapor pressure, 499
Equipartition Theorem, 506, 526
Equipotentials, **634–635**
in the brain, 635
Equipotential surfaces, 634
Equipotential volume, 634
Equivalence Principle, 414
Equivalent resistance, 693, 707
Escape speed (v_{esc}), **301**, 310
from surfaces of celestial bodies, 302
Ether, propagation of light by, 836
Euler, Leonhard, 130, 228, 250
European Council for Nuclear Research
(CERN), 992
Evaporation, 530
Exact measurement, 13–16
Excited states, 846
Exclusion Principle, 1093, 1105, 1110
Experimenta Nova (von Guericke), 363
Exponentials, 15
Exposition of the System of the World
(Laplace), 303
Extended bodies, equilibrium of, 184–194
External force, 108–109
weight as an, 132
External reflection, 868–869
Eye, structure of, 910–912
Eyeglasses, 918–923

$F = ma$, 126–130
Face-centered cubic structure, 327
Fahrenheit, G. D., 482
Fahrenheit temperature scale, 482
Farad (F), 639
Faraday, Michael, 499, 605
on diamagnetism, 728
electric generator and, 774
on electromagnetic fields, 603–604
on electromagnetic induction, 764–765
ice-pail experiment, 612
on lines-of-force, 603, 835
Royal institution lectures, 779
use of electrolysis, 1016
Faraday (F), 1016
Faraday cage, 610, 612
Faraday's Induction Law, **765–768**, 785
Far-field diffraction, 962
Far-point, 911–912, 921
Farsightedness, 919–920
Fatigue, 340–341
Fatigue limit, 340
Fat tank, 371
Fermat, Pierre, 1076
Fermi, Enrico, 1104, 1116, 1123

atomic bomb, 1128
creation of radioisotopes, 1122–1123
neutrino and, 1116
nuclear reactor and, 1126
theory of beta decay, 1118
Fermi (fm), 1104
Fermi level, 1110
Fermi National Accelerator Laboratory, 1142
Fermions, 1084, 1159
Ferrite-core inductor, 781
Ferrites, eddy currents and, 780
Ferromagnetic substances, 728
Ferromagnetism, 738
Feynman, Richard P., 278, 1092, 1094, 1095
on the atomic bomb, 1128
and Quantum Electrodynamics, 1095
Feynman diagram, **1142**, 1143
Fiberoptics, 880–882
cladding, 881
coherent bundle, 881
incoherent bundle, 881
Field lines, 229, 230, 609–610, 726
cutting, 771–773
for a positive charge, 634
Field of force, 602–603
Field particles, 230
Fields
conductors and, 610–613
currents and, 731–737
Field strength, 231
Field Theory, 228
Fine structure, 1082
Finite imagery
lenses and, 903–905
mirrors and, 929–931
Firing angle, of projectiles, 81
First Condition of Equilibrium, 164, 193
First harmonic, 446
First Law of Thermodynamics, 550–561, 579
First Postulate, 983
Fission, nuclear, 1123–1131
Fission reactor, 1126
FitzGerald, G., 981
Fizeau, 842
Flavors of quark, 1148
Fletcher, H., 1021
Floating, 370–371
Floating circuits, 679
Fluid dynamics, 354, 374–389
airfoils and, 384–385
Bernoulli's Equation and, 379–384
blood pressure and, 386–389
continuity equation and, 377–379
fluid flow and, 374–377
vortices and, 385–386
Fluids, 353–389
flow of, 613
force exerted by, 356
Fluid statics, 354–374
Boyle's Law and, 364–365
buoyant force and, 368–369
gases and, 354–355
hydrostatic pressure and, 355–364
liquids and, 354
Pascal's Principle and, 365–368
specific gravity and, 369–371
surface tension and, 371–374
Flux, defined, 613

Flux density, 767
Flux linkage, 768
Flux of the electric field, **613**–614
Flux of the magnetic field, 766, 785
Flying buttress, 192
FM transmissions, 850
f-number, 914
Focal length, 925, 932
Focal plane, 903
Focal points, 896, 901–903
Food, energy provided by, 519
Foot-pounds, 279
Force (**F**), **94**–99, 278
applied to bodies in motion, 104–105
buoyant, 368–369
center-seeking, 141–143
central, 221, 222
centripetal, 140
colinear, 131, 164–167
conservative, **296**, 309
coplanar, 164, 180
cosmic, 218–230
defined, **4**, 95–96, 278
elastic, 5, 450
electromagnetic, 5
erroneous ideas about, 99
external, 108, 114
frictional (F_f), 131, 146, 283
gravitational, 4–5, 95, 130–134, 139, 412–415
impulse of, 104
instantaneous, **106**–107
interaction and, 109–110
interatomic, 324
internal, 113
kinetic-friction, 152–153
line-of-action of, 126, 174
measuring, 96
moment-arm of, 175
moment of a, 174
nonworking, 289
normal, 131, 132, 143–144
reaction, 165
relationship to change in motion, 89
relationship to mass and acceleration, 126–139
restoring, 328, 408–412, 412–415
static-friction, 150–152
strong, 5
tensile, 126
varying, 106–108
varying with distance, 294–296
as a vector, **97**–99
weak, 5
on wires, 746–752
Force-carrying particles, 208
Force-displacement curve, 284
Forced oscillation, 416–418
Force equations, 183
Force field, 228
Force law, 596–597
Force-multiplier, 166–167
Force of friction, 131
Force per unit area (pressure), 155
Forces
mediation by field quanta, 1145
nonconcurrent, 180–183
Force-summation axis, 183

Force systems
concurrent, 164, 167–173
coplanar, 164
noncoplanar, 164
parallel and colinear, 164–167
Force-time curve, 107, 108
Force-time graph, 106
Force triangle, 168, 169, 170, 172
Force vectors, 98–99, 167, 169, 172, 603
Forward biased diode, 820
Fosbury Flop, 187
Four Forces, **4**–6, 95, 278, 617, 1108, 1155, 1158
energy transformation and, 551
field particles and, 617
isotropic nature of, 267
Fourier, Baron de, 446, 539, 671
on waveforms, 445
Fourier's Conduction Law, 539
Franklin, Benjamin, 588
on electrical matter, 1019
on electric force, 595
parallel plate capacitor and, 641
Franklin (unit of charge), 597
Fraunhofer diffraction, 962, 968, 973
Free-body diagram, **127**, 129, 134
Free charge, 843–844
Free-fall, **69**–83
acceleration in, 71–73
orbiting astronauts in, 226
Free particles, trajectory of, 743–746
Free space
permittivity of, 608, 619
propagation of light in, 984
Freeze drying, 501
Freezing, 528–530
Frequency (f), 402
angular (ω), 403
resonant (f_0), 416–418, 811
wave, 423, 433
Frequency modulation (FM), 850
Frequency spectrum, 447
Fresnel, Augustin Jean, 834
on polarization, 940
wave theory of, 969
Fresnel diffraction, 962
Friction, 5, **146**
air, 70, 82, 93
as a cause of rotation, 256
causes of, **154**–156
cold welding, 155
contact area, 148
deformable objects and, 282–283
as a driving force, 150–152
dry, 146
force of (F_f), 131, 146, 283
heating by, 515
kinetic, 146, **152**–153, 282
motion and, 146–156
Newton's experiments with, 90–91
rolling, 154
static, **146**–150
stick-slip, 419
Friction coefficients, 147, 152, 154
Friction electric machine, 639
Friction force, 283
Fringes of equal thickness, 960, 973
Frisch, Otto, 1124, 1125

Full-scale current, 699
Full-wave rectifier, 821, 822
Fundamental (frequency), 446, 461
Fusion, 528
  nuclear, 1123–1131

g (acceleration due to gravity), 72–73
  effect of Earth's rotation on, 217
  and free-fall, 73
  and weight, 96
G, determining, 212
g (vector acceleration due to gravity), 93, 130
Gabor, Dennis, 969
Gal, 229
Galaxy, 5, 18, 25
Gale, Henry G., 834
Galilei, Galileo, 13, 27–28, **70**, 73, 412–413, 419
  acceleration experiments, **71**–72
  ballistics and, 92
  discourse on falling bodies, 70
  elasticity, 327
  energy and, 278
  law of inertia and, **90**–91
  pendulum, 413
  projectile motion, 75, 224
  thermoscope of, 482
  thought experiment, 83
  waves, 461
Gallium, 529
Galvani, Luigi, 664, 667
Galvanometer, 699–700
Gamma decay, 1119
Gamma radiation, 857, 1025–1028
Gases, 324, 354–355
  diffusion of, 506
  ideal, 365
  light scattering of, 867
  molecule speed in, 504–506
  monatomic, 528
  Newton's experiments on, 109
  specific heat of, 528
Gas Laws, 494–506
  Ideal Gas Law, 496–498
  kinetic theory and, 502–506
  liquefaction, 498–500
  phase diagrams and, 500–501
Gauge blocks, 156
Gauge force field, 1146
Gauge forces, 618
Gauge pressure, 361–362
Gauge symmetry, 6, 13, 1145, 1147
Gauge Theory, 267, 1095, **1145**–1147
  importance of, 638
Gauss, Friedrich, 613, 904
  on surface tension, 373
Gaussian Lens Equation, 904, 910, 924
Gaussian surfaces, 613–614, 615–616
Gauss's Law, **613**–617, 619, 1066
Gay-Lussac, Joseph Louis, 495
Gay-Lussac's Law, 495, 507
Geiger, Hans, 1026, 1028–1029
Gell-Mann, Murray, 1141, 1148
Gemini spacecraft, 105
General Theory of Relativity, 3, 6, 13, 207–208, 229–230, 414
Generators, **774**–779
  ac generator, 774–776

dc generator, 776–778
  eddy currents and, 778–779
Genesis, 4
Geomagnetic poles, 726–727
Geometrical optics, **893**–932
  aspherical surfaces, 895
  finite imagery, 903
  focal points, 901
  Gaussian Lens Equation, 904
  lenses, **894**–925
  magnification (*M*$_T$), 905
  mirrors, 925–931
  sign convention, 905
  spherical thin lenses, 897
  thin lens combinations, 916
Geometric center, as the center-of-mass, 255
Geostationary (geosynchronous) orbit, 226–227
Gilbert, William, 726
  on gravity, 208, 209
  on magnetism, 724, 755
  the versorium, 620
Gillet, J. A., 836
Glashow, Sheldon L., 1150, 1154–1155, 1156
Global symmetry, 1146
Gluons, 1150
Goddard, Robert H., 112
Goeppert Mayer, Maria, 1110
Gold-leaf electroscope, 594–595
Gothic arches, 193
Goudsmit, S. A., 1083
Gram (g), 9, 11
Gramme dynamos, 778
Gram-molecular mass, 320
Grand Unification Theories (GUTs), 727, **1156**–1159
Graphs, **17**
  for representing motion, 65
Grating equation, 967
Gravimeter, 218
Gravitation,
  Law of Universal, 208–218
  Theory of, 3
Gravitational attraction, computing, 213
Gravitational constant (*G*), 212–213
Gravitational energy, 278
Gravitational field intensity, 228
Gravitational fields, 227–229, 231
  action upon planets, 263
Gravitational field strength, 228
Gravitational force (*F*$_G$), 4–5, 95, 130–134
  center-seeking, 139
Gravitational lenses, 3
Gravitational mass, 414
Gravitational-PE (PE$_G$), 292–294, 295, 298, 301, 380, 383, 551, 626
Gravitational restoring force, 412–415
Graviton, 230
Gravity
  acceleration due to (*g*), 71–73
  Aristotle's theory of, 208
  center of, 184–189, 194
  as the central force, 222
  effect on gases, 355
  effect on motion, 92
  hydrostatic pressure and, 356–358
  Kepler's Laws and, 220–224
  mass and, 100–101

negative, 230
  Newtonian theory of, 207–231
  oscillations and, 412
  relationship to planetary motion, 220
  relationship to time, 13
  specific, 369–371
  of a sphere, 215–216
  stability and, 188
  terrestrial, 213–214
  vector acceleration due to, 93
  weight and, 11
Gray, Stephen, 593, 620
Gregory, James, 931
Ground, 600
Ground state, 818, 846, 1057
Ground-state configurations, 1087
Ground wave, 845
Group velocity, 874
Guth, A. H., 230
Gyroscopic stability, 265–267

Hadrons, **1140**, 1149, 1159
  quark composition of, 1149
Hahn, Otto, 1123
Haldane, J. B. S., 72
Half-life (t$_{1/2}$), **1119**–1122
Half-wave rectifier, 821
Halley's comet, 221
Halogens, 1089
Hamilton, William Rowan, 1076
Hanger, 166
Hansen, H. M., 1057
Harmonic electromagnetic wave, 839, 840
Harmonic motion, 402–419
  damping and, 415–416
  elastic restoring force and, 408–412
  forcing and resonance and, 416–418
  gravitational restoring force and, 412–415
  self-excited vibrations and, 418–419
  simple, 402–408
Harmonic oscillator, 407, 408
Harmonics, 446
Harmonic waves, 424–425, 433, 857
Hauksbee, Francis, 620, 639, 1017
Hauksbee's Influence Machines, 639
Head-on collisions, 304
Hearing, 451–454
Heart activity, equipotentials due to, 637
Heat (*Q*), 513, **517**
  change of state and, 528–534
  defined, 517–518
  mechanical equivalent of, 520
  quantity of, 518–520
  specific, 526–528
  temperature and, 515–518
  thermal energy and, 513–540
  work and internal energy and, 552–553
Heat-as-matter theory, 515
Heat engine, 562, 579
Heat of fusion (*L*$_f$), 529, 541
Heat of vaporization (*L*$_v$), 529, 531, 541
Heat pumps, 566–568
Heats of combustion, 526
Heaviside, Oliver, 607
Heavy water, 1104
Heisenberg, Werner, 728, 1035, 1089, 1090
  and the atomic bomb, 1126
  on ferromagnetism, 728

Heisenberg Uncertainty Principle, 965, **1089**, 1096, 1143
and diffraction, 1091
Helium
discovery of, 1025
electron structure of, 1085–1086
liquid, 487, 677
Helium atom, nucleus of, 1026
Helmholtz, Hermann von, 550
on atoms, 1017
on sound, 457
Henry (H), 779
Henry, Joseph, 738, 779
on electromagnetic induction, 765
electromagnetism and, 737
Hermann, Jacob, 130
Herschel, William
on infrared radiation, 851
Hertz (Hz), 402–403
Hertz, Heinrich R., 402, 1018
generation of electromagnetic waves, 858, 860
on the Photoelectric Effect, 1050
Hexagonal close-packed structure, 327
Higgs boson, 1156
Higgs field, 1155
High-energy physics, 1137–1159
elementary particles, 1138–1142
quantum field theory, 1142–1159
High-temperature superconductors, 678
"High-tension" wires, 797
High-voltage ac, 796
Hip joints, rotational inertia of, 253
Hofstadter, R., 1104
Holography, 969–972
Homogeneity of space, 116, 267
Homogeneous waves, 840
Hooke, Robert, 140, 319
on elasticity, 327, 328
on gravity, 209, 210, 211
wave theory of, 834
Hooke's Law, 17, 318, 327–332, 346, 372
Horizontal balance, computing, 174
Horizontal force, computing, 129–130, 168–169
Horizontal reaction forces, 165, 193
Horizontal speed, of projectiles, 77–79
Horrocks, Jeremiah, 224
Horsepower (hp), 286
"Hot" wires, 816–817
Houtermans, Fritz, 1127
H-particles, 1033
H-sheet Polaroid, 945
Hubble Space Telescope, 535, 925, 926
Human body
center-of-gravity of, 186–187, 190
as a conducting antenna, 850
effect of electricity on, 817–818
as a fluid-dynamical system, 353
fluid pressures in, 361
levers and, 174
radiation by, 851
statics of, 164
temperature of, 482–483
Human bones, torques and breaking angles for, 345
Hume, David, 518
Hunt, F., 1055

Huygens, Christian, 13, 73, 141, 288, 297, 415
on collisions, 304
vs viva, 288
wave theory of, 940
Hydraulic machines, 366–368
Hydrodynamic systems, 374
Hydrogen
antiprotonic, 1094
charge-to-mass ratio of, 1017
electron structure of, 1085
formation of, 230
isotopes of, 1104–1105
liquid, 487
molecular, 319–320
structure of, 1082
Hydrogen atom
Bohr model of, 749
electrical and gravitational forces in, 597–598
energy levels of, 1061
Hydrogen visible spectra, 1032
Hydrostatic pressure, 355–364
Hyperbolic surfaces, 896
Hyperopia, 919–920
Hypotheses, 2

IC microprocessor chip, 818
Ideal fluids, 376
Ideal gases, 365
isothermal expansion of, 558
temperature of, 504
Ideal Gas Law, 496–498
Ideal liquids, 354
Ideal voltage source, 694
Ideal voltmeter, 692
Image distance (s_i), 872
Image focal length (f_i), 901
Image focal point, 901
Image reconstruction, 969
Impact, force-time graph for, 106
Impact time, 105, 108
Impedance (Z), 698–699, 802, 808–809, 825
Impedance triangle, 809
Impulse
defined, **104**
interaction and, 113
momentum change and, 102–106
Impulse-momentum, 108
Impurity semiconductors, 819
Incident angles, 879–880
Inclined plane, 133–134
Inclines
friction on, 149
rolling down, 256–257
Incoherent fiber bundle, 881
Index of refraction (n), 874–875, 885
table of, 875
"Indivisible moments," 66
Induced emf (electromotive force), 764–774
Faraday's Induction Law, 765–768
fields and, 773–774
Lenz's Law, 768–770
motional emf, 770–773
Induced radioactivity, 1122–1123
Inductance (L), **779**–780, 785
ac and, 800–803
Induction, electrostatic, 600–602
Induction motor, 798

Inductive reactance (X_L), 802, 825
Inductor, defined, 781
Inelastic collisions, 303–306
Inertia
defined, 92
moment of (I), 250–252, 258, 260
Newton's Law of, 89, 90–99, 117, 250
rotational, 250–254
Inertial mass (m), 250, 414
Inertial observer, 983
midway, 988
Inertial reference frame (system), 40, 41, 983
Inflationary era, 203
Infrared electromagnetic radiation, 535
Infrared (IR) pictures, 852
Infrared radiation, 851–852
Infrasound, 432–433
Inhomogeneous waves, 840
Initial phase, 406
Ink jet printer, E-field in, 617
Inner ear, 452
Input current, 824
Input transducers, 689
Instantaneous acceleration, 60–62, 128
Instantaneous angular acceleration, 246, 268
Instantaneous angular speed, 244, 268
Instantaneous current, 662, 798, 803
Instantaneous force, 106
versus time, 106–107
Instantaneous motion, 128–130
Instantaneous power, 286, 799
Instantaneous speed, 29–31, 50
Instantaneous velocity, 35, 36–39, 50
Instantaneous voltage, 798, 801, 803
Insulators, **592**–595, 674–675
Integral calculus, 215
Integrals, 65–66
Integrated circuits (IC), 818
Intensity (I)
of a sound wave, 473
wavefronts and, 447–449
Intensity-level (β), 473
of an acoustic wave, 454–457
Interacting systems, work and, 281–282
Interaction
Newton's Law of, 89, **108**–117, 134–136, 150–152, 211
in rockets, 111–112
Interaction pairs, 110, 128, 132, 179, 209
Interatomic binding, 488, 490
Interference, 458, 950–960, 972
coherence and, 951–952
Michelson Interferometer and, 959–960
scattering and, 865–866
in sound waves, 466
thin film, 956–959
Young's Experiment with, 952–956
Interference fringes, 951
Interferometers, 959
Intermediate vector bosons, **1152**, 1156, 1159
Internal combustion, 566
Internal compression engine, 567
Internal energy, 296–297, 527
as a measure of state, 557
work and heat and, 552–553
Internal forces, 113
Internal reflection, 868–869, 879
Internal resistance (r), 691

of a battery, 692
International Bureau of Weights and Measures, 12
International Prototype Meter, 9
Intervals, 65–66
Intrinsic semiconductor, 819
Invariance, 7
Invariant mass, 1002
Inverse (arc) tangent function, 44
Inverse relationships, 17
Inverse Square Law, 449
Ionization, 324
Ionization energy, 1060
Ionizing radiation, sources of, 1028
Ions, 593, 1016
Iron-core transformer, 813
Irradiance (*I*), 842–843, 858
  minimum and maximum in the, 951
Isentropy, 575
Isobaric process, 554, 555
Isolated systems, 550, 553
  momentum changes in, 113–114
Isothermal change, 557–559
Isothermal curves, 499
Isothermal process, 554, 559
Isothermals, 498
Isotopes, **1102**–1104
Isotropic medium, 870
Isovolumic (isochoric) process, 554

Jammer, Max, 101
Jensen, J. H. D., 1110
Joints, 171
Joliot-Curie, Jean Frédéric
  induced radioactivity, 1122
Joliot-Curie, Irène
  induced radioactivity, 1122
Joseph, Jan Baptiste, 445, 446
Joule, James Prescott, 279, 518, 520, 550, 680
Joule heat, 680
Joules, 279, 289, 520
Junction diodes, 819–821
Jupiter
  displacement of the Sun by, 223
  Great Red Spot on, 385

Kaons, 1140–1141
Kelvin, Lord (William Thomson), 289, 485, 486, 550, 569–570
  on atoms, 1028
  on ether, 981
  on lines-of-force, 835
  on permeability, 730
Kelvin (K), 8, 485, 507
Kelvin statement, 571, 572, 580
Kennedy, R., 982
Kepler, Johannes, 209
  astronomical telescope of, 924
  on gravity, 208
  Laws of Planetary Motion, **218**–224, 231, 263
Keplerian constant, computing, 220
Kevlar, 339–340
Kidney, 431
Kilocalorie (kcal), 519
Kilogram (kg), 8, 9, 12
Kilogram Number 20, 12

Kilomole, 320
Kinematic parameters, linear and angular, 246
Kinematics, **23**–50
  acceleration, 57–83
  relative motion, 39–50
  of rotation, 241–250
  speed, 24–31
  velocity, 31–39
Kinetic coefficient of friction (μ_k), 178
Kinetic energy (KE), 278, 288–291, 296, 310
  origin of concept, 288
  as a relative quantity, 290
  relativistic, 1003
Kinetic friction, 146, 152–153, 282
Kinetic Theory, 502–506, 552
  versus Caloric Theory, 514–515
Kinetic Theory of Gases, 365, 507
King Kong, 347
Kirchhoff, Gustav Robert, 702
  on thermal radiation, 1044
Kirchhoff's Radiation Law, 1044
Kirchhoff's Rules, **702**–707, 801
Knowlton, A. A., 982
Kohlrausch, 842
Kossel, W., 1065–1066
Kronig, R., 1083

Lamb, Willis, 1095
Lambert, J. H., 484
Lamb shift, 1095
Laminar flow, 374–375
Land, Edwin H., 945
Laplace, Pierre Simon de, 303
  speed of sound, 450
Larmor, J., 1028
Laser interferometry, 229
Lasers, 1062–1065
Latent heat of fusion, 529, 541
Lattice points, 326
Launch speed, computing, 225–226
Lavoisier, Antoine, 514
Law of Conservation of Charge, 590
Law of Conservation of Energy, 7, 277, 297, 310, 379, 550–561, 703
Law of Conservation of Momentum, 230, 306, 307
Law of the Constancy of c, 985
Law of Inertia, 89, 90–99, 117
  rotational equivalent of, 250
Law of Interaction, 89, 108–117, 211
  friction and, 150–152
  vertical acceleration and, 134–136
Law of the Lever, 173–174, 181
Law of Reflection, 869–871, 885
Law of Refraction, 876, 885
Law of Sines, 49, 99
Law of Universal Gravitation, 208–218, 231
  asphericity of Earth, 216–218
  confirmation of $1/r^2$, 210–211
  Earth's density, 214–215
  evolution of, 208–209
  gravitational constant, 212–213
  gravity of a sphere, 215–216
  product of the masses, 211
  terrestrial gravity, 213–214
Lawrence Livermore National Laboratory, 651
Laws, 19

defined, **2**–3
Laws of Planetary Motion, 218–224, 231, 263
L-C-R ac networks, 805–818
  domestic circuits and, 815–818
  series circuits, 805–812
  transformer and, 812–815
Leacock, Stephen, 589, 1020
Lead storage cell, 667
Leahy, William, 1129
Least Action, Principle of, 1076
Left-circular polarization, 940
Leibniz, Gottfried Wilhelm von, 31, 65, 288, 517
Lenard, P., 1018
Length, measurement of, **8**–11
Length contraction, 993–995
Lenoir, J., 566
Lenses, 894–925
  aspherical surfaces and, 895–897
  finite imagery and, 903–905
  focal points and planes and, 901–903
  magnification and, 905–907
  single, 907–916
  spherical thin lenses, 897–901
  thin-lense combinations, 916–925
Lensmaker's Formula, 900
Lenz, Heinrich Friedrich Emil, 768
Lenz's Law, **768**–770, 771, 773, 778, 785, 814
Lepton number (L), 1139
Leptons, 4, 1138, **1139**, 1159
  angular momentum of, 259
Leucippus, 319
Levers, 173–174
  bones and muscles as, 164
  end-pivoted, 188
Lewis, G. N., 1051
Leyden jar, 640–641, 664
Libration point, 233
Lift, 384–385
Light
  atoms and, 846–847
  behavior of, 1056, 1067
  elastic scattering of, 864
  electromagnetic-photon spectrum, 848–857
  electromagnetic radiation, 843–845
  electromagnetic waves, 835–837, 939
  energy quanta, 845–846
  granularity of, 1051
  irradiance and, 842–843
  as massless quanta, 845
  measurement of the speed of, 9–10
  natural, 940–942
  nature of, 834–848
  propagation of, 863–886
  propagation in free space, 984
  quasimonochromatic, 853
  scattering and absorption, 847–848
  speed of, 230, 840, 842, 981, 985
  transmission through dense media, 866–867
  as a transverse wave, 940
  unpolarized, 941
  vacuum speed of, 841
  waveforms and wavefronts, 837–840
  waves and particles, 834–835
  white, 852
Light-emitting diodes (LEDs), 820
Light-quanta, 1052
Light rays, 870

Light waves, sources of, 834
Light-year (ly), 21
Limits, concept of, 31
Linear acceleration, link to angular acceleration, 246
Linear charge density, 616
Linear equations, 17
Linear expansion, temperature coefficient of (α), 487–490
Linear mass-density (μ), 426
Linear momentum (**p**), 101–102
    conservation of, 7, 90, 112–117, 267, 308
    rotational equivalent of, 257
    symmetry and, **116**–117
Linear polarization, 940, 972
Linear polarizer, 942, 943, 944
Linear speed, determining, 245, 247
Line-of-action, 126, 164, 174, 175
    center-of-gravity and, 184
Line of stability, 1104, 1110
Lines-of-force, 603, 604
Line strengths, 169
Line voltage, 797
Liquefaction of gases, 498–500
Liquid-gas friction, 146
Liquid-hydrogen bubble chamber, 639
Liquids, 324, 354
    compression waves in, 431
    ideal, 354
    light scattering of, 867
    polar and nonpolar, 601
    surface tension of, 373
    volumetric expansion of, 490–494
Lithium, electron structure of, 1086
Load, 132–133
    deformation in proportion to, 327
    relationship to pressure, 155
    skeletal, 165–166
    supported by ropes, 166–167
Load resistor, 821
Local symmetry, 1146
Lodestone, 721, 722
Lodge, Sir Oliver, 1126
Logarithmic dependence, 558
Logarithms, 15
Longitudinal elastic wave, 429
Longitudinal magnification, 906
Longitudinal waves, 421–422
Loop, defined, 693
Loop equations, 705
Loop Rule, **703**, 707, 782
Lorentz-FitzGerald Contraction, 981–982
Loudness, sensation of, 454
Low-temperature superconductivity, 678
Low-voltage ac, 796
Lucretius, 728
Lyman series, 1032, 1061

Mach, E., 92, 1048
Machines
    hydraulic, 366–368
    perpetual-motion, 553, 571
    rotational motion in, 240
Mach's Principle, 92, 288
Magalotti, 591
Magdeburg sphere, 362–363
Magic numbers, 1109
Maglev (magnetically levitated train), 770

Magnetic bottle, 745
Magnetic dipole moment, 749–750
Magnetic domains, 728–730
Magnetic field (B)
    energy in, 783–784
    flux of (Φ_M), 766, 785
    induced, 773–774
    table of, 725
Magnetic flux (Φ_M), 766
Magnetic force (F_M), 741
Magnetic materials, permeability of, 733
Magnetic mirror, 745
Magnetic moment ($\boldsymbol{\mu}$), 749, 752
Magnetic monopole, 723
Magnetic-PE (PE_M), 296
Magnetic poles, **722**–724
Magnetic resonance imaging (MRI), 1107
Magnetic torques, 748
Magnetism, 208–209, **721**–752
    on the atomic level, 727–730
    of a current loop, 744
    diamagnetism, 728
    of the Earth, **726**–727
    electrodynamics, 730–740
    ferromagnetism, 728
    magnetic force, 741–752
    magnets and magnetic field, 722–730
    paramagnetism, 728
    of a solenoid, 735, 740
Magnetite, 721
Magnetite crystal, 723
Magnetotactic bacterium, 723
Magnification, 905–907
    angular, 915
Magnifying glasses, 909, 911, 915–916
Magnitude, 32
    of acceleration, 62
    of angular momentum, 257
    of average acceleration, 60
    of centripetal acceleration, 141
    of an electric field, 632
    of the electric field of a point-charge, 605
    of gravitational field strength, 229
    of the instantaneous velocity vector, 37
    of momentum, 102
    of normal force, 133
    of a reaction force, 181
    of velocity, 49
Malus, Étienne, 944
Malus's Law, 944, 972
Manhattan Project, 1128
Manometer, 362, 388
Maricourt, Peter de, 722
Mariner 2, 52
Marx, Karl, 132
Mass (m), 11, 100
    atomic standard for, 12
    center of, 254–255, 269
    conceptual origin of, 100
    determining, 105–106
    of an electron, 14
    gravitational, 414
    inertial, 100, 250, 414
    measurement of, 11–12
    Newton's Second Law and, 100–101
    operational definition of, 114–115
    relationship to force and acceleration, 126–139

Mass damper, 418
Mass defect, **1111**, 1131
Mass-energy, 1004
    conservation of, 1006
Masses
    comparison of, 100
    product of, 211
Mass flux, 378
Materials, elastic, 331–332
Mathematical Principles of Natural Philosophy (Newton), 91
Mathematical symbols, 16–17
Matter
    atoms and, 318–327
    center-to-center attraction in, 208
    changes of state of, 528–534
    defining, 1–2
    density of, 320–322
    electromagnetic interaction and, 587–588
    energy as a property of, 278
    as field, 1144
    interaction of, 4
    phases of, 324
    states of, 323–324
    tendency to sustain rotational motion, 260
    thermal properties of, 482–507
    wavelength of, 1074
Maupertuis, 1076
Maximum acceleration, computing, 68
Maximum current, 798
Maximum emf (electromotive force), calculating, 777
Maximum force, computing, 107
Maximum Power Transfer Theorem, 698–699, 707
Maximum range, 81–82
Maximum torque, determining, 178
Maximum voltage, 798
Maxwell, James Clerk, 502, 503, 835
    Electromagnetic Theory of, 876
    on magnetism, 738
    on measuring the speed of Earth, 980–981
Maxwell-Boltzmann distributions, 506
Maxwell's Equations, 836, 984
Mayer, Julius Robert, 520, 550
Mayer, Maria Goeppert, 1110
Mean free path, 505
Mean lifetime (τ), 1120–1121
Mean-Speed Theorem, 63–65, 67, 648
Measurement, 8–16
    exact, 13–16
    of force, 96
    of length, 8–11
    of mass, 100–101
    of mass and weight, 11–12
    significant figures in, 13–16
    of time, 13
Mechanical change in energy, 279
Mechanical energy, 288–297
    conservation of, 297–309
    defined, 298
    internal energy, 296–297
    kinetic energy, 288–291
    potential energy, 291–296
Mechanical equivalent of heat, 520
Mechanical waves, 420–433
    characteristics of, 420–425
    compression waves, 429–433

transverse waves, 425–429
Mechanics (Aristotle), 173
Medicine, use of X-rays n, 857
Megohms, 673
Meissner Effect, 730, 731
Meitner, Lise, 1123–1124, 1125
Melting, 528–530
Melting-point curve, 501
Meniscus lenses, 918
Mercury
    density of, 492
    superconducting state of, 677
Mercury barometer, 359, 360
Mercury cell, 666–667
Mercury manometer, 362, 388
Mersenne, Father, 458
    speed of sound, 449
    waves on a string, 461
Mesons, **1140**–1141, 1144, 1148, 1159
    kaon, 1140
    pion, 1140
    quark composition of, 1149
Messenger particles, 617
    spin-1, 618
Metals
    conductivity of, 537, 538, 593
    electromotive series for, 665
    electrostatic induction and, 601
    resistivity of, 676
    scattering by, 883
Meter (m), 8
    defined, 10
Method of mixing, 522
Metric System, **8**–11
Michelson, Albert Abraham, 981, 1090
Michelson Interferometer, 959–960, 970, 981,
    990
Michelson-Morley experiment, 980–982, 1008
Microcoulomb, 598
Microfarad, 639
Microscope, compound, 923–924
Microstates, 578–579
Microwave images, 851
Microwaves, 850–851
Middle ear, 452
"Middle third" rule, 191, 192
Milky Way Galaxy, 223
Milligals, 230
Millikan, Robert, 20, 834, 1020–1021, 1053
Millikan oil-drop experiment, 1020–1022
Mills, R. L., 1150
Minkowski, Herman, 997
Mirror formula, 928, 932
Mirrors, 925–931
    aspherical, 925–926
    finite imagery and, 929–931
    plane, 871–873
    spherical, 926–929
Mobile charge carriers, 663
Modern physics, 4
    nuclear atom, 1028–1036
    subatomic particles, 1016–1028
Moduli, 341–346
    shear and bulk, 343–346
    Young's modulus (Y), 341–343
Modulus of Elasticity, 341–343
Modulus of Rigidity, 343–344, 346
Molar specific heat capacity, 526

Mole, 320, 346
Molecular number-densities, 324
Molecules, 318
    mobility of, 356
    speed in gases, 504–506
Moment-arm, 175, 176, 177, 179
    for torques, 184
Moment of a force, 174, 175, 177, 257
Moment of inertia (I), 250, 258
    calculating, 251–252
    of a point-mass, 250
    spin rate and, 260
Momentum, 89–90, 101
    angular (**L**), 240, 257–268, 269
    Conservation of Linear, 112, 115, 230, 306, 307
    homogeneity of space and, **116**–117
    linear (**p**), **101**–102, 112–117, 257, 267
    measuring the magnitude of, 102
    rate of change of, 502
    relativistic, 1000–1003
    time-rate-of-change of, 103, 126
    transfer of, 303
Momentum change (Δ**p**)
    computing, 110–111
    impulse and, 102–106
    in systems, 113–114
Monatomic gases, 528, 561
Monochromatic light, 853
Monolithic IC, 824
Monopoles, magnetic, 723, 727
Moon
    average orbital speed of, 25
    centripetal acceleration of, 210
    diameter of, 241
    displacement of Earth caused by, 223
Morgan, J. P., 796
Morley, E. W., 981
Moseley, Henry, 1030, 1065
Motion, 23–50
    ballistic, 92–94
    changes in, 57
    circular, 143–146
    coupled, 138–139
    curvilinear, 139–146
    friction and, 146–156
    harmonic, 402–419
    instantaneous, 128–130
    nonuniform, 65
    1-dimensional, 39
    periodic, 401
    power and, 287–288
    projectile, 75–83
    relative, 39–50, 984
    repetitive, 401–402
    resistance to, 70
    resistance to a change in, 92
    rotational, 239–269
    straight-line, 36, 91
    translational versus rotational, 164
    uniform, 39, 91
    uniformly accelerated, 62–69
    vibrational, 402
Motional emf (electromotive force), 770–773,
    785
Motion-induced vortices, 419
Motions, coupled, 138–139
Motion variables, constant-a equations for,
    66–69

Motor, dc, 750
Moving bodies, center-of-mass of, 255
Moving coil galvanometer, 699
Multiple Photon Photoelectric Effect, 1053
Multiplication
    significant figures in, 14–15
    of a vector by a scalar, 34–35
Muons, 992–993
Muscles
    fibers, 282
    force exerted by, 187–188
    functioning of, 282
    tissue strength, 339
Musculoskeletal system
    structure of, 173
    support in, 192
mv (product of mass and speed), 101
Myopia, 921–922

National Aeronautic and Space
    Administration (NASA), 138
National Institute of Standards and
    Technology (NIST), 12, 987
Natural angular frequency, 409
Natural convection, 536
Natural frequency (f_0), 433
Natural light, 940–942, 972
Natural logarithm, 558
Natural radioactive series, 1114
Natural sciences, 2
Navier, Claude, 341
Near-point, 912, 919
Nearsightedness, 921–922
Necking, 337
Negative acceleration, 61, 67
    uniform, 68–69
Negative charge, 588–589
Negative gravity, 230
Negative lenses, 902
Negative magnification, 905
Net applied force, 97
Net force, 128–129
    computing, 98–99, 599–600
Net potential, determining, 636
Net torque, 179
Network, equivalent capacitance of, 647
Network analysis, **702**–707
Neutral-current process, 1156
Neutral equilibrium, 189
Neutrality, electrical, 588–589
Neutrinos (ν), 318, 1139
    beta decay and, 1116–1118
Neutrons (n), 1149
    angular momentum of, 259
    decay of, 1154
    interaction with protons, 1144
Neutron star, 260
Neutrophotos, 1034
Newton, Isaac, 65, 278–286, 423
    confirmation of $1/r^2$, 210–211
    curvilinear motion, 139
    death mask, 109
    definition of force, 95
    definition of quantity-of-motion, 102
    devising of calculus by, 30
    g as a function of altitude, 72
    Law of Inertia of, **90**–99
    light theory of, 834

momentum and, 89–90
speed of sound, 450
on viscosity, 354
on wave diffraction, 835
wave speed, 423
welding of dynamics and astronomy, 139
Newton (N), 96, 117
defined, 104
Newtonian Mechanics, 2–3
Newtonian reflecting telescope, 931
Newtonian Theory of Gravity, 207–231
cosmic force, 218–231
Law of Universal Gravitation, 208–218
Newton-meter, 279
Newton's First Law, 90
Newton's Second Law, 90, 100, **103**, 117, 211, 430
alternative formulation of ($F = ma$), 126–130
relationship of weight and mass in, 130
rotational equivalent of, 259
statics and, 164
Newtons per coulomb (N/C), 604
Newton's Third Law, 596
Nicholson, J. W., 1057, 1059
Nickel-cadmium cell, 667
Nitrogen, liquid, 487
Noble gases, 1085, 1088, 1089, 1110
Noddack, Ida, 1123
Node equations, 705
Node Rule, **704**, 707
Nodes, 459
defined, 693
electric circuits, 645
Noether, Amalie, 7, 8, 277, 308
Noether's Principle, **6–8**, 116, 267, 309, 638, 1146
linear momentum and, 116
Noise, 457
Nollet, Abbé, 641
Nonconcurrent forces, 180–183
Nonconductors, 592–595
Noncoplanar force systems, 164
Noninductive coils, 781
Nonohmic electrical devices, 672
Nonresonant elastic scattering, 848
Nonuniform motion, 65
Nonworking force, 289
Normal compressive strain (ε), 335, 346
Normal compressive stress (σ), 333
Normal force (F_N), 131, 132
relationship to centripetal force, 143–144
Normal tensile strain (ε), 335, 346
Normal tensile stress (σ), 333
Normal Zeeman Effect, 1083
Norman, Robert, 727
North pole, 723
npn transistor, 822, 823, 825
n-type semiconductor, 819
Nuclear atom, 1028–1036
atomic spectra, 1031–1032
neutron, 1034–1036
proton, 1032–1034
Rutherford scattering, 1028–1031
Nuclear Coulomb barrier, 1030
Nuclear electrons, uncertainty and, 1091–1092
Nuclear energy, 278
Nuclear fission, 1113, 1123–1131

Nuclear force, **1107**–1109
nucleus formation and, 1107
Nuclear fusion, 1113, 1123–1131
Nuclear magnetic resonance (NMR), **1106**
Nuclear-PE, 296
Nuclear physics, **1101**–1131
nuclear structure, 1102–1113
nuclear transformation, 1113–1131
Nuclear reactors, **1126**–1128
Nucleon-nucleon interactions, 1145
Nucleon number (A), 1102
Nucleons, 1035
Nucleus, 5, 6
composition of, 1035
Coulomb interaction with electron, 1059
radius of, 1105
shell model of, 1109–1111
size, shape, and spin of, **1104**–1107
size of, 323, 1029–1031
stability of, 1109–1113
Nuclides, **1102**
atomic mass of, 1103
binding-energy-per-nucleon for, 1112
Null result, 981
Nutation, 265

Object distance (s_o), 872
Object focal length (f_o), 901
Object focal point, 901
Objects
nonequilibrium behavior of, 125
point-masses in, 254
Object size, relationship to terminal speed, 70
Object-wave, 970
Oersted, Hans Christian, 722, 731, 732, 741
Ohm, Georg Simon, 671, 681
on resistance, 671–672
Ohmic circuit element, 672
Ohm's Law, 17, **672**, 680, 681, 694, 798, 800, 808
Ohm-meters, 674, 675
Oil-drop experiment, 1020–1022
Old Quantum Theory, 1043, 1044–1056
1-dimensional motion, 39
$1/r^2$, Newton's confirmation of, 210–211
Onnes, H. Kamerlingh, 677
"On the Electrodynamics of Moving Bodies" (Einstein), 983
On the Equilibrium of Liquids (Pascal), 365
Operational definitions, 2
Oppenheimer, J. Robert, 1130
on antimatter, 1093
and the atomic bomb, 1128
Optical center, 902
Optical density, 866
Optical path length (O.P.L.), 895
Optics, 833
geometrical, 893–932
physical, 939–972
Orbital, 1085
Orbital angular momentum
determining, 262–263
of planets, 263
Orbital angular momentum quantum number (l), 1082
Orbital magnetic quantum number (m_l), 1082
Orbital speed, relationship to orbit size, 225
Orbiting planet, gravitational field acting

on, 263
Orbits, of satellites, 224–227
Order, disorder and, 576–577
Order-of-magnitude, 25
calculations, 670–671
Oresme, Nicolas, 63
Oscillating dipole, 844
Oscillating spring, 408–412
Oscillation, 401–433
forced, 416–418
gravity and, 412
Oscillators, 798
maximum voltage of, 808
Otto, Nikolaus, 566
Ounce, 12
Outer ear, 452
Outlets, two- and three-prong, 817
Output current, 824
Overdamping, 415
Overtones, 442, 452
Ozone, 355

p (linear momentum), 101
Papin, Denis, 534
Parabola, 17, 18
conic section, 77
equation of, 77
Parallel-Axis Theorem, 275
Parallel circuit, 651
and capacitors, 643–645
and cells, 667
and resistors, 694
Parallel force systems, 164–167
Parallelogram method, 33–34, 44, 48
Parallel-plate capacitor, 617, **641**–643
charged, 642
Parallel resistance, 693–698
Parallel vectors, 35, 42
Parallel wires, current in, 750–751
Paramagnetic substances, 728
Paraxial rays, 898
Parent nucleus, 1114, 1131
Partial polarization, 941
Particle interactions, mediation by field quanta, 1145
Particles
angular momentum of, 257
determining the speed of, 1007
elementary, 4, 259
force-carrying, 208
masses and rest-energies for, 1005
messenger, 617
moment of inertia of, 251
relativistic kinetic energy of, 1003
spin and, 618
subatomic, 1016–1028
Particles of zero-mass, 1008
Particle theory of light, 834–835
Pascal, Blaise, 360, 366
Pascal (Pa), 333, 389
Pascal's Principle, 365–368, 389
Pascal's syringe, 366
Paschen series, 1032, 1061
Pauli, Wolfgang, 20, 1116–1117
beta decay and, 1116
Exclusion Principle, 1081, 1084, 1096
on spin, 1083
Peak altitude (s_p), 74–75, 80, 83

Peak-to-peak voltage, 798
Pendulum, 412–415, 433
    period of, 413
Pendulum clock, 13, 73
Perfect diamagnetism, 730
Perihelion, 219, 263
Periodic motion, 401
Periodic Table, 1065, 1066
    placement in, 1103
Periodic waves, 423
Period of a cycle (T), 402, 409, 433
Permanent magnet, 722
    magnetic field of, 726
Permeability (μ), 732
    of free space (μ₀), 732
    property of, 730
Permittivity (ε), **608**, 619
    table of values, 608
Permittivity of free space (ε₀), **608**, 619
Perpendicular components of vectors, 44–50
    summation of two vectors using, 47
Perpetual-motion machine, 553
Perpetual motion machine of the second kind, 571
Perrin, J., 1018, 1028
Phase, 406, 424
    initial, 406
Phase diagrams, 500–501
Phase differences, 800–801
Phases of matter, 324
Phase velocity, 425
Phasor, 806
Phasor addition, 806, 807
Philoponus, John, 70
Photocopy machine, 593
Photocurrent, 1050
Photoelectric effect, 1020, **1049**–1054
    work function (φ), 1052
Photoelectric Equation, Einstein's, 1053
Photoelectrons, 1050
Photon flux, 846
Photon flux density, 846
Photons, 4, 845, 858
    Einstein on, 1051–1054
    gamma-ray, 857
    as the quantum of the electromagnetic field, 1142
    as the quantum of the gauge field, 1147
    virtual, 617
Physical optics, 939–972
    diffraction, 960–972
    interference, 950–960
    polarization, 940–950
Physical quantities, symbols for, 16–17
Physics
    Classical, 3–4
    defined, **1**–2
    high-energy, 1137–1159
    language of, 16–19
    Modern, 4
    nuclear, 1101–1131
    quantification function of, 2
    testing of, 3
    thermal, 482–507
Picard, J., 1017
Picofarad, 639
Pioneer 10, 91, 302
Pions, 1140–1141, 1145

quark composition of, 1149
Pitch, 452
Pith balls, 592
Pivot, reaction force on, 182–183
Planck, Max, K. E. L., 1048, 1049
    on energy elements, 1048–1049
    on quantum theory, 1049
Planck's Constant (h), 259, 845, 1048, 1074, 1076, 1089, 1090
    quantum of action, 1059
Planck's Radiation Law, **1049**, 1067
Plane harmonic electromagnetic wave, 840
Plane mirrors, 871–873
Plane-of-incidence, 870, 871, 876, 885, 946
Plane of polarization, 940
Plane-polarized waves, 972
Planes, focal points and, 901–903
Planetary motion, **263**–265
    Kepler's laws of, 218–224, 231, 263
    investigations of, 139
    laws of, 231
    Newton's explanation of, 208
Planetary spin, effect on g, 72–73
Planetary wiggle, 223
Planets
    elliptical orbits of, 218
    gravitational field acting on, 263
    gravitational-PE and, 301
Planimeter, 284
Planté, 665
Plasma, 324, 485
Plato, 16, 502, 514
Plücker, J., 1031
Plutarch, 173
Plutonium, 1115, 1126–1128
Plutonium-239, production of, 1128
Plutonium Project, 1127
pn-junction, 819–821, 825
pnp transistor, 822, 823, 825
Poincaré, Jules Henri, 20, 279, 982, 984
Point-charges
    electric field of, 619
    force exerted on a test-charge, 603
    interacting, 633
    magnitude of the electric field of, 605
    potential of, 632–634
Pointed arches, 193
Point-mass, 184, 185
    angular momentum of, 258
    moment of inertia of a, 250
Poise, 175
Polaris, 266, 267
Polarity, of emf (electromotive force), 768
Polarization, of light, **940**–950
    Malus's Law, 944
    natural light and, 940–942
    polarizers and, 942–946
    processes of, 946–950
    by reflection, 946
Polarization angle (θₚ), 946, 972
Polarization of dielectrics, 643
Polarization of electric cells, 668
Polarized atom, 602
Polarized light, scattering of, 949
Polaroids, 945
Poles, 722–724
Pollio, Marcus Vitruvius, 442
Polycrystalline solids, 325

Polyphase ac, 797
Population inversion, 1063
Positive charge, 588–589
    direction of flow of, 663
    transfer of, 591–592
Positive lens, 902, 908, 909
Positive magnification, 905
Positive rays, 1033
Positron (e⁺), 1094
Positron-beta decay, 1131
Positron Emission Tomography (PET), 1095
Positronium, 1094
Potential, 627, 651
    of a point-charge, 632–634
    electric, 626–639
    in a uniform field, 631
Potential difference, **628**–631, 633, 644, 651
    in batteries, 632
    proton falling through, 630
    table of, 629
Potential energy (PE), 291–296, 310
    elastic (PEₑ), 329–331
    electrical (PE_E), 626
    gravitational (PE_G), 292
Pound (lb), 96
Potential-energy wells, 1110
Potentiometer, 679, 681
Power (P), **286**–288, 310
    act, 810
    apparent, 810, 825
    electrical energy and, 679–681
    equivalencies, 287
    factor, 810
    instantaneous, 799
    maximum, 698–699
    real dissipated, 810
Power, H., 364
Precession, 265, 266
Pressure, force per unit area (P), 155, 355, 389
    absolute, 361
    in a container, 358
    critical, 498
    gauge, 361
    in human body, 361
    hydrostatic, 355–364
Pressure conversions, 360
Pressure cooker, 534
Pressure-energy density, 380
Prevost, P., 535
Priestley, Joseph
    on electric force, 595
Primary charge distribution, 603
Primary colors, 883
Primary waves, 873, 874
Principal quantum number, 1060, 1082
Principia (Newton), 91
Principle of Complementarity, 1076
Principle of Least Action, 1076
Principle of Relativity, 983, 1008
Principle of the Constancy of the Speed of Neutrinos, 985
Product of the masses, 211
Profile of a wave, 838
Progressive (traveling) waves, 420, 459, 857
Progressive electromagnetic wave, 837
Projected horizontal distance, 81
Projectile motion, **75**–83
Projectiles

ballistic, 76–77
displacement of, 93
effect of air friction on, 82
Proof-planes, 595
Propagation, speed of (c), 840–842
Propagation number (k), 839, 857
Proper length, 993
Proper time, 992
Proton decay, 1158
Proton-electron pairs, 1034
Proton-proton chain, 1130
Protons (p)
angular momentum of, 259
charge on, 589–590
interaction with neutrons, 1144
quark composition of, 1149
PT-diagrams, 500–501
p-type semiconductor, 819
Pulleys, 166–167
Pulsars, 3, 260
Pure tone, 442, 452
Pythagorean Theorem, 32, 44, 46, 49, 225

Quadratic equations, 17
Quanta, 4, 1017
energy, 845–846
field, 1145
light, 845, 1052
Quantity
change in, 28–29
measuring, 2
Quantity-of-angular-motion, 257
Quantity-of-matter, 100
Quantity-of-motion, 101, 102, 112
Quantization, 4, 5
of angular momentum, 259
Quantized energy, 13
Quantum Chromodynamics (QCD),
1150–1152, 1159
Quantum Electrodynamics (QED), 1081,
1092–1095
Quantum Field Theory, 116, 230, 617–618,
1142–1159
creation of the Universe, 1156–1159
electroweak force, 1152–1156
gauge theory, 1145–1147
quarks, 1148–1150
Quantum jump, 1057
Quantum Mechanics, 4, 6, 617, 1073–1096
De Broglie waves, 1074–1076
principle of complementarity, 1076
Schrödinger wave equation, 1076–1081
Quantum numbers, 1096
Quantum of action (h), 1059
Quantum physics, 1081–1096
electron shells, 1084–1089
Pauli Exclusion Principle, 1084
quantum electrodynamics and antimatter,
1092–1095
quantum numbers, 1081–1082
spin, 1083–1084
Uncertainty Principle, 1089–1092
Zeeman effect, 1082–1083
Quantum randomness, 1081
Quantum state, 1078
Quantum Theory, 847, 1043–1066
atomic theory, 1056–1066
origins of, 1044–1056

Quark confinement, 1149
Quarks, 4, 590, 1137, 1148–1150, 1159
angular momentum of, 259
table of, 1148
transformation of, 5
Quasi-crystalline solids, 324
Quasimonochromatic light, 853

Radial acceleration, 141, 143
Radial force, 145
Radian (rad), 241
Radiant energy, 278, 318, 833–858
conceptions of light, 834–848
electromagnetic-photon spectrum, 848–857
quantized, 1049
Radiant flux density, 843
Radiation, thermal, 534–536
Radiation belts, 745–746
Radio
AM, 812
circuitry in, 674
Radioactive decay, 5, 1113–1118
measurement via, 1007
Radioactivity
discovery of, 1024–1028
induced, 1122–1123
Radioisotopes, half-lives of, 1122
Radio receiver, 589, 813
Radiowaves, 848–850
Radium, 1025
Radius of curvature, 141
Radon, 355, 1026, 1027, 1115
Raisin-pudding atom, 1028, 1029
Ramsay, William, 1025
Range, 81–82, 83
Rankine, W., 292
Rarefaction, 429, 462, 464
Rationalized units, 607–609
Rayleigh, Lord, 454, 506, 864, 968
Rayleigh scattering, 864–865
R-C circuits, 700–702
Reactance (X), 802
of a circuit, 808, 825
capacitive, 804, 825
inductive, 802, 825
Reaction force, 131, 165
determining, 182–183
magnitude of, 181
Reactor, nuclear, 1126
Real dissipated power, 810
Real gases, 498–500
Real images, 904
Real particles, 1
creation from virtual particles, 1144
Recoil speed, calculating, 114–115
Rectification, 820–821, 825
Rectilinear motion, 59
Rectilinear translation, 239
Reference circle, 404
Reference-wave, 970
Reflected waves, 427–428
Reflecting telescope, 930–931
Reflection
defined, 867
diffuse, 871
external, 868
internal, 868
Law of, 869–871, 885

of light waves, 867–873
polarization by, 946–947
of sound waves, 465–466
specular, 871
total internal, 879–882, 886
Reflection gratings, 967
Reflections on the Motive Power of Heat
(Carnot), 562
Refracted waves, 873
Refracting telescope, 924–925
Refraction, 873
index of (n), 874
Law of, 876, 885
of light waves, 873–882
of sound waves, 466
Refrigerants, 568
Refrigeration machines, 566–568, 579
Reines, F., 1117
Relative atomic masses, 319
Relative density, 369–370
Relative motion, 39–50
velocity and, 40
Relative permittivity (K_e), 608
Relativistic dynamics
relativistic energy, 1003–1008
relativistic momentum, 1000–1003
Relativistic linear momentum, 1001
Relativity, Theory of, 3, 6, 39, 90, 207–208,
229–230, 277, 414
Lorentz-FitzGerald Contraction, 981, 993
Michaelson-Morley Experiment, 980
postulates, the two, 983
relativistic dynamics, 1000–1008
simultaneity, 985
space-time, 997
table of, β and γ, 986
time dilation, 990
twin effect, 995
Renaldini, C., 482
Renormalization theory, 1095
Repetitive motion, 401–402
Resilience, 348
Resistance (R), 663, 671–681
ac and, 798–800
energy and power, 679–681
equivalent (R_e), 693, 707
internal (r), 691
movement against, 279
power dissipated in, 680, 681
resistivity, 673–677
superconductivity, 677–678
voltage drops and rises, 678–679
Resistivity (ρ), 673
table of, 675
temperature dependence of, 676–677
Resistors, 673, 674, 681
in series and parallel, 693–698
Resolution, angular limit of, 968
Resolving power, 968
Resonance, 416–418, 825
ac, series, 811
Resonance phenomenon, 417
Resonant cavity, 1064
Resonant frequency (f_0), 416–418, 811
Rest
objects at, 163
uniform motion and, 91
Rest energy (E_0), 1003, 1009

Restoring force, 328
    elastic, 408–412
    gravitational, 412–415
Resultant vector, 33, 44, 50
    computing the magnitude of, 46
    finding, 48
Reverse biased diode, 820
Reverse potential, 803
Reversible engines, 563
Reversible processes, 559, 579, 580
Revolving-armature machine, 776
Revolving-field machine, 776
Rey, J., 482
Reynolds, O., 374
Richer, J., 11
Right-circular polarization, 940
Right-Hand-Current Rule, 732, 734, 749, 838
Right-hand-screw rule, 177–178, 258, 742
Rigid bodies, 178–183
    center-of-mass in, 255
Ripple, 821
Ritter, J.
    discovery of ultraviolet radiation, 854
R-L circuit, **782–783**
rms current, 799
Roberval, 209
Robison, J., 457
    on the inverse-square law, 595
Rockets, 111–112, 313
Rolfe, W. J., 836
Rolling friction, 154
Roman balance, 174, 175
Romanus, Aegidius, 100
Röntgen, Wilhelm Konrad, 856, 1022
Ropes
    breaking strength, table of, 169
    determining tension on, 136, 138–139,
        168–169, 253–254
    pulleys and, 166–167
    weights on, 131
Rotating systems, dynamical equation of, 252,
    253
Rotation
    dynamics of, 250–269
    kinematics of, 241–250
Rotational equilibrium, 173–183, 179, 181
    rigid bodies, 178–183
    torque, 174–178
Rotational inertia, **250–254**
Rotational motion, 239–269
    resistance to change in, 251–252
    versus translational motion, 164
Rotational symmetry, 6–7
Rowland, H. A., 518
Royds, 1026
Rubbia, Carlo, 1156
Rubbing, charging by, 590–591
Rubbing machines, 637
Rumford, Count, 515, 516
Russell, Bertrand, 2, 1019
Rutherford, Ernest, 1025–1026
    on atomic power, 1126
    on protons, 1033–1034
    scattering, 1028

s (scalar value of displacement), 74
Saggita, 898
Salam, Abdus, 1155

Salt (NaCl), 326
Satellites
    energy formulation and, 301
    orbits of, 224–227
Saturated color, 884
Saturated vapor pressure, 499
Saturn, ring system of, 222
Savart, Félix, 732, 734
Scalar, multiplication of a vector by, 34–35
Scalar quantity, 24
Scalar components, 45, 46, 47, 49, 50
Scattering, 847–848, **863–886**, 885
    color and, 882–885
    dense media and, 866–867
    elastic, 864
    interference and, 865–866
    nonresonant elastic, 848
    polarization by, 948
    Rayleigh scattering, 864–865
    reflection and, 867–873
    refraction and, 873–882
    Rutherford, 1028–1031
Schrieffer, Robert, 678
Schrödinger, Erwin, 4, 1077
    on the quantum jump, 1080
Schrödinger Wave Equation, 1076–1081, 1096
Schwinger, Julian, 1095, 1152
Scientific calculator, 15–16
Scientific vocabulary, 16–19
Scotch yoke, 404
Scott, David R., 70
Search coil, 773–774
Sears Tower, 216
Secondary waves, 873, 874
Second Condition of Equilibrium, 179, 184,
    193
Second harmonic, 446
Second-Law equations, coupled, 138–139
Second Law of Thermodynamics, 570–579
    Clausius Formulation and, 571–572
    entropy and, 572–576, 577–579
    order and disorder and, 576–577
Seconds (s), 2, 8, 9, 13
Seismic waves, 422
Self-excited vibrations, 418–419
Self-field, 643
Self-induced emf, determining, 781
Self-induction, 779–784, 785
    back-emf and, 780–781
    energy in the magnetic field and, 783–784
    R-L circuit and, 782–783
Semiconductors, 675, 818–819
Series circuit, 651, 805–812
    and capacitors, 646–648
    and electric cells, 667
    and resistors, 694
Series limit, 1032
Series resonance, 811–812
Shakespeare, William, 211
    on time, 990
Sharks, 605
Shear Modulus, 343–344, 346
Shear strain (ε_s), 335, 346
Shear stress (σ_s), 334
Sheet polarizer, 945
Shell, defined, 1085
Shell model, 1109–1111
Shock absorbers, 330, 415

Short circuit, 649, 667
Shunt resistor, 699
SI (*Système International*), 8, 9
    units of force, mass, and acceleration in,
    127
Significant figures, **13–16**, 19
Silicon, doping of, 820
Silicon atoms, 819
Simon, Pierre, 1092
Simple cubic structure, 326
Simple harmonic motion (SHM), 402–408,
    433
Simple harmonic oscillator, 408, 410
    acceleration, 407
    displacement, 404
    velocity, 406
Simple pendulum, 413–415, 433
Sine function, 15, 44
Single-lens reflex camera, 914
Single-phase system, 797
Single Photon Photoelectric Effect, 1053
Single-slit diffraction, 963–965
Sink of a field, 609, 615
Sinusoidal sound wave, 445
Sinusoidal voltage, 799
Sinusoidal wave, 857
Sinusoidal waveforms, 447
Skin, human, 334
Sliding acceleration, 256
Slip, 331
Slope of a line, 28
Smith, Adam, 518
Snell's Law, 876–880, 885, 950
Socrates, 722
Soddy, F., 1025, 1102
Solar system, Brahe's model of, 218
Solenoids, 735–737
Solid-gas friction, 146
Solid-liquid friction, 146
Solids, 317–346
    atoms and matter, 318–327
    band structure of, 819
    classification of, 324–327
    crystalline, 324–327
    elasticity and, 327–341
    light scattering of, 867
    moduli and energy and, 341–346
    volumetric expansion of, 490–494
Solid-solid friction, 146
Solutions, checking, 19
Solvay Conference, 1048
Sommerfeld, Arnold, 1082
Sonograms, 433
Sound, 429, 441–474
    acoustics, 442–457
    hearing of, 451–454
    speed of, 449–451
    wavelengths, 444
Sound-level, 473
Sound production and propagation, 457–473
    beats, 458
    Doppler Effect, 469–473
    sound waves, 465–469
    standing waves, 458–465
Sound waves, 442–444, 465–469
Source of a field, 609, 615
South pole, 723
Space

defining, 1
homogeneity of, 116, 267
Space quantization, 1083
Spaceships, energy formulation and, 301
Space-time, 997–998
Spafford, H. G., 834
Spark, 593
Spatial period (wavelength), 423
Spatial symmetry, 116–117
Special Theory of Relativity, 3, 39, 90, 277, 979–1008
  addition of velocities, 998–1000
  effect on Classical Mechanics, 845
  groundwork for, 980–982
  length contraction, 993–995
  magnetism and, 731
  and Michelson-Morley experiment, 980
  postulates of, 983–985
  relativistic dynamics, 1000–1008
  simultaneity and time, 985–990
  space-time, 997–998
  time dilation, 990–993
  twin effect, 995–997
Specific gravity, 369–371
Specific heat capacity (*c*), **521**–528, 540
  atomic view of, 526–527
  of gases, 528
  of water, 527–528
Spectral lines, 1031
Spectroscopy, 1031
Specular reflection, 871
Speed (*v*), 24
  angular, 243–245, 248–249
  average, **24**–27, 50
  banking angle for, 144
  of compression waves, 429–431
  constant, 27–28
  instantaneous, **29**–31, 50
  mean, 63–65
  of sound, 449–451
  table of, 25
  terminal, 70
  of waves (*v*), 423
Speed conversion table, 26
Speed of efflux, 381, 382
Speed of light (c), 10, 842, 985
  characteristics of, 1003
  constancy of, 1008
Speedometer, 28
Speed-time graph, 28, 29, 61, 66, 69
Sphere, gravity of, 215–216
Spherical mirrors, 926–929
Spherical thin lenses, 897–901
Spherical waves, amplitude of, 449
Spin
  concept of, 1081, **1083**–1084
  defined, 727
Spin-1 particles, 618
Spin angular momentum quantum number, 1083
Spin axis, tendency to remained fixed, 265
Spin-down electrons, 1083
Spin magnetic quantum number, 1083
Spin rate, moment of inertia and, 260
Spin-up electrons, 1083
Split-ring commutator, 776
Spontaneous emission, 1062
Spontaneously broken symmetries, 8

Spring, oscillating, 408–412
  mass of, 412
Spring (elastic force) constant, 328
Spring scale, for measuring force, 96
Sputnik, 236
Stability
  balance and, 188–191
  gyroscopic, 265–267
Stable equilibrium, 188
  determining, 189
Stagnation point, 397
Standard, 19
Standard Model, **1158**
Standards of measure, 8
Standard temperature and pressure (STP), 319–320
  molecular parameters at, 505
Standing (stationary) waves, 458–465, 473
  in air columns, 462
  and the laser, 1064
  on strings, 459
Standing still, physics of, 131–133
Stanford Linear Accelerator Center (SLAC), 1149
Staphylococcus bacterium, 363
Stark, Johannes, 1053
Stars, thermonuclear fusion in, 1129–1131
State, defined, 553, 1085
State of polarization, 940
Static electric field, 617
Static equilibrium, 163, 611
Static friction, 146–150, 256
  moving via, 150–152
Static-friction coefficient (μ_s), 148–149
Statics, 90, 125, 163–205
  applied to hip dislocations, 170
  fluid, 354–374
  governing principles in, 164
Stationary states, 1057
Statistical mechanics, 1050
Steady-state current, 667
Steady-state motion, 374–375
Steam, 533
Steam engine, 562
Steelyard, 174, 175
Stefan, Joseph, 1045
Stefan-Boltzmann Law, **1045**–1046, 1066
Step-down transformer, 814
Step-up transformer, 814
Stevin, Simon, 358, 359
Stick-slip friction, 419
Stimulated emission, 1062–1065
  and the laser, 1063
Stoney, George, 1017, 1031
Stopping distance, calculating, **68**–69, 154
Stopping potential, 1050
Straight-line inertial path, 94
Straight-line motion
  in a fixed direction, 36
  infinite, 91
Strain (ε), 335–337, 346
Strain gauges, 690
Strange particles, **1141**
Strange quarks, 1148
Strato, 58, 71
Streamlines, 375
Strength, 337–340, 346
  breaking, 337

tensile, 337
ultimate, 337
yield, 337
Stress (σ), 332–335, 346
Stress birefringence, 950
Stress-strain diagram, 336, 339
Strings, standing waves on, 459–461
Strong force, 5, 95, 1006, 1107
  table of, 5
Strong interaction, 1154
Structures
  stability of, 188–191
  study of, 163–164
  total tension, 169
Sturgeon, W., 737
Subatomic matter, net charge of, 1021
Subatomic particles, 1015, 1016–1028
  cathode rays and, 1017–1022
  charges of, 589–590
  conservation of momentum in, 115–116
  discovery of radioactivity, 1024–1028
  quantum of charge, 1016–1017
  X-rays, 1022–1024
Sublimation, 501
Sublimation curve, 501
Subshell, defined, 1085
Subtraction
  significant digits in, 15
  of vectors, 35
Subtractive coloration, 884–885
Sum-of-the-forces, 179, 193
Sum-of-the-forces equals zero, 164
Sum-of-the-torques, 179, 181, 189, 193
Sum-of-the-torques equals zero, 164
Sun
  attractive action on planets, 209
  diameter of, 242
  distance to, 14
  galactic material acting on, 223–224
  thermonuclear fusion and, 1131
Superconducting elements, critical temperatures of, 677
Superconducting magnets, 738
Superconducting quantum interference devices (SQUIDS), 733
Superconductivity, 677–678, 681
Superconductors, 730
Supercooled state, 493
Supercritical mass, 1129
Superfluids, 487
Superheated steam, 533
Supernova 1987A, 91, 1118, 1158
Superposition Principle, 444–447
Surface charge density, 616
Surface-PE, 373
Surfaces of constant action, 1077
Surfaces of constant phase, 447
Surface tension (γ), 371–374, 389
Symbols, mathematical, 16–17
Symmetry, **6**
  angular momentum and, 267–268
  continuous and discontinuous, 7
  energy conservation and, 308–309
  global, 1146
  linear momentum and, 116–117
  local, 1146
  rotational, 6–7
  spontaneously broken, 8

Syncom II, 227
Systems
  binding energy in, 1111–1113
  concurrent force, 167–173
  coplanar force, 164
  equilibrium in, 125
  isolated, 113–114, 550, 553
  momentum changes in, 113–114
  nonconcurrent forces, 181
  noncoplanar force, 164
  order and disorder in, 576–577
  parallel and colinear force, 164–167
Système International, **8**, 9
  units of force, mass, and acceleration in, 127
Systolic pressure, 386
Szilard, Leo, 1125–1126

Tachyons, 1003
Tacoma Narrows Bridge, 419
Tait, P. G., 425
Tangent, 15
Tangent function, **44**
Tangential acceleration, 246, 247, 250
Tangential orbital speed, 225
Tau lepton, 1150
Telescope
  reflecting, 930–931
  refracting, 924–925
Temperature (*T*), 482–487, 540
  absolute, 485, 565, 568–570
  critical, 677, 681
  Curie, 730
  defined, 517
  heat and, 515–518
  range of, 485–487
  resistivity and, 676–677
  thermodynamic temperature and absolute zero, 484–485
Temperature coefficient of linear expansion (α), 488, 490, 507
  table of, 489
Temperature coefficient of resistivity (α_0), 676
Temperature coefficient of volume expansion (β), 490–494, 507
  table of, 491
Temperature gradient, 539
Temperature map, 514
Temporal coherence, 952
Temporal period (*T*), 423–424
Tendon, 339
Tensile force (F_T), 126
Tensile stress, normal, 333
Tensile (ultimate) strength, 337
Tension (F_T), 126
  on weighted ropes, 131
Terminal speed, **70**
Terminal voltage, 691–692
Terrestrial gravity, 213–214
Tesla, Nikola, 724, 796
  ac induction motor and, 796
  polyphase ac and, 797
Tesla (T), 724, 752
Test-charge, 603
Theories, 19
  defined, **2**–3
Theory of Everything (TOE), 6, 1159

Theory of Gravitation, 3
Thermal conductivity, 537–540, 541
Thermal energy, 278, 296, 513–540
  caloric versus kinetic, 514–515
  change of state and, 528–534
  heat and temperature, 515–518
  mechanical equivalent of heat, 520
  quantity of heat, 518–520
  specific heat capacity, 521–528
  Zeroth Law, 521
Thermal energy transfer, 534–540
  conduction, 537–540
  convection, 536
  radiation, 534–536
Thermal equilibrium, 521
Thermal neutrons, 1123
Thermal physics, 482–507
  gas laws, 494–506
  temperature, 482–487
  thermal expansion, 487–494
Thermal radiation, 534–536, 1044
Thermal reservoir, 557
Thermionic emission, 1024
Thermodynamic absolute temperature, 485
Thermodynamics, 3, 549–580
  cyclic processes and, 561–570
  First Law of, 550–561
  Second Law of, 570–579
  Zeroth Law of, 521
Thermodynamic temperature (*T*), 484–485, 507
Thermographs, 852
Thermometers, 482–483
Thermonuclear fusion, 1129–1131
Thermoscope, 482
Thilorier, M., 487
Thin film interference, 956–959
Thin-Lens Equation, 900
Thin lenses, combinations of, 916–925
Thin positive lens, 908
Third harmonic, 446
Thompson, Benjamin, 514–515
Thomson, G. P., 1075
Thomson, J. J., 1018–1019
  on electromagnetic waves, 1051
  separation of isotopes, 1102
Thomson, William, 550
  on lines-of-force, 835
  Lord Kelvin, 485
"Thomson atom," 1028
't Hooft, Gerhard, 1155
Thorndike, E., 982
Three-bar frames, 172
Three-coil dc generator, 777
Three Laws of Planetary Motion, 218–224, 231, 263
Three-phase ac, 797, 798
Thrust, 191
  in arches, 192
  calculating, 112
Tidal motions, 264
Timbre, 452–453
Time (*t*)
  defining, 1, **13**
  measurement of, 13
  momentum change and, 103
  with respect to an observer, 991
  simultaneity and, 985–990
  versus instantaneous force, 106–107

Time constant, 701, 782
Time dilation, 990–993
Time-rate-of-change
  of momentum, 103, 126
  of velocity, 58
Time-varying *E*-field, 773–774
Tip-to-tail method, 32–33, 46
Tokamak Fusion Reactor, 1133
Tomonaga, Sinitiro, 1095
Top quark, 1150
Torque (τ), **174**–178, 193
  center-of-gravity and, 184
  on a current loop, 748
  in a gravitational field, 254
  as the rotational equivalent of force, 257
  rotational motion and, 250
  scalar value of, 178
  as the time-rate-of-change of angular momentum, 259
Torque equations, 183
Torque vector (τ), 177
Torr, 359
Torricelli, Evangelista
  the barometer and, 358
Torricellian vacuum, 358, 362
Torricelli's Result, 381–383, 389
Torsion balance, Coulomb's, 595, 596
Total absorptivity, 1046
Total constructive interference, 866
Total destructive interference, 866
Total emissivity, 1045, 1046, 1067
Total energy, 1004
Total-flight time, 80, 83
Total impulse, 107
Total internal reflection, **879**–882, 886
Total tension structures, 169
Toughness, 331, 348
Towneley, R., 364
Transcendental functions, 15–16
Transformers, **812**–815
  turn ratio, 814
Transistors, 822–824
Translational equilibrium, 164–173, 179
  concurrent force systems, 167–173
  parallel and colinear force systems, 164–167
Translational motion, 239
Transmission axis, 942
Transmission gratings, 967
Transmission hologram, 970
Transmitted waves, 427–428
Transuranic element, 1123
Transverse electromagnetic (TEM) waves, 840
Transverse magnification (M_T), 905, 932
Transverse waves, 421–422, 425–429, 433
Traveling waves, 420
Treadmill scheme, 42
Triboelectric sequence, 591
Triple point, 501
Triple point of water, 485
Tritium, 1104
Trusses, 172, 173
Tube length, 923
Tuned-mass damper, 418, 419
Tungsten filament, 676
Turbulent flow, 375–377
  breathing, 377
Turn ratio, 814

Twin paradox, 996
Two-coil dc generator, 777
Two New Sciences (Galileo), 327
Tyndall, J., 1045

U.S. Customary units, 9, 26
    units of force, mass, and acceleration in, 127
U-235, fission of, 1124
Uhlenbeck, G. E., 1083
Ultimate (tensile) strength, 337
Ultrasonic levitation, 463
Ultrasound, 431–432
Ultraviolet radiation (UV), 842, **854**–855
Uncertainty, nuclear electrons and, 1091–1092
Uncertainty Principle, 1081, 1089–1092, 1096, 1110
Underdamped system, 415
Unequal-arm balance, 174, 175
Unified atomic mass units (u), 319, 1103, 1131
    table of, 1103
Uniform acceleration, 58, 60–69, 83
    equations of, 66–68
    of falling bodies, 72
    problems concerning, 67–68
Uniform electric field, 651
    potential in, 631–632
Uniform motion, rest and, 39, 91
Uniform negative acceleration, 68–69
Uniform speed, 27–28
Unit conversions, **26**
Unit prefixes, 9, 10
Units of charge, 597
Units of measure, 8, 19
    rationalized, 607–609
Universal gas constant, 497, 507
Universal Gravitation, Law of, 208–218, 231
Universal Gravitational Constant (*G*), 212–213
Universe
    age of, 303
    Aristotle's theory of, 69
    constant net charge of, 590
    creation of, 230, 727, 1156–1159
    expansion of, 303
    indeterministic nature of, 1081
    inner symmetry of, 240
    spatial homogeneity in, 267
    spatial symmetry in, 116–117
    temperature range in, 485
    total momentum of, 112–113
    understanding, 2
Unpolarized light, 972
Unstable equilibrium, 188
Up quarks, 1148
u-quark, 1138

Vacuum, 361–364
    propagation of electromagnetic waves in, 842
    speed of light in, 230, 984, 985
Vacuum pump, 364
Vacuum wavelengths, 878
Valence band, 818
Valence electrons, 818
    as the source of colored light, 855
Van Allen belts, 746
Van de Graaff, Robert J., 621

Van de Graaff generator, 612, 621, 637
Van der Meer, Simon, 1156
van Helmont, Jan Baptista, 355
van Musschenbroek, P., 640
Vapor compression cycle, 568
Vaporization, 530–532
    heats of, 529
Vapor pressure, equilibrium, 499
Vapors, condensation of, 499
Variable capacitor, 681
Variable dc voltage source, 681
Variable resistor, 681
Vector acceleration due to gravity, 93
Vector boson, 1150
Vector force field, 603
Vector product, 177
Vector quantity, 32, 50
Vectors, 31
    addition of, 32–33
    algebra of, 32–35
    colinear, 59
    components of, **44**–50
    coplanar, 177
    displacement vector (**s**), 32
    force, 97–99, 167, 169, 172
    magnitude, 33
    mathematics of, 31–35
    multiplication by scalars, 34–35
    reversing the direction of, 43
    scalar components of, 79
    subtraction of, 35
Velocity (**v**), 31–39
    addition and, 998–1000
    average, 35–36, 50
    instantaneous, 35, **36**–39, 50
    one-dimensional relativistic formula for the addition of, 998
    relative motion and, 40
    in simple harmonic motion, 406–407
    time-rate-of-change of, 58
    of waves, 423–424
Velocity field, 375
Velocity-time curve, 61, 65
Velocity triangle, 140
Velocity vectors, 37–38
    adding, 41
Venturi Effect, 383–384
Venus, ultraviolet photo of, 854
Versorium, 620
Vertical acceleration, 134–136
    gravity and, 75
Vertical displacement, 46
    work and, 285–286
Vertical reaction force, 165
Vertical speed, of projectiles, 77–79
Vibrational motion, 402
Vibrations, self-excited, 418–419
Vibratory periodic motion, 402
Villard, P., 1026
Virtual images, 871, 904, 907
Virtual mesons, 1144
Virtual particles, 6, 617, 618
    creation of real particles from, 1144
Virtual photons, 617, **1144**
Visco-elastic dampers, 416
Viscosity, 354
Viscous damping, 415
Visual perspective, 242

Vis viva, 288
Vocal tract, 464–465
Voices, frequency ranges of, 442, 453
Volta, Alessandro, 627, 640, 664
    on capacitance, 639
    on measuring potentials, 641
Voltage (*V*), 625, 627
    determining, 696–697
    drops and rises in, 678–679
    effective, 799, 825
    of household appliances, 698
    instantaneous, 798, 803
    line, 797
    maximum, 798
    peak-to-peak (*V*$_{pp}$), 798
    rms (*V*$_{eff}$), 799
Voltage divider, 679
Voltage gain, 824
Voltaic pile, 664, 668
Voltaic wet cell, 664, 665
Voltaire, 215
Voltmeter, 681, **699**–700
    placement of, 693
Volts per meter, 628
Volume (*V*), 11
Volume charge density (*ρ*), 616
Volume expansion, temperature coefficient of, 490–494
Volume flux (*J*), 378, 389, 613
Vomit Comet, 138
von Guericke, Otto, 363–364, 442
    electrostatic generator, 637
    vacuum and, 362
von Kármán vortex street, 385
von Kleist, G., 640
von Laue, Max, 1022–1023, 1053
Vonnegut, Kurt, Jr., 998
Vortices, 385–386
    motion-induced, 419

Wallis, John, 304
Walls, stability of, 188–191
Water
    absorption of microwave radiation by, 851
    expansion of, 491–494
    heavy, 1104
    phase diagram for, 501
    polarized molecules in, 601
    specific heat of, 524, 527–528
    tensile strength of, 357
Water calorimeter, 525
Water reactor system, 1128
Waterston, J. J., 506
Watson, J. D., 1024
Watt, James, 286, 518
Watt (W), 286, 310, 681
Watt-hour meter, ac, 798
Waveforms, 422–423, 445–446, 837–840
Wavefront, **447**, 837–840
    intensity and, 447–449
Wavefunction, 1077
    state of a system and, 1078
Wavelength, 423
    relationship to transmitting medium, 427
Wave motion, 3
Wave-particle duality, 1053, 1076
Wavepulse, 422

Waves, 401–433
  acoustic, 429, 454–457
  circularly polarized, 940
  compression, 429–433
  De Broglie, 1074–1076
  diffraction of, 960–972
  energy of, 424
  finding the velocity of, 423–424
  harmonic, 424–425, 433
  linearly polarized, 943
  mechanical, 420–433
  phase speed of, 874
  plane-polarized, 940, 972
  probability, 1078–1081
  reflection, absorption, and transmission of,
     427–428
  refracted, 873
  speed of, 423
  transverse, 425–429
  ultrasonic, 432
Wave theory of light, 834–835
Wavetrain, 423
Weak force, 5, 95, 1118–1119
  theory of, 6, 1152
Weak interaction, 1149, 1152–1156
Weber, Ernst, 459, 842
Weber, Wilhelm, 459, 766
Weber (Wb), 766, 785
Wedgwood, Thomas, 1044
Weight ($\mathbf{F}_W$), 11
  effective, 130, 135, 136, 137
  effect on equilibrium, 184–193
  gravity and, 100–101, 130–134
  measurement of, 11–12
  on ropes, 131

Weight-dependent fall, notion of, 70
Weightlessness, 97, 136–138
Weight-mass correspondence, 12
Weinberg, Steven, 1155
Weisskopf, Victor, 354
Wells, H. G., 999
Westinghouse, George, 796
White light, 852–854, 882
White noise, 457
Width of the central maximum, 965
Wien, Wilhelm, 1046
Wien's Displacement Law, 1046–1048, 1067
Wilson, C. T. R., 1020
Wind force, 165–166
Wind loads, 190
Wire-grid polarizers, 945
Wires, forces on, 746–752
Wollaston, 918
Wooster, W. A., 1116
Work (W), 278–286, 309
  change in kinetic energy and, 289–290
  changing forces and, 283–285
  defined, 517
  deformable objects and friction and,
     282–283
  First Law of Thermodynamics and,
     554–557
  in general, 280–282
  heat and internal energy and, 552–553
  internal energy sources and, 296–297
  power and, 286–288
  along a straight line, 279–280
  vertical displacement and, 285–286
Work diagram, 284
Work function (ϕ), 1052

"Work rule," 740
World-line, 997
World Trade Center, 165, 190, 403, 411
W-particles, 1152, 1154
Wrenches, 175, 176–177

X-rays, **856**–857
  Compton scattering of, 1055
X-ray spectrum, 1054–1055
X-ray tube, 1025

Yang, C. N., 1150
Yield point, 331
Yield strength, 337
Young, Thomas, 341, 343, 372, 444, 834
  on color, 883
  energy and, 288
  on light waves, 940
  superposition principle, 444
  on surface tension, 372
Young's Experiment, 952–956, 972, 1079
Young's Modulus, **341**–343, 346
Yukawa, Hideki, 1144, 1152

Z boson, 1154–1155
Zeeman, Pieter, 1082
Zeeman Effect, **1082**–1083
"Zero-g," 137
Zero-mass particles, determining energy and
     momentum of, 1007–1008
Zero-point energy, 487
Zeroth Law of Thermodynamics, 521, 540
Zweig, George, 1148

USEFUL PHYSICAL DATA

Standard Temperature and Pressure (STP)		$0°C = 273.15$ K 1 atm $\equiv 101.325$ kPa
Water		
Density, relative (4°C)	ρ_w	1.000×10^3 kg/m^3
Heat of fusion	L_f	333.7 kJ/kg
Heat of vaporization	L_v	2259 kJ/kg
Specific heat capacity	c	4.186 kJ/kg·K
Index of refraction	n_w	1.33
Standard acceleration due to Earth's gravity		$9.806\,65$ m/s^2
Speed of sound in air (20°C)		343 m/s
Speed of sound in air (STP)		331 m/s
Density of dry air (STP)		1.29 kg/m^3
Molecular mass of air		28.98 g/mol

ASTROPHYSICAL DATA

	Earth	Moon	Sun
Mass	5.975×10^{24} kg	7.35×10^{22} kg	1.987×10^{30} kg
Mean radius	6.371×10^6 m	1.74×10^6 m	6.96×10^8 m
Mean density	5.52×10^3 kg/m^3	3.33×10^3 kg/m^3	1.41×10^3 kg/m^3
Orbital period about galactic center			200×10^6 y
Orbital period	365 d 5 h 48 min	27.3 d	
Mean distance from Sun	1.50×10^{11} m		
Mean distance from Earth		3.85×10^8 m	
Surface gravitational acceleration	9.81 m/s^2	1.62 m/s^2	274 m/s^2
Surface pressure	1.013×10^5 Pa		
Magnetic moment	8.0×10^{22} A·m^2		
Surface temperature	≈ 287 K	125 K–375 K	5.8×10^3 K
Power output			3.85×10^{26} W
Period of rotation	23 h 56 min 4.1 s	27.3 d	~26 d

THE GREEK ALPHABET

Alpha	A	α	Nu	N	ν
Beta	B	β	Xi	Ξ	ξ
Gamma	Γ	γ	Omicron	O	o
Delta	Δ	δ	Pi	Π	π
Epsilon	E	ϵ	Rho	P	ρ
Zeta	Z	ζ	Sigma	Σ	σ
Eta	H	η	Tau	T	τ
Theta	Θ	θ	Upsilon	Y	υ
Iota	I	ι	Phi	Φ	ϕ
Kappa	K	κ	Chi	X	χ
Lambda	Λ	λ	Psi	Ψ	ψ
Mu	M	μ	Omega	Ω	ω